Reagents for Organic Synthesis

Fieser and Fieser's

Reagents for Organic Synthesis

VOLUME TWELVE

Mary Fieser

Harvard University

A WILEY-INTERSCIENCE PUBLICATION
JOHN WILEY & SONS
NEW YORK ● CHICHESTER ● BRISBANE ●
TORONTO ● SINGAPORE

Library of Congress Cataloging in Publication Data:

(Revised for vol. 12)

Fieser, Louis Frederick, 1899–
  Reagents for organic synthesis.

  Authors' names in reverse order in v. 2–7.
  Vols. 8, by Mary Fieser.
  Vol. 9, by Mary Fieser, Rick L. Danheiser, William
Roush.
  Vols. 8– have title: Fieser and Fieser's
Reagents for organic synthesis.
  Vols. 2–3 have imprint: New York, Wiley-Interscience.
  Vols. 4– A Wiley-Interscience publication.
  Includes bibliographical references and indexes.
  1. Chemistry, Organic—Synthesis. 2. Chemical tests
and reagents. I. Fieser, Mary, 1909– joint author.
II. Fieser and Fieser's Reagents for organic synthesis.
III. Title.

QD262.F5   547'.2   66-27894
ISBN 0-471-83469-6 (v. 12)

Printed in the United States of America

10 9 8 7 6 5 4

PREFACE

This volume of reagents includes for the most part references to papers published during 1983 and 1984. I am particularly grateful to Andrew Myers and William Siebel, who read the entire manuscript and offered many valuable suggestions. I am again indebted to my colleagues for careful proofreading, in particular to Daniel F. Lieberman, Joseph Tino, Theodore S. Widlanski, Robert W. Hahl, Stephen Wright, Erick M. Carreira, Jonathan Ellman, Stephen W. Kaldor, Felix Chau, Frank Hannon, Paul DaSilva Jardine, Wei Guo Su, Regan G. Shea, Andreas Spaltenstein, and in particular, Professor Paul B. Hopkins. The proofreading was coordinated by Richard T. Peterson and Marcello DiMare. The picture of some members of the group was taken by Julia D. Peterson.

MARY FIESER

Cambridge, Massachusetts
November 1985

CONTENTS

Reagents for Organic Synthesis

A

Acetic anhydride–Ferric chloride, 6, 260.

gem-Diacetates.[1]   This system converts aldehydes into *gem*-diacetates in moderate to high yield. PCl_3 can also be used as catalyst.[2] The products are stable to aqueous acid and to $NaHCO_3$ and Na_2CO_3 for 1–4 hours, but are cleaved after longer reaction periods.

[1] K. S. Kochhar, B. S. Bal, R. P. Deshpande, S. N. Rajadhyaksha, and H. W. Pinnick, *J. Org.,* **48,** 1765 (1983).
[2] J. K. Michie and J. A. Miller, *Synthesis,* 824 (1981).

Acetoacetic acid, CH_3COCH_2COOH (**1**). Preparation.[1]

Acetonylation.[2]   Alkylation of the dianion of **1** at 25° is accompanied by decarboxylation to provide methyl ketones (equation I).

$$(\text{I}) \quad \mathbf{1} \xrightarrow[\text{THF, } 0°]{n\text{-BuLi,}} \overset{\overset{\displaystyle OLi}{|}}{CH_3C}=CHCOOLi \xrightarrow[45-90\%]{\underset{0-25°}{RX,HMPT,}} \overset{\overset{\displaystyle O}{\|}}{CH_3CCH_2R} + CO_2$$

[1] R. C. Krueger, *Am. Soc.,* **74,** 5536 (1952).
[2] R. A. Kjonaas and D. D. Patel, *Tetrahedron Letters,* **25,** 5467 (1984).

Aceto(carbonyl)cyclopentadienyl(triphenylphosphine)iron, $C_5H_5Fe(CO)(COCH_3)$- $[P(C_6H_5)_3]$ (**1**). Orange crystals, m.p. 145°, air stable. The chiral iron complex is prepared as a racemate by reaction of $C_5H_5Fe(CO)_2CH_3$ with $P(C_6H_5)_3$.[1]

Stereoselective condensation with imines.[2]   The diethylaluminum enolate of **1** reacts with a wide variety of imines with high stereofacial bias, which is ascribed to a cyclic transition state in an aluminum chelate.

Example:

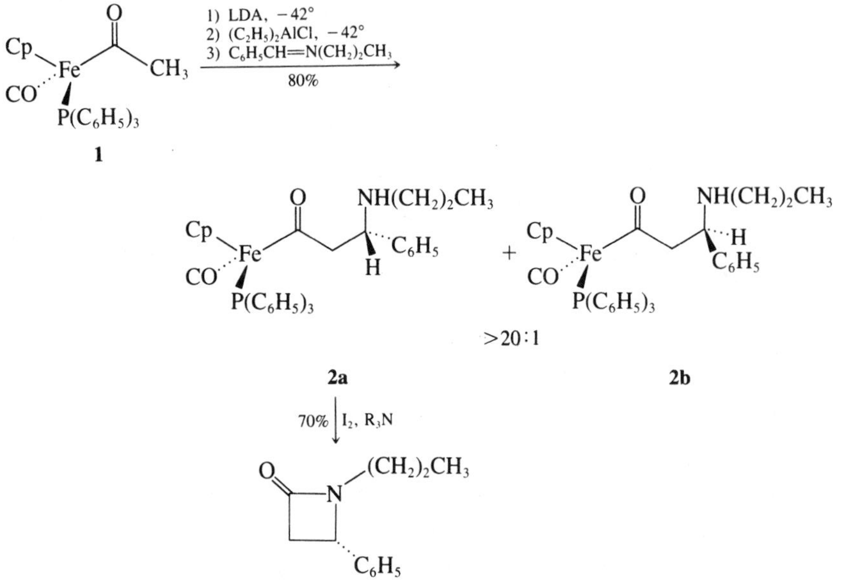

1

$> 20:1$

2a   **2b**

70% | I_2, R_3N

3

The products are converted into β-lactams (**3**) on oxidative removal of the metal. The two optically active forms of **1** are known,[3] and their use should permit enantioselective alkylation by imines.

Stereoselective aldol condensations.[4]   The stereoselectivity of aldol condensations of **1** can be controlled by the enolate counterion to give either one of the possible diastereomeric products predominantly.

Example:

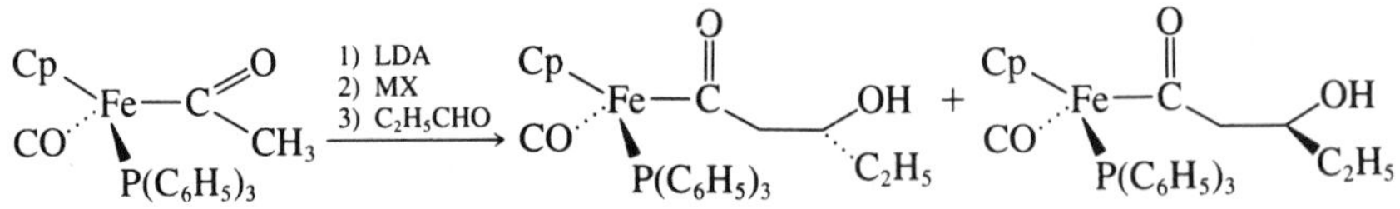

1

| MX | = i-Bu$_2$AlCl | 73% | 5.2:1 |
| | = SnCl$_2$ | 66% | 1:11.6 |

The products are converted into β-hydroxy carboxylic esters by oxidation with NBS in CH_2Cl_2/C_2H_5OH at 0°.

[1] J. P. Bibler and A. Wojcicki, *Inorg. Chem.*, **5**, 889 (1966).
[2] L. S. Liebeskind, M. E. Welker, and V. Goedken, *Am. Soc.*, **106**, 441 (1984).
[3] H. Brunner and E. Schmidt, *J. Organometal. Chem.*, **36**, C18 (1972).
[4] L. S. Liebeskind and M. E. Welker, *Tetrahedron Letters*, **25**, 4341 (1984).

(R)-2-Acetoxy-1,1,2-triphenylethanol,

$$HOC(C_6H_5)_2$$

$$CH_3CO \underset{O}{\overset{C \cdots H}{\diagup}} C_6H_5 \quad (1), \ \alpha_D \ +209°.$$

The reagent is prepared by addition of C_6H_5MgBr to (R)-mandelic acid methyl ester followed by acetylation of the resulting diol. Both (R)- and (S)-mandelic acid are available from Aldrich or Fluka.

Chiral β-hydroxy carboxylic acids.[1] The enolate of **1**, prepared by reaction with LDA followed by transmetallation with $MgBr_2$, reacts with aldehydes to provide the diastereomers **2** and **3** in the ratio 92–97:8–3. The adducts are cleaved by KOH in aqueous methanol to chiral β-hydroxy carboxylic acids (**4**) and the chiral diol **5**.

[1] M. Braun and R. Devant, *Tetrahedron Letters*, **25**, 5031 (1984).

Acetyl hypofluorite (1), 10, 1; 11, 5.

Fluorination of 1, 3-dicarbonyl compounds.[1] 1,3-Dicarbonyl compounds that exist partly in the enol form are converted by **1** into the 2-fluoro derivative in moderate yield. The yield is generally improved by reaction with the corresponding sodium enolate. When the enol content of the 1,3-dicarbonyl substrate is small, no reaction occurs with **1**, although the corresponding metal enolate does react.

Example:

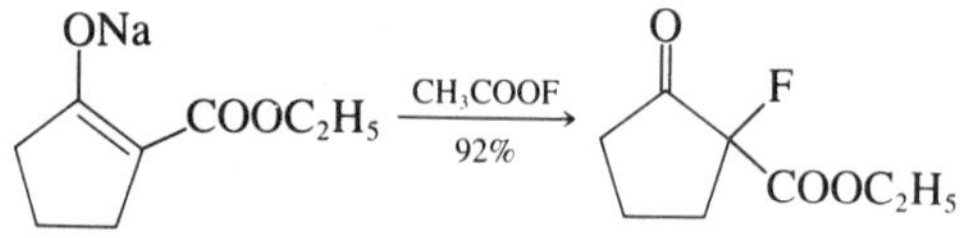

[1] O. Lerman and S. Rozen, *J. Org.*, **48**, 724 (1983).

Acyllithium reagents, RCOLi, **11**, 111. These unstable anions, when generated from RLi and CO at $-135°$ in the presence of a carbonyl compound (1 equiv.), can effect nucleophilic acylation to give acyloins. Since 1,2-addition of RLi is a competing reaction, acylation is favored when the carbonyl group is hindered and when R is secondary or tertiary. The anions also react with esters under the same conditions to give 1,2-diketones.[1]

[1] D. Seyferth, R. M. Weinstein, and W.-L. Wang, *J. Org.*, **48**, 1144 (1983); D. Seyferth, R. M. Weinstein, W.-L. Wang, and R. C. Hui, *Tetrahedron Letters*, **24**, 4907 (1983).

3-Acylthiazolidine-2-thiones, 11, 518–519.

Stereoselective aldol condensations.[1] Reaction of these substrates (**1**) with LDA results in decomposition, but the tin(II) enolates can be obtained by use of tin(II) triflate (**11**, 525) in the presence of N-ethylpiperidine. Addition of aldehydes to the enolates results in formation of β-hydroxy ketones (**2**) with high *syn*-selectivity.

Example:

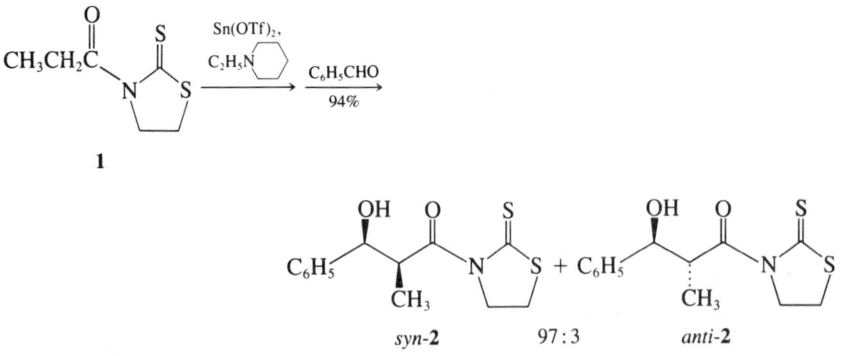

[1] T. Mukaiyama and N. Iwasawa, *Chem. Letters*, 1903 (1982).

B-1-Alkenyl-9-borabicyclo[3.3.1]nonanes (1), 8, 6.

trans,trans-*Dienones*. These boranes undergo conjugate addition/elimination to 4-methoxy-3-butene-2-one to furnish *trans,trans*-dienones in nearly quantitative yield.[1]
Example:

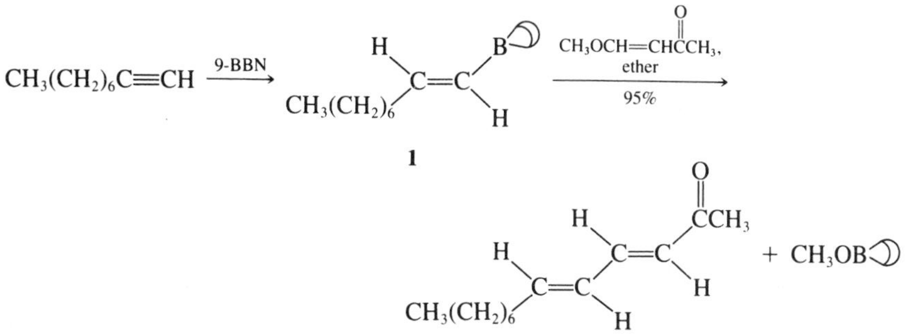

[1] G. A. Molander, B. Singaram, and H. C. Brown, *J. Org.*, **49**, 5024 (1984).

Alkylaluminum halides, 6, 251; 7, 146; 10, 177–181; 11, 7–12.

trans-*Hydrindenones*.[1] Cyclization of the methyl ketone **1** with CH_3AlCl_2 (2 equiv.) in CH_2Cl_2 leads to formation of the *trans*-hydrindanone **2**. Cyclization of the unstable dienone **3**, obtained in two steps from hydrocinnamic acid, is also possible with CH_3AlCl_2 under carefully controlled conditions (antioxidant, N_2, sealed tube, 90–93°). The *trans*-hydrindenone **4** is obtained in 53% yield. This product can be converted into intermediates in two routes to 11-oxo steroids.[2]

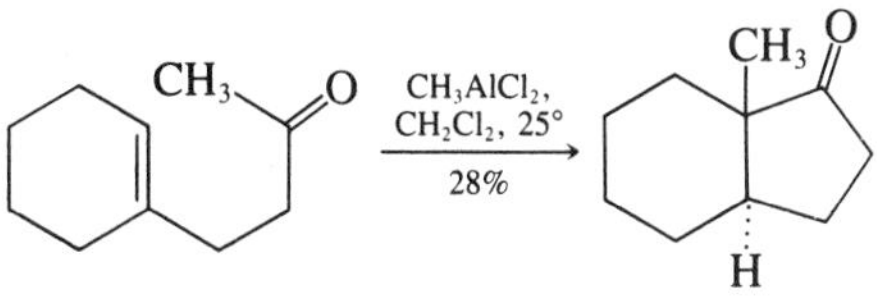

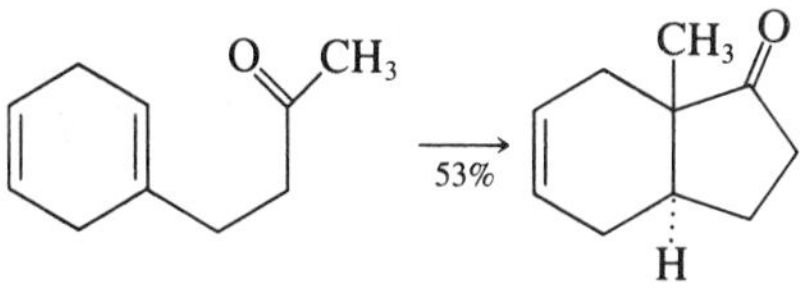

Aryltetralin lignans.[3]   This ring system can be prepared in high yield by a Prins reaction of a 1,4-diaryl-1-butene with paraformaldehyde catalyzed by an alkylaluminum halide prepared from CH_3AlCl_2 and $(CH_3)_2AlCl$ (1:1), which is more acidic than $(CH_3)_2AlCl$ alone.

Example:

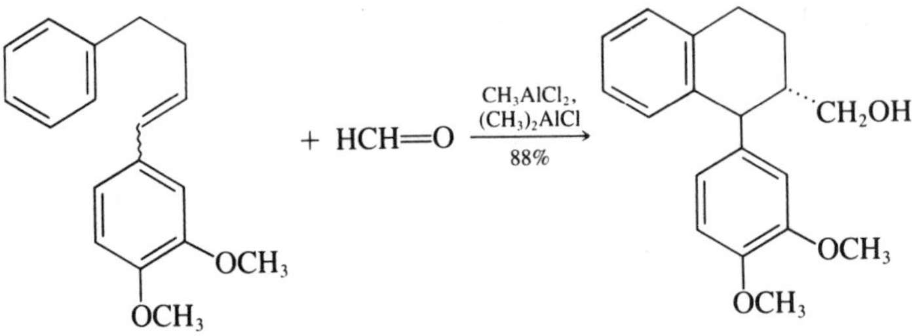

(E, 67%)   (92% *trans*)

Intramolecular cyclization of allylsilanes.[4]   An intramolecular version of the Sakurai reaction (**7**, 371; **11**, 529–530) provides a route to functionalized spiro[4,5]decanones. For cyclization of **1** to **2**, ethylaluminum dichloride is the preferred Lewis acid activator. The diastereoselectivity depends on the temperature and the solvent.

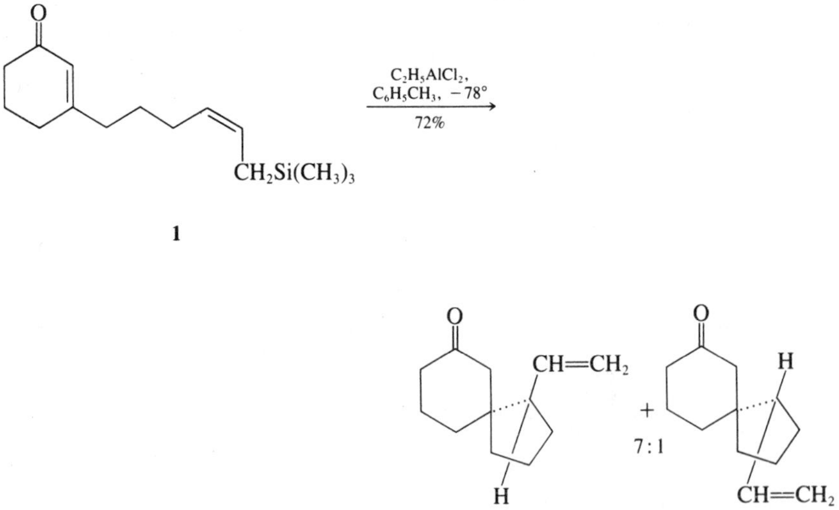

1

2a   **2b**

Imidoyl iodides.[5]   Reaction of oxime mesylates (or carbonates) of ketones with $(C_2H_5)_2AlI$ effects Beckmann rearrangement to imidoyl iodides, which can be directly alkylated with Grignard reagents and then reduced (DIBAH) to α-alkylated amines. Examples:

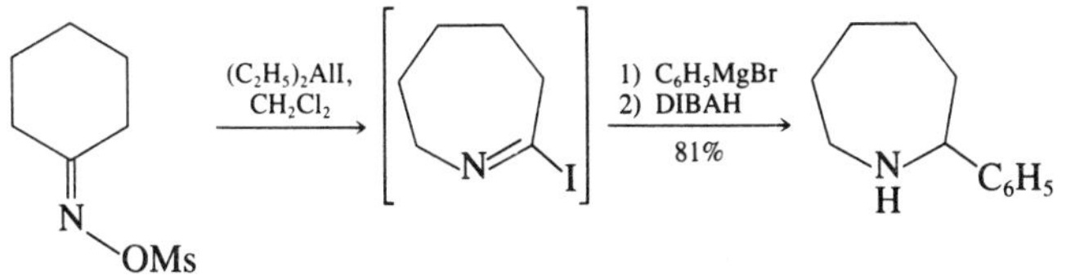

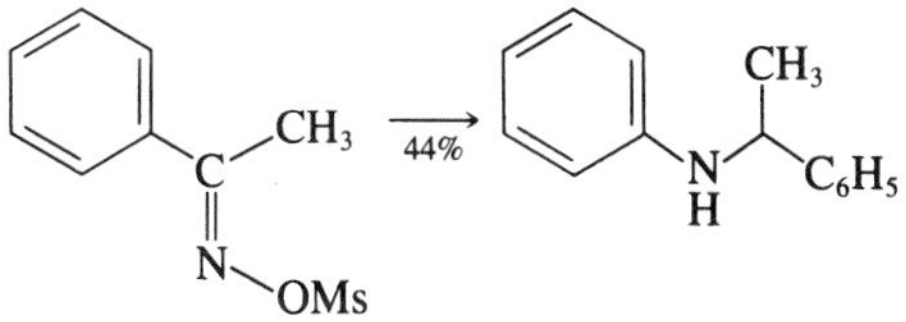

Enaminones.[6]   Coupling of enol silyl ethers with oxime mesylates can be effected via a Beckmann rearrangement with diethylaluminum chloride. Examples:

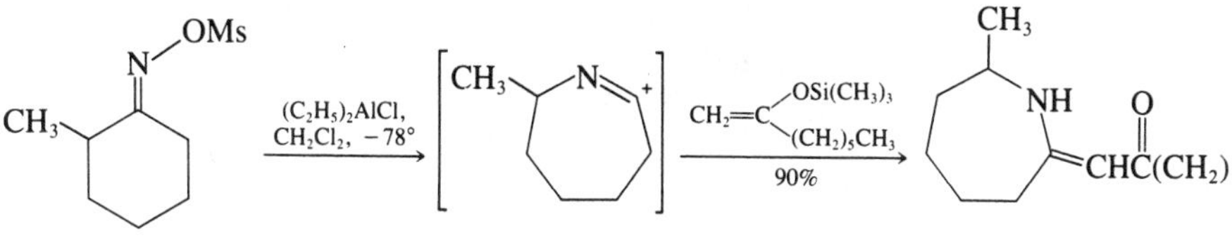

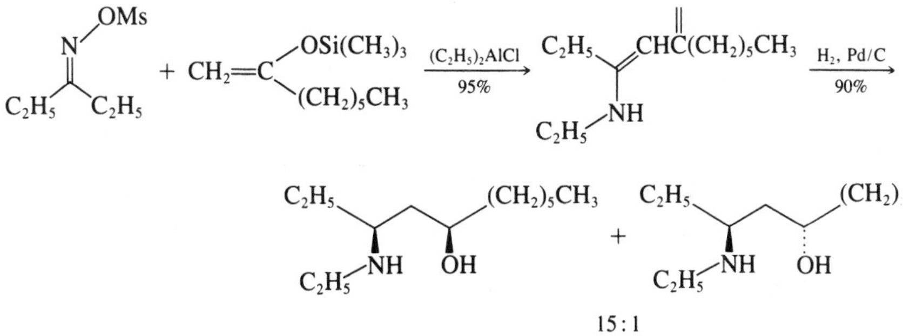

The enaminones are hydrogenated to γ-amino alcohols with marked stereoselectivity.
Ene reactions of aldehydes with 1-alkenes.[7]   Ethylaluminum dichloride in CH_2Cl_2

is the most satisfactory catalyst for this reaction. An example is formulated in equation (I). The product has been lactonized to the macrolide recifeiolide (**1**).

(I) $CH_3CHO + CH_2{=}CH(CH_2)_7COOH \xrightarrow[\substack{66\%}]{\substack{2C_2H_5AlCl_2, \\ CH_2Cl_2,\ 0°}}$

$$CH_3CHCH_2CH{=}CH(CH_2)_6COOH \longrightarrow$$
$$\underset{\displaystyle OH}{|}$$

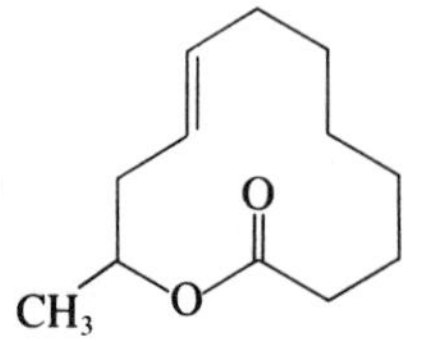

$(E/Z = 4:1)$

1

A similar ene reaction provides a short route to ricinelaidic acid, (E)-**2**, and ricinoleic acid, (Z)-**2**, in a 4:1 ratio.

$$C_6H_{13}CHO + CH_2{=}CH(CH_2)_8COOH \xrightarrow[\substack{41\%}]{\substack{2C_2H_5AlCl_2}}$$

$$C_6H_{13}CHCH_2CH{=}CH(CH_2)_7COOH$$
$$\underset{\displaystyle OH}{|}$$

2 $(E/Z = 4:1)$

Diels-Alder cyclization of enone silyloxyvinylallenes.[8]   The cyclization of the enone silyloxyvinylallene (**1**) can be effected thermally, but occurs in higher yield and with higher stereoselectivity with $(C_2H_5)_2AlCl$ (1.1 equiv.) at $-78{\rightarrow}0°$. The product **2** is converted into the sesquiterpene dehydrofukinone (**3**) in several steps.

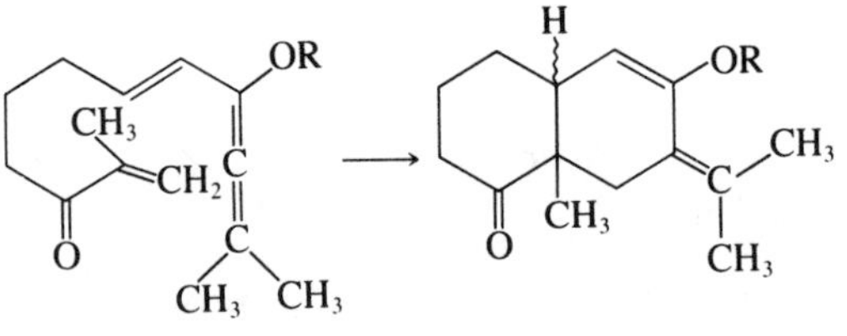

1, R = Si(CH$_3$)$_2$-t-Bu   **2**

$(C_2H_5)_2AlCl$	51%	*cis/trans* = 2:1
75°	42%	*cis/trans* = 1:1

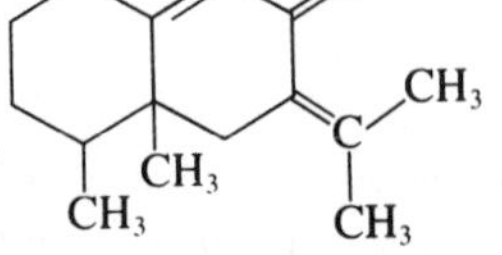

3

Intramolecular Diels-Alder reactions.   The Lewis acid-catalyzed intramolecular Diels-Alder reaction is useful for construction of *cis*-fused bicyclo[6.4.0]dodecanes. Thus

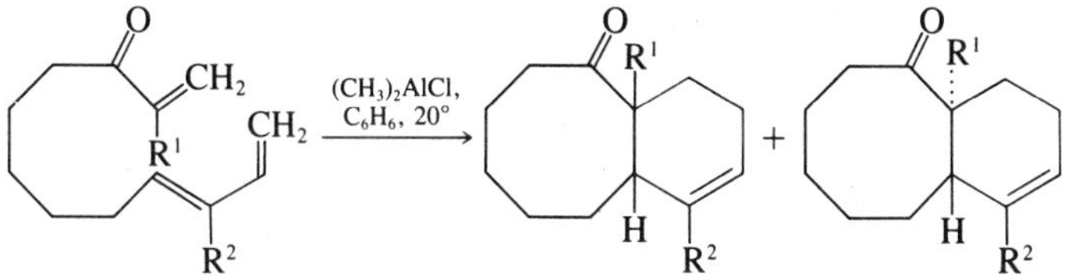

1a, R^1, R^2 = H	69%	100:0
1b, R^1, R^2 = CH$_3$	69%	74:26

in the presence of (CH$_3$)$_2$AlCl (1 equiv.), **1** cyclizes to the vicinally fused cyclooctane **2**, with preferential formation of the less stable *cis*-fused isomer.[9]

A related Diels-Alder reaction offers one approach to the taxane diterpenes. Thus in the presence of (CH$_3$)$_2$AlCl (3 equiv.) the model compound **3** cyclizes mainly to **4**. Even

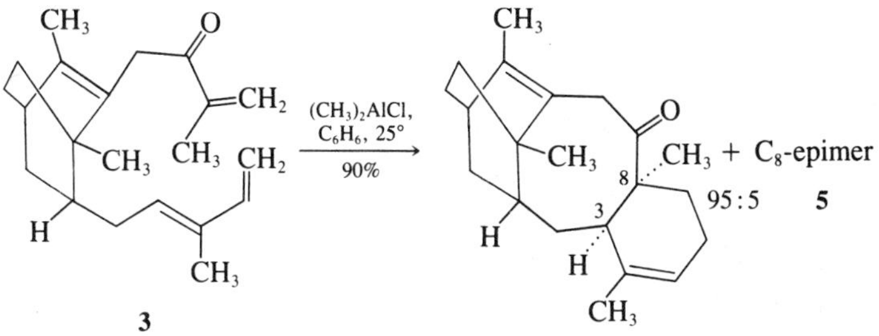

though the cycloaddition involves formation of an eight-membered ring, the yield of **4** and **5** is 90%. Surprisingly, the thermal cyclization of **3** gives **5** as the predominant product (70% yield).[10]

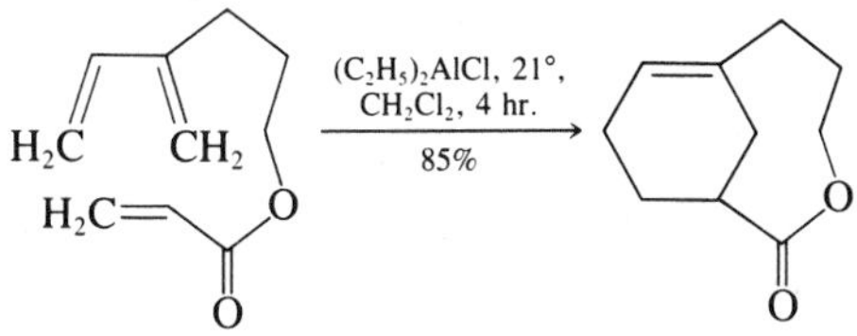

Intramolecular Diels-Alder reactions resulting in bridgehead alkenes usually require high reaction temperatures. Consequently catalysts that permit use of lower temperatures are useful. Of a number of Lewis acids studied, diethylaluminum chloride was found to be the most efficient for the intramolecular cycloaddition of the triene **7** to **8** at 21°. The thermal cycloaddition requires a temperature of 139°. For optimum yields, an equimolar amount of $C_2H_5AlCl_2$ is required.[11]

2,8,10-Dodecatrienals undergo intramolecular Diels-Alder cyclization in the presence of ethylaluminum dichloride or the milder Lewis acid diethylaluminum chloride at -78 to $-23°$.[12]

Example:

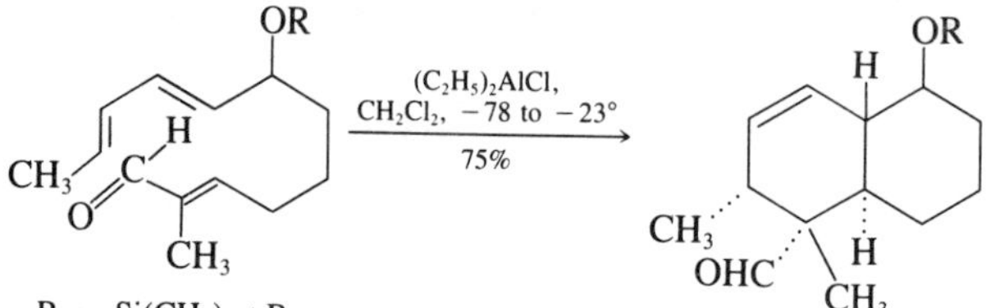

$$R = Si(CH_3)_2\text{-}t\text{-}Bu$$

Bisdehydrobromination of vic-dibromides.[13]   A route to N-acyl-1,2-dihydropyridines (**3**) involves bromination of an N-acyltetrahydropyridine followed by bisdehydrobromination of the *vic*-dibromide (**2**). Usual methods, including DABCO or DBU in DMF,

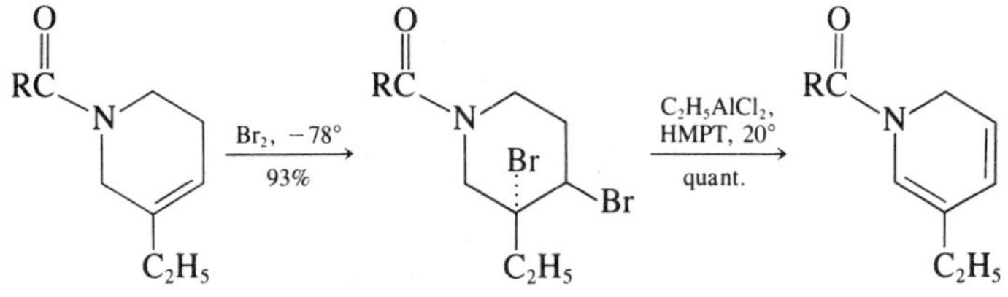

are ineffective or result in decomposition, but reaction with ethylaluminum dichloride in HMPT at 20° gives the dienes in almost quantitative yield. $AlCl_3$ is much less effective.

The products are reasonably stable, and are useful for Diels-Alder reactions. An example is the synthesis of an isoquinuclidine (**4**).[14]

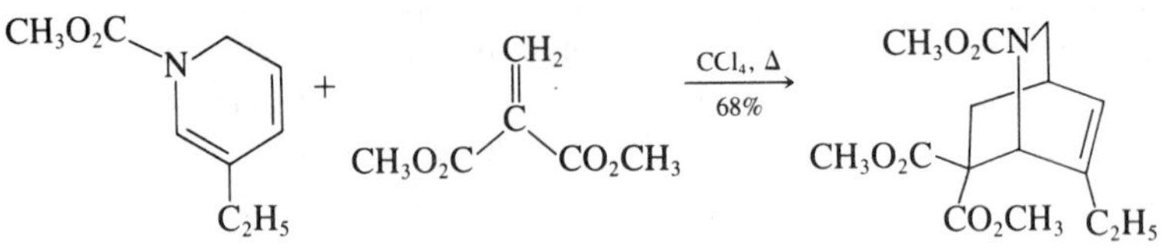

α-Trimethylsilyl-α,β-enones.[15]   These useful Michael acceptors (**2**) can be prepared by rearrangement of 1-(trimethylsilyl)-2-propynyl trimethylsilyl ethers (**1**) with methylaluminum dichloride.

$$RCOSi(CH_3)_3 \xrightarrow[\text{2) ClSi(CH}_3)_3]{\text{1) BrMgC}\equiv\text{CH}} \underset{\underset{Si(CH_3)_3}{|}}{RC}\overset{\overset{OSi(CH_3)_3}{|}}{-}C\equiv CH \xrightarrow[80-85\%]{CH_3AlCl_2} \underset{\underset{Si(CH_3)_3}{|}}{RC}\overset{\overset{O}{\|}}{-}C=CH_2$$

$$\qquad\qquad\qquad\qquad\qquad\qquad 1 \qquad\qquad\qquad\qquad 2$$

Cleavage of tetrahydropyranyl ethers.[16] Tetrahydropyranyl ethers are cleaved to the alcohol by dimethylaluminum chloride or methylaluminum dichloride in high yield at temperatures of -25 to $25°$, conditions that do not affect *t*-butyldimethylsilyl ethers. MOM and MEM ethers are converted into ethyl ethers by a methyl transfer reaction.

[1] B. B. Snider and T. C. Kirk, *Am. Soc.*, **105**, 2364 (1983).

[2] G. Stork, G. Clark, and C. S. Shiner, *Am. Soc.*, **103**, 4948 (1981); G. Stork and E. Logusch, *ibid.*, **102**, 1218 (1980).

[3] B. B. Snider and A. C. Jackson, *J. Org.*, **48**, 1471 (1983).

[4] D. Schinzer, *Angew. Chem., Int. Ed.*, **23**, 308 (1984).

[5] Y. Ishida, S. Sasatani, K. Maruoka, and H. Yamamoto, *Tetrahedron Letters*, **24**, 3255 (1983).

[6] Y. Matsumura, J. Fujiwara, K. Maruoka, and H. Yamamoto, *Am. Soc.*, **105**, 6312 (1983).

[7] B. B. Snider and G. B. Phillips, *J. Org.*, **48**, 464 (1983).

[8] H. J. Reich and E. K. Eisenhart, *ibid.*, **49**, 5282 (1984).

[9] K. Sakan and D. A. Smith, *Tetrahedron Letters*, **25**, 2081 (1984).

[10] K. Sakan and B. M. Craven, *Am. Soc.*, **105**, 3732 (1983).

[11] K. J. Shea and J. W. Gilman, *Tetrahedron Letters*, **24**, 657 (1983).

[12] J. A. Marshall, J. E. Audia, and J. Grote, *J. Org.*, **49**, 5277 (1984).

[13] S. Raucher and R. F. Lawrence, *Tetrahedron*, **39**, 3731 (1983).

[14] *idem, Tetrahedron Letters*, **24**, 2927 (1983).

[15] J. Ende and I. Kuwajima, *J.C.S. Chem. Comm.*, 1589 (1984).

[16] Y. Ogawa and M. Shibasaki, *Tetrahedron Letters*, **25**, 663 (1984).

N-Alkyl-β-aminoethylphosphonium bromides, $RHN(CH_2CH_2)\overset{+}{P}(C_6H_5)_3Br^-$ (**1**). These somewhat unstable salts can be obtained in high yield by reaction of primary amines and vinyltriphenylphosphonium bromide in CH_3CN.

Allyl amines.[1] Aldehydes and the dilithio anion (**2**) of **1** react via a Wittig reaction to form *sec*-allyl amines. *trans*-Selectivity is favored in reactions of aromatic, but not of aliphatic, aldehydes.

Example:

$$C_6H_5CHO + C_6H_5CH_2\overset{\overset{Li}{|}}{N}CH_2\overset{\overset{Li}{|}}{C}H\overset{+}{P}(C_6H_5)_3Br^- \xrightarrow[73\%]{} C_6H_5CH_2\overset{\overset{H}{|}}{N}CH_2CH=CHC_6H_5$$

$$\qquad\qquad\qquad\qquad\qquad 2 \qquad\qquad\qquad\qquad\qquad\qquad (E/Z = 71:29)$$

[1] R. J. Linderman and A. I. Meyers, *Tetrahedron Letters*, **24**, 3043 (1983).

Alkyl chloroformates, $ClCO_2R$.

Esterification.[1] Unhindered aliphatic carboxylic acids are esterified in high yield by reaction with an alkyl chloroformate, triethylamine, and a trace of 4-dimethylamino-pyridine in CH_2Cl_2 at $0°$. The reaction probably involves a mixed anhydride. Aromatic carboxylic acids are esterified by this method, but in lower yield because of formation of the acid anhydride.

[1] S. Kim, Y. C. Kim, and J. I. Lee, *J. Org.*, **50**, 560 (1985).

Alkyldimesitylboranes,   These boranes are prepared by reaction of fluorodimesitylborane with Grignard reagents. Unlike the usual trialkylboranes, which are attacked by a base on boron to form ate complexes, these new boranes can form carbanions adjacent to boron on treatment with moderately hindered bases such as lithium dicyclo-hexylamide. Mesityllithium (prepared from bromomesitylene and *t*-butyllithium or lithium dispersion) is even more useful for this purpose. Fortunately, less hindered bases permit removal of the boron when desired.[1] Thus the anion of methyl(dimesityl)borane is readily alkylated by primary alkyl iodides and bromides in moderate to high yield. It is even possible to prepare $Mes_2BCR^1R^2R^3$ directly from $Mes_2BCH_2CH_3$ in reasonable yields. As expected, the products are converted to carbinols on alkaline oxidation.[2]

Allyldimesitylborane, $Mes_2BCH_2CH=CH_2$ (**1**). M.p. $68°$. This borane is obtained in high yield by reaction of allylmagnesium bromide and Mes_2BF in ether. The anion (**2**) of **1** reacts with a variety of electrophiles to give vinyl boranes, which are of interest in themselves and as precursors to aldehydes (equation I).[3]

$$(I)\ \mathbf{1} \longrightarrow \underset{\mathbf{2}}{Mes_2BCH\dot{-}CH\dot{-}CH_2Li^+} \xrightarrow[90-95\%]{RX} \overset{(E)}{Mes_2BCH=CHCH_2R}$$

$$\xrightarrow[90-95\%]{\underset{THF}{H_2O_2,\ NaOH,}} RCH_2CH_2CHO$$

Boron-Wittig reactions.[4] The reaction of anions of Mes_2BR with nonenolizable aldehydes or ketones results in an adduct that spontaneously eliminates Mes_2BOLi to give an alkene in good yield.

Examples:

$$Mes_2BCH_2Li + (C_6H_5)_2C{=}O \xrightarrow[75\%]{} (C_6H_5)_2C{=}CH_2$$

$$\overset{Li}{\underset{|}{Mes_2BCH}}(CH_2)_6CH_3 + C_6H_5CHO \longrightarrow C_6H_5CH{=}CH(CH_2)_6CH_3 + C_6H_5CHOHCH_2(CH$$

$$\qquad\qquad (40\%, E > 99.9\%) \qquad\qquad\qquad (40\%)$$

Mes₂BCH₂MRₙ $(M = Si, Sn, Pb, S. Hg).$[5]   Most of these complex reagents can be prepared by the reaction of Mes_2BCH_2Li with $ClMR_n$. An interesting Peterson-type reaction observed with one of these boranes is shown in equation (I).

$$(I)\ Mes_2\overset{-}{B}CHSi(CH_3)_3Li^+ + C_6H_5CHO \longrightarrow adduct \xrightarrow[95\%]{\overset{H_2O_2,}{^-OH}} C_6H_5CH_2CHO$$

[1] A. Pelter, B. Singaram, L. Williams, and J. W. Wilson, *Tetrahedron Letters*, **24**, 623 (1983).
[2] A. Pelter, L. Williams, and J. W. Wilson, *ibid.*, **24**, 627 (1983).
[3] A. Pelter, B. Singaram, and J. W. Wilson, *ibid.*, **24**, 631 (1983).
[4] *Idem, ibid.*, **24**, 635 (1983).
[5] M. V. Garad, A. Pelter, B. Singaram, and J. W. Wilson, *ibid.*, **24**, 637 (1983).

N-Alkylhydroxylamines, RNHOH.

Intramolecular nitrone–alkene cycloaddition.[1]   Reaction of cycloalkanones substituted by a 3-(2-propenyl) or a 3-(3-butenyl) side chain with alkylhydroxylamines with azeotropic removal of water results in a bridged bicycloalkane fused to an isoxazolidine ring. The transformation involves formation of a nitrone that undergoes intramolecular cycloaddition with the unsaturated side chain.

   Example:

This methodology was used for synthesis of the sesquiterpene (±)-**3** from the unsaturated ketone **1** via **2**. The natural product (12S)-**3** is an antifertility agent in mice.

1 **2**

3

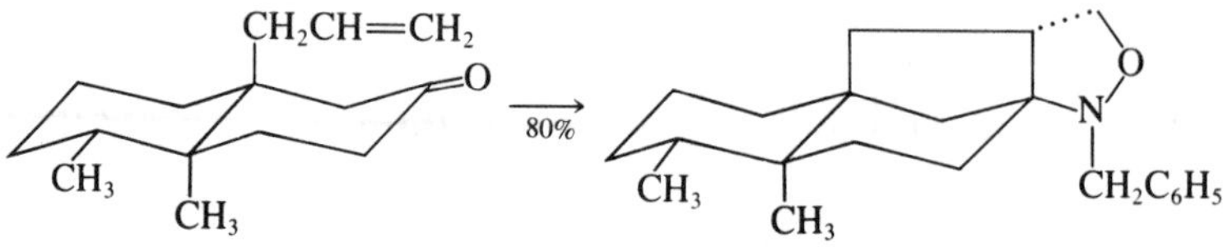

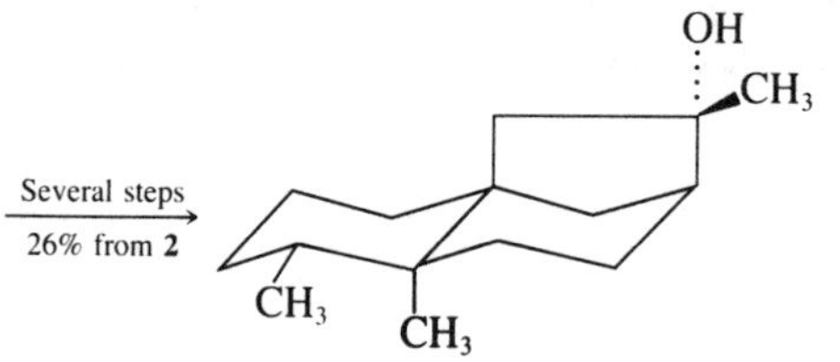

[1] R. L. Funk, L. H. M. Horcher II, J. U. Daggett, and M. M. Hansen, *J. Org.*, **48**, 2632 (1983).

Alkyllithium reagents.

Monoalkylation of primary amines.[1] A two-step sequence for this reaction involves conversion of the amine to the N-(cyanomethyl)amine (**2**) by reaction with chloroacetonitrile or with formaldehyde and KCN. Reaction of **2** with an organolithium or a Grignard reagent generates an unstable formaldehyde imine (**a**) that reacts with a second equivalent of the organometallic reagent to form a secondary amine (**3**).

$$\underset{\mathbf{1}}{R^1CH_2NH_2} \xrightarrow[75-85\%]{} \underset{\mathbf{2}}{R^1CH_2NHCH_2CN} \xrightarrow[-LiCN]{R^2Li} \underset{\mathbf{a}}{\left[R^1CH_2N{=}CH_2\right]} \xrightarrow[50-80\%]{R^2Li} \underset{\mathbf{3}}{R^1CH_2NHCH_2R^2}$$

[1] L. E. Overman and R. M. Burk, *Tetrahedron Letters*, **25**, 1635 (1984).

B-Allyl-9-borabicyclo[3.3.1]nonane (Allyl-9-BBN), $CH_2=CHCH_2B$⟩

syn-*Selective reaction with imines*.[1]   The reaction of B-allyl-9-BBN with chiral α-methyl aldehydes to form homoallylic alcohols shows only slight *syn*-selectivity (equation I). Similar low *syn*-selectivity is observed with other allylic organometallic compounds, presumably because of lack of chelation in a chair transition state. In contrast, the reaction of **1** with an imine of the aldehyde is highly or completely *syn*-selective (equation II).

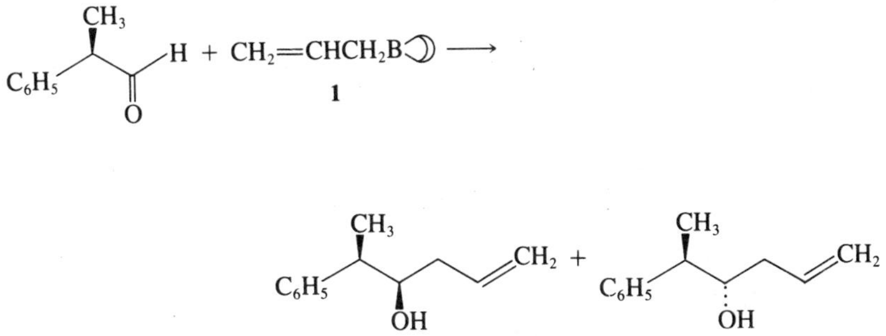

55:45

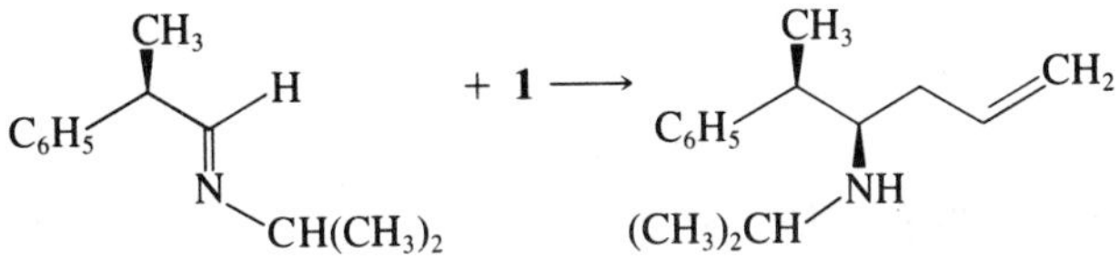

Even the reaction of allylmagnesium chloride with the imine shows moderate *syn*-selectivity.

[1] Y. Yamamoto, T. Komatsu, and K. Maruyama, *Am. Soc.*, **106**, 5031 (1984).

Allyl chloroformate, $ClCO_2CH_2CH=CH_2$ (1). Mol. wt. 120.54, b.p. 109–110°. Supplier: Aldrich.

α-*Allyl ketones*.[1]   Allyl enol carbonates, prepared as shown with **1**, rearrange in the presence of a Pd(0) catalyst to α-allyl ketones (or aldehydes).

Examples:

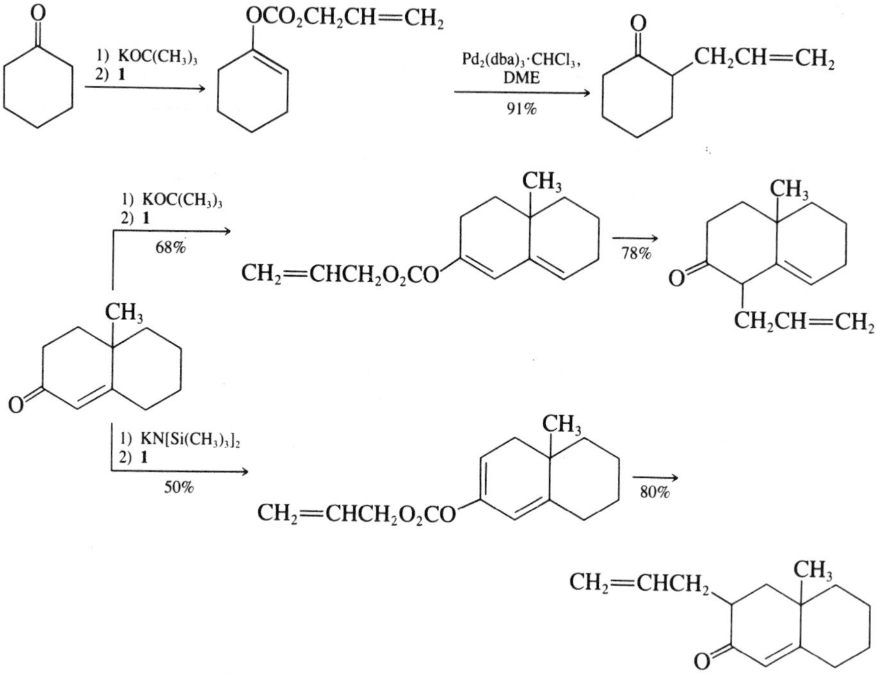

α,β-Enones from ketones.[2]   Allyl enol carbonates are converted in the presence of $Pd(OAc)_2$ and a phosphine, particularly dppe, into α,β-unsaturated ketones.
Examples:

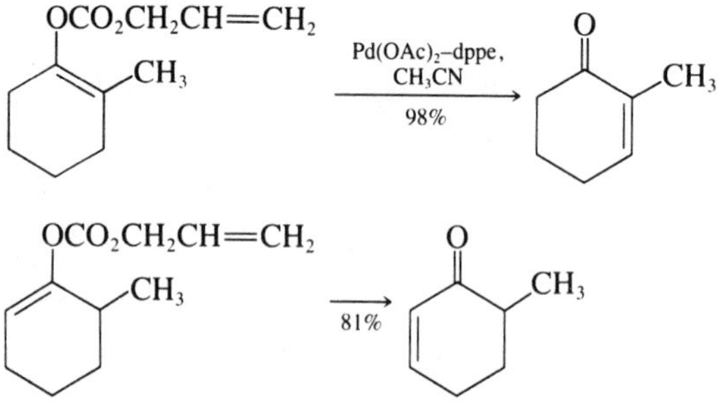

[1] J. Tsuji, I. Minami, and I. Shimizu, *Tetrahedron Letters*, **24**, 1793 (1983).
[2] I. Shimizu, I. Minami, and J. Tsuji, *ibid.*, **24**, 1797 (1983).

B-Allyldiisopinocampheylborane (1).

Preparation from (+)-α-pinene:

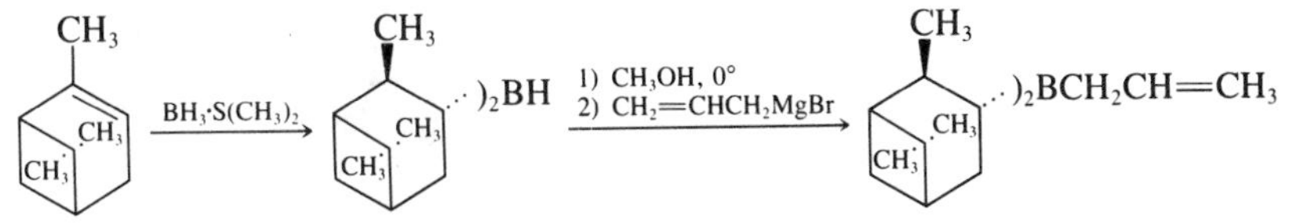

(91.3% ee)   (98.9% ee)   1 (98.9% ee)

Chiral homoallylic alcohols.[1]   The borane **1** reacts with aldehydes in ether at −78°
to furnish, after oxidation, chiral secondary homoallylic alcohols (**2**) with enantiomeric
excesses of 83–96%. The addition occurs in all six cases examined in the same stereochemical sense.

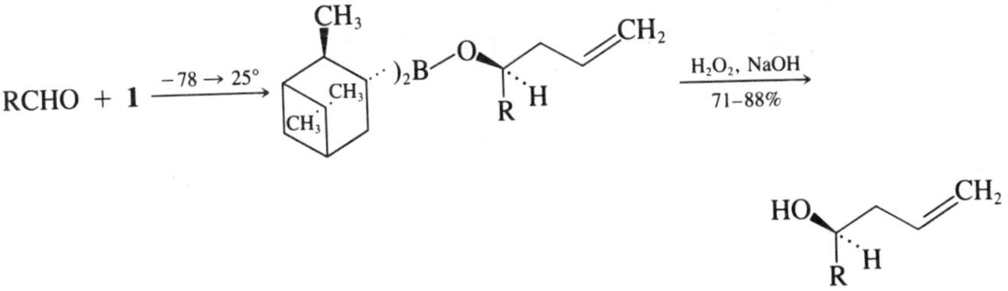

2 (83–96% ee)

B-3,3-Dimethylallyldiisopinocampheylborane (2).[2]   This related borane is prepared
by hydroboration of 3-methyl-1,2-butadiene with either (+)- or (−)-diisopinocampheyl-
borane. It reacts with aldehydes to give, after oxidation, products **3** with an irregular

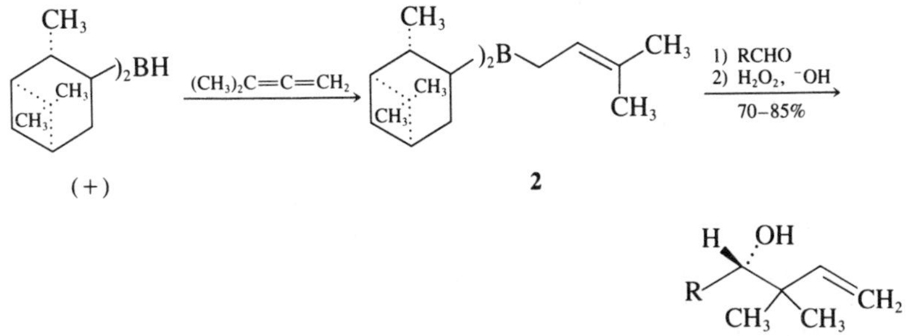

(S)-**3** (89–96% ee)

union of isoprene units (equation I). The reaction provides an asymmetric synthesis of rare terpenes of this type, such as artemesia alcohol (**4**), in both enantiomeric forms.

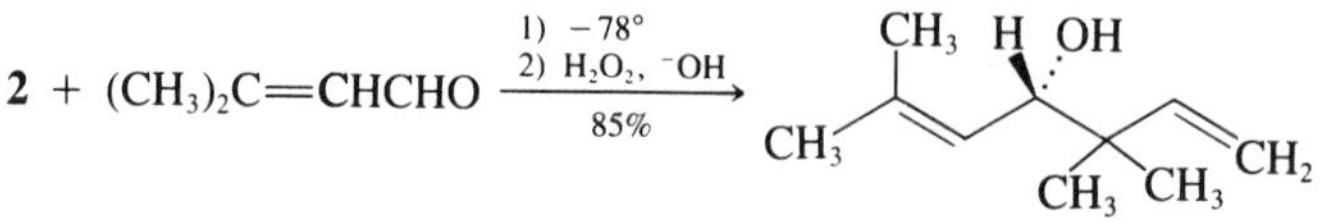

$$2 + (CH_3)_2C{=}CHCHO \xrightarrow[\substack{85\%}]{\substack{1)\ -78° \\ 2)\ H_2O_2,\ ^-OH}}$$

(−)-**4** (96% ee)

B-Allyldiisocaranylborane (**3**).   This compound, prepared from (+)-3-carene, undergoes even more enantioselective allylboration than does **1** to furnish chiral homoallylic alcohols in 86–99% ee (six examples).[3]

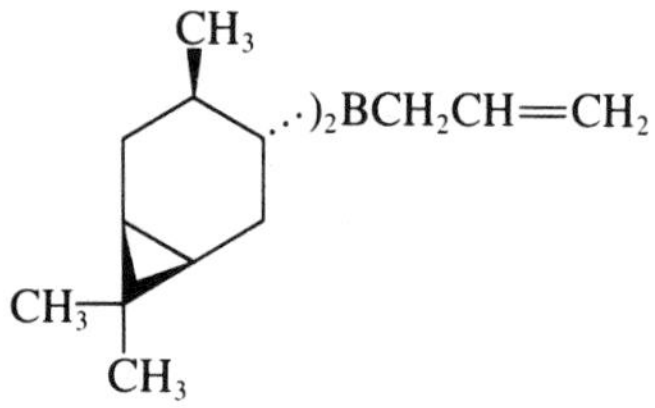

3

B-Methallyldiisopinocampheylborane (**4**).   This reagent (**4**), prepared by reaction of (+)-B-methoxydiisopinocampheylborane with methallyllithium, reacts with aldehydes to give, after oxidation, 2-methyl-1-alkene-4-ols (**5**) with >90% ee.

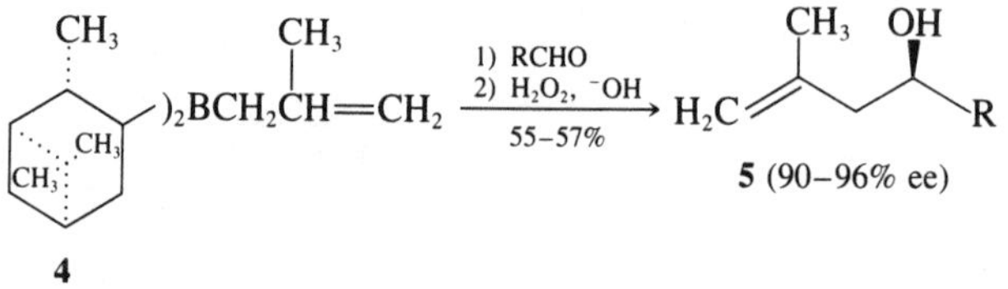

5 (90–96% ee)

4

[1] H. C. Brown and P. K. Jadhav, *Am. Soc.*, **105**, 2092 (1983).
[2] *Idem, Tetrahedron Letters*, **25**, 1215 (1984).
[3] *Idem, J. Org.*, **49**, 4089 (1984).
[4] H. C. Brown, P. K. Jadhav, and P. T. Perumal, *Tetrahedron Letters*, **25**, 5111 (1984).

Allyldiphenylphosphine (1).

Preparation:[1]

$$CH_2\!\!=\!\!CHCH_2OC_6H_5 + (C_6H_5)_2PH \xrightarrow[78\%]{\substack{n\text{-BuLi,}\\ \text{THF, }\Delta}} (C_6H_5)_2PCH_2CH\!\!=\!\!CH_2$$

1, b.p. 145°/0.9mm.

(E)- or (Z)-1,3-Dienes. Lithiation (*t*-BuLi) of **1** followed by addition of 1 equiv. of titanium(IV) isopropoxide results in a reagent that reacts with aldehydes to form an intermediate β-oxidophosphine, which is not isolated, but is converted into a (Z)-1,3-diene by alkylation with methyl iodide (equation I).[2]

(I)   **1** $\xrightarrow{\substack{1)\ t\text{-BuLi, THF}\\ 2)\ \text{Ti(O-}i\text{-Pr)}_4\\ 3)\ \text{RCHO}}}$

(E/Z ≈ 5:95)

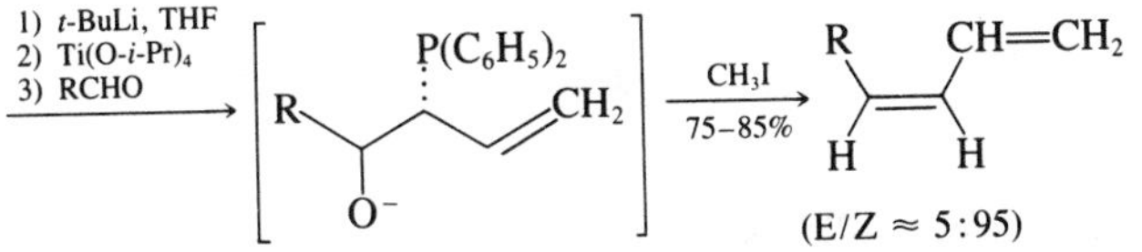

In contrast, the lithio derivative of allyldiphenylphosphine oxide (**2**)[3] reacts with aldehydes in the presence of HMPT to give (E)-1,3 dienes (equation II).

(II) $(C_6H_5)_2\overset{\displaystyle O}{\overset{\|}{P}}CH_2CH\!\!=\!\!CH_2$ $\xrightarrow[80-90\%]{\substack{1)\ n\text{-BuLi, THF, HMPT}\\ 2)\ \text{RCHO}}}$

2

(E/Z ≈ 95:5)

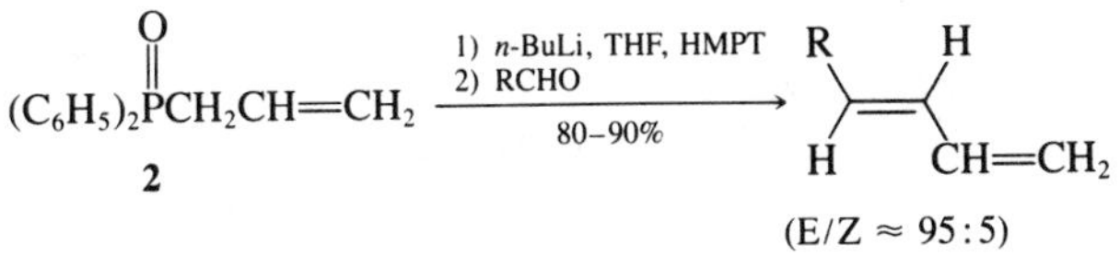

[1] F. G. Mann and M. J. Pragnell, *J. Chem. Soc.*, 4120 (1965).
[2] J. Ukai, Y. Ikeda, N. Ikeda, and H. Yamamoto, *Tetrahedron Letters*, **24**, 4029 (1983).
[3] Prepared by reaction of allyl alcohol with chlorodiphenylphosphine oxide and pyridine.

Allylidenecyclopropane (1), oxygen sensitive.

Preparation:

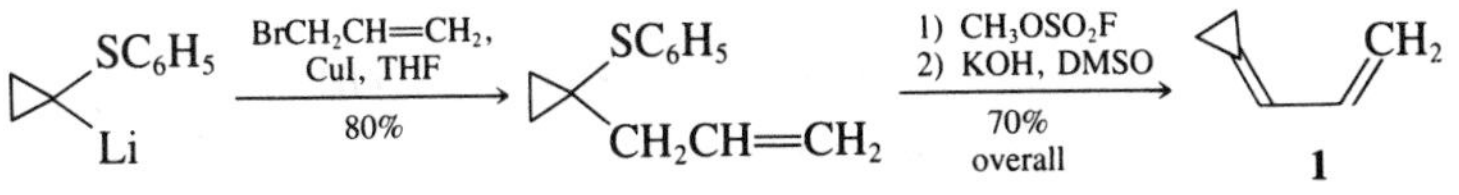

Diels-Alder reactions.[1] **1** is considerably more reactive in Diels-Alder reactions than the corresponding dimethyl-1,3-diene. It reacts readily with activated dienophiles (*e.g.*, maleic anhydride). It also reacts regioselectively with monosubstituted dienophiles to give the so-called "meta" adduct.

Example:

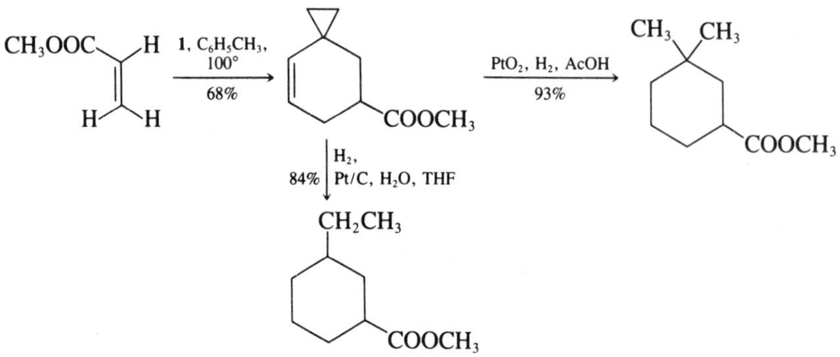

[1] Z. Zutterman and A. Krief, *J. Org.*, **48**, 1135 (1983).

Allyl phenyl selenide–Trialkylborane.

Homoallylic alcohols.[1]   The anion (LDA) of allyl phenyl selenide reacts with a trialkylborane in THF at −78° to form an ate complex (**a**) which rearranges at 0° to the complex **b**. An allylic rearrangement of **b** to **c** occurs at 25°. Reaction of **b** with an

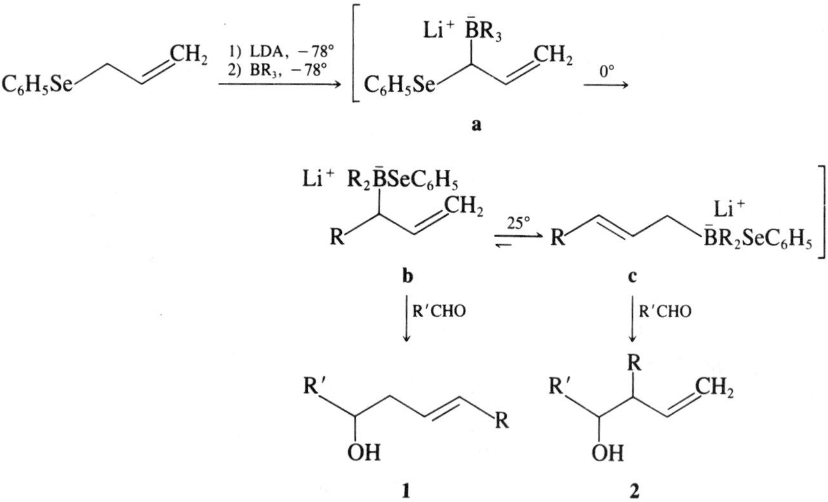

aldehyde results mainly in a linear homoallylic alcohol (**1**), predominantly with the (E)-configuration. Reaction of **c** with an aldehyde results in a branched homoallylic alcohol (**2**), with an *anti/syn* ratio of ~3:1.

[1] Y. Yamamoto, Y. Saito, and K. Maruyama, *J. Org.*, **48**, 5408 (1983).

Allyltri-*n*-butyltin (1), 11, 15–16.

Addition to α-hydroxy aldehydes.[1] The Lewis acid-catalyzed addition of **1** to aldehydes to afford homoallylic alcohols (**9, 8**) has been extended to the reaction with derivatives of a chiral α-hydroxy aldehyde (**2**), which can result in the monoderivative of a *syn*-diol (**3**) and/or an *anti*-diol (**4**). The diastereoselectivity can be controlled by the

2a, R = CH₂C₆H₅	MgBr₂	**3a**	250:1	**4a**	
	BF₃·O(C₂H₅)₂		39:61		
2b, R = SiMe₂-*t*-Bu	BF₃·O(C₂H₅)₂	**3b**	5:95	**4b**	
	MgBr₂		21:79		

choice of the Lewis acid and of the protecting group. MgBr₂, TiCl₄, and ZnI₂, all of which permit bidentate chelation, favor *syn*-selectivity. The effect is most marked with the benzyloxy derivative **2a**. In contrast, BF₃ etherate favors formation of *anti*-adducts; this tendency is most marked with the *t*-butyldimethylsilyl ether **2b**. The difference between the two protecting groups is ascribed to electronic factors rather than steric effects. In any case, by suitable choice of the protecting group and the catalyst, high stereoselectivity for *syn*- or *anti*-products is possible.

Addition of crotyltri-*n*-butyltin (**5**; **11**, 143) to chiral α-alkoxy aldehydes (**6**) presents a more complicated situation, since four products are possible.[2] Products **7** and **8** result from "chelation-controlled" diastereofacial selectivity; products **9** and **10** are products of Cram-Felkin control. In the reaction catalyzed by BF₃ etherate the major products are **7** and **9** in the ratio 67:33. Use of TiCl₄ or MgBr₂ results in formation of only **7** and **8**. With the former catalyst the **7/8** ratio is 63:37; with the latter, 92.5:7.5. The almost exclusive formation of **7** is consistent with the known *syn*-stereoselectivity in the reaction of **5** with achiral aldehydes.

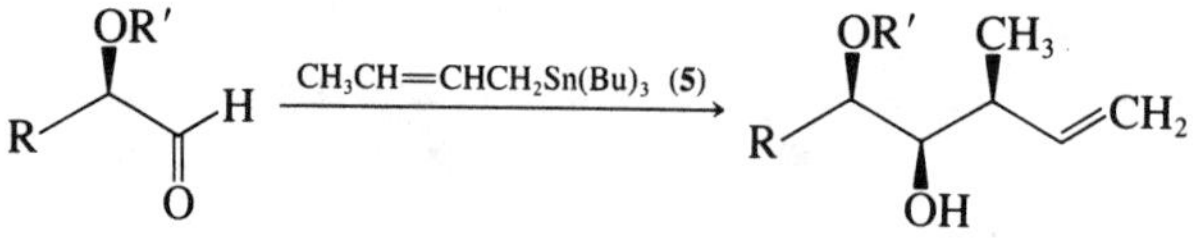

6, R' = CH₂C₆H₅

The sterochemistry of additions of allyltri-*n*-butyltin (**1**) and crotyltri-*n*-butyltin (**5**) with β-alkoxy aldehydes is more complex.[3] In the allyltin reactions, useful diastereofacial selectivity is observed only with $TiCl_4$ (4.6:1) or $SnCl_4$ (9:1). Optimum stereoselectivity in the reaction of the β-alkoxy aldehyde **11** with **5** is observed in the reaction catalyzed by BF₃ etherate, which results mainly in the product of Cram-Felkin control (equation I). The highest chelation-controlled selectivity in reactions of **5** with the β-benzyloxy aldehyde **12** is obtained with $MgBr_2$ (equation II).

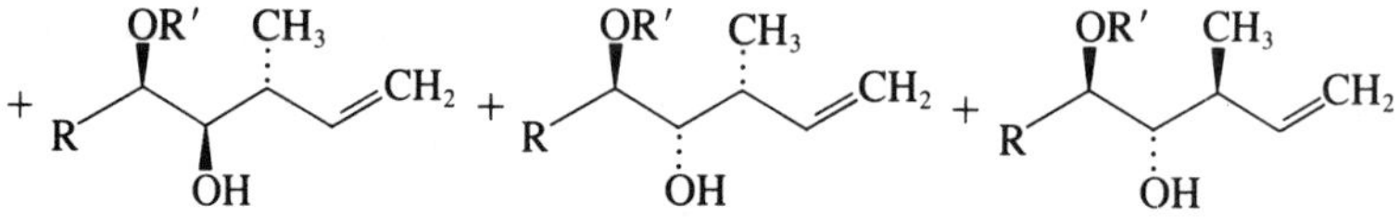

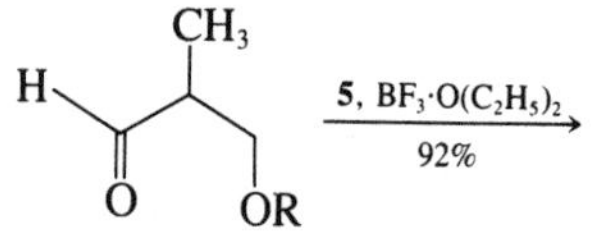

11 (R = Si(CH₃)₂-*t*-Bu)

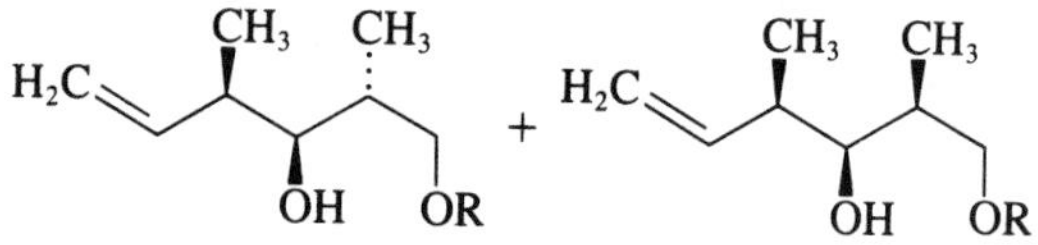

5:95

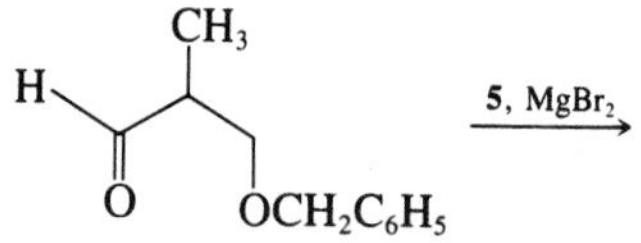

12

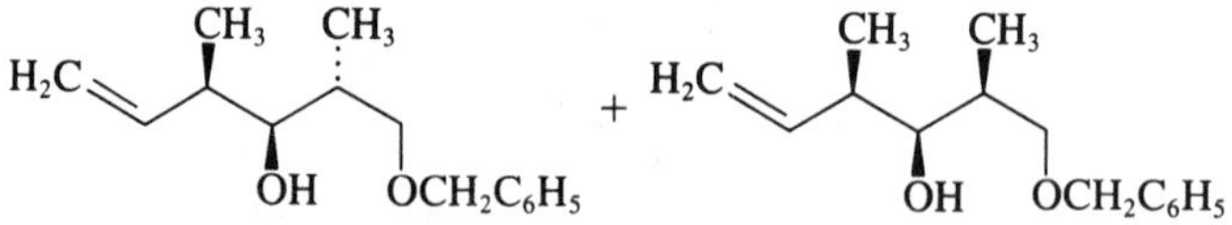

(81%)   (7%)

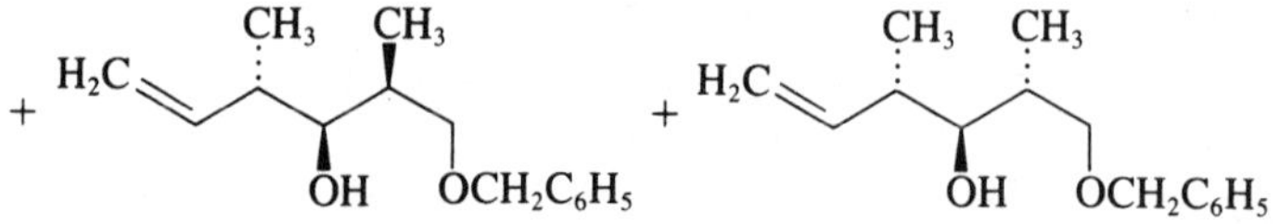

(10%)   (2%)

***C*-Glycosides.**[4] Thiophenyl glycosides undergo ready free-radical allylation with allyl- or methallyltri-*n*-butyltin. Thus the L-lyxose derivative **1** reacts with allyltri-*n*-butyltin under photochemical initiation to give **2** and **3** in the ratio 90:10. The reaction

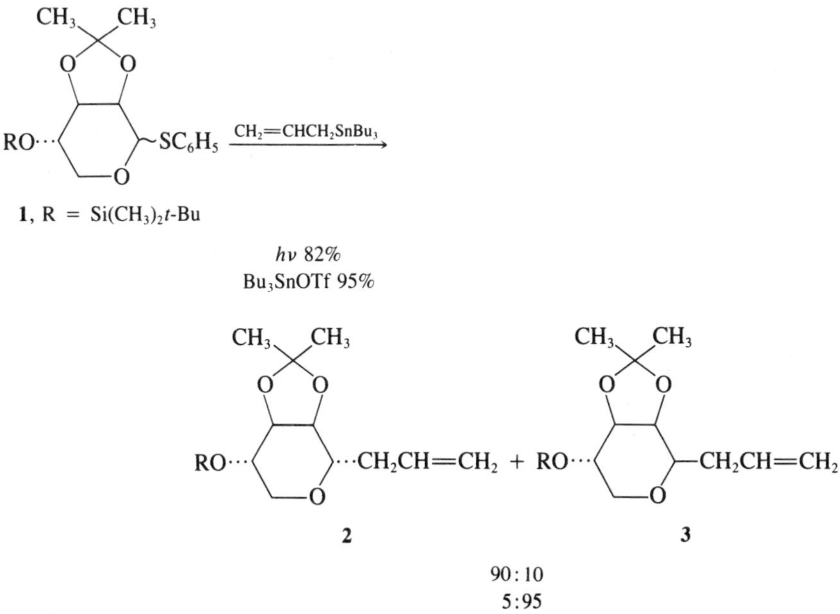

CH3 CH3
O O
RO···
SC6H5
O
CH2=CHCH2SnBu3
1, R = Si(CH3)2t-Bu
hν 82%
Bu3SnOTf 95%
···CH2CH=CH2 + RO···
—CH2CH=CH2
2
3
90:10
5:95

can be effected by catalysis with tri-*n*-butyltin triflate,[5] but in this case the β-anomer **3** is formed with high stereoselectivity. However, this marked difference in stereoselectivity between the two procedures is not general, but depends on steric factors and the protecting groups used.

[1] G. E. Keck and E. P. Boden, *Tetrahedron Letters*, **25**, 265 (1984).
[2] *Idem, ibid.*, 1879 (1984).
[3] G. E. Keck and D. E. Abbott, *ibid.*, 1883 (1984).
[4] G. E. Keck, E. J. Enholm, and D. F. Kachensky, *ibid.*, 1867 (1984).
[5] Prepared by reaction of tri-*n*-butyltin chloride with triflic acid.

Allyltrimethylsilane, 10, 6–8; **11**, 16–20.

Asymmetric addition to a chiral α-keto amide.[1] The Lewis acid-catalyzed addition of allyltrimethylsilane to a chiral α-keto amide (**1**) derived from methyl (S)-prolinate proceeds with good to high diastereoselectivity, probably because of chelation with the ester group of **1**. $SnBr_4$, $SnCl_4$, and $TiCl_4$ are the most effective catalysts.

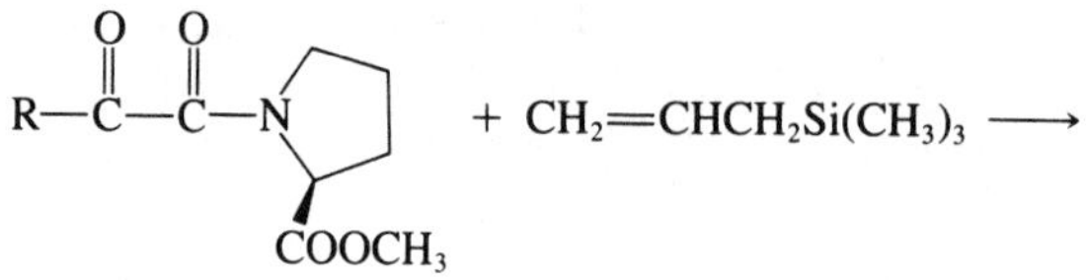

1 a) R = C$_6$H$_5$ SnBr$_4$ or SnCl$_4$
   b) R = CH$_3$ TiCl$_4$

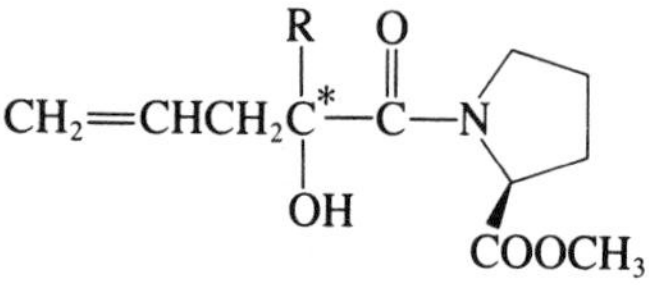

2

(87% de)
(56% de)

Methyl (E)-4,8-alkadienoates.[2]   γ-Alkenyl-γ-butyrolactones react with allyltrime-thylsilanes in the presence of trimethyloxonium tetrafluoroborate to afford methyl (E)-4,8-alkadienoates in 75–100% yield (equation I).

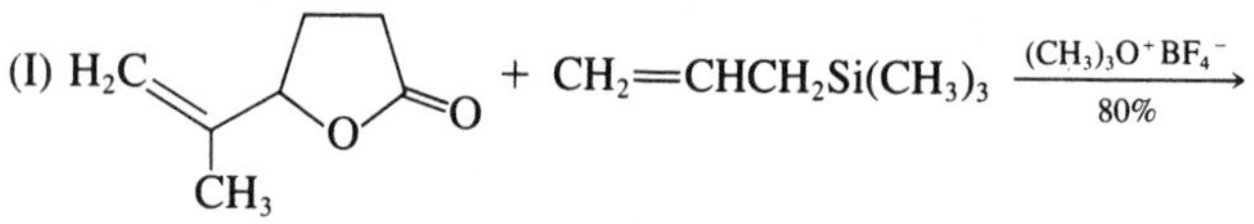

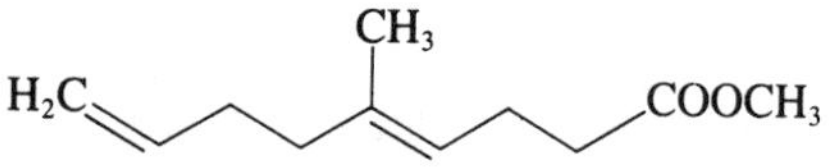

(I)

This regio- and stereoselective reaction was used to obtain the triunsaturated ester **1** in a short synthesis of the sesquiterpene β-sinensal (**2**).

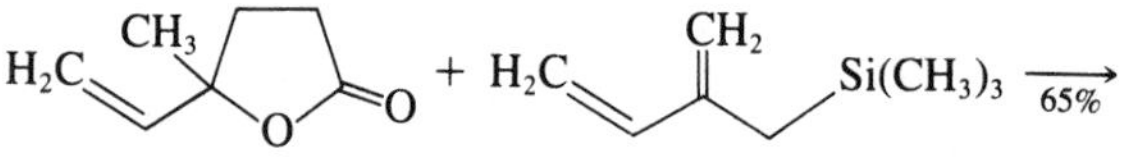

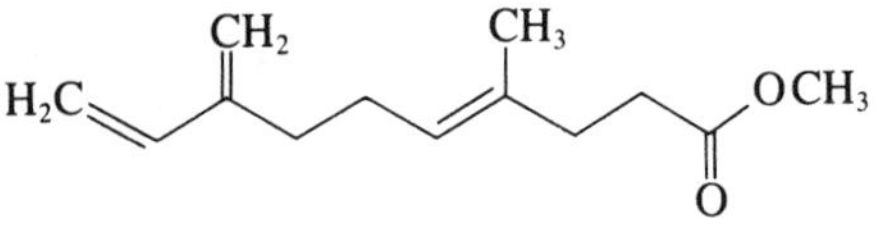

1

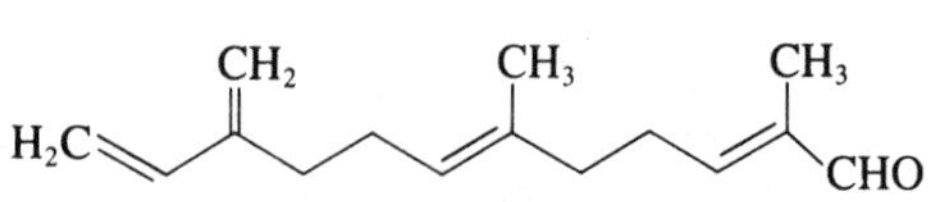

2

Conjugate addition to acyclic Michael acceptors.[3] Sakurai and Hosomi (**9**, 445–446) reported one example of the fluoride ion-catalyzed reaction of allyltrimethylsilane with an acyclic enone. In that case (reaction with $C_6H_5CH{=}CHCOCH_3$), 1,2- and 1,4-adducts are obtained in the ratio 2:1. 1,4-Addition is enhanced by use of DMF and HMPT as solvent and by increase in the size of the group adjacent to the carbonyl group. 1,4-Addition is the main or predominant reaction with α,β-unsaturated esters or nitriles. In this case, it is superior to or competitive with allylation with lithium diallylcuprate. Yields in 1,4-additions to α,β-enones can compare favorably with those obtained with reactions catalyzed by $TiCl_4$.

Examples:

$$C_6H_5CH{=}CHCOOCH_3 \xrightarrow[90\%]{(CH_3)_3SiCH_2CH{=}CH_2,\ Bu_4NF} C_6H_5\underset{\displaystyle CH_2CH{=}CH_2}{\underset{|}{C}}HCH_2COOCH_3$$

$$(CH_3)_3CCH{=}CHCN \xrightarrow[65\%]{} (CH_3)_3\underset{\displaystyle CH_2CH{=}CH_2}{\underset{|}{C}}HCH_2CN$$

Intramolecular conjugate additions.[4] This fluoride ion-catalyzed Michael addition is particularly useful for cyclization of suitably substituted allyltrimethylsilanes with a Michael acceptor. In the examples cited, BF_3 is not useful, because its use results in desilylation or isomerization.

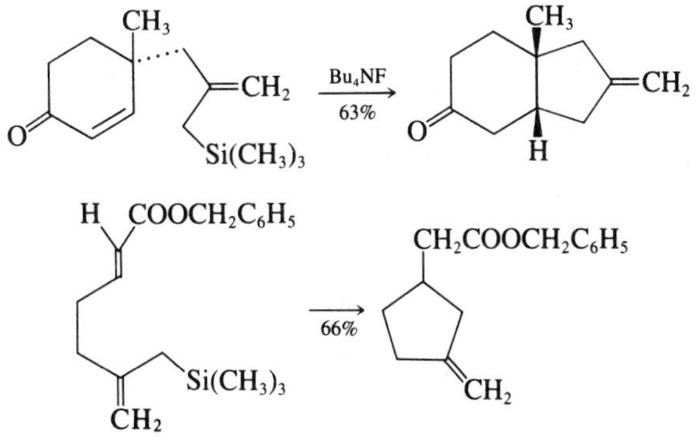

[1] K. Soai and M. Ishizaki, *J.C.S. Chem. Comm.*, 1016 (1984).
[2] T. Fujisawa, M. Kawashima, and S. Ando, *Tetrahedron Letters*, **25**, 3213 (1984).
[3] G. Majetich, A. M. Casares, D. Chapman, and M. Behnke, *ibid.*, **24**, 1909 (1983).
[4] G. Majetich, R. Desmond, and A. M. Casares, *ibid.*, **24**, 1913 (1983).

Aluminum chloride, 1, 24–34; **2**, 21–23; **3**, 7–9; **4**, 10–15; **5**, 10–13; **6**, 17–19; **7**, 7–9; **8**, 13–15; **9**, 11–13; **10**, 9–11; **11**, 25–28.

Intramolecular acylation of alkylsilanes.[1]   Cyclopentanones can be prepared by ring closure of 5-(trimethylsilyl)alkanoyl chlorides mediated by $AlCl_3$. The starting materials are readily available from alkylation of the dianion of a carboxylic acid with a 3-(trimethylsilyl)alkyl bromide or iodide (equation I).

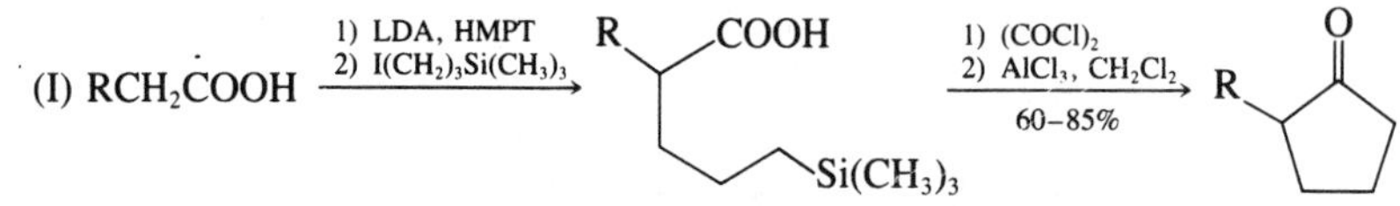

(I) RCH_2COOH

$\xrightarrow[\text{2) }I(CH_2)_3Si(CH_3)_3]{\text{1) LDA, HMPT}}$

$\xrightarrow[\text{60–85%}]{\substack{\text{1) }(COCl)_2 \\ \text{2) }AlCl_3,\ CH_2Cl_2}}$

1,4-Diene-3-ones.[2]   Acylation of vinyltrimethylsilane with α,β-unsaturated acyl chlorides catalyzed by $AlCl_3$ and conducted at low temperatures can lead to 1,4-diene-3-ones. Higher temperatures favor Nazarov cyclization to cyclopentenones. CH_2Cl_2, $CHCl_3$, and $ClCH_2CH_2Cl$ are suitable solvents.

Example:

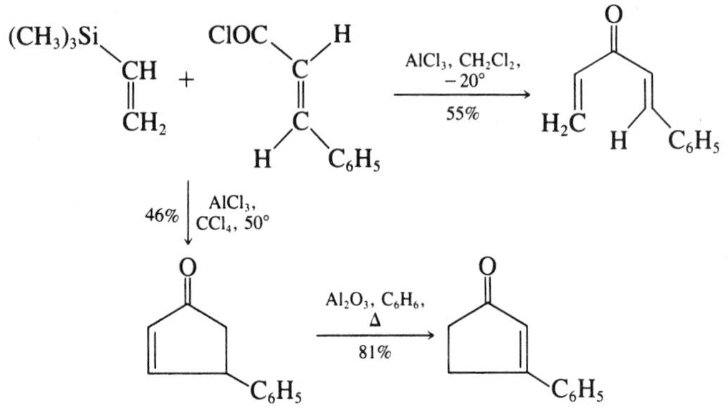

Cyclopentanones.[3]   Substituted cyclopentanones are formed on treatment of 1-acyl-1-(alkylthio)cyclobutanes with $AlCl_3$, $AlBr_3$, or $FeCl_3$. Yields range from 55 to 90%. A typical synthesis is shown in equation (I).

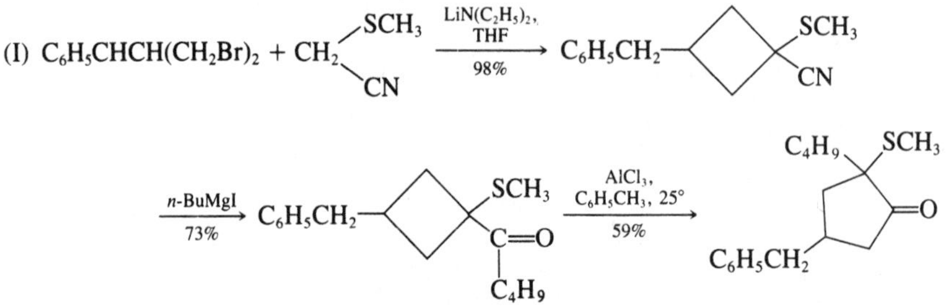

The cyclobutane derivative **1** is converted under the same conditions to the cyclopentenone **2**.

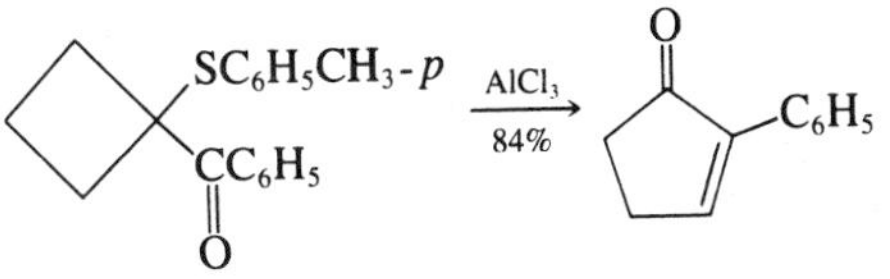

1   **2**

Cyclization of allylic sulfones.[4]   In the presence of aluminum chloride, allylic sulfones can undergo intramolecular cyclization of the Friedel-Crafts type with displacement of the sulfone group. Examples are the cyclization of **1** and **4** to **2** and **5**, respectively.

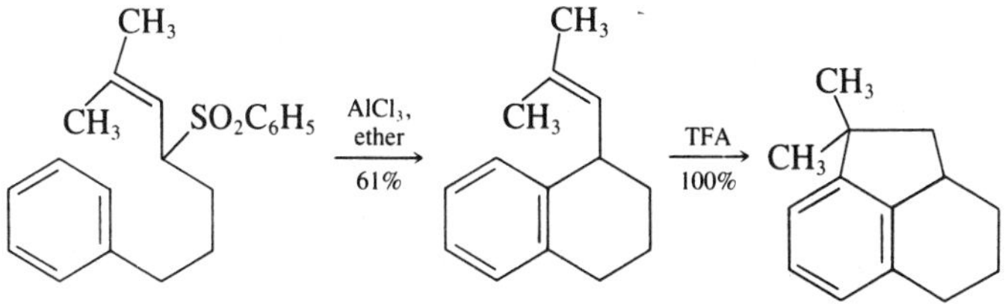

1   **2**   **3**

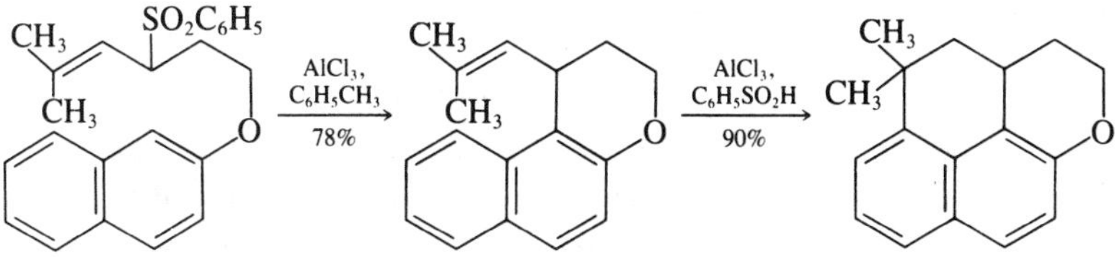

4   **5**   **6**

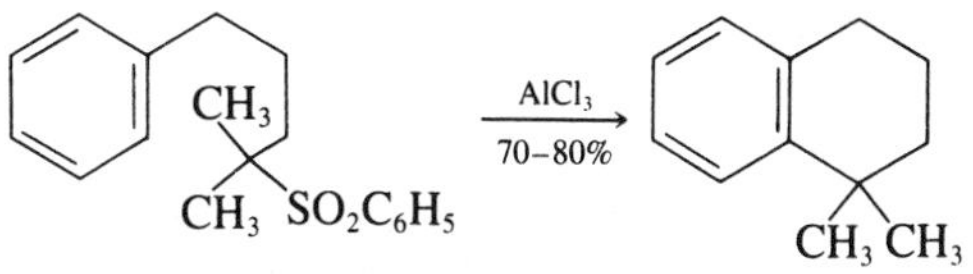

7   **8**

Rearrangement of **1**-*alkenylsilanes.*[5]   Alkenyl(chloromethyl)dimethylsilanes, readily available from hydrosilylation of alkynes with (chloromethyl)dimethylsilane, rearrange in the presence of aluminum chloride to allyl- and/or cyclopropylsilanes. (Z)-1-Alkenylsilanes rearrange exclusively to (Z)-allylsilanes (equation I). (E)-1-Alkenylsilanes rearrange to (E)-allylsilanes as the major product (equation II). Cyclopropylsilanes are formed as the major product from α-substituted vinylsilanes. Hydrogen peroxide oxidation of the rearranged products provides a stereoselective synthesis of (Z)- or (E)-allylic alcohols.

(I)

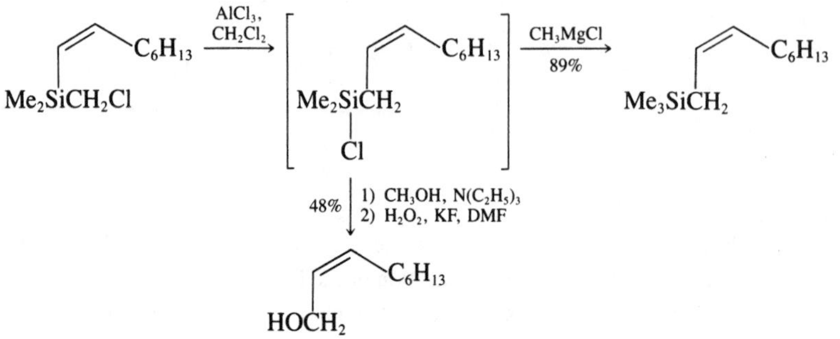

(II)

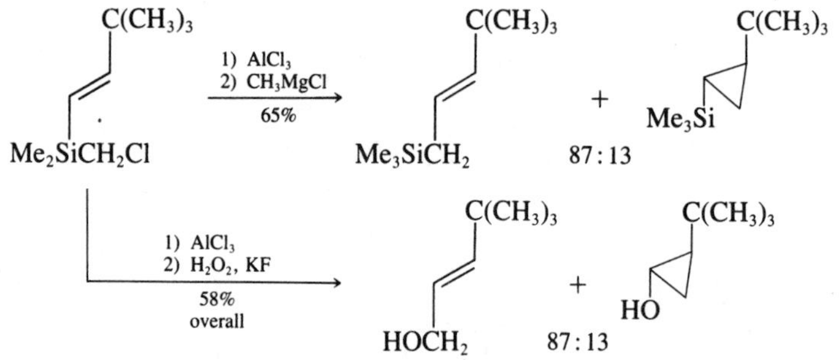

Crotylaluminum reagent.[6] The reaction of aldehydes with various crotylmetal reagents gives products in which the allylic group is attached at the more highly substituted position (**10**, 451). This regioselectivity can be reversed in reactions of crotylmagnesium chloride (**1**) or of crotyltributyltin[7] by addition of aluminum chloride, which results in predominant formation of the α-adduct. $C_2H_5AlCl_2$ and BF_3 exhibit similar but lower regioselectivity.

Examples:

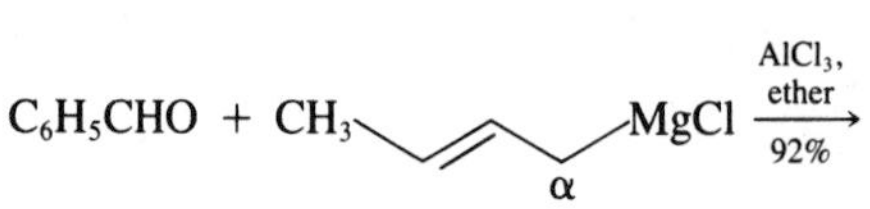

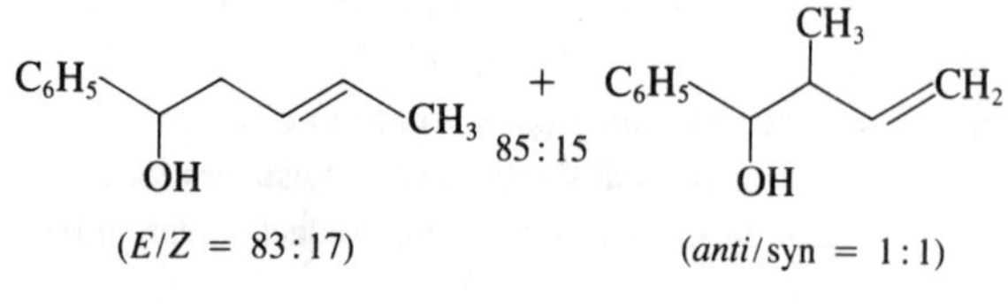

$$RCHO + CH_3\diagdown\diagup\diagdown\diagup\diagdown Sn(C_4H_9)_3 \xrightarrow[70-80\%]{\substack{AlCl_3, \\ (CH_3)_2CHOH}}$$

$$\sim 60-90 : 40-10$$

[1] H. Urabe and I. Kuwajima, *J. Org.*, **49**, 1140 (1984); *idem, Org. Syn.*, submitted (1984).

[2] G. Kjeldsen, J. S. Knudsen, L. S. Ravn-Petersen, and K. B. G. Torssell, *Tetrahedron*, **39**, 2237 (1983).

[3] M. Yamashita, J. Onozuka, G. Tsuchihashi, and K. Ogura, *Tetrahedron Letters*, **24**, 79 (1983).

[4] B. M. Trost and M. R. Ghadiri, *Am. Soc.*, **106**, 7260 (1984).

[5] K. Tamao, T. Nakajima, and M. Kumada, *Organometallics*, **3**, 1655 (1984).

[6] Y. Yamamoto and K. Maruyama, *J. Org.*, **48**, 1564 (1983).

[7] Y. Yamamoto, N. Maeda, and K. Maruyama, *J.C.S. Chem. Comm.*, 742 (1983).

Aluminum chloride–Ethanethiol, 9, 13; **10**, 11; **11**, 28.

Dehalogenation. This combination effects efficient dehalogenation of *o*- and *p*-bromo- and iodophenols in CH_2Cl_2 at 0°. The corresponding chlorides and fluorides are stable to these conditions. Simultaneous dealkylation can occur when the reaction is applied to halophenolic ethers.[1]

Dehalogenation of α-halo ketones.[2]   This combination converts α-halo ketones into dithioketals of the parent ketone. Although any halo ketone undergoes this reaction, yields are generally higher with α-bromo or α-iodo ketones.

[1] M. Node, T. Kawabata, K. Ohta, K. Watanabe, K. Fuji, and E. Fujita, *Chem. Pharm. Bull. Japan*, **31**, 749 (1983).

[2] K. Fuji, M. Node, T. Kawabata, and M. Fujimoto, *Chem. Letters*, 1153 (1984).

Aluminum chloride–Sodium iodide.

Demethylation. AlCl_3–NaI (1:1) in acetonitrile selectively demethylates aliphatic methyl ethers at room temperature in the presence of aromatic methyl ethers. Acetals are also cleaved, but esters and lactones are not affected. The methyl ether of a primary alcohol is cleaved somewhat more readily than the methyl ether of a secondary alcohol.[1]

2-Substituted tetrahydrofuranes are cleaved selectively by this system at the less hindered carbon atom to afford γ-iodo alcohols (equation I).[2]

$$(I)$$

$$(R = H, C_6H_5)$$

[1] M. Node, K. Ohta, T. Kajimoto, K. Nishide, E. Fujita, and K. Fuji, *Chem. Pharm. Bull. Japan,* **31**, 4178 (1983).

[2] M. Node, T. Kajimoto, K. Nishide, E. Fujita, and K. Fuji, *Tetrahedron Letters,* **25**, 219 (1984).

Aluminum iodide, AlI_3, **1**, 35. The reagent is prepared by reaction of aluminum foil (excess) with iodine in refluxing CS_2 or CH_3CN.

Cleavage of ethers.[1] AlI_3 cleaves alkyl phenyl ethers in refluxing CH_3CN or CS_2 selectively to phenols and alkyl iodides. The regioselectivity is the same as that shown by $AlCl_3$, but the reagent is more reactive. Cleavage of dialkyl ethers requires elevated temperatures and long reaction times.

[1] M. V. Bhatt and J. R. Babu, *Tetrahedron Letters,* **25**, 3497 (1984).

(S)-1-Amino-2-methoxymethyl-1-pyrrolidine (SAMP), **8**, 16–17; **9**, 17. Supplier of SAMP and RAMP: Merck-Schuchardt. An improved synthesis of SAMP (**1**) utilizes the Hofmann reaction to effect N-amination (equation I).[1]

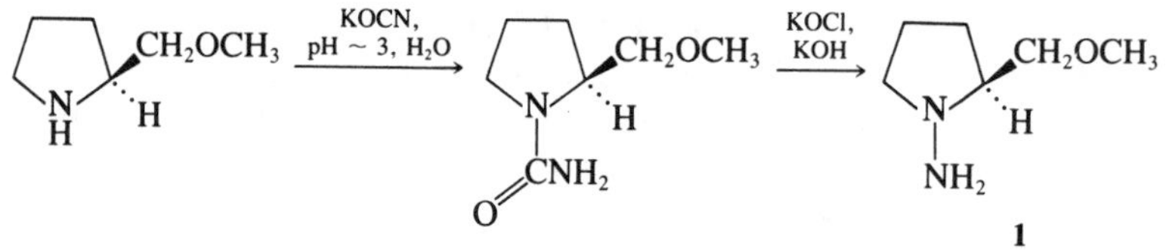

The report includes details for asymmetric α-alkylation of a symmetrical ketone (97% ee) via the SAMP hydrazone and references to other stereoselective reactions of SAMP (or RAMP) hydrazones. Under standard conditions (LDA, 0°), deprotonation takes place regioselectively at the less substituted α-position with uniform diastereoface differentiation.

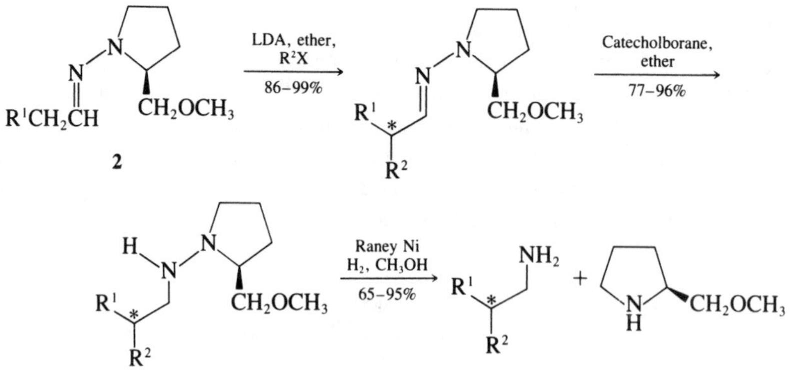

3 (>95% ee)

Enantioselective synthesis of primary amines.[2]   Chiral β-substituted primary amines (**3**) can be prepared in 95% ee from SAMP hydrazones (**2**) of primary aldehydes (equation I).

Asymmetric Michael additions.[3]   The anions of hydrazones (**2**), obtained from **1** and methyl ketones, undergo conjugate addition to α,β-unsaturated esters with virtually complete 1,6-asymmetric induction to give the adducts (S,R)-**3**. Ozonolysis converts these products into β-substituted δ-keto esters, (R)-**4**, obtained in optical yields of 96–100%, with recovery of the chiral auxiliary as the nitrosamine (S)-**5**. The overall chemical yields of **4** are 45–62%.

[1] D. Enders, P. Fey, and H. Kipphardt, *Org. Syn.*, submitted (1984).
[2] D. Enders and H. Schubert, *Angew. Chem., Int. Ed.*, **23**, 365 (1984).
[3] D. Enders and K. Papadopoulos, *Tetrahedron Letters*, **24**, 4967 (1983).

(S)-(−)-2-Amino-3-methyl-1,1-diphenylbutane-1-ol, $H{\cdots}\overset{\displaystyle CH(CH_3)_2}{\underset{\displaystyle H_2N}{C}}\!\!-\!\!\overset{}{\underset{\displaystyle OH}{C}}(C_6H_5)_2$. M.p. 94–95°,

$\alpha_D - 127.7°$. The alcohol is prepared by reaction of (S)-valine methyl ester hydrochloride with phenylmagnesium bromide.[1]

Asymmetric reduction of ketones.[2]   The reagent prepared from **1** and diborane reduces alkyl phenyl ketones to (R)-benzyl alcohols in 94–100% ee. Aliphatic methyl ketones can also be reduced to (R)-secondary alcohols in 55–78% ee. The enantioselectivity increases with the steric bulk of the alkyl group; it is highest with a *t*-alkyl group.

[1] S. Itsuno, K. Ito, A. Hirao, and S. Nakahama, *J. Org.*, **49**, 555 (1984).
[2] *Idem, J.C.S. Chem. Comm.*, 469 (1983).

(1S,2S)-2-Amino-1-phenyl-1,3-propanediol (1), **6**, 386. M.p. 109–113°, α_D +25.7°. Supplier: Aldrich.

(S)-1-Alkyl-1,2,3,4-tetrahydroisoquinolines.[1]   The bis(trimethylsilyl) ether (**2**) of **1** has been used as a chiral auxiliary to effect alkylation of tetrahydroisoquinoline (equation I) with high enantioselectivity. The chiral propanediol is much more effective than (R)-(−)-α-phenethylamine, and moreover leads consistently to (S)-1-alkyl derivatives.

(I)

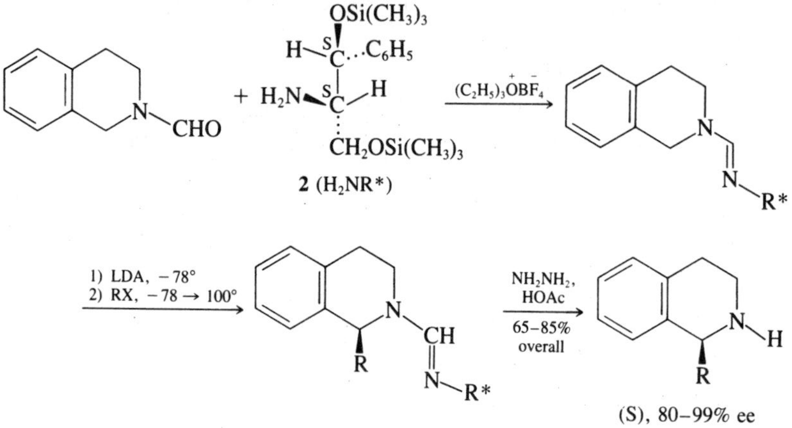

[1] A. I. Meyers and L. M. Fuentes, *Am. Soc.*, **105**, 117 (1983).

(S)-(−)-1-Amino-2-(silyloxymethyl)pyrrolidines (SASP),

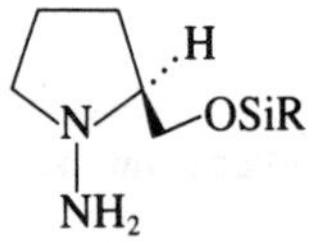

1a, R = *t*-BuMe$_2$
1b, R = *t*-BuPh$_2$

The enantiomerically pure hydrazines are prepared in four steps from (S)-proline.

   Resolution of α-substituted aldehydes.[1]   The SASP hydrazones of α-substituted aldehydes can be resolved by high-performance liquid chromatography. The separability factors are sufficient for analytical and preparative purposes. The (S,S)-isomer elutes consistently before the (S,R)-isomer. Both isomers can be cleaved to the enantiomerically pure aldehydes by ozonolysis or acid hydrolysis, with resolution yields of 35–70%.

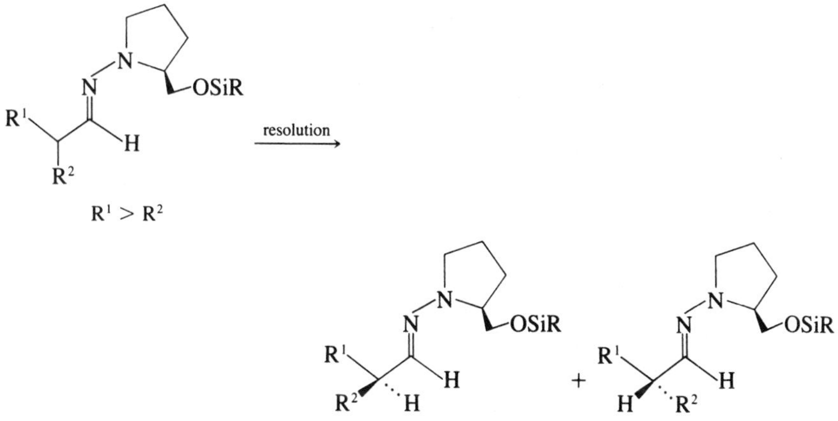

[1] D. Enders and W. Mies, *J.C.S. Chem. Comm.*, 1221 (1984).

Ammonium persulfate, 1, 952–954; **2**, 348; **3**, 238–239; **5**, 15–16; **6**, 20–21.

vic-*Diacetates*.[1] Reaction of mono- and disubstituted alkenes with ammonium per-sulfate (4 equiv.) in acetic acid catalyzed by $FeSO_4$ provides *trans*,*vic*-diacetates as the major products. *cis*-Alkenes can undergo partial isomerization to the *trans*-isomer. Oxidation of cyclohexene results in the *trans*-1,2-diacetate (79% yield).

[1] W. E. Fristad and J. R. Peterson, *Tetrahedron Letters*, **24**, 4547 (1983).

(S)-4-Anilino-3-methylamino-1-butanol (1), 11, 292–293.

Asymmetric reduction of α,β-*enones*.[1] Prochiral cyclic and acyclic α,β-enones are reduced by lithium aluminum hydride complexed with **1** to (S)-allylic alcohols in optical yields of 30–100% (equation I).

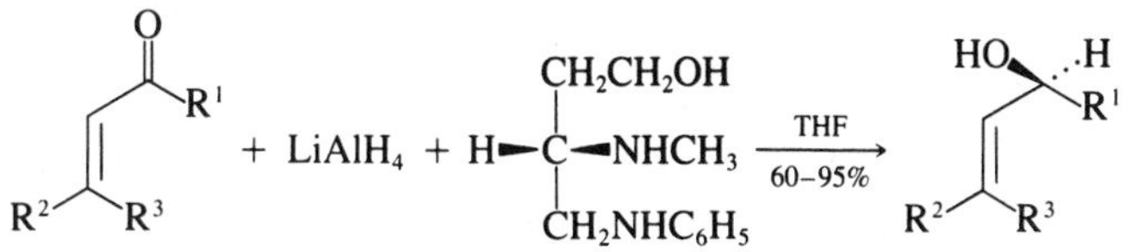

Surprisingly, when the phenyl group of **1** is replaced by a 2,6-xylyl group, the same reduction results in an (R)-allylic alcohol, but usually in lower optical yield.

[1] T. Sato, Y. Gotoh, Y. Wakabayashi, and T. Fujisawa, *Tetrahedron Letters*, **24**, 4123 (1983).

Arene(tricarbonyl)chromium complexes, 6, 27–28; **7**, 71–72; **8**, 21–22; **9**, 21; **10**, 13–14; **11**, 131–132.

Michael addition to a complexed styrene.[1]   A key step in a synthesis of 11-deoxy-anthracyclinone (**3**) involves the regioselective reaction of **1** with the carbanion of a protected acetaldehyde cyanohydrin to give **2** by stereospecific *exo*-addition.

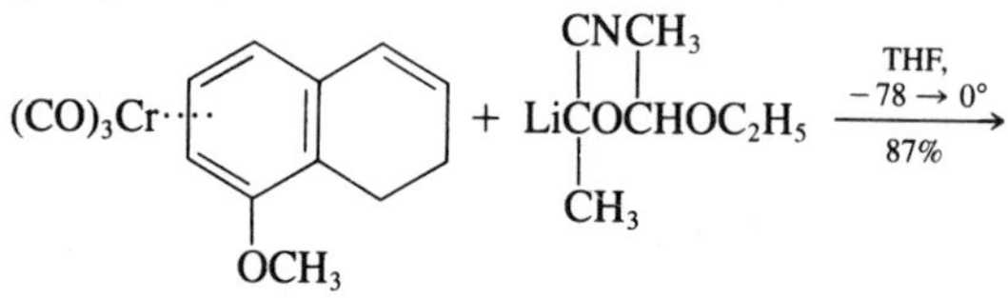

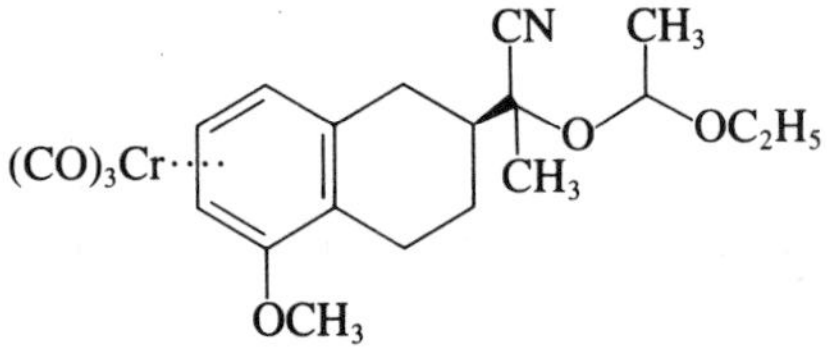

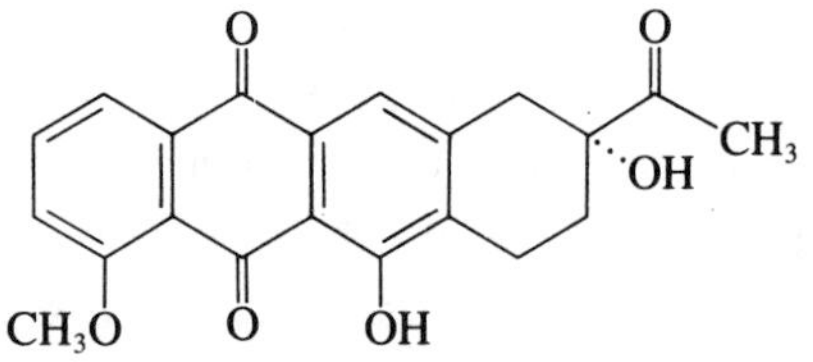

Benzylic functionalization.   The reaction of (toluene)Cr(CO)₃ (**1**) with KOC(CH₃)₃ and ethyl oxalate in DMSO results in the stable enol **2** in 78% yield. Similar products are formed from the reactions of (*m*-xylene)Cr(CO)₃ and of (indane)Cr(CO)₃.[2]

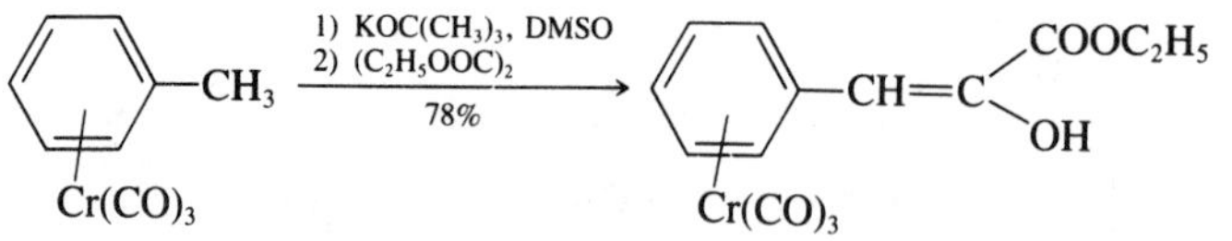

Alkylation of benzyl alkyl ethers.[3]   The anion of a benzyl alkyl ether is unstable and rearranges to the corresponding alkoxide (Wittig rearrangement). However, coordination of the ether to Cr(CO)₃ stabilizes the α-anion and permits alkylation to give an α-substituted benzyl ether. Complexation also stabilizes the anion of a benzyl alkyl sulfide. The products undergo decomplexation on exposure of ether solutions to air and sunlight (overall yields 65–80%).

Example:

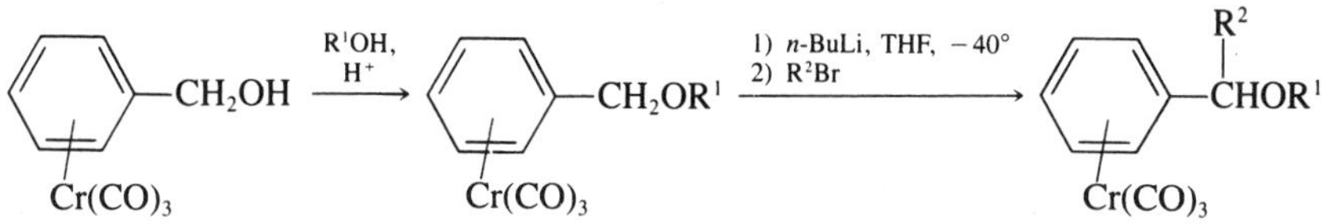

Alkylation of methoxyalkylbenzenes. Chromium tricarbonyl complexes of methoxyalkylbenzenes can undergo regiospecific alkylation. Thus the complex **1**, although it contains two benzylic sites, on deprotonation in the presence of formaldehyde undergoes attack only at the position *meta* to the methoxy group.[4]

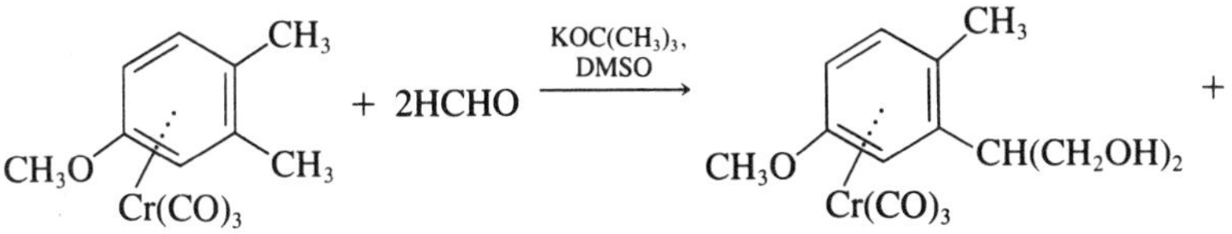

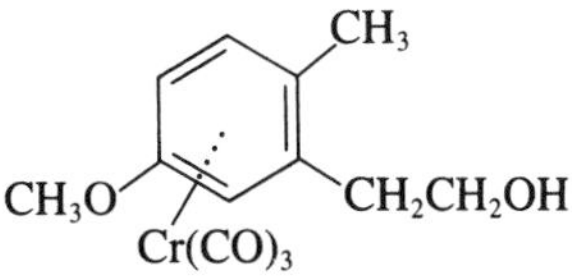

Diastereomeric complexes (**4**) of estradiol derivatives can be separated by chromatography. These differ in the location of the metal, which may be on the α- or β-side of the ring system. Both complexes undergo regiospecific alkylation at the *meta* C_6-position, but with opposite stereospecificities. Decomplexation by exposure to sunlight and air furnishes 6-substituted estradiol derivatives.

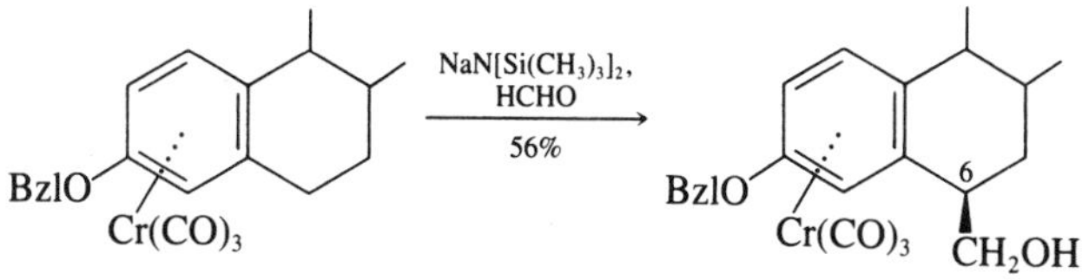

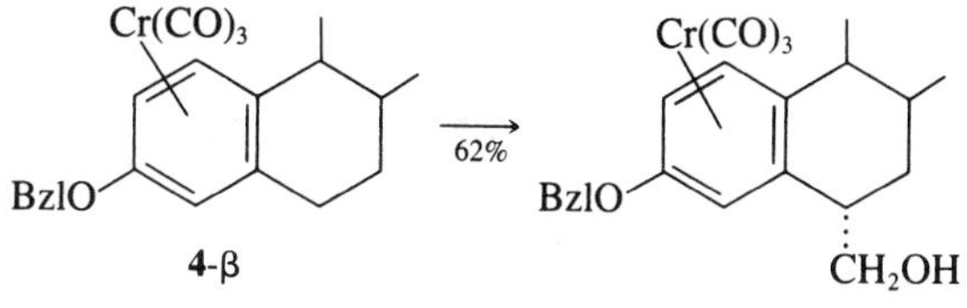

6

Stereoselective hydrogenation of **1,3-dienes.**   Hydrogenation of simple acyclic and cyclic 1,3-dienes catalyzed by (arene)Cr(CO)₃ complexes results in highly stereoselective 1,4-addition of hydrogen to produce (Z)-monoenes. Under these conditions 1,4-dienes are isomerized to 1,3-dienes and then reduced to (Z)-monoenes, but 1,5-dienes are not reduced.[5]

This reduction is useful for stereoselective preparation of exocyclic alkenes. Thus the alkaloid deplancheine (**2**) is obtained in essentially quantitative yield by hydrogenation of the diene **1** catalyzed by (toluene)Cr(CO)₃.[6]

1   **2**

This hydrogenation is particularly useful for stereoselective synthesis of the (E)-C₅-exocyclic trisubstituted double bond of carbacyclins.[7] Thus hydrogenation of the 1,3-diene **3** catalyzed by (methyl benzoate)Cr(CO)₃ can result in the desired product **4** in yields as

3 (E/Z = 1:2.2)   **4**

high as 95%. This intermediate, obtained in about 50% yield from the Corey lactone, has been converted into carbocyclin and several homologs.

[1] M. Uemura, T. Minami, and Y. Hayashi, *J.C.S. Chem. Comm.*, 1193 (1984).

[2] B. Caro, J.-Y. LeBihan, J.-P. Guillot, S. Top, and G. Jaouen, *ibid.*, 602 (1984).

[3] S. G. Davies, N. J. Holman, C. A. Laughton, and B. E. Mobbs, *ibid.*, 1316 (1983).

[4] G. Jaouen, S. Top, A. Laconi, D. Couturier, and J. Brocard, *Am. Soc.*, **106**, 2207 (1984); S. Top, A. Vessieres, J.-P. Abjean, and G. Jaouen, *J.C.S. Chem. Comm.*, 428 (1984).

[5] M. Cais, E. N. Frankel, and A. Rejoan, *Tetrahedron Letters*, 1919 (1968); E. N. Frankel and R. O. Butterfield, *J. Org.*, **34**, 3931 (1969); E. N. Frankel, E. Selke, and C. A. Glass, *ibid.*, **34**, 3936 (1969).

[6] P. Rosenmund and M. Casutt, *Tetrahedron Letters*, **24**, 1771 (1983).

[7] M. Shibasaki, M. Sodeoka, and Y. Ogawa, *J. Org.*, **49**, 4096 (1984).

Azidomethyl phenyl sulfide (1), 10, 14.

Amination of Grignard reagents.[1]   The addition of aliphatic Grignard reagents to **1** generates the thermodynamically less stable magnesium salt of the triazene, which can effect amination of RMgX.

Example:

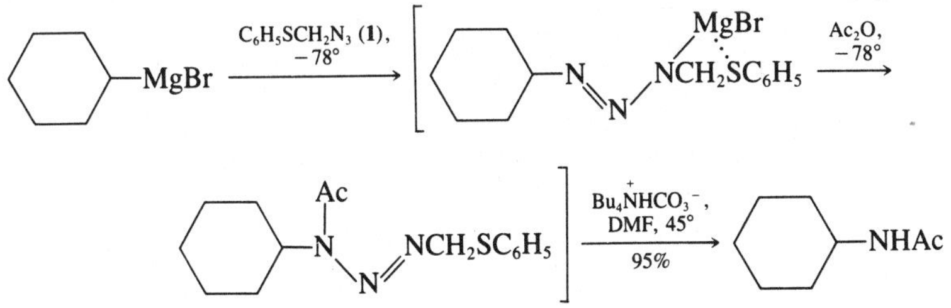

This azide is superior to other azides for amination; the ranking is as follows:

$$1 > CH_3SCH_2N_3 > CH_3OCH_2N_3 >> (CH_3)_3SiOCHN_3$$
$$\underset{CH(CH_3)_2}{|}$$

Aminopolyhydroxynaphthalenes.[2]   This $^+NH_2$ synthon was used to convert the protected bromopolyhydroxynaphthalene derivative **1** to the protected aminopolyhydroxynaphthalene derivative **2**, the aromatic nucleus of an ansamycin. One advantage of this amination is the satisfactory yield, which occurs in spite of the steric factors usually observed in the reaction of bromides with potassium azide.

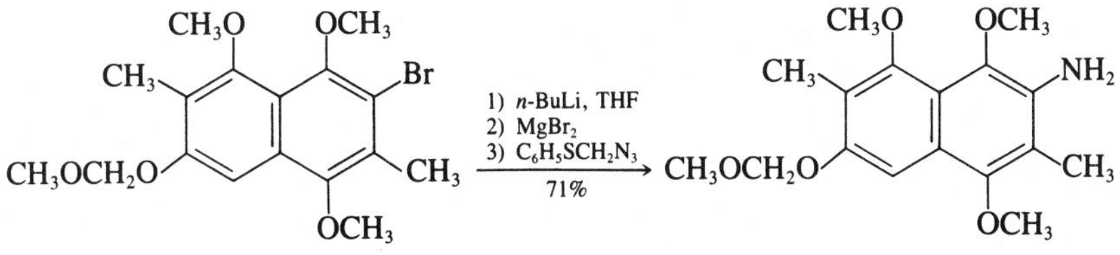

[1] B. M. Trost and W. H. Pearson, *Am. Soc.,* **105**, 1054 (1983).
[2] *Idem, Tetrahedron Letters,* **24**, 269 (1983).

Azidotrimethylsilane, 1, 1236; **3**, 316; **4**, 542; **5**, 719–720; **6**, 632; **9**, 21–22; **10**, 14–15; **11**, 32. *Caution.* West and Zigler[1] report the occurrence of a violent explosion during distillation of the product obtained by reaction of $ClSi(CH_3)_3$ and NaN_3 catalyzed by $AlCl_3$. They recommend use of reactions that do not require $AlCl_3$ as catalyst.

[1] R. West and S. Zigler, *Chem. Eng. News,* June 11, 4 (1984).

B

Barium hydroxide.

Diastereoselective Favorskii ring contraction.[1] Bicyclic α-chloro-δ-lactams can undergo highly diastereoselective ring contraction to proline derivatives when treated with base, preferably barium hydroxide in aqueous solution. Thus **1** is converted into **2** and **3** in the ratio 9:1. The ring contraction of **4** to **5** is stereospecific. In contrast, rearrangement

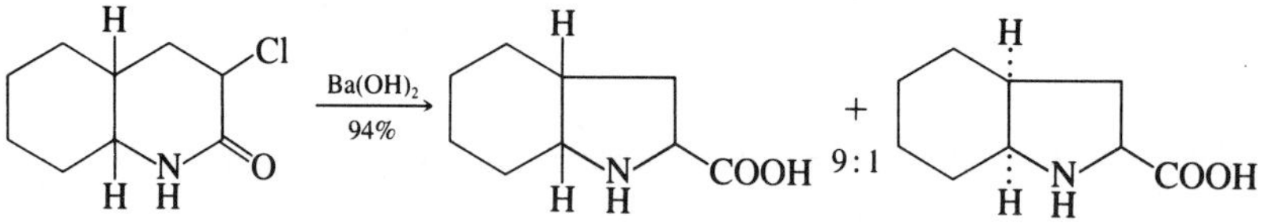

of the *trans*-isomer of **1** shows no diastereoselectivity. The high diastereoselectivity in the case of **1** and **4** evidently reflects the thermodynamic equilibrium of the respective isomers in the strongly basic medium.

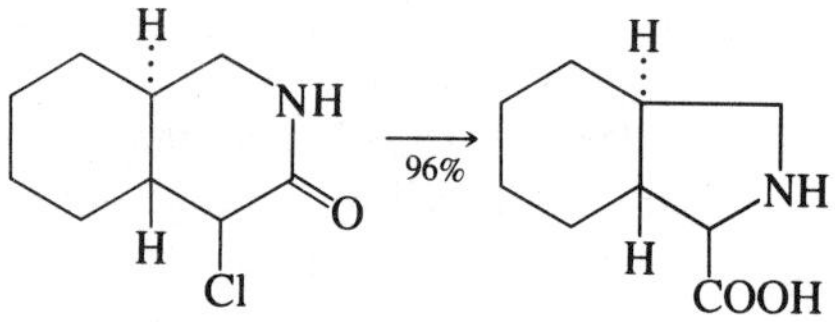

[1] R. Henning and H. Urbach, *Tetrahedron Letters*, **24**, 5339 (1983).

Barium manganate, 8, 21; **9**, 23; **10**, 16.

Selective oxidation.[1] The diol **1** is oxidized to the lactone **3** (cinnamolide) in 55% yield by Collins reagent. The same oxidation with Ag_2CO_3/Celite (Fetizon reagent) is essentially quantitative. The cheaper barium manganate is equally effective, and has the advantage that the intermediate lactol **2** can also be obtained if desired.

[1] D. M. Hollinshead, S. C. Howell, S. V. Ley, M. Mahon, N. M. Ratcliffe, and P. A. Worthington, *J.C.S. Perkin I*, 1579 (1983).

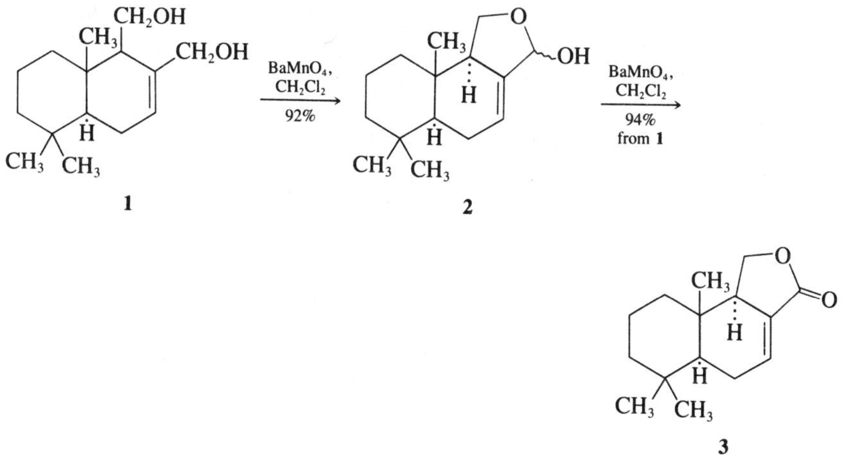

Benzeneselenenyl halides, 5, 518–522; **6**, 459–460; **7**, 286–287; **9**, 25–32; **10**, 16–21; **11**, 34–37.

cis-*Dichlorination of alkenes.*[1]   The *trans*-adducts of the reaction of benzeneselenenyl chloride with alkenes (**9**, 27) are converted to *cis*-dichlorides by reaction with chlorine and Bu$_4$NCl in CCl$_4$. The overall conversion can be conducted as a one-pot operation. In general only one stereoisomer is obtained, and yields are generally high.

Intramolecular amidoselenation; lactams.[2]   γ,δ-Unsaturated amides, on reaction with C$_6$H$_5$SeCl, can cyclize to γ- or δ-lactams. This amidoselenation is sensitive to the substituent on nitrogen (it fails with RCONHC$_6$H$_5$) and to substituents on the double bond.

Examples:

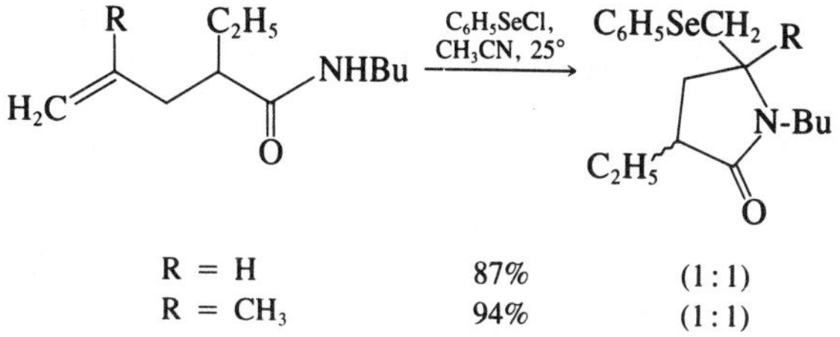

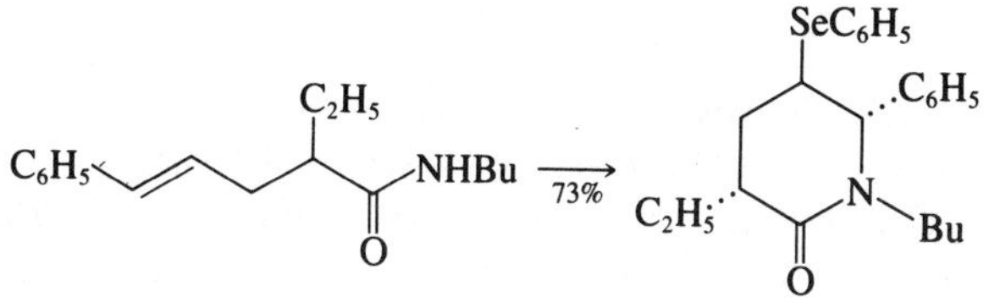

R = H	87%	(1:1)
R = CH$_3$	94%	(1:1)

C-Nucleosides.[3]   The first step in a new synthesis of C-nucleosides involves reaction[4] of ethoxycarbonylethylidenetriphenylphosphorane with the 2,3-di-O-isopropylidene-D-ribose derivative **1** to give the acyclic product **2**. On treatment with C_6H_5SeCl, **2** gives a single cyclic ribose derivative (**3**), shown to be the β-anomer by the conversion shown here to a derivative (**6**) of showdomycin. In addition, several other 1β-substituted 1-deoxyribose derivatives were prepared.

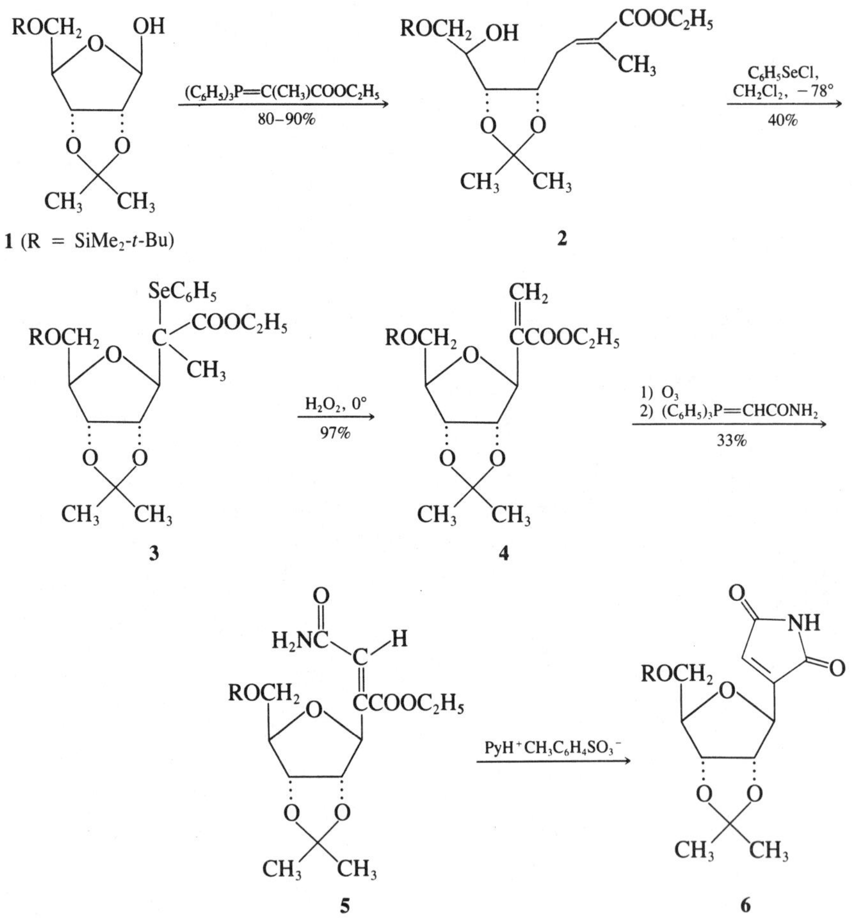

1 (R = SiMe$_2$-*t*-Bu)

2

3

4

5

6

Phenylselenoetherification.[5]   This reaction provides a short synthesis of the C-nucleosides showdomycin (**10**) and epishowdomycin (**11**) from D-ribose (**7**). Thus the stabilized ylide **8**, derived from maleimide, couples with **7** in acetic acid to form **9**. Cyclization with C_6H_5SeCl in refluxing trimethyl borate gives an unstable product, which on oxidation gives a mixture of **10** and **11**.

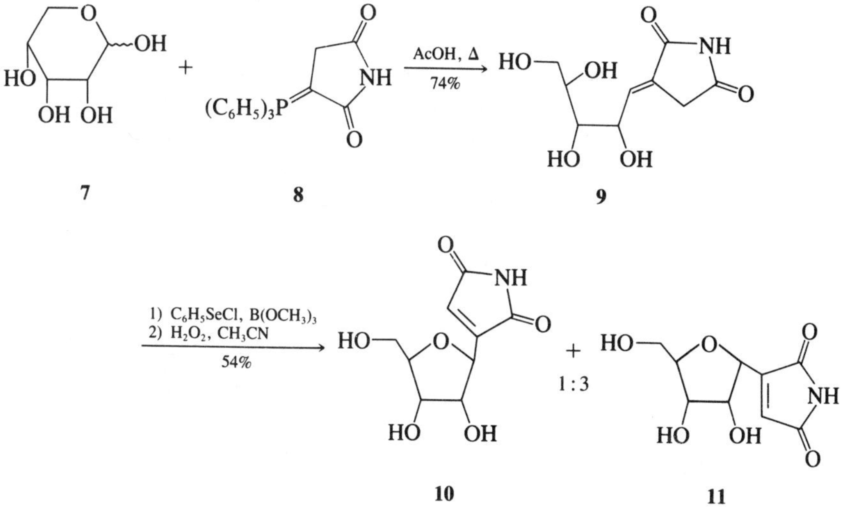

γ-Lactones.[6]   Successive treatment of alkenes with C_6H_5SeCl and silver crotonate affords β-phenylselenocrotonates. The products undergo ring closure to γ-lactones when treated with $(C_6H_5)_3SnH$ and a radical initiator. Application of this sequence to cyclohexene (equation I) or cyclopentene results in two bicyclo-γ-lactones. Both are *cis*-fused, and the major product results from *exo*-closure.

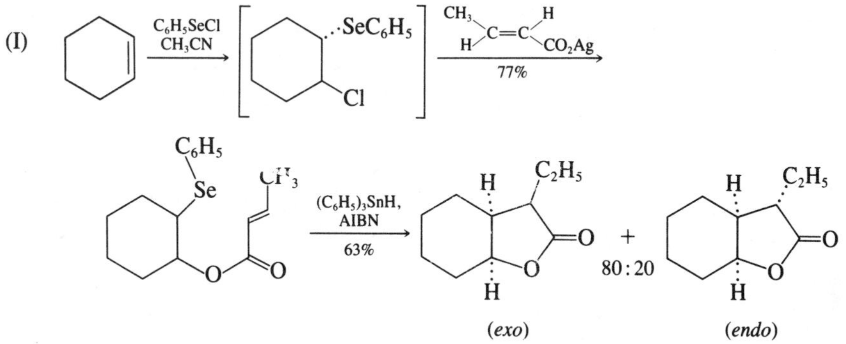

Application to (Z)- and (E)-2-butene results in essentially identical mixtures of four ethyldimethyl-γ-butyrolactones.

[1] A. M. Morella and A. D. Ward, *Tetrahedron Letters*, **25**, 1197 (1984).
[2] A. Toshimitsu, K. Terao, and S. Uemura, *ibid.*, **25**, 5917 (1984).
[3] P. D. Kane and J. Mann, *J.C.S. Chem. Comm.*, 224 (1983).
[4] H. Ohrui, G. H. Jones, J. G. Moffatt, M. L. Maddox, A. T. Christensen, and S. K. Byram, *Am. Soc.*, **97**, 4602 (1975).
[5] A. G. M. Barrett and H. B. Broughton, *J. Org.*, **49**, 3673 (1984).
[6] D. L. J. Clive and P. L. Beaulieu, *J. C. S. Chem. Comm.*, 307 (1983).

Benzenesulfenyl chloride, 5, 523–524; **6**, 30–32; **8**, 32–34; **9**, 35–38; **10**, 24; **11**, 39–40.

α-Phenylthiocycloalkenones; α-phenyl-α-(phenylthio)ketones. Some years ago Weygand and Bestmann[1] reported the reaction of C_6H_5SCl with diazoacetophenone to give the α-chloro-α-(phenylthio) adduct with evolution of N_2. This reaction has now been investigated with several cyclic and acyclic α-diazo ketones. It provides a useful route to a wide range of α-phenylthioalkenones and α-phenyl-α-(phenylthio)ketones.[2]

Examples:

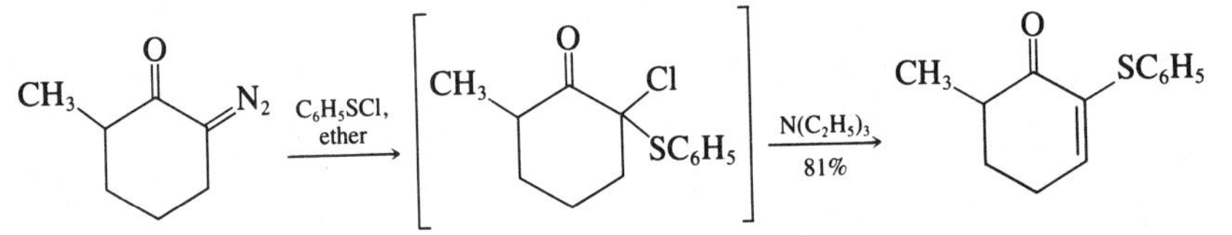

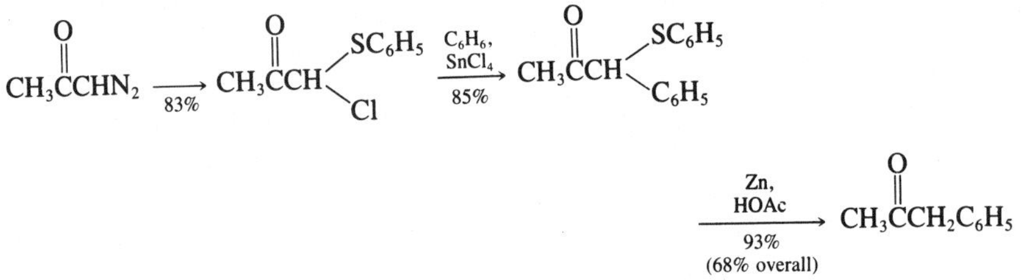

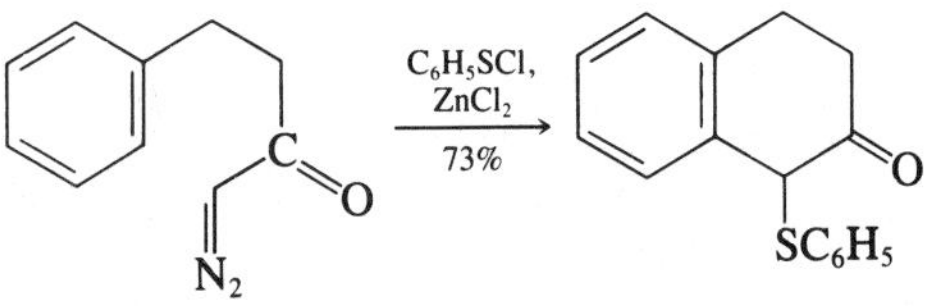

β-Lactams.[3] Phenylsulfenylation of an α,β-unsaturated amide (**1**) gives two regioisomeric adducts, which need not be separated, because on treatment with base both

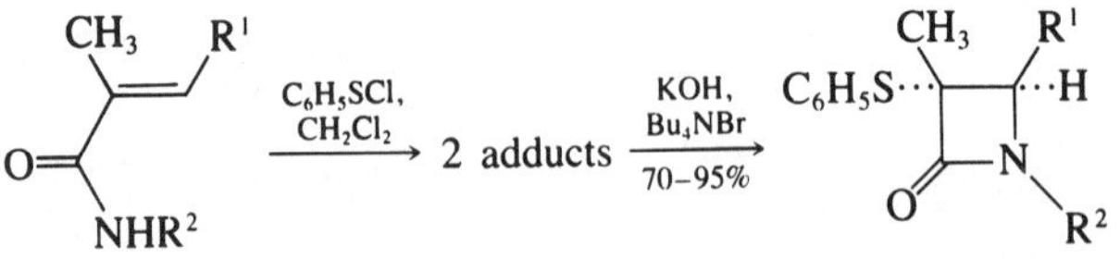

yield the same azetidinone (**2**) in 70–95% yield. An example is the synthesis of **3**, an intermediate to nocardicin antibiotics.

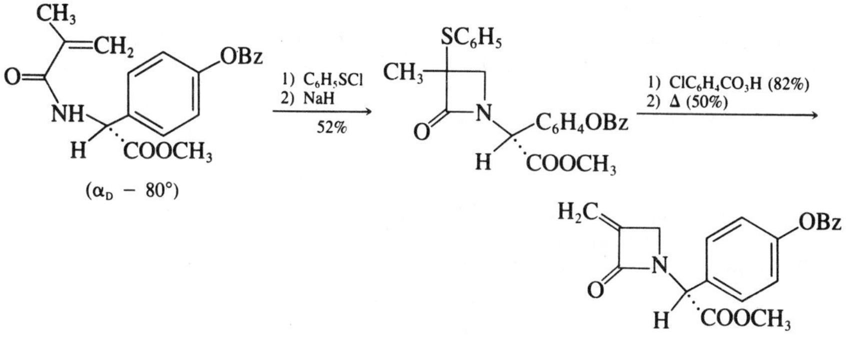

3

The reaction can also be used to construct the bicyclic carbapenam ring system (equation I).

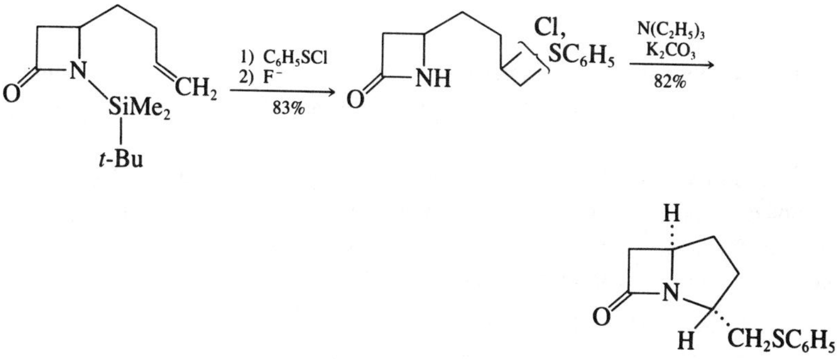

Sulfenocycloamination.[4]   Treatment of an ω-alkenylamine hydrochloride with C_6H_5SCl gives a mixture of two adducts that undergoes ring closure via an episulfonium ion to a single cyclic amine on treatment with base (K_2CO_3). The reaction is useful for synthesis of pyrrolidines and piperidines (equation I) and pyrrolizidines (equation II).

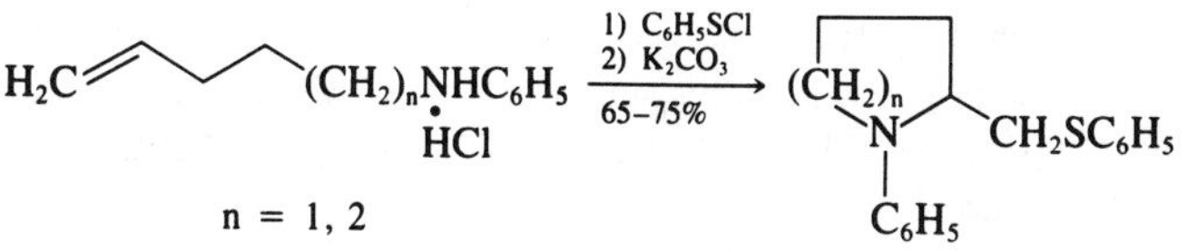

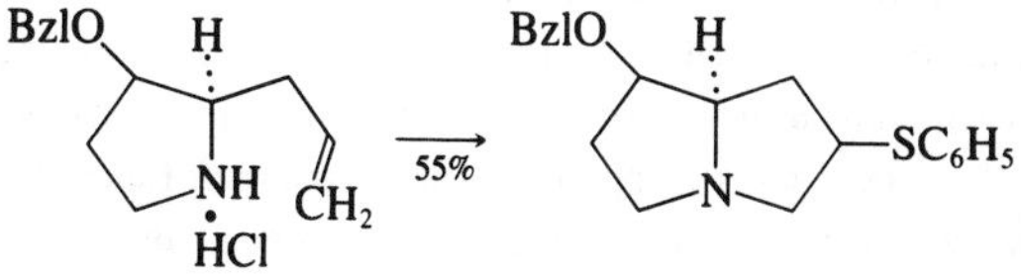

[1] F. Weygand and H. J. Bestmann, *Z. Naturforsch.*, **10b**, 296 (1955).
[2] M. A. McKervey and P. Ratananukul, *Tetrahedron Letters*, **24**, 117 (1983).
[3] M. Ihara, Y. Haga, M. Yonekura, T. Ohsawa, K. Fukumoto, and T. Kametani, *Am. Soc.*, **105**, 7345 (1983).
[4] T. Ohsawa, M. Ihara, K. Fukumoto, and T. Kametani, *J. Org.*, **48**, 3644 (1983).

O-Benzotriazolyl-N,N,N′,N′-tetramethyluronium hexafluorophosphate,

$[(CH_3)_2N]_2\overset{+}{C}$—O—N$\diagup^{N}\diagdown_{N}$ (1). The salt is prepared by reaction of hydroxybenzo-
$PF_6{}^-$ triazole with tetramethyluronium chloride hexafluoro-
phosphate.

Peptide synthesis.[1]   The reagent effects coupling of an N-protected amino acid with an amino acid ester or peptide in yields usually >90% and with low racemization.

[1] V. Dourtoglou, B. Gross, V. Lambropoulou, and C. Zioudrou, *Synthesis*, 572 (1984).

Benzoyl trifluoromethanesulfonate, $CF_3SO_2OCOC_6H_5$. B.p. 92–94°/2.2 mm. The reagent is prepared from benzoyl chloride and triflic acid.[1] It is extremely hygroscopic and unstable to air.

Benzoylation.[2]   This acyl triflate is particularly useful for benzoylation of hindered secondary hydroxyl groups and, in combination with pyridine, tertiary hydroxyl groups. Acetals and ketals, if present, are converted into carbonyl groups. Epoxides also react, generally to give a mixture of products.

[1] F. Effenberger and G. Epple, *Angew. Chem., Int. Ed.*, **11**, 299 (1972).
[2] M. Koreeda and L. Brown, *J.C.S. Chem. Comm.*, 1113 (1983); L. Brown and M. Koreeda, *J. Org.*, **49**, 3875 (1984).

Benzylchlorobis(triphenylphosphine)palladium(II), **8**, 35–36; **9**, 41–42; **10**, 26.

Coupling of ROCl with tetraorganotins (**9**, 41–42).[1]   An attractive feature of this ketone synthesis is that one group of the tin reagent is transferred rapidly, with formation of R_3SnCl, which undergoes a second coupling only slowly. Competitive reaction of various unsymmetrical organotin reagents with C_6H_5COCl indicates the following order reactivity: $C_6H_5C\equiv C \gg CH_2{=}CH > C_6H_5 > CH_3 > C_2H_5 > CH(CH_3)_2$. Vinyl groups couple with retention.

Coupling of allyl halides with vinyltin reagents.[2]   In the presence of a Pd(II) catalyst, allyl halides cross-couple with vinyltin reagents to give 1,4-dienes in high yield. The double-bond geometry of both partners is retained, and primary allyl halides couple without allylic transposition. The coupling is compatible with a variety of functional groups, such as ester, nitrile, and hydroxyl. The reaction involves inversion of configuration of the allyl halide. Aryltin reagents also undergo this coupling. Coupling in the presence of carbon monoxide results in insertion to furnish ketones, often in useful yields.

Examples:

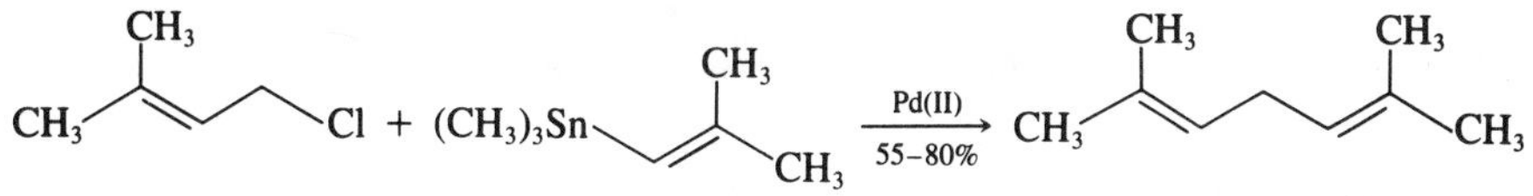

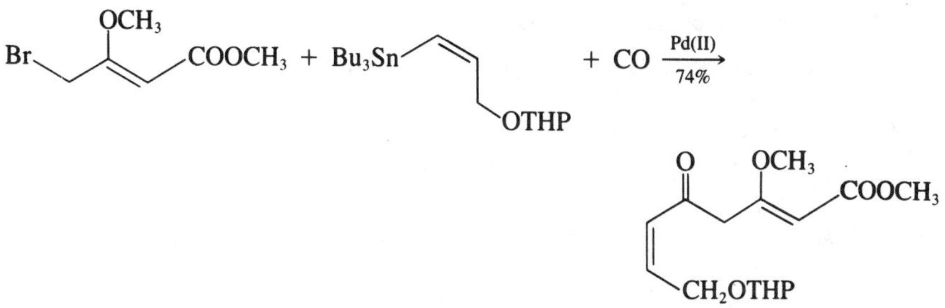

Divinyl ketones.[3]   An attractive route to unsymmetrical divinyl ketones is Pd(II)-catalyzed coupling of vinyl iodides with trimethyl- or tributylvinyltin reagents in the presence of carbon monoxide (15–50 psig). The reaction is sensitive to steric hindrance, particularly on the vinyltin partner. (E)-Geometry in both partners is retained, but (Z)-vinyltin derivatives undergo partial isomerization after coupling.

Example:

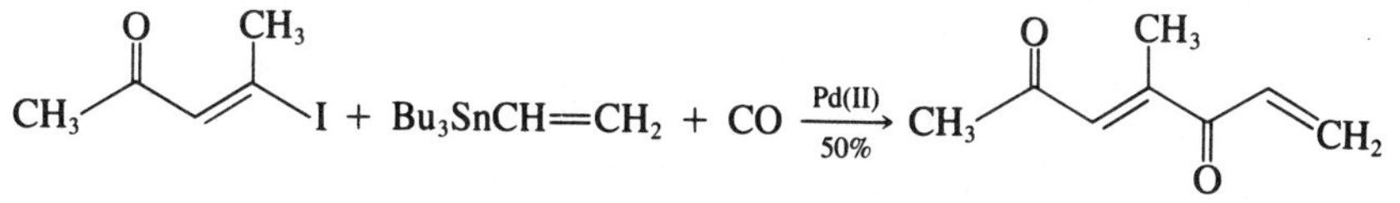

This coupling followed by Nazarov cyclization provides a useful synthesis of cyclopentenones.

Example:

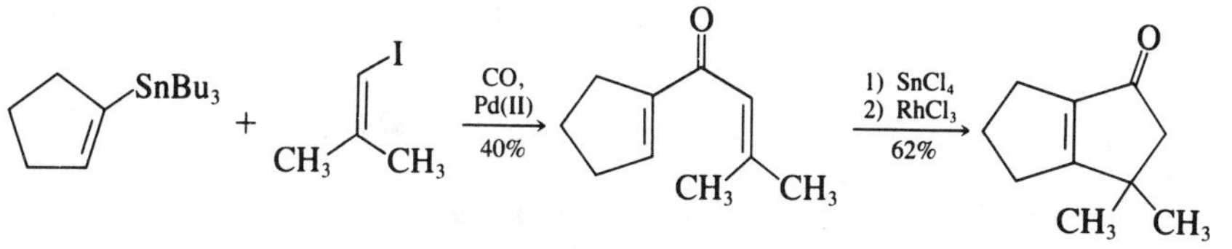

Ketone synthesis.[4]   Ketones can be prepared, generally in high yield, by a palladium-catalyzed reaction of acid chlorides with dialkylzinc compounds, prepared *in situ* by

$$\text{(I) } R^1COCl + R^2_2Zn \xrightarrow[75-98\%]{\substack{Pd(II), \\ THF, Et_2O, 23°}} R^1\overset{\overset{\displaystyle O}{\|}}{C}R^2$$

reaction of RMgCl with $ZnCl_2$ (equation I). Tetrakis(triphenylphosphine)palladium and (dppf)PdCl$_2$ are also effective catalysts.

[1] J. W. Labadie and J. K. Stille, *Am. Soc.*, **105**, 6129 (1983).
[2] F. K. Sheffy, J. P. Godschalx, and J. K. Stille, *ibid.*, **106**, 4833 (1984).
[3] W. F. Goure, M. E. Wright, P. D. Davis, S. S. Labadie, and J. K. Stille, *ibid.*, **106**, 6417 (1984).
[4] R. A. Grey, *J. Org.*, **49**, 2288 (1984).

1-Benzyl-1,2-dihydroisonicotinamide,

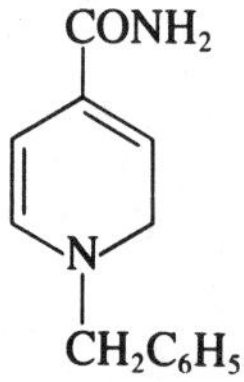

(1). Preparation.[1]

Reduction of pyridinium salts.[2]   Pyridinium salts are reduced by **1**, mainly to 1,4-dihydropyridines. The yield is generally higher than that obtained with NaBH$_4$ or Na$_2$S$_2$O$_4$. Examples:

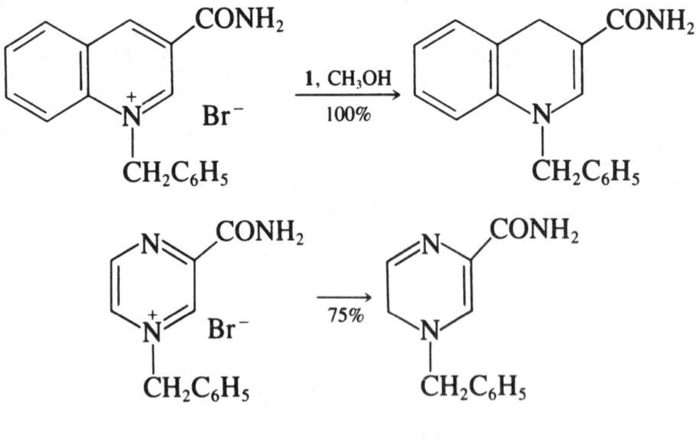

[1] G. Paglietti, P. Sanna, A. Nuvole, F. Soccolni, and R. M. Acheson, *J. Chem. Res. (M)*, 2326 (1983).
[2] A. Nuvole, G. Paglietti, P. Sanna, and R. M. Acheson, *ibid.*, 3245 (1984).

1-Benzyl-1,4-dihydronicotinamide (1), 6, 36–37; **7**, 15–16; **11**, 419.

Protiodenitration.[1] Simple nitroalkanes are not reduced by **1**, but α-nitro nitriles, esters, and ketones are selectively reduced by **1** under irradiation, with selective replacement of NO_2 by H. The reduction occurs via an electron-transfer chain mechanism.

Example:

1

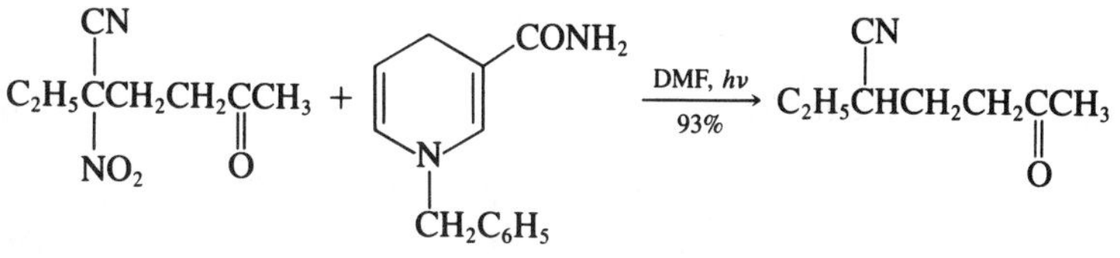

[1] N. Ono, R. Tamura, and A. Kaji, *Am. Soc.*, **105**, 4017 (1983).

9-(Benzyloxy)methoxyanthracene,

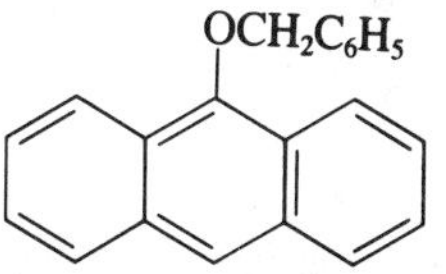

(1). The reagent is prepared by reaction of the sodium salt of anthrone with benzyloxymethyl chloride.

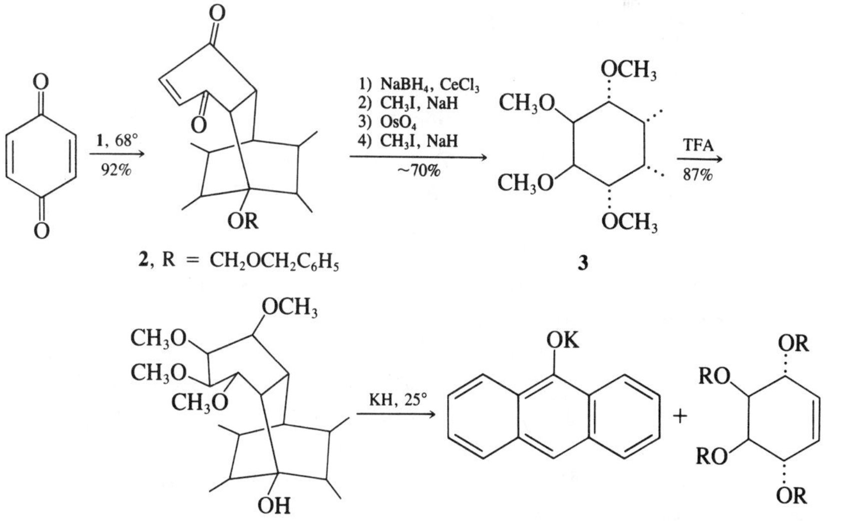

Cyclitols.[1] A novel synthesis of conduritol (**6**) from benzoquinone utilizes **1** for protection of one C═C bond and for differentiation of the carbonyl groups. Reaction of the quinone with **1** gives the Diels-Alder adduct **2**, which is converted selectively into **4**,

2, R = $CH_2OCH_2C_6H_5$

3

4

5 (R = CH_3, 99%)
6 (R = H)

formally the Diels-Alder adduct of **5** and 9-hydroxyanthracene. On treatment with potassium hydride, **4** undergoes a retro-Diels-Alder reaction at room temperature to give tetra-O-methyl conduritol (**5**).

[1] S. Knapp, R. M. Ornaf, and K. E. Rodriques, *Am. Soc.*, **105**, 5494 (1983).

2-Benzyloxy-1-propene, $CH_2{=}C{<}\begin{subarray}{l} OCH_2C_6H_5 \\ CH_3 \end{subarray}$ (**1**). The reagent is obtained from 2-benzyloxy-1-chloropropane by dehydrochlorination (t-BuOK in DMSO).

Protection of hydroxyl groups.[1]   An alcohol is converted into the acetal **2** by reaction of **1** catalyzed by dichloro(1,5-cyclooctadiene)palladium(II). POCl$_3$ or TsOH can be used

$$ROH + 1 \xrightarrow[96\%]{PdCl_2(COD)} (CH_3)_2C{<}\begin{subarray}{l} OCH_2C_6H_5 \\ OR \end{subarray} \xrightarrow[92-99\%]{\substack{H_2, \\ Pd/C}} ROH$$

2

as catalysts, but the yield is lower (60–90%). The alcohol can be regenerated by hydrogenation catalyzed by Pd/C (5%). The acetal is stable to hydride reagents and NaOH in THF, but is hydrolyzed under acidic conditions. Primary alcohols can be selectively acetalized with **1**. The particular advantage is that the protection and regeneration steps can be conducted under neutral conditions.

[1] T. Mukaiyama, M. Ohshima, and M. Murakami, *Chem. Letters,* 265 (1984).

Benzyl(triethyl)ammonium permanganate, 9, 43; **10**, 28–29; **11**, 44. *Caution:* The reagent when dry can ignite explosively, even at 20°.[1]

[1] J. Graefe and R. Rienäcker, *Angew. Chem., Int. Ed.,* **22**, 625 (1983).

Benzyltrimethylammonium hydroxide (Triton B), 1, 1252–1254; **5**, 29; **8**, 36–37.

Intramolecular Marschalk reaction.[1]   An intramolecular Marschalk reaction (**9**, 376) can be used to effect a synthesis of anthracyclinones from anthraquinones. Thus the α-hydroxy aldehyde **2**, formed on saponification of the α-hydroxydichloride **1**, on reduction of the quinone group cyclizes in the alkaline medium to the tetracyclic *trans-* and *cis*-diols (**3** and **4**) in about equal amounts. Cyclization under phase-transfer conditions results in improved yields and, more importantly, can alter the stereoselectivity. Triton B is the most effective catalyst for stereoselective cyclization to the desired natural *trans*-diol.

[1] K. Krohn and W. Priyono, *Tetrahedron,* **40**, 4609 (1984).

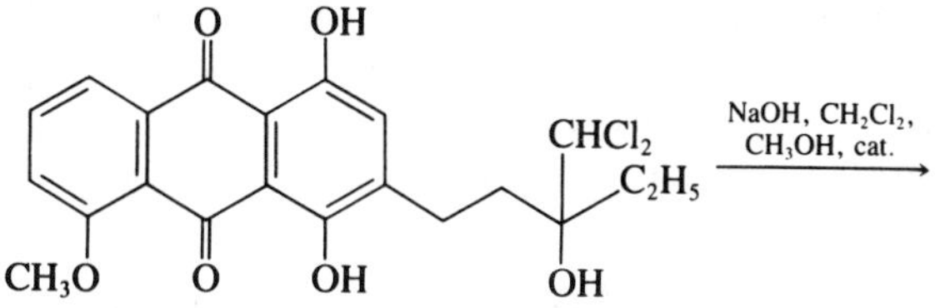

(S)-(+)- and R-(−)-1,1'-Binaphthyl-2,2'-diyl hydrogen phosphate (1).
Preparation:

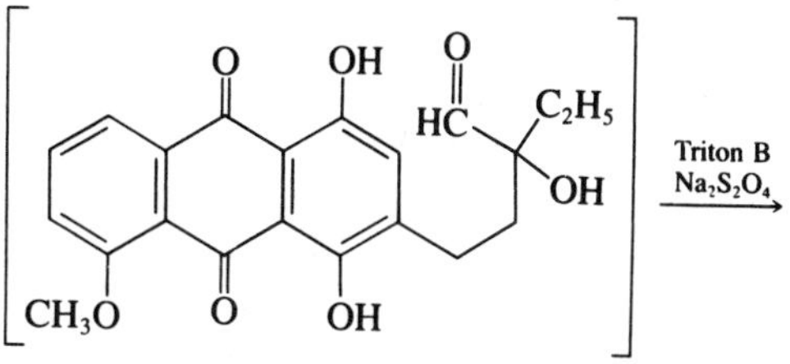

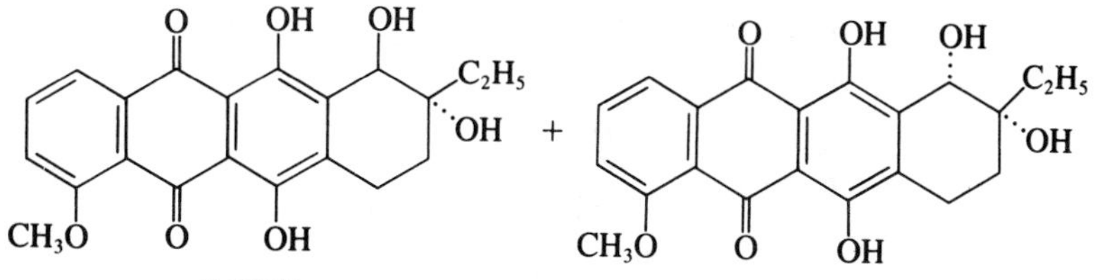

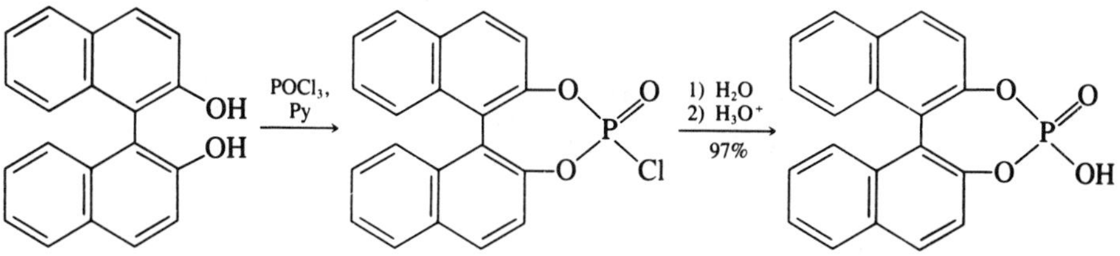

The racemic acid can be resolved by crystallization of the salt of (+)-**1** with cinchonine. The optically pure acids are useful for resolution of amines. Reduction of the corresponding methyl esters provides a particularly useful route to (S)-(−)- and (R)-(+)-binaphthol (**9**, 169–170).[1]

[1] J. Jacques and C. Fouquey, *Org. Syn.*, submitted (1984).

Bis(acetonitrile)dichloropalladium(II), 7, 21–22; **8**, 39; **9**, 44; **10**, 30–31; **11**, 46–47.

Cyclopentenones.[1]   1-Ethynyl-2-propenyl acetates (**1**)[2] in the presence of this Pd(II) salt and acetic acid (1 equiv.) cyclize to cyclopentenones (**2**) in 50–90% yield. The cyclopentadiene **a** has been identified as an intermediate by trapping with a dienophile.

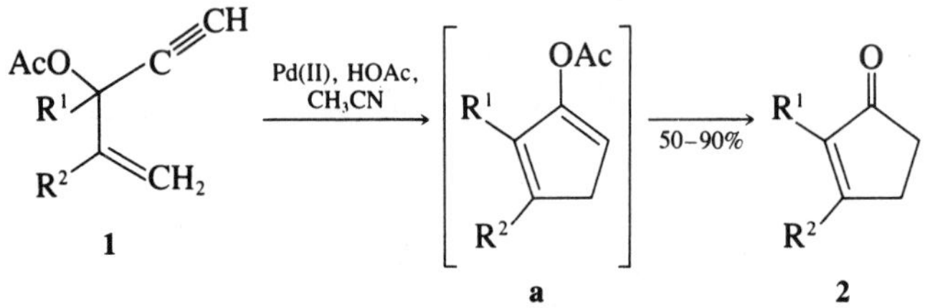

The R groups can be hydrogen or alkyl groups, but further substitution on the ethynyl or vinyl group suppresses the reaction, which can be regarded as a variation of the Nazarov cyclization.[3]

Cope rearrangement (**10**, 31).   Acyclic 1,5-dienes substituted at C_3 by an electron-withdrawing group undergo Pd(II)-catalyzed Cope rearrangement at 40°.[4]

Example:

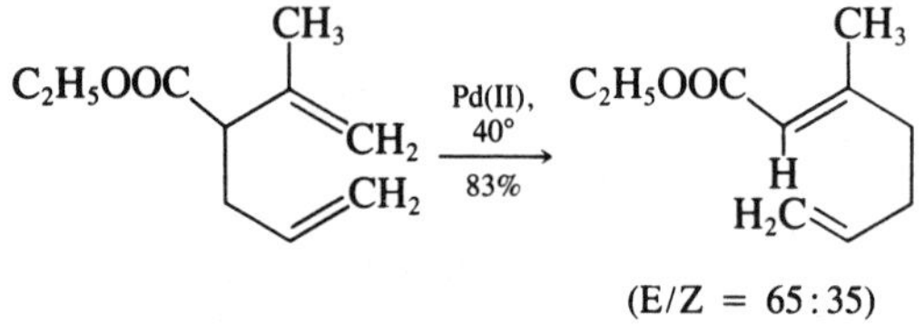

(E/Z = 65:35)

Stereoselective allylic rearrangement.[5]   2-Substituted 1-vinylcyclohexyl acetates (**1**) rearrange in the presence of this Pd catalyst selectively to the (E)-trisubstituted allylic acetates **2**.

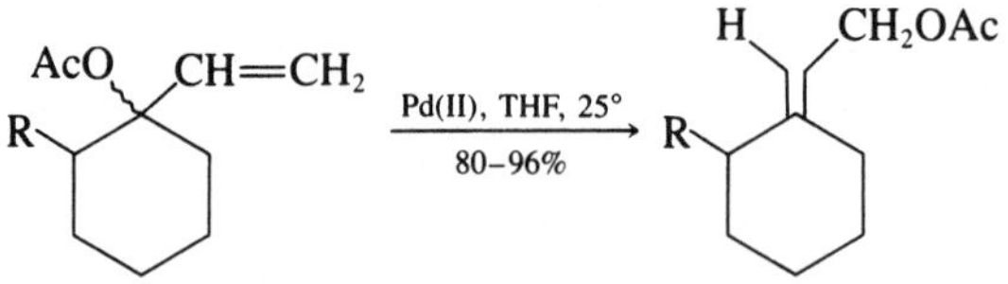

Enaminones.[6]   Tertiary amines substituted by a carbonyl group at the β-position undergo dehydrogenation to enaminones when treated with the Pd(II) complex (1 equiv.) and triethylamine (2 equiv.) in acetonitrile (equation I).

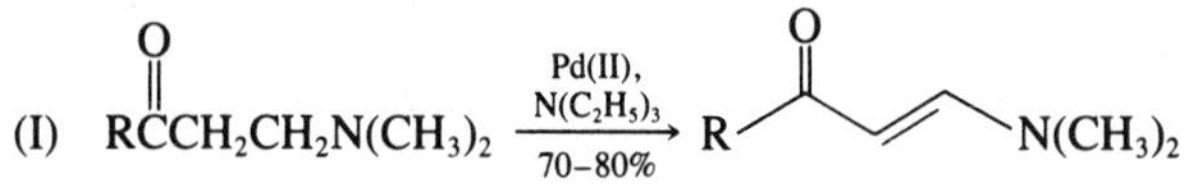

[1] V. Rautenstrauch, *J. Org.*, **49**, 950 (1984).
[2] M. L. Roumestant, M. Malacria, J. Goré, J. Grimaldi, and M. Bertrand, *Synthesis*, 755 (1976).
[3] Review: E. J. Santelli-Rouvier and M. Santelli, *ibid.*, 419 (1983).
[4] L. E. Overman and A. F. Renaldo, *Tetrahedron Letters*, **24**, 3757 (1983).
[5] Y. Tamaru, Y. Yamada, H. Ochiai, E. Nakajo, and Z. Yoshida, *Tetrahedron*, **40**, 1791 (1984).
[6] S.-I. Murahashi, T. Tsumiyama, and Y. Mitsue, *Chem. Letters*, 1419 (1984).

Bis(benzonitrile)dichloropalladium(II), 5, 31–32; **6**, 45–47; **9**, 44–45; **10**, 31–32; **11**, 48–49.

Cyclization of alkynoic acids.[1] In the presence of this Pd(II) catalyst and triethylamine, 3-, 4-, and 5-alkynoic acids cyclize to unsaturated lactones.

Examples:

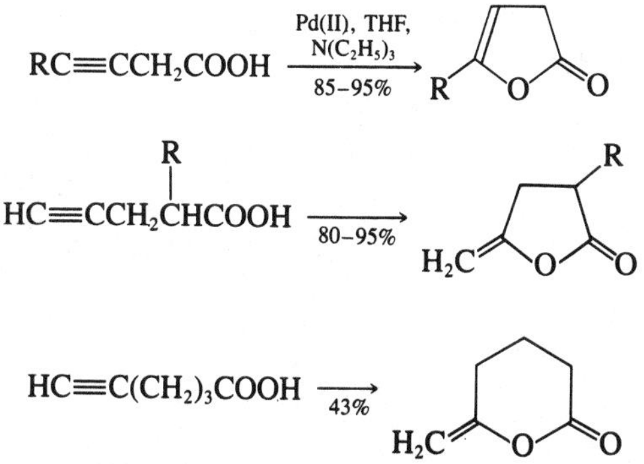

Oxy-Cope rearrangement (*cf.* **10**, 31).   Tertiary substituted 1,5-hexadiene-3-ols are isomerized by catalytic amounts of this Pd(II) complex to δ,ε-unsaturated ketones with a stereoselectivity >90%.[2]

Examples:

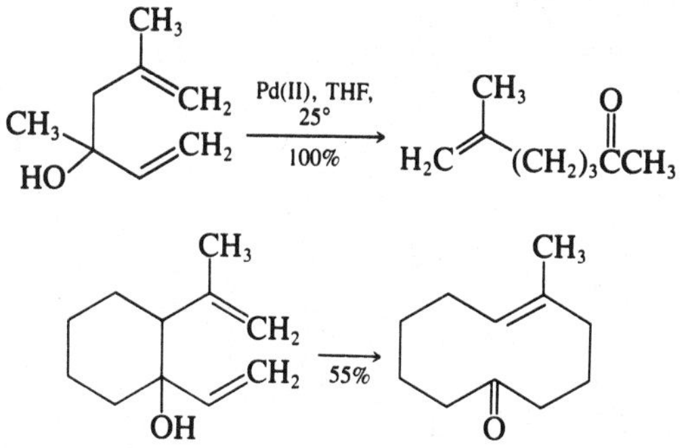

(E/Z = 9:1)

[1] C. Lambert, K. Utimoto, and H. Nozaki, *Tetrahedron Letters*, **25**, 5323 (1984).
[2] N. Bluthe, M. Malacria, and J. Gore, *ibid.*, **24**, 1157 (1983).

Bis(N-*t*-butylsalicyldimato)copper(II) (1). M.p. 184–185°. This copper chelate is prepared from salicylaldehyde by sequential reaction in CH_3OH with *t*-butylamine and $Cu(OAc)_2$.

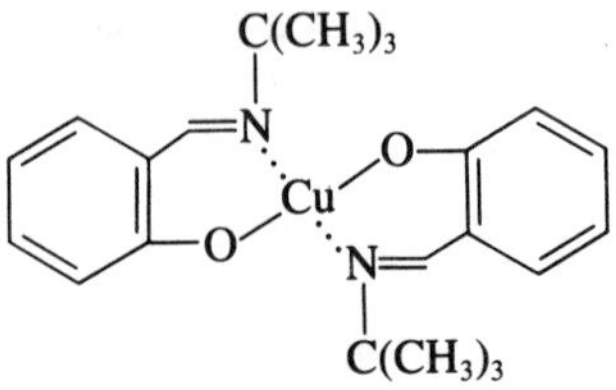

1

Cyclopropyl lactones. Copper-catalyzed decomposition of unsaturated esters of diazoacetic acid can provide cyclopropyl lactones (**2**, 83). The soluble copper chelate **1** is superior for this purpose to copper powder or copper oxide.[1] In a typical example, the lactone **3** is obtained from the diazo ester **2** in 92% yield.

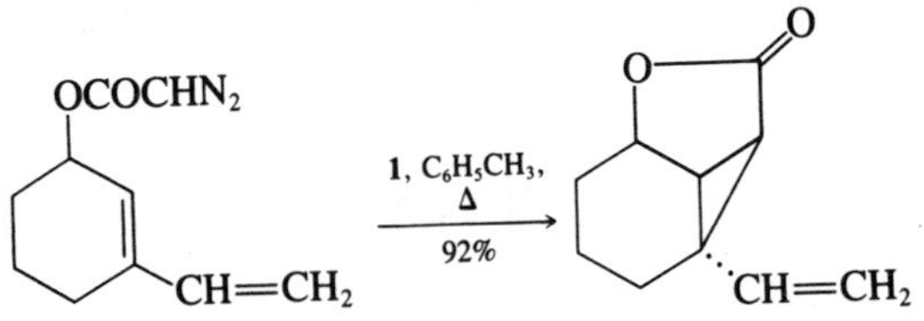

2   **3**

The esters are prepared by reaction of the allylic or homoallylic alcohol with the tosylhydrazone of glyoxylic acid chloride[2] and N,N-dimethylaniline as a base,[1] followed by elimination of *p*-toluenesulfinic acid (triethylamine). Yields of ~75% can be obtained.

[1] E. J. Corey and A. G. Myers, *Tetrahedron Letters,* **25**, 3559 (1984).
[2] C. J. Blankley, F. J. Sauter, and H. O. House, *Org. Syn. Coll. Vol.,* **5**, 258 (1973).

Bis(1,5-cyclooctadiene)nickel(0), 4, 33–35; **5,** 34–35; **7,** 428–429; **9,** 45–46; **10,** 33; **11,** 51.
Dimerization of functionalized 1,3-dienes; cyclooctadienes.[1] Butadienes carrying an electron-donating group dimerize in the presence of either Ni(COD)₂ or the Ni(0) obtained by reduction of Ni(acac)₂ with $(C_2H_5)_2AlOC_2H_5$ to 1,2-*trans*-disubstituted 3,7-cyclooctadienes in 70–90% yield.

Examples:

$$2CH_2\!=\!CHCH\!=\!CHOSi(CH_3)_3 \xrightarrow[90\%]{Ni(0)}$$

$$\left.\begin{array}{c} CH_2\!=\!CHCH\!=\!CHCOOCH_3 \\ + \\ CH_2\!=\!CHCH\!=\!CH_2 \end{array}\right\} \xrightarrow{33\%}$$

Dimerization of butadiene.[2]   Butadiene can be dimerized almost exclusively to a mixture of (Z,E)- and (E,E)-1,3,6-octatrienes in the presence of this Ni(0) catalyst complexed with an aminophosphinite such as **1** or **2**, obtained by reaction of ephedrine or prolinol, respectively, with $P(C_6H_5)_2N(CH_3)_2$. If the reaction is prolonged, the mixture isomerizes to (E,E,E)-2,4,6-octatriene.

$$CH_3NHCH(CH_3)CH(C_6H_5)OP(C_6H_5)_2$$

1

2

[1] P. Brun, A. Tenaglia, and B. Waegell, *Tetrahedron Letters*, **24**, 385 (1983).
[2] P. Denis, A. Mortreux, F. Petit, G. Buono, and G. Peiffer, *J. Org.*, **49**, 5274 (1984).

Bis(cyclopentadienyl)diiodozirconium, I_2ZrCp_2 (**1**). The complex can be obtained by reaction of Cl_2ZrCp_2 with BI_3 or generated *in situ* from Cl_2ZrCp_2 and NaI.

Carbozincation of alkynes.   A dialkylzinc in combination with I_2ZrCp_2 undergoes regioselective *cis*-addition to terminal alkynes to provide alkenylzinc derivatives (equation I). The reaction is catalytic in I_2ZrCp_2, but at least 1 equiv. of I_2ZrCp_2 is required for satisfactory results. Cl_2ZrCp_2 or Br_2ZrCp_2 is less effective.[1]

$$(I)\quad R'C\!\equiv\!CH + R_2Zn + \mathbf{1} \longrightarrow \underset{R}{\overset{R'}{>}}C\!=\!C\underset{ZnI}{\overset{H}{<}}$$

This reaction has been used to effect allylmetallation of internal alkynes for a novel synthesis of cyclopentenones.[2] Thus in the presence of I_2ZrCp_2, diallylzinc reacts with 2-butyne to provide, after iodinolysis, the diene **2** in 92% yield. The product was converted in one step into methylenemycin B (**3**) by reaction with carbon monoxide, $Pd[P(C_6H_5)_3]_4$ (1 equiv.), and triethylamine (1 equiv.).

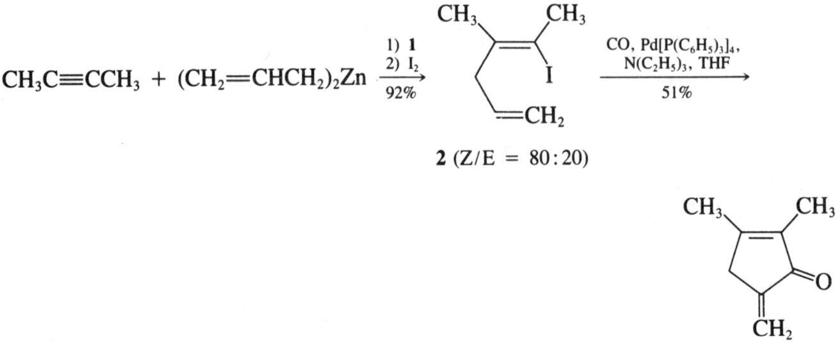

2 (Z/E = 80:20)

3

Allylzinc bromide can undergo regioselective addition to 1-(trimethylsilyl)-1-alkynes to give adducts in which ZnBr and $Si(CH_3)_3$ are attached to the same carbon atom. In the case of reactive alkynylsilanes, the reaction proceeds by *cis*-addition. However, if the reaction requires higher temperatures, products resulting from a net *trans*-addition are obtained (equation II).[3]

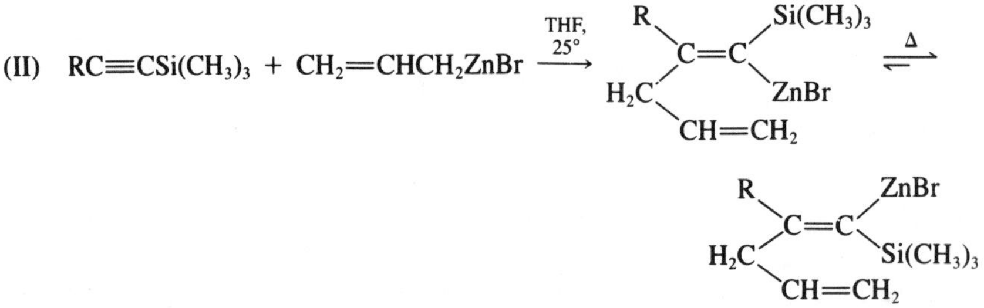

(II) $RC{\equiv}CSi(CH_3)_3$ + $CH_2{=}CHCH_2ZnBr$

[1] E. Negishi, D. E. Van Horn, T. Yoshida, and C. L. Rand, *Organometallics*, **2**, 563 (1983).
[2] E. Negishi and J. A. Miller, *Am. Soc.*, **105**, 6761 (1983).
[3] G. A. Molander, *J. Org.*, **48**, 5409 (1983).

Bis(cyclopentadienyl)titanacyclobutanes (1).
    Preparation from the Tebbe reagent:[1]

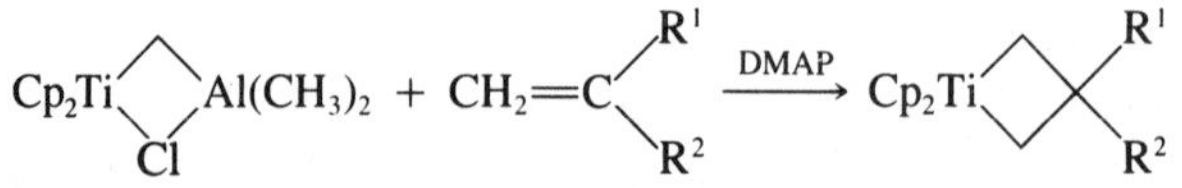

1

Titanium enolates.[2]   The labile **1a**, prepared from isobutylene, decomposes to **2**, which reacts with acid chlorides to give titanium enolates (**3**) of methyl ketones (equation I).

$$\text{(I)} \quad \textbf{1a} \quad \xrightarrow[-(CH_3)_2C=CH_2]{0°} \quad [Cp_2Ti=CH_2] \quad \xrightarrow{RCCl} \quad RC=CH_2 \quad \xrightarrow[75-95\%]{HCl} \quad RCCH_3$$

$$(R^1, R^2 = CH_3)$$

$$\qquad\qquad\qquad\qquad\qquad\qquad\qquad \textbf{2} \qquad\qquad\quad \textbf{3} \qquad\qquad\quad \textbf{4}$$

The enolates (**3**) undergo aldol condensation as expected (equation II).

$$\cdot \text{(II)} \qquad \begin{array}{c} Cp_2ClTiO \\ \diagdown \\ \qquad C=CH_2 \\ \diagup \\ C_6H_5CH_2 \end{array} + C_6H_5CHO \xrightarrow{69\%} C_6H_5CH_2\overset{O}{\overset{\|}{C}}CH_2\overset{OH}{\overset{|}{C}}HC_6H_5$$

Allenes.[3]   The reaction of **1** with a 1,1-disubstituted allene liberates a new titanacycle considered to have structure **2**. Addition of a ketone to **2** results in a new substituted allene **3**.

Examples:

$$(CH_3)_2C=C=CH_2 + 1 \xrightarrow{C_6H_6} \begin{array}{c} CH_3 \\ \diagdown \\ \quad C=\langle \text{TiCp}_2 \rangle \\ \diagup \\ CH_3 \end{array} + \begin{array}{c} R^1 \\ \diagdown \\ \quad C=CH_2 \\ \diagup \\ R^2 \end{array}$$

$$\textbf{2}$$

$$\xrightarrow[\text{overall}]{80\%} \Bigg| O=C(C_6H_5)_2$$

$$(CH_3)_2C=C=C(C_6H_5)_2 + Cp_2Ti(0)$$

$$\textbf{3}$$

$$(C_6H_5)_2C=C=CH_2 + 1 + O=\!\!\!\!\bigcirc \xrightarrow{72\%} (C_6H_5)_2C=C=\!\!\!\!\bigcirc$$

$$H_2C=C=CH_2 + 1 + O=C(C_6H_5)_2 \xrightarrow{58\%} CH_2=C=C(C_6H_5)_2$$

[1] T. R. Howard, J. B. Lee, and R. N. Grubbs, *Am. Soc.*, **102**, 6876 (1980).
[2] J. R. Stille and R. H. Grubbs, *ibid.*, **105**, 1664 (1983).
[3] S. L. Buchwald and R. H. Grubbs, *ibid.*, **105**, 5490 (1983).

Bis(dibenzylideneacetone)palladium, Pd(dba)$_2$, **9**, 46; **10**, 34; **11**, 53.

Cross-coupling of allyl halides with vinyltin reagents.[1]   Allyl halides couple with vinyl- or aryltin reagents in the presence of this catalyst in uniformly high yields (>80%). A wide variety of functional groups is tolerated. The geometry of the double bond in both partners is retained, but the carbon-to-carbon bond is formed with inversion at the allylic position.

Example:

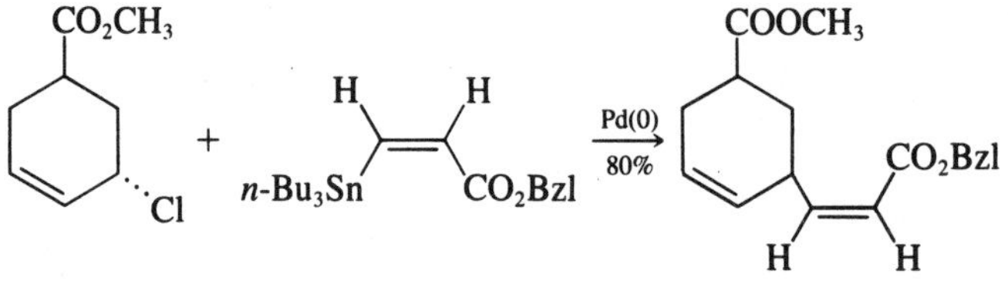

[1] F. K. Sheffy and J. K. Stille, *Am. Soc.*, **105**, 7173 (1983).

(R)-(+)- and (S)-(−)-2,2′-Bis(diphenylphosphino)-1,1′-binaphthyl (BINAP), 10, 36, **11**, 53–54. Details are available for the preparation of (±)-BINAP (**1**) and for the optical resolution with a chiral palladium complex (**2**), prepared from (S)-(−)-N,N-dimethyl-α-phenylethylamine. The complexes of **2** with (R)- and (S)-BINAP are obtained in about

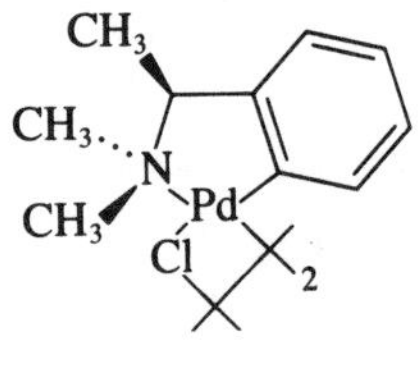

2

35% yield by fractional crystallization; they are converted to the optically pure phosphines by reduction with lithium aluminum hydride.[1]

Details are also available for isomerization of prochiral N,N-dialkylamines to the optically pure corresponding (E)-enamine (**11**, 53–54) using BINAP as the chiral catalyst.[2] This novel chemistry permits a short synthesis of optically pure (R)-(+)-citronellal (**4**) from isoprene utilizing the isomerization of N,N-diethylnerylamine (**2**) to R(−)-**3**.[3]

2

(−)-**3** (95% ee)   (+)-**4** (95.8% ee)

[1] H. Takaya and R. Noyori, *Org. Syn.*, submitted (1983).
[2] K. Tani, T. Yamagata, S. Akutagawa, H. Kumobayashi, T. Taketomi, H. Takaya, A. Miyashita, R. Noyori, and S. Otsuka, *Am. Soc.*, **106**, 5208 (1984); K. Tani, T. Yamagata, and S. Otsuka, *ibid.*, submitted (1983).
[3] K. Takabe, T. Yamada, T. Katagiri, and J. Tanaka, *ibid.*, submitted (1983).

Bis(isopropylthio)boron bromide, BrB[SCH(CH$_3$)$_2$]$_2$ (**1**). Mol. wt. 241.02, b.p. 75–80°/ 2 mm. The reagent is prepared in 90% yield by reaction of Pb[SCH(CH$_3$)$_2$]$_2$ with 1 equiv. of BBr$_3$.

Isopropylthiomethyl ethers,[1] ROCH$_2$SCH(CH$_3$)$_2$. Methoxyethoxymethyl (MEM) ethers on reaction with **1** and 4-dimethylaminopyridine in CH$_2$Cl$_2$ are converted into isopropylthiomethyl ethers. In the absence of DMAP, the MEM ether is cleaved to the free alcohol. Isopropylthiomethyl ethers can be cleaved to the alcohol under nonacidic conditions by silver nitrate and 2,6-lutidine.

The boron reagent **1** is recommended also for cleavage of mono MEM or methoxymethyl ethers of 1,2- or 1,3-diols, which are cleaved by acids mainly to cyclic diol formals.

[1] E. J. Corey, D. H. Hua, and S. P. Seitz, *Tetrahedron Letters*, **25**, 3 (1983).

Bis(methoxycarbonyl)sulfur diimide (1). Moisture-sensitive, viscous yellow oil.
  Preparation:

$$H_2NCOOCH_3 \xrightarrow[73\%]{Cl_2} Cl_2NCOOCH_3 \xrightarrow[quant.]{SCl_2,\ Py} CH_3OOCN{=}S{=}NCOOCH_3$$

1

Allylic amination.[1]   The reagent undergoes an ene reaction with a variety of alkenes to give products that undergo [2,3] sigmatropic rearrangement to N-(2-alkenyl)diaminosulfanes. The overall result is substitution of an allylic hydrogen by a nitrogen group. Mild alkaline hydrolysis results in carbamates, which can be hydrolyzed further to primary amines or reduced to methylamines by $LiAlH_4$.

Example:

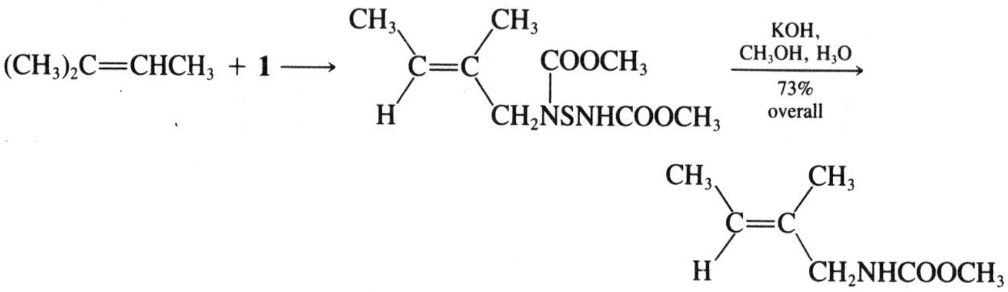

$$(CH_3)_2C=CHCH_3 + 1 \longrightarrow$$

with product structure and reaction conditions:

KOH, CH_3OH, H_2O
73% overall

[1] G. Kresze and H. Münsterer, *J. Org.*, **48**, 3561 (1983); G. Kresze, H. Braxmeier, and H. Münsterer, *Org. Syn.*, submitted (1984).

Bis(1-methoxy-2-methyl-1-propenyloxy)dimethylsilane, $(CH_3)_2Si\left[OC\begin{smallmatrix}OCH_3\\C(CH_3)_2\end{smallmatrix}\right]_2$ **(1).**

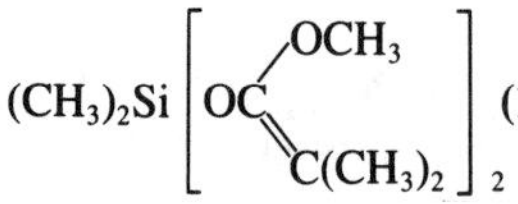

The divinyloxysilane is prepared by reaction of $(CH_3)_2SiCl_2$ with methyl lithioisobutyrate (2 equiv.) in THF (82% yield).

Bifunctional protection of **1,2-, 1,3-,** *and* **1,4-***diols, dithiols, and dicarboxylic acids.*   The reagent (**1**) converts these substrates into silylene derivatives in high yield.[1]

Examples:

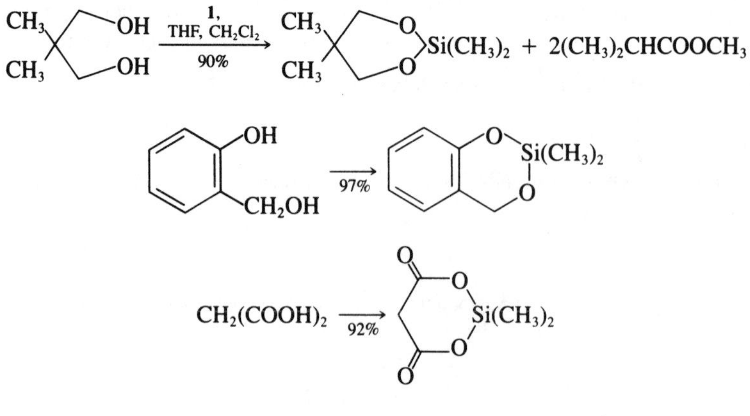

$CH_2(COOH)_2 \xrightarrow{92\%}$

[1] Y. Kita, H. Yasuda, Y. Sugiyama, F. Fukata, J. Haruta, and Y. Tamura, *Tetrahedron Letters*, **24**, 1273 (1983).

***trans*-2,5-Bis(methoxymethyl)pyrrolidine,** CH_3OCH_2 ⟨N-H⟩ CH_2OCH_3 (**1**).

(2R,5R)-(−)-**1** and (2S,5S)-(+)-**1** are obtained from (±)-*trans*-N-benzylpyrrolidine-2,5-dicarboxylic acid (resolution, reduction, etherification).

Asymmetric alkylation of amides.[1]   The enolates of amides (**2**) obtained by acylation of (+)- or (−)-**1** are alkylated consistently with >95% de. The stereoselectivity is independent of the base, solvent, or temperature. The products are converted without racemization into the corresponding acids by demethylation (BCl_3) followed by acid-catalyzed hydrolysis.

Example:

(2R,5R)-(−)-**2**   **3** (>95% de)

Similar results are obtained with the corresponding methoxymethoxymethyl ethers.

[1] Y. Kawanami, Y. Ito, T. Kitagawa, Y. Taniguchi, T. Katsuki, and M. Yamaguchi, *Tetrahedron Letters*, **25**, 857 (1984).

2,4-Bis(4-methoxyphenyl)-1,3,2,4-dithiaphosphetane-2,4-disulfide, 8, 327; **9**, 49–50; **10**, 39; **11**, 54.

Thiopeptides.   Lawesson's reagent is effective for transformation of protected dipeptides into the corresponding thiopeptides (**11**, 55), but not for higher peptides, possibly because of the limited solubility of the reagent. For monothionation of oligopeptides, Belleau *et al.*[1] recommend the related reagent **1**, prepared from diphenyl ether, which is

1, m.p. 187–190°

readily soluble in THF, CH_3CN, $CHCl_3$. An amide linkage involving glycine reacts particularly readily, permitting selective monothionation.

[1] G. Lajoie, F. Lépine, L. Maziak, and B. Belleau, *Tetrahedron Letters*, **24**, 3815 (1983).

$$CH_3$$

(S,S)- or (R,R)-N,N′-Bis(α-methylbenzyl)sulfamide, $SO_2(NH\overset{*}{C}HC_6H_5)_2$ **(1).** (S,S)-
1, m.p. 99°, α_D −80.1°, is obtained by reaction of (S)-(−)-α-methylbenzylamine with SO_2Cl_2 and triethylamine in CH_2Cl_2 (89% yield). (R,R)-**1** is similarly prepared from (R)-(+)-α-methylbenzylamine.

Asymmetric reduction of ketones.[1] Lithium aluminum hydride, after partial decomposition with 1 equiv. of **1** and an amine additive such as N-benzylmethylamine, can effect asymmetric reduction of prochiral ketones at temperatures of −20°. The highest selectivity is observed with aryl alkyl ketones (55–87% ee), but dialkyl ketones can be reduced stereoselectively if the two groups are sterically different. Thus cyclohexyl methyl ketone can be reduced with 71% ee.

[1] J. M. Hawkins and K. B. Sharpless, *J. Org.*, **49**, 3862 (1984).

Bis(N-methylpiperazinyl)aluminum hydride (1). This hydride was originally prepared from aluminum hydride and N-methylpiperazine, and was used to reduce carboxylic acids directly to aldehydes.[1] It can be prepared more conveniently from lithium aluminum hydride and the amine. It is useful for reduction of aliphatic and aromatic acids to aldehydes (80–95% yield). Significantly, it reduces α,β-unsaturated acids to aldehydes without reduction of the double bond (70–80% yield).[2]

[1] M. Muraki and T. Mukaiyama, *Chem. Letters,* 1447 (1974); *idem, ibid.,* 215, 875 (1975).
[2] T. D. Hubert, D. P. Eyman, and D. F. Wiemer, *J. Org.*, **49**, 2279 (1984).

2,5-Bis[(Z)-(2-nitrophenylsulfenyl)methylene]-3,6-dimethylene-7-oxabicyclo[2.2.1]heptane (1). This diene is obtained by double addition of 2-nitrobenzenesulfenyl chloride to **2** (**10**, 391–392).

1 (Ar = $C_6H_4NO_2$-o)

Diels-Alder reactions.[1] The diene **1** is similar to **2** in that it can undergo two successive Diels-Alder additions with different dienophiles with high regio- and stereoselectivity.

Example:

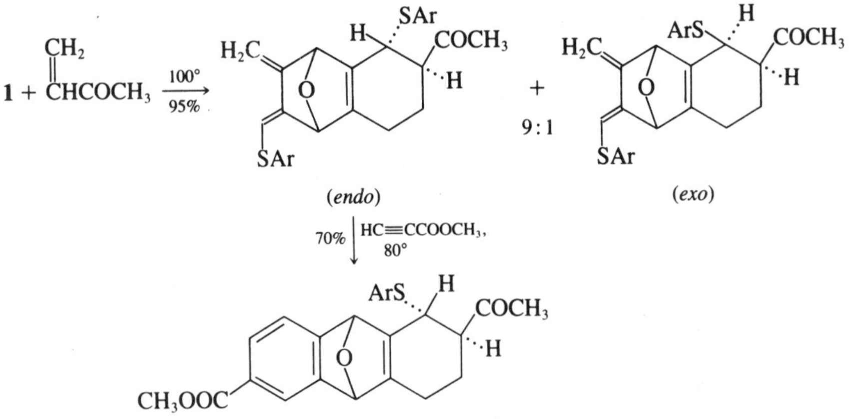

(endo)   (exo)

9 : 1

70% | HC≡CCOOCH₃, 80°

[1] J.-M. Tornare and P. Vogel, *J. Org.*, **49**, 2510 (1984).

1,1-Bis(phenylsulfonyl)cyclopropane (1). Mol. wt. 322.39, m.p. 145–146°.
  Preparation:[1]

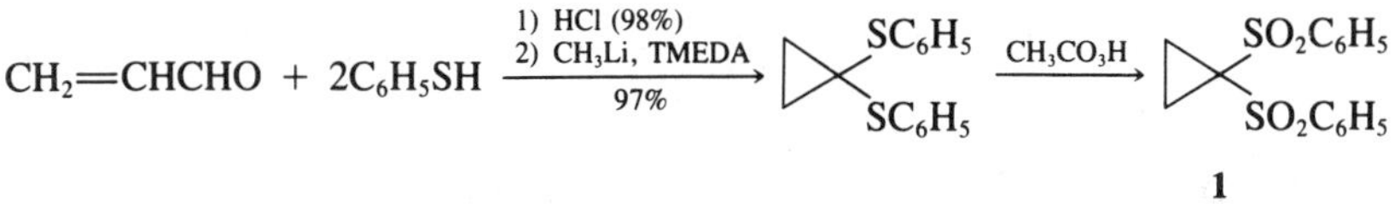

$$CH_2=CHCHO + 2C_6H_5SH \xrightarrow[\substack{97\%}]{\substack{1) \ HCl \ (98\%) \\ 2) \ CH_3Li, \ TMEDA}} \triangle\substack{SC_6H_5 \\ SC_6H_5} \xrightarrow{CH_3CO_3H} \triangle\substack{SO_2C_6H_5 \\ SO_2C_6H_5}$$

1

Propylene 1,3-dipole ($\overset{+}{C}H_2CH_2\overset{-}{C}H_2$).[2]   The ring of **1** is opened by various nucleophiles such as C_6H_5SH or organic cuprates.
Examples:

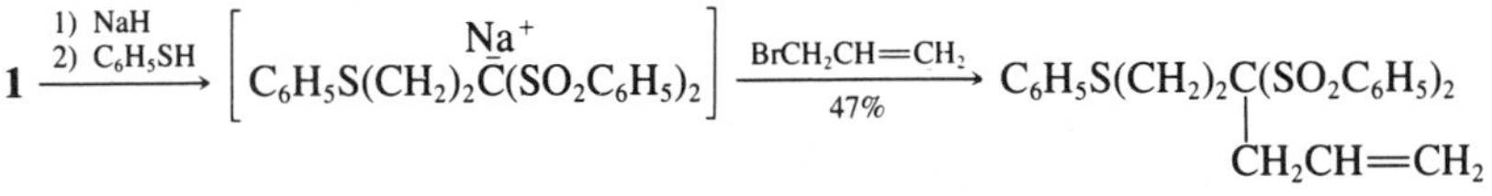

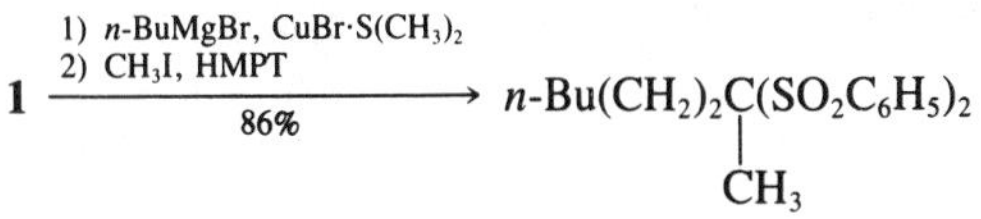

$$\mathbf{1} \xrightarrow{\substack{1) \ NaH \\ 2) \ C_6H_5SH}} \left[C_6H_5S(CH_2)_2\overset{Na^+}{\overset{|}{C}}(SO_2C_6H_5)_2 \right] \xrightarrow[\substack{47\%}]{BrCH_2CH=CH_2} C_6H_5S(CH_2)_2\underset{\substack{| \\ CH_2CH=CH_2}}{C}(SO_2C_6H_5)_2$$

$$\mathbf{1} \xrightarrow[\substack{86\%}]{\substack{1) \ n\text{-BuMgBr, CuBr·S(CH}_3)_2 \\ 2) \ CH_3I, \ HMPT}} n\text{-Bu}(CH_2)_2\underset{\substack{| \\ CH_3}}{C}(SO_2C_6H_5)_2$$

[1] T. Cohen, W. M. Daniewski, and R. B. Weisenfeld. *Tetrahedron Letters*, 4665 (1978).
[2] B. M. Trost, J. Cossy, and G. Burks, *Am. Soc.*, **105**, 1052 (1983).

Bispyridinesilver permanganate, 11, 61–62.

Deoximation.[1]   Oximes are converted into the corresponding carbonyl compounds by oxidation with this reagent in CH_2Cl_2 at room temperature. Yields range from 50–90%.

[1]H. Firouzabadi and A. Sardarian, *Syn. Comm.*, **13**, 863 (1983).

Bis(2,2,2-trichloroethyl) azodicarboxylate, $Cl_3CCH_2OOCN\!=\!NCOOCH_2CCl_3$(**1**). Mol. wt. 364.39, m.p. 108–110°.
    Preparation:

$$H_2NNH_2 \ + \ 2ClCO_2CH_2CCl_3 \ \xrightarrow[93\%]{} \ Cl_3CCH_2O_2CNHNHCO_2CH_2CCl_3 \ \xrightarrow[76\%]{HNO_3} \ 1$$

Diels-Alder reactions.[1]   This ester can be more useful than the corresponding methyl or ethyl ester because it is hydrolyzed under neutral conditions. Moreover, Diels-Alder reactions with **1** can proceed faster and at a lower temperature than those with other esters.

[1] R. D. Little and M. G. Venegas, *Org. Syn.*, **61**, 17 (1983).

N,N-Bis(trimethylsilyl)methoxymethylamine, $[(CH_3)_3Si]_2NCH_2OCH_3$ (**1**). The reagent is prepared by reaction of chloromethyl methyl ether with lithium bis(trimethylsilyl)amide (86% yield); b.p. 91–92°/86 mm.
    Aminomethylation.   The reagent reacts with Grignard reagents to form N,N-bis(trimethylsilyl)amines in 60–90% yield (equation I). A similar reaction with organo-lithium compounds requires added magnesium bromide for satisfactory yields.[1]

$$\text{(I)} \quad RMgX + 1 \ \xrightarrow[60-90\%]{ether} \ RCH_2N[Si(CH_3)_3]_2 \ \xrightarrow{CH_3OH} \ RCH_2NH_2$$

The reaction of ketene silyl acetals with **1** catalyzed by trimethylsilyl triflate affords N,N-bis(trimethylsilyl)-β-amino esters (**2**), which undergo desilylation to β-amino acids (**3**). These products are convertible into β-lactams on treatment with base.[2]

2

3

Zinc bromide catalysis effects a similar aminomethylation of silyl sulfides and phosphites (equation II).[3]

$$\text{(II)} \quad RSSi(CH_3)_3 + 1 \xrightarrow[\text{75-95\%}]{\text{ZnBr}_2} RSCH_2N[Si(CH_3)_3]_2 + CH_3OSi(CH_3)_3$$

[1] T. Morimoto, T. Takahashi, and M. Sekiya, *J.C.S. Chem. Comm.*, 794 (1984).
[2] K. Okano, T. Morimoto, and M. Sekiya, *ibid.*, 883 (1984).
[3] T. Morimoto, M. Aono, and M. Sekiya, *ibid.*, 1055 (1984).

Bis(trimethylsilyl)peroxide, $[(CH_3)_3SiO]_2$ (**1**), **11**, 67.
Oxidative desulfonylation.[1]   The α-anion of alkyl phenyl sulfones on reaction with **1** is converted into aldehydes or ketones in good to high yield (equation I).

Oxidation of alcohols.[2]   In the presence of catalytic amounts of pyridinium dichromate, this peroxide can oxidize primary and secondary alcohols to the corresponding carbonyl compounds in 70–100% yield. Reactions catalyzed by dichlorotris(triphenylphosphine)ruthenium are useful for highly selective oxidation of primary allylic and benzylic alcohols in the presence of secondary ones.

[1] J. R. Hwu, *J. Org.*, **48**, 4432 (1983).
[2] S. Kanemoto, K. Oshima, S. Matsubara, K. Takai, and H. Nozaki, *Tetrahedron Letters*, **24**, 2185 (1983).

Bis(trimethylsilyl)peroxide–Vanadyl acetylacetonate, $(CH_3)_2SiOOSi(CH_3)_3–VO(acac)_2$.
Isomerization of allylic alcohols.[1]   These reagents, in a 3:1 ratio, catalyze the rearrangement of primary allylic alcohols to tertiary allylic alcohols.

Examples:

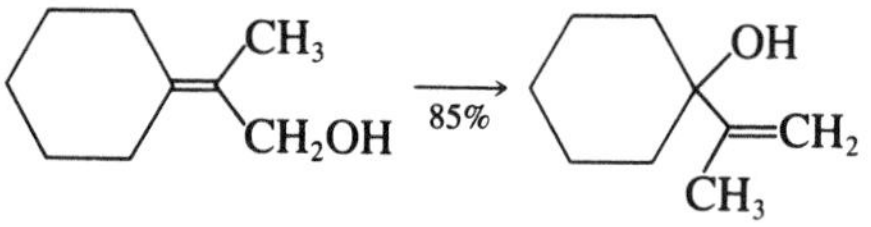

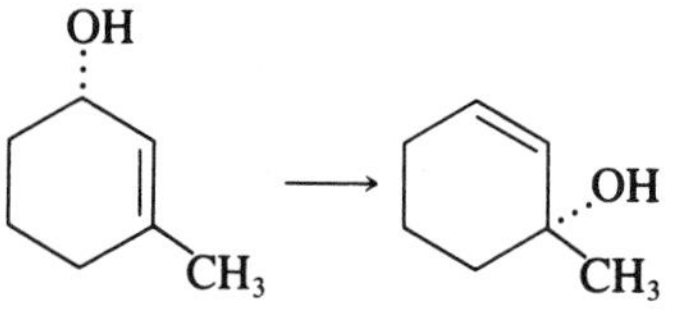

(S)-(−), 40% ee   (R)-(+), 29% ee

[1] S. Matsubara, K. Takai, and H. Nozaki, *Tetrahedron Letters*, **24**, 3741 (1983).

Borane–Dimethyl sulfide, 4, 124, 191; **5**, 47; **8**, 49; **10**, 49–50; **11**, 69.

Selective reduction of α-hydroxy esters.[1] Reduction of esters with BMS is slow, but can be facilitated by addition of catalytic amounts of sodium borohydride. This method of reduction can be used to effect regioselective reduction of α-hydroxy esters. Thus reduction of dimethyl (S)-(−)-malate (1) results in formation of methyl (3S)-3,4-dihydroxybutanoate (2).

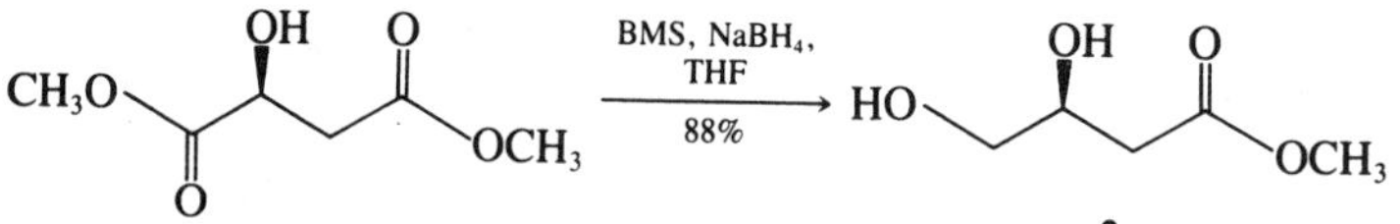

Another example:

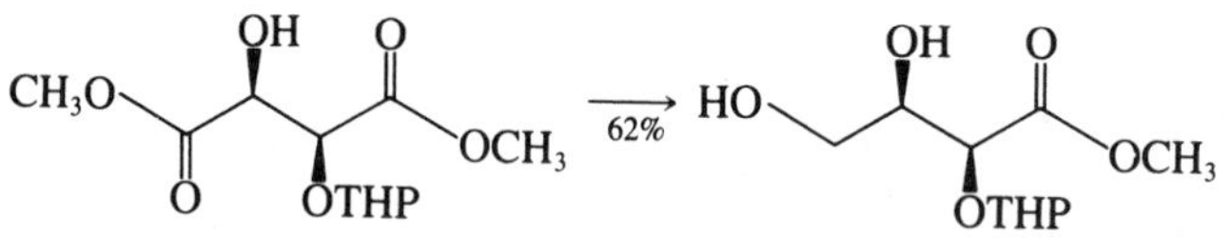

Reduction of amino acids.[2] The ethyl ester hydrochlorides of amino acids can be reduced to the corresponding amino alcohols wtihout racemization by borane–dimethyl sulfide in yields of about 50% (distilled). The rate is enhanced if the dimethyl sulfide is allowed to escape during the reaction.

[1] S. Saito, T. Hasegawa, M. Inaba, R. Nishida, T. Fujii, S. Nomizu, and T. Moriwake, *Chem. Letters*, 1389 (1984).
[2] G. A. Smith and R. E. Gawley, *Org. Syn.*, submitted (1983).

Borane–Pyridine, 1, 963–964; **8**, 50–51; **9**, 59; **11**, 69.

Reductive amination of carbonyl compounds.[1]   Sodium cyanoborohydride has been the reductant of choice for this reaction, even though it is highly toxic (**4**, 448–449). Borane–pyridine is recommended as a nontoxic substitute. The reactions are conducted in acetic acid, in which this borane is fairly stable, and a co-solvent such as CH_2Cl_2 or THF.

[1] A Pelter, R. M. Rosser, and S. Mills, *J.C.S. Perkin I,* 717 (1984).

Borane–Tetrahydrofurane.

Reduction of α,β-*unsaturated nitroalkenes.*[1]   These nitroalkenes are reduced by $BH_3·THF$ and a catalytic amount of $NaBH_4$ to alkylamines in 85–91% yield.

[1] M. S. Mourad, R. S. Varma, and G. W. Kabalka, *Syn. Comm.,* **14**, 1099 (1984).

Borane–Trimethylamine, 1, 1229–1230; **5**, 708.

Acylation of amines.[1]   Reaction of a primary or secondary amine, borane–trimethylamine, and a carboxylic acid in the molar ratio 1:1:3 in refluxing xylene results in formation of an amide in yields generally of 70–95% (equation I).

$$(I) \quad \begin{array}{c} R^1 \\ \diagdown \\ \diagup \\ R^2 \end{array} NH + R^3COOH \xrightarrow[\sim70-95\%]{BH_3·N(CH_3)_3} \begin{array}{c} R^1 \\ \diagdown \\ \diagup \\ R^2 \end{array} N\text{-}\overset{\overset{\displaystyle O}{\|}}{C}R^3$$

A similar reaction of a secondary amine, the borane complex, and a carboxylic acid in the molar ratio 1:2:2 results in N-alkylation to furnish a tertiary amine. The reaction is sluggish with aliphatic amines and acids, but proceeds in good yield with aromatic substrates.

[1] G. Trapani, A. Reho, and A. Latrofa, *Synthesis,* 1013 (1983).

Boron trichloride, 1, 67–68; **2**, 34–35; **3**, 31–32; **4**, 42–43; **5**, 50–51; **6**, 65; **9**, 62–63.

Selective demethylation of methoxyarenes.[1]   Boron trichloride is generally useful for selective cleavage of the more hindered methoxy group of polymethoxybenzenes and of *peri*-methoxynaphthalenes.

Example:

[1] C. F. Carvalho and M. V. Sargent, *J.C.S. Chem. Comm.,* 227 (1984).

Boron trifluoride etherate, 1, 70–72; **2,** 35–36; **3,** 33; **4,** 44–45; **5,** 54–55; **6,** 65–67; **7,** 31–32; **8,** 51–52; **9,** 64–65; **10,** 52–56; **11,** 172–75.

Selective demethylation.[1]   Treatment of 1,2-, 1,4-, 1,5-, and 1,8-dimethoxyanthra-quinone with BF_3 etherate in C_6H_6 (reflux) results in difluoroboron chelates, which are hydrolyzed to the corresponding 1-hydroxy-2-methoxy-, -4-methoxy-, -5-methoxy-, and -8-methoxyanthraquinones when heated in CH_3OH. This selective dealkylation can be used to convert 1,4,5-trimethoxyanthraquinone (**1**) into either 4-hydroxy-1,5-dimethoxy-anthraquinone (**4**) or 1,4-dihydroxy-5-methoxyanthraquinone (**5**).

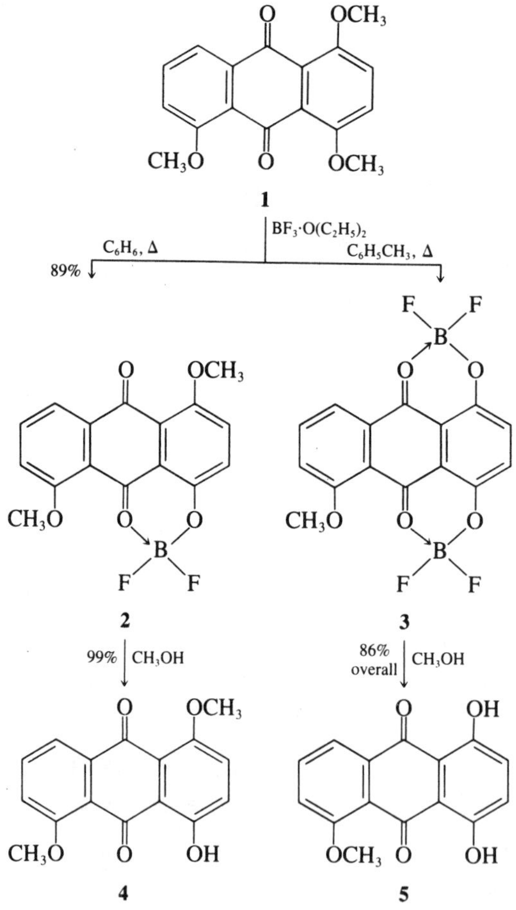

Nazarov cyclization.[2]   The key step in a recent synthesis[3] of the sesquiterpene tri-chodiene (**5**) is the stereospecific formation of the adjacent quaternary centers by a Nazarov cyclization of the dienone **1** to **2**, effected in high yield with BF_3 etherate (8 equiv.). In this case TFA is ineffective. Remaining steps involve cleavage of the central ring of **2** by Beckmann fragmentation of the corresponding oxime to give a mixture of **3** and **4**, both of which can be converted into **5** by suitable modification of the functional groups.

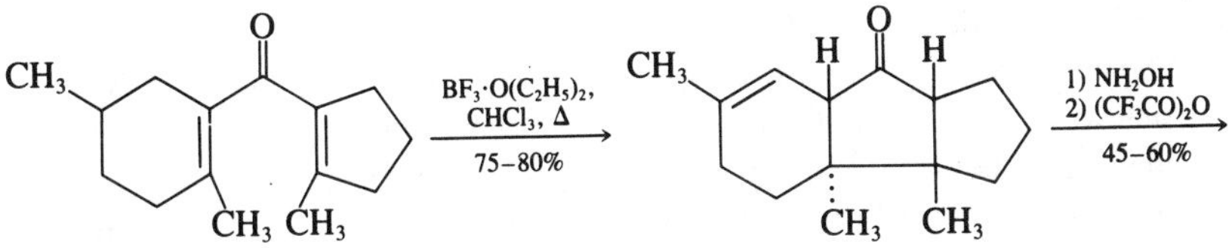

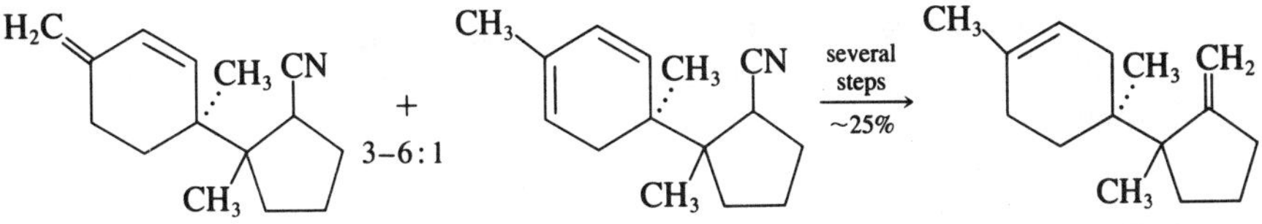

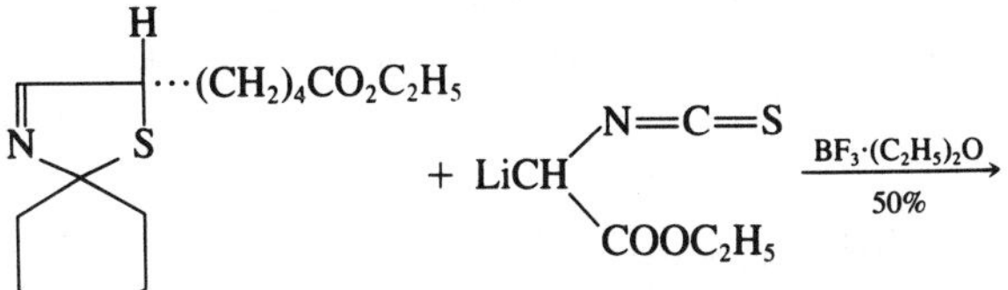

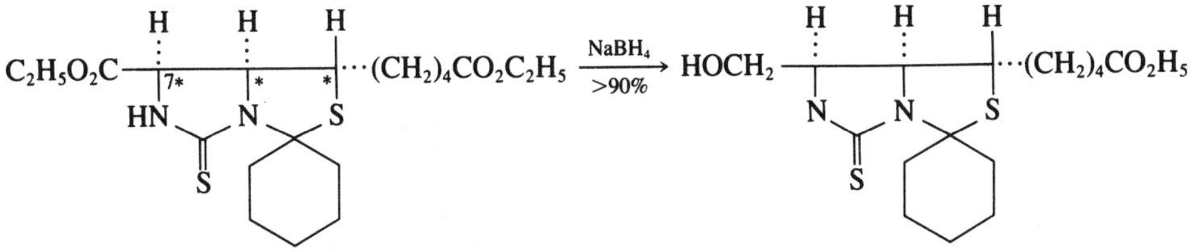

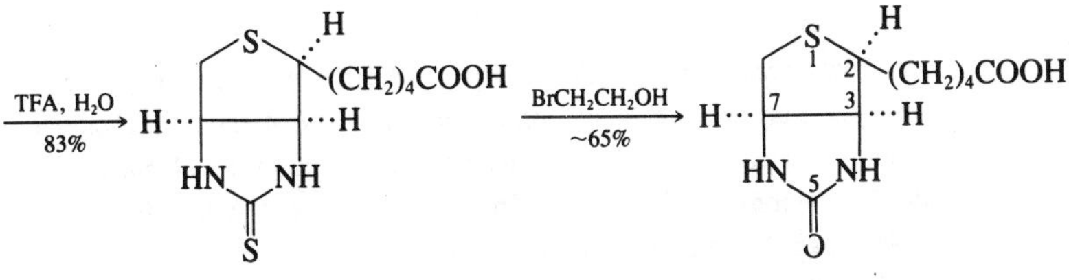

Activation of imines.[4]   A recent synthesis of (+)-biotin (**6**) is based on activation of the imine group of a 3-thiazoline (**1**) by BF$_3$ etherate to nucleophilic attack. Thus reaction of **1**, substituted by the biotin side chain, with the ester enolate **2** in the presence of 1 equiv. of the Lewis acid results in the thiazolidine **3** as the major product. The stereochemistry at the future C$_7$-center is determined to some extent by the ester group of **2**. Selective reduction of the C$_7$-ester group furnishes the alcohol **4**. Reaction of **4** with aqueous trifluoroacetic acid at 100° effects hydrolysis of the thiazolidine ring, formation of the thiophane ring, and ester hydrolysis to provide (±)-2-thiobiotin (**5**). However, conversion of **4** into the optically active (+)-d-camphorsulfonate (reaction with d-camphorsulfonyl chloride and chromatography of the resulting mixture) followed by reaction with TFA results in optically active (+)-**5**. The final step in the synthesis of biotin (**6**), conversion of a thiourea group to a urea, is effected with excess bromoethanol in N-methylpyrrolidine at 110°.

α,β-Acetylenic ketones.[5]   These ketones are readily available by reaction of lithium alkynyltrifluoroborates with acid anhydrides.

Example:

$$CH_3(CH_2)_5C{\equiv}CH \xrightarrow[\text{2) } BF_3 \cdot O(C_2H_5)_2, \ -78°]{\text{1) } n\text{-BuLi, THF, } -78°} \left[CH_3(CH_2)_5C{\equiv}C\bar{B}F_3 \right] Li^+$$

$$\xrightarrow[74\%]{(C_2H_5CO)_2O} CH_3(CH_2)_5C{\equiv}C{-}\overset{\displaystyle O}{\overset{\|}{C}}C_2H_5$$

Organolithium additions.[6]   Alkyl-, alkenyl-, and aryllithiums can react as potent nucleophiles in the presence of BF₃ etherate. Thus epoxides and oxetanes are rapidly alkylated by reaction with an organolithium and BF₃ etherate (2 equiv. of each) at −78°.

Examples:

$$\text{oxetane} + C_6H_5Li \xrightarrow{96\%} C_6H_5(CH_2)_3OH$$

Protiodesilylation.[7]   The complex of BF₃ and acetic acid serves as both a proton source and an efficient nucleophile for silicon (**9**, 492). The protonation of allylsilanes containing a carboxyl group or hydroxyl group is intramolecular and can result in 1,3- and 1,4-asymmetric induction. Thus treatment of **1** with BF₃·HOAc results in **2** with greater than 8:1 diastereoselectivity.

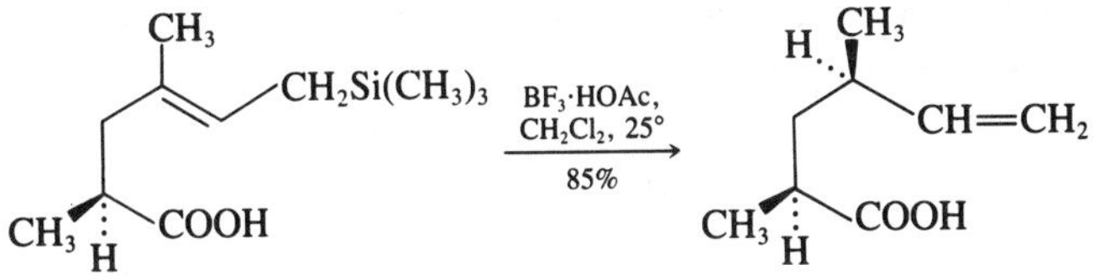

1 → 2

Under the same conditions, the alcohol **3a** is protonated selectively to give **4**, but the corresponding acetate (**3b**) is protonated with reverse selectivity.

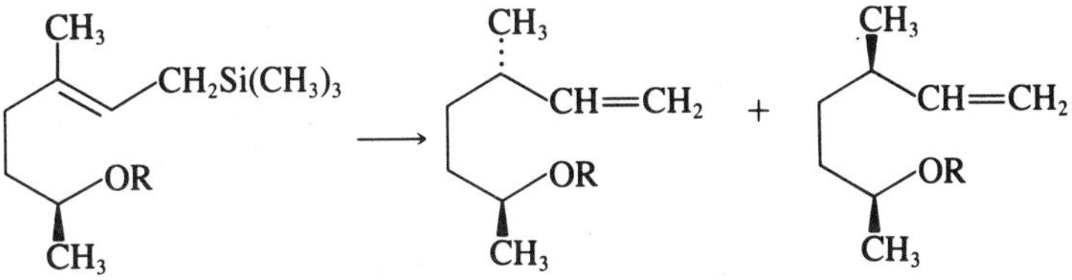

3a, R = H   **4**   66 : 34   **5**

3b, R = Ac   33 : 67

Diels-Alder catalyst.[8]   This Lewis acid is the most effective catalyst for the reaction of furane with methyl acrylate. The *endo/exo* ratio is 7:3. Use of ZnI_2 results in an *endo/exo* ratio of 1:2 (**11**, 605).

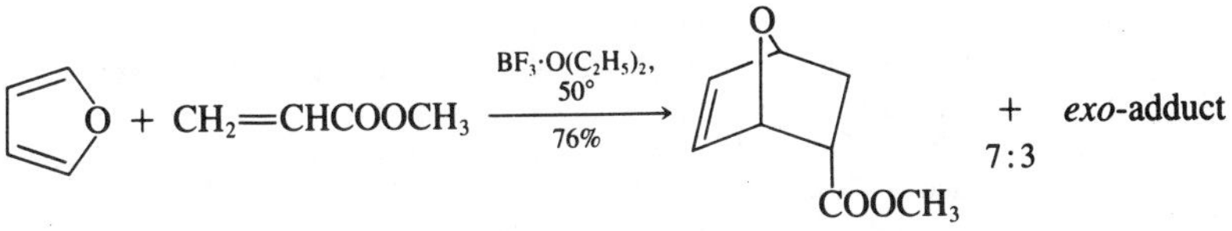

The reaction of furane with acryloyl chloride proceeds without a catalyst at room temperature, preferably in the presence of propylene oxide as a scavenger for hydrogen chloride (76% yield, *endo/exo* = 3:7).

Arylation with a homocuprate.[9]   A key step in a synthesis of the cannabis constituent **4** is the reaction of **1** with the homocuprate derived from olivetol dimethyl ether (**2**) by regiospecific lithiation followed by reaction with CuBr. The cuprate does not react with **1** in the absence of a Lewis acid, but in the presence of BF_3 etherate (3.5 equiv.), (−)-**3** is obtained in 78% yield with high regio- and stereospecificity. The dihydrobromide of

3 undergoes cyclization on treatment with PBr_3 to give a product that on demethylation and dehydrobromination is converted into **4**, in an overall yield from **1** of 59%.

(1S,4R)-**1**   **2**   (−)-**3**

4

Cleavage of oxetanes.[10]   Oxetanes can be cleaved in THF by lithium acetylides (excess) in the presence of BF_3 etherate (2–3 equiv.) to give γ-alkoxyalkynes. Other Lewis acids are much less effective.

[1] P. N. Preston, T. Winwick, and J. O. Morley, *J.C.S. Perkin I,* 1439 (1983).
[2] C. Santelli-Rouvier and M. Santelli, *Synthesis,* 429 (1983).
[3] K. E. Harding and K. S. Clement, *J. Org.,* **49**, 3870 (1984).
[4] R. A. Volkmann, J. T. Davis, and C. N. Meltz, *Am. Soc.,* **105**, 5946 (1983).
[5] H. C. Brown, U. S. Racherla, and S. M. Singh, *Tetrahedron Letters,* **25**, 2411 (1984).
[6] M. J. Eis, J. E. Wrobel, and B. Ganem, *Am. Soc.,* **106**, 3693 (1984).
[7] S. R. Wilson and M. F. Price, *ibid.,* **104**, 1124 (1982); *idem, Tetrahedron Letters,* **24**, 569 (1983).
[8] H. Kotsuki, K. Asao, and H. Ohnishi, *Bull. Chem. Soc. Japan,* **57**, 3339 (1984).
[9] R. W. Rickards and H. Rönneberg, *J. Org.,* **49**, 572 (1984).
[10] M. Yamaguchi, Y. Nobayashi, and I. Hirao, *Tetrahedron,* **40**, 4261 (1984).

Bromine, 3, 34; **4,** 46–47; **5,** 55–57; **6,** 70–73; **7,** 33–35; **8,** 52–53; **9,** 65–66; **10,** 56; **11,** 75–76.

Bromolactonization.   A key step in a synthesis of ramulosin (**4**) requires halolactonization of the γ,δ-unsaturated acid **1**, which can result in a γ- or a δ-lactone. Bromolactonization results mainly in the desired δ-lactone (**2**), which is converted in two steps into **4**.[1] Iodolactonization of **1** results mainly in the undesired γ-lactone and a mixture of isomeric δ-lactones.

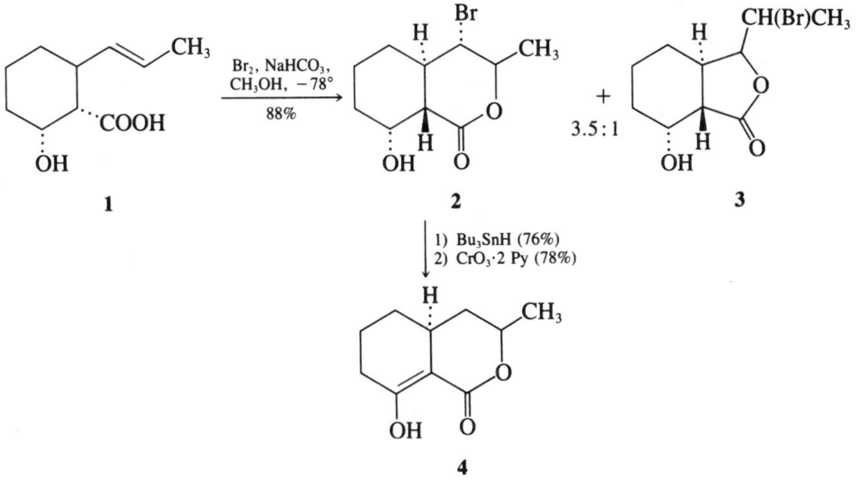

Protection of an anomeric center.[2]   The *p*-methoxybenzyl group is useful for protection of an anomeric center. *p*-Methoxybenzyl glycosides are prepared by Koenigs-Knorr glycosidation. Cleavage with DDQ (**11**, 166) is incomplete, but can be effected in high yield with bromine, NBS, or CAN in aqueous dichloromethane. Cleavage with bromine or NBS in CH_2Cl_2 results in the glycosyl bromide (60% yield).

6-Aryl-5-hydroxypyrones.[3]   Treatment of 5-(α-hydroxybenzyl)furoic acids (**1**) with Br_2 in CH_3OH containing Na_2CO_3 results in the hydroxypyrones **2** in about 40% yield. Only two intermediates have been identified in this multistep transformation.

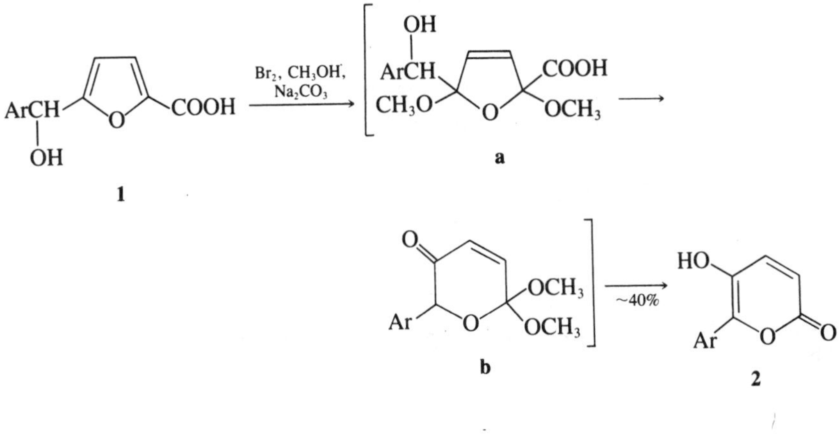

[1] R. Cordova and B. B. Snider, *Tetrahedron Letters*, **25**, 2945 (1984).
[2] B. Classon, P. J. Garegg, and B. Samuelsson, *Acta Chem. Scand.*, **38B**, 419 (1984).
[3] A. Pelter, R. S. Ward, D. C. James, and C. Kamakshi, *Tetrahedron Letters*, **24**, 3133 (1983).

Bromine–Nickel(II) alkanoates.

Regioselective oxidation of 2- and 4-alkyl-1,4-butanediols.[1]   2,2-Disubstituted 1,4-butanediols (**1**) are oxidized by this combination selectively to the γ-butyrolactones **2**. Cobalt(II) alkanoates and trityl tetrafluoroborate show comparable selectivity.

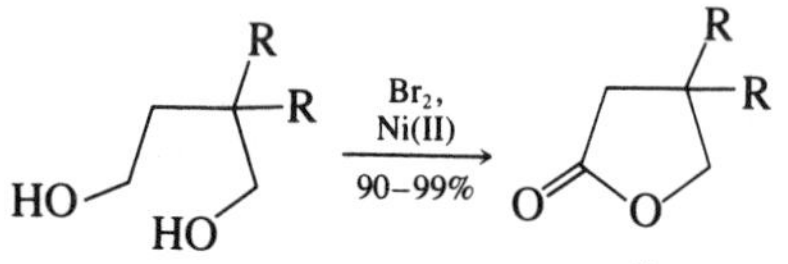

[1] M. P. Doyle, R. L. Dow, V. Bagheri, and W. J. Patrie, *J. Org.*, **48**, 476 (1983).

Bromine–Silver acetate.

Allylic acetoxylation.   Some years ago Petrow[1] reported that the reaction of dehydroisoandrosterone (**1**) with bromine and silver acetate resulted in acetoxylation at C_4 to give **2**. The reaction proceeds via the 5α,6β-dibromide (**a**). The same reaction is observed

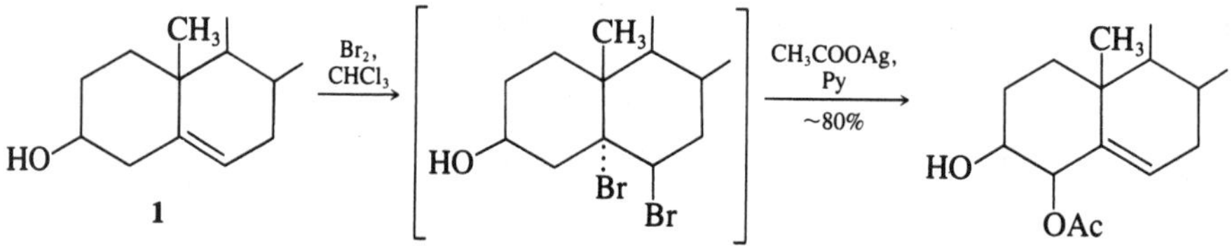

in the absence of the C_3-hydroxyl group, but in lower yield. The same reaction with Δ[5]-3β-chloro- or 3β-bromoandrostene-17-one results in formation of the 3α,6β-dibromides only.[2]

[1] V. A. Petrow, *J. Chem. Soc.*, 1077 (1937).
[2] J. R. Hanson, P. B. Reese, and H. J. Wadsworth, *J.C.S. Perkin I*, 2941 (1984).

1-Bromo-2-chlorocyclopropene (1). B.p. 45–48°/0.1 mm.
Preparation:

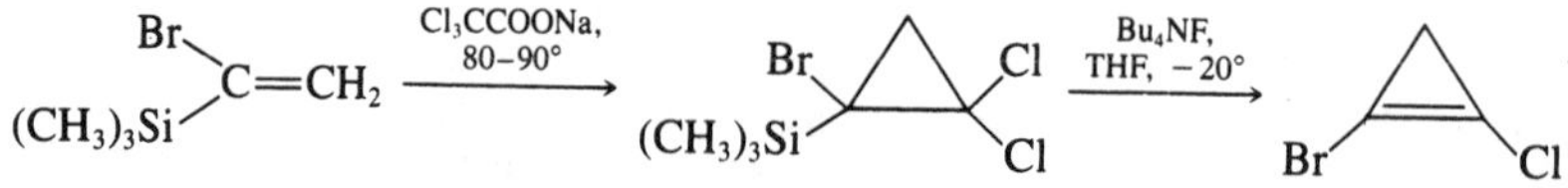

1

Diels-Alder reaction.[1]   The key step in a synthesis of the cycloproparene **4** involves cycloaddition of **1** to **2** to give the two regioisomers **3**. The mixture is converted into **4** by dehydrogenation and dehydrohalogenation.

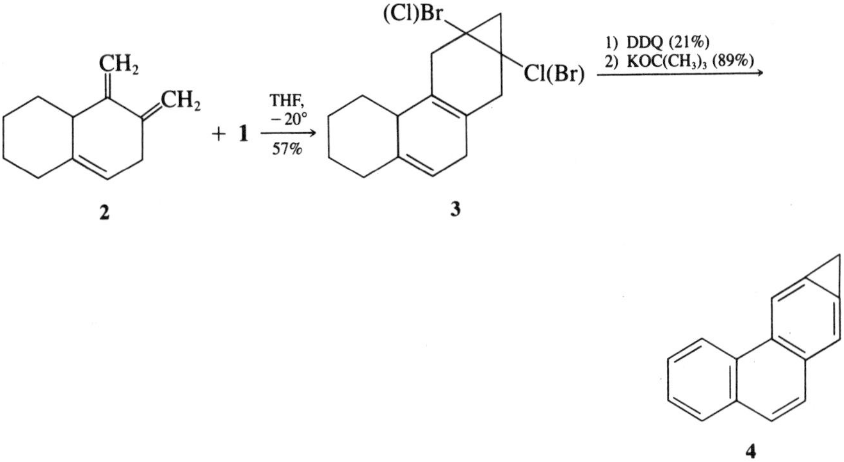

[1] W. E. Billups, L.-J. Lin, B. E. Arney, Jr., W. A. Rodin, and E. W. Casserly, *Tetrahedron Letters,* **25**, 3935 (1984).

1-Bromo-1-ethoxycyclopropane (1). B.p. 48–50°/25 mm.
Preparation:

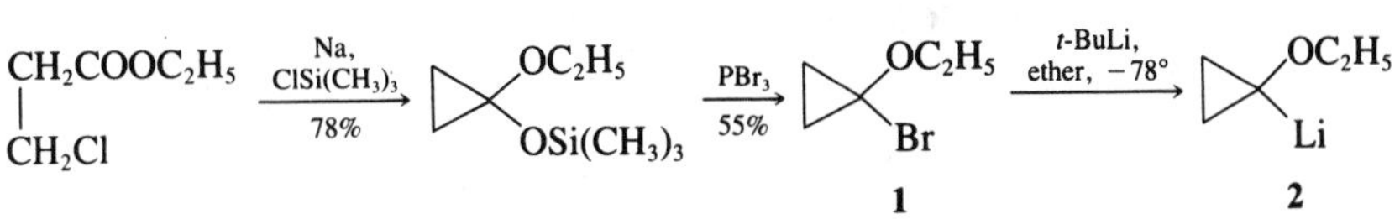

α-Substituted cyclobutanones.[1]   1-Ethoxycyclopropyllithium (**2**) reacts with a wide variety of aldehydes or ketones to form cyclopropylcarbinols (**3**), usually in >90% yield. These adducts rearrange in the presence of tetrafluoroboric acid (0.5–4 equiv.) at room temperature to cyclobutanones (**4**).

Example:

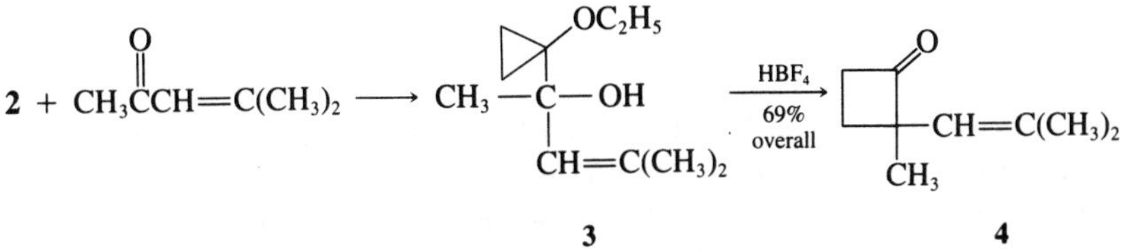

[1] R. C. Gadwood, *Tetrahedron Letters,* **25**, 5851 (1984).

Bromoform–Diethylzinc, 6, 194.

Cyclopropanecarboxylic esters.[1]   Reaction of ketene silyl acetals having an allylic hydrogen with a haloform and diethylzinc in pentane results in a cyclopropanecarboxylic ester as the major product.

Examples:

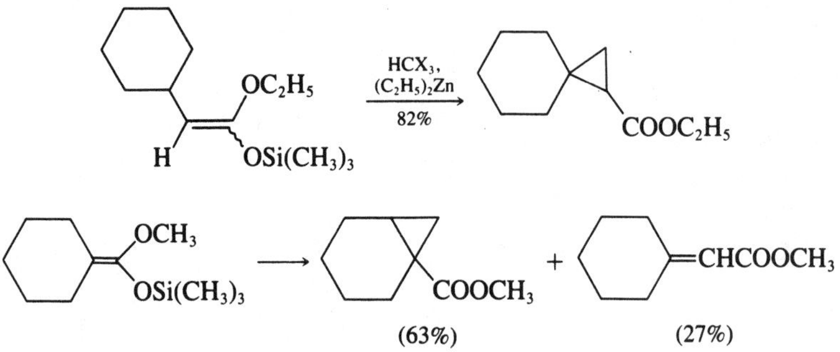

The same reaction with γ- or δ-unsaturated substrates results in intramolecular cyclopropanation.

Examples:

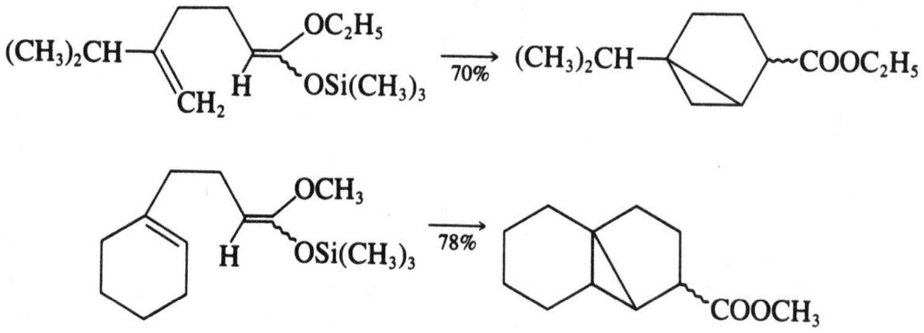

[1] G. Rousseau and N. Slougui, *Am. Soc.*, **106**, 7283 (1984).

Bromomagnesium diisopropylamide, $BrMgN[CH(CH_3)_2]_2$ (**1**). The base is prepared by reaction of CH_3MgBr and $HN[CH(CH_3)_2]_2$ in ether at 25°.

Thermodynamic magnesium enolates.[1]   Reaction of unsymmetrical cyclic ketones with **1** followed by chlorotrimethylsilane, triethylamine, and HMPT generates the more substituted silyl enol ether with high selectivity (95–97:5–3). Bromomagnesium hexamethyldisilazide is slightly less regioselective.

Example:

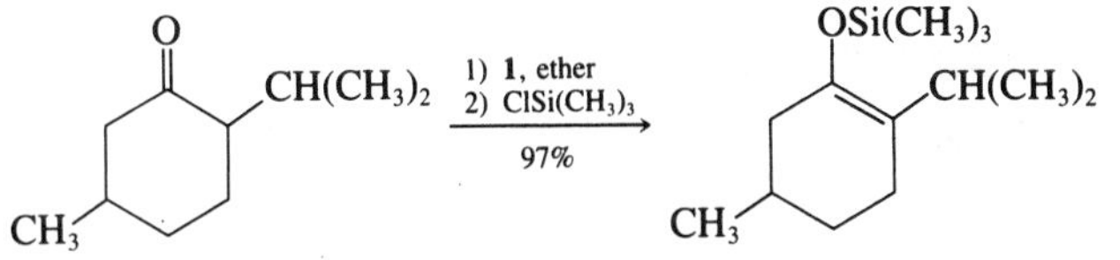

For another method, see Potassium hydride, this volume.

Aldol condensation.[2]   The final step in a synthesis of the tricyclic taxane ring system involves an intramolecular aldol condensation of **2**. Treatment with the usual bases results in a retro Michael reaction, but the desired cyclization to **3** can be effected in 90% yield by use of bromomagnesium diisopropylamide (**1**) or isopropylcyclohexylamide. The hydroxy ketone undergoes retroaldolization in the presence of mild acids or bases, but can be reduced and stored as the corresponding stable diol.

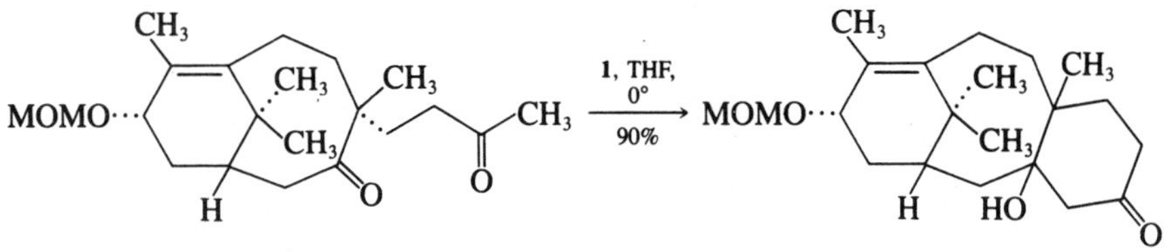

[1] M. E. Krafft and R. A. Holton, *Tetrahedron Letters*, **24**, 1345 (1983).
[2] R. A. Holton, *Am. Soc.*, **106**, 5731 (1984).

Bromomethanesulfonyl bromide, $BrCH_2SO_2Br$ (**1**). The reagent is a pale yellow liquid that is stable for several weeks at 25°. It is prepared by reaction of Br_2 with an aqueous slurry of 1,3,5-trithiane (46% yield).

1,3-Dienes.   The reagent undergoes a light-catalyzed regioselective addition to alkenes, particularly mono-, 1,1-di-, and trisubstituted alkenes. The adducts undergo dehydrobromination to α,β-unsaturated bromomethyl sulfones. On treatment with potassium *t*-butoxide these sulfones undergo a vinylogous Ramberg-Bäcklund reaction to furnish a 1,3-diene stereoselectively.[1]

Example:

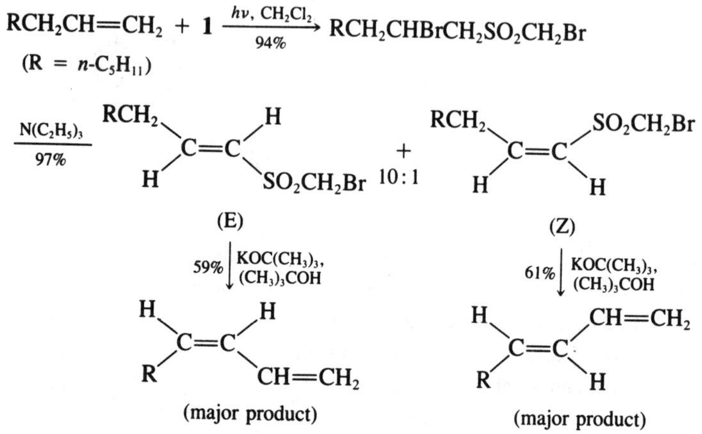

$$RCH_2CH{=}CH_2 + \mathbf{1} \xrightarrow[94\%]{h\nu,\ CH_2Cl_2} RCH_2CHBrCH_2SO_2CH_2Br$$
$$(R = n\text{-}C_5H_{11})$$

(E)   +   (Z)   10:1

(E) 59% ↓ KOC(CH₃)₃, (CH₃)₃COH

(Z) 61% ↓ KOC(CH₃)₃, (CH₃)₃COH

(major product)   (major product)

This process is generally applicable to cyclic and acyclic alkenes, and can be used to obtain branched, internal acyclic 1,3-dienes, 1,2-bismethylenecycloalkanes, and 3-methylene-1-cycloalkenes, as well as terminal 1,3-dienes.[2]

Examples:

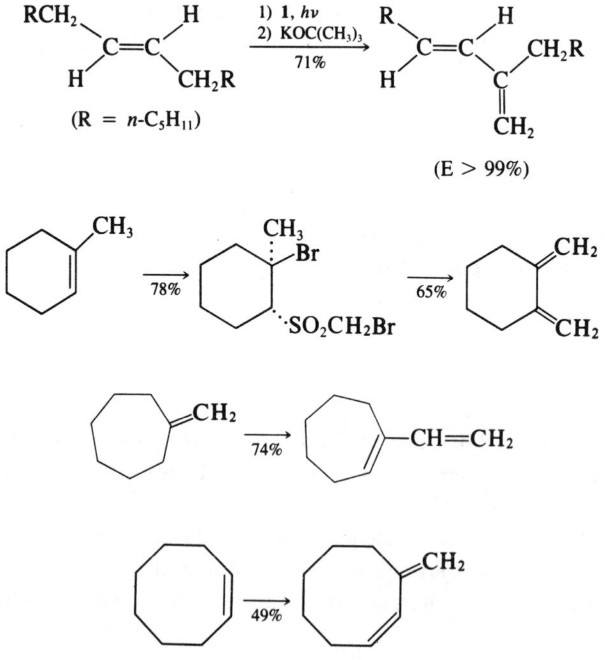

$(R = n\text{-}C_5H_{11})$

1) **1**, *hv*
2) KOC(CH₃)₃
71%

(E > 99%)

78%   65%

74%

49%

α-Methylene ketones.[3]   The reagent undergoes a light-catalyzed reaction with the silyl enol ether (2) of cycloheptanone to give 3. The product is converted by DBN into a mixture of the α-methylene ketone 4 and the 1,3-oxathiole 3,3-dioxide 5. Formation of these two products is fairly general, but the ratio varies widely with the substrate and the conditions. The reaction does not appear to be a useful general route to α-alkylidene ketones, since the initial reaction of the homologous reagent, $CH_3CHBrSO_2Br$, with 2 proceeds in modest yield (22%); the overall yield of α-ethylidene cycloheptanone (E/Z = 12:1) is 13%.

$$OSi(CH_3)_3 \quad \xrightarrow[77\%]{1, h\nu} \quad SO_2CH_2Br \quad \xrightarrow[-78 \to 23°]{DBN,} \quad CH_2 \quad + \quad SO_2$$

2   3   4 (77%)   5 (21%)

[1] E. Block and M. Aslam, *Am. Soc.*, **105**, 6164 (1983).
[2] E. Block, M. Aslam, V. Eswarakrishnan, and A. Wall, *ibid.*, **105**, 6165 (1983); E. Block and M. Aslam, *Org. Syn*, submitted (1984).
[3] E. Block, M. Aslam, R. Iyer, and J. Hutchinson, *J. Org.*, **49**, 3664 (1984).

Bromomethyllithium, $BrCH_2Li$; **Chloromethyllithium**, $ClCH_2Li$. These reagents are unstable when generated by bromine–lithium exchange with *n*-BuLi even at $-130°$. Stability is improved by coordination with lithium bromide. The most satisfactory results are obtained by generation from a bromohalomethane with *sec*-BuLi in the presence of 1 equiv. of LiBr in THF, ether, or pentane at $-100°$.

These reagents react with aldehydes and ketones to form epoxides or halohydrins, and with esters to form α-halomethyl ketones.[1]

Examples:

[1] T. Tarhouni, B. Kirschleger, M. Rambaud, and J. Villieras, *Tetrahedron Letters*, **25**, 835 (1984).

2-Bromomethyl-3-(trimethylsilylmethyl)butadiene,

(1).   The butadiene is obtained as an oil by reaction of 2,3-bis(trimethylsilylmethyl)butadiene with NBS.

Multiple ring annelations.[1] This diene can serve as a conjunctive reagent in the synthesis of multiple ring systems.

Example:

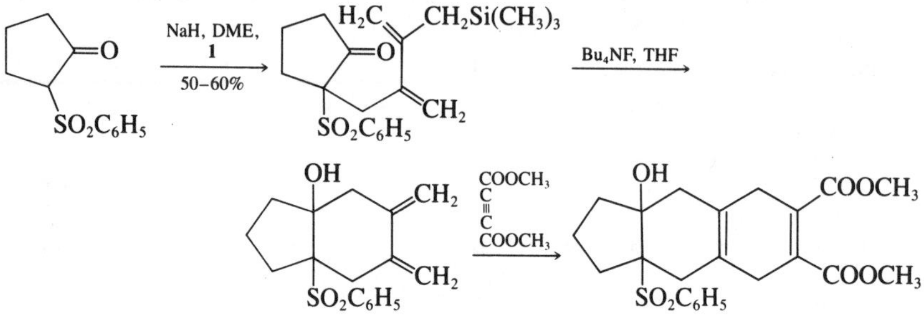

[1] B. M. Trost and R. Remuson, *Tetrahedron Letters,* **24**, 1129 (1983).

Bromonium dicollidine perchlorate, 10, 212–213.

Bromocyclization of allylic thiocarbamidates.[1]   Allylic thiocarbamidates (**1**), prepared as shown, on treatment with this reagent cyclize to bromo carbamates (**2**). This

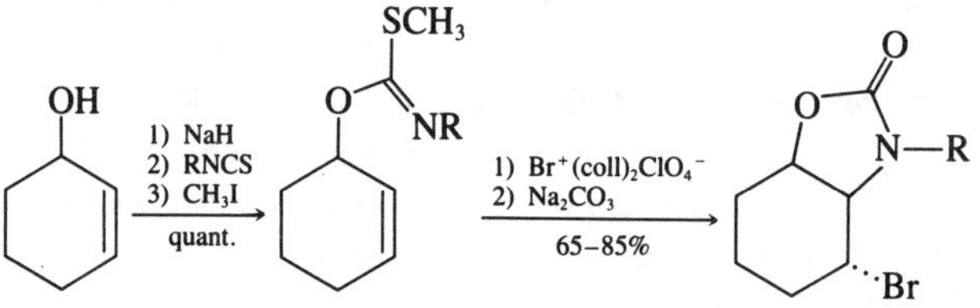

bromocyclization is applicable also to acyclic substrates. The products can be converted into *cis*-α-alkylamino alcohols by reduction of the bromine with tri-*n*-butyltin hydride followed by alkaline hydrolysis. When R is a *t*-butyl group it can be removed in high yield by treatment with trifluoroacetic acid at 25°.

These transformations have been used for a synthesis of the aminocyclitol sporamine (**4**).

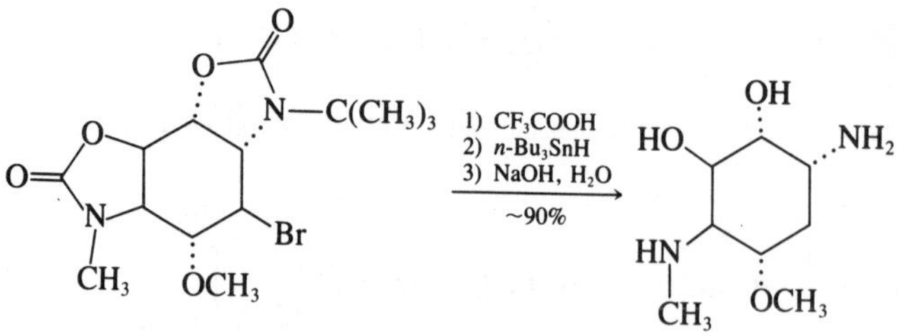

[1] S. Knapp and D. V. Patel, *Am. Soc.,* **105**, 6985 (1983).

N-Bromosuccinimide, 1, 78–80; **2,** 40–42; **3,** 34–36; **4,** 49–53; **5,** 65–66; **6,** 74–76; **7,** 37–40; **8,** 54–56; **9,** 70–72; **10,** 57–59; **11,** 79.

Cleavage of benzylidene acetals of hexoses.[1] 4,6-O-Benzylidene acetals of hexose derivatives on treatment with NBS undergo ring opening to 6-bromo-6-deoxy-4-benzoates. Internal benzylidene acetals are usually converted by this reaction into the two possible regioisomeric bromo benzoates.

Example:

α-Bromoethynyl steroids.[2] In the presence of silver nitrate, NBS or NIS reacts with 17α-ethynyl alcohols in acetone to form α-bromo- or α-iodoethynyl alcohols in 60–90% yield (equation I). Surprisingly, NCS does not undergo this reaction.

[1] S. Hanessian, *Carbohydrate Res.,* **1,** 86 (1976), and references cited therein; *idem, Org. Syn.,* submitted (1983).
[2] H. Hofmeister, K. Annen, H. Laurent, and R. Wiechert, *Angew. Chem., Int. Ed.,* **23,** 727 (1984).

Bromotrimethylsilane, 9, 73–74; **11,** 59. The silane can be prepared in high yield by reaction of triphenylphosphine dibromide and hexamethyldisiloxane catalyzed by powdered zinc (equation I).

$$(I)\ (C_6H_5)_3PBr_2\ +\ (CH_3)_3Si\text{-}O\text{-}Si(CH_3)_3\ \xrightarrow[96\%]{Zn,\ C_6H_4Cl_2\text{-}o}\ 2\ BrSi(CH_3)_3\ +\ (C_6H_5)_3P{=}O$$

Cyanotrimethylsilane can be prepared in 80% yield by reaction of freshly prepared bromotrimethylsilane with KCN and DMF at room temperature.[1]

Cleavage of MOM ethers. Methoxymethyl ethers are cleaved by this reagent at 0° in 80–97% yield. Acetals and THP, trityl, and *t*-butyldimethylsilyl ethers also are cleaved, but less readily.[2]

$ClSi(CH_3)_3/(C_2H_5)_4NBr$ is highly effective for selective cleavage of MOM ethers in the presence of an acetonide.[3]

[1] J. M. Aizpurua and C. Palomo, *Nouv. J. Chim.,* **8,** 51 (1984); C. Palomo, *Org. Syn.,* submitted (1983).
[2] S. Hanessian, D. Delorme, and Y. Dufresne, *Tetrahedron Letters,* 25, 2515 (1984).
[3] R. B. Woodward et al., *Am. Soc.,* **103,** 3213 (note 2) (1981).

D-(−)- and L-(+)-2,3-Butanediol (1), 11, 84–85.

Enantiomeric monoacetalization of diones.[1]   Acid-catalyzed monoacetalization of *cis*-9-methyldecalin-1,8-dione (**2**) with (2R,3R)-(−)-**1** forms a separable mixture of the diastereomeric monoacetals **3** and **4** in the ratio 9:1. The β-keto acetals **3** and **4** undergo

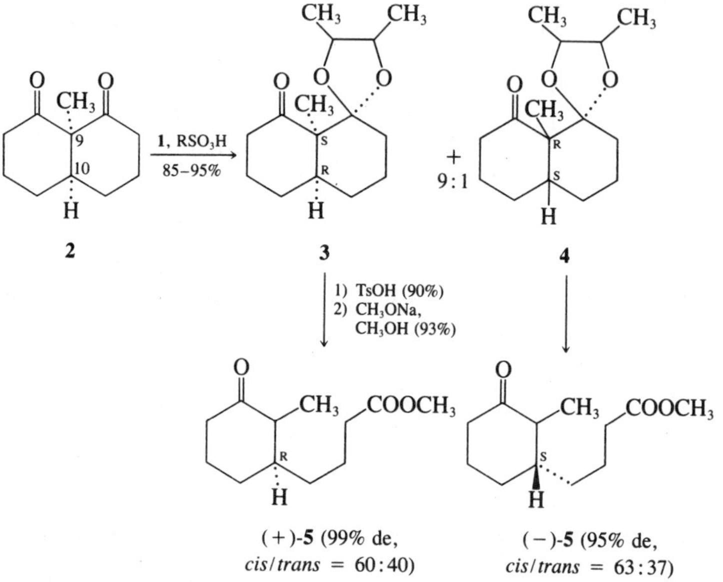

(+)-**5** (99% de,   (−)-**5** (95% de,
cis/trans = 60:40)   *cis/trans* = 63:37)

retro-Claisen fragmentation when treated with TsOH in refluxing benzene and then are transesterified with CH_3OH to give (+)- and (−)-**5**, respectively. Both are formed with >95% retention of configuration at C_{10} and as a mixture of *cis*- and *trans*-isomers. The other cleavage product is the monotosylate of meso-2,3-butanediol. The overall process provides a route to optically active 2,3-disubstituted cycloalkanones.

The reaction of **1** with the *trans*-isomer of **2** shows no enantiotopic differentiation.

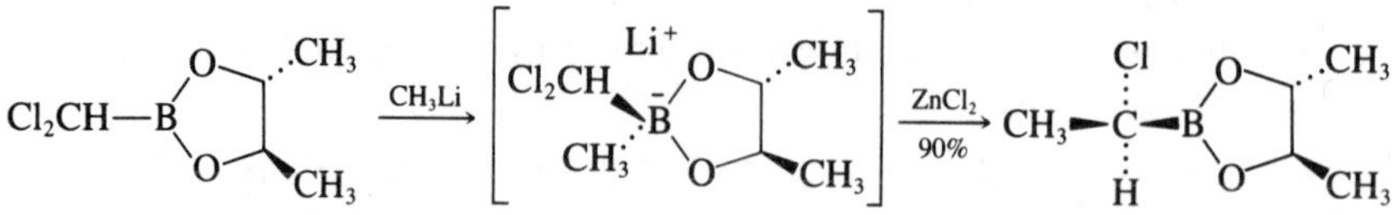

Boronic ester homologation.   (R,R)-2,3-Butanediol[2] and (+)-pinanediol[3] have been used as the chiral adjuncts in a diastereoselective homologation of dichloromethaneboronic esters (**1**) to the (αS)-α-chloroboronic esters (**2**). Reaction of **1** with an alkyllithium produces a borate complex (**a**), which rearranges diastereoselectively in the presence of $ZnCl_2$ to **2** with introduction of a chiral center adjacent to boron. The reaction permits sequential introduction of adjacent chiral centers, as shown by a synthesis of (3S,4S)-4-methyl-3-heptanol (3), a pheromone of the European elm bark beetle (equation I).

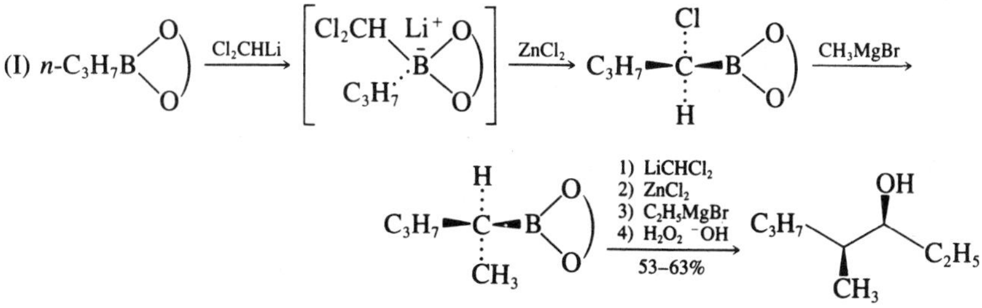

[1] R. O. Duthaler and P. Maienfisch, *Helv.*, **65**, 635 (1982); *idem, ibid.*, **67**, 832 (1984).
[2] K. M. Sadhu, D. S. Matteson, G. D. Hurst, and J. M. Kurosky, *Organometallics*, **3**, 804 (1984).
[3] D. S. Matteson and K. M. Sadhu, *Am. Soc.*, **105**, 2077 (1983).

2-Butenyl-9-borabicyclo[3.3.1]nonane (1).

Diastereoselective reactions with alkyl pyruvates.[1] The reagent (**1**) reacts with pyruvates to form the *anti*-substituted derivatives (**2**) selectively. The corresponding reaction with 2-butenyltributyltin in combination with BF_3 shows slight diastereoselectivity.

1

anti-**2** *syn*-**2**

R = CH_3 73:27
$CH_2C(CH_3)_3$ 85:15
C_6H_5 90:10

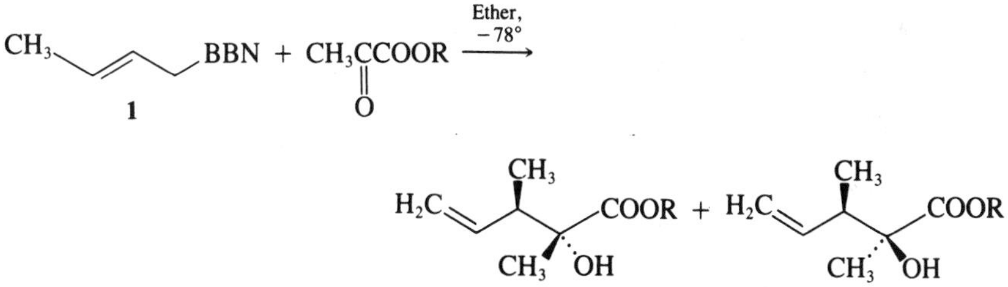

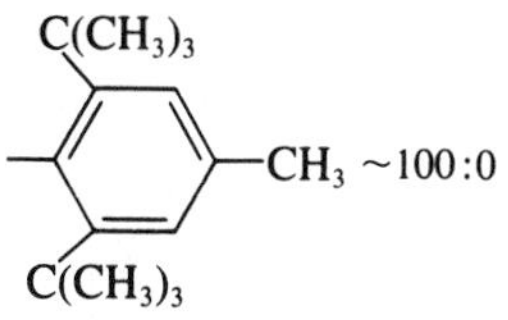

A similar condensation with the α-trimethylsilyl derivative (**3**) of **1** was used for a synthesis of crobarbatic acid (**4**).

CH₃—CH=CH—CH(Si(CH₃)₃)—BBN + CH₃CCOOCH₃ →[*n*-BuLi] (CH₃)₃Si...CH=CH—C(CH₃)H—C(OH)(CH₃)—COOCH₃

$$\text{CH}_3\underset{\textbf{3}}{\text{CH=CH-CH(Si(CH}_3)_3)\text{-BBN}} + \underset{\overset{\|}{\text{O}}}{\text{CH}_3\text{CCOOCH}_3} \xrightarrow{\textit{n}\text{-BuLi}}$$

(major product)

1) ClC₆H₄CO₃H (95%)
2) BF₃·O(C₂H₅)₂, CH₃OH (90%) →

CH₃O—(furanyl ring with CH₃, CH₃)—COOCH₃

1) CrO₃, H₂SO₄, O=C(CH₃)₂ (80%)
2) LiOH, H₂O, CH₃OH →

O=(lactone ring with CH₃, CH₃)—COOH

4

¹ Y. Yamamoto, T. Komatsu, and K. Maruyama, *J.C.S. Chem. Comm.*, 191 (1983).

2-Butenyl carbamates, CH₃CH=CHCH₂OCONR₂ (**1**). The (E)-2-butenyl carbamate is prepared by reaction of crotyl alcohol with N,N-diisopropylcarbamoyl chloride. The (Z)-isomer is obtained by hydrogenation of the 2-butynyl carbamate.

Diastereoselective homoaldol reaction.   The allylaluminum reagent (E)-**2**, obtained by deprotonation of (E)-**1** followed by transmetallation, undergoes a highly diastereoselective reaction with an aldehyde (or a ketone) to give (E)-*anti*-**3**. In contrast, (Z)-**2** reacts to give (E)-*syn*-**3**.¹

CH₃—CH=CH—CH₂—OCONR₂
(E)-**1**, R = *i*-Pr

1) LDA, TMEDA
2) ClAl(*i*-Bu)₂ →

CH₃—CH=CH—CH(Al(*i*-Bu)₂)—OCONR₂
(E)-**2**

R'CHO →

R'—CH(OH)—CH(CH₃)—CH=CH—OCONR₂
(E)-*anti*-**3** (*anti*/*syn* ≈90:10)

CH₃—CH=CH—CH(Al(*i*-Bu)₂)—OCONR₂
(Z)-**2**

R'CHO →

R'—CH(OH)—CH(CH₃)—CH=CH—OCONR₂
(E)-*syn*-**3** (*syn*/*anti* ≈90:10)

These enol carbamates are stable to acid, but undergo methanolysis in the presence of methanesulfonic acid and $Hg(OAc)_2$ to form γ-lactol methyl ethers, with retention of configuration at C_3 and C_4. These products can be oxidized (8, 97) to γ-lactones.[2,3] The overall process thus provides a route to protected γ-hydroxy aldehydes.
Examples:

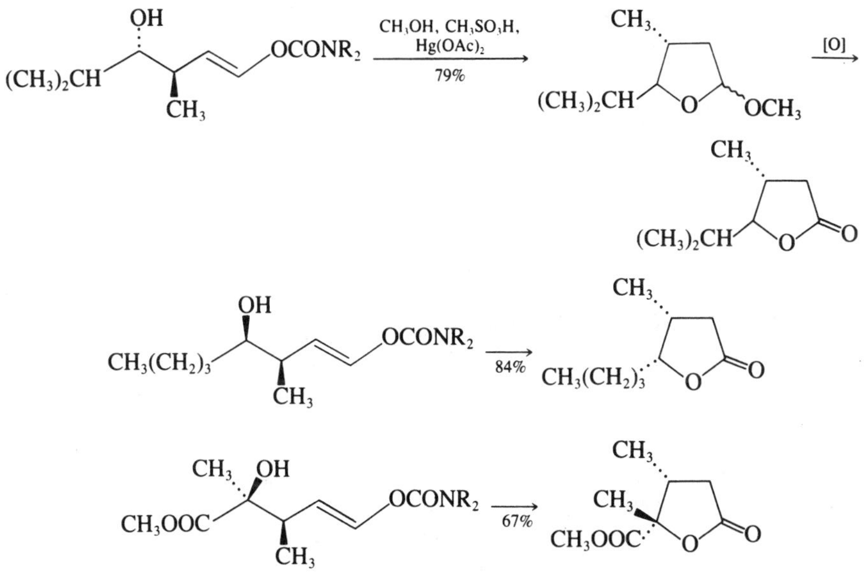

[1] D. Hoppe and F. Lichtenberg, *Angew. Chem., Int. Ed.*, **21**, 372 (1982).
[2] *Idem, ibid.*, **23**, 239 (1984).
[3] D. Hoppe and A. Brönneke, *Tetrahedron Letters*, **24**, 1687 (1983).

t-Butyl benzotriazol-1-yl carbonate,

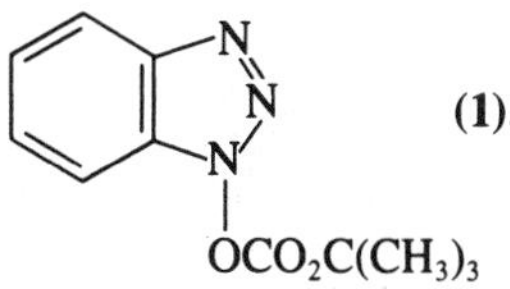

(1). The reagent is obtained in 85% yield by reaction of benzotriazol-1-yl chloroformate with *t*-butyl alcohol in methylene chloride in the presence of pyridine.

t-*Butoxycarbonylation of amino acids*.[1] Amino acids are converted by reaction with **1** and triethylamine in DMF at 25° into the N-Boc derivatives in >85% yield.

[1] S. Kim and H. Chang, *J.C.S. Chem. Comm.*, 1357 (1983).

t-**Butyldimethylchlorosilane, 4**, 57–58; 176–177; **5**, 74–75; **6**, 78–79; **7**, 59; **8**, 77; **9**, 77; **10**, 62; **11**, 88–90.

t-Butyldimethylsilyl (TBDMS) enol ethers.[1]   The thermodynamically more stable TBDMS enol ether can be obtained with high selectivity by addition of KH to a THF solution of the ketone and TBDMS chloride (excess) at $-78 \rightarrow 25°$. Addition of HMPT can improve the stereoselectivity marginally. This method is satisfactory for ketones that undergo self-condensation or degradation in the presence of bases or ketones that are hindered.

Examples:

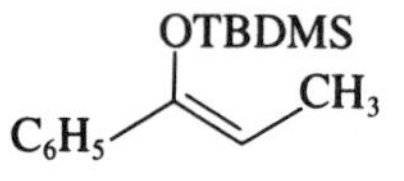

(100%)

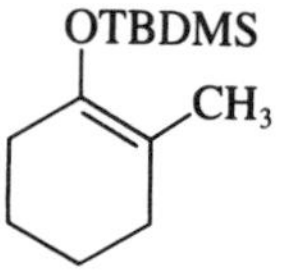

(98:2)

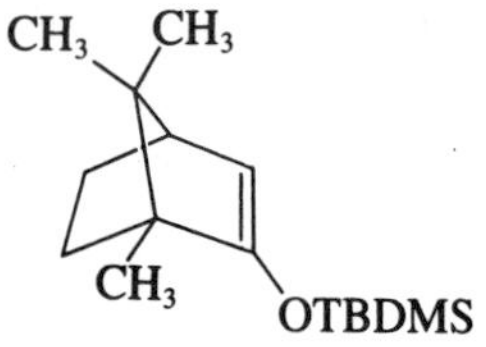

(87%)

[1] J. Orban, J. V. Turner, and B. Twitchin, *Tetrahedron Letters*, **25**, 5099 (1984).

t-Butyldimethylsilyl ethylnitronate,

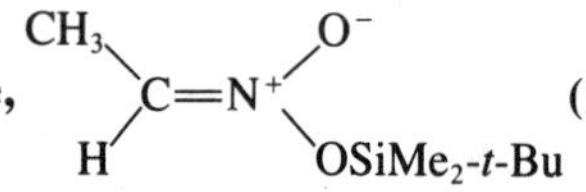

(1). The nitronate ester is prepared by lithiation (LDA) of nitroethane followed by reaction with the chlorosilane.[1]

Reaction with thiocarbonyl compounds.[2]   The thiocarbonyl compounds obtained by photochemical oxidation of phenacyl sulfides[3] can be trapped efficiently by a 1,3-dipolar cycloaddition with **1** to give **2**. This heterocycle can be cleaved to carbonyl compounds by $Bu_4N^+F^-$ or $(C_2H_5)_3N^+HF^-$. This process is more efficient and more general than photolysis of phenacyl sulfides in the presence of oxygen.

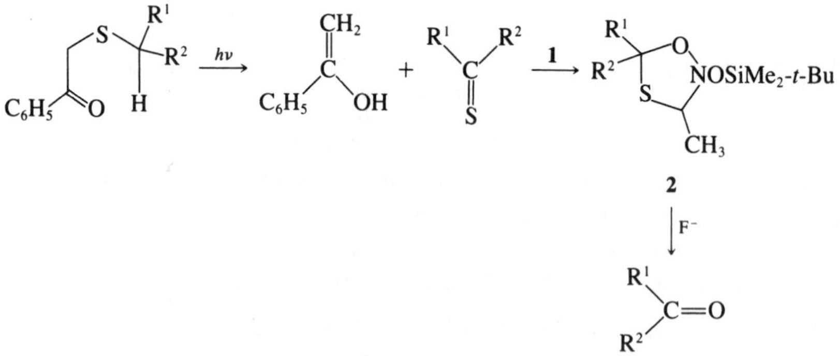

The substrates are available by reaction of mercaptans with phenacyl chloride, by Michael addition of phenacyl mercaptan to enones, or by alkylation of phenacyl mercaptan with alkyl halides.

[1] E. W. Calvin, A. K. Beck, B. Bastani, D. Seebach, Y. Kai, and J. D. Dunitz, *Helv.*, **63**, 697 (1980).
[2] E. Vedejs and D. A. Perry, *J. Org.*, **49**, 573 (1984).
[3] E. Vedejs, T. H. Eberlein, and D. L. Varie, *Am. Soc.*, **104**, 1445 (1982).

t-**Butyldimethylsilyl hydroperoxide–Mercury(II) trifluoroacetate.** The hydroperoxide is prepared by reaction of $(CH_3)_3C(CH_3)_2SiCl$ with imidazole and ethereal H_2O_2.

Epoxides.[1]   A new synthesis of epoxides involves peroxymercuration of an allylic alcohol with this combination followed by reduction of the peroxymercurial with aqueous $NaBH_4$.

Example:

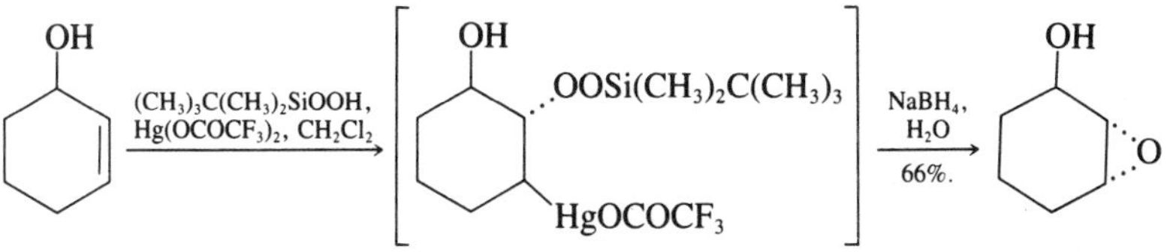

The reaction involves an intermediate β-peroxy carbon free radical (*cf.* Giese reaction, **11**, 315–316), which then displaces the silyloxy group to form an epoxide.

[1] E. J. Corey, G. Schmidt, and K. Shimoji, *Tetrahedron Letters*, **24**, 3169 (1983).

O-*t*-Butyldimethylsilylhydroxylamine, NH_2OSiMe_2-*t*-Bu (**1**). Preparation.[1] Carbonyl compounds are converted into silyloximes in high yield by reaction with this reagent in $CHCl_3$ in the presence of 4 Å molecular sieves. This oximination is useful for acid- or base-sensitive carbonyl compounds, since desilylation can be effected with fluoride ion.

Nitrosoalkenes.[2]   Nitrosoalkenes are usually generated by reaction of α-halo oximes with bases, and have been used for [4 + 2] cycloadditions. They are also generated efficiently from α-chloro silyloximes by fluoride ion. This elimination is used to effect an intramolecular cyclization of the nitrosoalkene generated from **1** to give a mixture of the dihydrooxazines **2** and **3**. The choice of the metal fluoride is critical for acceptable yields. Highest yields are obtained when the nitrosoalkene is generated slowly by a

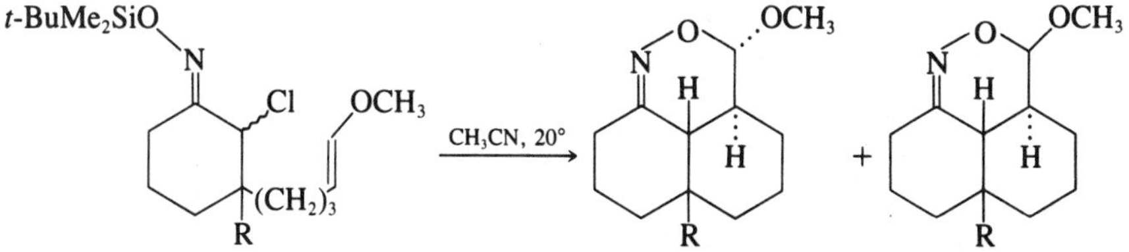

1a, R = H	CsF	**2a** (70%)	**3a** (13%)
1b, R = CH₃	KF	**2b** (65%)	**3b** (17%)

sparingly soluble metal fluoride such as CsF or KF. The geometry of the oxime and the orientation of the chlorine are not important, but the (E)-enol ether undergoes cyclization more rapidly than the (Z)-isomer.

[1] R. West and P. Boudjouk, *Am. Soc.*, **95**, 3983 (1973); F. Duboudin, E. Frainnet, G. Vinion, and F. Dabescat, *J. Organometal. Chem.*, **82**, 41 (1974).
[2] S. E. Denmark, M. S. Dappen, and J. A. Sternberg, *J. Org.*, **49**, 4741 (1984).

t-Butyldimethylsilyl trifluoromethanesulfonate (1), 10, 63; 11, 90–91.

Dealkylation[1] or transalkylation[2] of t-amines.   N-Oxides of *t*-amines on treatment with this triflate form silyloxyammonium salts (**2**), which rearrange in the presence of methyllithium to α-silyloxyamines (**3**). These can be converted into *sec*-amines (**4**).

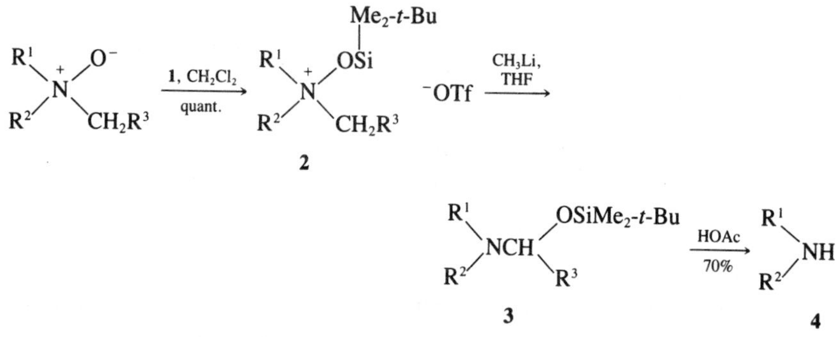

Alkylation of **3** in the presence of $Bu_4N^+F^-$ affords new tertiary amines in moderate yield (equation I).

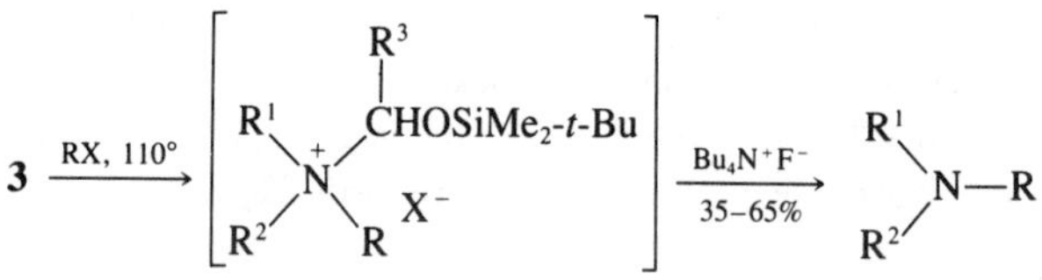

Enol ethers.[3]   Reaction of ketones, even highly hindered ones, with this reagent and triethylamine provide TBDMS enol ethers in 90–100% yield. The method is not useful for selective formation of the kinetic isomers.

[1] R. Okazaki and N. Tokitoh, *J.C.S. Chem. Comm.*, 192 (1984).
[2] *Idem, Chem. Letters*, 1937 (1984).
[3] L. N. Mander and S. P. Sethi, *Tetrahedron Letters*, **25**, 5953 (1984).

t-**Butyldiphenylchlorosilane, 6**, 81.

　　Selective protection of primary alcohols.[1]　Selective protection of primary hydroxyl groups of carbohydrates is possible by reaction of this chlorosilane and polyvinylpyridine in CH_2Cl_2 or THF in the presence of HMPT. Amberlite A-26 in the F^- form is recommended for desilylation.

[1] G. Cardillo, M. Orena, S. Sandri, and C. Tomasini, *Chem. Ind.*, 643 (1983).

t-**Butylhydrazine,** $(CH_3)_3CNHNH_2$. Suppliers: Aldrich, Fluka.

　　α-*Hydroxy ketones; ketones.*[1]　The *t*-butylhydrazones (**1**) of aldehydes on deprotonation can form an azo anion (**a**), which reacts with aldehydes or ketones to form the alkoxide (**b**). On treatment with *n*-BuLi and water, **b** is converted into the hydrazone **2** of an α-hydroxy ketone (**3**) (equation I).

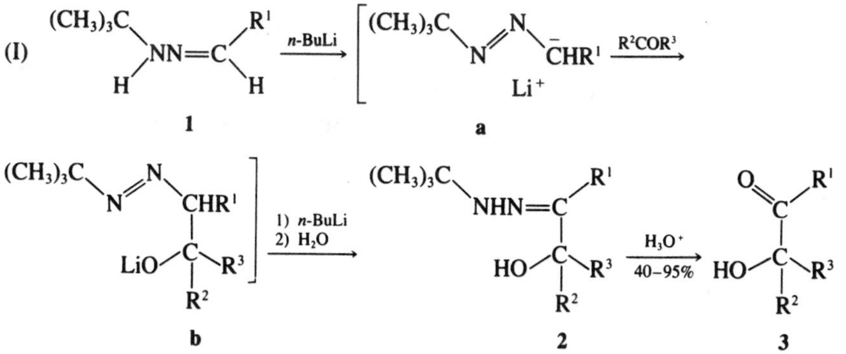

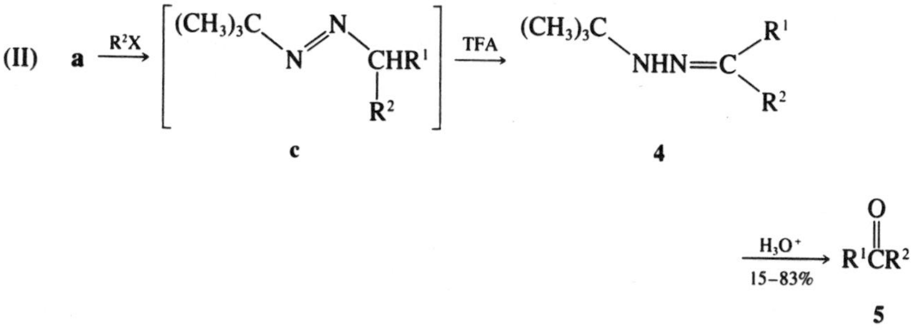

　　C-Alkylation of **a** results in a ketone (equation II), but N-alkylation is a competing reaction unless R^1 and R^2 are fairly bulky groups.

[1] R. M. Adlington, J. E. Baldwin, J. C. Bottaro, and M. W. D. Perry, *J.C.S. Chem. Comm.*, 1040 (1983).

t-Butyl hydroperoxide, 1, 88–89; **2**, 49–50; **3**, 37–38; **5**, 75–77; **6**, 81–82; **7**, 43–45; **8**, 62–64; **10**, 64–65.

α-Keto esters and nitriles.[1]   α-Hydroxy esters or nitriles can be dehydrogenated to α-keto esters or nitriles by *t*-butyl hydroperoxide in the presence of various ruthenium compounds such as $RuCl_3$, $RuCl_2[P(C_6H_5)_3]_3$, and $Ru(acac)_3$.

Examples:

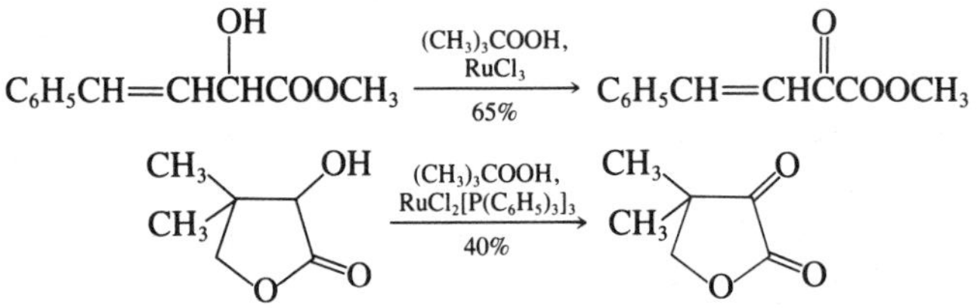

Cleavage of acetals.[2]   In the presence of a Pd(II) catalyst, *t*-butyl hydroperoxide oxidatively cleaves a five- or six-membered acetal to an ester of a diol. $Pd(OAc)_2$ and $PdCl_2$ can catalyze this reaction, but $CF_3CO_2PdOOC(CH_3)_3$ (**10**, 299) is most effective. The cleavage of acetals derived from unsymmetrical diols is not regioselective.

Example:

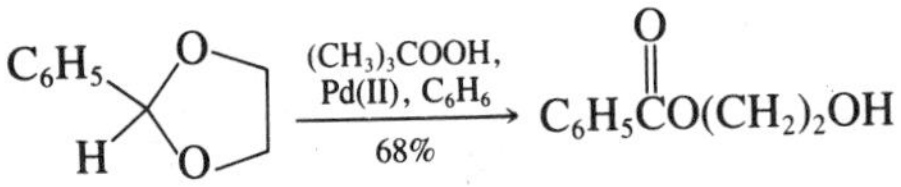

α′-Hydroxy-α,β-enones.[3]   2-Trimethylsilyloxy-1,3-cyclohexadienes (**1**) are oxidized to hydroxy enones (**2**) by *t*-butyl hydroperoxide (2 equiv.) in the presence of CuCl. *m*-Chloroperbenzoic acid has been used for this oxidation (**8**, 100–101).

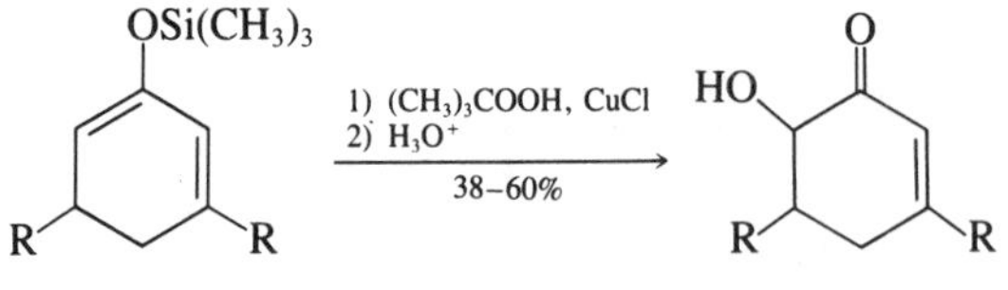

1 (R = H, CH₃)   **2**

Oxidation of α-isophorone (**3**) with *t*-butyl hydroperoxide catalyzed by $Pd(OAc)_2$ and $N(C_2H_5)_3$ results in **4** as the only oxidation product.

$$3 \xrightarrow[\text{50-55\%}]{\substack{(CH_3)_3COOH, \\ Pd(OAc)_2, \ N(C_2H_5)_3}} 4$$

[1] M. Tanaka, T. Kobayashi, and T. Sakakura, *Angew. Chem., Int. Ed.*, **23**, 518 (1984).
[2] T. Hosokawa, Y. Imada, and S.-I. Murahashi, *J.C.S. Chem. Comm.*, 1245 (1983).
[3] T. Hosokawa, S. Inui, and S.-I. Murahashi, *Chem. Letters*, 1081 (1983).

t-Butyl hydroperoxide–Benzyltrimethylammonium tetrabromooxomolybdate, $C_6H_5CH_2\overset{+}{N}(CH_3)_3{}^-OMoBr_4$. The Mo complex is prepared by reaction of MoO_3 with HBr (47%) and then $C_6H_5CH_2\overset{+}{N}(CH_3)_3Cl^-$ in HBr.[1]

Oxidation.[2] Oxidations with *t*-butyl hydroperoxide catalyzed with this Mo complex can be used to effect selective oxidations of secondary alcohols in the presence of primary ones in benzene at 60°. Primary alcohols are oxidized slowly to esters in methanol. Aldehydes are oxidized to carboxylic acids in benzene or to esters in methanol.

[1] J. F. Allen and H. M. Newman, *Inorg. Chem.*, **3**, 1612 (1964).
[2] Y. Masuyama, M. Takahashi, and Y. Kurusu, *Tetrahedron Letters*, **25**, 4417 (1984).

t-Butyl hydroperoxide–Bisoxobis(2,4-pentanedionate)molybdenum, **11**, 91.

Cleavage of allylic alcohols.[1] The system selectively cleaves the double bond and the adjacent bond bearing the hydroxyl group of acyclic and aromatic allylic alcohols. Cyclic allylic alcohols are oxidized in low yield to dicarboxylic acids.

Examples:

$$
\begin{array}{c}
RCH_2CH{=}CHCH_2OH \\
\text{or} \\
RCH_2\underset{|}{CHCH}{=}CH_2 \\
OH
\end{array}
\xrightarrow[\text{MoO}_2(\text{acac})_2]{t\text{-BuOOH,}}
\underset{(73-79\%)}{RCH_2COOH} + \underset{(6-11\%)}{RCOOH}
$$

$$
CH_3(CH_2)_2CH{=}\underset{\underset{CH_2CH_3}{|}}{C}CH_2OH \longrightarrow \underset{(70\%)}{CH_3(CH_2)_2COOH} + \underset{(7\%)}{CH_3CH_2COOH}
$$

[1] K. Jitsukawa, K. Kaneda, and S. Teranishi, *J. Org.*, **49**, 199 (1984).

t-Butyl hydroperoxide–Chromium carbonyl.

Allylic oxidation.[1] Reaction of alkenes with *t*-butyl hydroperoxide (90%) in the presence of catalytic amounts of $Cr(CO)_6$ results in oxidation to α,β-enones in 25–100%

yield. Use of acetonitrile as solvent can be advantageous, probably because $Cr(CO)_3(CH_3CN)_3$ is formed *in situ*. Oxidation of secondary alcohols to ketones under these conditions is a fairly slow reaction.

Examples:

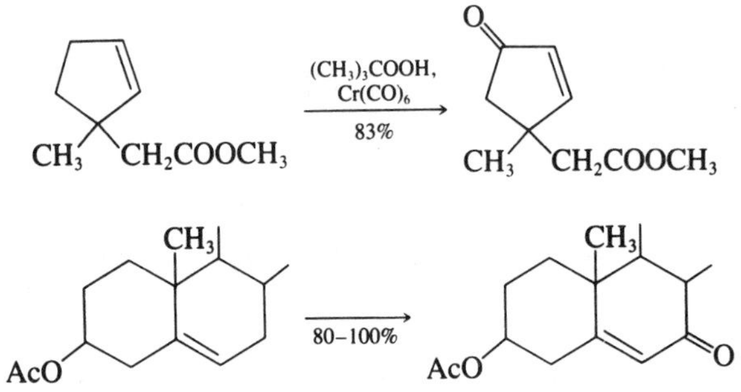

[1] A. J. Pearson, Y.-S. Chen, S.-Y. Hsu and T. Ray, *Tetrahedron Letters,* **25,** 1235 (1984); A. J. Pearson, Y.-S. Chen, G. R. Han, S.-Y. Hsu, and T. Ray, *J.C.S. Perkin I,* 267 (1985).

***t*-Butyl hydroperoxide–Dialkyl tartrate–Titanium(IV) isopropoxide, 10,** 64–65; **11,** 92–95.

Asymmetric epoxidation of allylic alcohols.[1] Details are available for enantiose-lective Sharpless epoxidation of the allylic alcohol **1**.

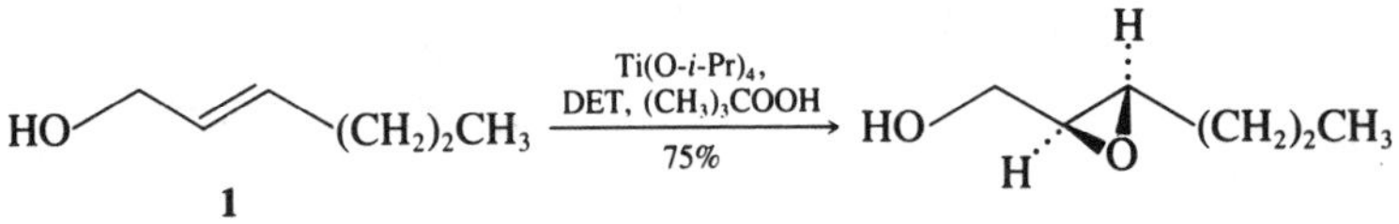

Asymmetric epoxidation of homoallylic alcohols.[2] Sharpless asymmetric epoxi-dation of primary homoallylic alcohols with L-(+)-diethyl tartrate proceeds with only moderate enantiomeric selectivity (23–55% ee) and opposite to that observed with allylic alcohols. Unfortunately, operation at low temperatures to improve the enantiomeric excess also retards the rate drastically. Even so, this epoxidation provides a useful synthesis of (+)-γ-amino-β(R)-hydroxybutyric acid (**1**).

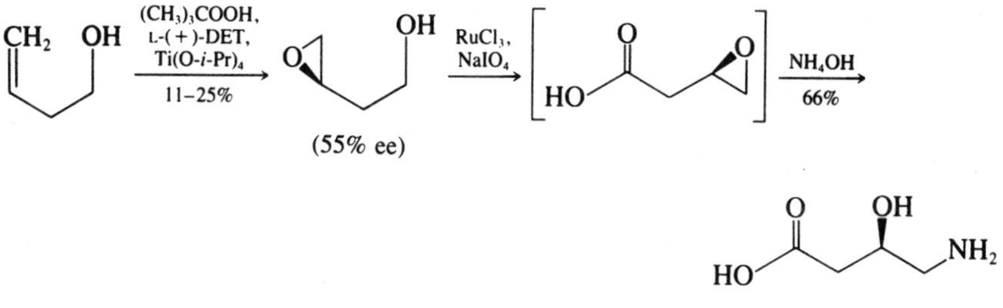

(55% ee)

1 (49% ee)

Epoxidation catalyzed by TiCl$_2$(O-i-Pr)$_2$.[3] Asymmetric epoxidation of 2-alkyl al-lylic alcohols such as **1** under the usual conditions results in a mixture of products formed by cleavage of the expected epoxide. Use of TiCl$_2$(O-i-Pr)$_2$[2] and (+)-DET in the ratio 2:1 results in the chloro diol **2a** in 76% yield and in 73% ee. On treatment with base, **2a** is converted into the oxide **3a**. This oxide is the enantiomer of the oxide **3b**, obtained in moderate yield with Ti(O-t-Bu)$_4$ and (+)-DET (2:2 ratio).

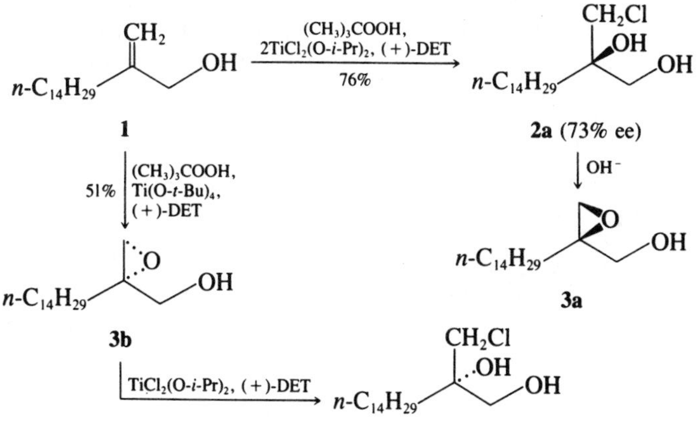

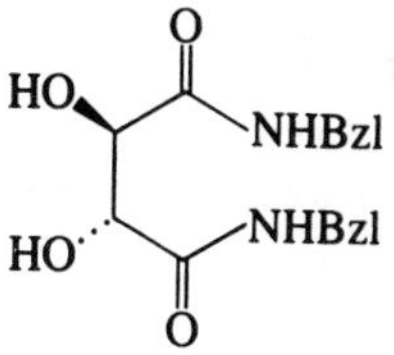

Epoxidation catalyzed by (+)-tartramides.[3] Epoxidation of (E)-α-phenylcinnamyl alcohol (**2**) catalyzed by Ti(O-i-Pr)$_4$ and the amide **1** derived from L-(+)-tartaric acid gives an unexpected result. When the usual ratio (2:2.4) of Ti(O-i-Pr)$_4$ to the amide **1** is used, epoxidation proceeds with the usual high enantiofacial selection (equation I). However, with a 2:1 Ti/**1** ratio, a strong inverse enantiofacial selection occurs (equation II). Although the 2:1 system is generally less enantioselective than the standard one, it should be useful in cases where the desired enantioselection requires use of the expensive unnatural tartrate.

(I)

2

3a (96% ee)

(II)   **2**

3b (82% ee)

The 2:2 and 2:1 systems obviously have different structures; this is confirmed by measurement of the molecular weights. The proposed structures for the 2:2 and 2:1 catalysts are shown in formulas **4a** and **4b**.

4a

4b

Chiral sulfoxides.[4]   The Sharpless reagent for asymmetric epoxidation also effects asymmetric oxidation of prochiral sulfides to sulfoxides. The most satisfactory results are obtained for the stoichiometry $Ti(O-i-Pr)_4/L\text{-}DET/H_2O/(CH_3)_3COOH = 1:2:1:2$ for 1 equiv. of sulfide. In the series of alkyl *p*-tolyl sulfides, the (R)-sulfoxide is obtained in 41–90% ee; the enantioselectivity is highest when the alkyl group is methyl. Methyl phenyl sulfide is oxidized to the (R)-sulfoxide in 81% ee. Even optically active dialkyl sulfoxides can be prepared in 50–71% ee; the enantioselectivity is highest for methyl octyl sulfoxide.

Resolution of β-hydroxy t-amines.[5]   These amines can undergo kinetic resolution by partial oxidation of one enantiomer to the N-oxide by $(CH_3)_3COOH$, $Ti(O-i-Pr)_4$, and either $(+)$- or $(-)$-diisopropyl tartrate in the ratio $0.6:2:1.2$. The enantiomeric excess of the recovered enantiomer often exceeds 90%. The choice of $(+)$- or $(-)$-tartrate determines the unreactive enantiomer. The N-oxides are easily converted to the optically active amines by reduction with $LiAlH_4$. The substituents on the nitrogen are important. When the nitrogen is substituted by two benzyl groups, only slight resolution is observed. The process is highly predictable. The slow-reacting enantiomer when $(+)$-DIPT is used is always the one in which the hydroxyl group is up and the amine to the right as drawn in the first example.

Examples:

(37%, 95% ee)   (59%, 63% ee)

(95% ee)

Review. Sharpless[6] has discussed the general problems of achieving diastereoselectivity through reagent control in reactions with chiral substrates, which result from the fact that double asymmetric synthesis is involved. Matching and mismatching asymmetric induction and powerful asymmetric reagents are required. In addition, asymmetric reagents are useless in reactions with racemates, except for kinetic resolution. However, chiral reagents have one advantage over enzymes, in that they are effective with both enantiomers of a given substrate.

Masamune *et al.*[7] have reviewed in detail the effects of double asymmetric induction not only for epoxidation, but also for the aldol, Diels-Alder, and catalytic hydrogenation reactions. The merits of this strategy are illustrated by an analysis of Woodward's synthesis[8] of erythromycin A (**1**), which has 10 chiral centers.

1

[1] J. G. Hill, K. B. Sharpless, C. M. Exon, and R. Regenye, *Org. Syn.*, **63**, 66 (1985).

[2] B. E. Rossiter and K. B. Sharpless, *J. Org.*, **49**, 3707 (1984).

[3] L. D.-L. Lu, R. A. Johnson, M. G. Finn, and K. B. Sharpless, *ibid.*, 728 (1984).

[4] P. Pitchen, E. Duñach, M. N. Deshmukh, and H. B. Kagan, *Am. Soc.*, **106**, 8188 (1984); *see also* F. DiFuria, G. Modena, and R. Seraglia, *Synthesis*, 325 (1984).

[5] S. M. Viti, S. Miyano, L. D.-L. Lu, Y. Gao, and K. B. Sharpless, *J. Org.*, in press (1985).

[6] K. B. Sharpless, *Chemica Scripta*, **25**, in press (1985).

[7] S. Masamune, W. Choy, J. S. Peterson, and L. R. Sita, *Angew. Chem., Int. Ed.*, **24**, 1 (1985).

[8] R. B. Woodward *et al.*, *Am. Soc.*, **103**, 3210, 3213, 3215 (1981).

t-Butyl hydroperoxide–Titanium(IV) chloride.

Chlorohydroxylation of alkenes.[1]   Anhydrous *t*-butyl hydroperoxide or di-*t*-butyl peroxide (**1**, 211–212) in the presence of $TiCl_4$ effects chlorohydroxylation of alkenes. The reaction shows moderate to high regioselectivity as well as fair to good diastereoselectivity when the substrate is substituted in the allylic or homoallylic position.

Examples:

$$CH_3(CH_2)_7CH{=}CH_2 \xrightarrow[62\%]{\substack{(CH_3)_3COOH, \\ TiCl_4, \ CH_2Cl_2}} CH_3(CH_2)_7CHClCH_2OH \quad + \quad CH_3(CH_2)_7CHOHCH_2Cl$$

$$96:4$$

(74%)

(*cis*/*trans* = 82:18)

(20%)

(*syn*/*anti* = 80:20)

The epoxides obtained from the chlorohydrins are the opposite diastereomers to those formed by vanadium-catalyzed epoxidations (**9**, 81–82).

[1] J. M. Klunder, M. Caron, M. Uchiyama, and K. B. Sharpless, *J. Org.*, **50**, 912 (1985).

t-Butyl hydroperoxide–Vanadyl acetylacetonate, **2**, 287; **4**, 346; **5**, 75–76; **7**, 43–44; **9**, 81–82; **10**, 66; **11**, 97–99.

Selective oxidation of secondary alcohols.[1]   This system exhibits high selectivity for oxidation of a secondary hydroxyl group in the presence of a primary hydroxyl group.

Example:

$$HOCH_2CH_2\underset{\underset{OH}{|}}{C}HCH_3 \xrightarrow[96\%]{\substack{(CH_3)_3COOH, \\ VO(acac)_2, \ C_6H_6}} HOCH_2CH_2\underset{\underset{O}{\|}}{C}CH_3$$

[1] K. Kaneda, Y. Kawanishi, K. Jitsukawa, and S. Teranishi, *Tetrahedron Letters*, **24**, 5009 (1983).

N-*t*-Butylhydroxylamine, (CH$_3$)$_3$CNHOH (**1**). Mol. wt. 89.14, m.p. 64–65°. Preparation.[1]

α-*Acyloxy aldehydes*.[2]   The N-*t*-butylnitrones (**2**) of aldehydes react with acid chlorides in the presence of triethylamine to form α-acyloxy imines (**3**) via a [3.3] sigmatropic rearrangement. The imines are hydrolyzed by buffered aqueous acetic acid to α-acyloxy aldehydes (**4**).

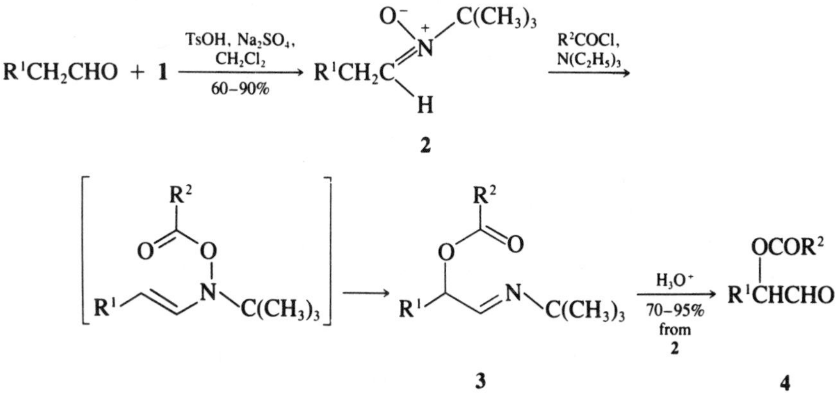

[1] A. Calder, A. R. Forrester, and S. P. Hepburn, *Org. Syn.*, **52**, 77 (1972).
[2] C. H. Cummins and R. M. Coates, *J. Org.*, **48**, 2070 (1983).

_t_-Butyl isocyanide, 2, 50–51; **9**, 82; **10**, 67; **11**, 99–100.

Cyanation of acetals.[1]   Reaction of acetals with *t*-butyl isocyanide catalyzed by TiCl$_4$ results in cyanohydrin ethers in high yield. A similar reaction with ethylene acetals results in spirolactones.

Examples:

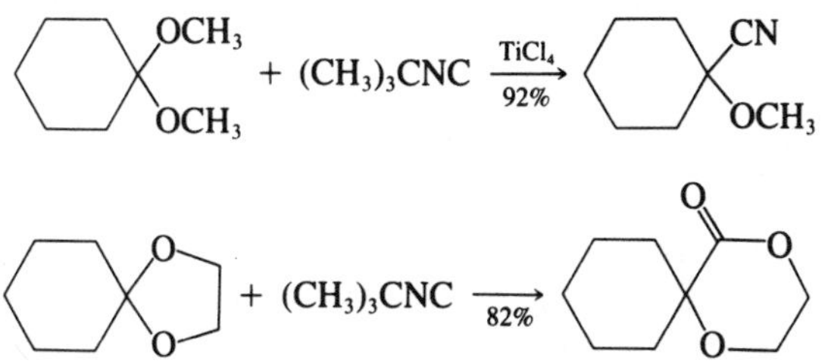

[1] Y. Ito, H. Imai, K. Segoe, and T. Saegusa, *Chem. Letters*, 937 (1984).

n-**Butyllithium, 1,** 95–96; **2,** 51–53; **4,** 60–63; **5,** 78; **6,** 85–91; **7,** 45–47; **8,** 65–66; **9,** 83–87; **10,** 68–71; **11,** 101–103.

Asymmetric Wittig rearrangement.   Three laboratories[1-3] have reported enantio- and *syn*-selective [2,3] Wittig rearrangements. Thus rearrangement of the optically active allylic ether **1** provides the allylic alcohol (3R,4R)-**2** as the major product, with complete

chirality transfer and with high (E)- and *syn*-selectivity.[2] Moderate enantioselectivity (65%) is observed in the rearrangement of a chiral tertiary allylic ether.[3]

Intramolecular conjugate additions of ω-iodo-α,β-unsaturated esters.[4]   Reaction of the ester **1a** with *n*-butyllithium at −100° results in *t*-butyl cyclopentylacetate (**2a**) in 82% yield. The cyclization presumably involves a rapid lithium–iodine exchange to form a nucleophilic center that undergoes an intramolecular conjugate addition to form a cy-

1a, n = 4, R = H	82%	**2**
b, n = 4, R = CH₃	86%	
c, n = 3, R = H	78%	
d, n = 5, R = H	14%	

clopentyl derivative. A similar reaction occurs with **1b**, and a similar cyclization to *t*-butyl cyclobutylacetate (**2c**) with **1c**. However, extension to **1d** results in a low yield of the desired cyclohexyl derivative.

Michael acceptor groups other than α,β-unsaturated esters give only traces of cyclized products under similar conditions. The major products result from 1,2- and 1,4-additions of *n*-butyllithium in the case of ω-iodo-α,β-unsaturated ketones or amides.

[1] N. Sayo, K. Azuma, K. Mikami, and T. Nakai, *Tetrahedron Letters*, **25,** 565 (1984).
[2] D. J.-S. Tsai and M. M. Midland, *J. Org.*, **49,** 1842 (1984).
[3] J. A. Marshall and T. M. Jenson, *ibid.*, **49,** 1707 (1984).
[4] M. P. Cooke, Jr., *J. Org.*, **49,** 1144 (1984).

n-Butyllithium–Diisobutylaluminum hydride.

Reduction of diynols to enynols.[1]   Conjugated diynols (**1**) are selectively reduced to (E)-enynols (**2**) by conversion to lithium alkynoxides with *n*-BuLi followed by reduction with DIBAH.

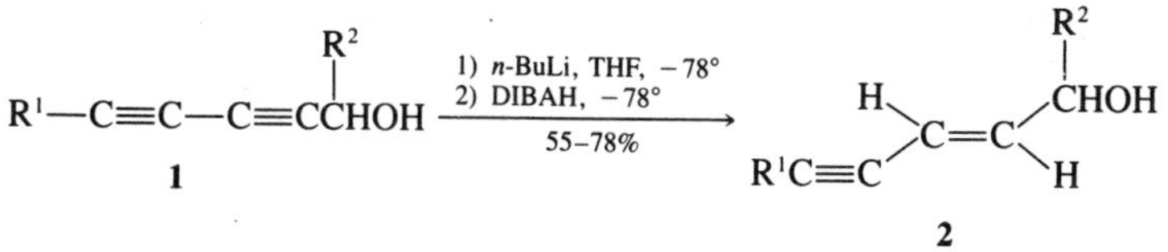

$$R^1—C\equiv C—C\equiv CCHOH \xrightarrow[\text{55-78\%}]{\begin{array}{l}1)\ \textit{n-}\text{BuLi, THF, }-78°\\2)\ \text{DIBAH, }-78°\end{array}} \underset{\mathbf{2}}{R^1C\equiv C}\overset{H}{\diagup}C=C\overset{CHOH}{\diagdown}_H$$

1

[1] R. E. Doolittle, *Synthesis*, 730 (1984).

***n*-Butyllithium–Potassium *t*-butoxide, 5**, 552; **8**, 67; **9**, 87; **10**, 72–73; **11**, 103.

α-Silyl ketones.[1]   Silyl enol ethers with sterically hindered silyl groups rearrange to α-silyl ketones in the presence of *n*-BuLi (2 equiv.) and KO-*t*-Bu (2.5 equiv.). Trimethylsilyl ethers do not undergo this rearrangement, but triisopropylsilyl (TIPS) and diisopropylmethylsilyl (DIMS) ethers do if they contain an allylic α-proton. The silyl group rearranges preferentially to the less hindered terminus of the intermediate allyl anion. The rearrangement is less useful with acyclic substrates because of side reactions.
    Example:

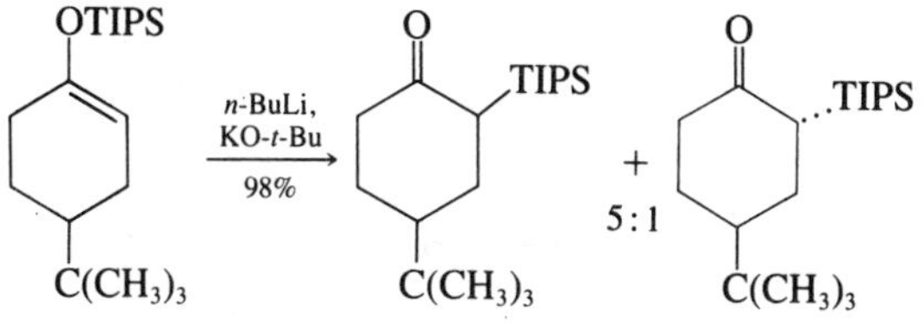

[1] E. J. Corey and C. Rücker, *Tetrahedron Letters*, 4345 (1984).

***sec*-Butyllithium, 5**, 78–79; **9**, 87–88; **10**, 75.

Metallation of benzamides.[1]   The *ortho*-metallation of tertiary amides (**10**, 75; **11**, 367–368) permits a novel epoxy cyclialkylation route to benzofuranes and -pyranes.[2]
    Examples:

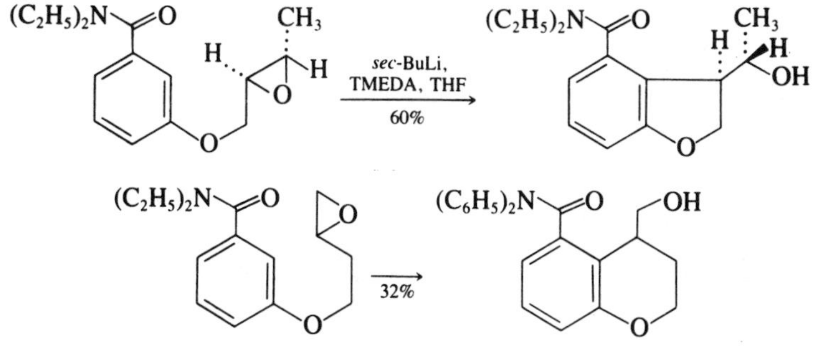

Metallation of O-aryl carbamates.[3]   O-Aryl carbamates are metallated by *sec*-BuLi–
TMEDA at the *ortho*-position. Metallation followed by reaction with an electrophile offers
an attractive route to *o*-substituted carbamates and phenols.
    Example:

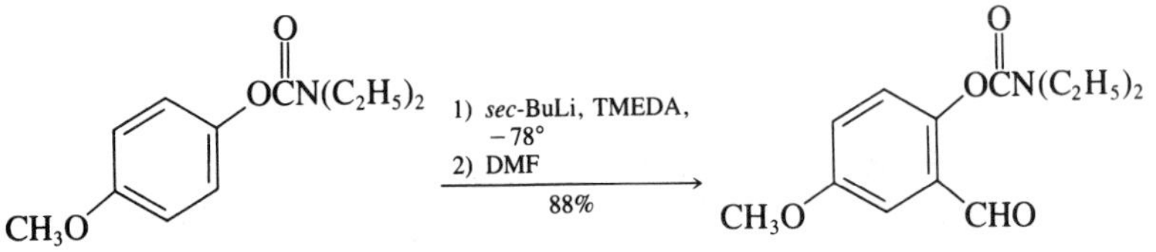

Lithiation of the *o*-methyl carbamate **1** under the same conditions followed by $(CH_3)_3SiCl$
quench results in **2** and **3** in the ratio 2:1. However, lithiation with LDA and reaction
with $(CH_3)_3SiCl$ gives **3** as the major product (66% yield).

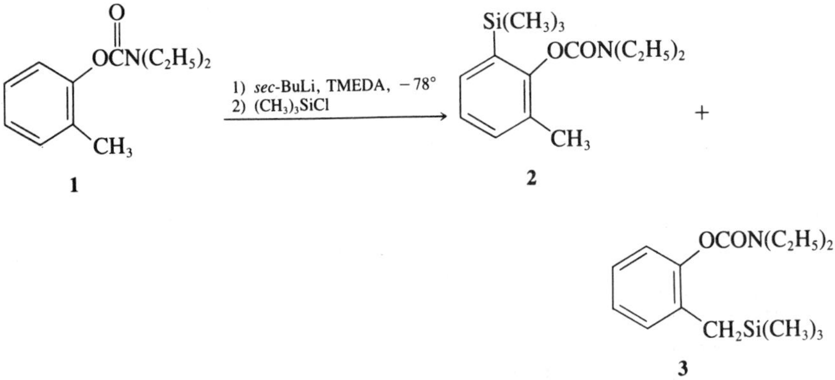

The *o*-lithio intermediates undergo an unexpected O → C 1,3-carbonyl migration to
salicylamides at 25°. This migration is an equivalent of the Fries rearrangement.
    Example:

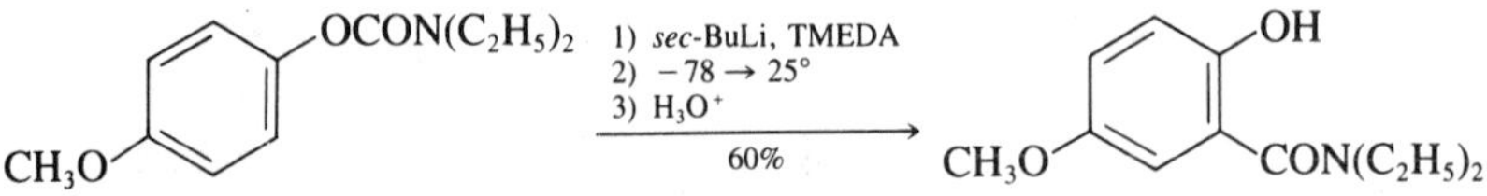

β-*Deprotonation of a vinyl ether*.[4]   The enol ether **1** is deprotonated selectively at the β-position by *sec*-BuLi at −78°. The anion reacts with alkyl iodides under CuI and HMPT catalysis and also with aldehydes and ketones. Only one geometrical isomer, presumed to be (Z), is isolated in all cases.

Examples:

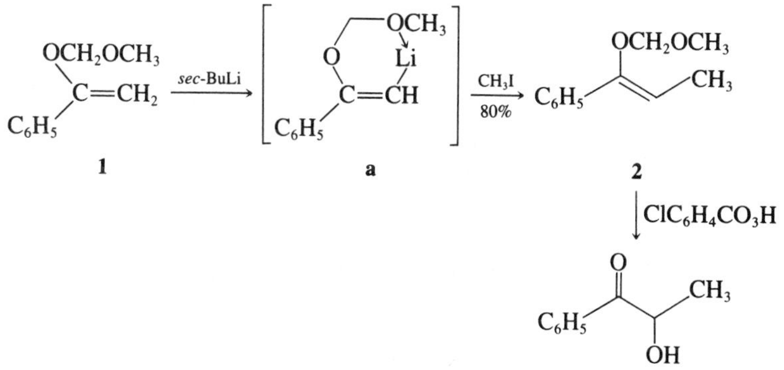

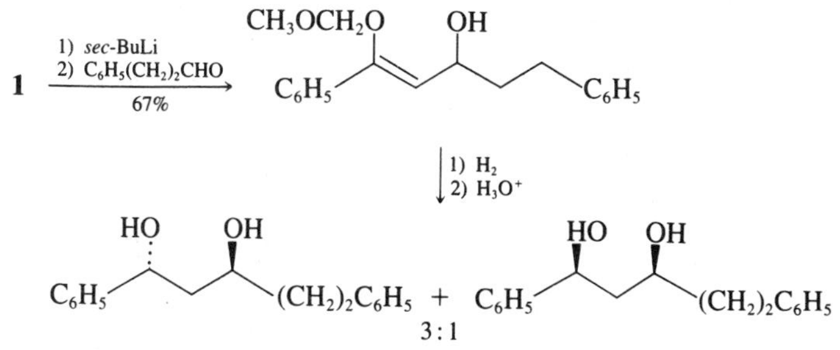

[1] P. Beak and V. Snieckus, *Accts. Chem. Res.*, **15**, 306 (1982).
[2] K. Shankaran and V. Snieckus, *J. Org.*, **49**, 5022 (1984).
[3] M. P. Sibi and V. Snieckus, *ibid.*, **48**, 1935 (1983).
[4] P. G. McDougal and J. G. Rico, *Tetrahedron Letters*, **25**, 5977 (1984).

sec-Butyllithium–Potassium *t*-butoxide.

Deprotonation of N-methylamines.[1]   Aliphatic N-methylamines are metallated exclusively on the methyl group by *sec*-butyllithium and potassium *t*-butoxide when excess amine is used as solvent. Deprotonation cannot be effected with *n*-BuLi/KOC(CH$_3$)$_3$ or *t*-BuLi/KOC(CH$_3$)$_3$. The resulting carbanion reacts readily with alkyl halides, but gives low yields in reactions with enolizable carbonyl compounds. Metal exchange with lithium bromide gives the lithium reagent, which reacts normally with carbonyl compounds.

Examples:

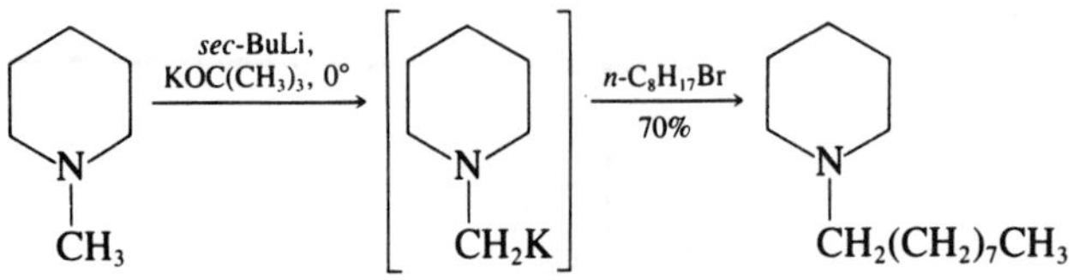

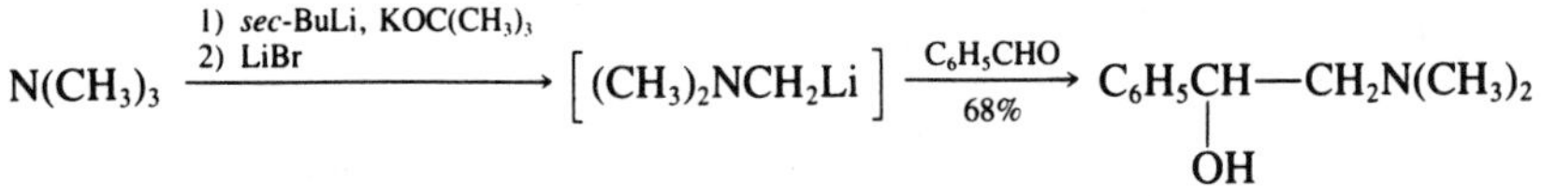

[1] H. Ahlbrecht and H. Dollinger, *Tetrahedron Letters,* **25**, 1353 (1984).

t-Butyllithium, **1**, 96–97; **5**, 79–80; **7**, 47; **8**, 70–72; **9**, 89; **10**, 76–77; **11**, 103–105.

α-Methoxyvinyllithium.[1] Details are available for deprotonation of methyl vinyl ether by *t*-butyllithium to provide α-methoxyvinyllithium (**a**), a useful acyl anion equivalent. Thus **a** reacts with ClSi(CH₃)₃ to provide 1-(methoxyvinyl)trimethylsilane (**1**), a useful precursor to acetotrimethylsilane (**2**).

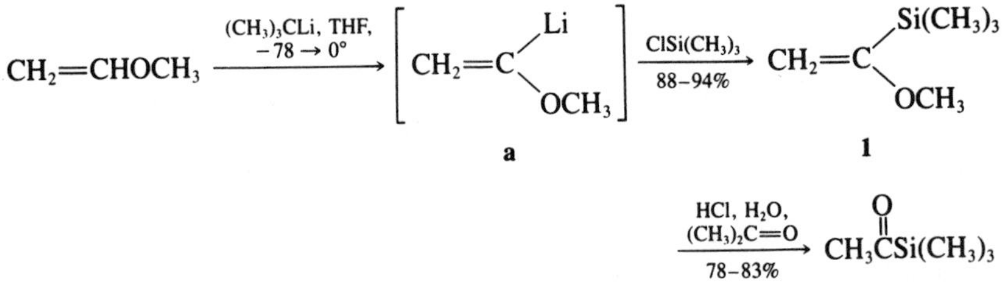

[1] J. A. Soderquist and A. Hassner, *Am. Soc.,* **102**, 1577 (1980); J. A. Soderquist, *Org. Syn,* submitted (1984).

t-Butylmethoxyphenylsilyl bromide, $C_6H_5\underset{\underset{\displaystyle C(CH_3)_3}{|}}{\overset{\overset{\displaystyle OCH_3}{|}}{Si}}\!-\!Br$. This silane is prepared by reaction of *t*-butyldiphenylmethoxysilane with bromine in ethylene dichloride (65% yield).

t-*Butylmethoxyphenylsilyl ethers* (*t*-BMPSi ethers).[1] In DMF in the presence of $N(C_2H_5)_3$, this bromosilane reacts with primary, secondary, and tertiary alcohols to form silyl ethers in good yield, and also with some enolizable ketones to form enol silyl acetals. Selective silylation of primary alcohols is possible by use of CH_2Cl_2 as solvent. The hydrolytic stability of these ethers is intermediate between that of *t*-butyldimethylsilyl ethers and that of *t*-butyldiphenylsilyl ethers. The most useful feature of this new protecting group is the selective cleavage by fluoride ion in the presence of other silyl ethers.

[1] Y. Guindon, R. Fortin, C. Yoakim, and J. W. Gillard, *Tetrahedron Letters*, **25**, 4717 (1984).

2-*t*-Butylperoxy-1,3,2-dioxaborolane (1).
Preparation:

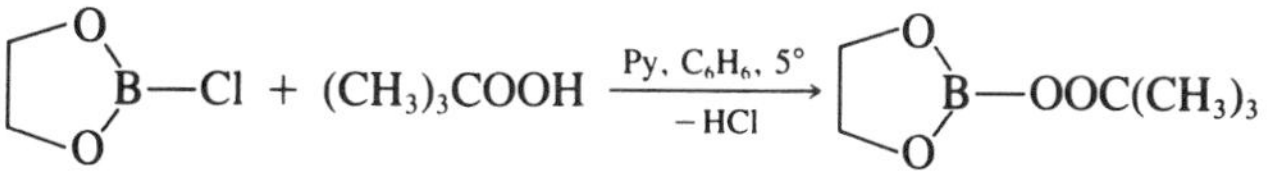

1 (impure oil)

ArMgBr → ArOH.[1]   The reagent converts an arylmagnesium bromide or an aryllithium into an ate complex (**a**), which rearranges to **b**. Phenols are obtained on acid-catalyzed hydrolysis of **b**. Yields are usually higher from Grignard reagents; they are poor with sterically hindered substrates.

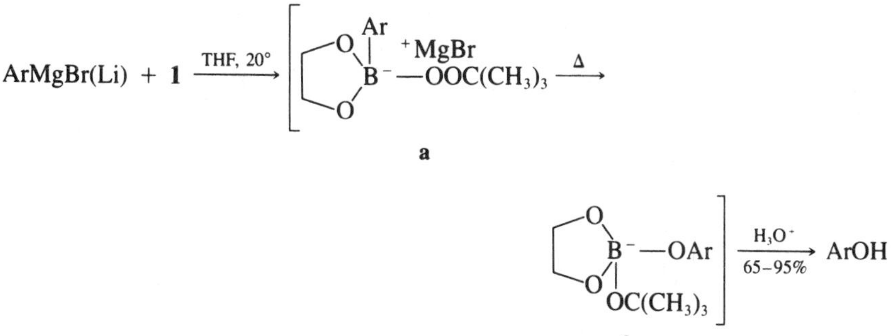

[1] R. W. Hoffmann and K. Ditrich, *Synthesis*, 107 (1983).

t-**Butyl 2-pyridyl carbonate (1).** M.p. 48–49°.

Preparation:

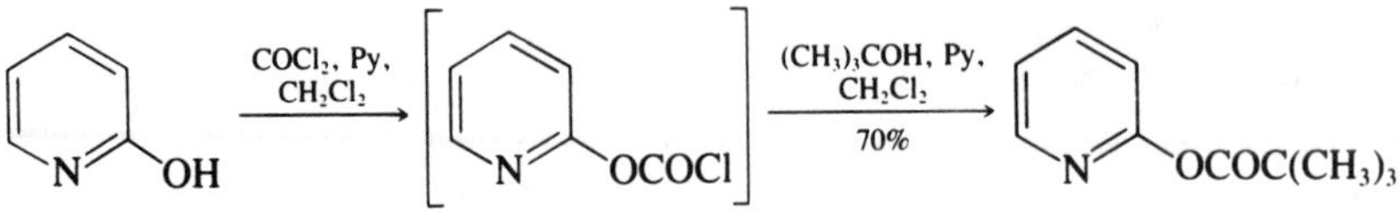

N-Boc Amino acids.[1]   *t*-Butoxycarbonylation of amino acids is effected in 85–98% yield by reaction with **1** and triethylamine in aqueous DMF at room temperature for 4–10 hours followed by acidification with oxalic acid.

[1] S. Kim and J. I. Lee, *Chem. Letters,* 237 (1984).

N''-(*t*-Butyl)-N,N,N',N'-tetramethylguanidinium *m*-iodylbenzoate (1).

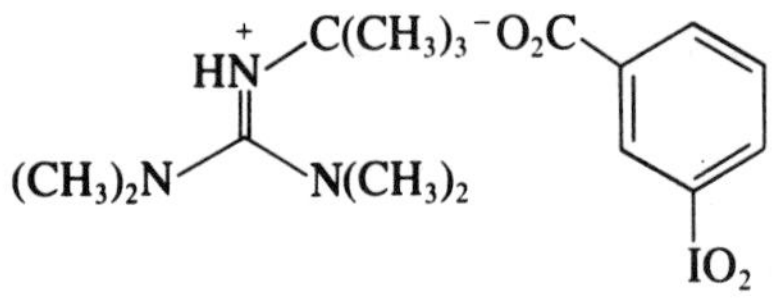

The salt is obtained by mixing *m*-iodylbenzoic acid (**11**, 275) with the hindered base (**11**, 105–106). The salt is soluble in CH_2Cl_2.

Oxidation.   The salt effects rapid cleavage of glycols in high yield.[1] It can be used also for the oxidation of secondary nitro compounds to ketones (Nef reaction).[2] An excess of the guanidine base is used to form the nitronate, which is then oxidized to the ketone by the salt. Yields are ~80–95%, and are higher than those obtained with iodosylbenzene, C_6H_5IO. Oxidation of primary nitro compounds by this method gives only low to moderate yields of aldehydes.

[1] D. H. R. Barton, C. R. A. Godfrey, J. W. Morzycki, W. B. Motherwell, and A. Stobie, *Tetrahedron Letters,* **23**, 957 (1982).
[2] D. H. R. Barton, W. B. Motherwell, and S. Z. Zard, *ibid.,* **24**, 5227 (1983).

C

Calcium–Amines, 9, 94.

Reduction of arenes. Calcium in combination with amines, generally methylamine or ethylenediamine, reduces aromatic hydrocarbons to monoenes in generally high yield.[1] Addition of *t*-butyl alcohol to this system effects reduction to nonconjugated dienes (Birch-type products). Ethylenediamine is essential in this case; addition of a second amine is generally desirable for solubility reasons.[2]

Examples:

| | Ca, amine | — | 84% |
| | + (CH₃)₃COH | 88% | 11% |

| | Ca, amine | — | 80% | 9% |
| | + (CH₃)₃OH | 89% | 6% | 2% |

[1] R. A. Benkeser, F. G. Balmonte, and J. Kang, *J. Org.*, **48**, 2796 (1983).
[2] R. A. Benkeser, J. A. Laugal, and A. Rappa, *Tetrahedron Letters*, **25**, 2089 (1984).

Camphor-10-sulfonic acid (1), 1, 108–109; **2,** 58–59; **4,** 68; **11,** 107–108.

Asymmetric Diels-Alder reactions. This chiral reagent, which is accessible as the (+)- or (−)-isomer, is useful for effecting highly diastereoselective Diels-Alder reactions. Thus (+)-**1** has been converted in two steps into the crystalline chiral alcohol **2** with a bulky shielding group at C₁₀. The acrylate (**3**) of **2** undergoes a TiCl₂(O-*i*-Pr)₂-promoted Diels-Alder addition to cyclopentadiene to give after two crystallizations the pure (2R)-adduct (**4**) in 83% yield and in 99% de. Lithium aluminum hydride reduction of **4** furnishes pure (2R)-**5** with regeneration of **2**. One advantage of this process is that all the intermediates are highly crystalline and easily purified.

(+)-1 2 (R*OH) 3

4 (99% de) (2R)-5 (99% ee)

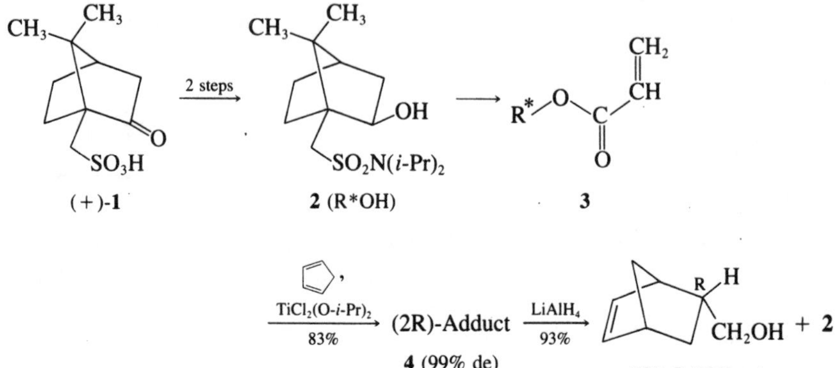

The $TiCl_4$-catalyzed addition of 1,3-butadiene to **3** proceeds with lower chiral efficiency, but an almost pure adduct (**6**) was obtained after two crystallizations in 60% yield and in 86% de. Hydrolysis of the adduct furnishes the important chiral cyclohexene **7**.[1]

6 (86% de) 7

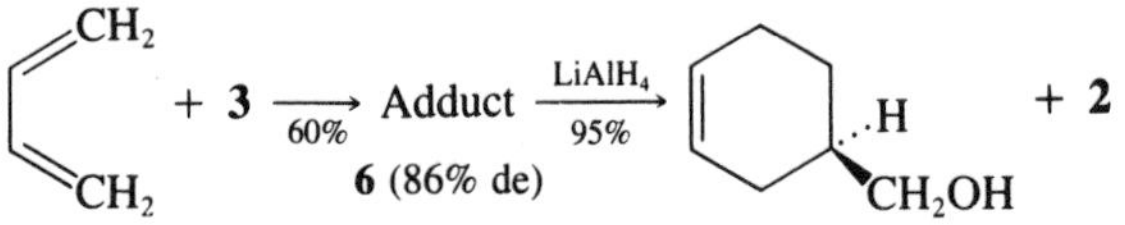

Acryloyl and crotonoyl amides (**9**) derived from the sultam (**8**), available from (+)-**1**, are somewhat more reactive dienophiles than the corresponding esters (**3**) derived from **2**. Diels-Alder reactions of **9** with cyclopentadiene catalyzed by $TiCl_4$ or $C_2H_5AlCl_2$ proceed in high yield, with high *endo* selectivity (~99%) and diastereoselectivity of ~95% de. The chiral auxiliary is removed by reduction with $LiAlH_4$ to furnish **8** and the chiral alcohol in 89–95% yield.[2]

8

9a, R = H
b, R = Ch3

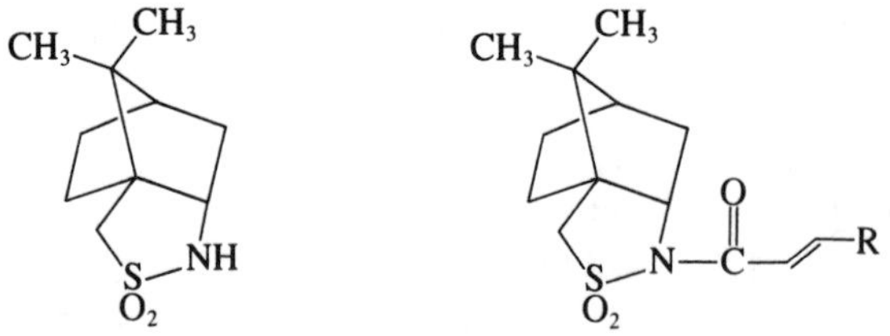

[1] W. Oppolzer, C. Chapuis, and M. J. Kelly, *Helv.*, **66**, 2358 (1983).
[2] W. Oppolzer, C. Chapuis, and G. Bernardinelli, *ibid.*, **67**, 1397 (1984).

1-Carbomethoxyethyl(diphenyl)phosphine oxide, $(C_6H_5)_2\overset{\overset{O}{\|}}{P}CH\overset{\diagup COOCH_3}{\diagdown CH_3}$ **(1).** The phosphine oxide is prepared by reaction of ethyl diphenylphosphinite with methyl 2-bromopropionate at 100° (m.p. 107–108°, 85% yield).[1]

Wittig-Horner reaction.[2] The reagent reacts with a wide range of aldehydes, aliphatic, aromatic, and α,β-unsaturated, in the presence of a base to form (E)-α-methyl-α,β-unsaturated esters stereoselectively. The reaction can be carried out either with potassium *t*-butoxide in DMF or with K_2CO_3 in C_6H_6/H_2O and Aliquat 336 as phase-transfer agent.

Example:

(E/Z $\geqslant$ 95:5)

[1] T. Bottin-Strzalko, G. Etemad-Moghadam, M. J. Pouet, J. Seyden-Penne, and M. P. Simonnin, *Nouv. J. Chim.*, **7**, 155 (1983).
[2] G. Etemad-Moghadam and J. Seyden-Penne, *Tetrahedron,* **40**, 5153 (1984).

Carbon monoxide, 2, 60, 204; **3,** 41–43; **5,** 96; **7,** 53; **8,** 76–77; **9,** 95; **11,** 111.

Acylsilanes.[1] Carbonylation of trimethylsilylmethyllithium (**1**) in ether at 15° followed by quenching with $ClSi(CH_3)_3$ results in the trimethylsilyl enolate (**2**) of acetotrimethylsilane. The reaction evidently involves insertion of CO to give an acyllithium (**a**), which undergoes a 1,2-silicon shift to give the lithium enolate (**b**) of an acylsilane.

This reaction is general for α-silylalkyllithium compounds and, by quenching with H_2O or $ClSi(CH_3)_3$, can be used to obtain acylsilanes or the enol silyl ethers, which are obtained as (E)-isomers.

Examples:

$$(syn/anti = 93:7)$$

[1] S. Murai, I. Ryu, J. Iriguchi, and N. Sonoda, *Am. Soc.*, **106**, 2440 (1984).

Carbonyl difluoride, COF_2. Pungent hygroscopic gas. Preparation.[1] Supplier: PCR Research Chemicals.

   Fluorination.[2]   The reagent reacts with organophosphines or phosphites to give the corresponding difluoro compounds in 70–80% yield. It reacts with dialkylamines in the presence of triethylamine to give N-fluoro derivatives. A similar reaction is observed with tertiary hydrocarbons.

[1] M. M. Farlow, E. H. Man, and C. W. Tulloch, *Inorg. Syn.*, **6**, 155 (1960).
[2] O. D. Gupta and J. M. Shreeve, *J.C.S. Chem. Comm.*, 416 (1984).

N,N'-Carbonyldiimidazole, 1, 114–116; **2,** 61; **5,** 97–98; **6,** 97; **8,** 77; **9,** 96.

   Alkyl halides.[1]   Alcohols can be converted into alkyl bromides or iodides by reaction with N,N'-carbonyldiimidazole and an activated halide. Any halide more reactive than the resultant halide can be used, but in practice allyl bromide and methyl iodide are preferred because they are effective and are easily removed after reaction. At least 3 equivalents are necessary for satisfactory yields. Acetonitrile is preferred as solvent. The yields are generally >80%.

[1] T. Kamijo, H. Harada, and K. Iizuka, *Chem. Pharm. Bull. Japan*, **31**, 4189 (1983).

(E)-(Carboxyvinyl)trimethylammonium betaine, $\overset{+}{(CH_3)_3 N}$ **(1)**. Mol. wt. 129.16, m.p. 240–245° (dec.). The betaine is prepared by reaction of ethyl propiolate with trimethylamine in CH_2Cl_2–H_2O (59% yield).[1]

   γ,δ-Unsaturated aldehydes.[2]   The reaction of the sodium salt of allylic alcohols with **1** results in *trans*-alkoxyacrylic acids (**2**), which undergo Claisen rearrangement with loss of carbon dioxide at 150–200° to give γ,δ-unsaturated aldehydes (**3**) (equation I). One drawback of this method is that tertiary alcohols do not react with **1**.

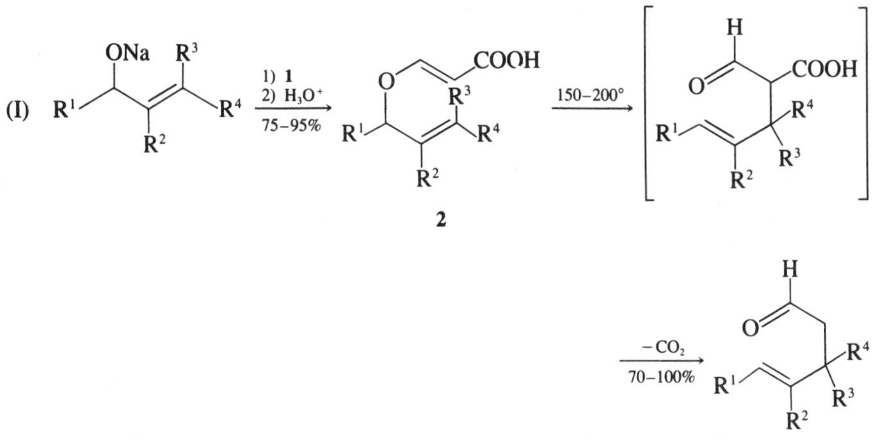

(I)

2

3

[1] A. W. McCulloch and A. G. McInnes, *Can. J. Chem.*, **52**, 3569 (1974).
[2] G. Büchi and D. E. Vogel, *J. Org.*, **48**, 5406 (1983).

Cerium(IV) ammonium sulfate (CAS), 7, 57; 8, 80–81.

Oxidation of arenes[1] (**8**, 80–81).   This oxidation can be carried out in a two-phase system with sodium dodecyl sulfate as a micellar catalyst. CAS can be used in catalytic amounts if ammonium persulfate is used in excess to convert Ce(III) as formed to Ce(IV). This oxidation is slow in the absence of a Ag(I) salt. This catalytic two-phase oxidation is very useful for preparation of polycyclic quinones from hydrocarbons, but is ineffective for other substrates.

[1] J. Skarzewski, *Tetrahedron,* **40**, 4997 (1984).

Cerium(III) chloride, 9, 99.

Aldol reaction.   Quantitative yields of 1,2-adducts of alkyllithiums to ketones can be obtained at $-65°$ in the presence of CeI_3.[1] Cerium enolates, formed by reaction of $CeCl_3$ with lithium enolates, also show enhanced reactivity in reactions with carbonyl compounds, particularly ketones. Yields of aldols are increased, but the stereoselectivity remains moderate.[2]

Example:

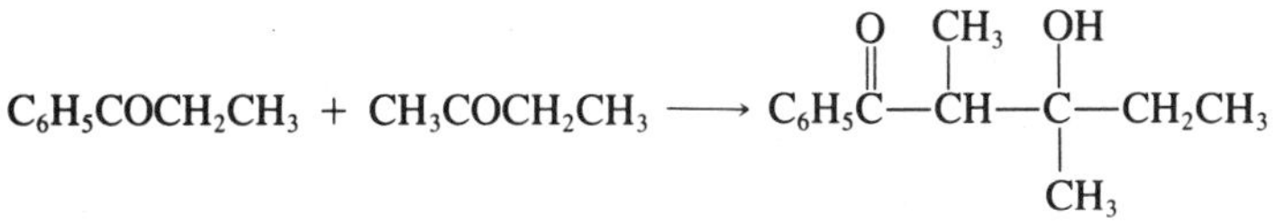

$$C_6H_5COCH_2CH_3 + CH_3COCH_2CH_3 \longrightarrow C_6H_5\overset{O}{\overset{\|}{C}}-\overset{CH_3}{\overset{|}{CH}}-\overset{OH}{\overset{|}{\underset{CH_3}{C}}}-CH_2CH_3$$

LDA	11%
LDA, CeCl₃	62%

[1] T. Imamoto, T. Kusumoto, and M. Yokoyama, *J.C.S. Chem. Comm.*, 1042 (1982).
[2] *Idem, Tetrahedron Letters*, **24**, 5233 (1983).

Cesium carbonate, 7, 57; **8**, 81; **9**, 99–100; **11**, 114–115.

Macrocyclic sulfides.[1]   Macrocyclic sulfides are prepared conveniently by reaction of the cesium salts of 1,ω-dithiols, prepared *in situ* from Cs_2CO_3 in DMF, with 1,ω-dibromoalkanes.

Example:

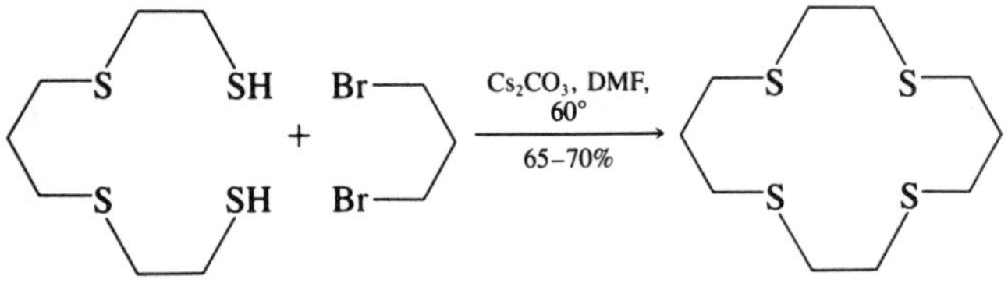

Rubidium carbonate is less effective than cesium carbonate for this cyclization; other metal carbonates are either less effective or inactive.

[1] J. Buter and R. M. Kellogg, *J. Org.*, **46**, 4481 (1981); *idem, Org. Syn.*, submitted (1984).

Cesium fluoride, 7, 57–58; **8**, 81–82; **9**, 100; **10**, 81–84; **11**, 115–117.

o-Quinone methides.[1]   *o*-Quinone methides can be generated by fluoride-induced desilylation of disilyl derivatives of *o*-hydroxybenzyl alcohols. This 1,4-elimination followed by an intramolecular Diels-Alder reaction provides a synthesis of either (+)- or (−)-hexahydrocannabinol methyl ether (**1**).

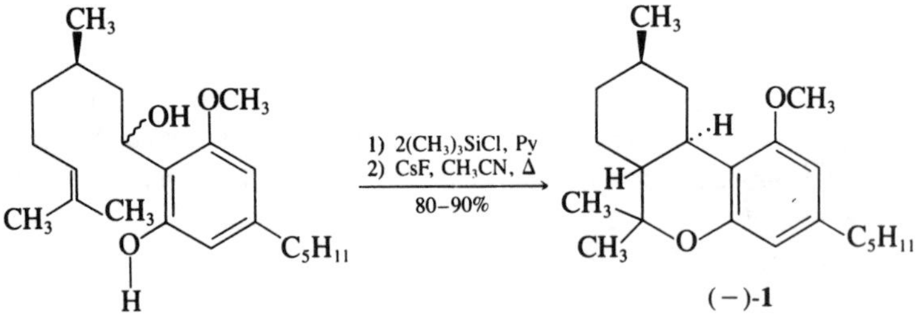

[1] J. P. Marino and S. L. Dax, *J. Org.*, **49**, 3671 (1984).

Cesium fluoride–Tetraalkoxysilanes.

Michael reactions (**10**, 82).   In the presence of equimolar amounts of CsF and $Si(OCH_3)_4$ or $Si(OC_2H_5)_4$, ketones and aryl acetonitriles undergo Michael addition to α,β-enones and α,β-unsaturated esters and nitriles. The reaction fails with highly hindered ketones or acceptors, and also with aldehydes because of competing aldolization.[1]

Examples:

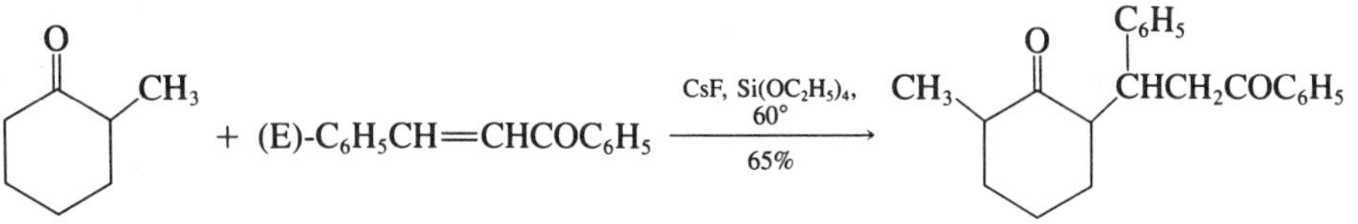

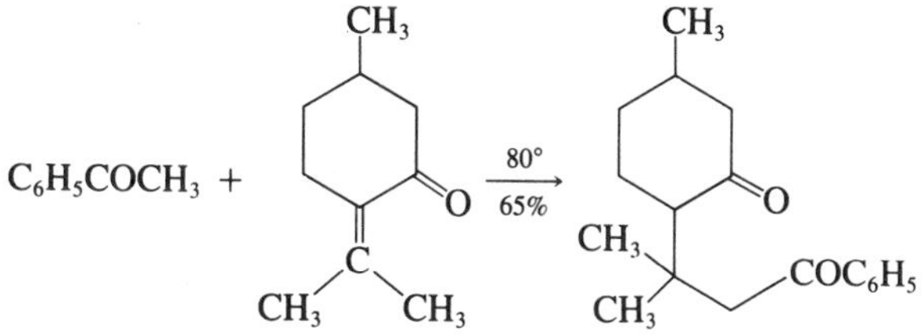

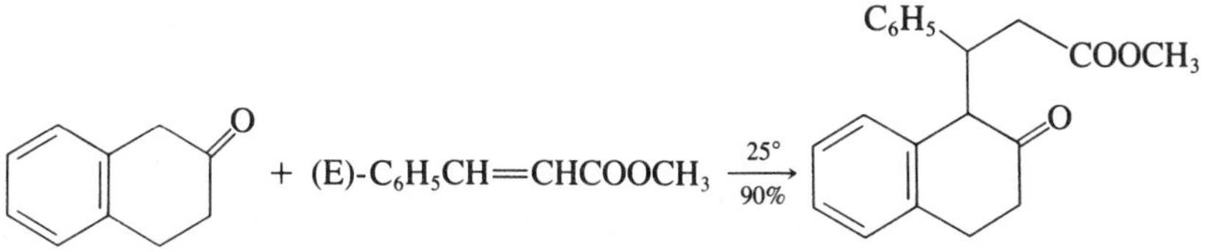

[1] J. Boyer, R. J. P. Corriu, R. Perz, and C. Reye, *Tetrahedron*, **39**, 117 (1983).

Cesium propionate, 11, 118.

Alcohol inversion.   Elimination competes with S_N2 substitution in the inversion of secondary alcohols by the Mitsunobu reaction or by reaction of mesylates with cesium propionate or cesium acetate.[1] Elimination in the inversion of cyclopentyl and cyclohexyl alcohols can be largely suppressed by reaction of the mesylate with cesium acetate (excess) and a catalytic amount of 18-crown-6 in refluxing benzene. Even inversion of an allylic alcohol can be effected in moderate yield under these conditions (equation I).[2]

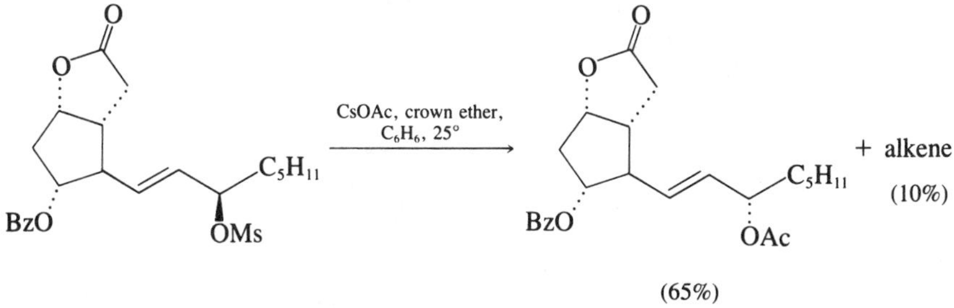

[1] Y. Torisawa, H. Okabe, and S. Ikegami, *Chem. Letters*, 1555 (1984).
[2] J. W. Huffman and R. C. Desai, *Syn. Comm.*, **13**, 553 (1983).

Chlorine oxide, 11, 119.

Ene chlorination.[1] Trisubstituted alkenes undergo an ene-type chlorination with Cl_2O (0.5 equiv.) to form an allylic chloride in high yield. Terminal double bonds are more reactive than inner double bonds; mono- and 1,2-disubstituted double bonds react slowly to give mainly products of decomposition.

Examples:

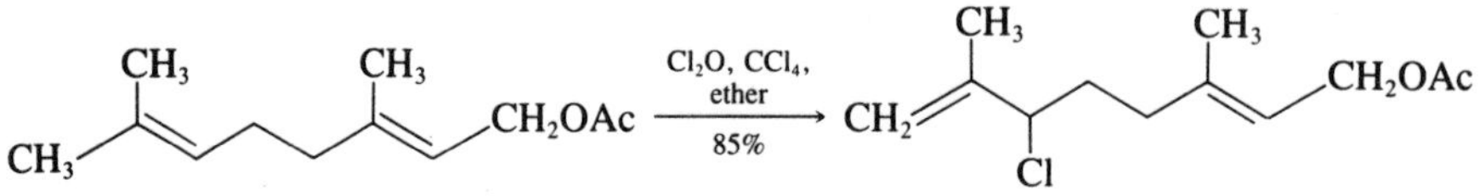

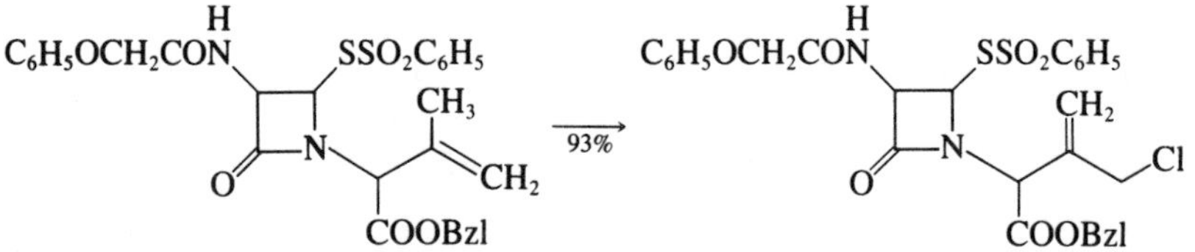

[1] S. Torii, H. Tanaka, N. Tada, S. Nagao, and M. Sasaoka, *Chem. Letters,* 877 (1984).

μ-Chlorobis(cyclopentadienyl)(dimethylaluminum)-μ-methylenetitanium (Tebbe reagent), **8,** 83–84; **10,** 87–88.

Methyl ketones.[1] The Tebbe reagent (**1**) reacts with acyl chlorides to form methyl ketones in 35–65% yield (equation I). That an intermediate titanium enolate may be involved is indicated because addition of methyl iodide before aqueous work-up results in formation of an ethyl ketone.

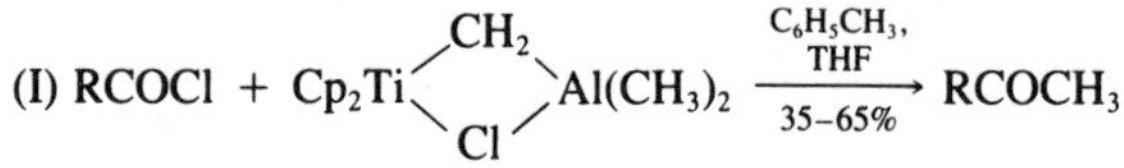

1

Methylenation of ketones.[2] The Tebbe reagent (**1**) or the unstable β,β-disubstituted titanacyclobutane **2** methylenate even easily enolizable ketones rapidly at room temperature. Yields are generally high except in reactions with a very hindered ketone.

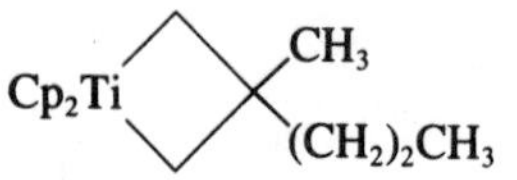

2

However, both reagents react with α,α-disubstituted ketones to form titanium enolates. These enolates are thermally stable and do not undergo aldol condensations.

Example:

[1] T.-S. Chou and S.-B. Huang, *Tetrahedron Letters*, **24**, 2169 (1983).
[2] L. Clawson, S. L. Buchwald, and R. H. Grubbs, *ibid.*, **25**, 5733(1984).

Chlorobis[1,3-bis(diphenylphosphine)propane]rhodium, $Rh(dppp)_2Cl$ (**1**). The complex is prepared by heating $COClRh[P(C_6H_5)_3]_2$ with the phosphine in xylene.[1]

Decarbonylation; decarboxylation. At 110–170° this complex in catalytic amounts effects decarbonylation of simple aldehydes in quantitative yield.[2] Decarbonylation of typical indole-2-carboxaldehydes with *in situ*-generated catalyst proceeds in 82–95% yield. In fact, decarboxylation of some indole-2-carboxylic acids can be effected in higher overall yield by conversion to the aldehyde ($LiAlH_4$; MnO_2) followed by decarbonylation than by copper-catalyzed decarboxylation.[3]

[1] J. Chatt and B. L. Shaw, *J. Chem. Soc. A*, 1437 (1966).
[2] D. H. Doughty and L. H. Pignolet, *Am. Soc.*, **100**, 7083 (1978).
[3] M. D. Meyer and L. I. Kruse, *J. Org.*, **49**, 3195 (1984).

Chlorocyanoketene (**1**), **8**, 88; **9**, 103. The pseudoisopropyl ester of the azidofuranone (**2**) is now preferred over the pseudomethyl ester used originally as the precursor because it is less prone to detonation. The ketene can be generated by dehydrohalogenation of the

unstable chlorocyanoacetyl chloride, but this material gives only moderate yields in cycloaddition to imines.[1]

Cycloaddition to alkenes.[2] This ketene adds regioselectively to alkenes, even tetrasubstituted ones, with preservation of the stereochemistry of the alkene to afford 2-chloro-2-cyanocyclobutanones in good yield.

Example:

$$C_6H_5CH{=}CH_2 \; + \; 1 \xrightarrow[86\%]{}$$

[1] P. L. Fishbein and H. W. Moore, *Org. Syn.*, submitted (1984).
[2] *Idem, J. Org.*, **49**, 2190 (1984).

Chlorodicarbonylrhodium(I) dimer, $[RhCl(CO)_2]_2$. Mol. wt. 388.76 m.p. 124–125°, air sensitive. Suppliers: Alfa, Strem.

Expansion of aziridines to β-lactams.[1] This rhodium complex catalyzes a regio-specific carbonylation of N-*t*-butyl-2-arylaziridines (**1**) to form lactams (**2**). The reaction fails if the alkyl group on nitrogen contains acidic hydrogens adjacent to nitrogen.

[1] H. Alper, F. Urso, and D. J. H. Smith, *Am. Soc.*, **105**, 6737 (1983).

2-Chloro-1,3,2-dithioborolane, Mol. wt. 138.45, b.p. 24–25°/0.5 mm.

The reagent is prepared by reaction of BCl_3 with 1,2-ethanedithiol at $-78°$.[1]

Cleavage of methoxyethoxymethyl ethers.[2] The reagent selectively cleaves MEM ethers in the presence of benzyl, silyl, allyl, and methyl ethers as well as acetals, acetates, and benzoates. However, methoxymethyl ethers and acetonides are cleaved at about the same rate as MEM ethers.

[1] A. Finch and J. Pearn, *Tetrahedron*, **20**, 173 (1964).
[2] D. R. Williams and S. Sakdarat, *Tetrahedron Letters*, **24**, 3965 (1983).

α-Chloroethyl chloroformate, $Cl\overset{\text{O}}{\overset{\|}{C}}OCHClCH_3$ (**1**). The reagent is prepared by reaction of phosgene with acetaldehyde in the presence of benzyltributylammonium chloride (96% yield).

Dealkylation of tertiary amines.[1] This reagent is preferred over vinyl chloroformate (**8**, 530) for dealkylation of tertiary amines. The conditions are milder and the yields are somewhat higher. A typical process is outlined in equation (I). Dealkylation of aromatic

(I)

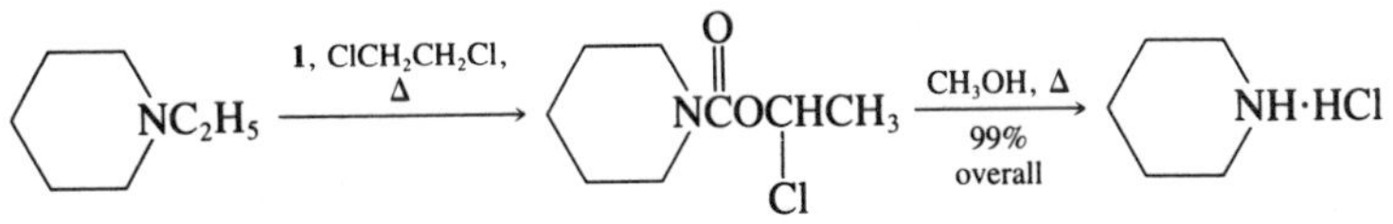

amines is also possible but requires higher temperatures. Carboalkoxy and acetoxy groups are stable; however, benzyl cleavage occurs more readily then alkyl cleavage.

[1] R. A. Olafson, J. T. Martz, J.-P. Senet, M. Piteau, and T. Malfroot, *J. Org.*, **49**, 2081 (1984).

2-Chloroethyl(dimethyl)sulfonium iodide, $ClCH_2CH_2\overset{+}{S}(CH_3)_2I^-$ (**1**). The salt, m.p. 83–85°, is prepared by reaction of 2-chloroethyl methyl sulfide with CH_3I.[1]

α-Cyclopropyl ketones.[2] The potassium enolate of ketones reacts with the reagent in *t*-butyl alcohol to form α-cyclopropyl ketones in 40–85% yield. The reagent is rapidly destroyed by base in other solvents. The actual reagent is probably 2-iodoethyl(dimethyl)sulfonium chloride, and indeed addition of NaI can enhance the yields.

Example:

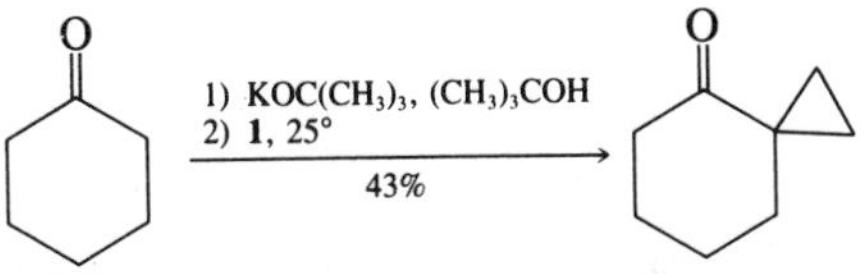

[1] W. von E. Doering and K. Schreiber, *Am. Soc.*, **77**, 514 (1955).
[2] S. M. Ruder and R. C. Ronald, *Tetrahedron Letters*, **25**, 5501 (1984).

4-Chloro-2-lithio-1-butene, $ClCH_2CH_2C(Li){=}CH_2$ (**1**).

Preparation:[1]

$$(CH_3)_3SnCu\cdot S(CH_3)_2 + HC{\equiv}C(CH_2)_2Cl \xrightarrow[59\%]{THF,\ CH_3OH} H_2C{=}C\begin{matrix} Sn(CH_3)_3 \\ (CH_2)_2Cl \end{matrix}$$

$$\xrightarrow{CH_3Li} \mathbf{1}$$

Methylenecyclopentane annelation.[2] The cuprates derived from **1** by addition of C_6H_5SCu or CuCN are sufficiently stable to permit conjugate addition to cyclic enones.

The adducts cyclize in the presence of KH to methylenecyclopentanes. The overall annelation can be effected in one step by addition of HMPT (1.5 equiv.); this gives about the same overall yield.

Example:

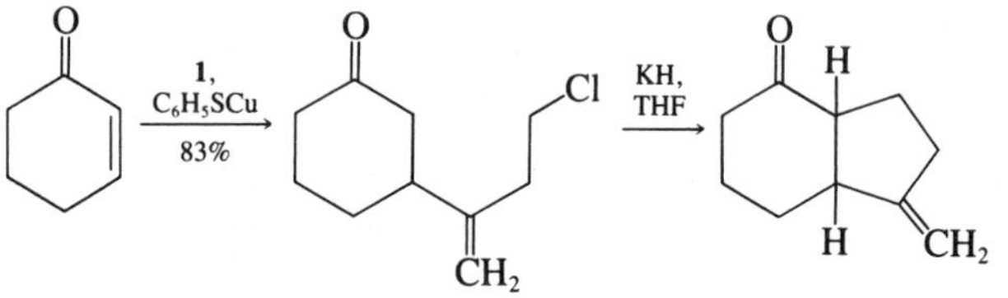

This annelation provides a key step in a synthesis of the sesquiterpene pentalenene (**2**, equation I).[3]

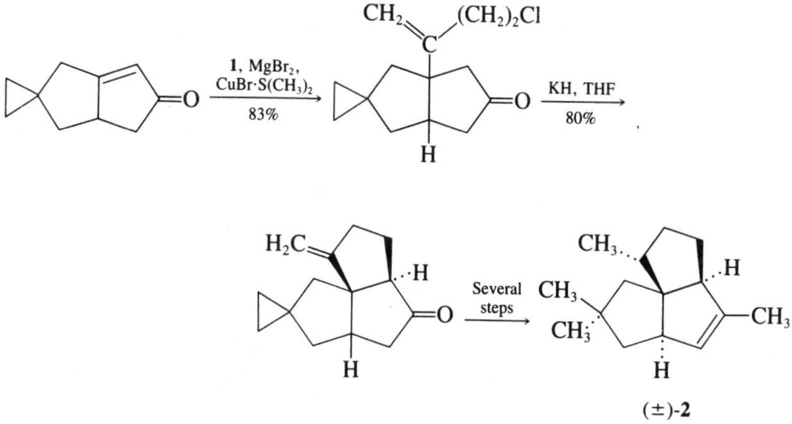

The Grignard reagent **3**, obtained by reaction of **1** with anhydrous MgBr$_2$, undergoes conjugate addition to 2-methyl-2-cyclopentene-1-one (**4**) in the presence of CuBr·S(CH$_3$)$_2$ and BF$_3$ etherate to afford the chloro ketone **5**. This procedure is more efficient than use of the corresponding cuprate reagent. The product on treatment with potassium hydride in THF provides the annelated product **6**. A similar conjugate addition of **3** to **7** provides **8**. This product is converted into the marine sesquiterpenoid $\Delta^{9(12)}$-capnellene (**9**) by deoxygenation of the methyl xanthate of the corresponding alcohol.[4]

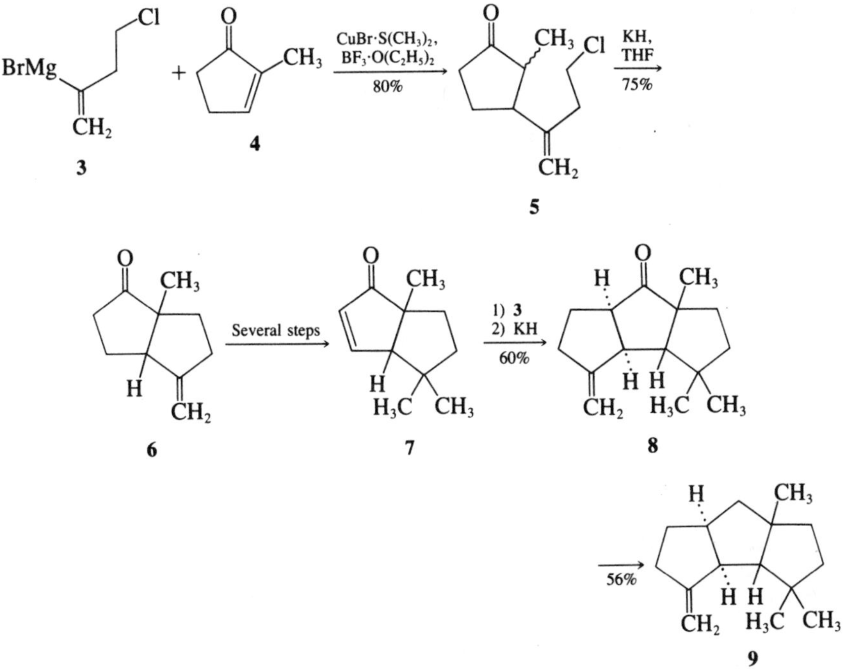

BrMg + **4** $\xrightarrow[80\%]{\text{CuBr·S(CH}_3)_2,\ \text{BF}_3\text{·O(C}_2\text{H}_5)_2}$ **5** $\xrightarrow[75\%]{\text{KH, THF}}$

6 $\xrightarrow{\text{Several steps}}$ **7** $\xrightarrow[60\%]{\text{1) 3 2) KH}}$ **8** $\xrightarrow{56\%}$ **9**

3-Methylenetetrahydrofuranes.[5] The adducts (**2**) of aldehydes or ketones with **1** cyclize to 3-methylenetetrahydrofuranes (**3**) when treated with KH in THF. Alternatively, addition of HMPT to a solution of the initial reaction mixture effects cyclization directly to **3**.

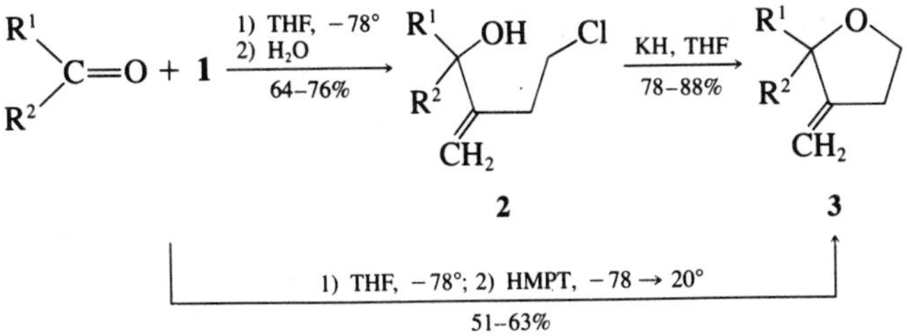

$\underset{R^2}{\overset{R^1}{>}}C{=}O + 1$ $\xrightarrow[64-76\%]{\text{1) THF, }-78°\ \ \text{2) H}_2\text{O}}$ **2** $\xrightarrow[78-88\%]{\text{KH, THF}}$ **3**

1) THF, −78°; 2) HMPT, −78 → 20°
51–63%

5-Chloro-2-lithio-1-pentene, $CH_2{=}C(Li)CH_2CH_2CH_2Cl$ (**2**).[6] Both the Grignard and the cuprate reagents derived from **2** undergo conjugate addition to enones in moderate yield. In some cases use of BF_3 etherate as catalyst improves yields significantly. The products are cyclized by base to annelated methylenecyclohexanes. Overall yields vary from 40 to 70%.

Example:

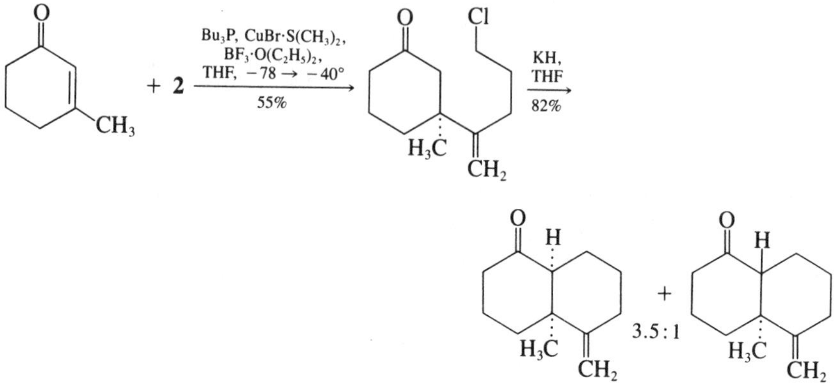

[1] E. Piers and J. M. Chong, *J.C.S. Chem. Comm.*, 934 (1983).
[2] E. Piers and V. Karunaratne, *ibid.*, 935 (1983).
[3] *Idem, ibid.*, 959 (1984).
[4] E. Piers and V. Karunaratne, *Can. J. Chem.*, **62**, 629 (1984).
[5] *Idem, J. Org.*, **48**, 1774 (1983).
[6] E. Piers and B. W. A. Yeung, *ibid.*, **49**, 4567 (1984).

Chloromethyl methyl ether, 1, 132–135; **5,** 120; **7,** 61–62; **9,** 107. This ether can be prepared by heating methoxyacetic acid (Aldrich) with thionyl chloride under reflux for 2 hours followed by fractional distillation (81% yield).[1]

[1] M. Jones, *Synthesis*, 727 (1984).

2-Chloro-1-methylpyridinium iodide (1), 8, 95–96. Review.[1]

β-*Lactams*.[2]   β-Lactams are obtained in high yield by treatment of γ-amino acids with the reagent and triethylamine in CH_2Cl_2 at room temperature.

Example:

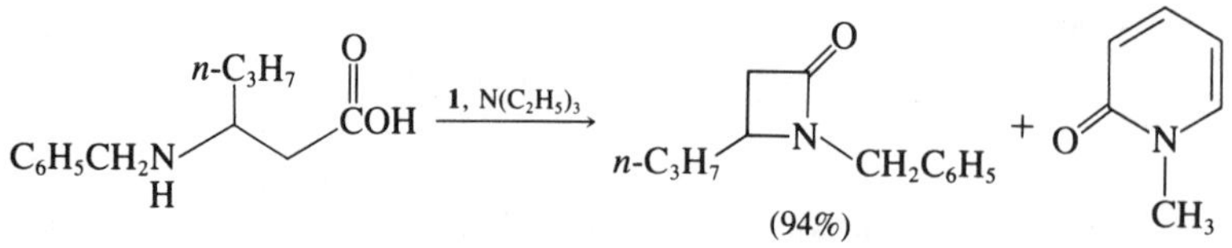

Lactonization.[3]   Lactonization of γ-hydroxy acids to *trans*-fused bicyclic γ-lactones generally requires acidic conditions or DCC. Cyclization of hydroxy acids of structure **2** is difficult because of dehydration, but is effected in >95% yield by treatment with excess **1** and $N(C_2H_5)_3$ in refluxing CH_2Cl_2.

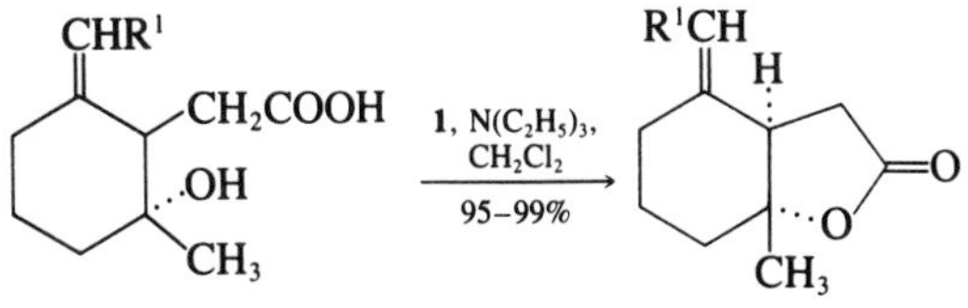

2 → **3**

1, $N(C_2H_5)_3$,
CH_2Cl_2
95–99%

[1] T. Mukaiyama, *Angew. Chem., Int. Ed.*, **18**, 707 (1979).
[2] H. Huang, N. Iwasawa, and T. Mukaiyama, *Chem. Letters*, 1465 (1984).
[3] L. Strekowski, M. Visnick, and M. A. Battiste, *Synthesis*, 493 (1983).

Chloromethyl methyl sulfide, 6, 109–110; 8, 94.

Macrolactonization.[1]   Lactonization to the 11- and 12-membered pyrrolizidine di-lactones by the usual methods employing activation of the carboxyl group is generally unsuccessful or proceeds in low yield. A new method uses a (methylthio)methyl ester[2] for protection of the carboxyl group and then for lactonization. On oxidation it is converted into a (methylsulfonyl)methyl group, which is labile to base (**9**, 107). Thus conversion of the hydroxyl group of **1** to the lithium alkoxide (triphenylmethyllithium, THF, $-78 \rightarrow -60°$) is accompanied by lactonization to **2** (40% yield). Reduction of the N-oxide group with zinc and deprotection of the MOM ether furnishes the pyrrolizidine alkaloid ($\pm$)-integerrimine (**3**).

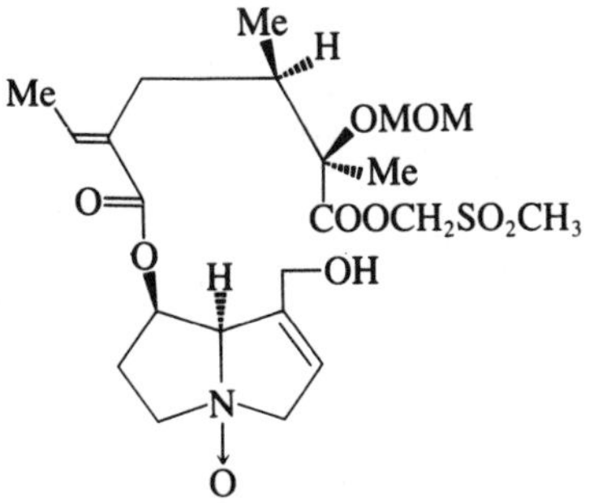

1

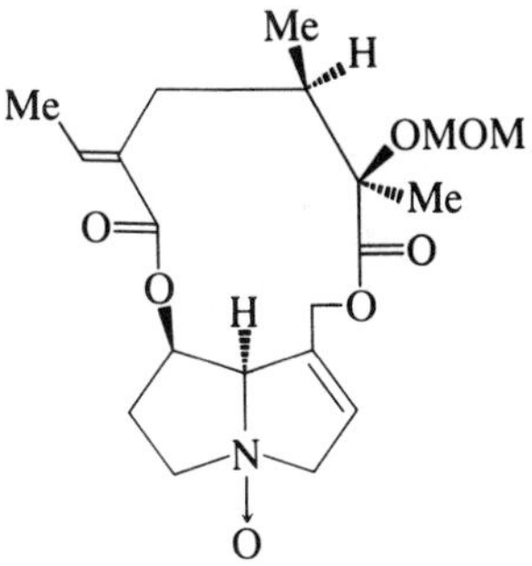

2

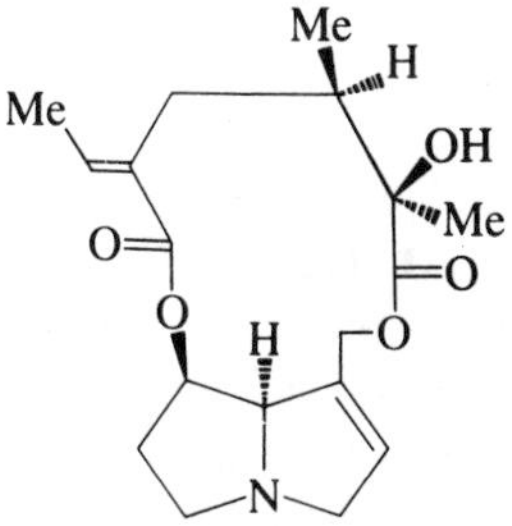

3

[1] K. Narasaka, T. Sakakura, T. Uchimaru, and D. Guédin-Vuong, *Am. Soc.*, **106**, 2954 (1984).
[2] MTM esters are obtained in high yield by reaction of the acid with MTMCl, NaI, and diisopropylethylamine in DME.

Chloromethyltrimethylsilane, $(CH_3)_3SiCH_2Cl$. Mol. wt. 122.67, b.p. 98–99°. Supplier: Aldrich.

   α,β-*Epoxytrimethylsilanes*.[1]   Deprotonation of this silane with *sec*-BuLi/TMEDA affords chloromethyl(trimethylsilyl)lithium (**1**), which reacts with aldehydes or ketones to form α,β-epoxytrimethylsilanes (**2**). Yields are low with sterically hindered or readily

$$(CH_3)_3SiCH_2Cl \xrightarrow[\substack{\text{95\%}}]{\substack{\text{\textit{sec}-BuLi, TMEDA,} \\ \text{THF, } -55°}} (CH_3)_3SiCHCl \underset{\underset{Li}{|}}{} \xrightarrow[\substack{\text{20--95\%}}]{\substack{R^1 \\ R^2}C=O} \underset{R^2}{\overset{R^1}{}}\underset{H}{\overset{O}{\triangle}}Si(CH_3)_3$$

1   2

enolizable carbonyl compounds. The products are converted by mild acid hydrolysis into homologated aldehydes (70–98% yield). Magnus *et al.*[1] report a new transformation of an epoxysilane into an enol formate (equation I).

[1] C. Burford, F. Cooke, G. Roy, and P. Magnus, *Tetrahedron*, **39**, 867 (1983).

m-**Chloroperbenzoic acid, 1**, 135–139, **2**, 68–69; **3**, 49–50; **6**, 110–114; **7**, 62–64; **8**, 97–102; **9**, 108–110; **10**, 92–93; **11**, 122–124.

   Oxidation of allylic iodides to rearranged allylic alcohols.[1]   The oxidation of RI to ROH by *m*-chloroperbenzoic acid (**8**, 98–99) has been extended to oxidation of allylic iodides. In this case the presence of a base ($NaHCO_3$) is necessary for satisfactory yields, and at least 3 equiv. of the peracid is necessary. The oxidation involves a [2,3] sigmatropic rearrangement.

Examples:

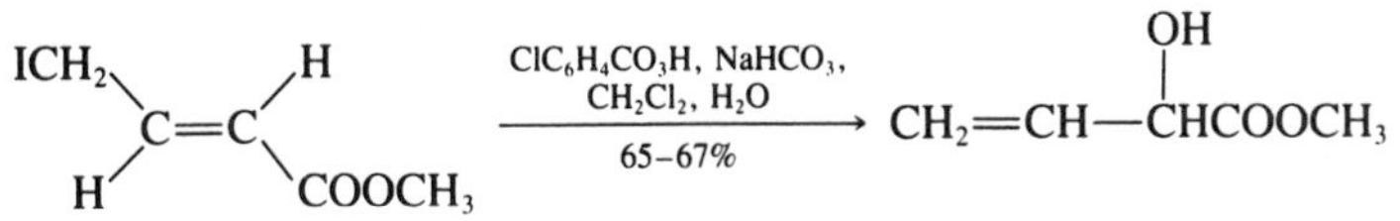

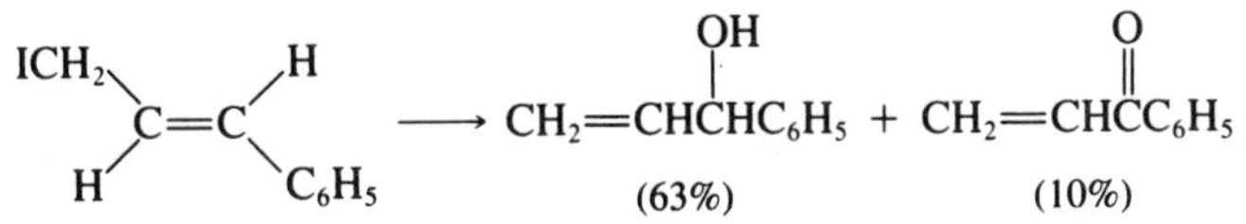

$$ICH_2, \quad H$$
structure with $CIC_6H_4CO_3H, NaHCO_3,$ CH_2Cl_2, H_2O over arrow, 65–67%, giving $CH_2=CH-\overset{OH}{\underset{|}{C}}HCOOCH_3$

structure giving $CH_2=CH\overset{OH}{\underset{|}{C}}HC_6H_5$ + $CH_2=CH\overset{O}{\underset{||}{C}}C_6H_5$

(63%)   (10%)

Oxidative cleavage of N,N-dimethylhydrazones.[2]   *m*-Chloroperbenzoic acid is as efficient and more economical than $NaIO_4$ (**7**, 126) for cleavage of these hydrazones. The cleavage of the hydrazone **1** was effected without epimerization of the axial methyl group.

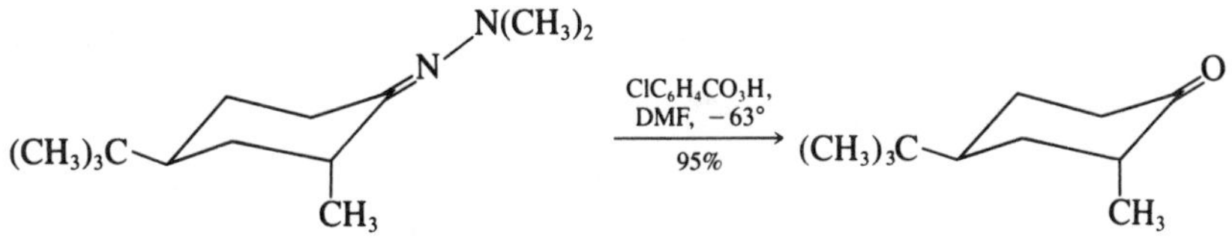

$CIC_6H_4CO_3H$, DMF, $-63°$ over arrow, 95%

1   2

Hydroxylation of a β-keto ester.[3]   Reaction of the unsaturated β-keto ester **1** with *m*-chloroperbenzoic acid gives the expected epoxide in 88% yield. Reaction with excess peracid unexpectedly also effects hydroxylation to provide **2** in high yield.

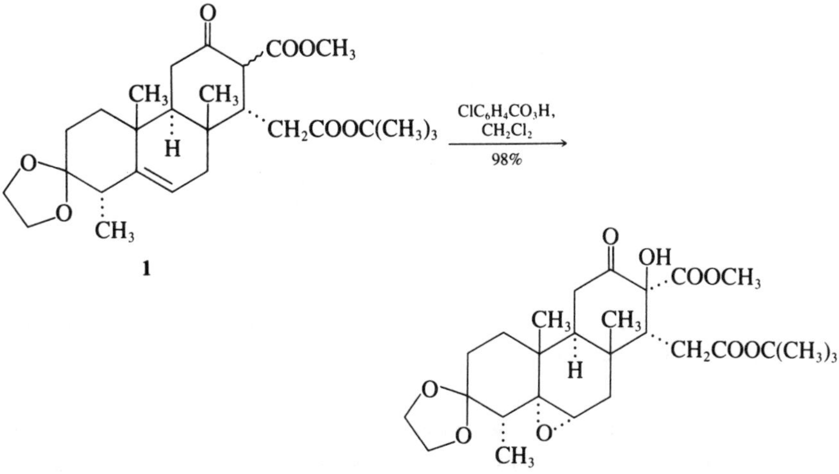

$CIC_6H_4CO_3H$, CH_2Cl_2 over arrow, 98%

1

2

Hydroxylation of resorcinol ethers.[4]   Methyl or THP ethers of resorcinol are oxidized regioselectively by *m*-chloroperbenzoic acid in CH_2Cl_2 at 0° to derivatives of 1,2,4-trihydroxybenzene in yields of 55–80%.

Example:

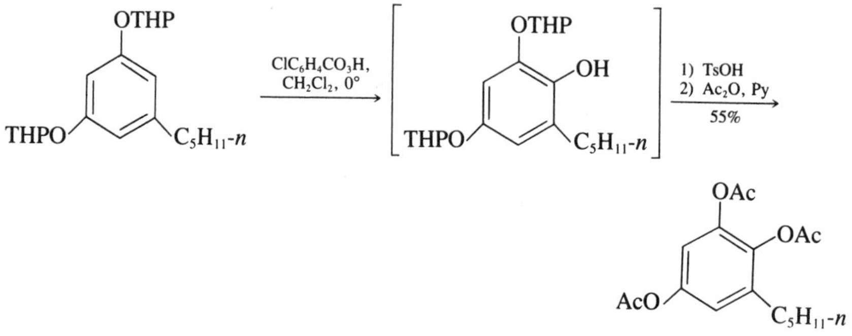

$RSeC_6H_5 \rightarrow ROCH_3$.[5]   Oxidation of an alkyl phenyl selenide with *m*-chloroperbenzoic acid (2–5 equiv.) in methanol affords the corresponding alkyl methyl ethers in high yield. Oxidation of selenides with a vicinal phenyl group is accompanied by rearrangement of the phenyl group. *vic*-Methoxy selenides derived from cycloalkenes are oxidized under these conditions to dimethyl acetals of ring-contracted aldehydes.

Examples:

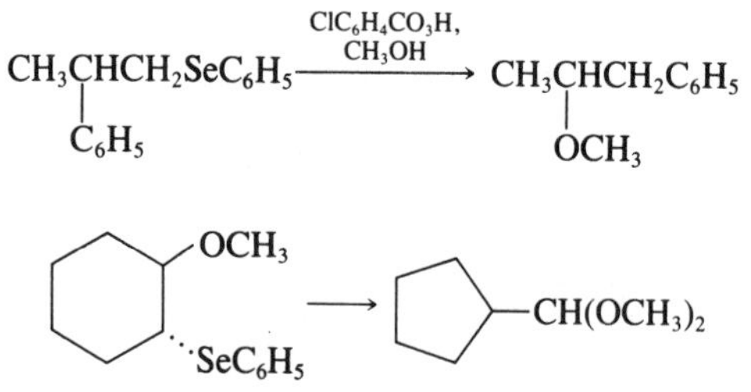

This oxidation is believed to involve a selenone, which can be cleaved by methanol either directly or with rearrangement.

$$-\overset{\overset{\displaystyle S}{\|}}{N}HCR \rightarrow -\overset{\overset{\displaystyle O}{\|}}{N}HCR.$$[6]   Thioamides and thiolactams are converted into amides and lactams, respectively, by oxidation with $ClC_6H_4CO_3H$ in CH_2Cl_2 at 25° (exothermic reaction). This oxidation occurs more readily than epoxidation of double bonds, if these are present.

Examples:

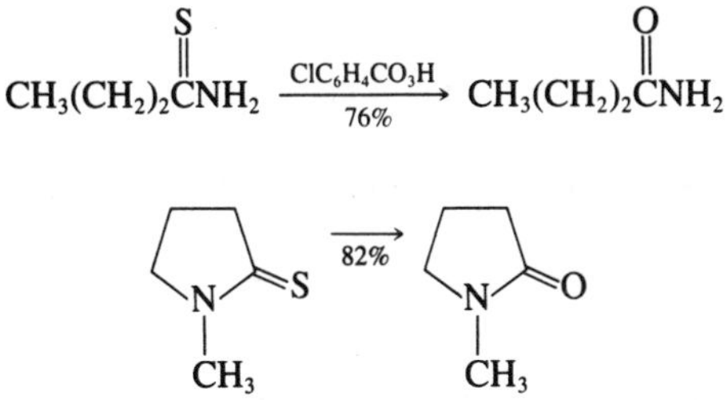

$$\underset{CH_3(CH_2)_2\overset{\overset{\displaystyle S}{\|}}{C}NH_2}{} \xrightarrow[76\%]{ClC_6H_4CO_3H} \underset{CH_3(CH_2)_2\overset{\overset{\displaystyle O}{\|}}{C}NH_2}{}$$

[1] S. Yamamoto, H. Itani, T. Tsuji, and W. Nagata, *Am. Soc.*, **105**, 2908 (1983).
[2] M. Duraisamy and H. M. Walborsky, *J. Org.*, **49**, 3411 (1984).
[3] C. H. Heathcock, C. Mahaim, M. F. Schlecht, and T. Utawanit, *ibid.*, **49**, 3264 (1984).
[4] M. Srebnik and R. Mechoulam, *Synthesis,* 1046 (1983).
[5] S. Uemura, S. Fukuzawa, and A. Toshimitsu, *J.C.S. Chem. Comm.,* 1501 (1983).
[6] K. S. Kochhar, D. A. Cottrell, and H. W. Pinnick, *Tetrahedron Letters,* **24**, 1323 (1983).

N-Chlorosuccinimide, 1, 139; **2,** 69–70; **5,** 127–129; **6,** 115–118, **7,** 65; **8,** 103–105; **9,** 111; **10,** 94.

Allylic amines.[1]   Allylic phenyl selenides on reaction with NCS (excess), an alkyl carbamate, and triethylamine at room temperature are converted into the corresponding rearranged carbamate-protected primary allylic amines.

Examples:

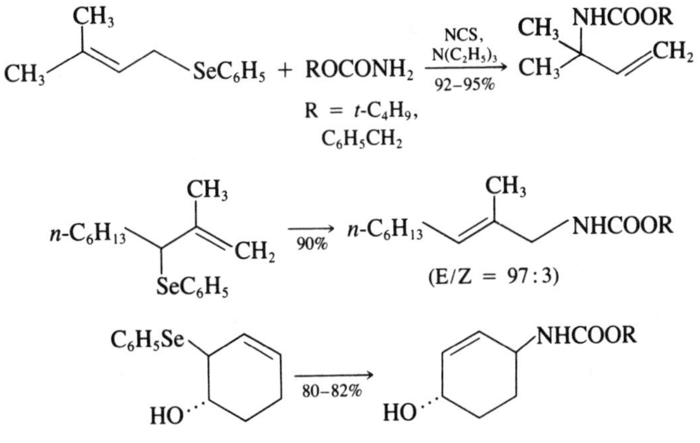

The same rearrangement with anhydrous chloromine T[2] is less attractive, because the resulting *p*-toluenesulfonamides are less useful derivatives of amines.

[1] R. G. Shea, J. N. Fitzner, J. E. Fankhauser, and P. B. Hopkins, *J. Org.*, **49**, 3647 (1984).
[2] J. E. Fankhauser, R. M. Peevey, and P. B. Hopkins, *Tetrahedron Letters*, **25**, 15 (1984).

1-Chlorosulfinyl-4-dimethylaminopyridinium chloride, $(CH_3)_2N$—⟨pyridinium⟩—$\overset{+}{N}$—$\overset{\overset{O}{\|}}{S}Cl\,Cl^-$

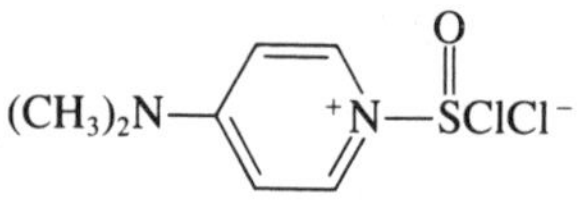

The reagent is prepared by reaction of 4-dimethylaminopyridine with thionyl chloride in CH_2Cl_2 at $-10°$.

Nitriles.[1]   Aldoximes are dehydrated to nitriles in 10–30 minutes by reaction with **1** (slight excess) (equation I).

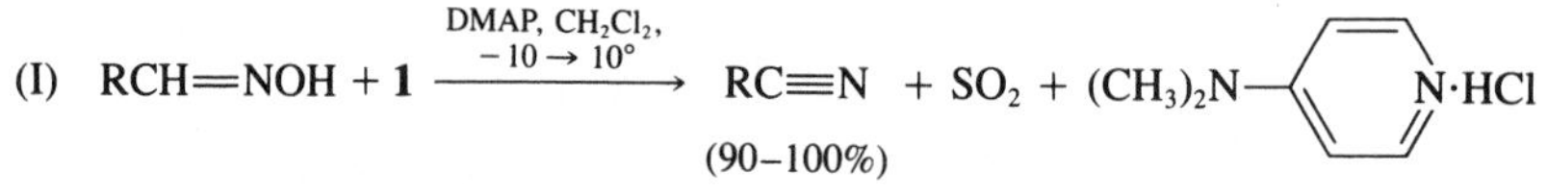

(I) $RCH{=}NOH + \mathbf{1}$ $\xrightarrow[\;-10 \to 10°\;]{\text{DMAP, } CH_2Cl_2,}$ $RC{\equiv}N + SO_2 + (CH_3)_2N$—⟨pyridine⟩—$N{\cdot}HCl$

(90–100%)

[1] A. Arrieta and C. Palomo, *Synthesis*, 472 (1983).

Chlorosulfonyl isocyanate (CSI, **1**), **1**, 117–118; **2**, 70; **3**, 51–53; **4**, 90–94; **5**, 132–136; **6**, 122; **7**, 65–66; **8**, 105–106; **10**, 94–95; **11**, 125.

3-(Methylethylidene)azetidinones.[1]   CSI reacts with the allenyl acetate **2** to give, after reduction, the unsaturated azetidinone **3**, which was converted into the stable sulfide **4** in 44% overall yield. This product was converted in several steps into a carbapenem (**5**). Compounds of this type are potent β-lactamase inhibitors.

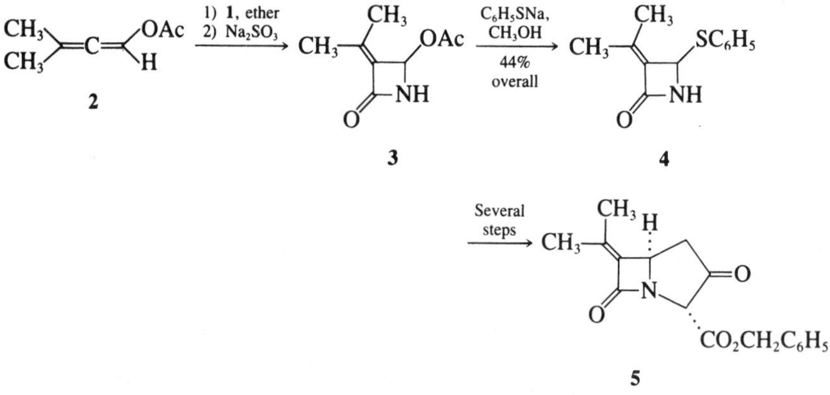

[1] J. D. Buynak, H. Pajouhesh, D. L. Lively, and Y. Ramalakshmi, *J.C.S. Chem. Comm.*, 948 (1984).

N-Chlorotriethylammonium chloride, $(C_2H_5)_3\overset{+}{N}Cl\,Cl^-$ (**1**). The salt is prepared by addition of the tertiary amine to CCl_4 saturated with Cl_2 at $-20°$. The salt is soluble in TFA.[1]

Regioselective chlorination.[2]   This N-chloroammonium salt (1) as well as N-chloropiperidine, chlorinates aromatic compounds substituted with an electron-donating group with high *para*-selectivity. Benzene and toluene are attacked only slowly.

Example:

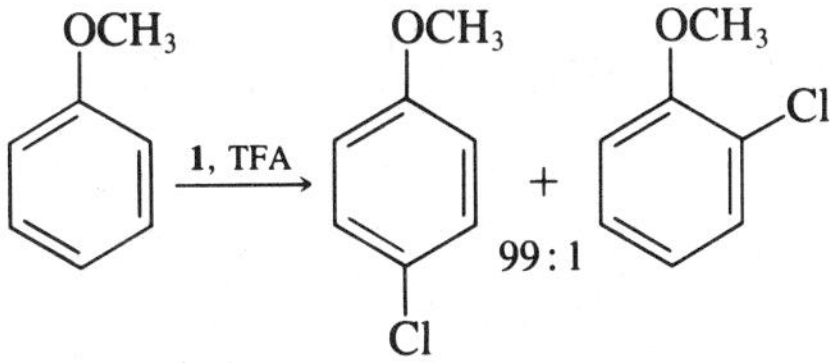

[1] S. E. Fuller, J. R. L. Smith, R. O. C. Norman, and R. Higgins, *J.C.S. Perkin II*, 545 (1981).
[2] J. R. L. Smith and L. C. McKeer, *Tetrahedron Letters*, **24**, 3117 (1983).

1-Chloro-N,N,2-trimethylpropenylamine (1), 4, 94–95; **5**, 136–138; **6**, 122–123; **7**, 66. Preparation.[1]

Allylic chlorides.[2]   The last steps in a synthesis of *cis*-maneonene-A (**4**), a natural product isolated from a Hawaiian alga, require conversion of the allylic alcohol **2** into the chloro compound **3**. Ghosez's chloroenamine (**1**) proved more effective than the Mitsunobu reagent, $P(C_6H_5)_3$–NCS, or CH_3SO_2Cl–LiCl. The reaction with **1** occurs with inversion.[3]

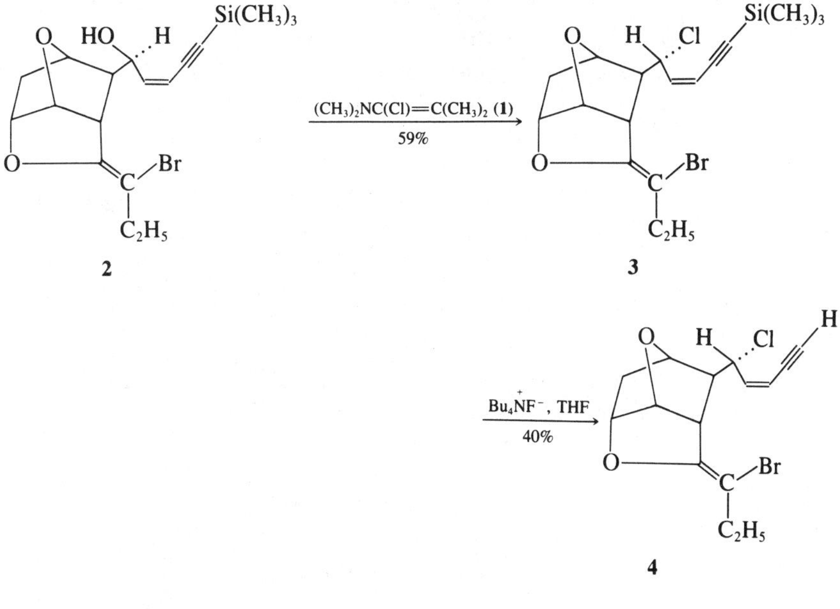

Ketone synthesis.[3]   Reaction of this α-chloroenamine (**1**) with carboxylic acids provides an activated intermediate (**a**) that reacts with Grignard reagents in the presence of catalytic amounts of copper(I) iodide to form ketones in generally good yield. This reaction is selective for carboxylic acid groups in the presence of various functional groups (cyano, ester, and even carbonyl). This reaction fails with vinyl or allyl Grignard reagents, and yields are only moderate with aryl Grignard reagents.

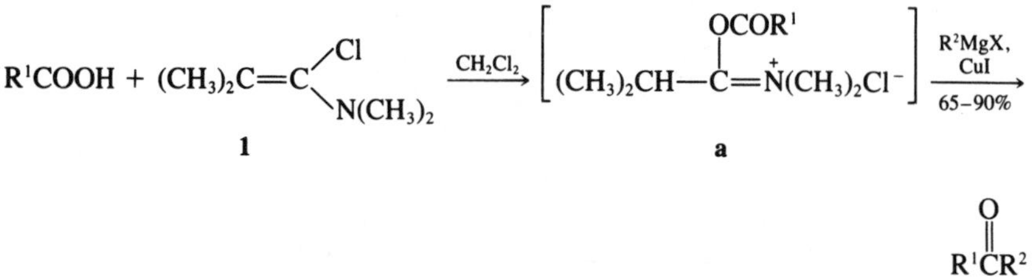

Activation of allylic[4] and propargylic alcohols.[5]   These alcohols can couple with Grignard reagents in the presence of an α-chloroenamine. For this purpose the related reagent 1-chloro-2-methyl-N,N-tetramethylenepropenylamine (**2**) is more effective than **1**.

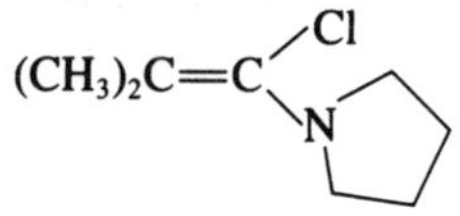

2

Reaction of allylic alcohols with Grignard reagents can occur at the α- or γ-position. The site depends in part on the structures of the alcohol and the Grignard reagent; γ-substitution can be enhanced by addition of HMPT and CuI.

Example:

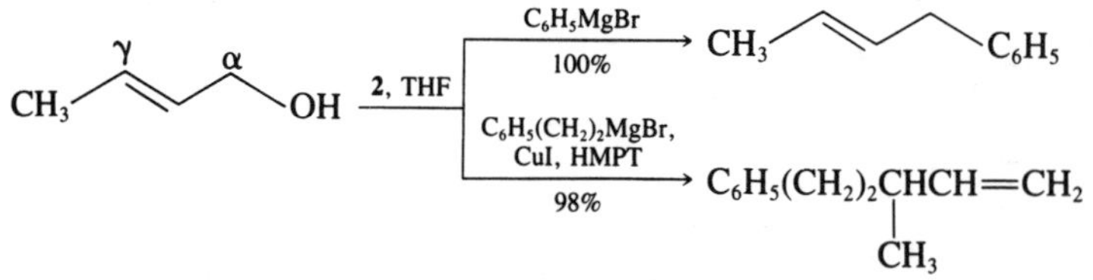

The coupling of propargyl alcohols with Grignard reagents in the presence of HMPT and CuI provides a useful synthesis of allenes in yields of 60–95%.

Example:

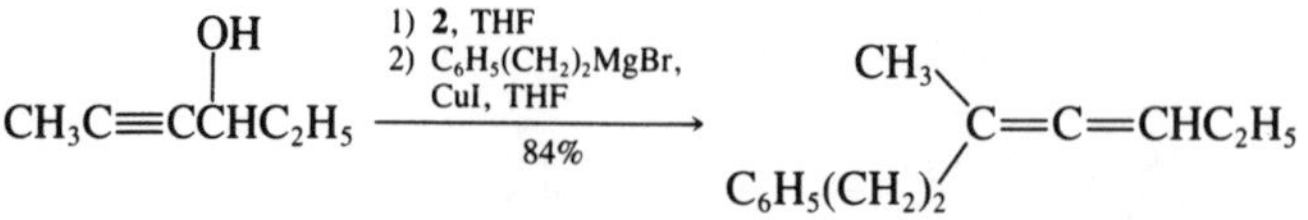

Of particular interest is that reaction of optically active propargyl alcohols results in optically active allenes. An example is the synthesis of both enantiomers of **5** from (S)- or (R)-**3**.

$$CH_3(CH_2)_7\underset{\overset{|}{OH}}{CH}C\equiv CH + CH_2{=}CHCH_2MgBr \xrightarrow[84\%]{2}$$

(S)-(−)-**3**

(R)-(−)-**4**

$$\xrightarrow{\text{Several steps}}$$

(R)-(−)-**5** (86% ee)

Butenolides and furanes.[6] The [2 + 2] cycloadducts (**3**) of alkynes with keten-iminium salts (**2**), readily formed from **1** (**11**, 560–561), undergo Baeyer-Villiger oxidation to give $\Delta^{\alpha,\beta}$-butenolides **4** exclusively. The products are readily reduced to furanes (**5**) by DIBAH.

$$(CH_3)_2C{=}C{=}\overset{+}{N}(CH_3)_2CF_3SO_3{}^{-} \xrightarrow{60{-}80\%}$$

2

+

$$R^1{-}C\equiv C{-}R^2$$

3

$$\xrightarrow[50{-}65\%]{ClC_6H_4CO_3H, \ CH_2Cl_2}$$

4

$$\xrightarrow[55{-}75\%]{DIBAH, \ THF}$$

5

[1] B. Haveaux, A. Dekoker, M. Rens, A. R. Sidani, J. Toye, and L. Ghosez, *Org. Syn.*, **59**, 26 (1980).

[2] A. B. Holmes, C. L. D. Jennings-White, and D. A. Kendrick, *J.C.S. Chem. Comm.*, 415 (1983).

[3] T. Fujisawa, T. Mori, K. Higuchi, and T. Sato, *Chem. Letters*, 1791 (1983).

[4] T. Fujisawa, S. Iida, H. Yukizaki, and T. Sato, *Tetrahedron Letters*, **24**, 5745 (1983).

[5] T. Fujisawa, S. Iida, and T. Sato, *ibid.*, **25**, 4007 (1984).

[6] C. Schmit, S. Sahraoui-Taleb, E. Differding, C. G. Dehasse-DeLombaert, and L. Ghosez, *ibid.*, **25**, 5043 (1984).

Chlorotrimethylsilane, 1, 1232; **2**, 435–438; **3**, 310–312; **4**, 537–539; **5**, 709–713; **6**, 626–628; **7**, 66–67; **8**, 107–109; **9**, 112–113; **10**, 96; **11**, 125–127.

Reformatsky reaction.[1]   In the presence of catalytic amounts of $ClSi(CH_3)_3$, ethyl formate or orthoformate undergoes a tandem Reformatsky reaction with α-bromopropionates (equation I). Three diastereomers, *syn–syn, syn–anti,* and *anti–anti,* are formed, typically in the ratio 49:42:9.

$$\text{(I)}\quad HCOOC_2H_5 + CH_3CHBrCOOR \xrightarrow[78-91\%]{\substack{Zn,\; ClSi(CH_3)_3,\\ ether}} ROOC \diagdown\diagup\overset{OH}{\diagdown}\diagup COOR$$

(with CH_3 and CH_3 substituents shown)

$R^1COOH \rightarrow R^1COR^2$.[2]   Conversion of carboxylic acids into methyl ketones by reaction with CH_3Li has the disadvantage that the products can react further with CH_3Li to form tertiary alcohols. Various expedients to limit the secondary reaction have been suggested. One recent solution is sequential reaction of the acid with CH_3Li and $ClSi(CH_3)_3$, which should effectively remove any excess CH_3Li.

Example:

$$p\text{-}HOC_6H_4COOH \xrightarrow[87\%]{\substack{1)\; CH_3Li,\; THF\\ 2)\; ClSi(CH_3)_3}} p\text{-}HOC_6H_4COCH_3$$

[1] J. K. Gawrónski, *Tetrahedron Letters,* **25**, 2605 (1984).
[2] G. M. Rubottom and C. Kim, *J. Org.,* **48**, 1550 (1983).

Chlorotrimethylsilane–Acetic anhydride.

Cleavage of methyl ethers.[1]   Alkyl methyl ethers are converted into acetates in high yield on contact with these two reagents. Both alkyl and aryl methylthiomethyl ethers are cleaved in 60–70% yield by this system. Tertiary alcohols can be acetylated by this reaction (two examples, 90% yield)

[1] N. C. Barus, R. P. Sharma, and J. N. Baruah, *Tetrahedron Letters,* **24**, 1189 (1983).

Chlorotrimethylsilane–Sodium, **2**, 435–436; **3**, 311–312; **4**, 537; **5**, 711–712; **11**, 127.

Vinylsilanes.[1]   Vinyl halides are converted into vinyltrimethylsilanes on reaction with Na (2.5 equiv.) and chlorotrimethylsilane in ether (25°, 2 hours). Yields are 65–85%. The configuration of the original bond is almost completely retained. *cis*-Vinyl halides can give rise to 1-trimethylsilylalkynes, formed by dehydrohalogenation followed by silylation, as by-products. This Wurtz-Fittig reaction is a convenient route to vinylsilanes when the vinyl halide is readily available.

[1] P. F. Hudrlik, A. K. Kulharni, S. Jain, and A. M. Hudrlik, *Tetrahedron,* **39**, 877 (1983).

Chlorotrimethylsilane–Sodium iodide, 9, 251–252; **10**, 97; **11**, 127–128.

Cleavage of MEM ethers.[1] 2-Methoxyethoxymethyl (MEM) ethers are cleaved by $NaI/ClSi(CH_3)_3$ in CH_3CN at -20 or 25° in moderate to high yield. Iodotrimethylsilane (commercial) is less effective. Fairly selective cleavage of MEM ethers is possible in the presence of lactones, methyl or benzyl ethers, and methyl esters.

Enol silyl ethers (**10**, 97).[2] Iodotrimethylsilane, generated *in situ,* is recommended for preparation of the enol silyl ether of acetone. The yield (60% based on chlorotrimethylsilane) is comparable to that obtained using the more expensive trimethylsilyl triflate.

[1] J. H. Rigby and J. Z. Wilson, *Am. Soc.*, **106**, 8217 (1984); *idem, Tetrahedron Letters,* **25**, 1429 (1984).
[2] N. D. A. Walshe, G. T. Goodwin, G. C. Smith, and F. E. Woodward, *Org. Syn.*, submitted (1984).

Chloro[(trimethylsilyl)methyl]ketene, (**1**).

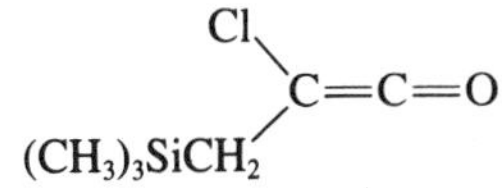

Preparation *in situ:*

$$(CH_3)_3SiCH_2CH_2COOH \xrightarrow[56\%]{1)\ SOCl_2\quad 2)\ NCS} (CH_3)_3SiCH_2CH(Cl)COCl \xrightarrow{N(C_2H_5)_3} \mathbf{1}$$

α-Methylenecyclobutanones.[1] The reagent reacts regioselectively with activated alkenes (vinyl ethers, silyl enol ethers) to give cyclobutanones. These products undergo ring expansion with diazomethane to cyclopentanones. Both products undergo desilylative elimination in the presence of fluoride ion to form α-methylene ketones.

Example:

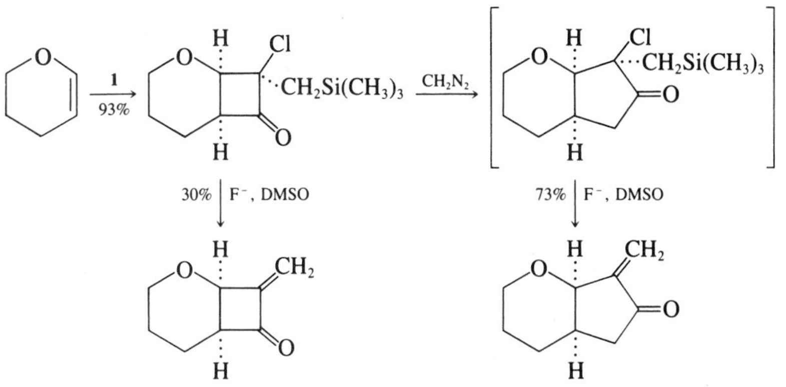

[1] L. A. Paquette, R. S. Valpey, and G. D. Annis, *J. Org.*, **49**, 1317 (1984).

Chlorotris(triphenylphosphine)cobalt, $CoCl[P(C_6H_5)_3]_3$ (**1**). This bright green, moderately air-stable complex is obtained by reaction of $CoCl_2[P(C_6H_5)_3]_2$ with $P(C_6H_5)_3$ and Zn in CH_3CN.[1] The reagent reacts with benzocyclobutenediones[2] to insert the metal between the two carbonyl groups.

Pyranonaphthoquinones.[3]   The reaction of phthaloylmetal complexes with alkynes to give naphthoquinones (**10**, 221–222) has been extended to a synthesis of pyrano-naphthoquinones. An intramolecular version of the reaction is used to control the regio-selectivity. The method is outlined in Scheme (I) for a synthesis of nanaomycin (**6**) and the *cis*-epimer (**7**).

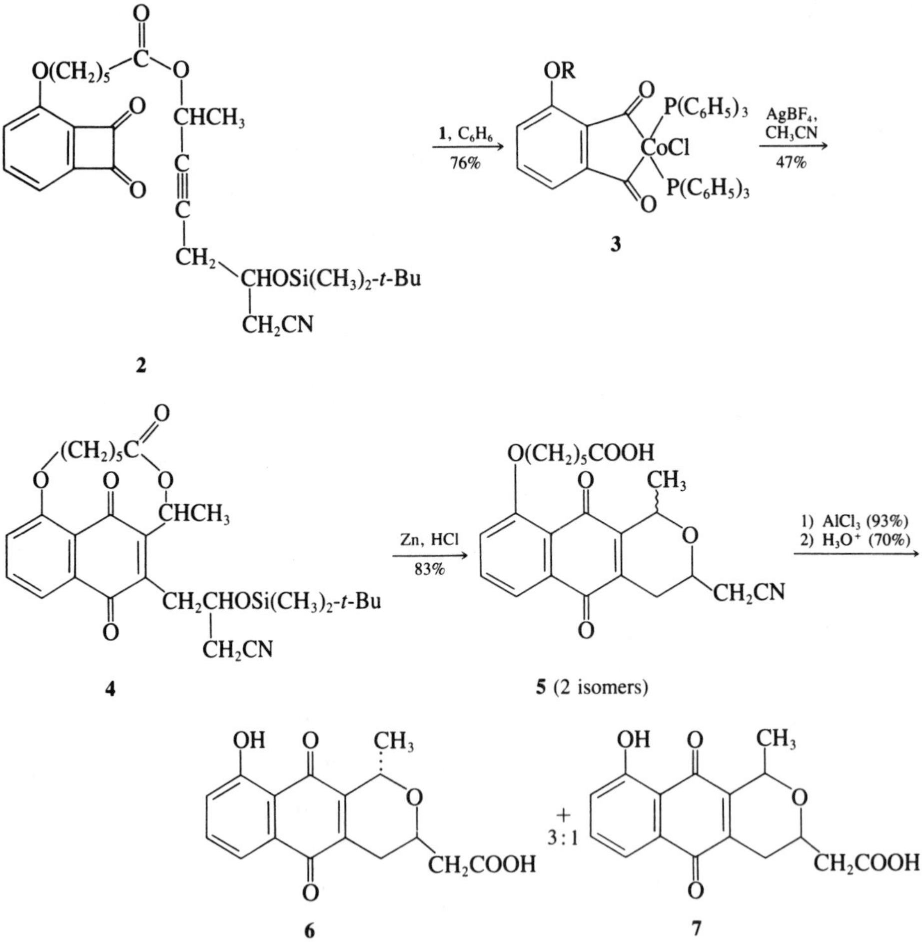

Scheme (I)

Reductive coupling of allylic halides.   This cobalt complex (1 equiv.) effects reductive coupling of allylic halides to form 1,5-dienes with preservation of the geometry of the double bonds.[4] The major product from coupling of terpenoid allylic halides is that formed by head-to-head coupling. The triphenylphosphine liberated during the reaction is removed as methyltriphenylphosphonium iodide, obtained by reaction with methyl

(I)

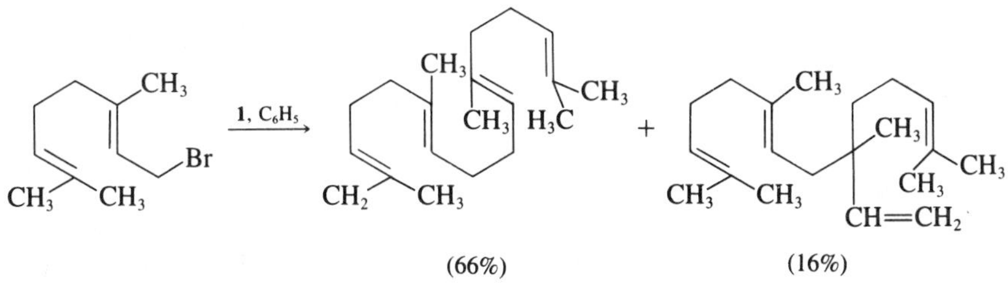

(66%)   (16%)

iodide. The coupling of geranyl bromide is typical (equation I). This coupling reaction converts (E,E)-farnesyl bromide to (E,E,E)-squalene in 66% yield.[5]

Naphthoquinones.[1]   The phthaloylcobalt complex (2) obtained by reaction of benzocyclobutenedione with 1 (92% yield) reacts with alkynes in the presence of AgBF₄ (1–

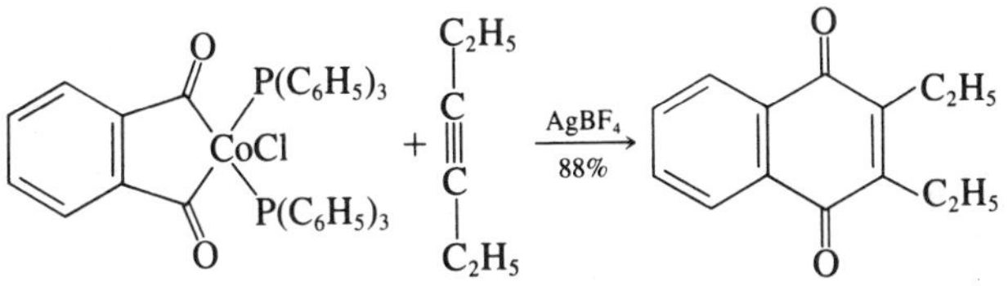

2

2 equiv.) to give naphthoquinones in 30–90% yield. Ligand exchange with diphos in acetonitrile provides a cationic complex (3), which reacts directly with functionalized alkynes to provide naphthoquinones in good yield.

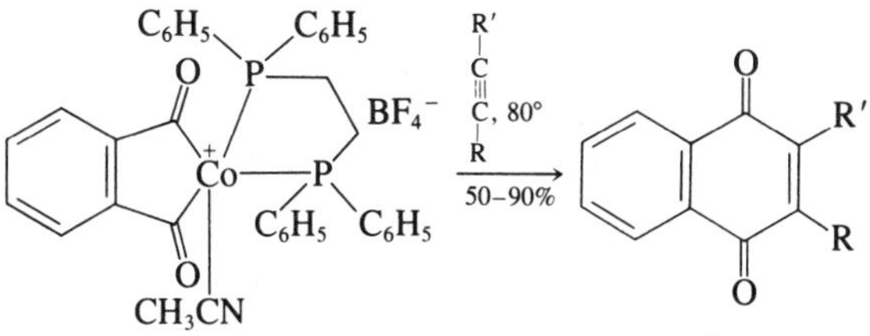

3   4

Benzoquinones.[6]   Reaction of a cyclobutenedione such as **2** with **1** provides the maleoylcobalt complex **3**, which also reacts with alkynes in the presence of AgBF$_4$ to afford benzoquinones. Ligand exchange with dimethylglyoxime in pyridine affords the complex **4**, which reacts directly with alkynes to produce benzoquinones in high yield, especially when CoCl$_2 \cdot$6H$_2$O is added. The unsymmetrical complex **4** shows some regiochemical preference, the highest being with 1-ethoxypropyne.

Example:

[1] S. L. Baysdon and L. S. Liebeskind, *Organometallics*, **1**, 771 (1982).
[2] M. S. South and L. S. Liebeskind, *J. Org.*, **47**, 3815 (1982).
[3] *Idem, Am. Soc.*, **106**, 4181 (1984).
[4] Y. Yamada and D. Momose, *Chem. Letters*, 1277 (1981).
[5] D. Momose, K. Iguchi, T. Sugiyama, and Y. Yamada, *Tetrahedron Letters*, **24**, 921 (1983).
[6] L. S. Liebeskind, J. P. Leeds, S. L. Baysden, and S. Iyer, *Am. Soc.*, **106**, 6451 (1984).

Chlorotris(triphenylphosphine)rhodium, 1, 140, 1252; **2,** 248–253; **3,** 325–329; **4,** 559–562; **5,** 736–740; **6,** 562–563; **7,** 68; **9,** 113–114; **10,** 98–99; **11,** 130.

3,4-Disubstituted cyclopentanones.   3,4-Disubstituted-4-pentenals (**1**) cyclize to *cis*-3,4-disubstituted cyclopentanones (**2**) in 50–85% yield in the presence of the Wilkinson complex.[1] In contrast, 2,3-disubstituted-4-pentenals (**3**) under the same conditions cyclize to *trans*-2,3-disubstituted cyclopentanones (**4**) in low yield because of concomitant formation of cyclopropanes (**5**).[2]

This cyclization can be used to convert the aldehyde **6** into the bicyclic ketone **7**, an intermediate for the synthesis of carbacyclin (**8**). In this case, decarbonylation competes with cyclization, and the yield is only moderate.[3]

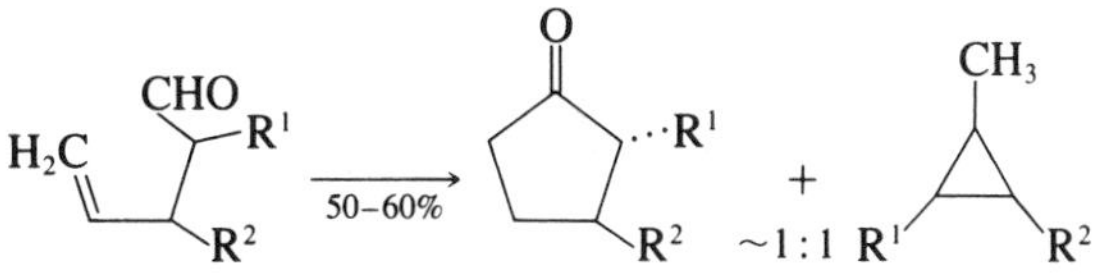

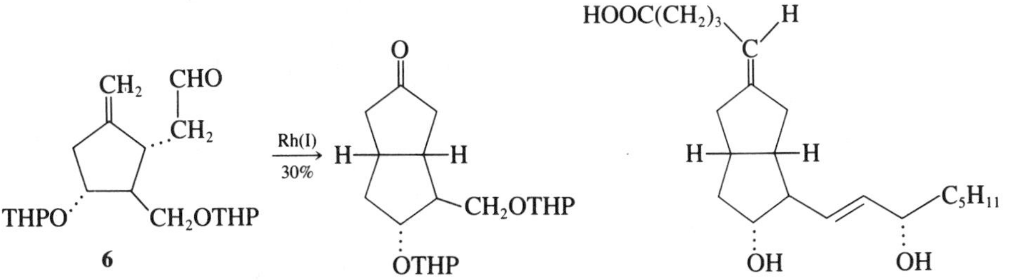

[1] K. Sakai, Y. Ishiguro, K. Funakoshi, K. Ueno, and H. Suemune, *Tetrahedron Letters,* **25**, 961 (1984).
[2] K. Sakai, J. Ide, O. Oda, and N. Nakamura, *ibid.,* **25**, 1287 (1972).
[3] K. Ueno, H. Suemune, and K. Sakai, *Chem. Pharm. Bull. Japan,* **32**, 3768 (1984).

Chromic acid, 1, 142–144; **2**, 70–72; **3**, 54; **4**, 95–96; **5**, 138–140; **6**, 123–124; **7**, 68–69; **9**, 114–115.

Jones reagent (**1**, 142–143).[1]   Phenols substituted by at least one alkyl group in the *ortho*-position can be oxidized to *p*-quinones by a two-phase (ether/aqueous CrO_3) Jones oxidation. Yields range from 30 to 85%, but the process is simple and more economical than use of Fremy's salt or thallium(III) nitrate.

[1] D. Liotta, J. Arbiser, J. W. Short, and M. Saindane, *J. Org.,* **48**, 2932 (1983).

Chromium carbene complexes, 11, 397–401.

Alkenyl complexes.[1]   These complexes are prepared by reaction of an alkenyllithium with $Cr(CO)_6$ followed by methylation. These α,β-unsaturated complexes react with terminal alkynes to give, after oxidative work-up, phenols or quinones in which the acetylene group is *ortho* to the phenolic hydroxyl group derived from carbon monoxide.

Examples:

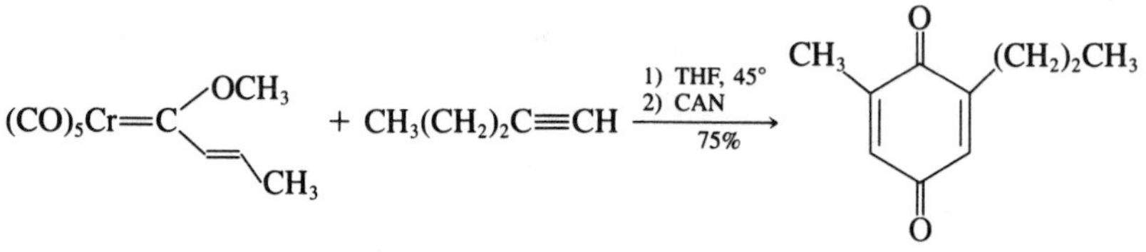

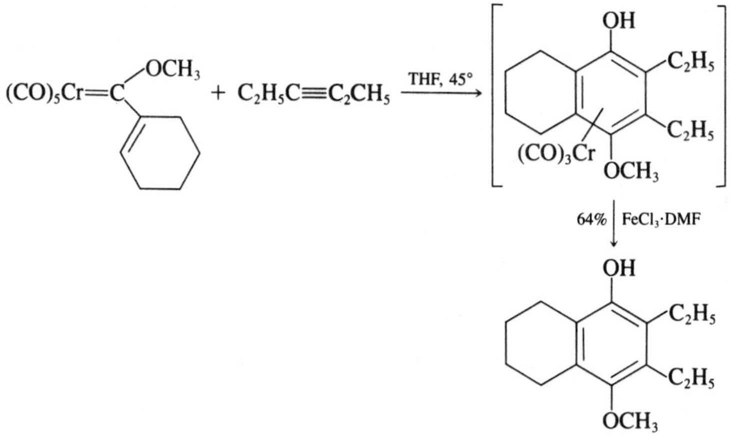

When this reaction is carried out on a β,β-disubstituted α,β-unsaturated complex, cyclohexa-2,4-dienones are obtained, evidently because tautomerization to a phenol is

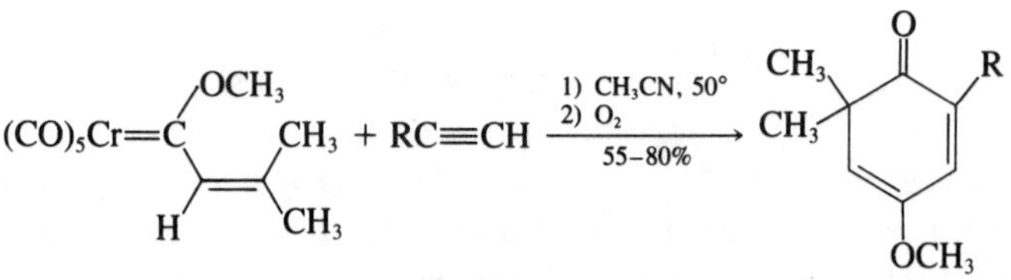

blocked (equation I). This regioselective reaction is also stereoselective in the case of a chiral carbene complex such as **1** (equation II).[2]

(II)

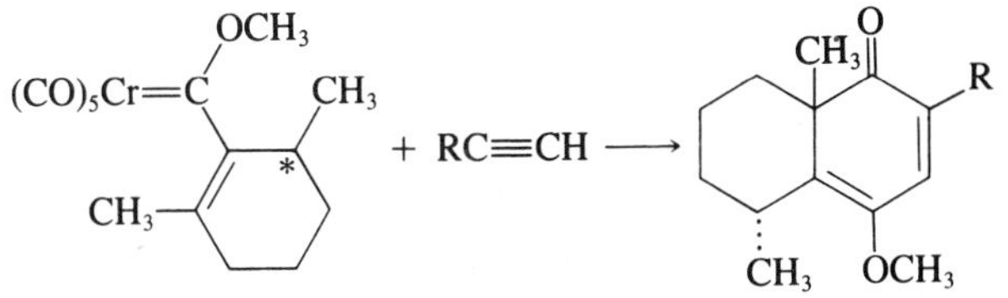

1

(*trans/cis* =
90–95 : 10–5)

This benzannelation provides the key step in a synthesis of the anthracycline **5**. Thus reaction of the complex **2** with the acetylene **3** provides, after oxidation, the tetrahydronaphthol **4**. The fourth ring of **5** is formed by ring closure to an anthrone followed by air oxidation (Triton B) to an anthraquinone.[3]

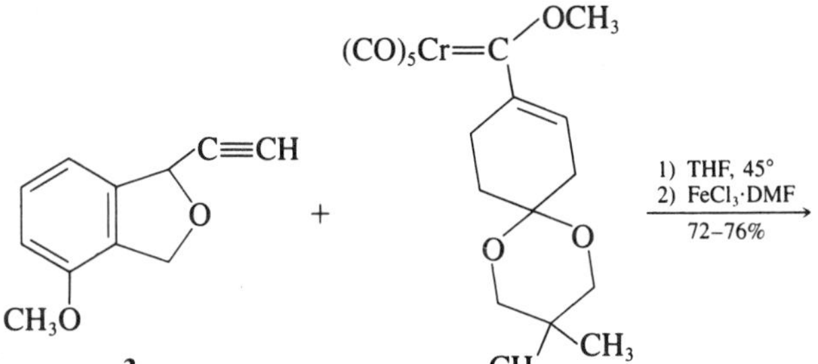

1) THF, 45°
2) FeCl₃·DMF
72–76%

3   2

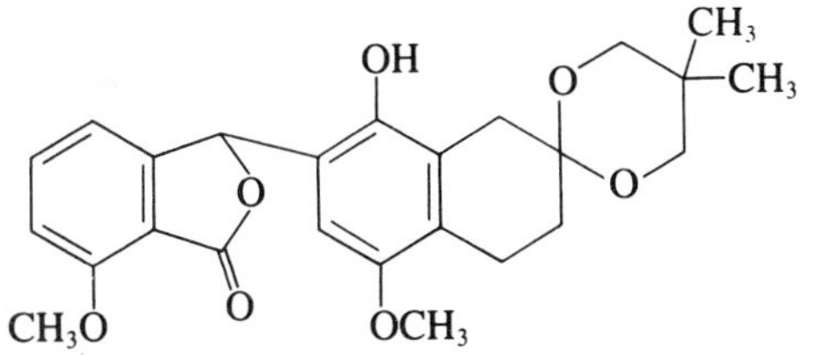

4

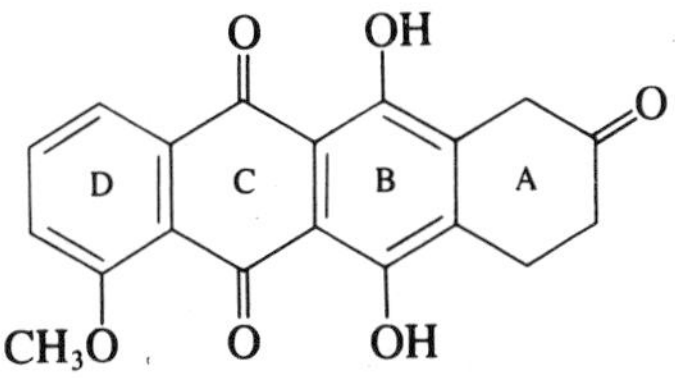

5

Alkenyl complexes can undergo cycloaddition with dienes.[4] Thus the complex **6**, prepared from vinyllithium, can be used as an alternative to the less reactive methyl acrylate. The cycloadducts can undergo various useful transformations (Scheme I). The reaction of **6** with cyclopentadiene gives a mixture of the *endo-* and *exo*-adducts in the ratio 94:6 in 78% yield.

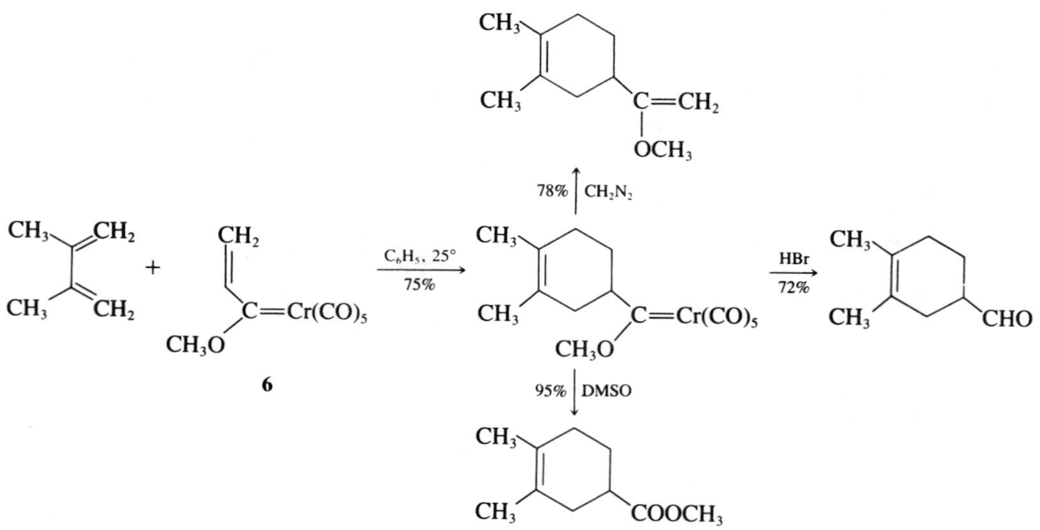

Scheme (I)

CHO

$\xrightarrow[\text{72\%}]{\text{HBr}}$

CH_3 CH_3

$C=CH_2$ | OCH_3

$\xleftarrow[\text{78\%}]{CH_2N_2}$

CH_3 CH_3

$C=Cr(CO)_5$ | CH_3O

$\xleftarrow[\text{75\%}]{C_6H_5,\ 25°}$

CH_3 CH_3

6

$\xrightarrow[\text{95\%}]{\text{DMSO}}$

$COOCH_3$

CH_3 CH_3

CH_2 | $C=Cr(CO)_5$ | CH_3O

+

CH_3 CH_3 | CH_2 CH_2

Alkynyl complexes.[5]   These complexes, which are readily available by reaction of an alkynyllithium with chromium carbonyl followed by methylation with CH_3OSO_2F, undergo facile Diels-Alder reactions with dienes to provide a general route to α,β-unsaturated chromium carbene complexes. The latter undergo benzannelation or cyclohexadienone annelation on reaction with an alkyne. A (trimethylsilyl)ethynyl complex is useful because benzannelation results in migration of silicon from carbon to oxygen to provide a protected phenol.

Examples:

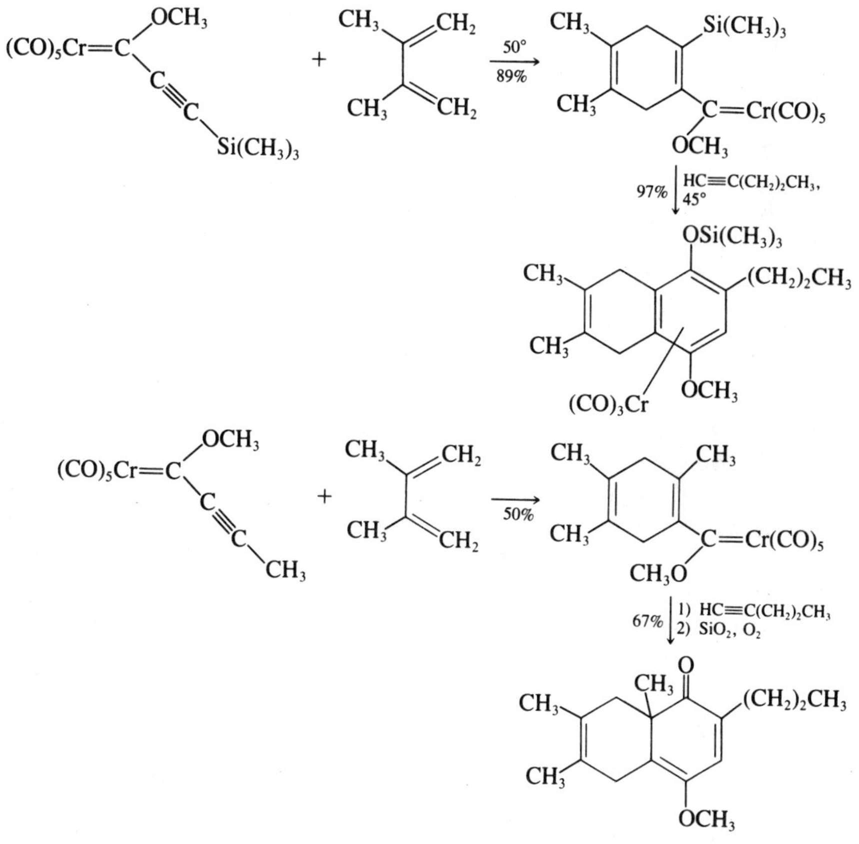

The cycloaddition and annelation reactions can be conducted concurrently in a one-pot operation to provide a regioselective synthesis of complex bicyclic systems. This method is an attractive alternative to preparation of α,β-alkenyl carbene complexes from cyclohexenyllithiums.

Example:

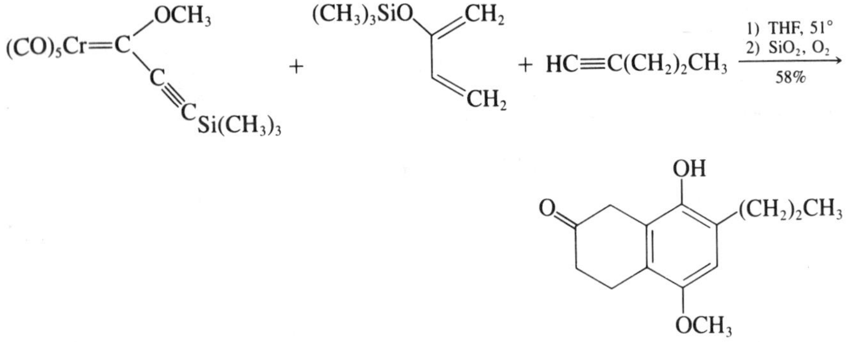

Reviews.[6]

[1] W. D. Wulff, K.-S. Chan, and P.-C. Tang, *J. Org.*, **49**, 2293 (1984).
[2] P.-C. Tang and W. D. Wulff, *Am. Soc.*, **106**, 1132 (1984).
[3] W. D. Wulff and P.-C. Tang, *ibid.*, **106**, 434 (1984).
[4] *Idem, ibid.*, **105**, 6726 (1983).
[5] *Idem, ibid.*, **106**, 7565 (1984).
[6] K. H. Dötz, *Pure Appl. Chem.*, **55**, 1689 (1983); *Angew. Chem., Int. Ed.*, **23**, 587 (1984).

Chromium(II) chloride, 8, 110–112; **9**, 119; **10**, 101; **11**, 132–134.

Diastereoselective macrocyclization.[1] A key step in a synthesis of the 14-membered cembranoid asperdiol (**4**) involves intramolecular cyclization of the aldehydo allylic bromide (**1**) with chromium(II) chloride. The intermolecular version of this reaction is known to be *anti*-selective (**8**,112). Treatment of racemic **1** with $CrCl_2$ (5 equiv., THF) results in a 4:1 mixture of the two *anti*-diastereomers **2** and **3** in 64% combined yield. The stereochemistry of this cyclization is evidently controlled by the remote epoxide group. The natural product was obtained by deprotection of **2** (Na/NH$_3$, 51% yield).

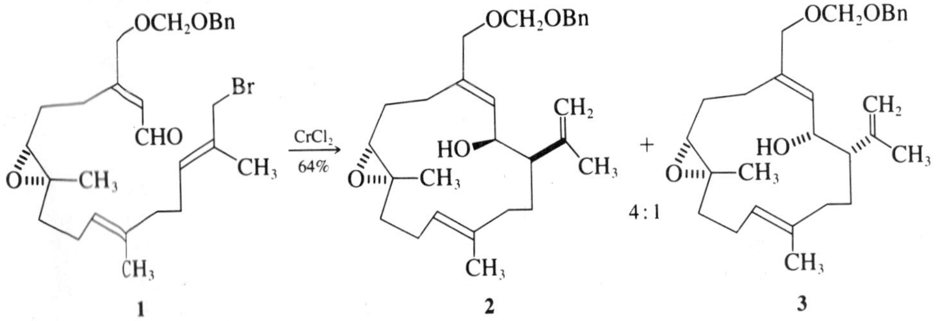

OH

(structure 4)

4

o-Quinodimethides.[2]   Reaction of an α,α'-dibromo-*o*-xylene with $CrCl_2$ (3 equiv.) in THF results in reduction to an *o*-quinodimethide, which can be trapped by a dienophile. Example:

OCH₃ ... (reaction scheme)

$$\text{OCH}_3 \xrightarrow[\text{THF, HMPT}]{\text{CrCl}_2,} \quad \left[\text{OCH}_3 \right] \xrightarrow[85\%]{\text{CH}_2=\text{CHCCH}_3} \text{OCH}_3 \cdots \text{COCH}_3$$

Alkenylchromium compounds.[3]   Addition of $CrCl_2$ to a vinyl iodide or bromide in DMF forms a vinylchromium compound that adds selectively to aldehydes to form an allylic alcohol. $CrCl_2$ also effects addition of iodobenzene to aldehydes to form benzyl alcohols.

Examples:

$$\text{CH}_2=\text{C}\begin{smallmatrix}\text{CH}_3\\ \\ \text{I}\end{smallmatrix} + \text{RCHO} \xrightarrow[\text{quant.}]{\text{CrCl}_2,\ \text{DMF}} \text{CH}_2=\text{C}\begin{smallmatrix}\text{CH}_3\\ \\ \text{CHR}\\ \\ \text{OH}\end{smallmatrix}$$

$$\text{OHC(CH}_2)_4\text{C(CH}_2)_3\text{CH}_3 + \text{C}_6\text{H}_5\text{I} \xrightarrow[81\%]{\text{CrCl}_2,\ \text{DMF}} \text{C}_6\text{H}_5\text{CH(CH}_2)_4\text{C(CH}_2)_3\text{CH}_3$$

[1] W. C. Still and D. Mobilio, *J. Org.*, **48**, 4785 (1983).

[2] D. Stephan, A. Gorgues, and A. Le Coq, *Tetrahedron Letters*, **25**, 5649 (1984).

[3] K. Takai, K. Kimura, T. Kuroda, T. Hiyama, and H. Nozaki, *ibid.*, **24**, 5281 (1983).

Cobalt *meso*-tetraphenylporphine (CoTPP, **1**). The cobalt salt is prepared as maroon crystals by reaction of TPP with Co(OAc)$_2$ in CHCl$_3$–HOAc. The salt is soluble in benzene, chloroform, and pyridine.[1]

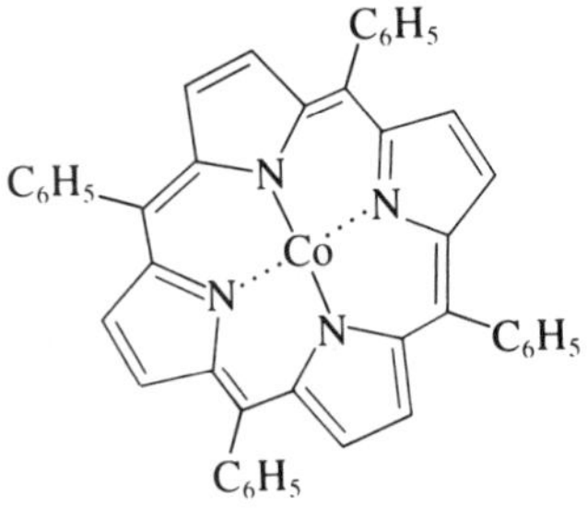

1

Rearrangement of **1,4-endoperoxides**.[2] CoTPP catalyzes the rearrangement of 1,4-endoperoxides to *syn*-1,2; 3,4-diepoxides at $-78°$. Neither *meso*-tetraphenylporphine nor zinc *meso*-tetraphenylporphine catalyzes this rearrangement. N,N,N′,N′-Tetramethylethylenediamine catalyzes the reaction, but only very slowly. The yields from the catalyzed reaction are much higher than those obtained by thermolysis.

Examples:

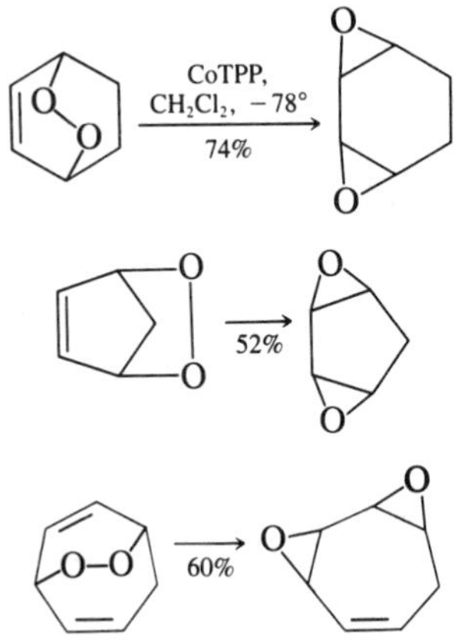

CoTPP also cleaves dioxetanes such as **2** to dicarbonyl compounds.[3]

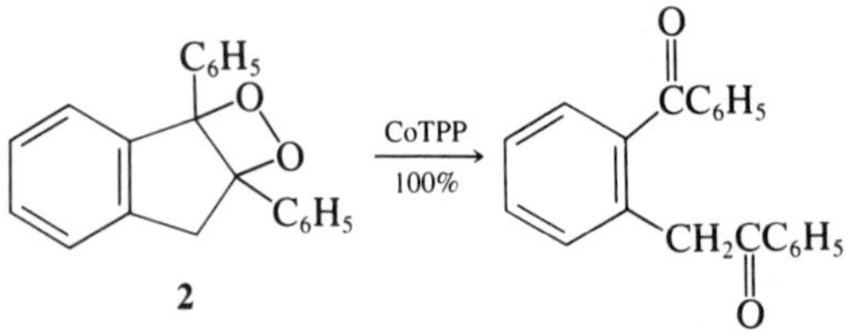

[1] P. Rothemund and A. R. Menotti, *Am. Soc.*, **70**, 1808 (1948).
[2] J. D. Boyd, C. S. Foote, and D. K. Imagawa, *ibid.*, **102**, 3641 (1980).
[3] M. Balci and Y. Sütbeyaz, *Tetrahedron Letters*, **24**, 311 (1983).

2,4,6-Collidinium *p*-toluenesulfonate,

$$\text{CH}_3\text{C}_6\text{H}_4\text{SO}_3^-$$

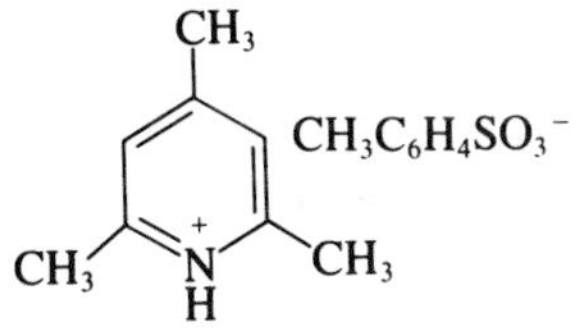

Selective enone ketalization.[1]   The reaction of ethylene glycol catalyzed by pyridinium *p*-toluenesulfonate (PPTS) does not discriminate between saturated and α,β-unsaturated ketones. In contrast, this hindered pyridinium salt (**1**) permits selective ketalization of enones in the presence of saturated keto groups. 2,6-Lutidinium *p*-toluenesulfonate (**2**) is as effective.

Example:

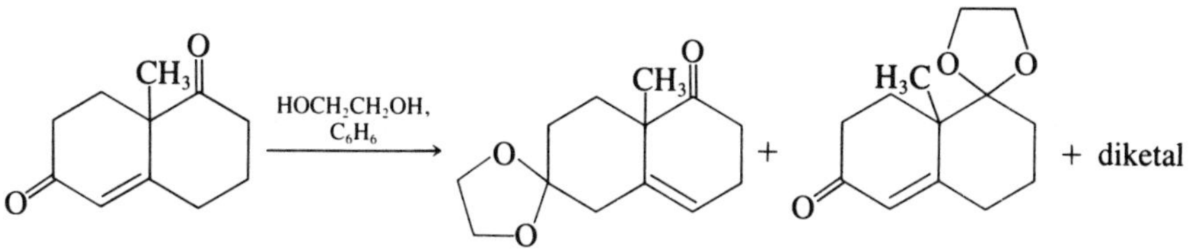

1	(67%)	(12%)	(3%)
2	(76%)	(5%)	(11%)
PPTS	(35%)	(34%)	(14%)

[1] T. J. Nitz and L. A. Paquette, *Tetrahedron Letters*, **25**, 3047 (1984).

Collins reagent, **2**, 74–75; **3**, 55–56; **4**, 216–217; **9**, 121; **11**, 139.

RCH₂OH → RCOOC(CH₃)₃.[1]   This transformation can be effected in sugars with the Collins reagent (4 equiv.) in CH_2Cl_2/DMF followed by addition of acetic anhydride (**9**,121) and a large excess of *t*-butyl alcohol (equation I). This conversion is probably general except for aromatic aldehydes.

[1] E. J. Corey and B. Samuelsson, *J. Org.*, **49**, 4735 (1984).

Copper, 1, 157–158; **2**, 82–84; **3**, 63–65; **4**, 102–103; **5**, 146–148; **7**, 73–74; **8**, 113–114; **9**, 122.

Reactive Cu(0).   A highly reactive zerovalent copper can be prepared by reduction of CuI·P(C₂H₅)₃ (**4**, 516–517; **5**, 244; **8**, 122–123) with lithium naphthalenide under argon. This material reacts with alkyl, aryl, and vinyl halides at 0–25° to form organocopper compounds. Generally the alkylcopper species formed undergoes self-coupling almost immediately to give RR in very high yield. Reagents derived from aryl and alkynyl halides are generally stable at 25° and on addition of water are reduced to ArH. These arylcopper reagents can be used for cross-coupling with acid chlorides and allylic and benzylic bromides in moderate to high yield.[1]

Examples:

$$C_6H_5Cu + C_6H_5COCl \xrightarrow[70\%]{} C_6H_5COC_6H_5$$

$$C_6H_5CH_2Cu + CH_2{=}CHCH_2Br \xrightarrow[50\%]{} C_6H_5CH_2CH_2CH{=}CH_2$$

[1] G. W. Ebert and R. D. Reike, *J. Org.*, **49**, 5280 (1984).

Copper(II) acetate, 1, 159–160; **2**, 84; **3**, 65; **4**, 105; **5**, 156–157; **6**, 138; **10**, 103.

Vinyl sulfones.[1]   Anions of primary sulfones are oxidized by Cu(OAc)₂ (2 equiv.) to (E)-vinyl sulfones in yields of 60–85% (equation I). Secondary sulfones are more difficult to oxidize.

$$(I)\ RCH_2CH_2SO_2C_6H_5 \xrightarrow[60-85\%]{\substack{1)\ n\text{-BuLi} \\ 2)\ Cu(OAc)_2}} R \diagup\!\!\!\diagdown\!\!\!\diagup SO_2C_6H_5$$

[1] J.-B. Baudin, M. Julia, C. Rolando, and J.-N. Verpeaux, *Tetrahedron Letters*, **25**, 3203 (1984).

Copper bronze, 4, 100.

Furane synthesis.[1]   Copper-catalyzed decomposition of ethyl diazopyruvate (**1**) in the presence of enol ethers or of alkynes provides a useful route to ethyl α-furoates.

Examples:

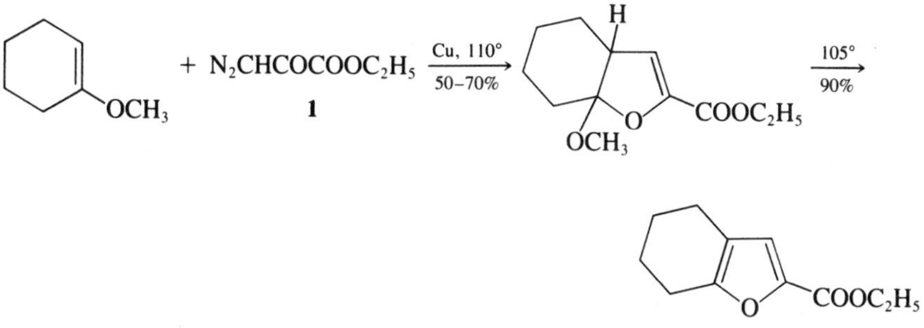

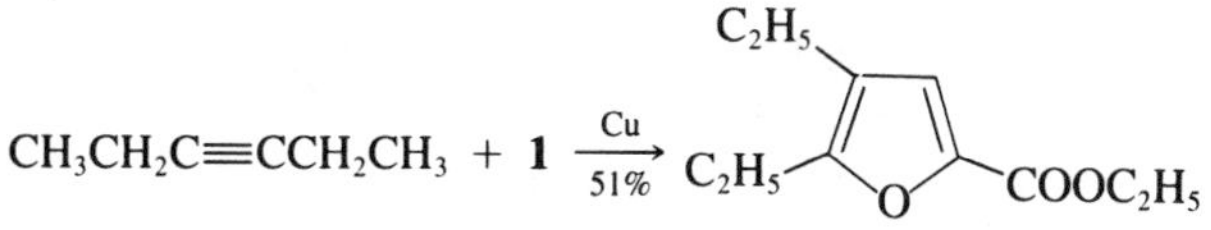

$$\text{CH}_3\text{CH}_2\text{C}\equiv\text{CCH}_2\text{CH}_3 + \mathbf{1} \xrightarrow[51\%]{\text{Cu}}$$

[1] E. Wenkert, M. E. Alonso, B. L. Buckwalter, and E. L. Sanchez, *Am. Soc.*, **105**, 2021 (1983).

Copper(I) chloride, 1, 166–169; **2**, 91–92; **3**, 67–69; **4**, 109–110; **5**, 164–165; **6**, 145–146; **7**, 80–81; **8**, 118–119; **9**, 123; **11**, 140–141.

   1,4-cis-*Addition to cyclic allylic epoxides*.[1]   In the presence of CuCl, sodium carboxylates react with cyclic allylic epoxides by a *syn*-1,4-addition to give mono esters of ene-1,4-diols.

Examples:

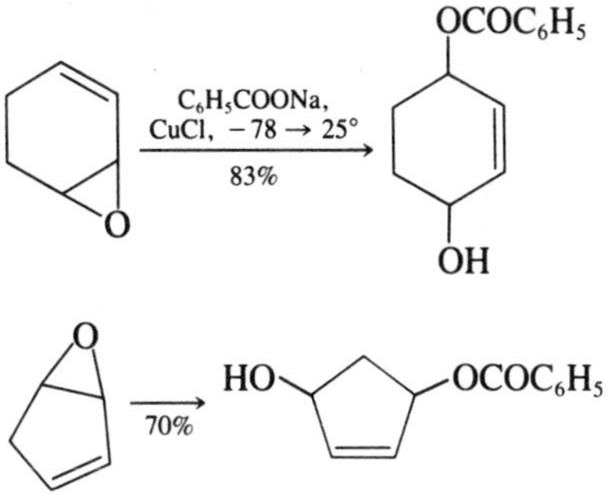

[1] J. P. Marino and J. C. Jaén, *Tetrahedron Letters*, **24**, 441 (1983).

Copper(I) iodide, 1, 169; **2**, 92; **3**, 69–71; **5**, 167–168; **6**, 147; **8**, 81–83; **9**, 124–125; **10**, 107; **11**, 141.

Optically active allylic alcohols and homoallylic epoxides.[1]   The reaction of optically active 2,3-epoxy halides (**1**), available by Sharpless asymmetric epoxidation of allylic
alcohols, with a preformed mixture of vinylmagnesium bromide and CuI (2 equiv. each)
in THF at $-23°$ does not result as expected (**7**,81–82) in a homoallylic epoxide, but

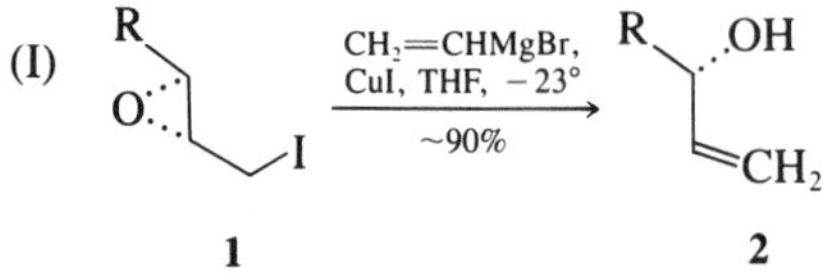

(I)

1   **2**

rather in an allylic alcohol (**2**, equation I). This transformation can be carried out more
conveniently with zinc dust and NaI (2.5 equiv.) in refluxing CH_3OH or with *n*-BuLi (2.5
equiv.) in THF at $-23°$. Yields by either method are 85–100%.

The desired substitution reaction can be effected in 74–85% yield by addition of
vinylmagnesium bromide (2 equiv.) to a mixture of **1**, CuI (0.1 equiv.), and HMPT (4
equiv.) in THF at $-23°$.

Example:

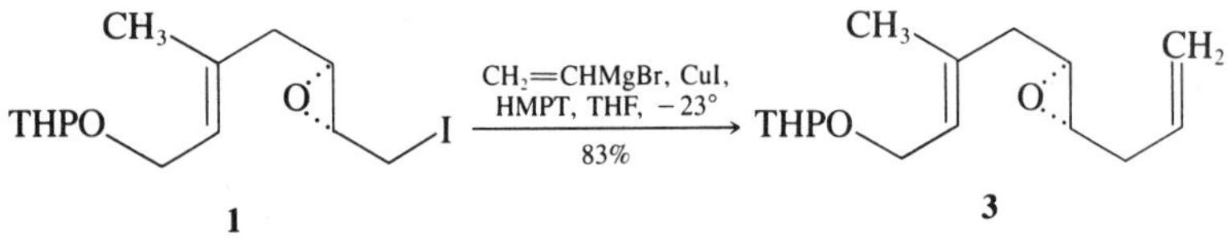

1   **3**

Indoles; isoquinolines.   Anions[2] derived from the enaminones **1** cyclize to indoles
(**2**) when heated with CuI (1.5–2.0 equiv.) in HMPT at 100–170°. The related enaminones
3 under the same conditions cyclize to dihydroisoquinolines (**4**).

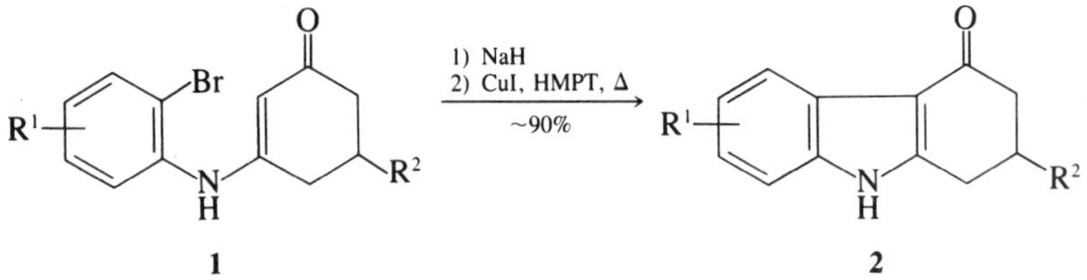

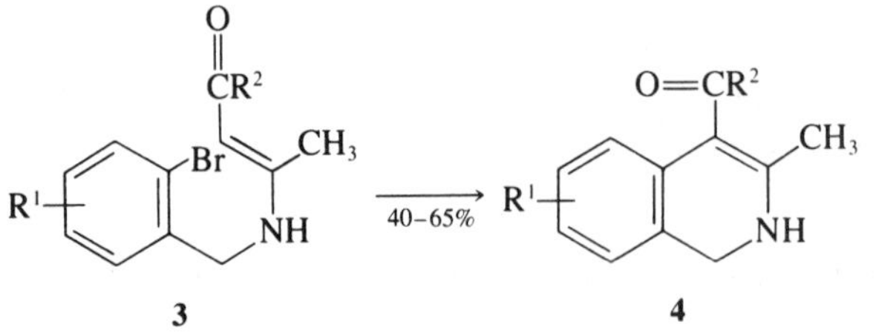

1   **2**

3   **4**

Related cyclization to an indoline:

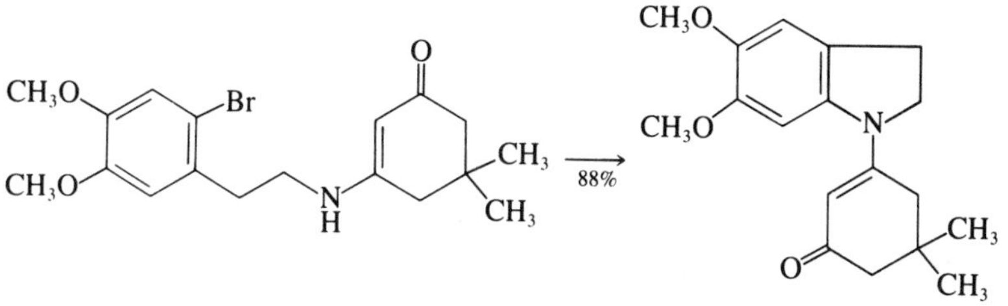

***CuI-catalyzed reactions of RLi and RMgX.*[3]** CuI-catalyzed reactions of these re-
agents with unsaturated substrates, carbonyl compounds, and epoxides reported since 1970
have been reviewed (89 references). Where possible, use of stoichiometric and catalytic
copper reagents was compared.

[1] K. C. Nicolaou, M. E. Duggan, and T. Ladduwahetty, *Tetrahedron Letters,* **25**, 2069 (1984).
[2] A. Osuka, Y. Mori, and H. Suzuki, *Chem. Letters,* 2031 (1982).
[3] E. Erdik, *Tetrahedron,* **40**, 641 (1984).

Copper(I) iodide–Triethyl phosphite, 8, 122–123.

Alkylation of alkenylaluminates.[1] 1-Trimethylsilylvinylaluminates (*e.g.,* **2**), obtained by *cis*-addition of DIBAH to 1-trialkylsilylalkynes (**1**) followed by alkyllithium
addition, react only with highly reactive alkyl halides. However, in the presence of 25
mole % of CuI·P(OC$_2$H$_5$)$_3$, alkylation can be effected with primary alkyl iodides in 75–
85% yield. The products are useful intermediates to trisubstituted alkenes (**3**).
Example:

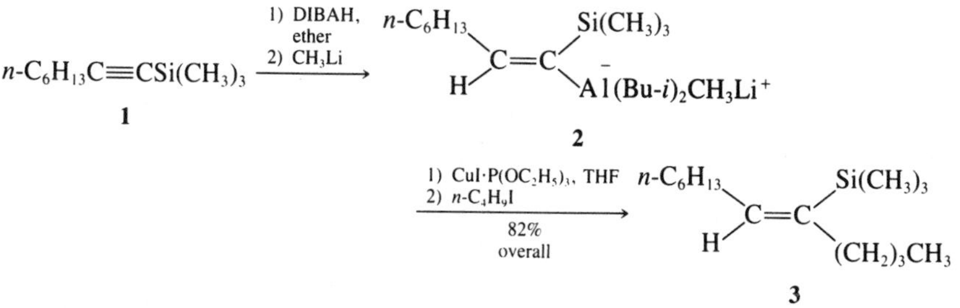

[1] F. E. Ziegler and K. Mikami, *Tetrahedron Letters,* **25**, 131 (1984).

Copper(I) phenylacetylide, C$_6$H$_5$C≡CCu (1).

The bright yellow reagent is prepared by
reaction of phenylacetylene with 1 equiv. of CuI in aqueous ammonia (77% yield).[1]

Ullmann diaryl ether synthesis. [2]   This copper(I) derivative is recommended as the condensing reagent in the Ullmann synthesis of diaryl ethers from phenols and bromoarenes in refluxing pyridine (equation I).

$$(I) \ Ar^1OH \xrightarrow{\ 1. \ Py, \ \Delta\ } Ar^1OCu \xrightarrow[40-75\%]{Ar^2Br, \ \Delta} Ar^1OAr^2$$

[1] R. D. Stephens and C. E. Castro, *J. Org.*, **28**, 3313 (1963).
[2] A. Afzali, H. Firouzabadi, and A. Khalafi-nejad, *Syn. Comm.*, **13**, 335 (1983).

Copper(II) tetrafluoroborate, $Cu(BF_4)_2$. Supplier: Alfa.

1,6-*Diketones*. [1]   Reaction of the silyloxycyclopropane **1** with $Cu(BF_4)_2$ (0.5 equiv., ether, 15°) affords the 1,6-diketone **2** as a mixture of racemic and *meso*-isomers in 80% isolated yield. A β-copper ketone (**a**) is postulated as intermediate. The same reaction

can be effected with $AgBF_4$, but in lower yield (42%). This desilylative dimerization is general. Attack of the metal always involves scission of bond *a* rather than the more hindered *b* bond. The reaction is also applicable to monocyclic silyloxycyclopropanes.

Example:

$AgBF_4$	42%
$Cu(BF_4)_2$	78%

[1] I. Ryu, M. Ando, A. Ogawa, S. Murai, and N. Sonoda, *Am. Soc.*, **105**, 7192 (1983).

Copper(I) trifluoromethanesulfonate, **5**, 151–152; **6**, 130–133; **7**, 75–76; **8**, 125–126; **10**, 108–110; **11**, 142–143.

***Photobicyclization of* 1,6-*dienes*.** [1]   Irradiation of the triene myrcene (**1**) affords three of the five possible [2 + 2] cycloadducts. Triplet-sensitized irradiation is more selective and results in the vinylcyclobutane **2** (75% yield) as the major product. Surprisingly, CuOTf-catalyzed photolysis results in the previously unknown vinylcyclobutane **3** (20% yield) in addition to the monocyclobutene **4** (35% yield).

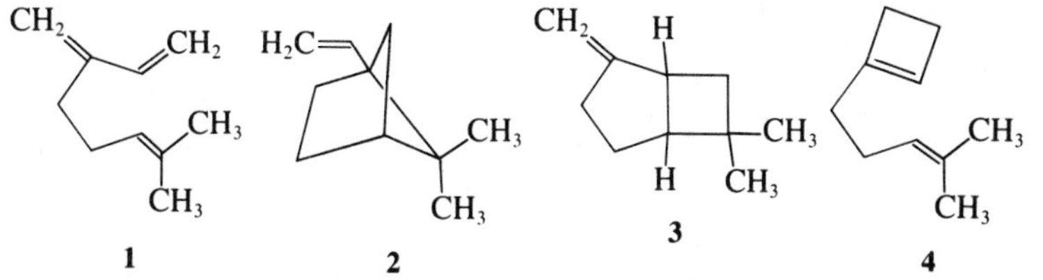

1 **2** **3** **4**

This catalyzed photocycloaddition of a conjugated diene to a double bond provides a useful route to bicyclic vinylcyclobutanes.[2]

Examples:

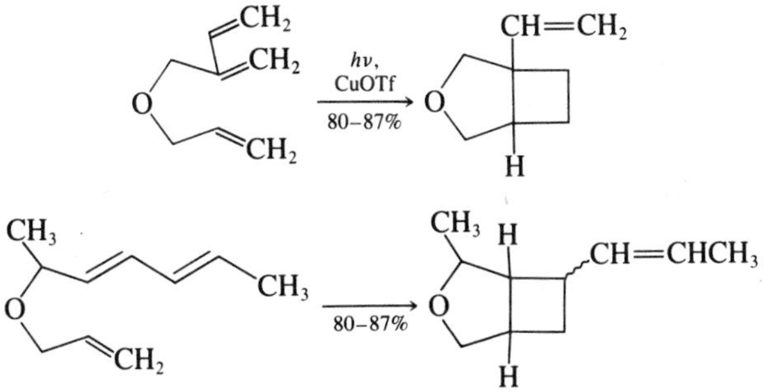

(*exo/endo* = 87 : 13)

Pyrrolidines.[3]   Ethyl N,N-diallylcarbamates, when irradiated in the presence of CuOTf, undergo bicyclization to bicyclic pyrrolidines (equation I).

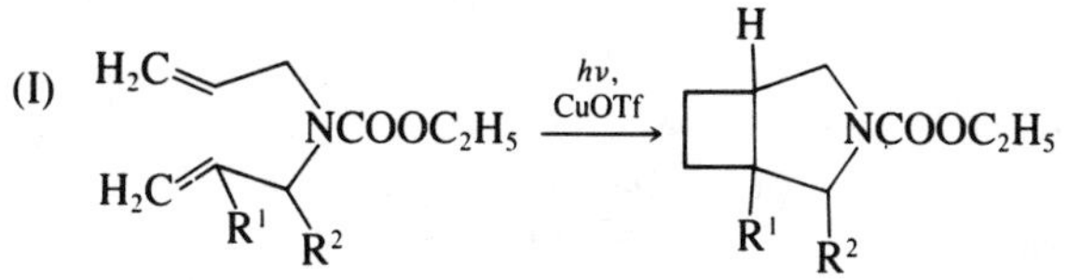

$R^1 = H, R^2 = CH_3$ 60%
$R^1 = CH_3, R^2 = H$ 76%

Another example:

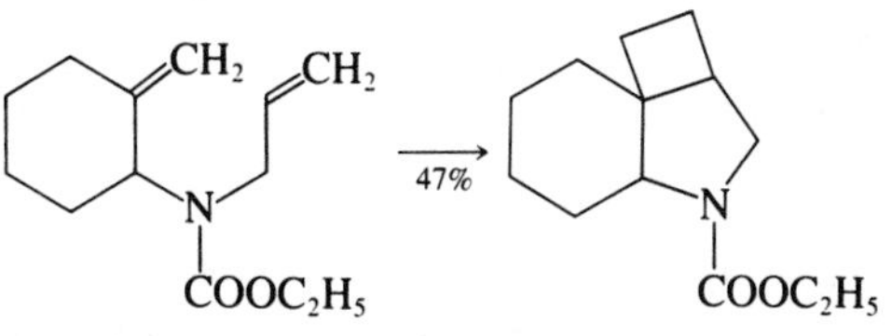

[1] Review: R. G. Salomon, *Tetrahedron*, **39**, 485 (1983).
[2] K. Avasthi, S. R. Raychaudhuri, and R. G. Salomon, *J. Org.*, **49**, 4322 (1984).
[3] R. G. Salomon, S. Ghosh, S. R. Raychaudhuri, and T. S. Miranti, *Tetrahedron Letters*, **25**, 3167 (1984).

Crotyltri-*n*-butyltin, 11, 143.

   Addition to aldehydes.[1]   A reinvestigation of this reaction catalyzed by BF_3 etherate indicates that the *syn*-selectivity is considerably enhanced (25:1 versus 9:1) by use of 2 equiv. of the reagent rather than 1 equiv. This effect probably results from the greater reactivity and *syn*-selectivity of the *trans*-isomer. Other Lewis acid catalysts give varying amounts of products arising from the alternate mode of addition to the aldehyde.

   The $TiCl_4$-catalyzed reaction can be controlled to give either the *syn*- or the *anti*-adduct (equation I). Normal addition of the reagent last results in *syn*-selectivity, but addition of

(I)   RCHO $\xrightarrow{\quad TiCl_4 \quad}$

R = $c\text{-}C_6H_{11}$

	(90.5)	(7.0)	(2.6)
normal	(90.5)	(7.0)	(2.6)
inverse	(4.4)	(90.8)	(4.9)

the aldehyde to a mixture of $TiCl_4$ and the tin reagent results in high *anti*-selectivity, possibly owing to formation of a crotyltitanium species.

[1] G. E. Keck, D. E. Abbott, E. P. Boden, and E. J. Enholm, *Tetrahedron Letters,* **25,** 3927 (1984).

(E)- and (Z)-Crotyltrimethylsilane, $CH_3CH{=}CHCH_2Si(CH_3)_3$ **(1).** The silanes are obtained by coupling of (trimethylsilyl)methylmagnesium chloride with (E)- and (Z)-bromopropene, with $NiCl_2[(C_6H_5)_2PCH_2CH_2CH_2P(C_6H_5)_2]$ as catalyst.

   syn-*Homoallylic alcohols.*[1]   Aldehydes react with (E)-**1** in the presence of $TiCl_4$ to form *syn*-homoallylic alcohols with >93% selectivity (equation I). (E)-Cinnamyltri-

(I) CH_3 $Si(CH_3)_3$ + RCHO $\xrightarrow[90-98\%]{TiCl_4,\ CH_2Cl_2,\ -78°}$

(E)-**1**

(95–99% *syn*)

methylsilane, $C_6H_5CH{=}CHCH_2Si(CH_3)_3$, shows similar *syn*-selectivity in reactions with aldehydes. Unexpectedly, the corresponding (Z)-allylsilanes show only slight *syn*-selectivity (65–75:35–25).

[1] T. Hayashi, K. Kabeta, I. Hamachi, and M. Kumada, *Tetrahedron Letters,* **24,** 2865 (1983).

Crown ethers, 4, 142–145; **5,** 152–155; **6,** 133–137; **7,** 76–79; **8,** 128–130; **9,** 126–127; **10,** 110–112; **11,** 143–145.

Diaza-crown ethers.[1]   4,13-Diaza-18-crown ethers (**1**) can be prepared in one step by reaction of primary amines with triethylene glycol diiodide (equation I). These crown

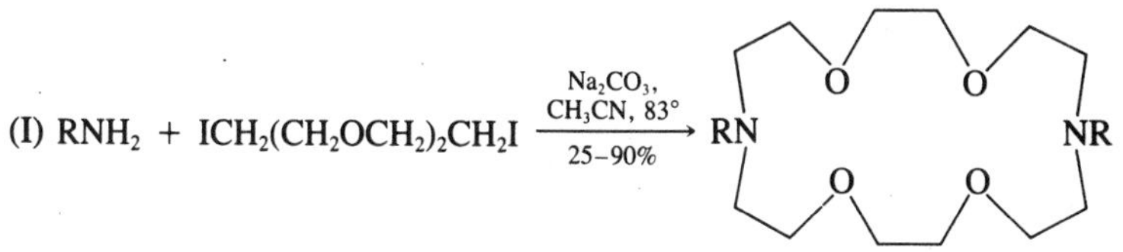

$$(I) \quad RNH_2 \; + \; ICH_2(CH_2OCH_2)_2CH_2I \xrightarrow[\substack{25-90\%}]{\substack{Na_2CO_3, \\ CH_3CN,\ 83°}}$$

1a, R = $HOCH_2CH_2$
1b, R = $C_2H_5OCOCH_2$

ethers are of interest because when R is a polar donor group, as in **1a** and **1b**, they can show selective binding affinity for Ca^{2+} ions over Na^+ or K^+ ions.

[1] V. J. Gatto and G. W. Gokel, *Am. Soc.,* **106,** 8240 (1984).

Cyanamide, 1, 170; **3,** 71; **5,** 168.

vic-Diamines.[1]   The first step in a recent conversion of an alkene such as *trans*-2-butene (**1**) into a *vic*-diamine involves addition of cyanamide and NBS to form a bromo-cyanamide adduct (**2**). Selective hydrogenation of **2** furnishes a bromoformamidine (**3**), which cyclizes to an imidazoline (**4**) when treated with sodium methoxide. The imidazoline is cleaved to a diamine (**5**) under more basic conditions (50% aqueous KOH). Overall yields are higher (50–70%) if the intermediates **3** and **4** are not isolated.

Example:

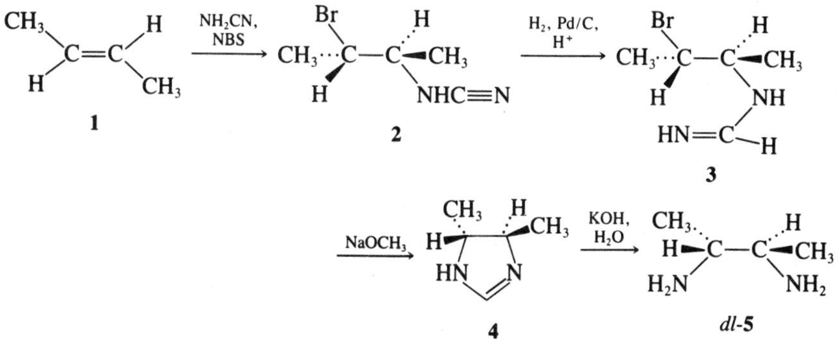

[1] H. Kohn and S.-H. Jung, *Am. Soc.,* **105,** 4106 (1983); S.-H. Jung and H. Kohn, *Tetrahedron Letters,* **25,** 399 (1984).

Cyanotrimethylsilane, 4, 542–543; **5**, 720–722; **6**, 632–633; **7**, 397–399; **8**, 133; **9**, 127–129; **10**, 112–114; **11**, 147–150.

α-Chloro nitriles.[1]   Aryl aldehydes or alkyl aryl ketones react with this reagent and $TiCl_4$ (1 equiv.) in CH_2Cl_2 to give α-chloro nitriles in modest to high yield. Cyanohydrins are formed as by-products or major products when the aryl group is substituted by electron-attracting groups.

Example:

$$C_6H_5CHO + (CH_3)_3SiCN + TiCl_4 \xrightarrow[85\%]{CH_2Cl_2} C_6H_5\overset{\overset{\displaystyle Cl}{|}}{C}HCN$$

α-Alkylthio nitriles.   The reagent in the presence of $SnCl_4$ or $SnCl_2$ reacts with thioacetals or thioketals to form α-alkylthio nitriles. The product can be converted into primary or secondary nitriles by desulfuration with Raney nickel.[2]

Example:

$$(C_2H_5)_2C(SC_2H_5)_2 + (CH_3)_3SiCN \xrightarrow[80\%]{\substack{SnCl_4, \\ CH_2Cl_2}} (C_2H_5)_2C\overset{\displaystyle SC_2H_5}{\underset{\displaystyle CN}{\big\langle}} \xrightarrow[81\%]{Raney\ Ni} (C_2H_5)_2CHCN$$

Tertiary nitriles can be prepared by a similar reaction with tertiary chlorides.[3]
Example:

$$(CH_3)_3CCl + (CH_3)_3SiCN \xrightarrow[75\%]{SnCl_4} (CH_3)_3CCN$$

Addition to ketones.[4]   Potassium cyanide in combination with 18-crown-6 is generally superior to ZnI_2 as the catalyst for addition of cyanotrimethylsilane to ketones substituted at the α-position with an electron-withdrawing group.

α-Amino nitriles.[5]   The adducts of cyanotrimethylsilane with aldehydes (**4**, 542–543; **5**, 721) are converted *in situ* into α-amino nitriles by reaction with ammonia or an amine in an alcohol (equation I).

$$(I)\ RCHO \longrightarrow \left[\underset{\displaystyle CN}{\overset{\displaystyle |}{RCHOSi(CH_3)_3}} \right] \xrightarrow[85-97\%]{\substack{HNR^1R^2, \\ CH_3OH}} \underset{\displaystyle CN}{\overset{\displaystyle |}{RCHNR^1R^2}}$$

Reaction with α,γ-diketo esters.[6]   The reaction of $CNSi(CH_3)_3$ with the α,γ-diketo ester **1** can be controlled to give one of the two possible adducts.

Example:

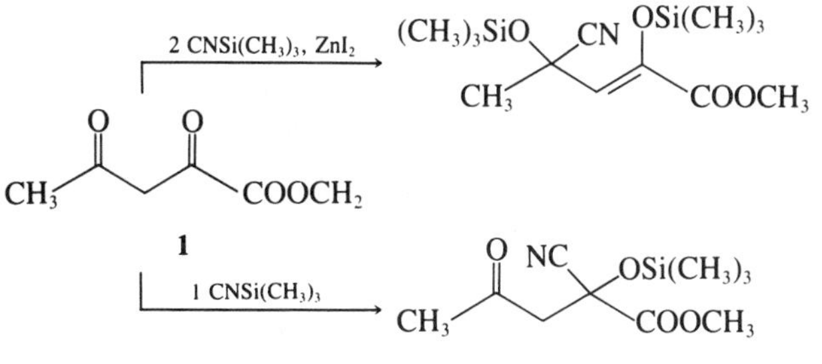

β,γ-Unsaturated amino acids.[7] These unsaturated amino acids can be prepared in a one-pot reaction from imines of α,β-unsaturated aldehydes by a variation of the Strecker reaction (equation I). N-Unsubstituted β,γ-unsaturated amino acids are prepared most conveniently by use of 4,4′-dimethoxybenzhydrylamine as the amine, since the dimethoxybenzhydryl group is removed by mild acid treatment (**9**,174–175).

$$\text{(I)}\ R^1CH{=}CHCHO + R^2NH_2 \longrightarrow \left[R^1CH{=}CHCH{=}NR^2 \xrightarrow{(CH_3)_3SiCN} R^1CH{=}CH\underset{\displaystyle CN}{\underset{|}{C}}HNHR^2 \right] \xrightarrow{H_3O^+,\ \Delta} R^1CH{=}CH\underset{\displaystyle COOH}{\underset{|}{C}}HNHR^2$$

Cleavage of epoxides (11,147). Cleavage of the epoxide (**1**), or the acetate or the trimethylsilyl ether of (R)-cyclohexenol with cyanotrimethylsilane (excess) catalyzed by zinc iodide proceeds regio- and stereoselectively to give **2** in 71% yield. This product can be converted into the aminodiol **3** with three contiguous chiral centers.[8]

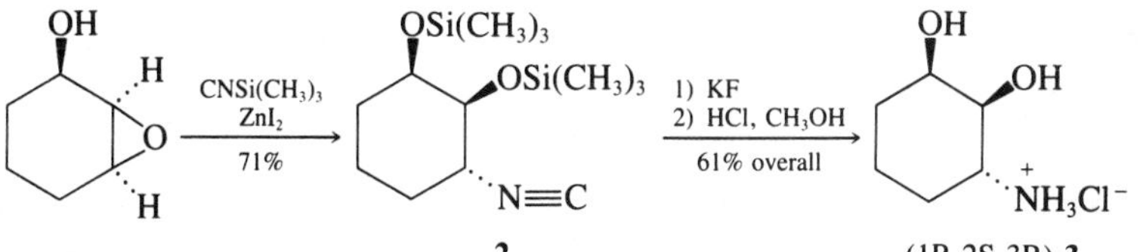

(R)-**1** **2** (1R,2S,3R)-**3**

α-Keto esters.[9]   The cyanohydrin silyl ether (1) of methyl glyoxylate (4, 542–543; 5, 720) can be converted by alkylation of the anion into the enol acetate (2) of α-keto esters.

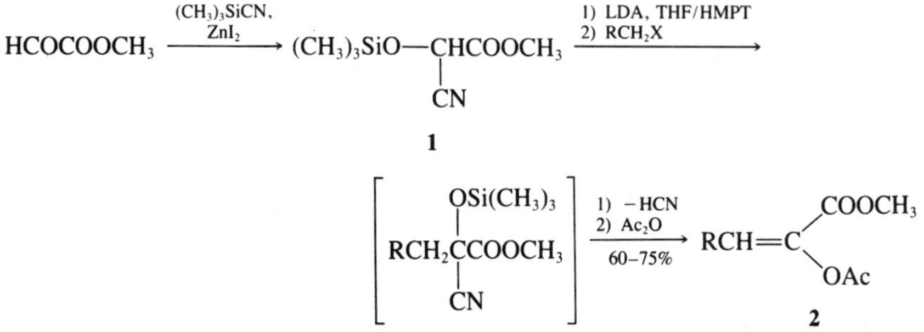

α-Cyano-N-heterocycles.[10]   In the presence of triethylamine, cyanotrimethylsilane (excess) undergoes selective addition to pyridine and quinoline N-oxides in refluxing acetonitrile. In the presence of catalytic amounts of n-Bu$_4$NF the reaction proceeds at 5°. The cyanotrimethylsilane is conveniently generated from ClSi(CH$_3$)$_3$, NaCN, and N(C$_2$H$_5$)$_3$ in DMF. The initial adducts lose trimethylsilanol to give α-cyano derivatives of pyridines and quinolines.

Example:

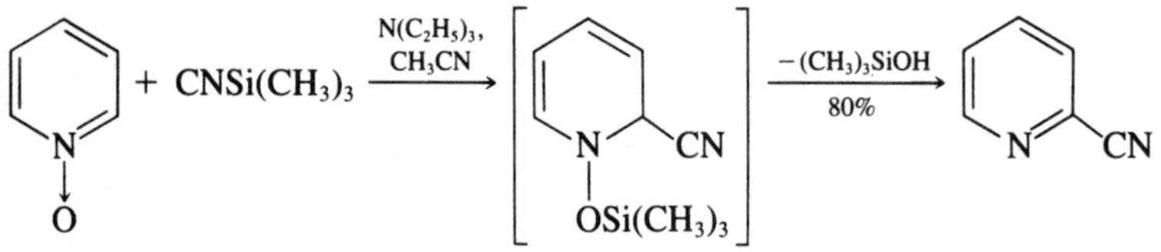

[1] S. Kiyooka, R. Fujiyama, and K. Kawaguchi, *Chem. Letters,* 1979 (1984).
[2] M. T. Reetz and H. Müller-Starke, *Tetrahedron Letters,* **25**, 3301 (1984).
[3] M. T. Reetz, I. Chatziiosifidis, H. Kunzer, and H. Müller-Starke, *Tetrahedron,* **39**, 961 (1983).
[4] W. J. Greenlee and D. G. Hangauer, *Tetrahedron Letters,* **24**, 4559 (1983).
[5] K. Mai and G. Patil, *ibid.,* **25**, 4583 (1984).
[6] L. H. Foley, *Syn. Comm.,* **14**, 1291 (1984).
[7] W. J. Greenlee, *J. Org.,* **49**, 2632 (1984).
[8] P. G. Gassman and R. S. Gremban, *Tetrahedron Letters,* **25**, 3259 (1984).
[9] T. Mukaiyama, T. Oriyama, and M. Murakami, *Chem. Letters,* 985 (1983).
[10] H. Vorbrüggen and K. Krolikiewicz, *Synthesis,* 316 (1983).

Cyclobutylidene, $\overset{\cdot\cdot}{\Box}$ (1). This carbenoid can be generated *in situ* by reaction of 1,1-dibromocyclobutane in ether with CH$_3$Li/LiBr at $-70°$.

[1 + 2] *Coupling*.   Generation of **1** in the presence of alkenes results in stereo-specific coupling to form spiro[2,3]hexanes.
Examples:

[1] U. H. Brinker and M. Boxberger, *Angew. Chem., Int. Ed.*, **23**, 974 (1984).

β-Cyclodextrin, 6, 151–152; **8**, 133–135; **9**, 129; **11**, 150–151.

Regioselective reaction of phenols.   Reaction of phenols with chloroform in aqueous alkaline solution catalyzed by β-cyclodextrin results in virtually complete attack at the *para*-position by dichlorocarbene to give, after hydrolysis, 4-hydroxybenzaldehydes. If the *para*-position is substituted, 4-(dichloromethyl)-2,5-cyclohexadienones are obtained as the major product. The selectivity results from formation of a ternary complex from β-cyclodextrin, chloroform, and the phenol.[1]

The β-cyclodextrin-catalyzed reaction of phenols and carbon tetrachloride in an alkaline medium in the presence of copper powder also results in almost exclusive attack at the *para*-position to give 4-hydroxybenzoic acids. 2-Methylphenol also undergoes almost exclusive *para*-carboxylation. β-Cyclodextrin has only a negligible effect on the carboxylation of 3-methylphenol.[2]

[1] M. Komiyama and H. Hirai, *Am. Soc.*, **105**, 2018 (1983).
[2] *Idem, ibid.*, **106**, 174 (1984).

(1,5-Cyclooctadiene)(pyridine)(tricyclohexylphosphine)iridium(I)   hexafluorophosphate (1, cf. 10, 116).[1]

Stereoselective hydrogenation.[2]   A suitably placed hydroxyl group can exert control over the hydrogenation catalyzed by **1** of various cyclohexenols with allylic or homoallylic double bonds.

Examples:

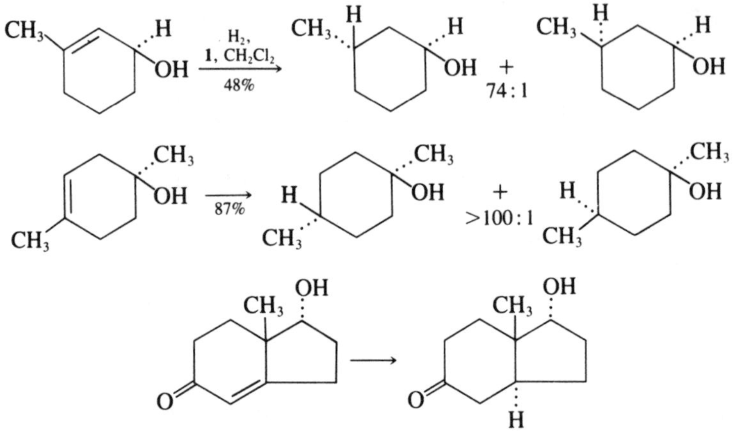

The last example is particularly striking because hydrogenation of indenones ordinarily results in exclusive formation of *cis*-indanones when the α-hydroxy group is absent or replaced by a β-hydroxyl group or carbonyl group.

[1] R. H. Crabtree, H. Felkin, T. Fellebun-Khan, G. E. Morrise, *J. Organometal. Chem.*, **168**, 183 (1979).
[2] G. Stork and D. E. Kahne, *Am. Soc.*, **105**, 1072 (1983).

Cyclopropenone 1, 3-propanediyl ketal (1).
Preparation (**8**, 398).[1]

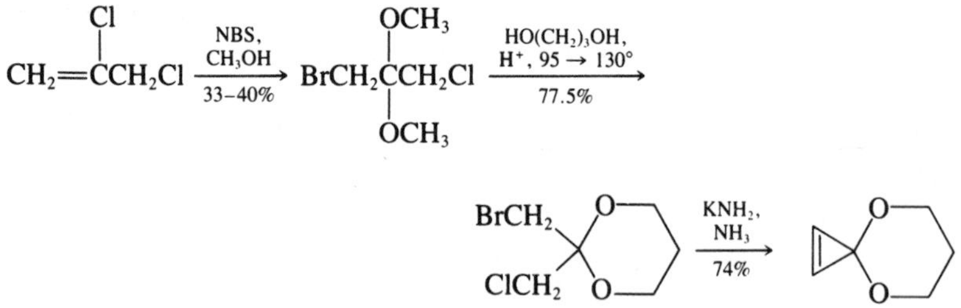

1
b.p. 35–40°/0.35 mm

[3 + 2] Cycloaddition. The cyclopropenone ketal (**1**) (or 3,3-dimethoxycyclopropene) undergoes a thermal [3 + 2] cycloaddition with alkenes substituted by two electron-withdrawing groups to form cyclopentenone ketals. The ketal serves as an equivalent of a 1,3-dipole that rearranges to an allyl cation. In all cases, the addition is regiospecific.[2,3]

Examples:

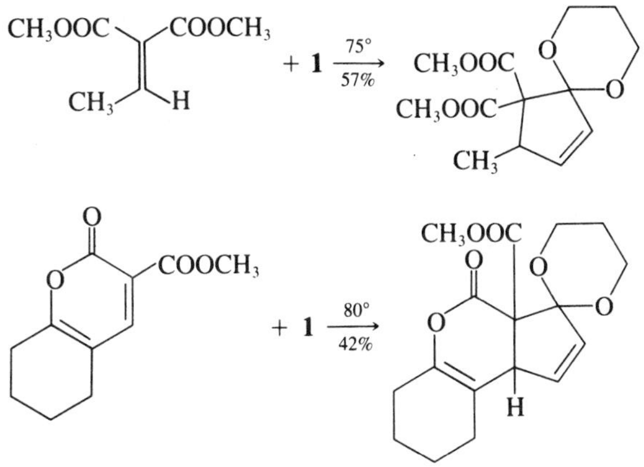

Cyclopropenone ketals can undergo a similar [3 + 2] cycloaddition with aldehydes or ketones to provide butenolide ortho esters, which in turn can be converted into butenolides, furanes, or γ-keto esters.[4]

Example:

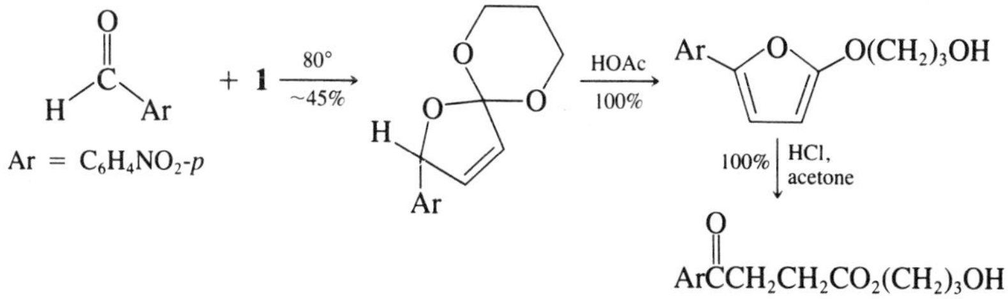

[1 + 2] *Cycloaddition.*[5]   Reaction of cyclopropenone ketals with alkenes bearing only one electron-withdrawing group results in unstable cyclopropane ketene ketals, which are not isolated, but rather are converted into *cis*-disubstituted cyclopropanes by acid hydrolysis.

Example:

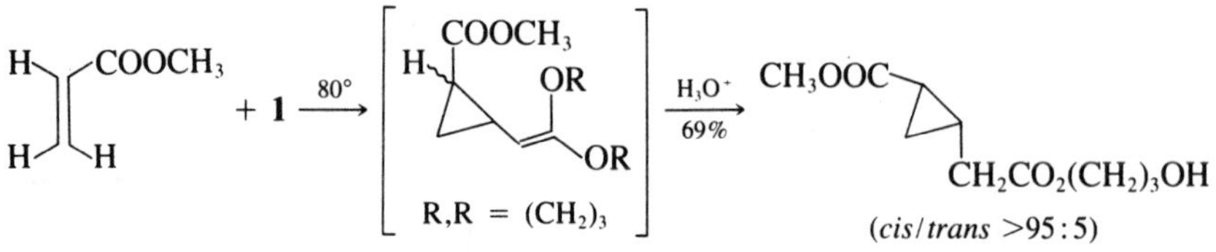

$$(cis/trans >95:5)$$

[1] G. B. Butler, K. H. Herring, P. L. Lewis, V. V. Sharpe III, and R. L. Veazey, *J. Org.*, **42**, 679 (1977).

[2] D. L. Boger and C. E. Brotherton, *Am. Soc.*, **106**, 805 (1984).

[3] D. L. Boger, C. E. Brotherton, and G. I. Georg, *Org. Syn.*, submitted (1984).

[4] *Idem, Tetrahedron Letters*, **25**, 5615 (1984).

[5] D. L. Boger and C. E. Brotherton, *ibid.*, **25**, 5611 (1984).

D

Dialkylboryl trifluoromethanesulfonates, 7, 91–92; **9**, 131, 140; **10**, 118; **11**, 159

Macrolactonization.[1] Trimethylsilyl ω-trimethylsilyloxycarboxylates cyclize to macrolides in the presence of 1 equiv. of a dialkylboryl triflate. Dipropylboryl triflate is more effective than diethyl- or dibutylboryl triflate (equation I).

$$
\text{(I)} \quad
\begin{array}{c}
\text{OSi(CH}_3)_3 \\
| \\
\text{(CH}_2)_n \\
| \\
\text{CO}_2\text{Si(CH}_3)_3
\end{array}
\xrightarrow[\substack{83-94\%}]{\substack{\text{Pr}_2\text{BOTf,} \\ \text{C}_6\text{H}_5\text{CH}_3, \ \Delta}}
\begin{array}{c}
\text{C}=\text{O} \\
\text{(CH}_2)_n \quad \text{O}
\end{array}
$$

$$(n = 11,12,14,15)$$

[1] N. Taniguchi, H. Kinoshita, K. Inomata, and H. Kotake, *Chem. Letters*, 1347 (1984).

1,4-Diazabicyclo[2.2.2]octane (DABCO), **2**, 99–101; **4**, 119; **5**, 176–177; **6**, 157; **7**, 86–87; **11**, 153–154.

Coupling of aldehydes and acrylates.[1] In the presence of DABCO as catalyst, aldehydes couple with acrylates at the α-position to form 2-(1-hydroxyalkyl)acrylates (equation I). These products are readily converted into α-methylene-β,γ-unsaturated es-

$$
\text{(I)} \quad \text{R}^1\text{CHO} + \text{H}_2\text{C} \diagup\diagdown \text{COOR}^2
\xrightarrow[\substack{\sim 60-95\%}]{\substack{\text{DABCO,} \\ 25°}}
\begin{array}{c}
\text{R}^1\text{CHOH} \\
| \\
\text{H}_2\text{C} \diagup\diagdown \text{COOR}^2
\end{array}
$$

ters. This sequence furnishes a short synthesis of the 2-methylidene-3-butenoic ester **a**, which dimerizes to the diester (**3**) of mikanecic acid, a terpenoid dicarboxylic acid.[2]

$$
\text{CH}_3\text{CHO} + \text{H}_2\text{C} \diagup\diagdown \text{COOC(CH}_3)_3
\xrightarrow[92\%]{\text{DABCO}}
\begin{array}{c}
\text{CH}_3\text{CHOH} \\
| \\
\text{H}_2\text{C} \diagup\diagdown \text{COOC(CH}_3)_3
\end{array}
\xrightarrow[93\%]{\substack{\text{NBS,} \\ \text{S(CH}_3)_2}}
$$

1

$$
\begin{array}{c}
\text{CH}_2\text{Br} \\
\text{CH}_3 \diagdown \diagup \text{COOC(CH}_3)_3 \\
\text{H}
\end{array}
\xrightarrow[\text{HOC(CH}_3)_3]{\text{KOC(CH}_3)_3}
\left[
\begin{array}{c}
\text{CH}_2 \\
\\
(\text{CH}_3)_3\text{COOC} \diagdown \diagup \\
\text{CH}_2
\end{array}
\right]
\xrightarrow{40\%}
$$

(Z)-**2**

a

$$
(\text{CH}_3)_3\text{CO}_2\text{C} \diagdown
\begin{array}{c}
\text{CH}=\text{CH}_2 \\
\diagup\diagdown \text{CO}_2\text{C(CH}_3)_3
\end{array}
$$

3

The adducts can also be converted into (E)-α-methyl-α,β-unsaturated esters (equation II).

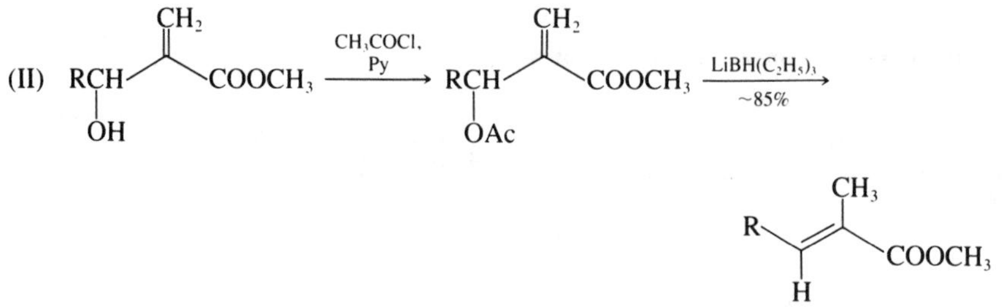

(II)

[1] J. Rabe and H. M. R. Hoffmann, *Angew. Chem., Int. Ed.*, **22**, 795, 796 (1983); H. M. R. Hoffmann and J. Rabe, *Helv.*, **67**, 413 (1984).
[2] O. Goldberg and A. S. Drieding, *ibid.*, **59**, 1904 (1976).

1,5-Diazabicyclo[5.4.0]undecene-5 (DBU), **2**, 101; **4**, 16–18; **5**, 177–178; **6**, 158; **7**, 87–88; **8**, 141; **9**, 132–133; **11**, 155.

Biphenylenes.[1] A new route to polycyclic biphenylenes involves base-catalyzed condensation of benzocyclobutane-1,2-diones with *o*-bis(cyanomethyl)arenes. DBN is useful, but DBU is the base of choice. Yields can be improved by addition of calcium hydride as a dehydrating agent.

Example:

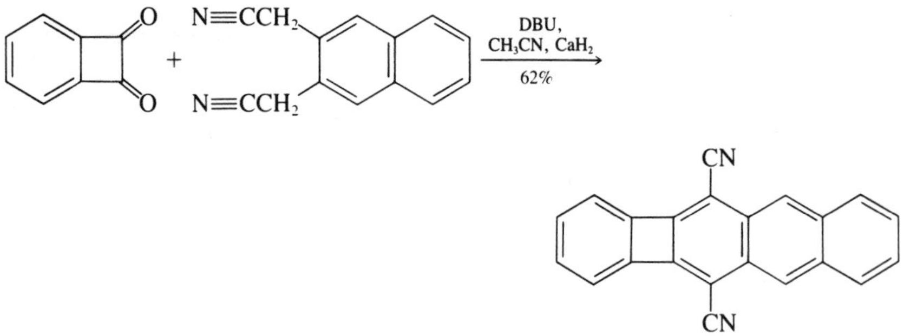

[1] P. R. Buckland, N. P. Hacker, and J. F. W. McOmie, *J.C.S. Perkin I*, 1443 (1983).

Diazadieneiron(0) complexes. The complexes are prepared by reaction of N,N'-dialkyl-1,4-diaza-1,3-dienes (dad) with $FeCl_2$ to form Fe(II) complexes (**1**), which are reduced by $Al(C_2H_5)_3$ or Grignard reagents to give (dad)Fe(0).

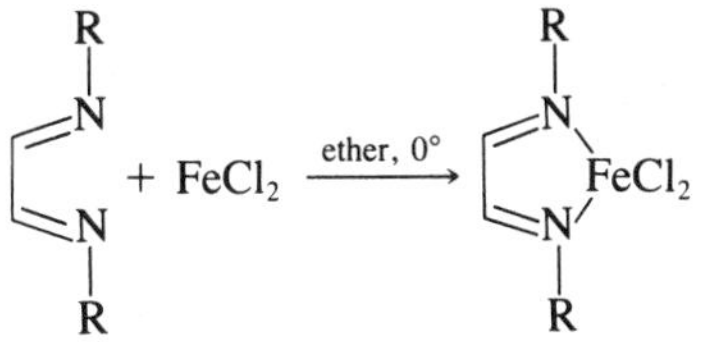

1

Diels-Alder catalyst.[1] The usually sluggish reaction of dienes with internal un-strained alkynes proceeds at 20–90° in the presence of **1** and an activator to give 1,4-cyclo-hexadienes in yields of 13–80%.

Example:

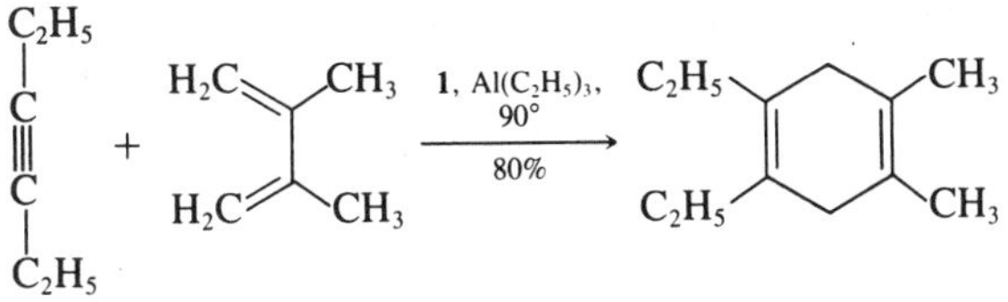

[1] H. tom Dieck and R. Diercks, *Angew. Chem., Int. Ed.,* **22,** 778 (1983).

Diazomethane, 1, 191–195; **2,** 102–104; **3,** 74; **4,** 120–122; **5,** 179–182; **6,** 158; **7,** 88–89; **9,** 133–136.

Review. Black[1] has reviewed recent work on the reactions of diazomethane (110 references).

[1] T. H. Black, *Aldrichim. Acta,* **16,** 36 (1983).

Dibenzoyl peroxide, 1, 196–198; **4,** 122–123; **5,** 182–183; **6,** 160–161; **7,** 89.

$R_2NH \rightarrow R_2NOH$.[1] This reaction has generally been conducted by reaction of the amine with dibenzoyl peroxide followed by saponification of the intermediate O-benzoyl-hydroxylamine. Yields can be markedly improved by addition of Na_2HPO_4 to trap the benzoic acid formed and by debenzoylation with $KOCH_3/CH_3OH$ (equation I).

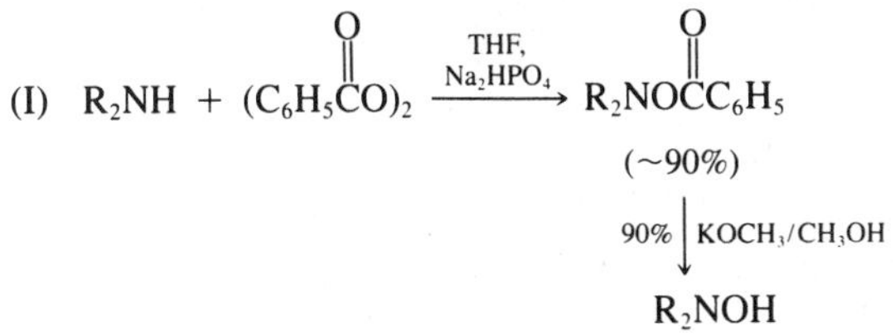

Enolate hydroxylation.[2]   α-Hydroxylation of the ketone **1** is best effected by reaction of the enolate with dibenzoyl peroxide. In this case Vedejs oxidation fails, and *m*-chlo-

roperbenzoic acid oxidation of the enol silyl ether (**8**, 100–101) gives **2** in only 22% overall yield. Surprisingly, the product (**2**) is the 1β-isomer (axial), and is subject to two 1,3-diaxial interactions (C_5 and C_9). Even so, it is not epimerized by acid or base because of stabilization by hydrogen bonding.

[1] A. J. Biloski and B. Ganem, *Synthesis*, 537 (1983).
[2] J. W. Huffman, R. C. Desai, and G. F. Hillenbrand, *J. Org.*, **49**, 982 (1984).

Dibromodifluoromethane, CF_2Br_2. Mol. wt. 209.83, b.p. 24–25°. Suppliers: Armageddon Chem. Co., Fluka.

α-Trifluoromethyl esters.[1]   These esters can be prepared by a modified malonic ester synthesis (equation I). Thus the anion of a malonic ester reacts with CF_2Br_2 by an

$$(I) \quad RCH(COOC_2H_5)_2 \xrightarrow[\substack{\text{2) } CF_2Br_2 \\ 50-80\%}]{\substack{\text{1) NaH, THF}}} \underset{\underset{CF_2Br}{|}}{RC(COOC_2H_5)_2} \xrightarrow[\substack{\Delta \\ 35-60\%}]{\substack{\text{KF, DMSO,}}} \underset{\underset{CF_3}{|}}{RCHCOOC_2H_5}$$

ionic chain reaction involving difluorocarbene to give a bromodifluoromethyl-substituted malonic ester. On reaction with strictly anhydrous KF in DMSO at 150–170°, these products undergo deethoxycarbonylation followed by halogen exchange to form α-trifluoromethyl esters.

[1] T. S. Everett, S. T. Purrington, and C. L. Bumgardner, *J. Org.*, **49**, 3702 (1984).

1,3-Dibromo-5,5-dimethylhydantoin (1), 1, 208; **2,** 108.

Regioselective bromination.   A procedure for a convenient large-scale preparation of 4-demethoxydaunomycinone (**6**) involves introduction of a hydroxyl group at C_7 of the tetracyclic precursor **2**. With an early method involving bromination followed by methanolysis, yields are low with Br_2, $(CH_3)_4\overset{+}{N}Br_3{}^-$, and NBS, because of low reactivity and

2, R = H
3, R = Ac

4

1) CH$_3$COOAg, HOAc
2) (CH$_3$CO)$_2$O, H$^+$
$\xrightarrow{61\%}$ Tetraacetates $\xrightarrow[\text{85\%}]{\text{HCl, CH}_3\text{OH}}$

5

6, R^1 = OH, R^2 = H
7, R^1 = H, R^2 = OH

aromatization. Bromination of the triacetate **3**, however, with **1** proceeds in high yield to give a single bromo derivative (**4**). The 9-acetoxy group is necessary for regioselectivity. The product is converted into a mixture of two tetraacetates (**5**), which is hydrolyzed to **6** and **7** in the ratio 2:1. The unnatural epimer **7** is converted into **6** by treatment with TFA.[1]

[1] D. Dominguez, R. J. Ardecky, and M. P. Cava, *Am. Soc.*, **105**, 1608 (1983).

Di-*t*-butyl dicarbonate, (Boc)$_2$O, **4**, 128; **7**, 91; **8**, 145; **10**, 122.

1-Boc derivatives of pyrroles.[1]   Base-sensitive pyrroles can be converted into the 1-Boc derivative by reaction with (Boc)$_2$O in CH$_3$CN or CH$_2$Cl$_2$ catalyzed by 4-dimethyl-aminopyridine. Triethylamine can be used as a base, but is not usually required. Yields are generally >80%.

Protection of tryptophan.[2]   The Boc group can also be used to protect the indole in tryptophan. The methyl ester of di-Boc-Trp can be prepared in almost quantitative yield and used for peptide synthesis. Both Boc groups are cleaved with TFA in a reasonable yield. The N$^\alpha$-Boc group can be cleaved selectively with 2.7 M HCl in dioxane at 25°.

[1] L. Grehn and U. Ragnarsson, *Angew. Chem., Int. Ed.*, **23**, 296 (1984).
[2] H. Franzén, L. Grehn, and U. Ragnarsson, *J.C.S. Chem. Comm.*, 1699 (1984).

Di-*n*-butyltelluronium carboethoxymethylide, (C$_4$H$_9$)$_2$$\overset{+}{\text{Te}}$—$\overset{-}{\text{C}}$HCOOC$_2H_5$ (**1**). The reagent is prepared[1] by treatment of carboethoxymethyldibutyltelluronium bromide, (C$_4$H$_9$)$_2$Te(Br)CH$_2$COOC$_2$H$_5$, m.p. 63°,[2] with potassium *t*-butoxide in THF.

α,β-Unsaturated esters.[1]   The reagent reacts with a variety of aldehydes and ketones to give α,β-unsaturated esters in ~50–90% yield. The (E/Z)-ratio varies from 1:1 to >50:1.

Example:

(E/Z >50:1)

[1] A. Osuka, Y. Mori, H. Shimizu, and H. Suzuki, *Tetrahedron Letters*, **24**, 2599 (1983).
[2] M. P. Balfe, C. A. Chaplin, and H. Phillips, *J. Chem. Soc.*, 341 (1958).

Di-*n*-butyltin oxide, 5, 188; 9, 141; 10, 123–125.

Macrolides (**10**, 124).[1]   The tin-mediated lactonization of ω-hydroxy carboxylic acids is particularly useful for formation of 13- to 17-membered macrolides. Highest yields are obtained when di-*n*-butyltin oxide is used in stoichiometric amounts. The method is not useful for macrocyclization of ω-bromo or ω-mercapto carboxylic acids.

The method is useful for synthesis of five-, six-, and seven-membered lactams, but not of β-lactams or macrolactams. The bridged lactam **2** was obtained from **1** in 77% yield by reaction with di-*n*-butyltin oxide.

[1] K. Steliou and M.-A. Poupart, *Am. Soc.*, **105**, 7130 (1983).

Dicarbonylcyclopentadienylcobalt, 5, 172–173; 6, 153–154; 7, 84, 94–95; 8, 146–147; 9, 142–143; 10, 126–127; 11, 161–162.

Tetrahydroprotoberberines.[1]   A novel entry into this ring synthesis involves the cobalt-catalyzed cocyclization of a tetrahydroisoquinoline such as **1** with trimethylsilyl-methoxyethyne (**2**). The reaction proceeds regioselectively in favor of the more hindered isomer (**3**). In the absence of the trimethylsilyl group of **1**, the condensation results in two isomers in a 1:1 ratio.

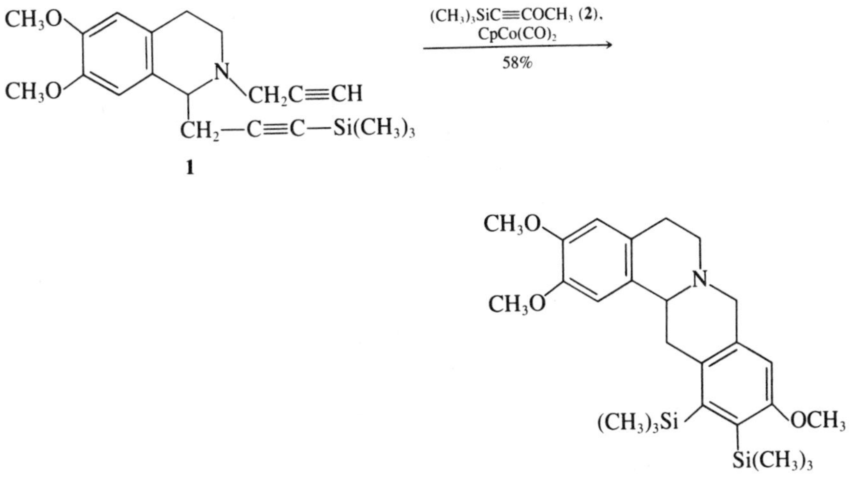

Estrones.[2]   An A → BCD approach to estrones involves a cobalt-catalyzed intra-molecular [2 + 2 + 2] cycloaddition of the enediyne **1** to form the B, C, and D rings in one step. The high stereoselectivity suggests that the cyclization involves a Diels-Alder-type reaction of the vinyl group with a cobaltacyclopentadiene formed by coupling of the two alkyne units. The homoannular diene obtained on demetalation is isomerized easily to the diene **3**.

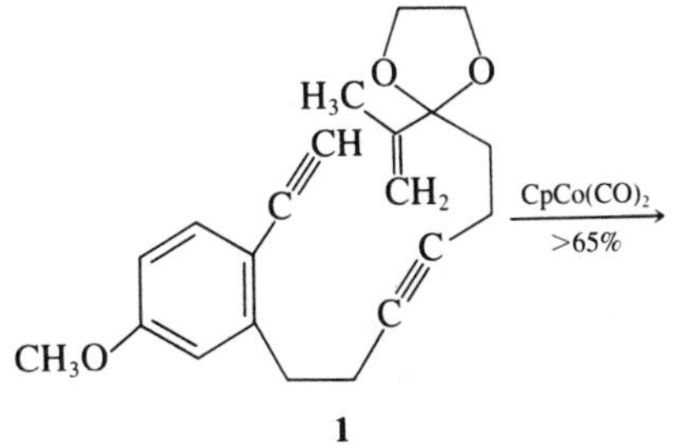

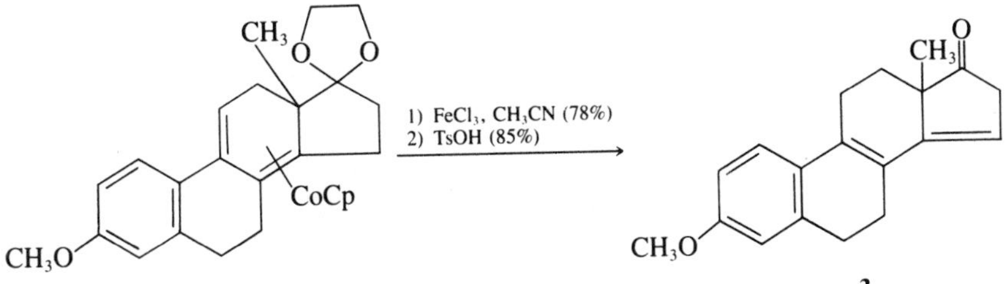

Benzocyclobutadienes.[3]   Cobalt-catalyzed cotrimerization provides a short synthesis of the tetrasilylbenzo[3.4]cyclobuta[1,2-*b*]biphenylene **1**, which can be desilylated to the

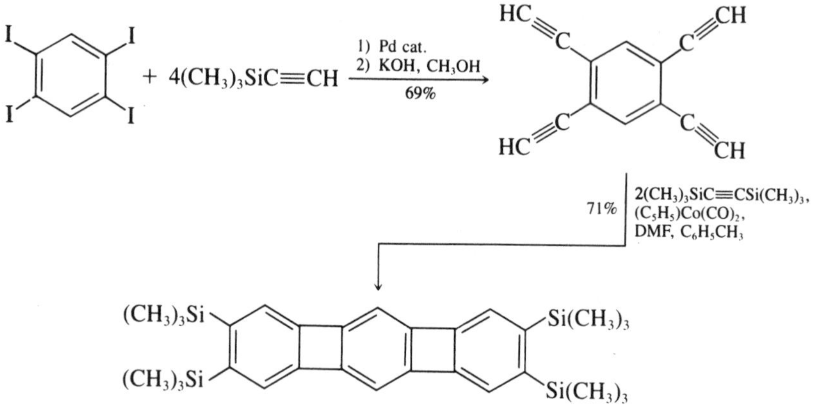

parent hydrocarbon [KOC(CH$_3$)$_3$, HOC(CH$_3$)$_3$, DMSO]. The outer benzene rings have little aromatic character and undergo very ready hydrogenation.

Pyridones.[4]   This catalyst effects [2 + 2 + 2] cycloadditions of 5-isocyanatoalkynes (**1**) with a variety of alkynes to form 2-pyridones. 1-Trimethylsilylalkynes (**2**) react re-

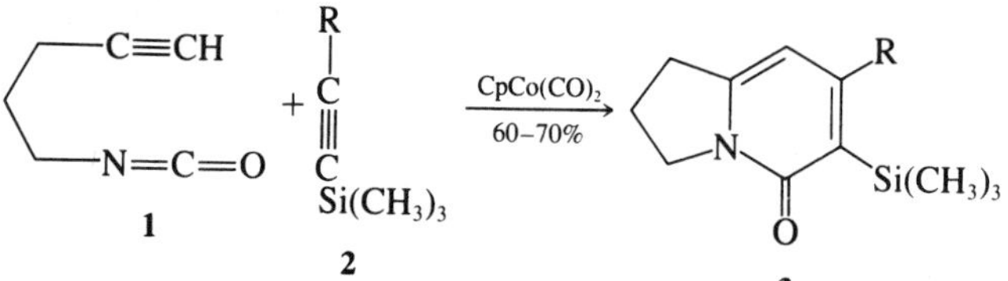

gioselectively to form **3** as the major or sole product. The pyridone **4** was obtained by this cycloaddition (60% yield) and used to prepare an intermediate in a synthesis of camptothecin (**5**).

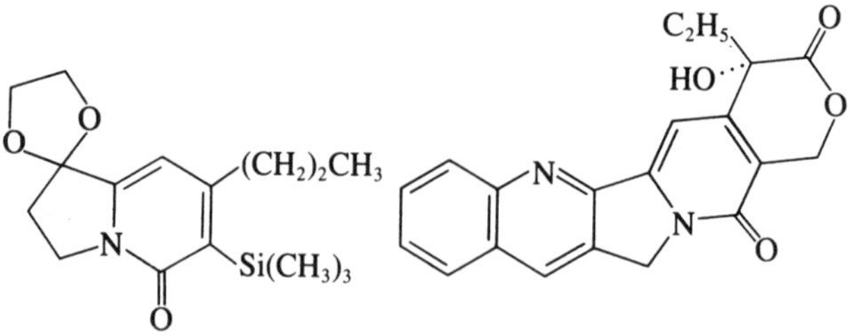

Decahydro-4a,5b-dimethylphenanthrene (**3**).[5]   This very air-sensitive hydrophen-anthrene nucleus is obtained by a [2 + 2 + 2] cycloaddition of enediyne (**1**) mediated by CpCo(CO)$_2$. This cyclization is noteworthy because it involves cyclization of a tetra-

2 (60%)   **3** (20–25%)

substituted double bond and results in stereospecific formation of two adjacent quaternary centers. The product decomposes rapidly in air, probably via an endoperoxide.

Review.   Vollhardt[6] has reviewed the use of cobalt-catalyzed [2 + 2 + 2] cycload-ditions for synthesis of annelated benzenes, pyridines, and other heterocycles, as well as of complex natural products.

[1] R. L. Hillard III, C. A. Parnell, and K. P. C. Vollhardt, *Tetrahedron*, **39**, 905 (1983).
[2] E. D. Sternberg and K. P. C. Vollhardt, *J. Org.*, **49**, 1574 (1984).
[3] B. C. Berris, G. H. Hovakeemian, and K. P. C. Vollhardt, *J.C.S. Chem. Comm.*, 502 (1983).
[4] R. A. Earl and K. P. C. Vollhardt, *Am. Soc.*, **105**, 6991 (1983).
[5] M. Malacria and K. P. C. Vollhardt, *J. Org.*, **49**, 5010 (1984).
[6] K. P. C. Vollhardt, *Angew. Chem., Int. Ed.*, **23**, 539 (1984).

Di-μ-carbonylhexacarbonyldicobalt, 1, 224–225; **3**, 89; **4**, 139; **5**, 204–205; **6**, 172; **7**, 99–100; **8**, 148–150; **9**, 144–145; **10**, 129–130; **11**, 162–163.

Fused cyclopentenones.   An intramolecular alkene–alkyne cyclization mediated by Co$_2$(CO)$_8$ provides a route to bicyclic [3.3.0]enones (equation I).[1] This reaction provides a key step in a synthesis of **1**, a linearly fused triquinane,[2] which is a precursor to the antitumor sesquiterpene coriolin (**2**).

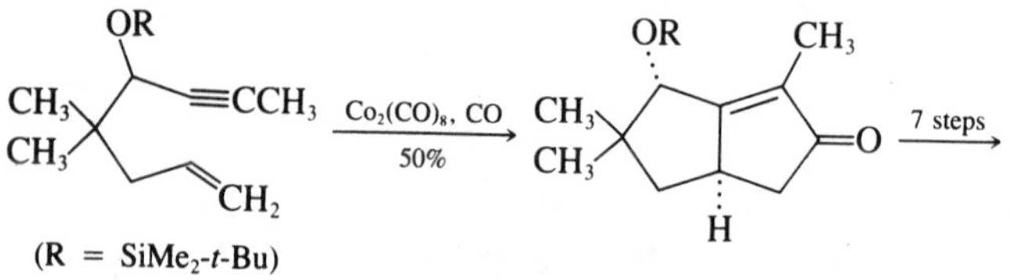

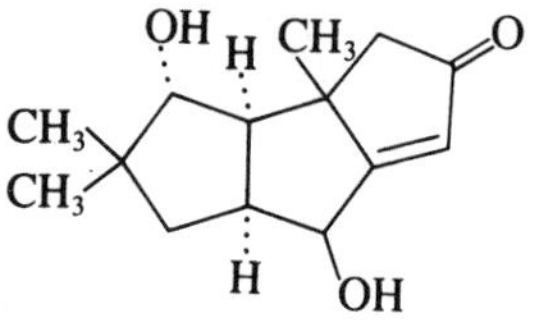

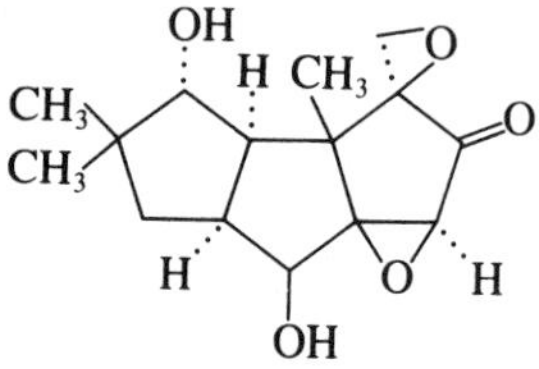

The cyclization provides a route to the angularly fused triquinane system of **3**, which lacks two of the methyl groups present in the natural sesquiterpene isocomene (**4**). Cyclization of a suitably substituted enyne for synthesis of **4**, however, proceeds in minute yield.[3]

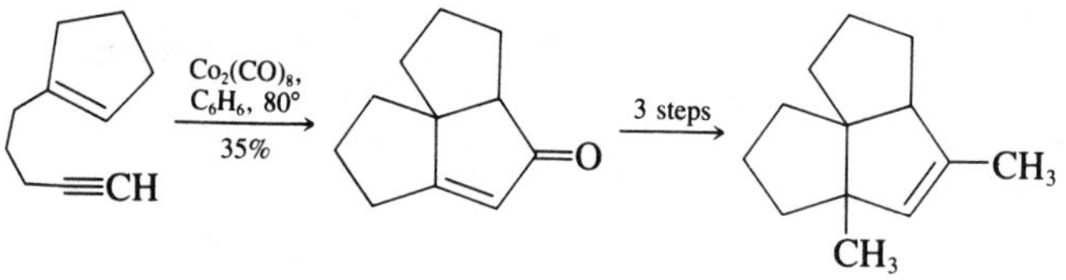

Alkyne hexacarbonyldicobalt complexes. The complexes of alkynes with $Co_2(CO)_8$[4] react with 2,5-dihydrofuranes to give 3-oxabicyclo[3.3.0]-7-octene-6-ones in high yield (equation I) when the reaction is conducted under a CO atmosphere.

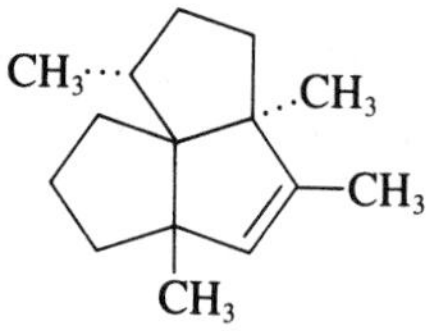

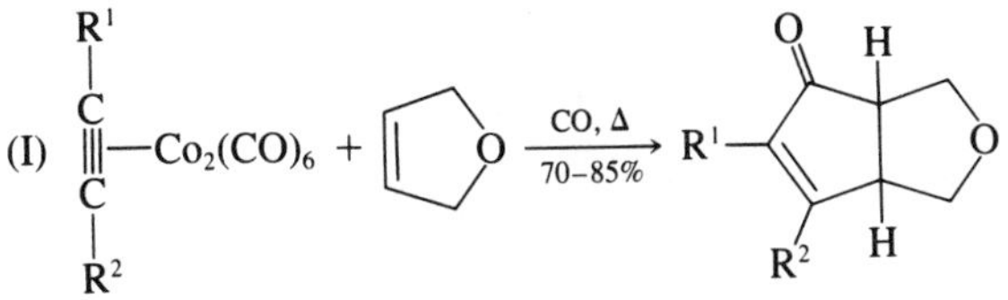

This reaction has been used for short total syntheses of cyclomethylenomycin A (**1**) and cyclosarkomycin (**2**).[5]

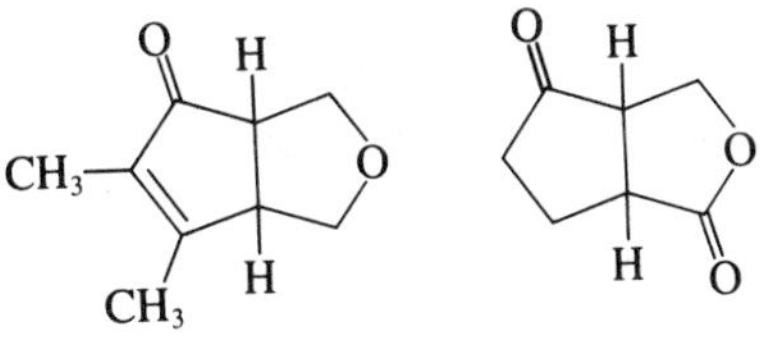

1,4-Diynes.[6]   The (propargyl acetate)$Co_2(CO)_6$ complex **1** couples with the trialkynylalane **2**[7] in CH_2Cl_2 at 0° (or better, $-78°$) to give, after aqueous work-up, the diyne complex **3**. The free diyne (**4**) is obtained by demetallation with CAN.

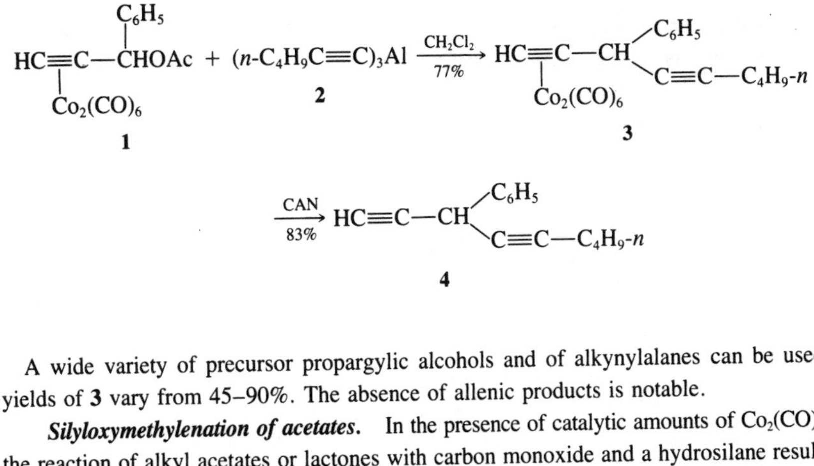

A wide variety of precursor propargylic alcohols and of alkynylalanes can be used; yields of **3** vary from 45–90%. The absence of allenic products is notable.

Silyloxymethylenation of acetates.   In the presence of catalytic amounts of $Co_2(CO)_8$, the reaction of alkyl acetates or lactones with carbon monoxide and a hydrosilane results in a [(trialkylsilyloxy)methylene]alkane in moderate to high yield. A series of plausible intermediates in this transformation have been suggested. The reaction is conducted in C_6H_6 or $C_6H_5CH_3$ in an autoclave (50 atm.) at 200°.[8]

Examples:

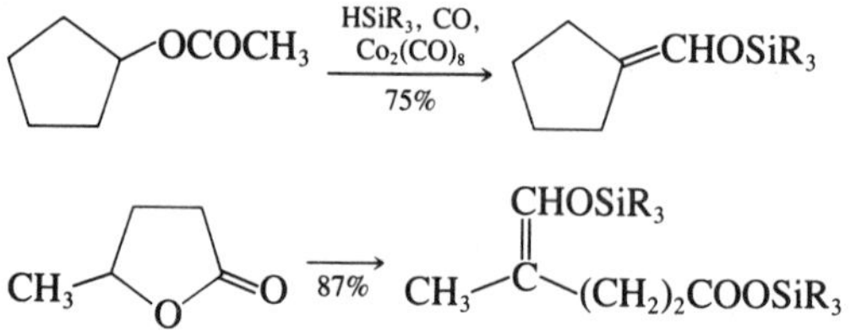

Carbonylation of ketones.[9] Cyclobutanones react with carbon monoxide and a hydrosilane in the presence of $Co_2(CO)_8$ and triphenylphosphine to give disilyloxycyclopentenes. Under these conditions, presumably $R_3SiCo(CO)_4$ is generated as the active species, which then reacts with the C=O bond with insertion of carbon monoxide. This reaction is limited to strained cyclic ketones; simple ketones are merely converted into enol silyl ethers.

Example:

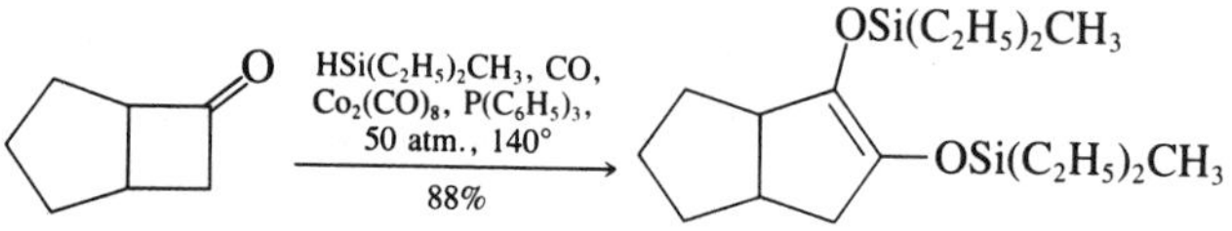

Oxymethylation of epoxides.[10]   The reaction of a hydrosilane and carbon monoxide with an epoxide catalyzed by $Co_2(CO)_8$ results in a regio- and stereoselective ring opening to give disilyl ethers of a 1,3-diol in generally high yield.

Examples:

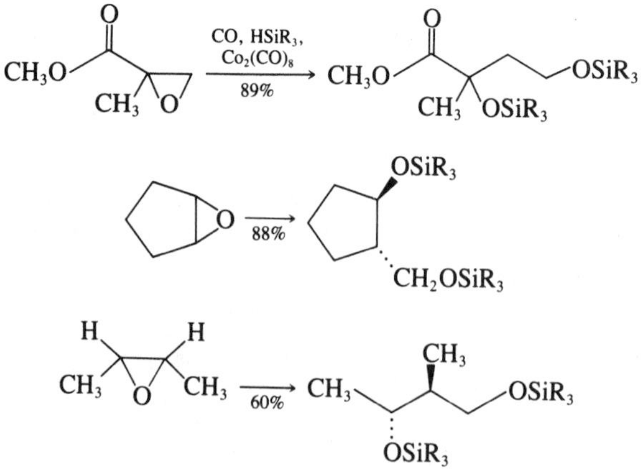

*Separation of diastereomeric **13,14-dehydroprostaglandins**.*[11]   Diastereomeric acetylenic prostaglandin analogs such as **1** cannot be separated by chromatography, but the

1 (13,14-dehydro PGF$_{2\alpha}$)

complexes with $Co_2(CO)_8$ can be separated by chromatography on silica gel plates into
two individual complexes (66% for each diastereomer). The dehydroprostaglandins are
regenerated with iron(III) nitrate.

This separation should be widely applicable.

***Hydrocarbonylation of* 1,4-dienes.**   1,4-Dienes bearing two substituents at C_3 undergo
hydrocarbonylation catalyzed by $Co_2(CO)_8$ to give 2-methylcyclopentanones as the main
product (equation I).[12]

This reaction has been used for a synthesis of 2-cuparenone (**1**).[13]

[1] N. E. Schore and M. C. Croudace, *J. Org.*, **46**, 5436 (1981).
[2] C. Exon and P. Magnus, *Am. Soc.*, **105**, 2477 (1983).
[3] M. T. Knudsen and N. E. Schore, *J. Org.*, **49**, 5025 (1984).
[4] I. U. Khand, G. R. Knox, P. L. Pauson, and W. E. Watts, *J.C.S. Perkin I*, 977 (1973).
[5] D. C. Billington, *Tetrahedron Letters*, **24**, 2905 (1983).
[6] S. Padmanabhan and K. M. Nicholas, *ibid.*, 2239 (1983).
[7] R. E. Connor and K. M. Nicholas, *J. Organometal. Chem.*, **125**, C45 (1977).
[8] N. Chatani, S. Murai, and N. Sonoda, *Am. Soc.*, **105**, 1370 (1983).
[9] N. Chatani, H. Furukawa, T. Kato, S. Murai, and N. Sonoda, *ibid.*, **106**, 430 (1984).
[10] T. Murai, S. Kato, S. Murai, T. Toki, S. Suzuki, and N. Sonoda, *ibid.*, 6093 (1984).
[11] C. O.-Yang and J. Fried, *Tetrahedron Letters*, **24**, 2533 (1983).

[12] P. Eilbracht, M. Acker, and W. Totzauer, *Ber.*, **116**, 238 (1983).
[13] P. Eilbracht, E. Balss, and M. Acker, *Tetrahedron Letters*, **25**, 1731 (1984).

Dichlorobis(cyclopentadienyl)titanium, 10, 130–131; **11,** 163–164.

π-Allyltitanium complexes (**10,** 131).[1]   The π-allyltitanium complexes (**a**) of cyclopentadiene and cyclohexadiene react stereoselectively with aldehydes to introduce a 1′-hydroxyalkyl side chain (equation I).

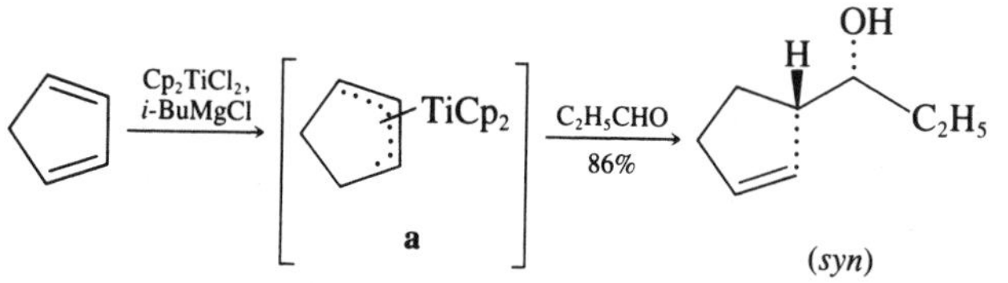

The reaction of the π-allyltitanium complex of 1-vinyl-1-cyclopentene with aldehydes is both regio- and stereoselective (equation II). The regioselectivity of this reaction decreases with an increase in the size of the ring.

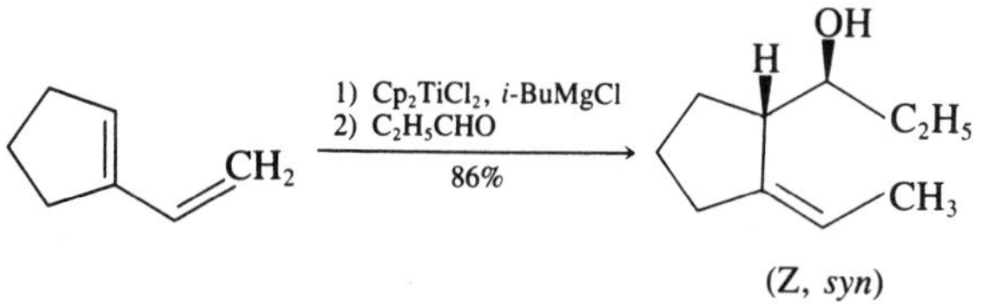

Hydromagnesiation; vinylsilanes.[2]   The hydromagnesiation (**10,** 130–131) of 1-trimethylsilyl-1-alkynes with isobutylmagnesium bromide followed by alkylation affords (Z)-1,2-dialkylvinylsilanes in high yield.

Example:

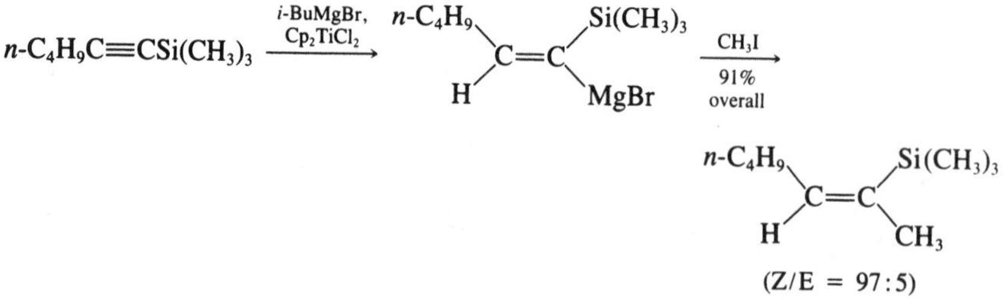

This reaction was used in a synthesis of the sex pheromone (**1**) of the false codling moth.

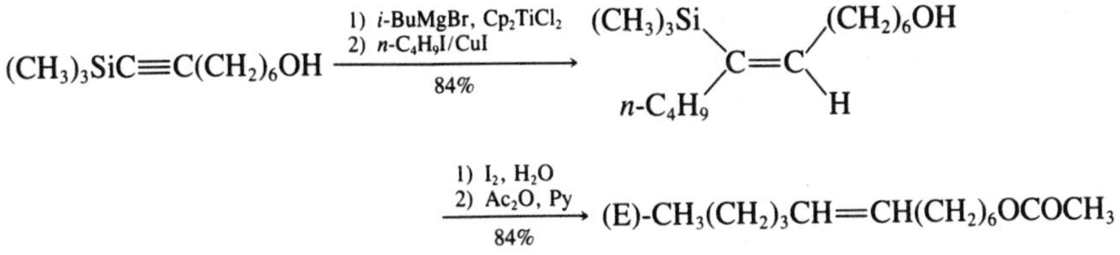

$$(CH_3)_3SiC\equiv C(CH_2)_6OH \xrightarrow[\substack{84\%}]{\substack{1)\ i\text{-BuMgBr, Cp}_2TiCl_2 \\ 2)\ n\text{-C}_4H_9I/CuI}}$$

$$\xrightarrow[\substack{84\%}]{\substack{1)\ I_2,\ H_2O \\ 2)\ Ac_2O,\ Py}} (E)\text{-}CH_3(CH_2)_3CH{=}CH(CH_2)_6OCOCH_3$$

1 (E/Z = 96 : 4)

Furanes.[3]   Hydromagnesiation of 3-trimethylsilyl-2-propyne-1-ol (**1**) (**11**, 163) followed by reaction with an aldehyde or ketone provides E-3-trimethylsilyl-2-alkene-1,4-diols (**2**). These are converted into 3-trimethylsilyl-2,5-dihydrofuranes (**3**) on dehydration

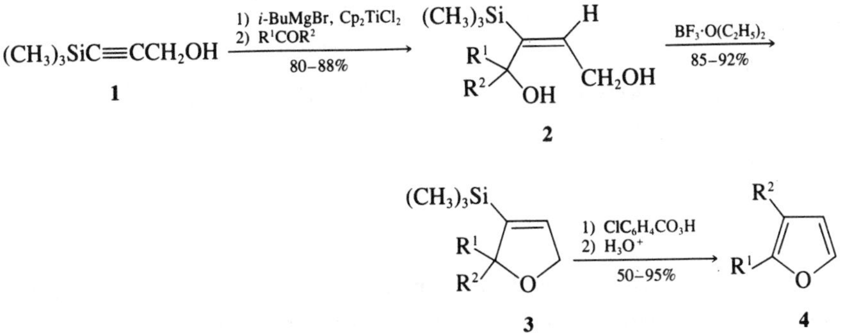

$$(CH_3)_3SiC\equiv CCH_2OH \xrightarrow[\substack{80-88\%}]{\substack{1)\ i\text{-BuMgBr, Cp}_2TiCl_2 \\ 2)\ R^1COR^2}}$$

1

$$\xrightarrow[\substack{85-92\%}]{BF_3 \cdot O(C_2H_5)_2}$$

with BF$_3$ etherate. Epoxidation of **3** followed by treatment with 20% H$_2$SO$_4$ affords furanes (**4**). Yields of **4** are high when R^1 = R^2. When R^1 = H, 2-substituted furanes are formed predominantly, because of preferential migration of hydrogen.

Allylalumination of alkynes.[4]   The reaction of allyldiisobutylaluminum, prepared from allylmagnesium bromide and *i*-Bu$_2$AlCl, with 1-octyne in the presence of Cl$_2$ZrCp$_2$ results in *cis*-allylalumination to provide a mixture of two regioisomers, **1** and **2**. Internal

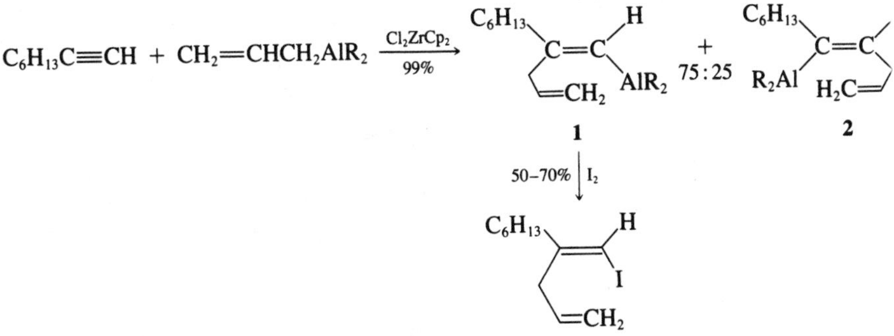

$$C_6H_{13}C\equiv CH + CH_2{=}CHCH_2AlR_2 \xrightarrow[\substack{99\%}]{Cl_2ZrCp_2}$$

alkynes react in a similar manner, but 1-silyl-1-alkynes do not react. An allylic rearrangement is not observed in the reaction with (E)-crotyldimethylalane (equation I).

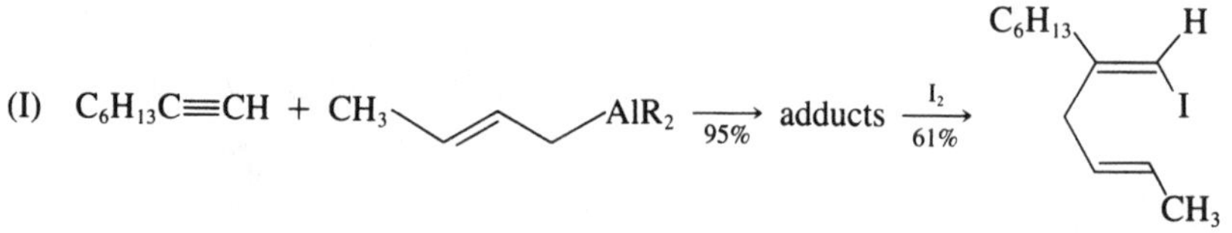

(I) $C_6H_{13}C{\equiv}CH$ + CH_3
AlR_2
95%
adducts
I_2
61%
C_6H_{13}
H
I
CH_3

***E,E*-Exocyclic dienes.**[5] A titanium reagent (**1**) consisting of $(C_5H_5)_2TiCl_2$, $CH_3P(C_6H_5)_2$, and sodium amalgam in the ratio 1:1:2 effects cyclization of diynes (**2**) in which n = 3, 4, or 5 to (E,E)-exocyclic dienes (**3**) (equation I). Highest yields (80%) are obtained when

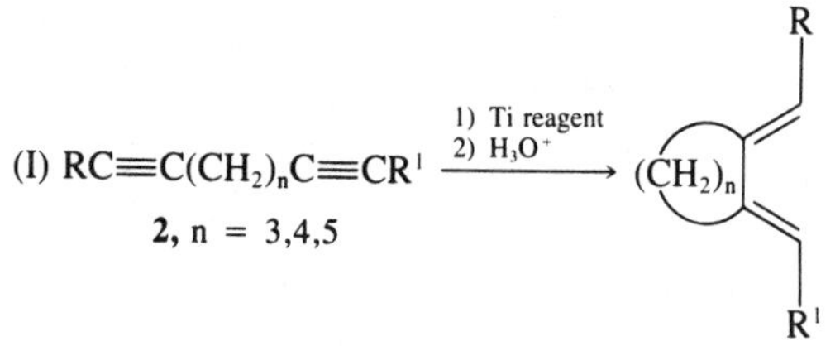

(I) $RC{\equiv}C(CH_2)_nC{\equiv}CR^1$
2, n = 3,4,5
1) Ti reagent
2) H_3O^+
R
$(CH_2)_n$
R^1

3

n = 4 and when R and R^1 are methyl, but larger alkyl and phenyl groups are tolerated. Yields are low for trimethylsilylalkynes, and the reaction fails with terminal alkynes. As expected, the products undergo cycloaddition with activated dienophiles with the usual *cis-endo*-stereochemistry.

Use of this cyclization in combination with an intramolecular Diels-Alder reaction is illustrated by a synthesis of the tricyclic lactone **4**.

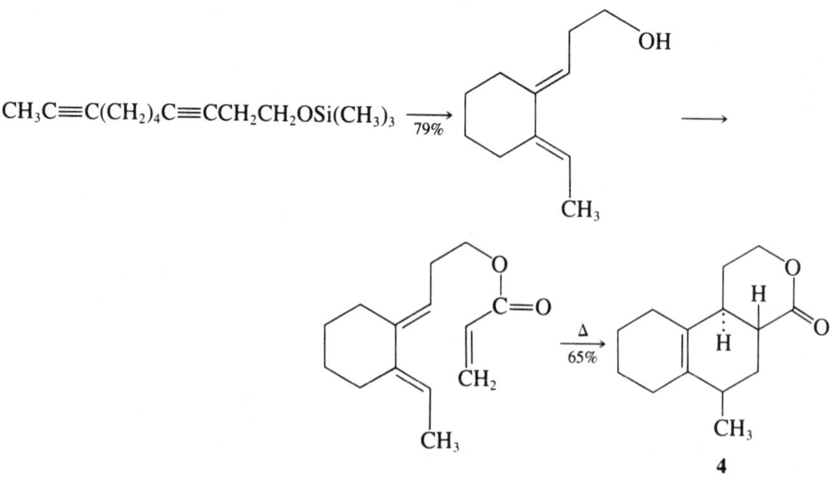

$CH_3C{\equiv}C(CH_2)_4C{\equiv}CCH_2CH_2OSi(CH_3)_3$
79%
OH
CH_3
O
C=O
CH_2
CH_3
Δ
65%
O
O
H
H
CH_3

4

[1] Y. Kobayashi, K. Umeyama, and F. Sato, *J.C.S. Chem. Comm.*, 621 (1984).
[2] F. Sato, H. Watanabe, Y. Tanaka, T. Yamaji, and M. Sato, *Tetrahedron Letters*, **24**, 1041 (1983).
[3] F. Sato, H. Kanbara, and Y. Tanaka, *ibid.*, **25**, 5063 (1984).
[4] J. A. Miller and E. Negishi, *ibid.*, **25**, 5863 (1984).
[5] W. A. Nugent and J. C. Calabrese, *Am. Soc.*, **106**, 6422 (1984).

Dichlorobis(cyclopentadienyl)zirconium–*t*-Butylmagnesium chloride, Cl_2ZrCp_2–$(CH_3)_3CMgCl$.

Hydrozirconation.[1]   The combination of zirconocene dichloride and *t*-butylmagnesium chloride in C_6H_6–ether effects hydrozirconation of monosubstituted alkenes. The actual reagent may be $HZrCp_2Cl$ (Schwartz reagent, **6**, 175–177) or *t*-$BuZrCp_2Cl$. The reaction produces monoalkylzirconium derivatives with the metal attached to the terminal position of the alkene.

The reaction with styrene or 1,3-butadiene under the same conditions (benzene–ether) results in *t*-butylzirconation (equation I). However, in reactions conducted in benzene alone, hydrozirconation is observed.

[1] E. Negishi, J. A. Miller, and T. Yoshida, *Tetrahedron Letters*, **25**, 3407 (1984).

Dichloro[bis(1,2-diphenylphosphine)ethane]nickel(II),

$[(C_6H_5)_2PCH_2CH_2P(C_6H_5)_2]NiCl_2(1)$. Mol. wt. 528.04, air-stable, orange solid. Supplier: Strem.

Grignard synthesis of ketones.[1]   The major product of the reaction of carboxylic acids with Grignard reagents is the tertiary alcohol obtained by addition of 2 equiv. of the reagent. The ketone becomes the major product if **1** is present as catalyst. In the case of diaryl or alkyl aryl ketones, yields are in the range 60–75%. Yields are only moderate in the case of dialkyl ketones, partly because of incomplete consumption of the carboxylic acid, even in reactions with a large excess of Grignard reagent.

[1] V. Fiandanese, G. Marchese, and L. Ronzini, *Tetrahedron Letters*, **24**, 3677 (1983).

Dichloro[1,1'-bis(diphenylphosphine)ferrocene]palladium(II), $PdCl_2(dppf)$, **9**, 147.

Cross-coupling of RMgX with organic halides (**9**, 147).   This complex is the most active and selective known catalyst for coupling of *n*- and *sec*-butylmagnesium chloride or the corresponding alkylzinc reagents with aryl and vinyl bromides.[1]

Alkylation of allylic ethers.[2]   PdCl$_2$(dppf) is the most effective catalyst for alkylation of allyl trialkylsilyl ethers with Grignard reagents to give an internal alkene. NiCl$_2$(dppf) also catalyzes this reaction, but with the opposite regioselectivity, thus forming a terminal alkene.

Example:

PdCl$_2$(dppf)	96 : 4
NiCl$_2$(dppf)	12 : 88

Acetylenic ketones.[3]   PdCl$_2$(dppf) is the preferred catalyst for the carbonylation of vinylic and aryl halides in the presence of terminal alkynes to form acetylenic ketones (equation I).

$$\text{(I) } R^1X + CO + HC\equiv CR^2 \xrightarrow[55-93\%]{\substack{Pd(II),\\N(C_2H_5)_3}} R^1COC\equiv CR^2$$

[1] T. Hayashi, M. Konishi, Y. Kobori, M. Kumada, T. Higuchi, and K. Hirotsu, *Am. Soc.*, **106**, 158 (1984).
[2] T. Hayashi, M. Konishi, K. Yokota, and M. Kumada, *J.C.S. Chem. Comm.*, 313 (1981).
[3] T. Kobayashi and M. Tanaka, *ibid.*, 333 (1981).

Di-μ-chlorobis(1,5-hexadiene)dirhodium, [RhCl(1,5-C$_6$H$_{10}$)]$_2$(**1**). The complex is prepared by reaction of the diene with RhCl$_3$ in aqueous C$_2$H$_5$OH.[1]

Hydrogenation of arenes.[2]   Hydrogenation of benzene rings occurs selectively and stereospecifically at 25° and atmospheric pressure in the presence of this rhodium catalyst under phase-transfer conditions (tetrabutylammonium hydrogen sulfate, buffered aqueous hexane, pH 7.4–7.6).

Examples:

(77%)   (23%)

(80%)   (20%)

[1] G. Winkhaus and H. Singer, *Ber.*, **99**, 3602 (1966).
[2] K. R. Januszkiewicz and H. Alper, *Organometallics*, **2**, 1055 (1983).

Dichlorobis(trifluoromethanesulfonato)titanium(IV), $TiCl_2(OTf)_2$.

Preparation:[1]

$$2CF_3SO_3H + TiCl_4 \xrightarrow{-18 \rightarrow 25°} TiCl_2(SO_3CF_3)_2 + 2HCl$$

Ester condensations.[2]   Esters undergo Claisen condensation in the presence of $TiCl_2(OTf)_2$ (1.5 equiv.) and triethylamine (2.2 equiv.) and 4Å molecular sieves at 0–20°.

Example:

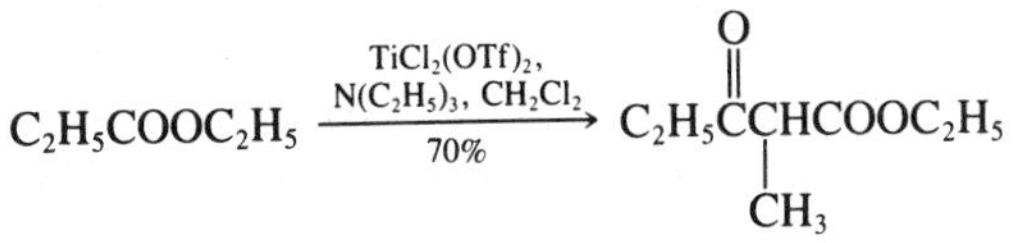

$$C_2H_5COOC_2H_5 \xrightarrow[70\%]{\substack{TiCl_2(OTf)_2, \\ N(C_2H_5)_3, CH_2Cl_2}} C_2H_5\overset{\displaystyle O}{\overset{\displaystyle \|}{C}}\underset{\displaystyle CH_3}{CH}COOC_2H_5$$

Under these conditions dicarboxylic esters undergo Dieckmann condensation.
Example:

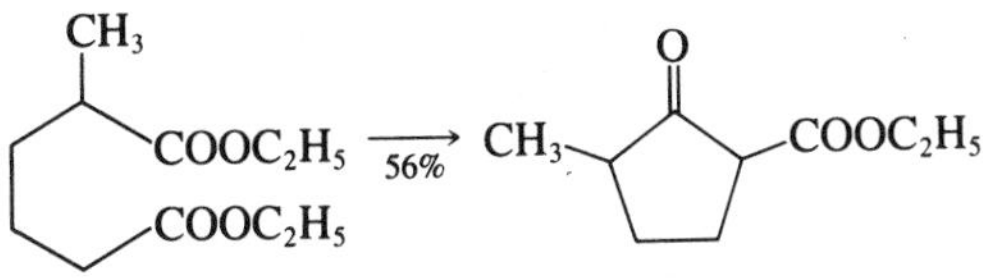

Condensation of lactones with esters (large excess) can be effected.
Example:

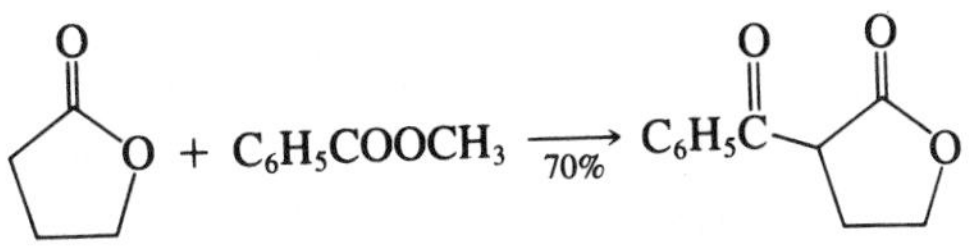

[1] R. E. Noftle and G. H. Cady, *Inorg. Chem.*, **5**, 2182 (1966).
[2] Y. Tanabe and T. Mukaiyama, *Chem. Letters*, 1867 (1984).

Dichlorobis(tri-*o*-tolylphosphine)palladium(II).

β,γ-Unsaturated ketones.[1]   Tri-*n*-butyltin enolates, prepared from enol acetates *in situ*, couple with vinyl bromides in the presence of a Pd(II) catalyst to provide β,γ-unsaturated ketones in moderate to high yield.

Example:

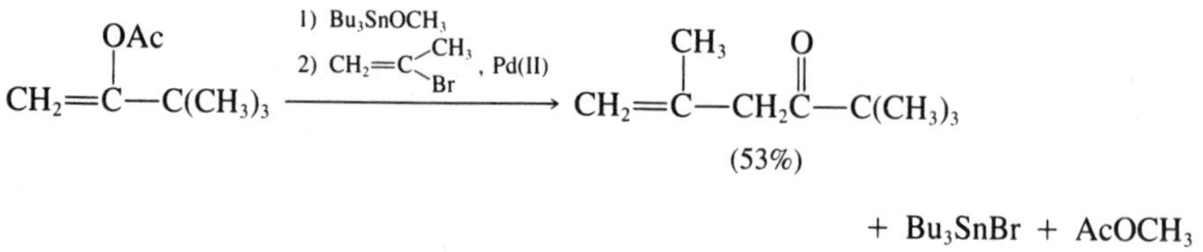

(53%)

+ Bu_3SnBr + $AcOCH_3$

[1] M. Kosugi, I. Hagiwara, and T. Migita, *Chem. Letters,* 839 (1983).

2,3-Dichloro-5,6-dicyanobenzoquinone, 1, 215–219; **2,** 112–117; **3,** 83–84; **4,** 120–134; **5,** 193–194; **6,** 168–170; **7,** 96–97; **8,** 153–156; **9,** 148–151; **10,** 135–136; **11,** 166–167.

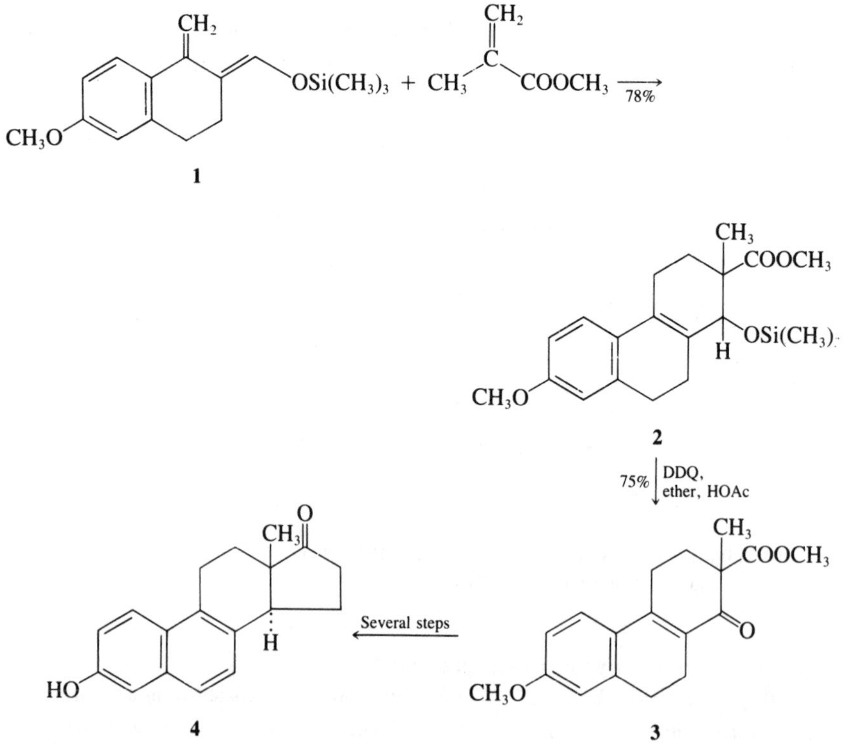

Allylic oxidation.[1]   A new synthesis of equilenin (**4**) involves a Diels-Alder reaction to provide ring C. The silyl ether (**2**) obtained in this way can be oxidized directly to the desired ketone (**3**) by DDQ in acetic acid. The oxidation may involve prior hydrolysis to the alcohol, which is also oxidized by DDQ in 80% yield. Other oxidants are inefficient for oxidation of the allylic alcohol: MnO_2 (29%), PCC (13–15%), DMSO–Ac_2O. The ketone (**3**) is converted into **4** by dehydrogenation of ring B followed by a Reformatsky reaction.

Protection of hydroxyl groups (**11**, 166).[2]   3,4-Dimethoxybenzyl ethers are oxidized by DDQ more readily than *p*-methoxybenzyl ethers. Moreover, the dimethoxybenzyl ethers of secondary alcohols can be selectively oxidized in the presence of the corresponding ethers of primary alcohols. Benzyl, *p*-methoxybenzyl, and 3,4-dimethoxybenzyl ethers all undergo hydrogenolysis catalyzed by Pt/C or Pd/C, but selective hydrogenolysis of benzyl ethers is possible with W-2 Raney Ni.

[1] N. S. Narasimhan and C. P. Bapat, *J.C.S. Perkin I,* 1435 (1984).
[2] Y. Oikawa, T. Tanaka, K. Horita, T. Yoshioka, and O. Yonemitsu, *Tetrahedron Letters,* **25**, 5393 (1984); Y. Oikawa, T. Tanaka, K. Horita, and O. Yonemitsu, *ibid.,* 5397 (1984).

1,4-Dichloro-1,4-dimethoxybutane, $CH_3O \diagdown Cl \quad Cl \diagdown OCH_3$ The butane is obtained

by reaction of 2,5-dimethoxytetrahydrofurane with $ClSi(CH_3)_3$ (86% yield).[1]

N-Acylpyrroles.[2]   The reagent converts primary amides into acylpyrroles (**2**) on reaction in CH_3CN or $CHCl_3$ in the presence of Amberlyst A-21 resin (equation I).

$$\text{(I)} \quad \underset{\text{O}}{\overset{\text{O}}{RCNH_2}} + \mathbf{1} \xrightarrow[\text{55–85\%}]{\text{Resin}} RC\!-\!N$$

2

These products undergo cleavage of the C—N bond on reaction with various nucleophiles. Thus they are converted in high yield to methyl esters on reaction with $NaOCH_3$ in CH_3OH. They are reduced to aldehydes by $NaBH_4$ in about 80% yield. They react with R^1Li or R^1MgCl to form tertiary alcohols, $RC(OH)R^1_2$, in 55–75% yield. Reaction with primary amines, R^1NH_2, results in secondary amides, $RCONHR^1$, in ~85% yield.

Thus this method is useful for conversion of primary amides to a variety of products arising from ready cleavage of the C—N bond.

[1] T. H. Chan and S. D. Lee, *Tetrahedron Letters,* **24**, 1225 (1983).
[2] S. D. Lee, M. A. Brook, and T. H. Chan, *ibid.,* 1569 (1983).

(E)-1,2-Dichloroethylene, $\underset{H}{\overset{Cl}{\diagdown}}C\!=\!C\underset{Cl}{\overset{H}{\diagup}}$ Mol. wt. 96.94, b.p. 78°. Suppliers: Aldrich, Fluka.

(E,E)-1-Chloro-1,3-dienes.[1]   In the presence of $Ni[P(C_6H_5)_3]_4$ or $Pd[P(C_6H_5)_3]_4$, **1**

reacts with an (E)-alkenylalane, prepared by hydroalumination of a 1-alkyne, to give (E,E)-1-chloro-1,3-dienes, which are useful precursors to dienynes.

Example:

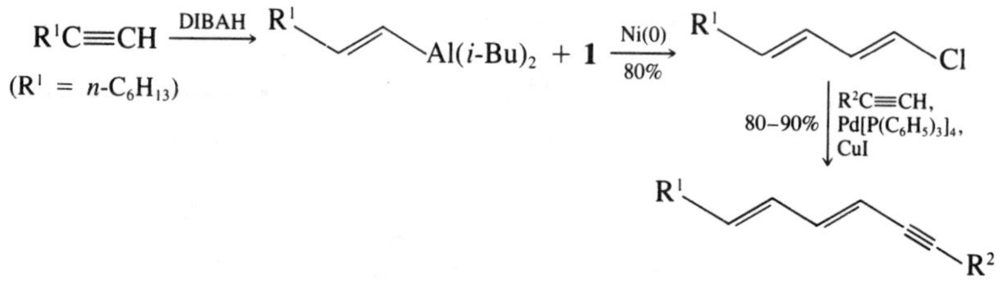

[1] V. Ratovelomanana and G. Linstrumele, *Tetrahedron Letters*, **25**, 6001 (1984).

Dichloroketene, 1, 221–222; **2,** 118; **3,** 87–88; **4,** 134–135; **8,** 156; **9,** 152–154; **10,** 139–140; **11,** 168–170.

Cyclobutenediones.[1]   A general route to these diones involves the regiospecific cycloaddition of dichloroketene to the phenylthio enol ether (**1**) of a ketone. The adduct (**2**) on treatment with triethylamine eliminates C_6H_5SCl and rearranges to **3**. Peracid oxidation of **3** results directly in a cyclobutenedione (**4**).

Example:

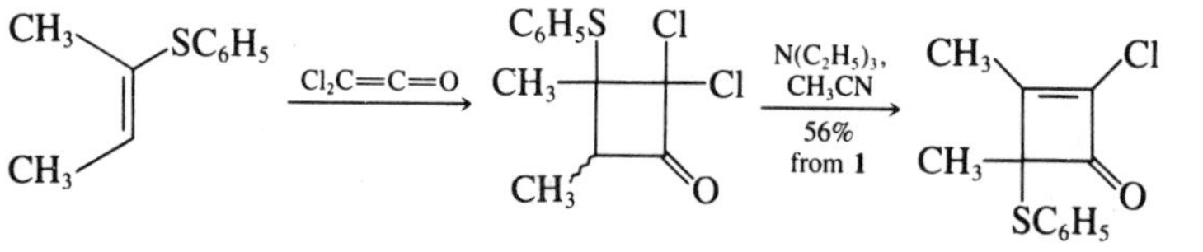

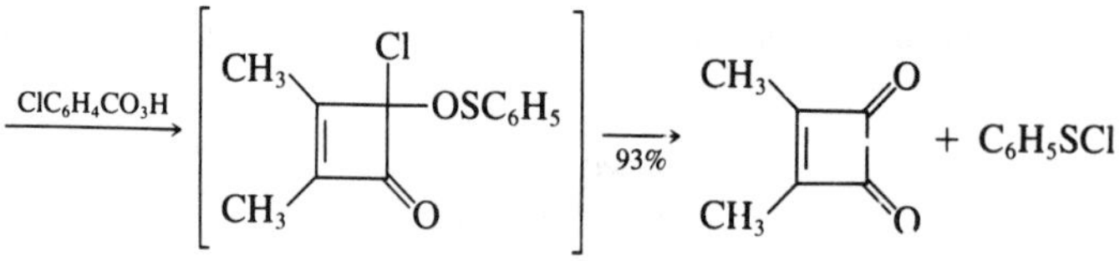

Cyclopentanone annelation (**10,** 139–140).[2]   The iterative cyclopentenone anne-lation has been extended to a synthesis of hirsutic acid (**7**) starting with the ester **1**. This alkene surprisingly does not react with chloromethylketene, but does react stereoselectively with dichloroketene to give **2** as the major product. One of the chlorine atoms was replaced by methyl by treatment with $(CH_3)_2CuLi$ (3 equiv.) and then with CH_3I and HMPT. Ring expansion of this product was effected with CH_2N_2 to give the cyclopentanone **3**. Reduction of the ketone followed by reduction of the resulting chlorohydrin gave **4** in 38–42% overall yield. Repetition of the annelation followed by Zn/HOAc reduction gave **5** in 68% yield. The ketone **5** was dehydrogenated to the enone **6** by Pd(II) oxidation, a method that proved superior to selenium-based methods and to dehydrotrimethylsilylation (**8**, 378). Concluding steps to **7** followed usual methods.

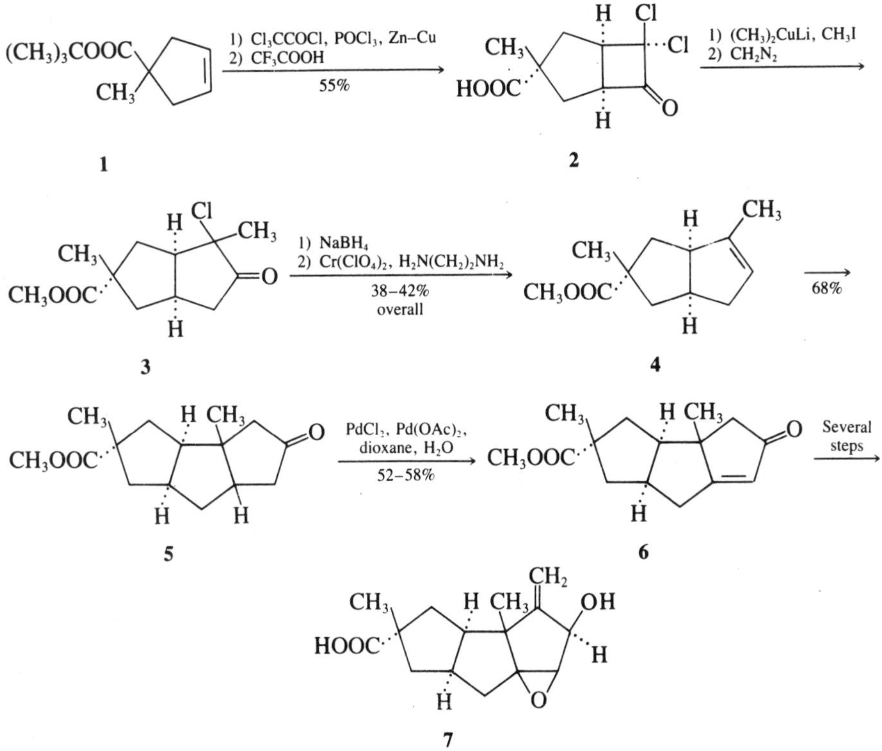

Enantiospecific addition to chiral vinyl sulfoxides.[3] The cycloaddition of dichloroketene (**11**, 169–170) to optically pure (R)- or (S)-1-cyclohexenyl tolyl sulfoxides (**1**) leads to optically pure γ-butyrolactones (**2**) with complete enantiospecificity at the two chiral centers. A third chiral *exo*-center is introduced in the addition of monochloroketene.

Example:

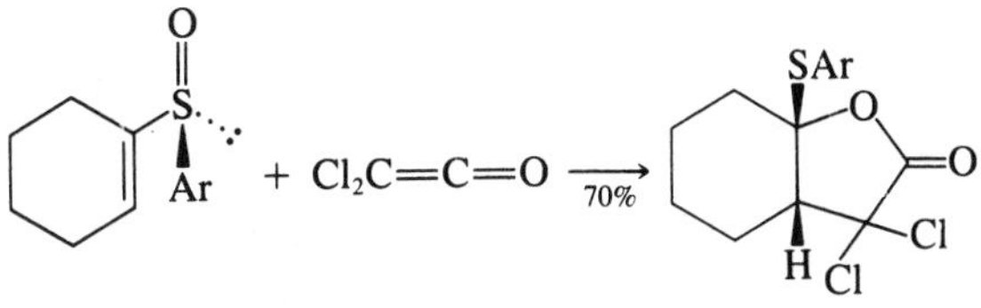

Macrolides.[4]   Reaction of α-alkenyl cyclic sulfide (**1**) with dichloroketene is accompanied by a [3.3] sigmatropic rearrangement to give the 10-membered thiolactone (**2**) (*cf.* **9**, 153–154). After dechlorination and deprotection, reaction with camphorsulfonic acid (CSA) effects an S-to-O acyl transfer to give the mercapto lactone (**3**).

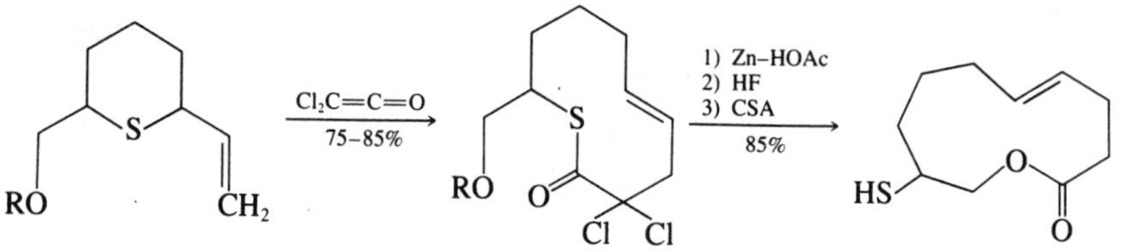

1, R = Si(CH$_3$)$_2$-*t*-Bu   **2**   **3**

[1] L. S. Liebeskind and S. L. Baysdon, *Tetrahedron Letters*, **25**, 1747 (1984).
[2] A. E. Greene, M.-J. Luche, and J.-P. Deprés, *Am. Soc.*, **105**, 2435 (1983).
[3] J. P. Marino and A. D. Perez, *ibid.*, **106**, 7643 (1984).
[4] E. Vedejs and R. A. Buchanan, *J. Org.*, **49**, 1840 (1984).

Dichlorophenylborane, C$_6$H$_5$BCl$_2$. The borane is prepared by reaction of Sn(C$_6$H$_5$)$_4$ with BCl$_3$ in refluxing benzene.[1]

syn-Selective aldol reactions.[2]   The boron enolate of ethyl ketones, obtained by reaction of C$_6$H$_5$BCl$_2$ and ethyldiisopropylamine, undergoes aldol reaction with aldehydes with high *syn*-selectivity.

Example:

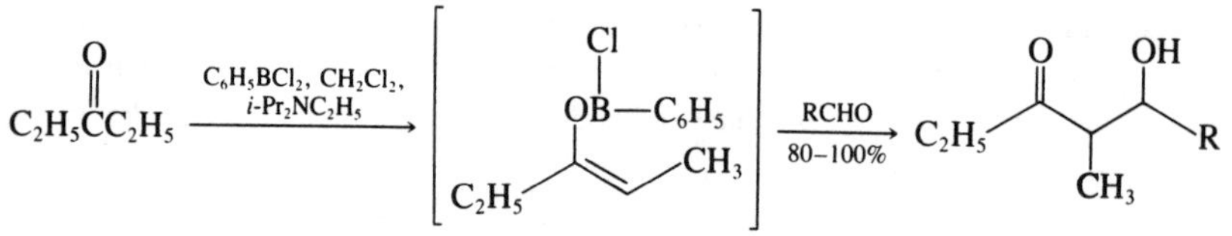

(*syn/anti* = >99:1)

2-Aminobenzhydrols.[3]   Reaction of anilines with benzaldehydes mediated by C$_6$H$_5$BCl$_2$ (1 equiv.) and N(C$_2$H$_5$)$_3$ (excess) results in 2-aminobenzhydrols as the major product.

Example:

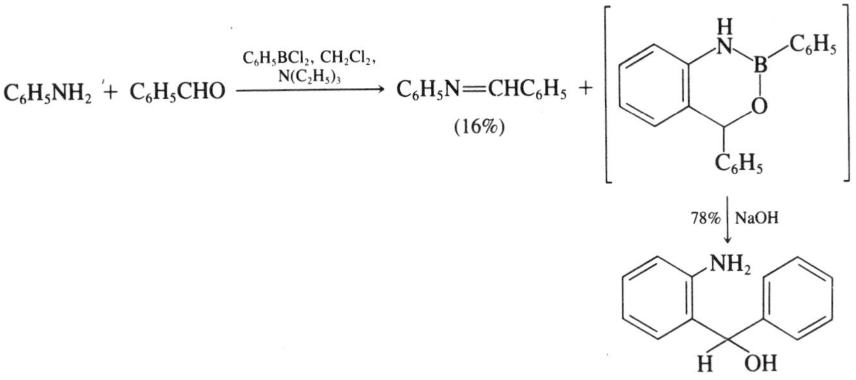

$$C_6H_5NH_2 + C_6H_5CHO \xrightarrow[\text{N}(C_2H_5)_3]{C_6H_5BCl_2, \ CH_2Cl_2,} C_6H_5N{=}CHC_6H_5 + \left[\ \right]$$

(16%)

78% | NaOH

[structure diagrams]

1. K. Niedenzu and J. W. Dawson, *Am. Soc.*, **82**, 4223 (1960).
2. H. Hamana, K. Sasakura, and T. Sugasawa, *Chem. Letters*, 1729 (1984).
3. T. Toyoda, K. Sasakura, and T. Sugasawa, *Tetrahedron Letters*, **21**, 173 (1980).

1,4-Dichloro-1,1,4,4-tetramethyldisilylethylene (1), 10, 140–141.

Pyrrolidines.[1]   The Grignard reagent **2**, obtained from 3-bromopropylamine protected as the stabase adduct with **1**, reacts with the N-methoxy-N-methyl amides **3** (**11**, 201–202) to form an intermediate ketone that cyclizes to an imine on liberation of the free primary amino group. Reduction of the imine results in a 2-substituted pyrrolidine (**4**).

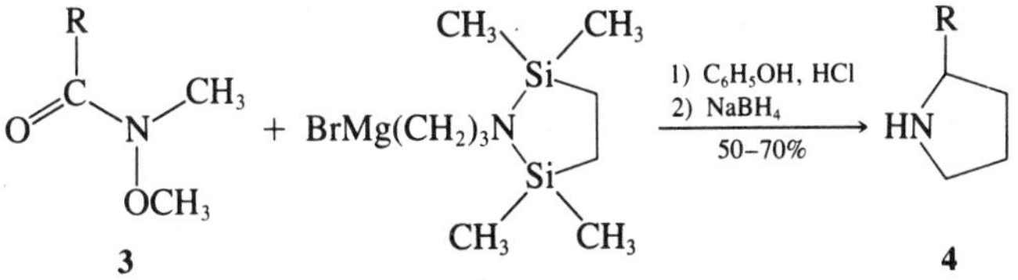

1. F. Z. Basha and J. F. DeBernardis, *Tetrahedron Letters*, **25**, 5271 (1984).

Dichlorotris(triphenylphosphine)ruthenium(II), 4, 564; 5, 740–741; 6, 654–655; 7, 99; 8, 159–161; 10, 141–142; 11, 171–172.

Reduction of nitroarenes.[1]   Nitrobenzenes substituted with chloro, methyl, or methoxy groups can be reduced to the corresponding anilines by formic acid and this ruthenium(II) complex in yields generally greater than 90%. Alcoholic solvents increase the rate. A base, in particular triethylamine, is necessary for a high degree of conversion. The same system also reduces azaarenes such as quinoline and indole.

Examples:

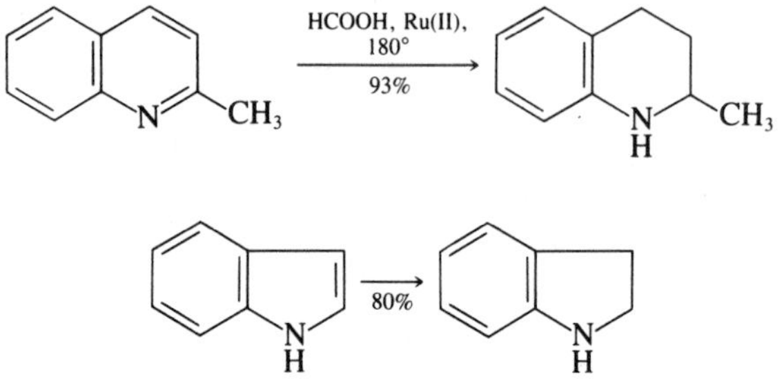

[1] Y. Watanabe, T. Ohta, Y. Tsuji, T. Hiyoshi, and Y. Tsuji, *Bull. Chem. Soc. Japan*, **57**, 2440 (1984).

Dicyclopentadienylcobalt (cobaltocene), $(C_5H_5)_2Co$ (**1**). Supplier: Strem.

Cycloaddition of nitriles with dipropargyl ethers.[1]   A new synthesis of pyridoxine (**4**, vitamin B_6) is based on the [2 + 2 + 2] cycloaddition of acetonitrile with dipropargyl ethers catalyzed by Cp_2Co or $Cp(CO)_2Co$ to form the pyridine derivative **1**. Subsequent steps involve rearrangement of the N-oxide of **1** to the 3-hydroxypyridine **2** with acetic anhydride. The final step involves the known cleavage of the dihydrofurane ring.

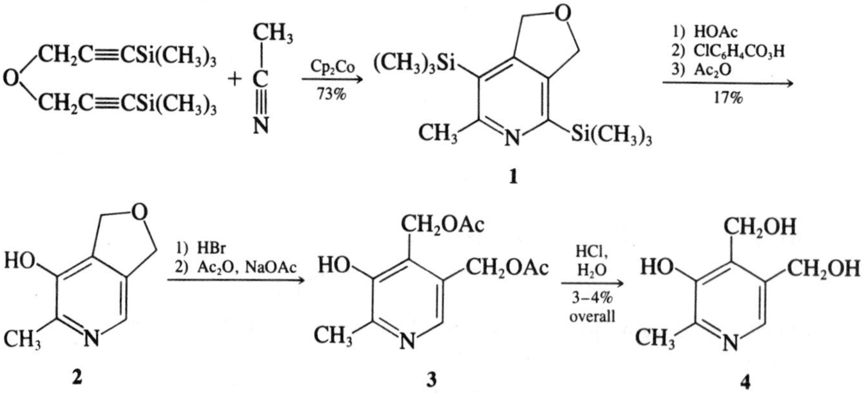

[1] R. E. Greiger, M. Lalonde, H. Stoller, and K. Schleich, *Helv.*, **67**, 1274 (1984).

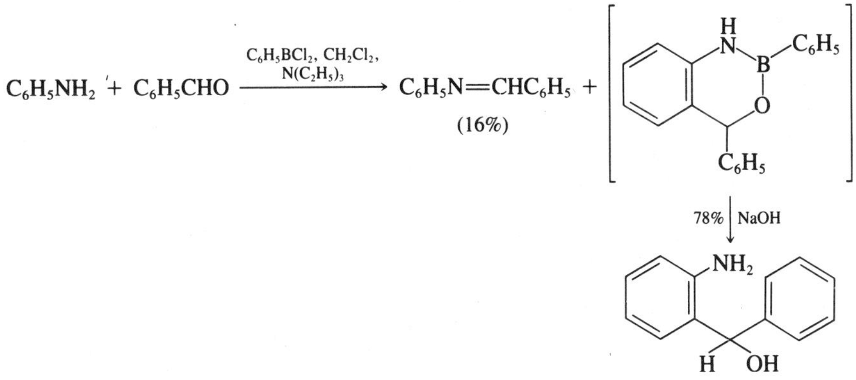

$$78\% \downarrow \text{NaOH}$$

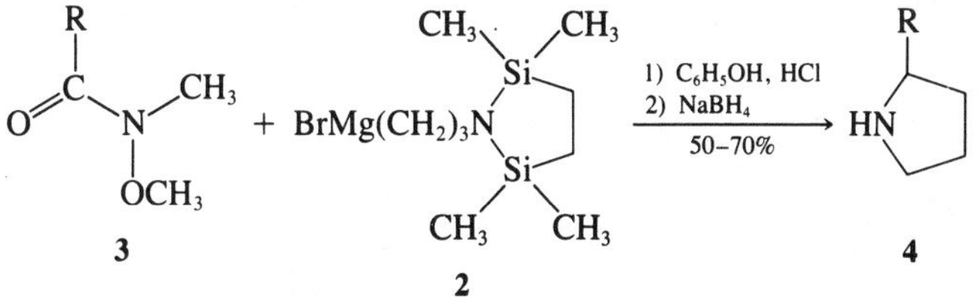

1,2-Diethoxy-1,2-disilyloxyethene,

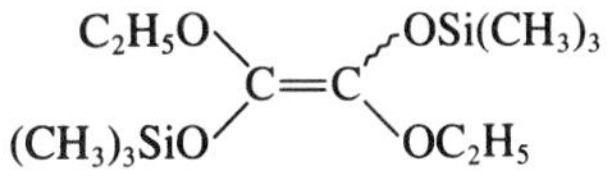

$$C_2H_5O \diagdown \diagup OSi(CH_3)_3$$
$$ C=C$$
$$(CH_3)_3SiO \diagup \diagdown OC_2H_5$$

(1). The reagent is prepared as a (Z/E)-mixture by reaction of diethyl oxalate with $ClSi(CH_3)_3$ and Na/K (25% yield).[1]

α-*Keto esters.*   The reagent in the presence of $ZnCl_2$ reacts with aldehydes or ketones to give a protected form of α-keto esters (equation I).

(I)

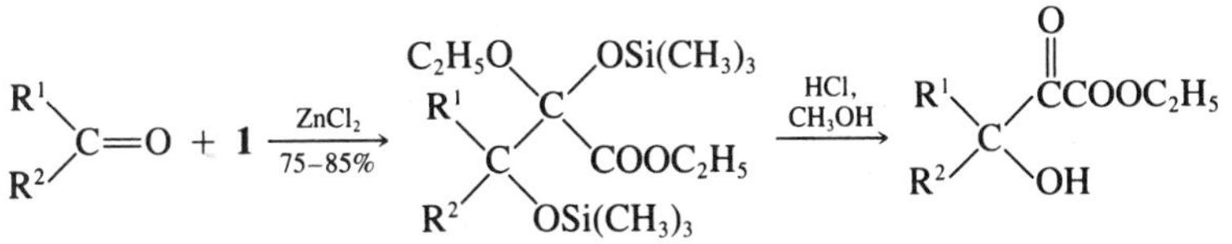

The reagent undergoes 1,4-addition to α,β-enones (equation II).[2]

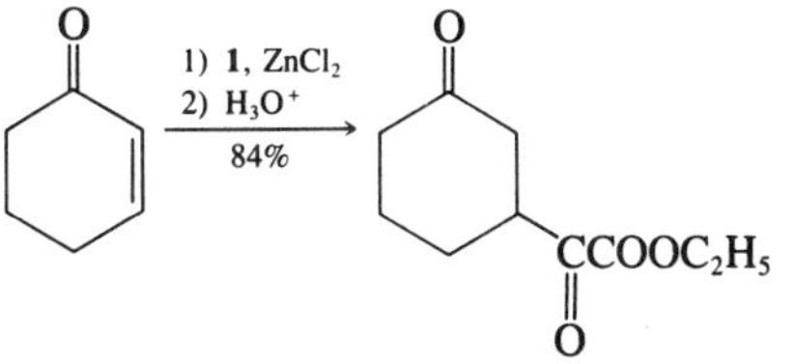

[1] Y. N. Kuo, F. Chen, C. Ainsworth, and J. J. Bloomfield, *J.C.S. Chem. Comm.*, 136 (1971).
[2] M. T. Reetz, H. Heimbach, and K. Schwellnus, *Tetrahedron Letters*, **25**, 511 (1984).

Diethoxymethyldiphenylphosphine oxide, $(C_6H_5)_2\overset{\overset{O}{\|}}{P}-CH(OC_2H_5)_2$ (1). Mol. wt. 304.4 m.p. 77–79°. The oxide is obtained in quantitative yield by reaction of chlorodiphenylphosphine with ethyl orthoformate (110°).[1]

Ketene diethyl ketals.   These useful intermediates can be obtained by a Wittig-Horner reaction of the anion of **1** with aldehydes and ketones. Higher yields are obtained when the adducts are isolated and then treated with potassium *t*-butoxide.[2]

Examples:

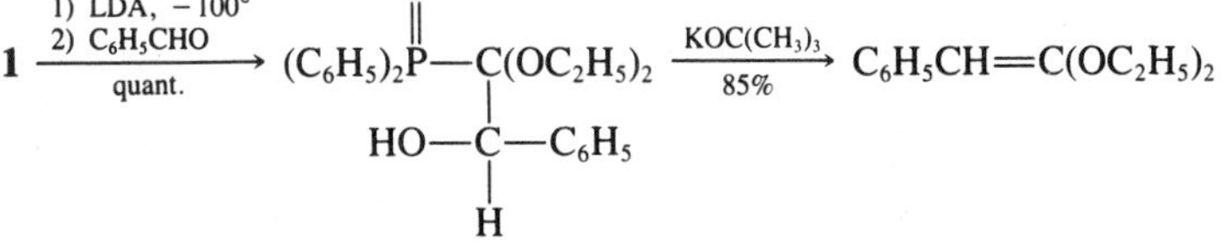

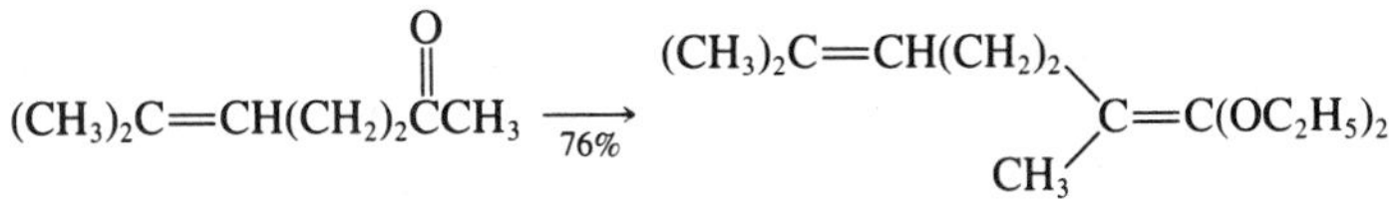

[1] W. Dietsche, *Ann.*, **712**, 21 (1968).
[2] T. A. M. van Schaik, A. V. Henzen, and A. van der Gen, *Tetrahedron Letters*, **24**, 1303 (1983).

Diethoxymethylsilane, $CH_3SiH(OC_2H_5)_2$ (**1**). Mol. wt. 134.25, b.p. 94.5°. Supplier: Petrarch.

Hydrosilylation.[1]   This silane is useful for hydrosilylation of alkenes and alkynes because the C—Si bond of the adducts is oxidized by 30% H_2O_2 in the presence of KF, KHF_2, or $NaHCO_3$ with formation of the corresponding alcohol. The oxidation occurs with retention of configuration at carbon. At least one alkoxy group on silicon is necessary for this oxidation. Hydrosilylation followed by oxidation permits conversion of 1-alkenes to *anti*-Markownikoff alcohols (equation I) and of internal alkynes to ketones (equation II).

(I) $CH_3OOC(CH_2)_8CH{=}CH_2 \xrightarrow{\textbf{1}, \text{ RhCl[P(C}_6\text{H}_5)_3]_3} CH_3OOC(CH_2)_{10}SiCH_3(OC_2H_5)_2$

$$\xrightarrow[\substack{79\% \\ \text{overall}}]{\substack{H_2O_2, \text{ KHF}_2, \\ \text{DMF}}} CH_3OOC(CH_2)_{10}OH$$

(II) $n\text{-}C_4H_9C{\equiv}CC_4H_9\text{-}n \xrightarrow[82\%]{\substack{1) \textbf{ 1}, \text{ H}_2\text{PtCl}_6\cdot6\text{H}_2\text{O} \\ 2) \text{ H}_2\text{O}_2, \text{ KHF}_2}} n\text{-}C_4H_9CH_2\overset{\displaystyle O}{\overset{\|}{C}}C_4H_9\text{-}n$

[1] K. Tamao, N. Ishida, T. Tanaka, and M. Kumada, *Organometallics*, **2**, 1694 (1983).

Diethylaluminum cyanide, $(C_2H_5)_2AlCN$. Mol. wt. 111.12. Supplier: Alfa.

Cyanomethyl ethers.[1]   Methoxyethoxymethyl (MEM) ethers are converted into cyanomethyl ethers by reaction with excess diethylaluminum cyanide in toluene at 100° (equation I). The same reaction with methoxymethyl (MOM) ethers is considerably slower.

(I) $ROCH_2OCH_2CH_2OCH_3 + (C_2H_5)_2AlCN \xrightarrow[65-80\%]{} ROCH_2CN$

[1] E. J. Corey, D. H. Hua, and S. P. Seitz, *Tetrahedron Letters*, **25**, 3 (1984).

2-Diethylamino-4-phenylsulfonyl-2-butenenitrile (1), 11, 176.

Saturated and α,β-*unsaturated esters* (*cf.* **11**, 176).[1]   The reagent is alkylated under phase-transfer conditions selectively at the γ-position by primary or secondary alkyl iodides and by benzylic or allylic bromides. The products are convertible into saturated and unsaturated esters.

Example:

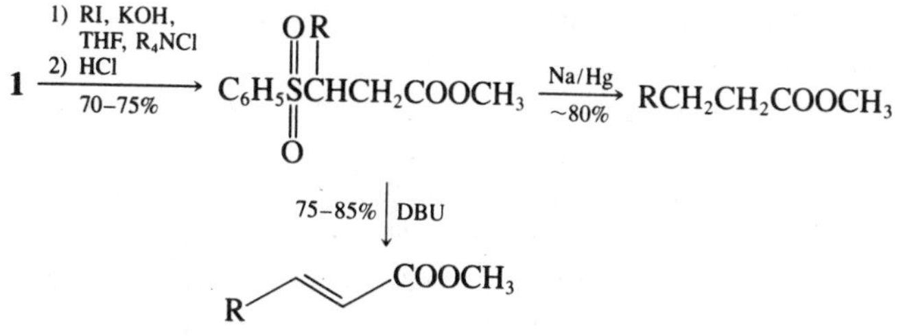

$$1 \xrightarrow[\substack{70-75\%}]{\substack{1)\ RI,\ KOH,\\THF,\ R_4NCl\\2)\ HCl}} \underset{\overset{\|}{O}}{C_6H_5\overset{\overset{O}{\|}}{S}}CHCH_2COOCH_3 \xrightarrow[\sim80\%]{Na/Hg} RCH_2CH_2COOCH_3$$

75–85% | DBU

R⌒⌒COOCH₃

¹ S. De Lombaert and L. Ghosez, *Tetrahedron Letters*, **25**, 3475 (1984).

N,N-Diethylaminopropyne, 2, 133–134; **3**, 98; **5**, 217–219; **7**, 107–108; **9**, 164–165.

*Stereoselective addition to an α,β-enone.*¹   4-Methylcyclopentenone (**1**) reacts with this reagent to give the adduct **2** with 97% stereoselectivity. The adduct can be converted into 3,4-disubstituted cyclopentanones with high stereochemical control at three centers.

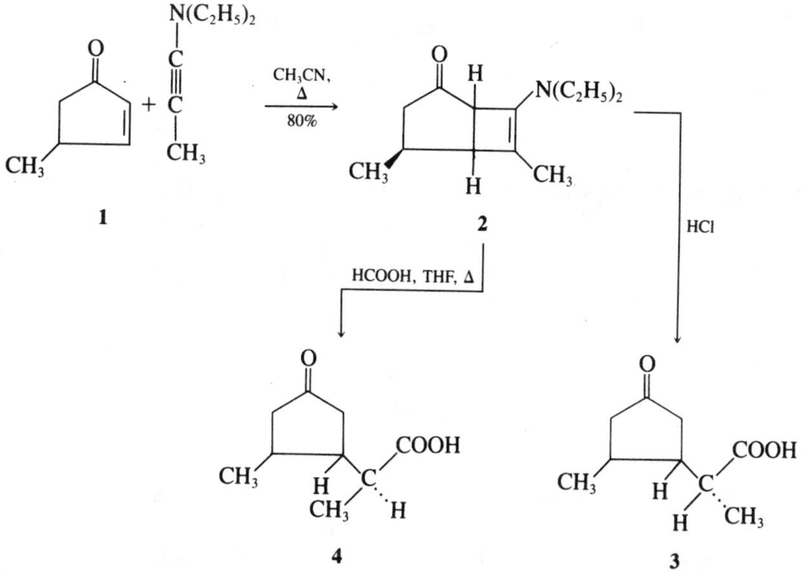

¹ J. Ficini, D. Desmaële, and A.-M. Touzin, *Tetrahedron Letters*, **24**, 1025 (1983).

(Diethylamino)sulfur trifluoride (DAST), **6**, 183–184; **8**, 166–167; **10**, 142; **11**, 177.

*Glycosyl fluorides.*¹   Phenylthioglycosides are converted into glycosyl fluorides under mild conditions by reaction with DAST and NBS or with pyridinium poly(hydrogen fluoride) and NBS. Yields range from 70 to 90%, being somewhat higher with DAST. The method is compatible with many functional groups, including O-glycoside bonds. An anomeric mixture is usually obtained, but both anomers form mainly α-glycosides on glycosidation with $SnCl_2$ or $AgClO_4$ (**10**, 374).

Oligosaccharides.[2]   Glycosyl fluorides are useful substrates for synthesis of oligosaccharides. The *t*-butyldiphenylsilyl group is recommended for protection of the hydroxyl group, since it is removed selectively by fluoride ion. This method is effective for preparation of the $C_1 \rightarrow C_6$ linked hexasaccharide **2** from the glucose derivative **1**. A disaccharide is obtained by coupling the glycosyl fluoride formed from **1** with the alcohol obtained by deprotection of **1** with fluoride ion. Similar coupling of the disaccharide results in a tetrasaccharide, which is then coupled with the disaccharide to give **2**. The yields decrease somewhat as the chain is lengthened, being about 65% for coupling to the hexasaccharide.

1

2

C-Glycosides.[3]   Glycosyl fluorides react with a number of nucleophiles to provide C-glycosides. Lewis acid catalysts are not essential for reaction with $Al(CH_3)_3$ or $Al(CH_3)_2CN$, but $MgBr_2$ is useful in reactions with Michael acceptors such as $CH_2{=}CHCN$. The products are reduced by alane to tetrahydropyranes. They also react with S- and N-nucleophiles to afford the corresponding glycosides.

α,β-Dehydroamino acids.[4]   N-Protected (Cbo and Boc) β-hydroxy-α-amino acid esters (ethyl, benzyl) are converted into the corresponding α,β-dehydroamino acid derivatives by DAST and pyridine in CH_2Cl_2 at 0° (65–90% yield). The hydroxy group is probably converted into the —$OSF_2N(C_2H_5)_2$ derivative, which undergoes *trans*-elimination. Thus *threo*-substrates are converted into (Z)-alkenes and *erythro*-substrates into (E)-alkenes.

[1] K. C. Nicolaou, R. E. Dolle, D. P. Papahatjis, and J. L. Randall, *Am. Soc.*, **106**, 4189 (1984).
[2] K. C. Nicolaou, S. P. Seitz, and D. P. Papahatjis, *ibid.*, **105**, 2430 (1983).
[3] K. C. Nicolaou, R. E. Dolle, A. Chucholowski, and J. L. Randall, *J.C.S. Chem. Comm.*, 1153, 1155 (1984).
[4] L. Somekh and A. Shanzer, *J. Org.*, **48**, 907 (1983).

Diethyl N-benzylideneaminomethylphosphonate (1), 9, 161–162; **11**, 178–179.

gem-*Acylation–alkylation of carbonyl compounds.*[1] This reaction provides a key step in a synthesis of the Amaryllidaceae alkaloid haemanthidine (**6**) from piperonal. Thus addition of *n*-butyllithium to the azadiene **3**, obtained by reaction of the anion of **1** with the ketone **2**, followed by geminal hydroxyalkylation and acylation furnishes the aldehyde

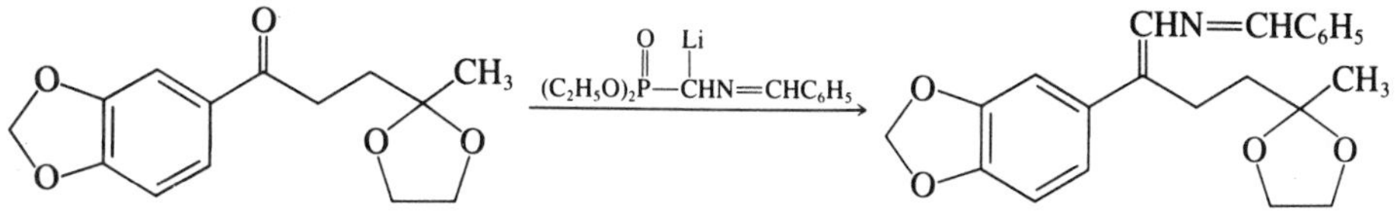
O
O
O
CH3
O
O
O Li
(C2H5O)2P—CHN=CHC6H5
CHN=CHC6H5
CH3

2 **3**

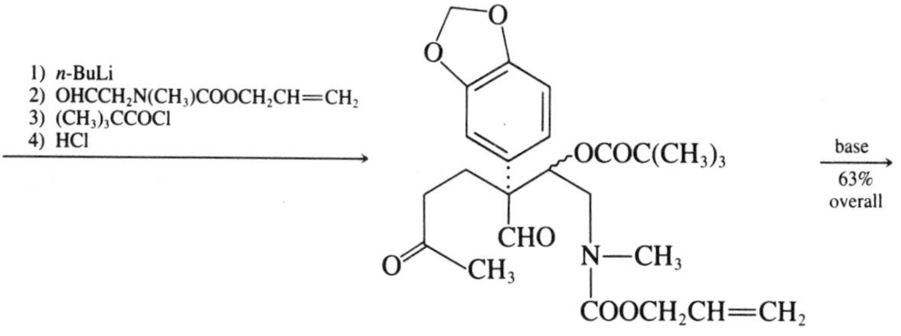
1) n-BuLi
2) OHCCH2N(CH3)COOCH2CH=CH2
3) (CH3)3CCOCl
4) HCl
OCOC(CH3)3
CHO
CH3
N—CH3
COOCH2CH=CH2
base
63%
overall

4

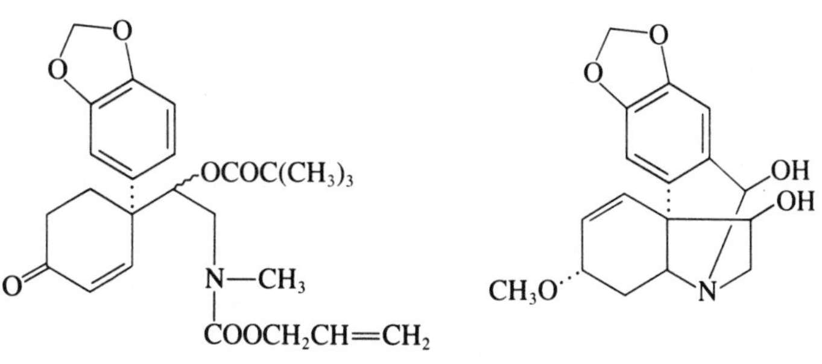
OCOC(CH3)3
N—CH3
COOCH2CH=CH2
OH
OH
CH3O
N

5 **6**

4 with a quaternary center. Base-catalyzed cyclization of **4** provides a 4,4-disubstituted cyclohexenone **5**. Remaining steps to **6** include cyclization to a hydroindole and a Bischler-Napieralski cyclization.

[1] S. F. Martin and S. K. Davidsen, *Am. Soc.*, **106**, 6431 (1984).

Diethyl (diazomethyl)phosphonate (1), 9, 181.

Cyclopentyne. The anion (**2**) of **1**, generated with KH at $-78°$, reacts with cyclobutanone to generate cyclopentyne (**b**), which can be trapped by [2 + 2] cycloaddition with an alkene.[1]

(I) (structure) $=O + (C_2H_5O)_2\overset{O}{\overset{\|}{P}}C^-N_2 \overset{K^+}{\longrightarrow}$ (structure) $\longrightarrow$ [(structure) $=CN_2 \longrightarrow$ (structure)] $\xrightarrow[28\%]{}$ (structure **3**)

2   **a**   **b**

3

Cyclopentyne has also been generated with comparable efficiency from dibromomethylenecyclobutane (**4**, equation II).[2] These two routes are more efficient than the previously preferred route, debromination of 1,2-dibromocyclopentene with *n*-butyllithium.[3]

(II) (structure) $=O \xrightarrow[82\%]{P(C_6H_5)_3,\ Br_4}$ (structure) $=CBr_2 \xrightarrow[-LiBr]{C_6H_5Li,}$ [(structure)] $\xrightarrow{32\%}$ **3**

4   **b**

The cycloaddition of cyclopentyne is stereospecifically *cis* to both *cis*- and *trans*-alkenes.[1,2] Only [2 + 2] cycloaddition is observed in the reaction of cyclopentyne with 1,3-butadiene (equation III).

(III) **4** $\longrightarrow$ [(structure)] $\xrightarrow[36\%]{-4°}$ (structure with $CH=CH_2$) $\xrightarrow[100\%]{180°}$ (structure with CH_2)

b

[1] J. C. Gilbert and M. E. Baze, *Am. Soc.*, **105**, 664 (1983); *idem, ibid.*, **106**, 1885 (1984).
[2] L. Fitjer, U. Kliebisch, D. Wehle, and S. Modaressi, *Tetrahedron Letters*, **23**, 1661 (1982); L. Fitjer and S. Modaressi, *ibid.*, **24**, 5495 (1983).
[3] G. Wittig and J. Heyn, *Ann.*, **726**, 57 (1969).

Diethyl phosphite (Diethyl phosphonite), 1, 251–253; 2, 132–133.

Reduction of bromides.   gem-Dibromocyclopropanes and gem-dibromoalkenes are reduced to the monobromides by $(C_2H_5O)_2P(O)H$ and $N(C_2H_5)_3$.[1] Under these conditions gem-bromochlorocyclopropanes are reduced to chlorocyclopropanes.[2] 1,1-Dibromo-2-tri-methylsilyloxycyclopropanes or α-bromo-α,β-enones are reduced in this way to β,γ-enones.[3] Other examples indicate that only activated halogen atoms are subject to this reduction.

Examples:

$$C_6H_5CH{=}CBrCOOCH_3 \xrightarrow[69\%]{} C_6H_5CH{=}CHCOOCH_3$$

[1] T. Hirao, T. Masunaga, Y. Ohshiro, and T. Agawa, *J. Org.*, **46**, 3745 (1981).
[2] T. Hirao, S. Kohno, Y. Ohshiro, and T. Agawa, *Bull. Chem. Soc. Japan*, **56**, 1881 (1983).
[3] T. Hirao, T. Masunaga, K. Hayashi, Y. Ohshiro, and T. Agawa, *Tetrahedron Letters*, **24**, 399 (1983).

Diethyl phosphorocyanidate, 5, 217; 6, 192–193; 7, 107; 9, 163–164; 10, 145; 11, 181.

α,β-Unsaturated nitriles.[1]   The reagent reacts with aromatic ketones in the presence of lithium cyanide to give a cyanophosphate, which is converted into an α,β-unsaturated nitrile by treatment with boron trifluoride etherate. The adducts of aliphatic ketones are stable under the same conditions.

Examples:

Aryl acetonitriles.[2] The aryl cyanophosphates obtained by reaction of aryl ketones or aldehydes (80–100% yield) are converted into aryl acetonitriles by hydrogenolysis using Pd/C. This method was used to convert the ketone **1** to ibuprofen (**2**) in 81% overall yield.

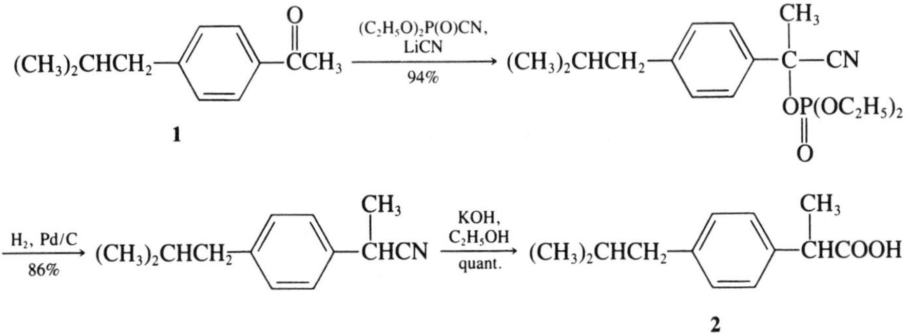

[1] S. Harusawa, R. Yoneda, T. Kurihara, Y. Hamada, and T. Shioiri, *Tetrahedron Letters*, **25**, 427 (1984).
[2] S. Harusawa, S. Nakamura, S. Yagi, T. Kurihara, Y. Hamada, and T. Shioiri, *Syn. Comm.*, **14**, 1365 (1984).

Diethyl [(2-tetrahydropyranyloxy)methyl]phosphonate (1), 11, 181–182.

Homologation of ketones to aldehydes (*cf.* **9**, 162–163).[1] The reagent is particularly useful for conversion of an α,β-unsaturated ketone into the homologated β,γ-unsaturated aldehyde (equation I).

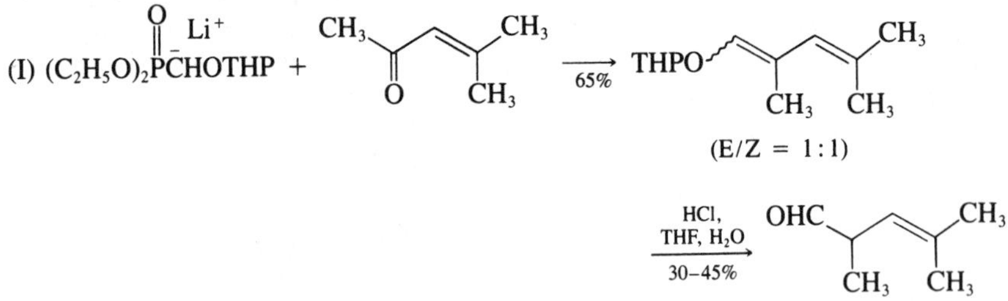

[1] S. D. Young and C. H. Heathcock, *Org. Syn.*, submitted (1982).

***gem*-Difluoroallyllithium,** $LiCF_2CH{=}CH_2$ (**1**), **10**, 188. The reagent is prepared *in situ*, preferably by lithium–halogen exchange (equation I) rather than by transmetallation.

$$\text{(I)}\ CF_2Br_2 + CH_2{=}CH_2 \xrightarrow[78\%]{Bz_2O_2} BrCH_2CH_2CF_2Br \xrightarrow{n\text{-BuLi, THF, ether, } -95^\circ} \mathbf{1}$$

Difluoroallylation.[1]   The reagent reacts with aldehydes, ketones, and esters regioselectively to give products in which the new C—C bond is formed at the CF_2 terminus of the allyl system.

Examples:

$$(C_2H_5)_2C{=}O \xrightarrow[70\%]{1} (C_2H_5)_2\underset{\underset{OH}{|}}{C}CF_2CH{=}CH_2$$

$$(C_2H_5)_3SiCl \xrightarrow[51\%]{} (C_2H_5)_3SiCF_2CH{=}CH_2$$

[1] D. Seyferth, R. M. Simon, D. J. Sepelak, and H. A. Klein, *Am. Soc.*, **105**, 4634 (1983).

2,3-Dihydrofurane,  **(1).**   Mol. wt. 70.09, b.p. 55°, flammable.

4-Hydroxybutylation of hydroquinones.[1]   Reaction of hydroquinones with **1** in the presence of D-camphor-10-sulfonic acid gives monotetrafuryl ethers (**2**), which rearrange to 2-tetrahydrofurylhydroquinones (**3**) in the presence of additional amounts of the acid

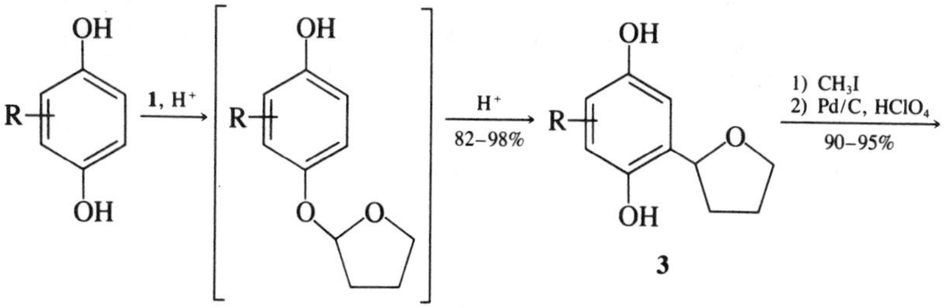

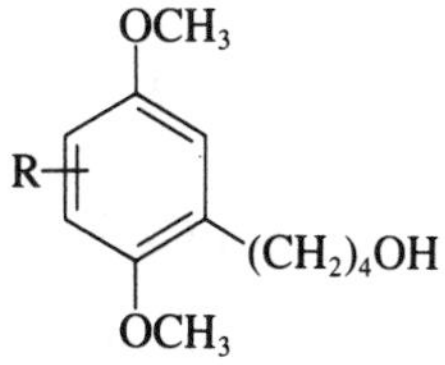

catalyst. Methylation followed by hydrogenolysis provides 4-hydroxybutylhydroquinone ethers (**4**). These products can be converted into a variety of quinones substituted by alkenyl or alkynyl groups, some of which are of interest because they can inhibit formation of leukotrienes. One of the most potent is the quinone **5**.

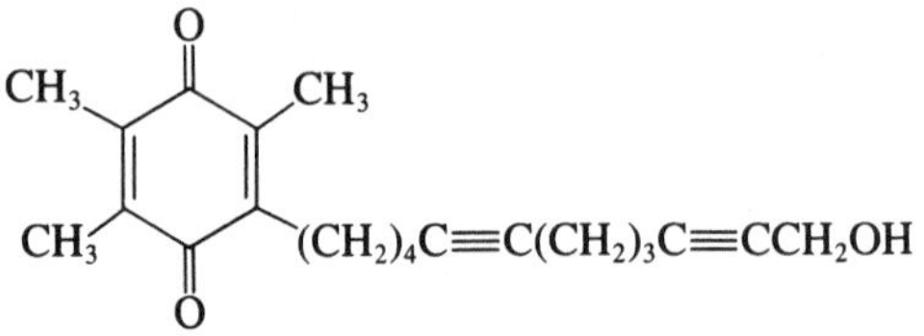

5

[1] M. Shiraishi and S. Terao, *J.C.S. Perkin I*, 1591 (1983).

2,2'-Dihydroxy-1,1'-binaphthyl (1), **9**, 169–170; **10**, 148–149. The (S)-(−)-diol (**1**) can be obtained in high optical purity (~95% ee) by dimerization of β-naphthol with the complex of (S)-(+)-amphetamine with $Cu(NO_3)_2$. The chemical yield is 85%.[1]

Enantioselective reduction of ketones.[2]   Full details are available for high enantio-selective reduction of prochiral aryl, alkynyl, and alkenyl ketones with the complex of lithium aluminum hydride, (R)- or (S)-**1**, and ethanol, (R)- or (S)-BINAL-H (**2**). The enantioface differentiation depends more on electronic than on steric differences between the substituents attached to the carbonyl group. Thus these reducing agents can be more effective than those dependent on steric effects. An example is the reduction of the prostaglandin intermediate **3** to the desired allylic alcohol **4** in 100% ee with (S)-**2**. An optical yield of 92% had been obtained previously by using a bulky blocking group at C_{11} and a bulky trialkylborohydride (**4**, 104).

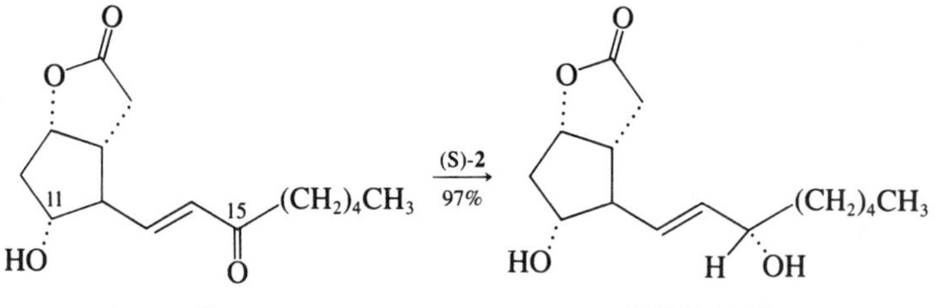

3   4 (15S, 100% ee)

This reduction is useful for synthesis of optically active styrene oxide (equation I).

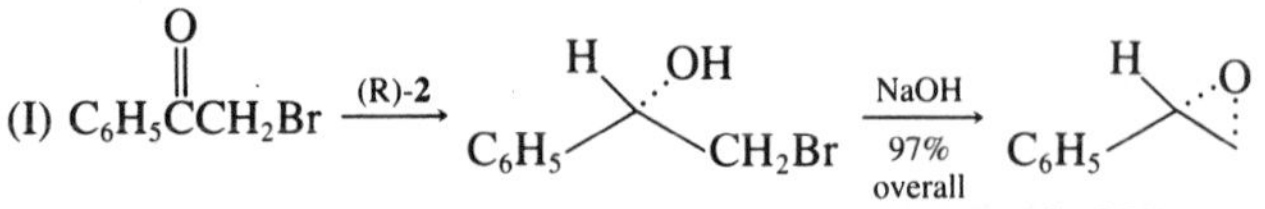

(S, 95% ee)

[1] J. Brussee and A. C. A. Jansen, *Tetrahedron Letters*, **24**, 3261 (1983).
[2] R. Noyori, I. Tomino, Y. Tanimoto, and M. Nishizawa, *Am. Soc.*, **106**, 6709 (1984); R. Noyori, I. Tomino, M. Yamada, and M. Nishizawa, *ibid.*, 6717 (1984).

(S)-(−)-10,10′-Dihydroxy-9,9′-biphenanthryl (1). The reagent is obtained by oxidative coupling of 9-phenanthrol in the presence (R)-1,2-diphenylethylamine with $Cu(NO_3)_2 \cdot 3H_2O$.

(S)-**1**

Asymmetric hydride reduction.[1]   The complex (**2**) prepared *in situ* from **1**, $LiAlH_4$, and ethanol (1:1:1 ratio) reduces alkyl aryl ketones to the corresponding (S)-alcohols in 85–98% ee. Optical yields are only moderate (20–35%) in reductions of aliphatic ketones. The complex is comparable to the related complex obtained using optically active 2,2′-dihydroxy-1,1′-binaphthyl (**9**, 169–170).

[1] K. Yamamoto, H. Fukushima, and M. Nakazaki, *J.C.S. Chem. Comm.*, 1490 (1984).

Diisobutylaluminum hydride, 1, 260–262; **2,** 140–142; **3,** 101–102; **4,** 158–161; **5,** 224–225; **6,** 198–201; **7,** 111–113; **8,** 173–174; **9,** 171–172; **10,** 149; **11,** 185–188.

Ate complex with n-BuLi (2).   This complex (**2**) reduces ketones, esters, acid chlorides, and acid anhydrides readily, even at −78°. Consequently, it is useful for selective reduction of these groups in the presence of halide, amide, and nitrile groups, which are inert at low temperatures. It is even possible to reduce a ketone selectively in the presence of an ester group by use of 1 equiv. of reagent at −78°.

Reduction of oximes.[2]   Oximes of aliphatic ketones are reduced by DIBAH to rearranged secondary amines in 70–90% yield.

Example:

Reaction with chiral acetals. The chiral ketals derived from $(2R,4R)-(-)-2,4-$pentanediol (**1**) can be cleaved with high diastereoselectivity by aluminum hydride reagents, in particular DIBAH, Cl_2AlH, and Br_2AlH. Oxidative removal of the chiral auxiliary affords optically active alcohols. This process provides a useful method for highly asymmetric reduction of dialkyl ketones.[3]

Example:

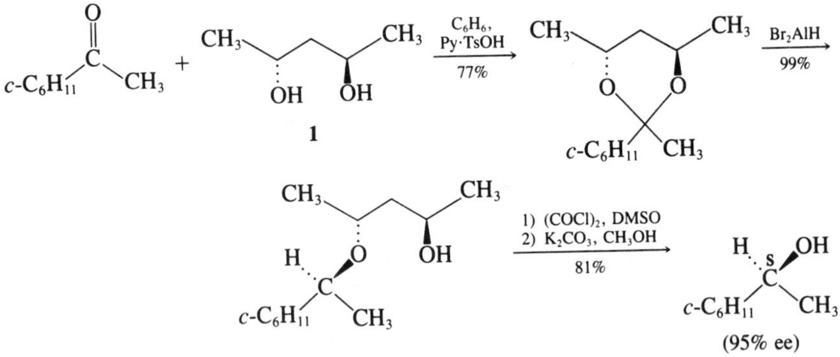

The chiral acetals of α,β-enals derived from $(R,R)-(+)-N,N,N',N'$-tetramethyltartaric acid diamide (**9**, 47–48)[4] undergo either 1,4- or 1,2-addition of R_3Al with high asymmetric induction. The course of reaction can be controlled by the choice of solvent. 1,4-Addition is favored in 1,2-dichloroethane (or toluene); 1,2-addition is the main or only reaction in chloroform. The adducts can be converted into optically active β-alkyl aldehydes or allylic alcohols (Chart I).[5]

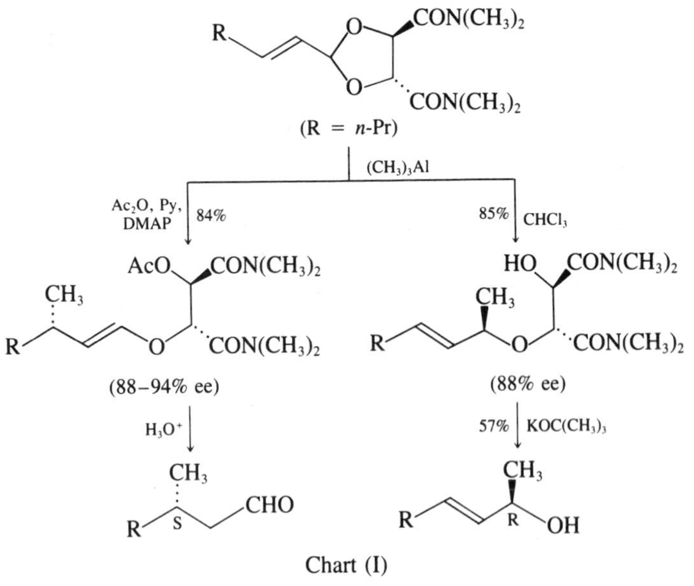

Chart (I)

Δ¹-*Pyrrolines*; Δ¹-*piperideines*.[6] A convenient, general synthesis of Δ^1-pyrrolines is based on the reduction of suitable nitriles with DIBAH (equation I).

Δ^1-Piperideines can also be prepared by this method (equation II).

[1] S. Kim and K. H. Ahn, *J. Org.*, **49**, 1717 (1984).
[2] S. Sasatani, T. Miyazaki, K. Maruoka, and H. Yamamoto, *Tetrahedron Letters*, **24**, 4711 (1983).
[3] A. Mori, J. Fujiwara, K. Maruoka, and H. Yamamoto, *ibid.*, 4581 (1983).
[4] D. Seebach, H.-O. Kalinowski, W. Langer, G. Crass, and E.-M. Wilka, *Org. Syn.*, **61**, 24 (1983).
[5] J. Fujiwara, Y. Fukutani, M. Hasegawa, K. Maruoka, and H. Yamamoto, *Am. Soc.*, **106**, 5004 (1984).
[6] L. E. Overman and R. M. Burk, *Tetrahedron Letters*, **25**, 5737 (1984).

Diisopinocampheylborane (Ipc₂BH) (1), **1**, 262–263; **4**, 161–162; **6**, 202; **8**, 174; **11**, 188. Two methods have been reported for preparation of either (+)- or (−)-Ipc₂BH of high optical purity from commercially available (−)- or (+)-α-pinene of lower optical purity. Essentially pure material (99% ee) selectively crystallizes from the reaction of α-pinene with borane–dimethyl sulfide in THF–diethyl ether after storage at 0° for several hours. Somewhat purer reagent (99.9% ee) can be prepared by hydroboration of (+)-α-pinene (84 or 91.6% ee) with optically pure (−)-monoisocampheylborane (**8**, 267).[1]

Asymmetric hydroboration.[2] Hydroboration of the chiral cyclopentene **2** with a number of hindered boranes followed by alkaline hydrolysis usually results in three hydroxy ketones. The most regio- and diastereoselective reaction is observed with (+)-diisopi-nocampheylborane, use of which results in only two products in a ratio of 95:5. In

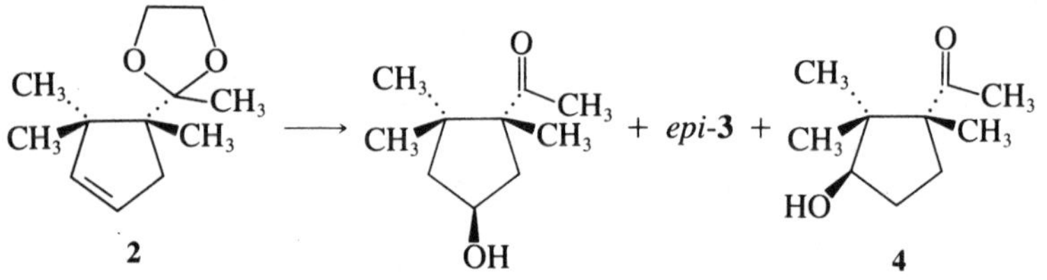

Diborane	59%	48:11:4
Disiamylborane	84%	75:5:20
9-BBN	17%	71:29:0
(−)-1	95%	50:9:41
(+)-1	97%	95:0:5

contrast, (−)-1 shows slight regioselectivity. The product 3 was used for synthesis of some optically active carotenoids found in red pepper, such as capsorubin (5).

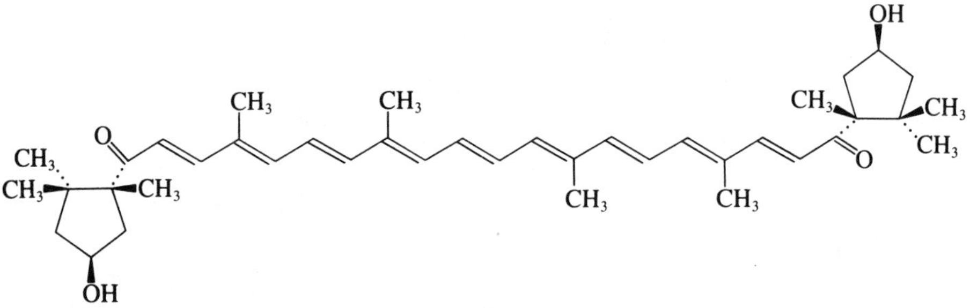

5

[1] H. C. Brown and B. Singaram, *J. Org.*, **49**, 945 (1984).
[2] A. Rüttimann, G. Englert, H. Mayer, G. P. Moss, and B. C. L. Weedon, *Helv.*, **66**, 1939 (1983).

Diisopropyl peroxydicarbonate (1), 1, 263–264; 2, 142.

Quinoline synthesis.[1]   The arylimidoyl radical (a) generated by treatment of N-benzylideneaniline (2) with 1 reacts with 1-alkynes to form 4-alkylquinolines (3). This reaction provides a useful general route to substituted quinolines (yields 15–85%).

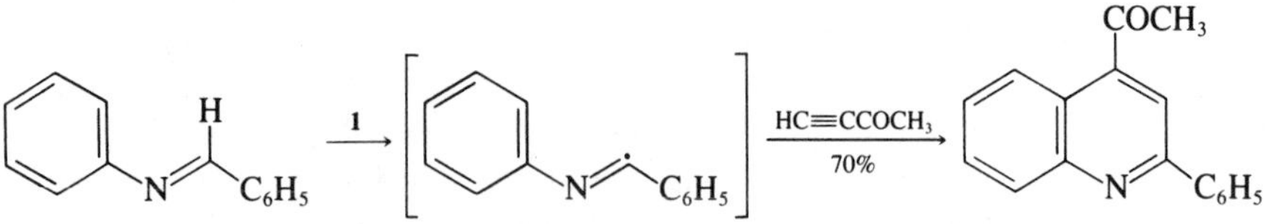

[1] R. Leardini, G. F. Pedulli, A. Tundo, and G. Zanardi, *J.C.S. Chem. Comm.*, 1320 (1984).

Diisopropyl sulfide–N-Chlorosuccinimide.

Oxidation of primary and secondary alcohols.[1]   In contrast to dimethyl sulfide–NCS (**4**, 88–90), which oxidizes both primary and secondary alcohols, diisopropyl sulfide–NCS can effect selective oxidation of these substrates. At 0° it oxidizes only primary alcohols to aldehydes, but at −78° it oxidizes only secondary alcohols to ketones. However, allylic or benzylic alcohols are oxidized at either temperature.

[1] K. S. Kim, I. H. Cho, B. K. Yoo, Y. H. Song, and C. S. Hahn, *J.C.S. Chem. Comm.*, 762 (1984).

Dilithium tetrabromonickelate(II), Li_2NiBr_4. The combination of LiBr (2 equiv.) and $NiBr_2$ (1 equiv.) dissolves in THF to form a dark green solution of Li_2NiBr_4.

Bromohydrins.[1]   The reagent reacts with epoxides regioselectively to yield a *trans*-bromohydrin with bromine predominantly at the less hindered position. Acid- and base-sensitive functional groups are stable to the reagent.

Examples:

[1] R. D. Dawe, T. F. Molinski, and J. V. Turner, *Tetrahedron Letters*, **25**, 2061 (1984).

Dilithium tetrachlorocuprate, 4, 163–164; **5**, 226; **6**, 203; **7**, 114; **8**, 176; **11**, 190–191.

2-Substituted 1,3-butadienes.[1]   These useful dienes are easily prepared by cross-coupling of 1,3-butadiene-2-ylmagnesium chloride with primary alkyl iodides or bromides catalyzed by Li_2CuCl_4. Similar coupling is possible with aryl iodides, but yields are lower.

[1] S. Nunomoto, Y. Kawakami, and Y. Yamashita, *J. Org.*, **48**, 1912 (1983).

Di-2-mesitylborane, $[1,3,5-(CH_3)_3C_6H_2]_2BH$ (**1**). Mol. wt. 250.20, m.p. 68°, air stable. The reagent is obtained in 70% yield by reduction of fluorodimesitylborane with $LiAlH_4$.

Hydroboration.[1]   This borane is recommended for hydroboration of alkynes, particularly for regioselective hydroboration of unsymmetrical alkynes (equation I). 1-Alkynes are converted into aldehydes in high yield. Since alkenes react only slowly with this borane, selective hydroboration of alkynes in the presence of alkenes is possible.

[1] A. Pelter, S. Singaram, and H. C. Brown, *Tetrahedron Letters*, **24**, 1433 (1983).

1,3-Dimethoxy-1-trimethylsilyloxy-1,3-butadiene, (1).

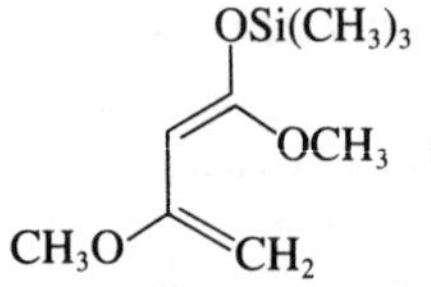

Preparation.[1]

Cyclocondensation with carbonyl compounds.[2] This diene undergoes condensation with both aldehydes and ketones in the presence of 1 equiv. of zinc chloride or BF$_3$ etherate to afford unstable orthoester adducts, which are hydrolyzed to δ-lactones (equation I). The europium(III) catalysts Eu(fod)$_3$ and the chiral Eu(hfc)$_3$[3] are also excellent catalysts,

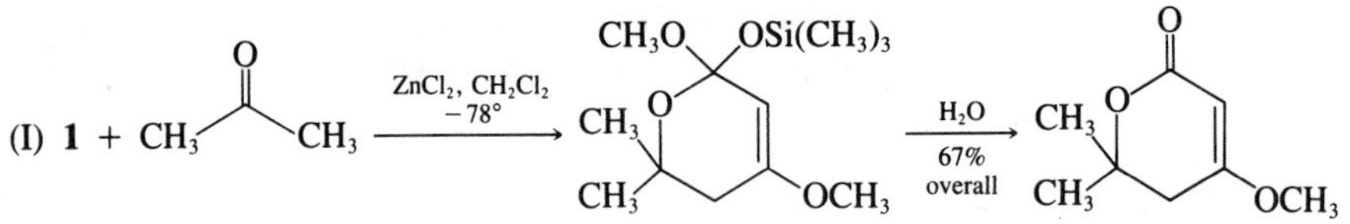

(I) 1 +

and the latter catalyst is particularly effective for diastereoselective reactions of 1 with chiral α-alkoxy aldehydes. In such reactions, diastereoselection owing to chelation control is as high as 60:1. The usual Lewis acid catalysts, ZnCl$_2$ or MgBr$_2$, effect diastereoselectivities of only 2:1. This diastereoselective control was effective in a synthesis of optically pure (−)-pestalotin (**2**, equation II).

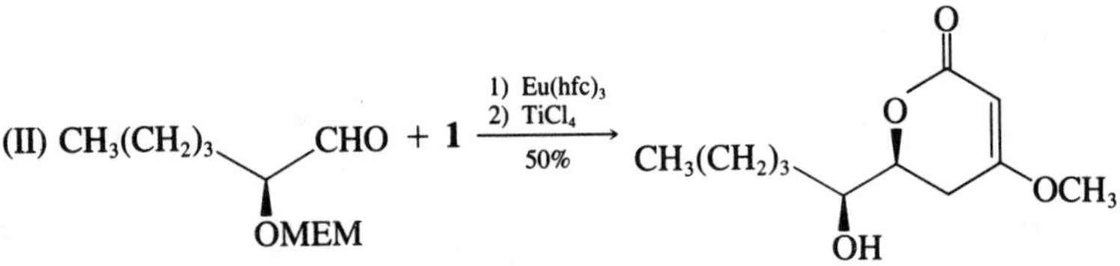

(II)

A related cyclocondensation using bis-1,1-dimethoxy-3-trimethylsilyloxy-1,3-butadiene provides a synthesis of 3-keto-δ-lactones (equation III).[4]

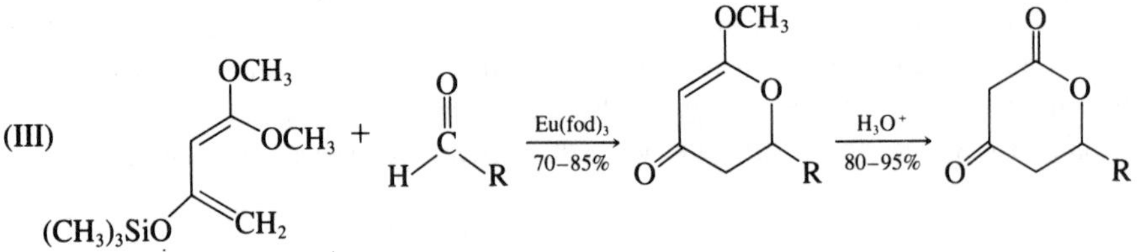

(III)

[1] P. Brassard and J. Savard, *Tetrahedron Letters*, 4911 (1979).
[2] M. M. Midland and R. S. Graham, *Am. Soc.*, **106**, 4294 (1984).

[3] Fod = 6,6,7,7,8,8,8-heptafluoro-2,2-dimethyl-3,5-octanedionata; hfc = [3-(heptafluoropropyl)-hydroxymethylene]-d-camphorate.
[4] S. Castellino and J. J. Simms, *Tetrahedron Letters*, **25**, 2307 (1984).

1,2-Dimethoxy-1-trimethylsilyloxy-1,3-pentadiene,
(E,E and Z,E)-**1**.

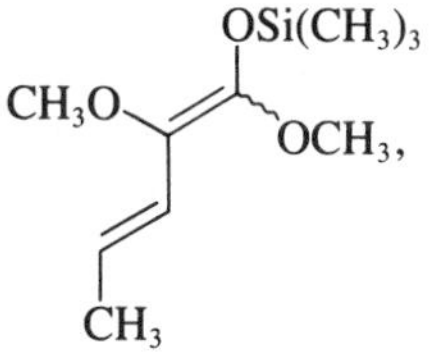

Preparation:

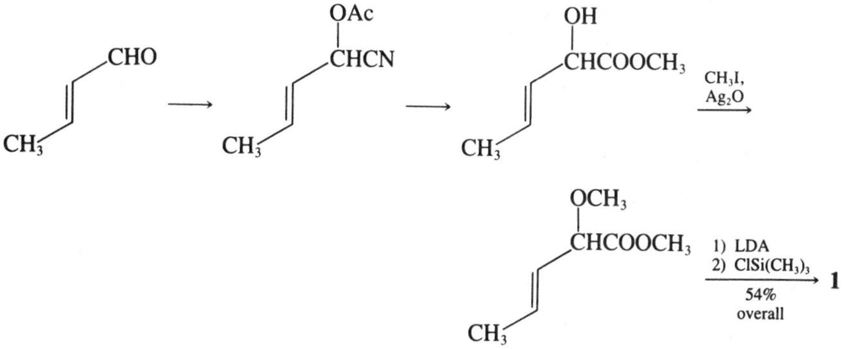

Diels-Alder reactions.[1] This diene reacts with 2-chloro-1,4-quinones to give adducts that undergo aromatization when heated. Yields tend to be low in reactions with benzoquinones, but are usually satisfactory with naphthoquinones. The reagent was developed to provide a synthesis of the trimethyl ether (**2**) of erythrolaccin.

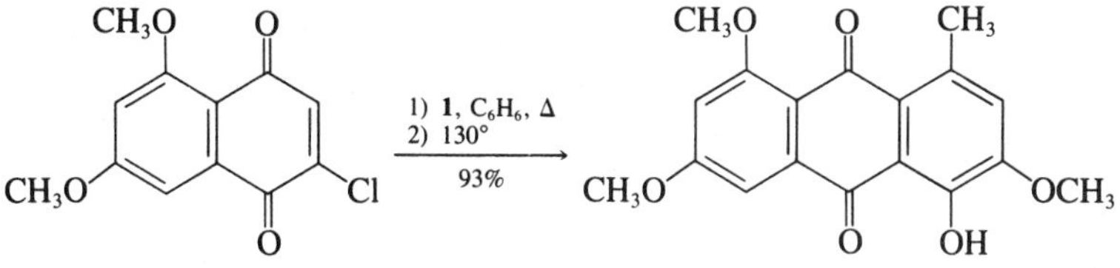

2

[1] V. Guay and P. Brassard, *Tetrahedron*, **40**, 5039 (1984).

Dimethylaluminum methyl selenolate, 8, 182–183.

Ketones from esters.[1] Methylselenol esters, readily available by reaction of carboxylic esters with $(CH_3)_2AlSeCH_3$ (**8**, 182), react with organocuprates to give ketones in generally high yield.

Examples:

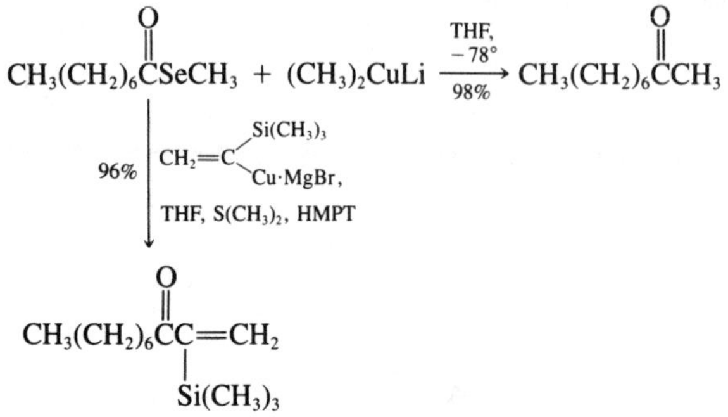

$$CH_3(CH_2)_6\overset{O}{\overset{\|}{C}}SeCH_3 \ + \ (CH_3)_2CuLi \ \xrightarrow[98\%]{\substack{THF, \\ -78°}} \ CH_3(CH_2)_6\overset{O}{\overset{\|}{C}}CH_3$$

The yields are generally higher than those obtained by use of the corresponding acyl chlorides.

[1] A. F. Sviridov, M. S. Ermolenko, D. V. Yashunsky, and N. K. Kochetkov, *Tetrahedron Letters*, **24**, 4355, 4359 (1983).

4-Dimethylamino-3-butyne-2-one (1), **8**, 183–184; **9**, 177–178.

Macrolactonization. This acetylene reacts selectively with the carboxyl group of an ω-hydroxy carboxylic acid (**2**) to form a stable hydroxy enol ester **3**. Addition of 10-camphorsulfonic acid or mercury(II) trifluoroacetate results in lactonization to macrolactones (**4**).

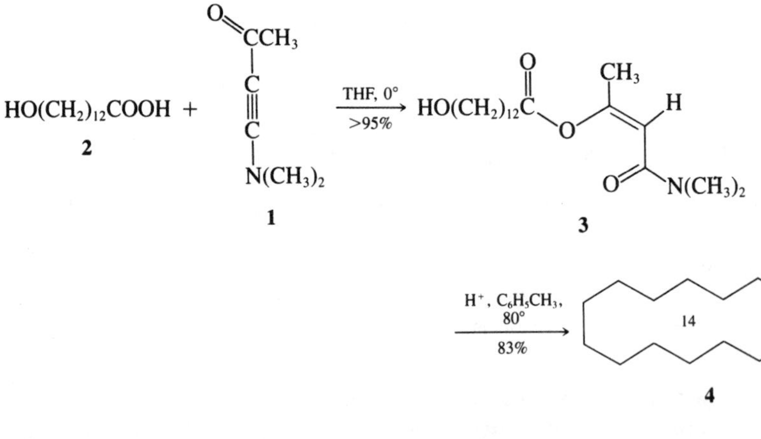

This method was used to effect lactonization of the protected seco acids of 7-epibrefeldine A (**5**) in 74% yield.

5

¹ H.-J. Gais, *Tetrahedron Letters*, **25**, 273 (1984).

1-(3-Dimethylaminopropyl)-3-ethylcarbodiimide, $(CH_3)_2N(CH_2)_3N{=}C{=}NC_2H_5$ (1); **1,** 274. Suppliers: Aldrich, Fluka.

Peptide synthesis. This water-soluble diimide is useful for synthesis of peptides, particularly if it is used in combination with 1-hydroxybenzotriazole.¹ The method was used extensively in a total synthesis of urogastrone, a polypeptide with 53 amino acid residues and three disulfide bonds, which controls human epidermal growth.²

¹ T. Kimura, M. Takai, Y. Masui, T. Morikawa, and S. Sakakibara, *Biopolymers*, **20**, 1823 (1981).
² D. Hagiwara, M. Neya, Y. Miyazaki, K. Hemmi, and M. Hashimoto, *J.C.S. Chem. Comm.*, 1676 (1984).

4-Dimethylaminopyridine (DMAP). Use of 4-dialkylaminopyridines for catalysis of acylation and alkylation has been reviewed (133 references).¹

¹ E. F. V. Scriven, *Chem. Soc. Rev.*, **12**, 129 (1983).

4-(N,N-Dimethylamino)pyridinium chlorosulfite chloride,

Mol. wt. 241.14, m.p. 155–157°. The reagent is prepared by reaction of thionyl chloride with DMAP in CH_2Cl_2 at $-20°$ (83% yield).

*Esterification.*¹ This reaction can be carried out by addition of the acid to a solution of **1**, prepared *in situ*, in CH_2Cl_2. After 15–60 minutes, when the acid has been converted into the acid chloride by **1**, an alcohol and DMAP (1 equiv.) are added and the reaction allowed to proceed for 30–90 minutes. Yields are generally 80–92%.

¹ A. Arrieta, T. García, and C. Palomo, *Syn. Comm.*, **12**, 1139 (1982).

Dimethylboron bromide, $(CH_3)_2BBr$ (1). Mol. wt. 120.79 b.p. 30–32°. Supplier: Alfa.

*Cleavage of ethers¹ and acetals.*² This reagent cleaves cyclic and acyclic acetals and ketals, including β-methoxyethoxymethyl (MEM), methoxymethyl (MOM), and methylthiomethyl (MTM) ethers, at $-78°$ in high yield. It also cleaves alkyl methyl and aryl methyl ethers to the alcohols at $20°$ in high yield. Methyl glycosides are converted efficiently into glycosyl bromides.

Examples:

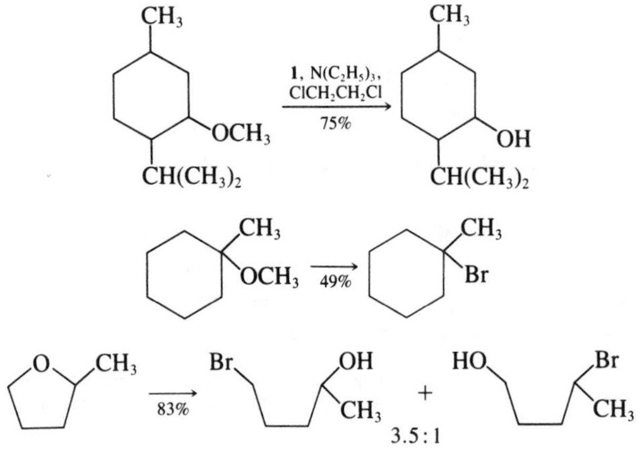

Deoxygenation of sulfoxides.[3] Dimethylboron bromide and 9-borabicy-clo[3.3.1]nonyl bromide (Aldrich) convert alkyl and aryl sulfoxides to sulfides at low temperature and in high yield.

[1] Y. Guindon, C. Yoakim, and H. E. Morton, *Tetrahedron Letters*, **24**, 2969 (1983).
[2] *Idem, J. Org.*, **49**, 3912 (1984).
[3] Y. Guindon, J. G. Atkinson, and H. E. Morton, *ibid.*, **49**, 4538 (1984).

Dimethyl bromomalonate, $BrCH(COOCH_3)$ **(1)**. Mol. wt. 211.03, b.p. 105–108°. Supplier: Fluka.

Bromination–dehydrobromination.[1] Reaction of "Red Salt" (**2**) with dimethyl bromomalonate in $CH_3OH/NaOCH_3$ produces **3** in 75–80% yield. Hydrolytic decarboxylation of **3** in refluxing aqueous HCl leads to bicyclo[3.3.0]octadiene-3,7-dione (**4**).

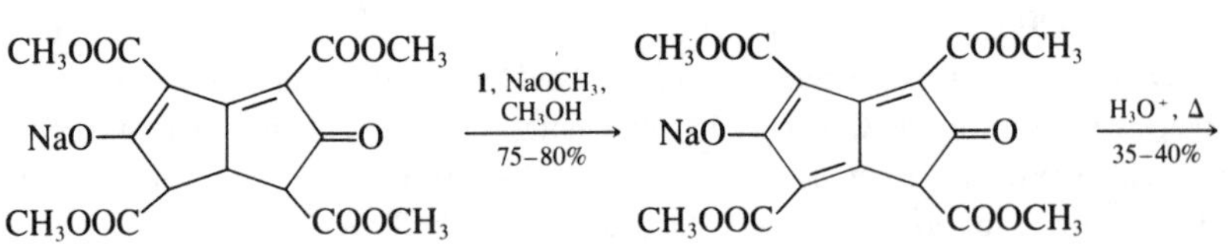

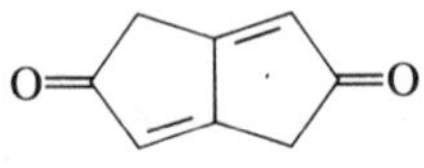

[1] A. M. Docken, *J. Org.*, **46**, 4096 (1981).

Dimethyl carbonate, 2, 149–150; **5**, 234.

Methoxycarbonylation–methylation of esters.[1] Reaction of a primary ester with dimethyl carbonate in the presence of sodium hydride results in α-methylation as well as methoxycarbonylation.

Example:

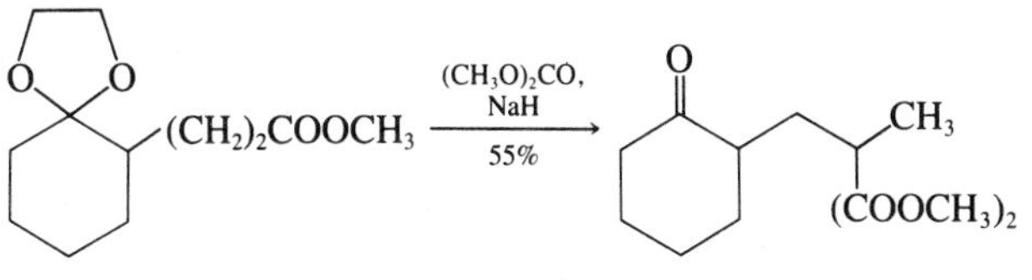

[1] D. Sengupta and R. V. Venkatesawaran, *J. Chem. Res. (S)*, 372 (1984).

N,N-Dimethylchloromethyleniminium chloride–Lithium *t*-butoxyaluminum hydride.

—COOH → —CHO.[1] This conversion can be effected in one flask by treatment of the carboxylic acid with N,N-dimethylchloromethyleniminium chloride (prepared from DMF and oxalyl chloride) and pyridine to form a betaine, which is then reduced to the aldehyde by the hydride in the presence of a catalytic amount of CuI at $-78°$. The reaction is applicable to aliphatic and aromatic acids, and can tolerate halo, cyano, ester, and even carbonyl groups. Yields are usually ~65–80%.

Examples:

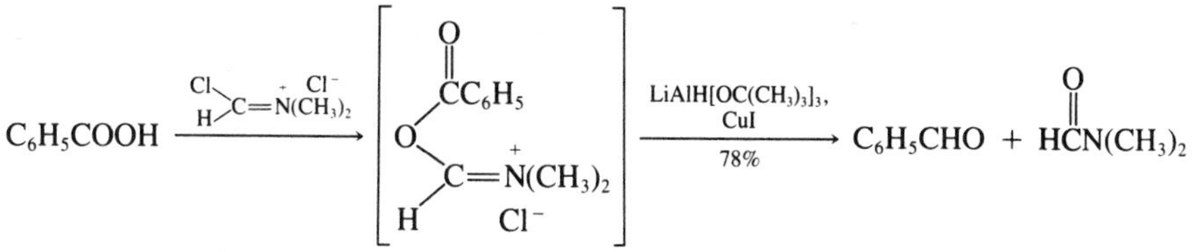

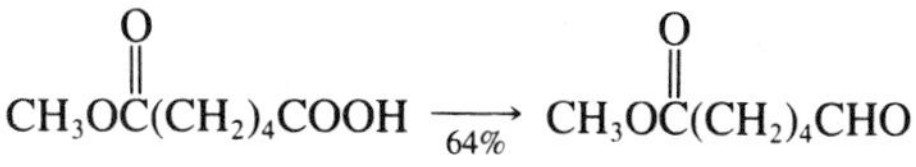

[1] T. Fujisawa, T. Mori, S. Tsuge, and T. Sato, *Tetrahedron Letters*, **24**, 1543 (1983).

1,1'-Di(methylcyclopentadienyl)tin(II), Sn (1).

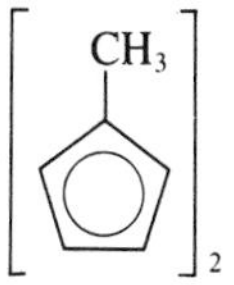

The reagent is prepared by reaction of methylcyclopentadienyllithium or sodium methylcyclopentadienide and anhydrous $SnCl_2$ in THF or DMF; it is a pale yellow liquid, readily oxidized by air, and sensitive to water.[1]

Acylation of alcohols and amides.[2]   Tin(II) alkoxides are formed *in situ* by reaction of **1** with alcohols in toluene at room temperature.[3] On addition of an acid chloride, esters are formed in good to excellent yields, particularly if HMPT is also present. Since conditions are nearly neutral, this method has wide application. The reaction is carried out at 100° in the case of a hindered alcohol or acid chloride.

Examples:

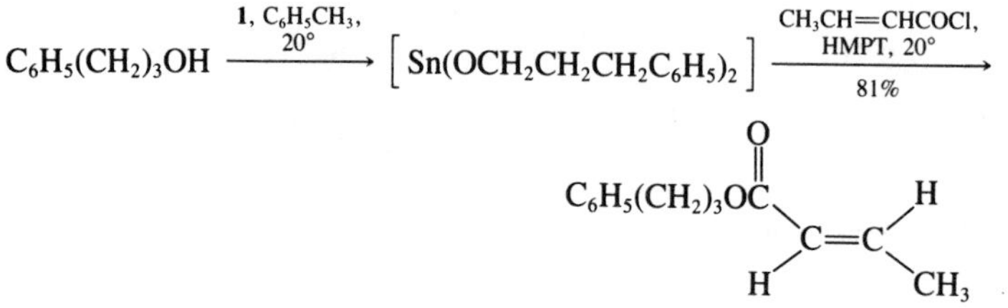

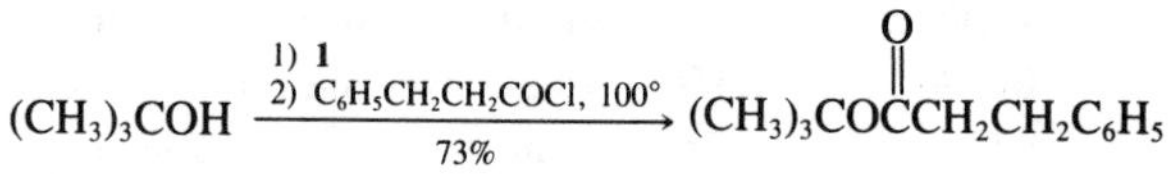

$$C_6H_5(CH_2)_3OH \xrightarrow[20°]{1,\ C_6H_5CH_3,} \left[Sn(OCH_2CH_2CH_2C_6H_5)_2 \right] \xrightarrow[81\%]{CH_3CH=CHCOCl,\ HMPT,\ 20°}$$

$$(CH_3)_3COH \xrightarrow[73\%]{1)\ \textbf{1} \atop 2)\ C_6H_5CH_2CH_2COCl,\ 100°} (CH_3)_3COCCH_2CH_2C_6H_5$$

A similar reaction can be used for acylation of amides with acid chlorides. Example:

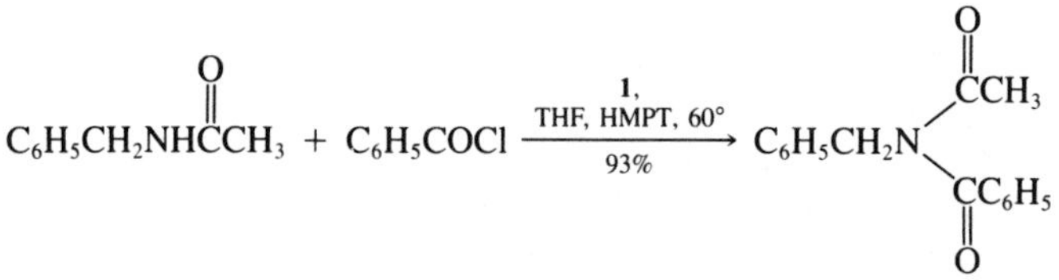

N-Alkylimides.[4]   Acid anhydrides react with primary amines in the presence of **1** (1.5 equiv.) in refluxing *p*-xylene to afford N-alkylimides in 60–90% yield (equation I).

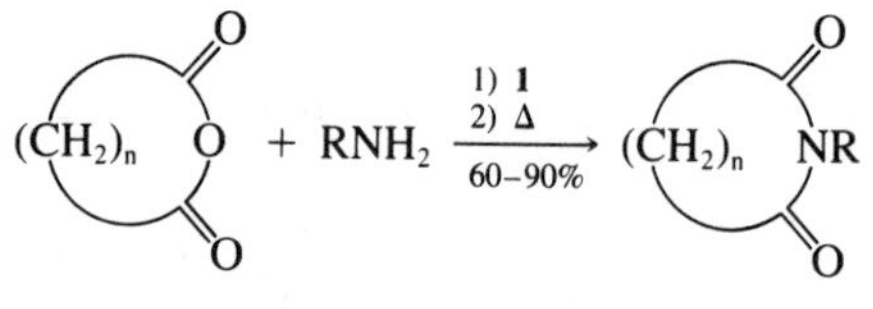

(n = 2, 3)

[1] L. D. Dave, D. F. Evans, and G. Wilkinson, *J. Chem. Soc.*, 3684 (1959).

[2] T. Mukaiyama, J. Ichikawa, and M. Asami, *Chem. Letters*, 293, 683 (1983).

[3] P. G. Harrison and S. R. Stobart, *J.C.S. Dalton, Trans.*, 940 (1973).

[4] T. Mukaiyama, J. Ichikawa, M. Toba, and M. Asami, *Chem. Letters*, 879 (1983).

Dimethyl diazomalonate, 5, 244; **8**, 187.

Deoxygenation of epoxides.[1]   In the presence of rhodium(II) acetate, dimethyl diazomalonate converts epoxides into the corresponding alkenes with formation of dimethyl oxomalonate. Alkene isomerization or cyclopropanation is not observed. Yields are generally >80%.

Examples:

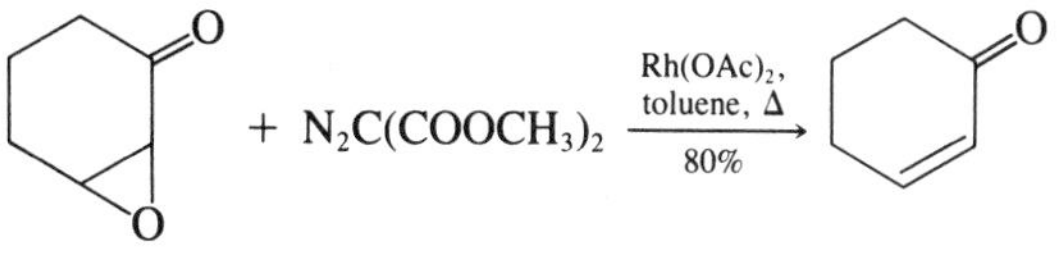

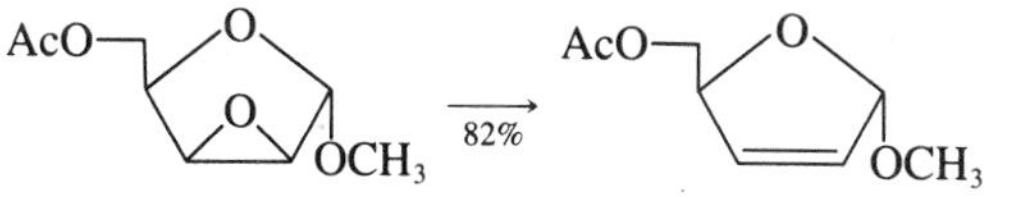

[1] M. G. Martin and B. Ganem, *Tetrahedron Letters*, **25**, 251 (1984).

Dimethylformamide, 1, 278–281; **2**, 153–154; **3**, 115; **4**, 184; **5**, 247–249; **7**, 124; **8**, 189–190; **9**, 182; **11**, 198.

Epoxymethano bridging.   Reaction of the bromoketone **1** with LiBr and Li$_2$CO$_3$ in DMF at 140° (**5**, 395) unexpectedly results in the epoxymethano-bridged compound **2** (79% yield). Actually, **2** is obtained in somewhat higher yield when **1** is heated with

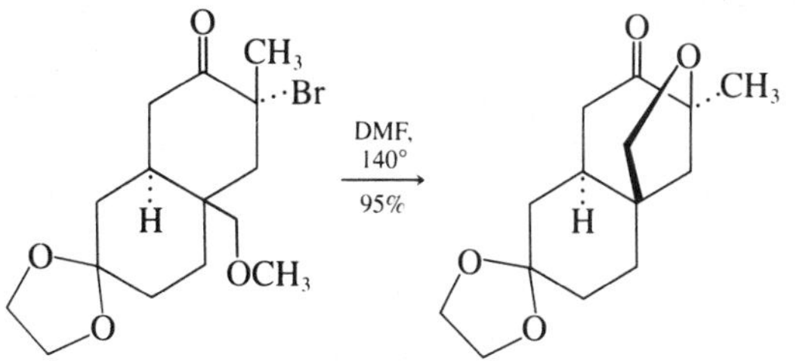

DMF with no additives at 140°. Bridges of this type are a common feature of several quassinoids.[1]

—CH$_2$OH → —CH$_2$Cl.   Some years ago Bredereck *et al.*[2] reported that ethyl chloroformate, ClCOOC$_2$H$_5$, is converted in the presence of DMF at room temperature into ethyl chloride. This decarboxylation has since been shown to provide a general synthesis of primary alkyl chlorides via the alkyl chloroformates, which are readily prepared by reaction of primary alcohols with phosgene in CH$_2$Cl$_2$ or ClCH$_2$CH$_2$Cl. Overall yields are

generally >90%.[3] N,N'-Tetramethylurea also catalyzes this decarboxylation, but higher temperatures are required and yields are lower.

[1] K. Kanai, R. E. Zelle, H.-L. Sham, P. A. Grieco, and P. Callant, *J. Org.*, **49**, 3867 (1984).
[2] H. Bredereck, F. Effenberger, and G. Simchen, *Ber.*, **96**, 1350 (1963).
[3] R. Richter and B. Tucker, *J. Org.*, **48**, 2625 (1983).

Dimethylformamide–Thionyl chloride, $(CH_3)_2\overset{+}{N}\!\!=\!\!CHOS\overset{\displaystyle O}{\underset{\displaystyle Cl}{\diagup}}$ Cl$^-$ 1, 286–289; 9, 465.

Dehydration.[1]   This reagent is useful for synthesis of acyl azides from carboxylic acids, NaCN, and pyridine, with tetrabutylammonium bromide as catalyst (75–95% yield). In combination with pyridine, it effects dehydration of oximes to nitriles in 80–90% yield.

β-Lactams can be prepared directly in 40–75% yield from carboxylic acids and imines with the reagent (1 equiv.) and triethylamine (excess). In general, a mixture of *cis*- and *trans*-azetidinones is formed.

Example:

$$PhtCH_2COOH + p\text{-}CH_3OC_6H_4CH\!\!=\!\!NC_6H_5 \xrightarrow[\substack{70\%}]{\substack{1,\ N(C_2H_5)_3,\\ CH_2Cl_2,\ 25°}}$$

[1] A. Arrieta, J. M. Aizpurua, and C. Palomo, *Tetrahedron Letters,* **25**, 3365 (1984).

N,N-Dimethylformamide dineopentyl acetal (1), 1, 283; **6,** 222; **9,** 183.

Dehydrative decarboxylation.[1]   The conversion of β-hydroxy carboxylic acids to alkenes by reaction with a DMF dialkyl acetal involves an *anti*-elimination, and thus is complementary to the known *syn*-elimination of these hydroxy acids via a β-lactone (**5,** 22; **9,** 504). These reactions were used to obtain both the (E)- and the (Z)-1-alkoxy-1,3-

2, R = C$_6$H$_5$

(Z)

butadienes from the isomeric γ,δ-unsaturated α-alkoxy-β-hydroxy carboxylic acids (**2**) obtained by reaction of methacrolein with the dianion of an alkoxyacetic acid.

[1] J. I. Luengo and M. Koreeda, *Tetrahedron Letters*, **25**, 4881 (1984); M. Koreeda and J. I. Luengo, *J. Org.*, **49**, 2079 (1984).

N,N-Dimethylhydrazine, 1, 289–290; **2**, 154–155; **3**, 117; **5**, 254; **6**, 223; **7**, 126–130; **8**, 192–193; **9**, 184–185; **11**, 200–201.

Spiroacetals. Enders *et al.*[1] have reported a novel synthesis of 1,6-dioxo-spiro[4.4]nonane-spiroacetals from acetone and epoxides via the dimethylhydrazone.

Example:

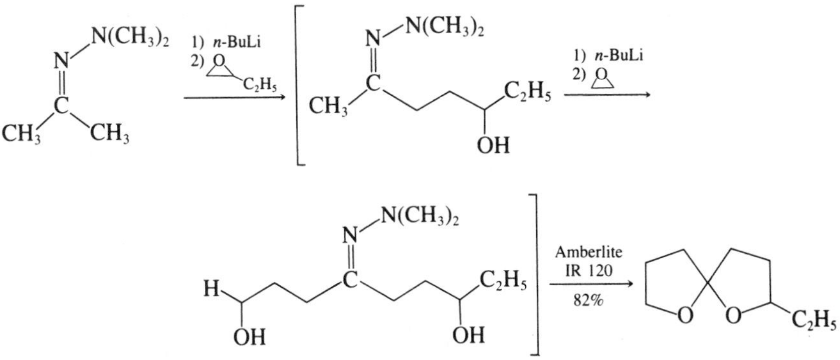

[1] D. Enders, W. Dahmen, E. Dedericks, and P. Weuster, *Syn. Comm.*, **13**, 1235 (1983).

N,O-Dimethylhydroxylamine, CH_3ONHCH_3 (**1**), **11**, 201–202.

Amination.[1] This amine in combination with methyllithium reacts with alkyllithiums to provide alkylmethylamines, which are isolated as benzamides.

Example:

$$n\text{-BuLi} + CH_3Li + \mathbf{1} \longrightarrow n\text{-BuNHCH}_3 \xrightarrow[\text{72\% overall}]{C_6H_5COCl} C_6H_5CON(CH_3)(\text{Bu-}n)$$

An intramolecular amination has been effected (equation I).

(I)

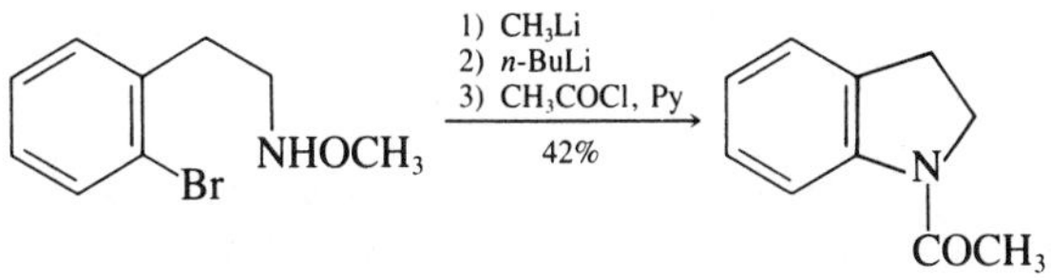

[1] B. J. Kokko and P. Beak, *Tetrahedron Letters*, **24**, 561 (1983).

Dimethyl methoxymethylenemalonate, $CH_3OCH\!=\!C(COOCH_3)_2$ **(1).**
Preparation.[1] Supplier: Fluka.

   3-Carbomethoxy-2-pyrones.[2]   Ketone enolates (NaH or LDA) react with **1** with loss of methanol to give 3-carbomethoxy-2-pyrones (**2**) in generally good yield. The final ring closure can be promoted if necessary by addition of acetic anhydride or TsOH and elevated temperatures.
   Example:

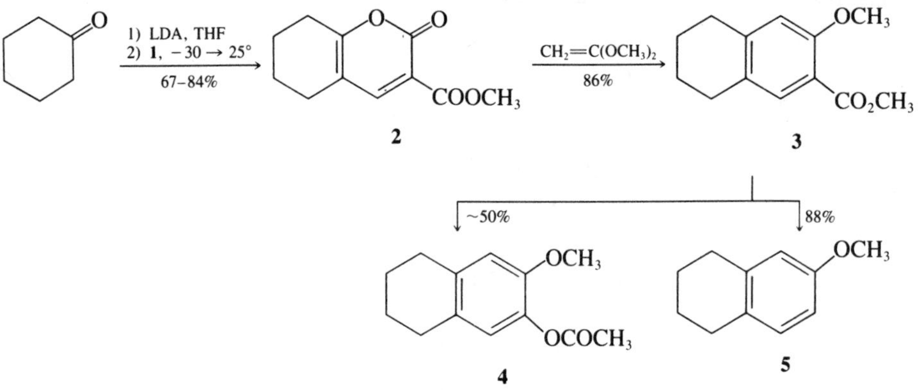

These pyrones undergo Diels-Alder reactions with electron-deficient alkenes with loss of CO_2 and formation of an aromatic ring. Thus reaction with 1,1-dimethoxyethylene (**11**, 279–281) provides a regiospecific route to methyl salicylates (*e.g.*, **3**), which are convertible in several steps into catechol derivatives (**4**) or by decarbomethoxylation into phenol ethers (**5**). Similar reactions with vinylene carbonate or 1,1,2-trimethoxyethylene provide regiospecific routes to phenols, differentially protected derivatives of catechol, resorcinol, or pyrogallol (equation II).

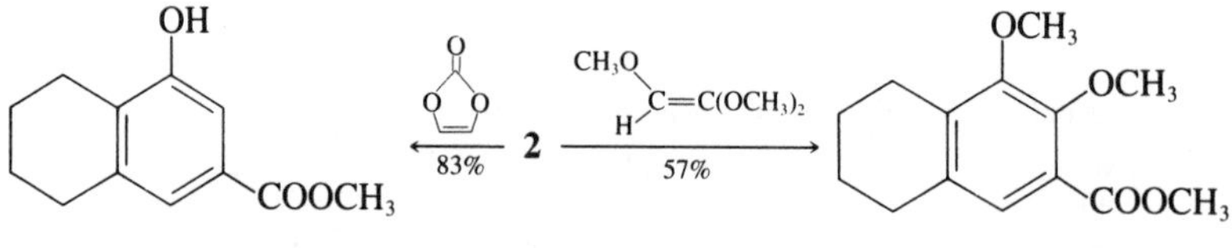

An example is the conversion of the 3-carbomethoxy-2-pyrone **6** into either one of the alkaloids rufercine (**7**) or imeluteine (**8**) by a Diels-Alder reaction followed by ester hydrolysis and decarboxylation promoted by pentafluorophenylcopper.[3]

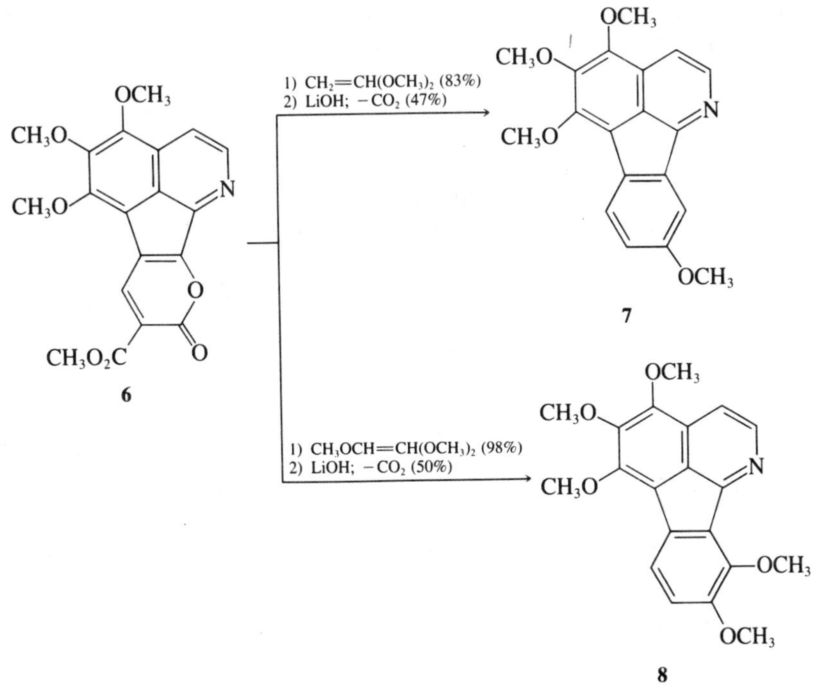
1) $CH_2{=}CH(OCH_3)_2$ (83%)
2) LiOH; $-CO_2$ (47%)
1) $CH_3OCH{=}CH(OCH_3)_2$ (98%)
2) LiOH; $-CO_2$ (50%)
6
7
8

[1] R. C. Fuson, W. E. Parkham, L. J. Reed, *J. Org.*, **11**, 194 (1946).
[2] D. L. Boger and M. D. Mullican, *ibid.*, **49**, 4033 (1984); *Org. Syn.*, submitted (1984).
[3] D. L. Boger and C. E. Brotherton, *J. Org.*, **49**, 4050 (1984).

Dimethyl(methylthio)sulfonium tetrafluoroborate, $CH_3S\overset{+}{S}(CH_3)_2BF_4^-$ (**1**), **11**, 204–206.

Alkynyl sulfenylation.[1] Reaction of a lithium acetylide with the adduct (**a**) of the reagent (**1**) with an alkene results in a complex mixture or an allylic sulfide. However, alkynyl sulfenylation can be effected with the ate complexes of a lithium acetylide with a thiophile. Two complexes can be used: the 2:1 complex with $(C_2H_5)_2AlCl$ or the 1:1 complex with $Al(C_2H_5)_3$.

Example:

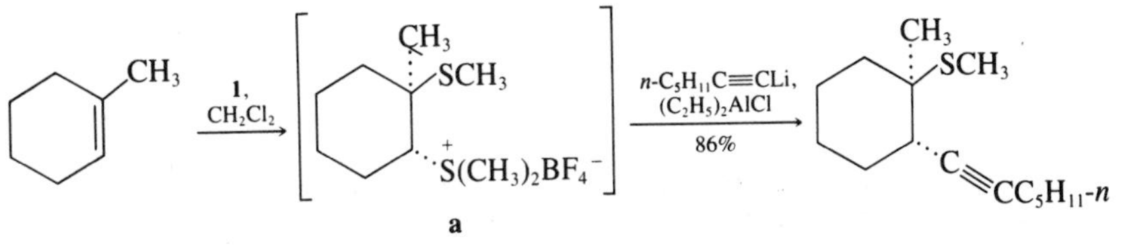
1,
CH_2Cl_2
n-$C_5H_{11}C{\equiv}CLi$,
$(C_2H_5)_2AlCl$
86%
a

The products can be converted into 1,3-enynes by sulfoxide elimination (equation I) or by alkylation–elimination (equation II; **7**, 149–150).

(I)

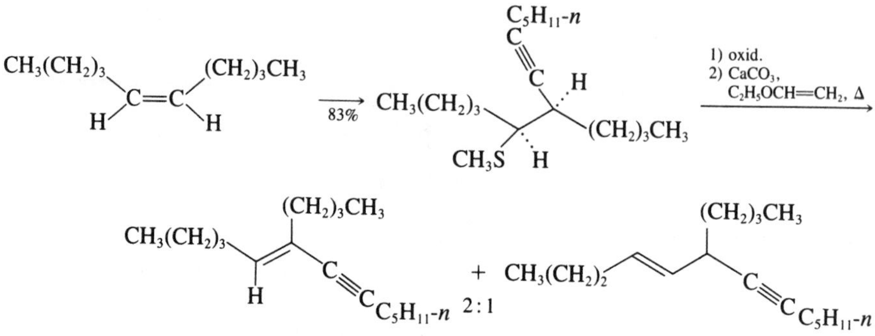

(II)

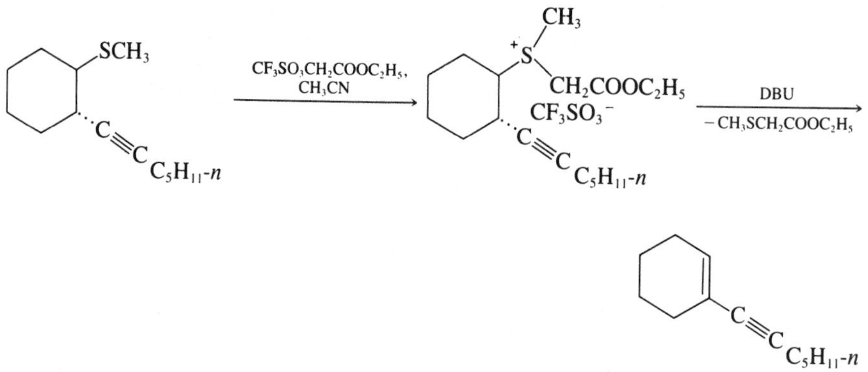

Formylation.[2] Reaction of $(C_6H_5S)_3CH$ (**2**) with **1** generates the carbocation $(C_6H_5S)_2C^+H$, which reacts with phenols or aromatic ethers to give the thioacetals (**3**), which are readily hydrolyzed to aldehydes. The formylation shows high *para*-preference. Example:

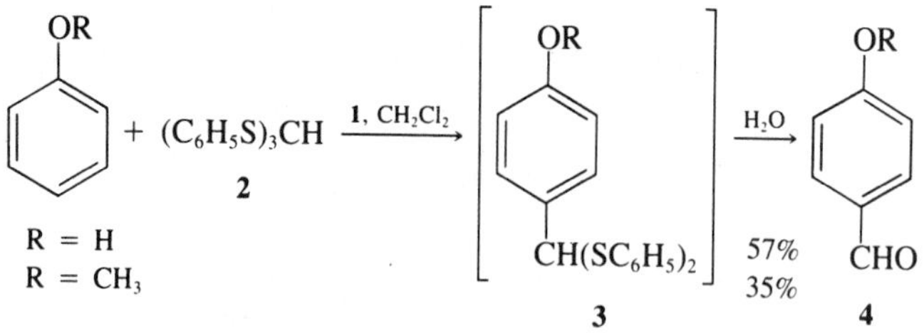

R = H
R = CH_3

[1] B. M. Trost and S. J. Martin, *Am. Soc.*, **106**, 4263 (1984).
[2] R. A. J. Smith and A. R. bin Manas, *Synthesis*, 166 (1984).

Dimethylphenylsilane, $C_6H_5(CH_3)_2SiH$. (**1**). Mol. wt. 136.27, b.p. 157°/744 mm. Supplier: Aldrich.

Diastereoselective reduction of ketones.[1]   Trialkylsilanes can reduce aldehydes and ketones in the presence of a catalytic amount of tetrabutylammonium fluoride in HMPT (*cf.* reductions with activated hydrosilanes, **11**, 554). Under these conditions esters and nitriles are not reduced. The reaction is stereoselective. Thus 2-methylcyclohexanone is reduced selectively to *cis*-2-methylcyclohexanol; the *cis*-selectivity depends on the bulkiness of the hydrosilane, being highest (95%) with triphenylsilane.

This hydrosilylation is particularly useful for highly *anti*-selective reduction of α-amino or α-hydroxy ketones with no detectable racemization.

Examples:

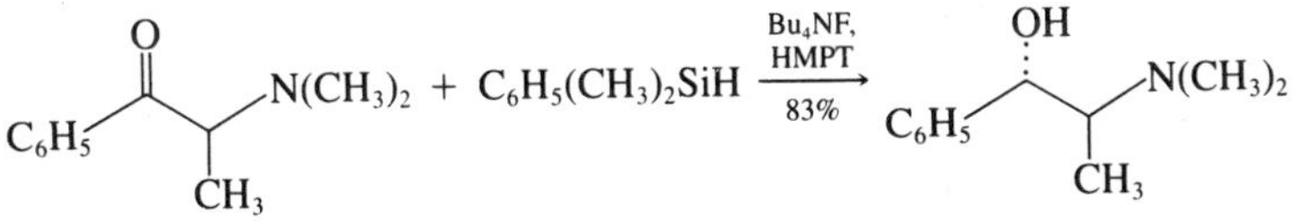

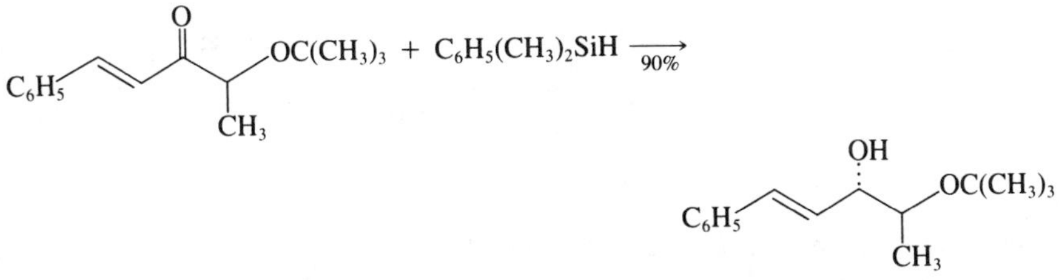

In contrast, hydrosilylation under acidic conditions (TFA) (**5**, 695) is highly *syn*-selective.

Examples:

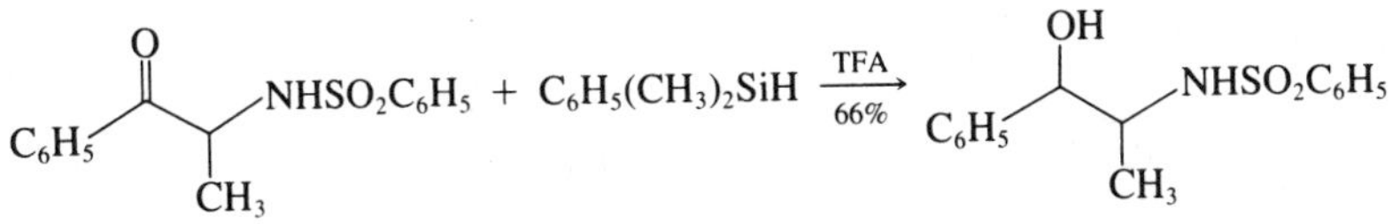

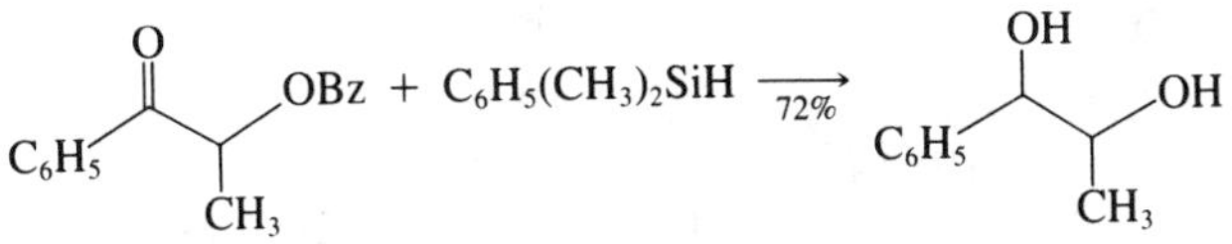

[1] M. Fujita and T. Hiyama, *Am. Soc.*, **106**, 4629 (1984).

Dimethylphenylsilyllithium, 7, 133; **8**, 196–197; **10**, 162–163; **11**, 209.

Conjugate additions. The silyl cuprate reagent $(PhMe_2Si)_2CuLi$ (**1**) undergoes conjugate addition to α,β-unsaturated aldehydes, ketones, or esters to form β-silyl carbonyl compounds.[1] Methylation of the intermediate enolate in acyclic systems is highly diastereoselective in favor of the isomer in which the silyl and methyl groups have the *anti-*

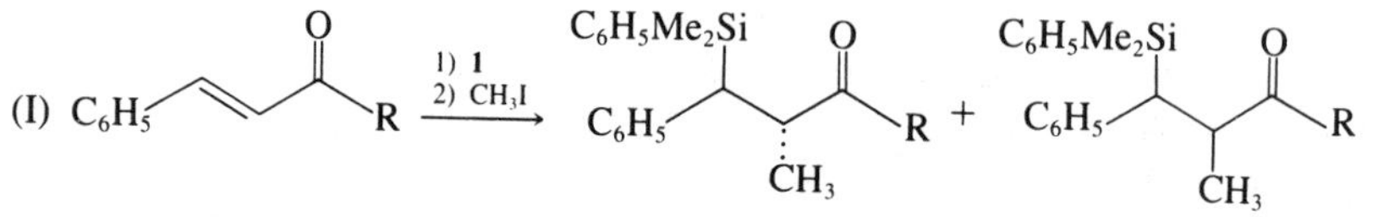

R = H	74%	
R = CH₃	57%	92:8
R = OCH₃	88%	98:2
		97:3

85:15

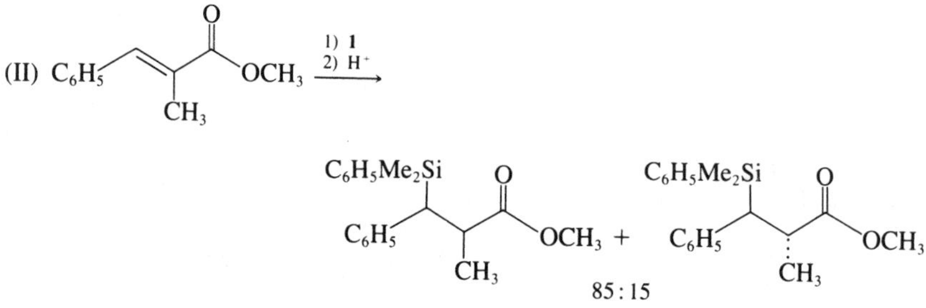

orientation (equation I).[2] The reverse diastereoselectivity is observed on protonation of the adduct of **1** with the α-substituted enone (equation II).

Conversion of $C_6H_5(CH_3)_2Si-$ to $-OH$.[3] This useful conversion can be effected in two steps. Reaction with tetrafluoroboric acid complexed with diethyl ether results in conversion to a dimethylfluorosilyl group, which on oxidation with *m*-chloroperbenzoic acid (excess) in the presence of triethylamine or KF is converted into a hydroxyl group with retention of the original configuration.

Examples:

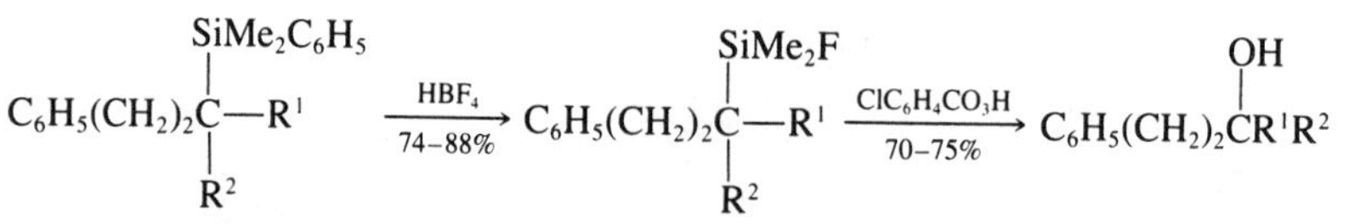

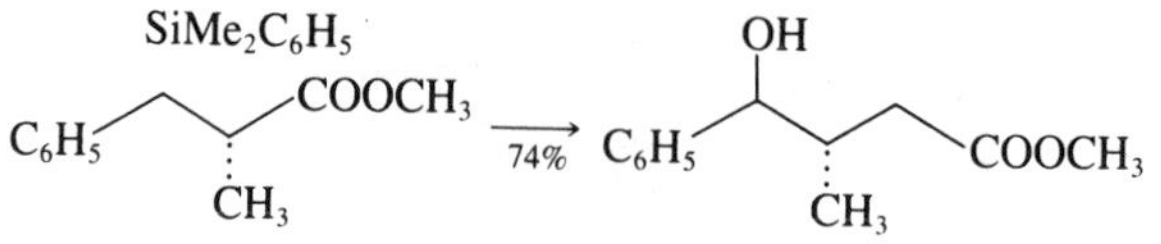

Conjugate addition of a phenyldimethylsilyl group followed by this unmasking transformation constitutes a novel overall aldol reaction.

Allylsilanes.   The S_E2' reaction of the cuprate (2) derived from 1 with tertiary allylic acetates results in stereospecific *anti*-reaction to provide allylsilanes. Similar stereospecificity is observed in reaction of 2 with propargyl acetates to form allenylsilanes.

Examples:

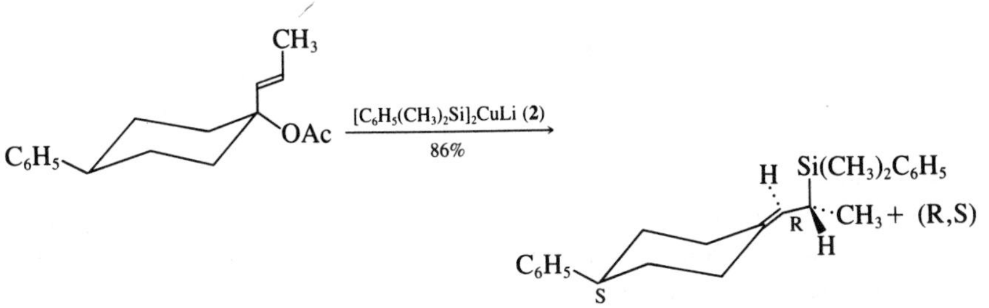

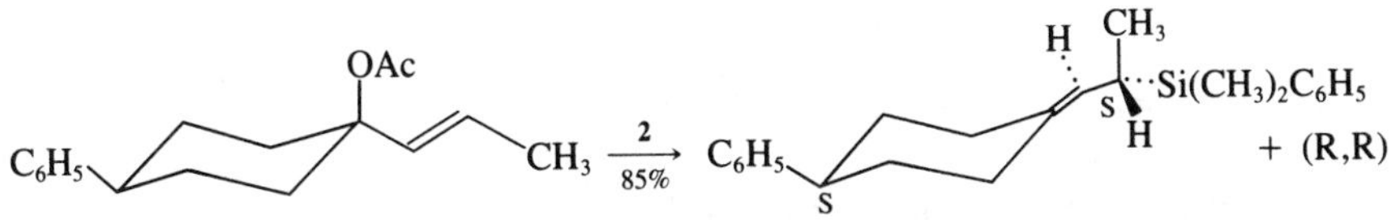

These allylsilanes undergo *anti*-reactions with bridging electrophiles such as osmium tetroxide or *m*-chloroperbenzoic acid.[4]

These two *anti*-reactions have been used for a stereospecific synthesis of dihydronepetalactone (7) from norbornenone (3).[5] Addition of propynylmagnesium bromide, followed by Lindlar reduction of the triple bond, acetylation, and reaction with 2, provides the allylsilane 4. Epoxidation of 4 and desilylation results in 5. Oxy-Cope rearrangement of 5 results in a single ketone (6) with the desired four chiral centers. The remaining steps to 7 involve conventional reactions.

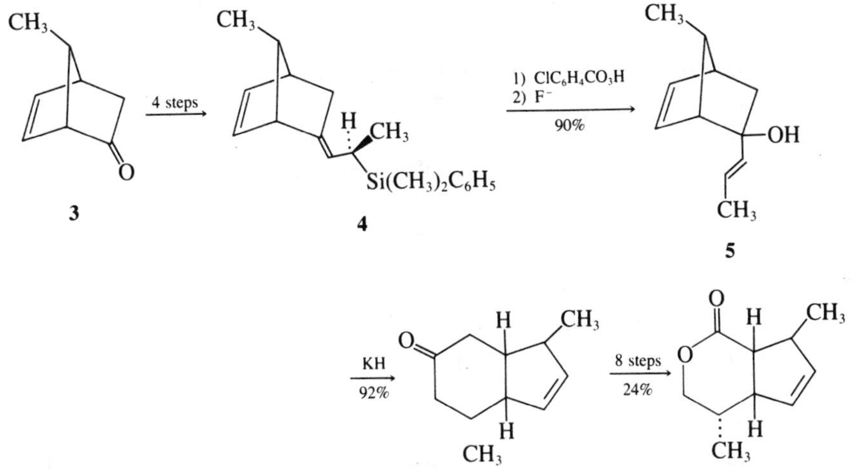

[1] D. J. Ager, I. Fleming, and S. K. Patel, *J.C.S. Perkin I,* 2520 (1981); I. Fleming, T. W. Newton, and F. Roessler, *ibid.,* 2527 (1981).
[2] W. Bernhard, I. Fleming, and D. Waterson, *J.C.S. Chem. Comm.,* 28 (1984).
[3] I. Fleming, R. Henning, and H. Plaut, *ibid.,* 29 (1984).
[4] I. Fleming and N. K. Terrett, *J. Organometal Chem.,* **264**, 99 (1984).
[5] *Idem, Tetrahedron Letters,* **25**, 5103 (1984).

Dimethyl sulfide–Methanesulfonic acid.

Selective demethylation.[1]   Alkyl methyl phosphates undergo demethylation when treated with dimethyl sulfide or ethanethiol and methanesulfonic acid (equation I). The products are isolated as the aniline salts.

$$\text{(I)}\quad \underset{\text{ROP(OCH}_3)_2}{\overset{\displaystyle\text{O}}{\|}} + 2(CH_3)_2S + 2CH_3SO_3H \xrightarrow[65-85\%]{25°}$$

$$\underset{\text{ROP(OH)}_2}{\overset{\displaystyle\text{O}}{\|}} + 2CH_3\overset{+}{S}(CH_3)_2CH_3SO_3^-$$

[1] L. Jacob, M. Julia, B. Pfeiffer, and C. Rolando, *Synthesis,* 451 (1983).

Dimethyl sulfoxide, 1, 296–310; 2, 157–158; 3, 119–123; 4, 192–194; 5, 263–266; 7, 133–135; 8, 198–199; 9, 189; 10, 166–167; 11, 214–215.

Disulfides.[1]   Thiols are oxidized to disulfides by dimethyl sulfoxide. This oxidative coupling is particularly useful for coupling of aryl thiols (equation I).

$$\text{(I)}\quad 2ArSH + \underset{\text{CH}_3\text{SCH}_3}{\overset{\displaystyle\text{O}}{\|}} \xrightarrow[80-97\%]{25°} ArSSAr + S(CH_3)_2 + H_2O$$

[1] W. E. Fristad and J. R. Peterson, *Org. Syn.,* submitted (1984).

Dimethyl sulfoxide–Thionyl chloride.

ortho-*Methylthiomethylation of phenols*.[1]   The complex (**1**) formed from these two reagents reacts with phenols to form an unstable sulfonium salt that rearranges in the presence of triethylamine to *ortho*-methylthiomethylphenols (**2**). Selective *ortho*-substi-

tution is observed with *o*- and *p*-substituted phenols, but *meta* electron-donating groups prevent this substitution.

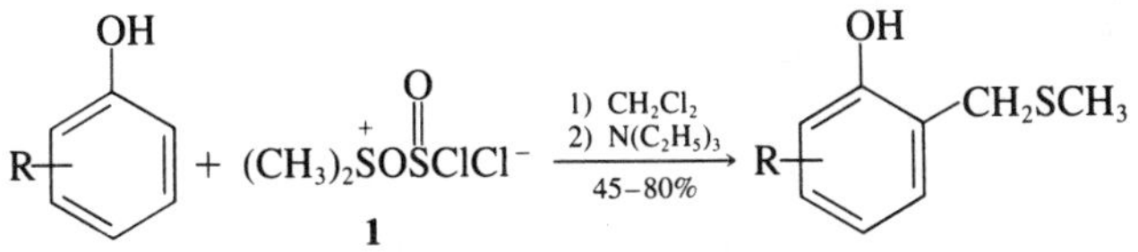

[1] K. Sato, S. Inoue, K. Ozawa, and M. Tazaki, *J.C.S. Perkin I*, 2715 (1984).

Dimethylsulfoxonium methylide, 1, 315–318; **2,** 171–173; **3,** 125–127; **4,** 197–199; **5,** 254–257; **7,** 133; **8,** 194–196; **9,** 186–187; **10,** 168–169; **11,** 213–214.

Isocoumarins.[1]   2-Acylbenzoates react with this reagent (1 equiv.) to give isocoumarins.

Example:

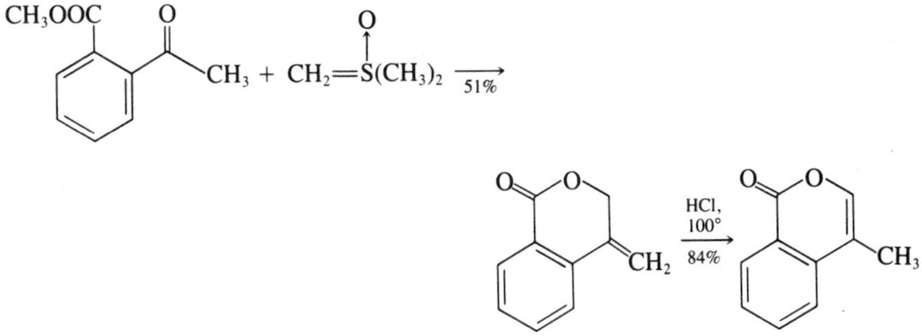

Oxetanes.[2]   Oxetanes are obtained in high yield in one step by reaction of aldehydes or ketones with 2 equiv. of this reagent (**1**) at 50° (3 days).

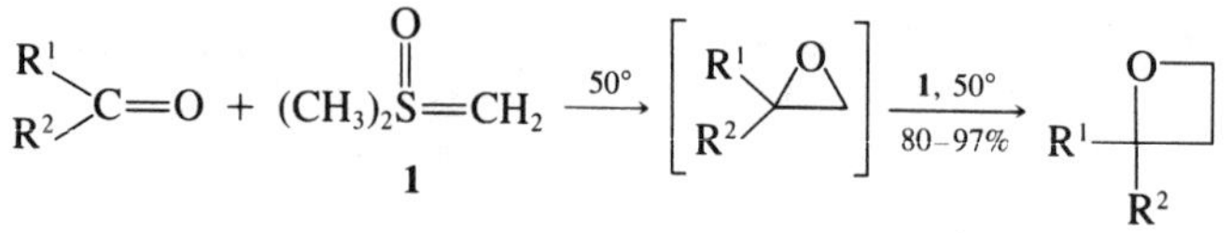

[1] K. Beautement and J. M. Clough, *Tetrahedron Letters*, **25**, 3025 (1984).
[2] K. Okuma, Y. Tanaka, S. Kaji, and H. Ohta, *J. Org.*, **48**, 5133 (1983).

Dimethyl 1,2,4,5-tetrazine-3,6-dicarboxylate, 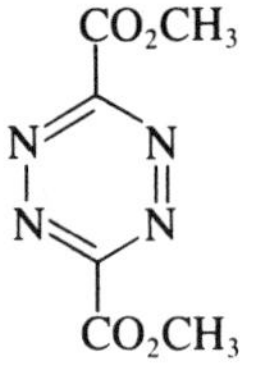 (**1**). Preparation.[1]

Diels-Alder reactions; 1,2,4-triazines.[2] The electron-deficient azadiene system present in **1** can undergo Diels-Alder reactions with electron-rich dienophiles to give an adduct that loses nitrogen to provide 1,2-diazines. Reactions with imidates ($>$C$=$NH) substituted with an active leaving group such as SCH_3 proceed at moderate temperatures to afford 1,2,4-triazines in high yield (equation I).

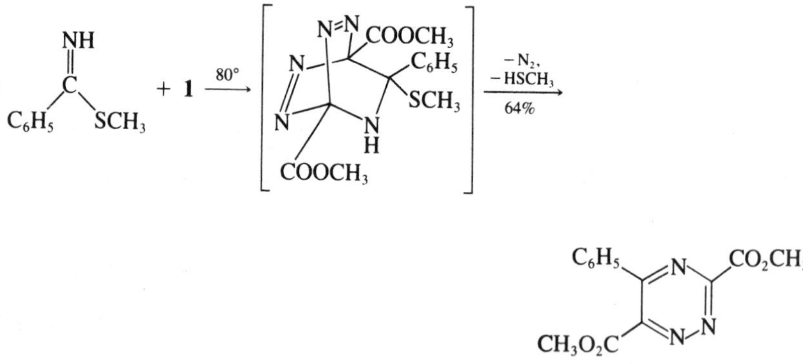

A synthesis of streptonigrin (**5**), an antitumor antibiotic, depends on two sequential Diels-Alder reactions of azadienes, reaction of **1** with an S-methyl thioimidate to give the 1,2,4-triazine **2**, and reaction of **2** with a morpholinoenamine (**3**) to afford **4** (Chart I).[3]

Chart (I)

The intramolecular Diels-Alder reaction of 1,2-diazines with an alkyne side chain provides a general synthesis of indolines. The reaction requires a high temperature (200–230°). The most satisfactory solvent is 1,3,5-triisopropylbenzene (b.p. 232–236°).[4]
Example:

[1] D. L. Boger and J. S. Panek, *Tetrahedron Letters*, **24**, 4511 (1983); D. L. Boger, R. S. Coleman, and J. S. Panek, *Org. Syn.*, submitted (1984).
[2] D. L. Boger, *Tetrahedron*, **39**, 2869 (1983).
[3] D. L. Boger and J. S. Panek, *J. Org.*, **48**, 621 (1983).
[4] D. L. Boger and R. S. Coleman, *ibid.*, **49**, 2240 (1984).

Dimethyl N-(p-tosyl)sulfoximine (1), 9, 193–194.

Oxetanes.[1] The reaction of ketones with 3 equiv. of the sodium anion (2) of 1 results in exclusive or predominant formation of the thermodynamically more stable oxetane. For example, axially C—O-bonded oxetanes such as 3 are formed from unhindered cyclohexanones. The reagent 1 can therefore serve as a $\overset{+}{C}H_2CH_2$ synthon. Thus reaction of 3 with aqueous HCl in THF yields a single homoallylic alcohol (4). It (3) is converted into the alcohol 5 in 93% yield on hydroboration followed by oxidation. The oxetane ring is opened by a few other nucleophiles [C_6H_5SNa, HMPT, $(CH_3)_2CuLi$], but is inert to several other nucleophiles, such as $C_6H_5SCH_2Li$ and $LiCH_2CO_2C(CH_3)_3$.

On the other hand, the epoxides obtained by reaction of ketones with 1 equiv. of 2 react with a large variety of nucleophiles. Thus this method was used to synthesize the diterpene lactone isomarrubiin (6, equation I).

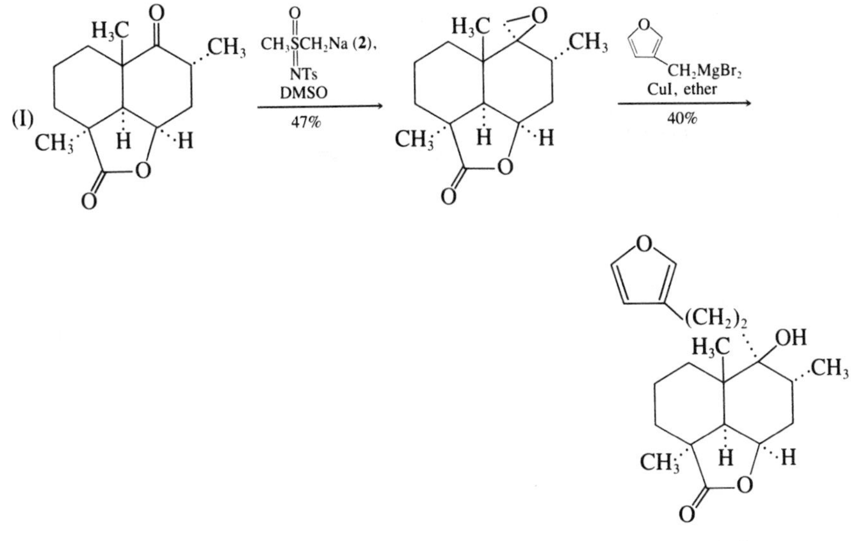

6

[1] S. C. Welch, A. S. C. P. Rao, J. T. Lyon, and J.-M. Assercq, *Am. Soc.*, **105**, 252 (1983).

Diphenyl phosphoroazidate, 4, 210–211; **5,** 280; **6,** 193; **7,** 138; **8,** 211–212; **10,** 173; **11,** 222.

Amination.[1] Aryllithium or aryl Grignard reagents react with this reagent to form phosphoryltriazenes, which are reduced by sodium bis(2-methoxyethoxy)aluminum hydride to amines (45–88% yield).

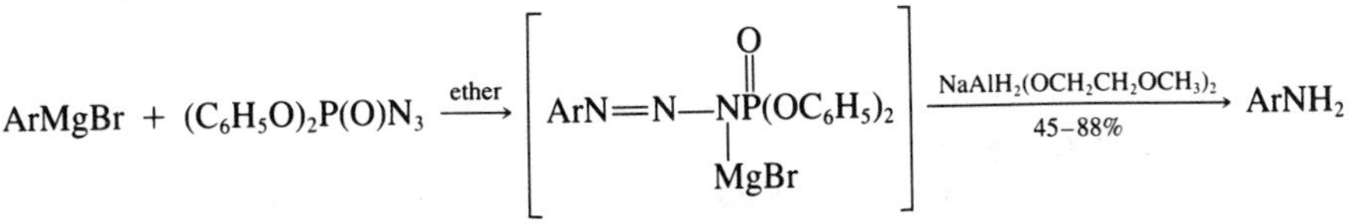

$$ArMgBr + (C_6H_5O)_2P(O)N_3 \xrightarrow{ether} \left[ArN{=}N{-}\overset{\overset{O}{\|}}{\underset{\underset{MgBr}{|}}{N}}P(OC_6H_5)_2 \right] \xrightarrow[45-88\%]{NaAlH_2(OCH_2CH_2OCH_3)_2} ArNH_2$$

[1] S. Mori, T. Aoyama, and T. Shioiri, *Tetrahedron Letters,* **25,** 429 (1984).

Diphenylsulfonium cyclopropylide, 4, 211–214; **5,** 281.

Trialkylation of ketones.[1] The silyl ether of a vinylcyclopropanol such as **1,** obtained by reaction of a ketone with diphenylsulfonium cyclopropylide followed by silylation, reacts with acetals in the presence of trimethylsilyl triflate (**10**, 438) to give a mixture of four products (**2**) of ring (Z)-alkylation.

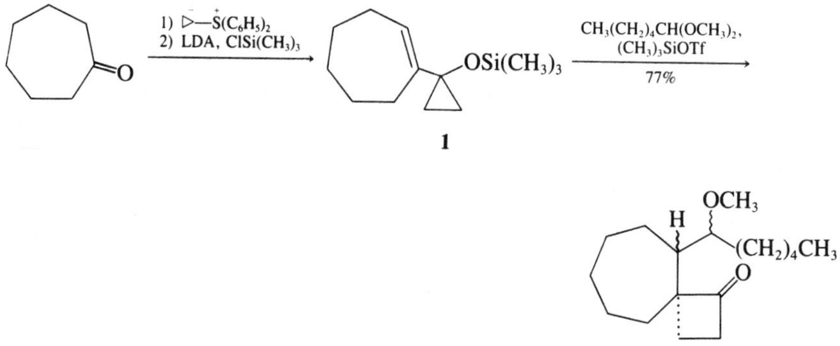

1) ▷—$\overset{+}{S}(C_6H_5)_2$
2) LDA, $ClSi(CH_3)_3$
$OSi(CH_3)_3$
$CH_3(CH_2)_4CH(OCH_3)_2$,
$(CH_3)_3SiOTf$
77%
1
OCH_3
H
$(CH_2)_4CH_3$
O

2

An intramolecular version of this reaction has been reported (equation I).

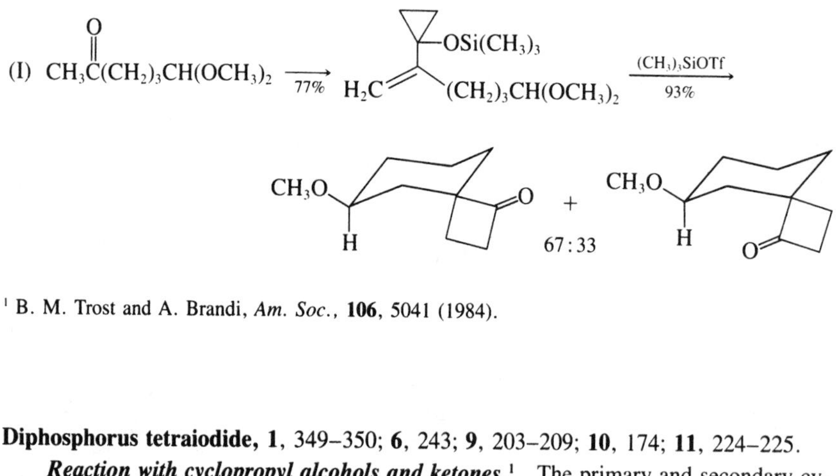

(I) $CH_3\overset{O}{\overset{\|}{C}}(CH_2)_3CH(OCH_3)_2$
77%
H_2C
$OSi(CH_3)_3$
$(CH_2)_3CH(OCH_3)_2$
$(CH_3)_3SiOTf$
93%
CH_3O
H
O
+
67:33
CH_3O
H
O

[1] B. M. Trost and A. Brandi, *Am. Soc.*, **106**, 5041 (1984).

Diphosphorus tetraiodide, 1, 349–350; **6**, 243; **9**, 203–209; **10**, 174; **11**, 224–225.

Reaction with cyclopropyl alcohols and ketones.[1] The primary and secondary cyclopropyl alcohols **1** and **2** are converted by reaction with P_2I_4 or PI_3 mainly into the corresponding iodides. In contrast, the tertiary α-cyclopropyl alcohol **3** is converted into the homoallylic iodide **4** under very mild conditions. α-Cyclopropyl aldehydes and ketones are also cleaved by P_2I_4 to γ-iodo carbonyl compounds in high yield, particularly when the solvent is acetone.

Reduction of thiiranes to alkenes.[2] P_2I_4 (DMF, 80°) converts a variety of episulfides into the corresponding alkenes (65–95% yield). With α,β-disubstituted episulfides the stereochemistry is retained. A number of other reagents, including Raney Ni in ethanol

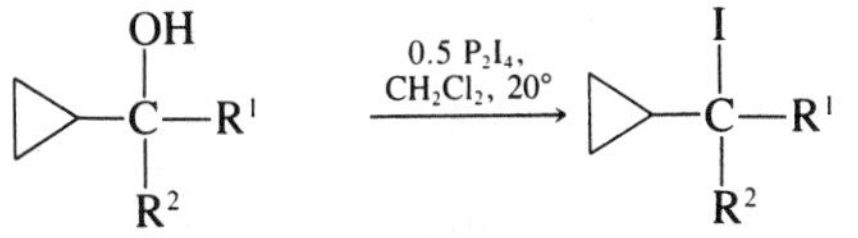

1, R^1, R^2 = H   20 hr.   71%
2, R^1 = H, R^2 = CH_3 1.5 hr.   76%

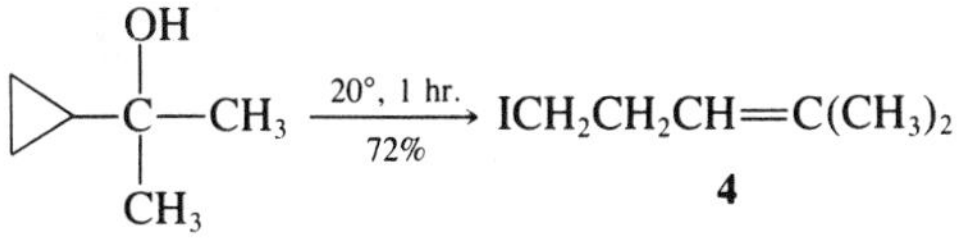

at 90°, Li in $C_2H_5NH_2$ at −15°, and Bu_3SnH at 110°, effect this reduction, but without complete retention of stereochemistry.

[1] J. N. Denis and A. Krief, *J.C.S. Chem. Comm.*, 229 (1983).
[2] J. R. Schauder, J. N. Denis, and A. Krief, *Tetrahedron Letters*, **24**, 1657 (1983).

Di(N-succinimidyl)oxalate, 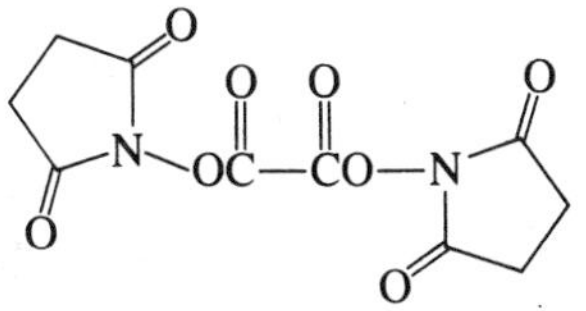 (**1**). Mol. wt. 284.18, m.p. 245°

(dec.). The reagent is prepared in 80% yield by reaction of N-hydroxysuccinimide with oxalyl chloride and pyridine. Supplier: Fluka.

Peptide synthesis.[1] This reagent, in contrast to N-hydroxysuccinimide (**1**, 487), reacts directly with N-protected amino acids in the presence of pyridine to form an active ester, which can be used for peptide synthesis.

Example:

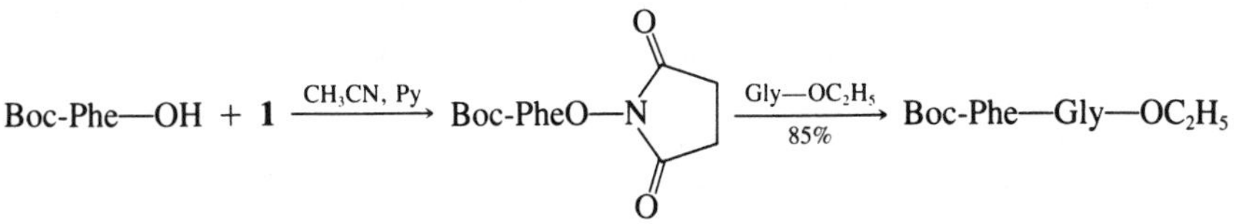

[1] K. Takeda, I. Sawada, A. Suzuki, and H. Ogura, *Tetrahedron Letters*, **24**, 4451 (1983).

Dodecamethylcyclohexasilane, $[(CH_3)_2Si]_6$ (**1**). Mol. wt. 348.94, m.p. 250°. The reagent is obtained in high yield by reaction of $(CH_3)_2SiCl_2$ with lithium (2 equiv.) in THF.[1]

Allylic dimethylmethoxysilanes.[2]   Dimethylsilylene generated by photolysis of **1** reacts with allylic methyl ethers to form an intermediate that undergoes a [2,3] sigmatropic rearrangement to form an allylic methoxydimethylsilane in modest yield. The products are converted in quantitative yield to the corresponding allylic trimethylsilanes by reaction with methyllithium. Reactions with dimethylsilylene generated by reaction of dichloro-dimethylsilane with sodium or lithium are not always regioselective, but yields are generally higher.

Example:

$$CH_2=CHC\underset{\underset{CH_3}{|}}{\overset{\overset{CH_3}{|}}{-}}OCH_3 \xrightarrow[50\%]{(CH_3)_2Si:, \ C_6H_{12}} (CH_3)_2CH=CHCH_2\underset{\underset{CH_3}{|}}{\overset{\overset{CH_3}{|}}{Si}}OCH_3 \xrightarrow[90\%]{CH_3Li, \ ether} (CH_3)_2C=CHCH_2Si(CH_3)_3$$

This reaction was used for a synthesis of the seven-membered cyclic terpene karahanaenol (**3**) from the methyl ether (**2**) of geraniol by an electrophilic cyclization of an allylsilane.[3]

[1] M. Laguerre, J. Donogues, and R. Calas, *J.C.S. Chem. Comm.*, 272 (1978).
[2] D. Tzeng and W. P. Weber, *J. Org.*, **46**, 693 (1981).
[3] D. Wang and T.-H. Chan, *J.C.S. Chem. Comm.*, 1273 (1984).

***p*-Dodecylbenzenesulfonyl azide, 11**, 535–536. This relatively safe azide is prepared from the commercially available mixture of isomeric *p*-dodecylbenzenesulfonic acids (K and K). The reaction of the sulfonyl chloride mixture with sodium azide is carried out in aqueous hexane or in acetone with Aliquat 336 as the phase-transfer catalyst (90–95% yield).[1]

[1] G. G. Hazen, F. W. Bollinger, F. E. Roberts, W. K. Russ, J. J. Seman, and S. Staskiewicz, *Org. Syn.*, submitted (1984).

E

N-(Ethoxycarbonyl)phthalimide (1), 1, 111–112. Material prepared by the original method is contaminated with phthalimide, which can be removed by fractional crystallization from chloroform. Pure reagent, m.p. 83°, can be obtained in a yield of 95%.[1]

N-Phthaloyl derivatives of α-amino acids and alcohols can be obtained without racemization by reaction with **1** in THF (addition of triethylamine can be helpful). The $H_2NCOOC_2H_5$ also formed can be removed by evaporation under reduced pressure. Yields are generally 70–80%.[2]

[1] P. M. Worster, C. C. Leznoff, and C. R. McArthur, *J. Org.*, **45**, 174 (1980).
[2] C. R. McArthur, P. M. Worster, J.-L. Jiang, and C. C. Leznoff, *Can. J. Chem.*, **60**, 1836 (1982); C. R. McArthur, P. M. Worster, and A. U. Okon, *Syn. Comm.*, **13**, 311 (1983).

Ethoxy(trimethylsilyl)acetylene, $(CH_3)_3SiC{\equiv}COC_2H_5$ **(1).** The reagent is prepared by silylation[1] of ethoxyacetylene.

Carboxylic anhydrides. Ethoxyacetylene has seen limited use for dehydration of carboxylic acids because of its instability to heat and high volatility.[2] This derivative (**1**) converts carboxylic acids, even ones that are inert to ethoxyacetylene, into anhydrides in 90–100% yield in CH_2Cl_2 at 25° to reflux temperatures.[3]

Example:

$$(E)\text{-}C_6H_5CH{=}CHCOOH \xrightarrow[100\%]{1,\ CH_2Cl_2,\ \Delta}$$

$$[(E)\text{-}C_6H_5CH{=}CHCO]_2O\ +\ (CH_3)_3SiCH_2COOC_2H_5$$

[1] R. A. Ruden, *J. Org.*, **39**, 3607 (1974).
[2] J. R. Edman and H. E. Simmons, *ibid.*, **33**, 3808 (1968).
[3] Y. Kita, S. Akai, M. Yoshigi, Y. Nakajima, H. Yasuda, and Y. Tamura, *Tetrahedron Letters*, **25**, 6027 (1984).

1-Ethoxy-1-trimethylsilyloxycyclopropane (1), 8, 219–220.

4- *and* 6-Keto esters.[1] Reaction of **1** with $ZnCl_2$ in ether (sonication) generates $ClSi(CH_3)_3$ and the zinc homoenolate **2**. In the presence of HMPT (1 equiv.) and a catalytic amount of $CuBr \cdot S(CH_3)_2$, the mixture undergoes conjugate addition to enones, enals, and ynones to give the trimethylsilyl enol ether of 6-keto esters. No reaction occurs in the absence of $ClSi(CH_3)_3$, which may be necessary for activation of the Michael acceptor. HMPT is probably required for transmetallation to copper.

Example:

$$1 \xrightarrow{\text{ZnCl}_2, \text{ ether}} \text{Zn(CH}_2\text{CH}_2\text{COOC}_2\text{H}_5)_2 + \text{ClSi(CH}_3)_3$$

1   **2**

76% CuBr·S(CH₃)₂, HMPT,

This homoenolate anion also acylates acid chlorides readily to give γ-keto esters (equation I), but does not react with aldehydes or epoxides.

(I) **1** $\xrightarrow[\text{3) HMPT}]{\substack{\text{1) ZnCl}_2, \text{ ether} \\ \text{2) CuI}}} \xrightarrow[89\%]{\text{C}_6\text{H}_5(\text{CH}_2)_2\text{COCl}}$ $\text{C}_6\text{H}_5(\text{CH}_2)_2\overset{\overset{\displaystyle O}{\|}}{\text{C}}(\text{CH}_2)_2\text{COOC}_2\text{H}_5$

¹ E. Nakamura and I. Kuwajima, *Am. Soc.*, **106**, 3368 (1984); *idem, Org. Syn.*, submitted (1984).

Ethyl acetoacetate, $CH_3COCH_2COOC_2H_5$.

Esterification of amino acids. The amino group of an amino acid can be protected as the enamine **1**, formed by reaction with ethyl acetoacetate¹ and a base in benzene/DMSO.² After alkylation the protective group is removed by addition of an acid, particularly TsOH, to provide amino acid ester salts (equation I).

(I) $H_2NCHCOOH + CH_3COCH_2COOC_2H_5 \xrightarrow[\text{C}_6\text{H}_6, \text{ DMSO}]{\text{KOH, K}_2\text{CO}_3,}$

R^1 (above first structure)

1

$\xrightarrow[75-90\%]{\substack{\text{1) R}^2\text{X} \\ \text{2) TsOH}}}$ $\underset{NH_3^+TsO^-}{R^1CHCOOR^2}$

¹ J. A. MacLaren, *Aust. J. Chem.*, **25**, 1293 (1972).
² A. M. Kotodziejczyk and M. Slebioda, *Synthesis*, 865 (1984).

Ethyl chloroformate, 1, 364–367; **2**, 193; **4**, 228–230; **5**, 294–295; **7**, 147; **9**, 214–215.

Primary allylic amines.[1] Aziridines can be converted into allyl carbamates by reaction with ethyl chloroformate to form N-ethoxycarbonylaziridines followed by thermolysis in benzene at 200–250°. Rearrangement of trisubstituted ethoxycarbonylaziridines is regiospecific.

Examples:

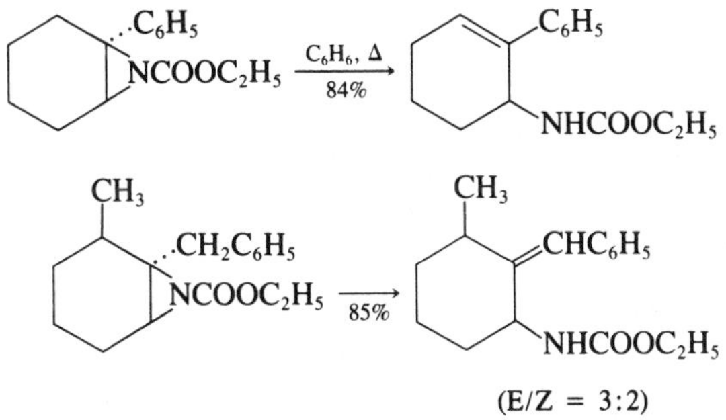

$(E/Z = 3:2)$

Azlactones.[2] N-Acyl-2-amino acids are converted into 2-oxazoline-5-ones in high yield by dehydration with ethyl chloroformate.

Example:

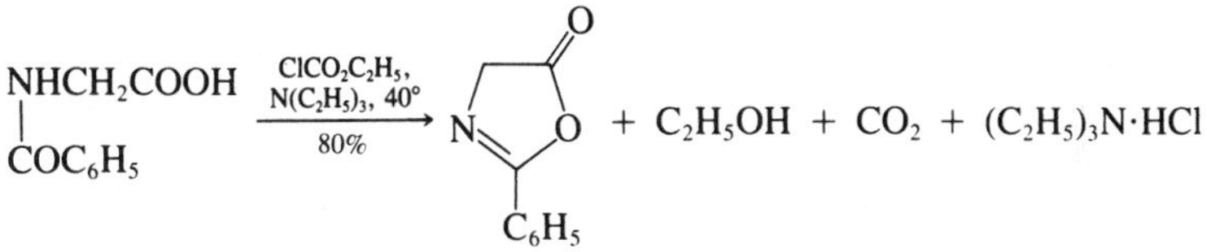

[1] A Laurent, P. Mison, A. Nafti, R. B. Cheikh, and R. Chaabouni, *J. Chem. Res. (M)*, 354 (1984).
[2] L. D. Taylor and T. E. Platt, *Org. Prep. Proc.*, **1**, 217 (1969); *idem, Org. Syn.*, submitted (1984).

Ethyl diazoacetate, 1, 367–370; **2**, 193–195; **3**, 138–139; **4**, 228–230; **5**, 295–300; **6**, 252–253; **8**, 222.

Homologation of ketones (**1**, 369–370; **6**, 252–253; **8**, 222).[1] Ethyl diazoacetate is recommended as the most useful diazoalkane for monohomologation of cyclic and acyclic ketones without formation of epoxides as by-products. One advantage is that the usually slow reaction can be catalyzed by BF_3 etherate (or triethyloxonium tetrafluoroborate).

However, as with other diazoalkanes, the homologation is not regiospecific. This drawback can be overcome by homologation of an α- (or α'-) halo ketone followed by removal of the halogen by reduction with zinc. The products are readily decarboxylated either at 230° or with $CaCl_2 \cdot 2H_2O$ in DMSO at 150°. This method is regiospecific because the unsubstituted α-methylene group migrates owing to steric and electronic factors. One limitation is that hindered ketones fail to react.

Examples:

$$C_6H_5COCH_2Br \xrightarrow[\text{overall}]{98\%} C_6H_5CH_2COCH_3$$

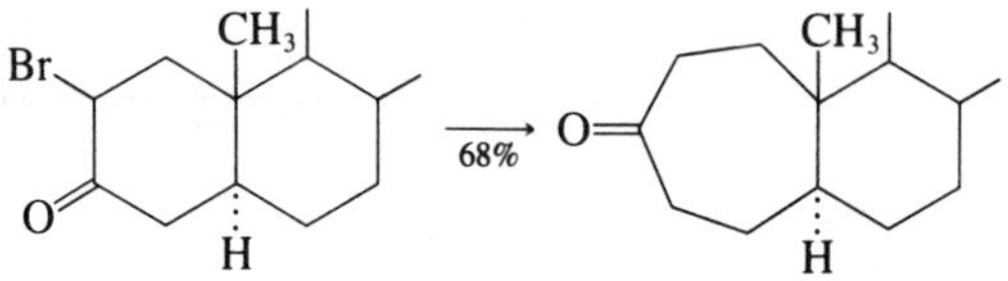

β,γ-Unsaturated acetals.[2]   Reaction of ethyl diazoacetate with dimethyl acetals of α,β-unsaturated aldehydes catalyzed by BF_3 etherate gives as the main product acetals of β,γ-unsaturated aldehydes by a carbon–carbon insertion. A similar reaction with ketals gives a complex mixture. The reaction is less selective with diethyl acetals.

Examples:

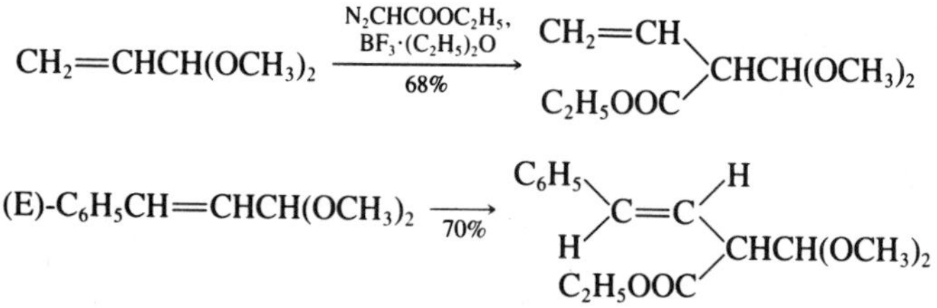

[1] V. Dave and E. W. Warnhoff, *J. Org.*, **48**, 2590 (1983).
[2] M. P. Doyle, M. L. Trudell, and J. W. Terpstra, *ibid.*, **48**, 5146 (1983).

Ethylene chloroboronate, B—Cl (1). The reagent is obtained by reaction of ethylene glycol with BCl_3 in CH_2Cl_2 (85% yield).[1]

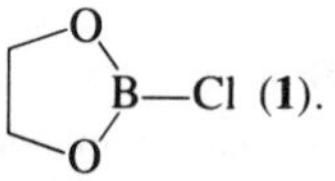

Regio- and stereoselective aldol condensations.   The enol boronates of ketones, obtained by reaction with 1 and diisopropylethylamine (1 equiv.), react with both aliphatic and aromatic ketones at −78° to −15° to form β-hydroxy ketones with high *syn*-diastereoselectivity.[2]

Example:

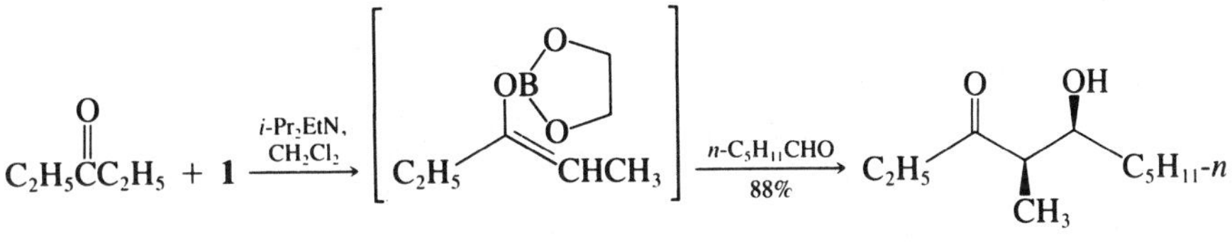

$$(syn/anti = 97:3)$$

The reaction of ketones (with the exception of methyl ethyl ketone) with **1** and diisopropylethyl amine (DPEA) leads to regioselective formation of alkenyloxy dialkoxy boranes. A methyl group is deprotonated with high selectivity. A C_2H_5 group is deprotonated almost exclusively in the presence of a CHR_2 group to form the (Z)-enolate (equation I).[3]

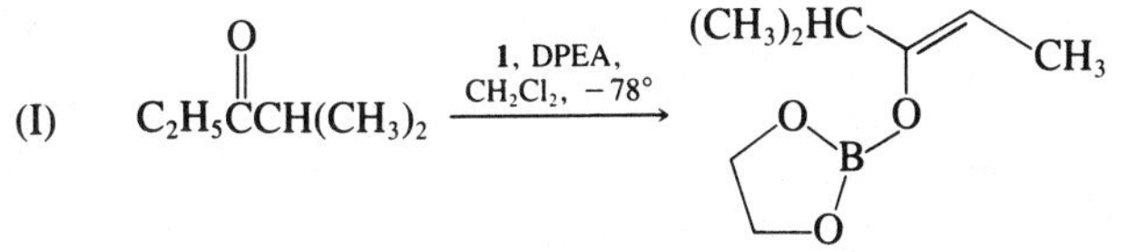

(I)

$$(Z/E \geqslant 95:5)$$

Both the (Z)-enolates from acyclic ketones and the (E)-enolates from cyclic ketones react with aldehydes to form β-hydroxy ketones with high *syn*-diastereoselectivity.
Examples:

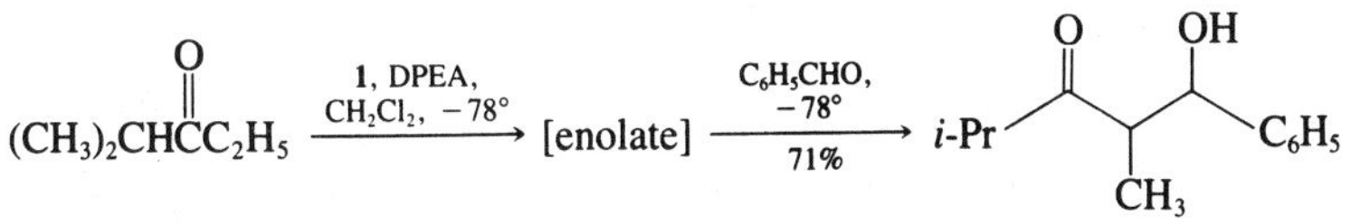

$$(syn/anti = 99:1)$$

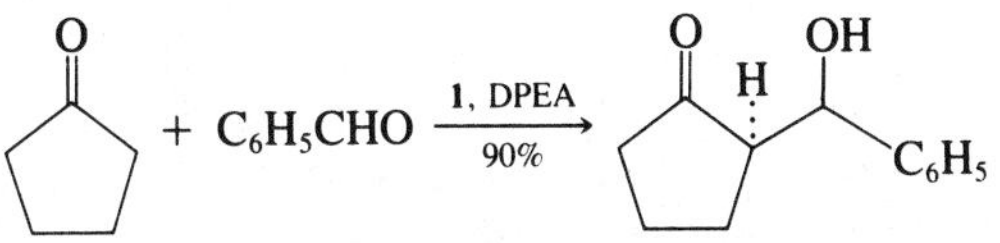

$$(syn/anti = 96:4)$$

Aldol condensations with phenyl thiopropionate (2).[4]   Reaction of **2** with **1** and a tertiary amine at $-78°$ results in a 70:30 mixture of (Z)- and (E)-enolates, whereas reaction at 25° furnishes an 18:82 mixture of these enolates. However, both enolates react with aldehydes with high *syn*-diastereoselectivity.

Example:

$$C_6H_5SCC_2H_5 \xrightarrow[\text{CH}_2\text{Cl}_2]{\text{1, DPEA,}} [\text{enolates}] \xrightarrow[55\%]{n\text{-}C_5H_{11}CHO} C_6H_5S$$

2

$$(syn/anti = 92:8)$$

[1] C. Gennari, L. Colombo, and G. Poli, *Tetrahedron Letters*, **25**, 2279 (1984).
[2] C. Gennari, S. Cardani, L. Colombo, and C. Scolastico, *ibid.*, 2283 (1984).
[3] C. Gennari, L. Colombo, C. Scolastico, and R. Todeschini, *Tetrahedron*, **40**, 4051 (1984).
[4] C. Gennari, A. Bernardi, S. Cardani, and C. Scolastico, *ibid.*, 4059 (1984).

Ethyl formate, 1, 380; **2**, 197; **4**, 233

α-*Alkylidenelactones*.[1] Formylation of γ-butyrolactone or δ-valerolactone followed by reaction with an aldehyde results in α-alkylidene lactones (equation I). α-Methylene lactones are obtained in high yield by this method, but mixtures of isomeric alkylidene lactones are usually formed.

(I)

n = 1,2

[1] A. W. Murray and R. G. Reid, *J.C.S. Chem. Comm.*, 132 (1984).

Ethyl (S)-lactate, $CH_3CH(OH)CO_2C_2H_5$ (**1**). Suppliers: Aldrich, Fluka.

Asymmetric Diels-Alder reactions.[1] Diels-Alder reaction of the acrylate (**2**) of ethyl (S)-lactate with cyclopentadiene gives four adducts with the usual preference for *endo*- rather than *exo*-addition, which is increased by use of a Lewis acid catalyst (*endo/exo* = 12–39:1). The Lewis acid catalyst can also control the diastereoface selectivity of *endo*-addition but not that of *exo*-addition (equation I). In the noncatalyzed reaction or in the presence of BF_3 etherate or $AlCl_3$, (1S,2S)-**3** is formed in excess. With $SnCl_4$ or $TiCl_4$, (1R,2R)-**3** is formed in excess. The stereoselectivity can be increased and polymerization decreased by addition of *n*-hexane. The adducts (**3**) can be hydrolyzed with LiOH in THF/H_2O at 20° without epimerization.

(I) 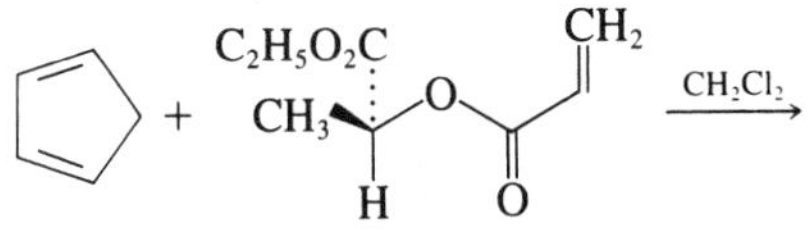   +   **2** $\xrightarrow{CH_2Cl_2}$

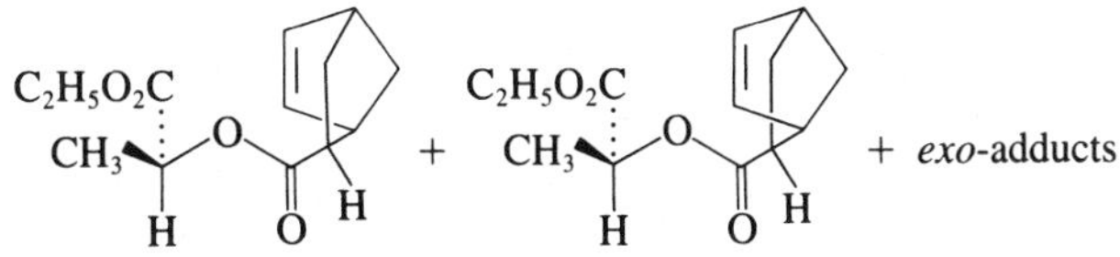

(1R,2R)-**3**   +   (1S,2S)-**3**   + *exo*-adducts

no cat.	42:58	
$BF_3 \cdot OC_2H_5$	34:66	12:1
$ZrCl_4$	48:52	13:1
$SnCl_4$	84:16	18:1
$TiCl_4$	91:9	31:1

[1] T. Poll, G. Helmchen, and B. Bauer, *Tetrahedron Letters,* **25,** 2191 (1984).

Ethyl lithio(trimethylsilyl)acetate, $(CH_3)_3SiCHCO_2C_2H_5$ **(1), 5,** 373–374; **6,** 255.

Indole alkaloids.[1]   The ester enolate **1** undergoes 1,4-addition to the pyridinium salt **2** to afford, after acid-promoted cyclization, the tetracycle **3** as an epimeric mixture. The

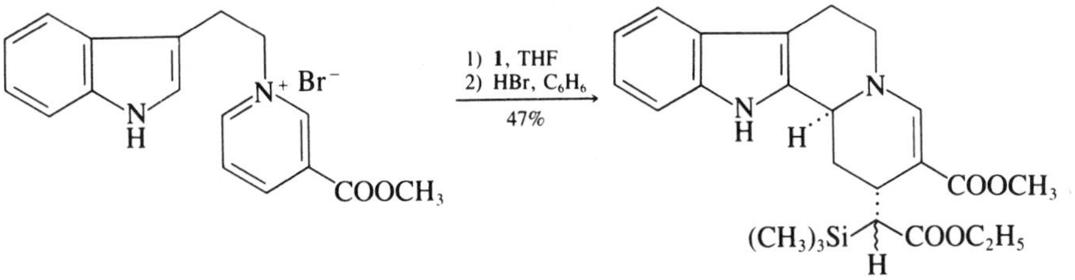

1) **1,** THF
2) HBr, C_6H_6

47%

2

3

Several steps

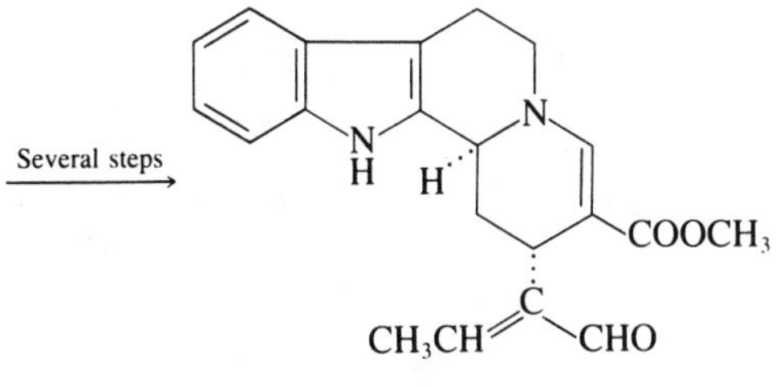

4 (Z/E = 3:1)

product was converted in a few steps into the alkaloid vallesiachotamine [(Z)-**4**] and the (E)-isomer.

[1] D. Spitzner and E. Wenkert, *Angew. Chem., Int. Ed.*, **23**, 984 (1984).

(R)-Ethyl p-tolylsulfinylmethylenepropionate, $p\text{-}CH_3C_6H_4$

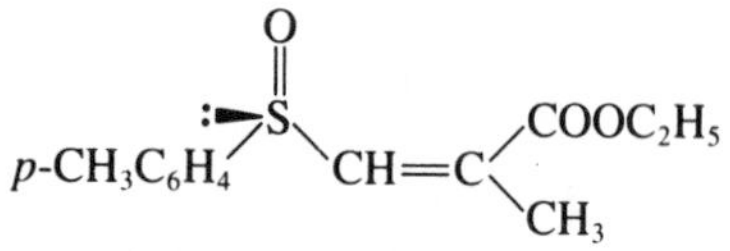

The reagent is obtained as a mixture of (E)- and (Z)-isomers from optically active diethyl p-tolylsulfinylmethanephosphonate.

Diels-Alder reactions.[1]   (R)-(E)- and (R)-(Z)-**1** react with cyclopentadiene to give as the major products the *endo*- and *exo*-adducts shown (Scheme I). The high diastereoselectivity is attributed to attachment of the chiral auxiliary directly to the double bond involved in the cycloaddition and to the difference in size between the lone-paired electrons and the p-tolyl group.

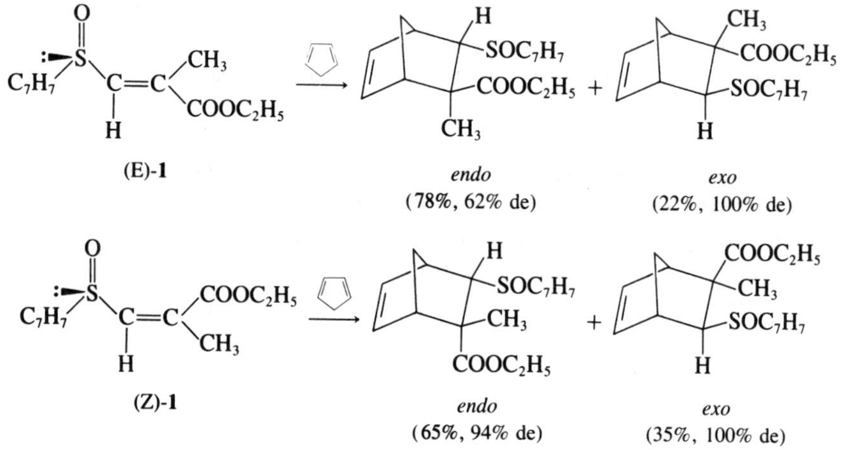

Scheme (I)

[1] T. Koizumi, I. Hakamada, and E. Yoshii, *Tetrahedron Letters*, **25**, 87 (1984).

Ethyl trimethylsilylacetate, 7, 150–152; **8,** 226–227; **11,** 234–235.

Peterson olefination.[1]   This reaction usually shows little or no selectivity. In contrast, the reaction of ethyl lithio(trimethylsilyl)acetate with α-substituted cyclohexanones can show moderate to high (Z)-selectivity. (Z)-Selectivity is somewhat higher in reaction with ethyl potassio(trimethylsilyl)acetate or t-butyl lithio(trimethylsilyl)acetate, but yields are lower.[2] (E)-Selectivity is favored in reaction of the same substrates with the Wittig-Horner reagent triethyl sodiophosphonoacetate.

Example:

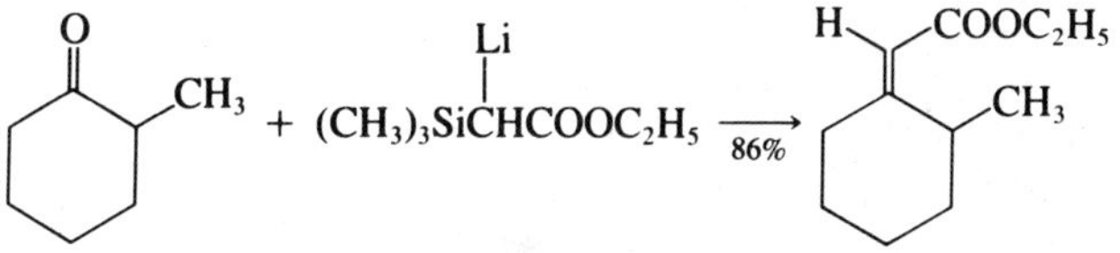

$$(Z/E \ = \ 89:11)$$

[1] Review: D. J. Ager, *Synthesis,* 384 (1984).
[2] L. Strekowski, M. Visnick, and M. A. Battiste, *Tetrahedron Letters,* **25,** 5603 (1984).

F

Ferric chloride, 1, 390–392; **2**, 199; **3**, 145; **4**, 236; **5**, 307–308; **6**, 260; **7**, 153–155; **8**, 229; **9**, 222; **10**, 185.

Photooxidation of cycloalkenes.[1] The FeCl$_3$-catalyzed photooxidation of cycloalkenes in pyridine can lead to three types of products, depending on the substitution pattern of the cycloalkenes. The three general reactions are shown in equations (I)–(III) for the oxidation of cyclohexenes.

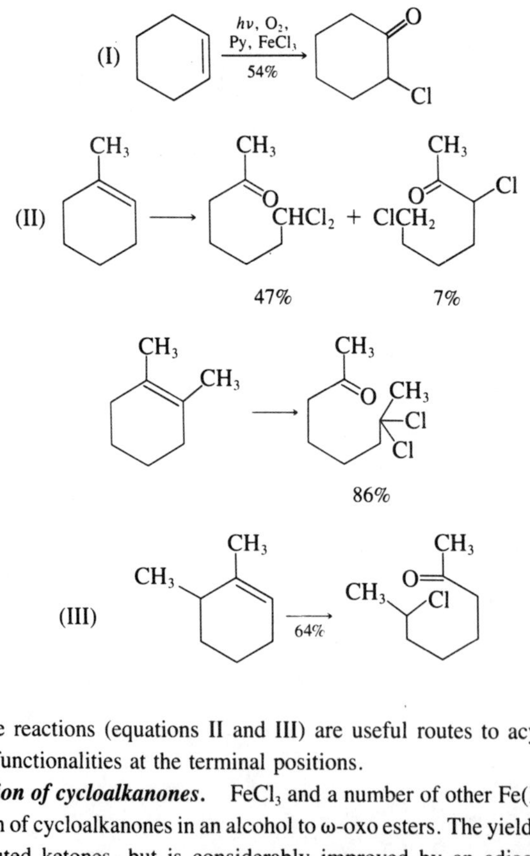

The cleavage reactions (equations II and III) are useful routes to acyclic compounds with different functionalities at the terminal positions.

Oxygenation of cycloalkanones. FeCl$_3$ and a number of other Fe(III) salts catalyze the O$_2$ oxidation of cycloalkanones in an alcohol to ω-oxo esters. The yield is only moderate with unsubstituted ketones, but is considerably improved by an adjacent alkyl group. Trifluoromethyl or methoxy groups inhibit oxygenation. Enol acetates are easily oxidized, but lactones are merely hydrolyzed.[2]

Example:

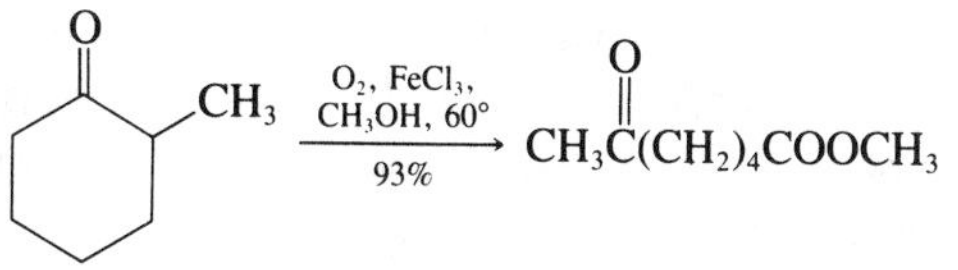

CH$_3$ $\xrightarrow[\text{93\%}]{\begin{array}{c}\text{O}_2,\ \text{FeCl}_3,\\\text{CH}_3\text{OH},\ 60°\end{array}}$ CH$_3$C(CH$_2$)$_4$COOCH$_3$

---

[1] A. Kohda, K. Nagayashi, K. Maemoto, and T. Sato, *J. Org.*, **48**, 425 (1983).
[2] S. Ito and M. Matsumoto, *ibid.*, **48**, 1133 (1983).

Ferric nitrate/K10 Bentonite (Clayfen), **11**, 237.

Cleavage of N,N-dimethylhydrazones.[1]   These hydrazones are cleaved to ketones by clay-supported Fe(NO$_3$)$_3$ in 70–90% yield.

Cleavage of thioacetals.[2]   Fe(NO$_3$)$_3$ or Cu(NO$_3$)$_2$ supported on the clay cleaves thioacetals to the carbonyl compound in essentially quantitative yield. They are particularly efficient for cleavage of dithianes and dithiolanes.

Coupling of thiols.[3]   Reaction of thiols with Fe(NO$_3$)$_3$ supported on Clayfen results in coupling to disulfides via a thionitrite, RSNO. Yields are ~40–95%.

Oxidation of 1,4-dihydropyridines.[4]   Clay-supported Fe(NO$_3$)$_3$ or Cu(NO$_3$)$_2$ (Claycop) oxidizes 1,4-dihydropyridines to pyridines at 25°. Reactions are generally slower with Claycop than with Clayfen, but yields are usually higher.

Example:

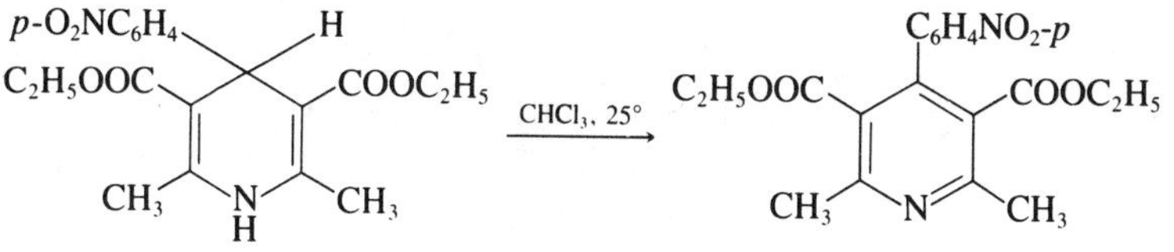

p-O$_2$NC$_6$H$_4$ H
C$_2$H$_5$OOC   COOC$_2$H$_5$ $\xrightarrow{\text{CHCl}_3,\ 25°}$ C$_6$H$_4$NO$_2$-*p*
CH$_3$ N CH$_3$ C$_2$H$_5$OOC   COOC$_2$H$_5$
     H   CH$_3$ N CH$_3$

Fe(NO$_3$)$_3$	1 hr.	87.4%
Cu(NO$_3$)$_2$	6 hr.	92.8%

---

[1] P. Laszlo and E. Polla, *Tetrahedron Letters*, **25**, 3309 (1984).
[2] M. Balogh, A. Cornélis, and P. Laszlo, *ibid.*, **25**, 3313 (1984).
[3] A. Cornélis, N. Depaye, A. Gerstmans, and P. Laszlo, *ibid.*, **24**, 3103 (1983).
[4] M. Balogh, I. Hermecz, Z. Meszaros, and P. Laszlo, *Helv.*, **67**, 2270 (1984).

N-Fluoro-N-alkyl-*p*-toluenesulfonamides, *p*-CH$_3$C$_6$H$_4$SO$_2$NFR (1).

These generally stable fluorosulfonamides are prepared by treatment of N-alkyl-*p*-toluenesulfonamides in CFCl$_3$/CHCl$_3$ at −78° with fluorine diluted in nitrogen (~50% yield).

Fluorination of carbanions.[1]   These reagents selectively fluorinate a broad variety of carbanions in fair to good yield. Oxygen and nitrogen anions are not affected. The base can be NaH, KH, or *n*-BuLi, although lithium enolates are less reactive than potassium enolates. Nonpolar solvents are preferable to DMF, THF, or ether. Strongly basic anions can effect β-elimination of HF from **1** when R = CH_3. Elimination is suppressed when R = neopentyl or norbornyl.

Examples:

$$C_6H_5CH(COOC_2H_5)_2 \xrightarrow[81\%]{\substack{1)\ NaH \\ 2)\ \mathbf{1},\ 20°}} C_6H_5CF(COOC_2H_5)_2$$

$$C_6H_5MgBr \xrightarrow[50\%]{\mathbf{1},\ 20°} C_6H_5F$$

[1] W. E. Barnette, *Am. Soc.*, **106**, 452 (1984).

Formaldehyde, 1, 397–402; **2**, 200–201; **4**, 238–239; **5**, 312; **6**, 264–267; **7**, 158–160; **8**, 231–232; **9**, 224–225; **10**, 186; **11**, 240–241.

Monomeric HCHO.[1]   Details are available for preparation of ethereal solutions of HCHO by pyrolysis of paraformaldehyde at 150° in a special apparatus. These solutions can be stored for an extended time at −78° under argon.

Anhydrous reagent is particularly advantageous in the trapping of regiospecifically generated enolates in the preparation of α-hydroxymethyl ketones (**6**, 264–267).

α-Methylene ketones or esters.[2]   The first step in a three-step conversion of α-nitro ketones (**1**)[3] into α-methylene ketones (**4**) is the reaction of formaldehyde (37%) catalyzed by triphenylphosphine to give, after acetylation, the acetoxymethylated product (**2**). The remaining steps involve denitration followed by loss of acetic acid.

Example:

$$CH_3(CH_2)_2\overset{\overset{\displaystyle O}{\|}}{C}\underset{\underset{\displaystyle NO_2}{|}}{C}HC_2H_5 \xrightarrow[76\%]{\substack{1)\ HCHO,\ P(C_6H_5)_3,\ (CH_3)_2CHOH \\ 2)\ Ac_2O,\ Py}} CH_3(CH_2)_2\overset{\overset{\displaystyle O}{\|}}{C}-\underset{\underset{\displaystyle NO_2}{|}}{\overset{\overset{\displaystyle C_2H_5}{|}}{C}}CH_2OAc$$

$$\textbf{1} \qquad\qquad\qquad\qquad\qquad \textbf{2}$$

$$\xrightarrow[74\%]{\substack{Bu_3SnH, \\ AIBN.\ C_6H_6}} CH_3(CH_2)_2\overset{\overset{\displaystyle O}{\|}}{C}-\underset{\underset{\displaystyle H}{|}}{\overset{\overset{\displaystyle C_2H_5}{|}}{C}}CH_2OAc \xrightarrow[90\%]{\substack{DBU. \\ C_6H_6}} CH_3(CH_2)_2\overset{\overset{\displaystyle O}{\|}}{C}-\overset{\overset{\displaystyle C_2H_5}{|}}{C}=CH_2$$

$$\textbf{3} \qquad\qquad\qquad\qquad\qquad\qquad \textbf{4}$$

The same method can be used to convert α-nitro esters into α-methylene esters, except that the hydroxymethylation step is catalyzed with sodium hydroxide rather than triphenylphosphine.

Methylation of arylamines.[4]   Reaction of $ArNH_2$ with paraformaldehyde, CH_3ONa, and CH_3OH results in N-(methoxymethyl)-N-arylamines (**2**), which slowly decompose to form a triazinane via an imine, $ArN{=}CH_2$ (**a**). The imine can be reduced by $NaBH_4$ in CH_3OH to $ArNHCH_3$ (**3**) in high yield.

Example:

$$C_6H_5NH_2 \xrightarrow{\substack{(CH_2O)_n, \\ CH_3ONa,\ CH_3OH}} C_6H_5NHCH_2OCH_3 \xrightarrow{\substack{CH_3OH, \\ \Delta}} [C_6H_5N{=}CH_2]$$

$$\textbf{1} \qquad\qquad\qquad \textbf{2} \qquad\qquad\qquad\qquad \textbf{a}$$

$$\xrightarrow{NaBH_4} C_6H_5NHCH_3$$

$$\textbf{3}$$

The reaction is conveniently carried out without isolation of the intermediate **2** in 82% yield.

Primary and secondary amines also undergo reductive methylation in >80% yield on reaction with aqueous formaldehyde and a salt of phosphorous acid (NaH_2PO_3) in dioxane.[5]

[1] A. Mitra, *Org. Syn.*, submitted (1983).
[2] N. Ono, H. Miyake, M. Fujii, and A. Kaji, *Tetrahedron Letters*, **24**, 3477 (1983).
[3] Preparation by acylation of RCH_2NO_2: R. L. Crumbie, J. S. Nimitz, and H. S. Mosher, *J. Org.*, **47**, 4040 (1982).
[4] J. Barluenga, A. M. Bayón, and G. Asensio, *J.C.S. Chem. Comm.*, 1334 (1984).
[5] H. Loibner, A. Pruckner, and A. Stutz, *Tetrahedron Letters*, **25**, 2535 (1984).

(α-Formylethylidene)triphenylphosphorane, $(C_6H_5)_3P{=}C\begin{smallmatrix}CH_3\\CHO\end{smallmatrix}$ (**1**). Mol. wt. 318.34, m.p. 220–222°. The Wittig reagent is prepared by reaction of ethyl formate with ethylidenetriphenylphosphorane.[1]

Furanes.[2] The reagent is alkylated by an α-bromo ketone on oxygen to give an oxophosphorane, which is converted regioselectively to a furane. An example is the synthesis of menthofurane (**2**).

2

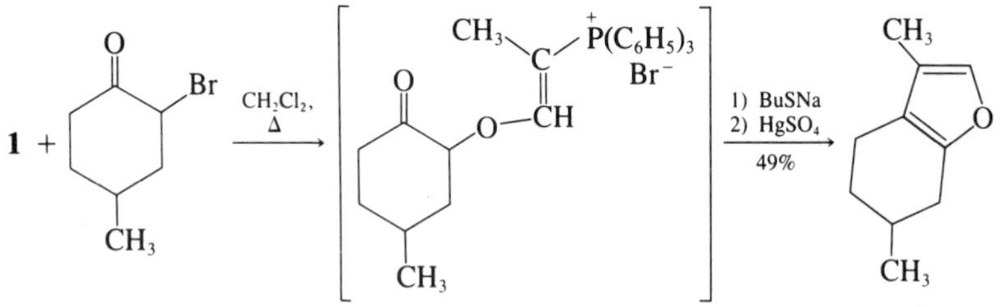

[1] S. Trippett and D. M. Walker, *J. Chem. Soc.*, 1266 (1961).
[2] M. E. Garst and T. A. Spencer, *J. Org.*, **48**, 2442 (1983).

G

Grignard reagents, 1, 415–424; **2**, 205; **5**, 321; **6**, 269–270; **7**, 163–164; **8**, 235–238; **9**, 229–233; **10**, 189–194; **11**, 245–249.

Acyloins.[1] Acyloins, both aliphatic and aromatic, can be prepared in good yield by reaction of Grignard reagents with O-trimethylsilyl cyanohydrins, which are readily available by reaction of cyanotrimethylsilane with aldehydes and ketones (**4**, 542–543).

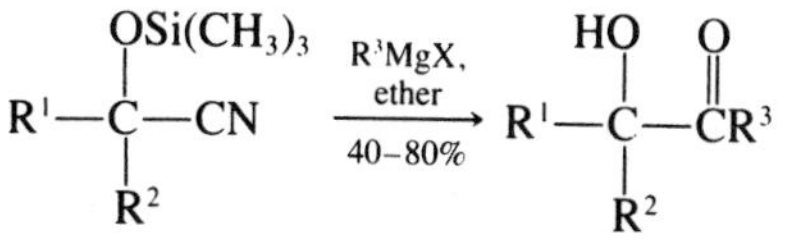

$$R^1\!-\!\underset{\underset{R^2}{\displaystyle |}}{\overset{\overset{\displaystyle OSi(CH_3)_3}{\displaystyle |}}{C}}\!-\!CN \quad \xrightarrow[\text{40-80\%}]{\substack{R^1MgX, \\ \text{ether}}} \quad R^1\!-\!\underset{\underset{R^2}{\displaystyle |}}{\overset{\overset{\displaystyle HO}{\displaystyle |}}{C}}\!-\!\overset{\overset{\displaystyle O}{\displaystyle \|}}{C}R^3$$

[1] L. R. Krepski, S. M. Heilmann, and J. K. Rasmussen, *Tetrahedron Letters,* **24**, 4075 (1983).

H

B-Halo-9-borabicyclo[3.3.1]nonene (B-X-9-BBN, X = Cl, Br, I). These derivatives can be prepared most simply by reaction of 9-BBN with the halogens (75–80% yield) or with the hydrogen halides (80–85% yield). They are all liquids, but form solid complexes with $S(CH_3)_2$.[1]

Haloboration. The reaction of B-Br- and B-I-9-BBN with 1-alkynes leads, after protonolysis, to 2-halo-1-alkenes in high yield (equation I).[2]

$$(I)\ RC{\equiv}CH + XB\bigcirc \longrightarrow \underset{X}{\overset{R}{\diagdown}}C{=}C\underset{B\bigcirc}{\overset{H}{\diagup}} \xrightarrow[90-100\%]{HOAc} \underset{X}{\overset{R}{\diagdown}}C{=}C\underset{H}{\overset{H}{\diagup}}$$

$$(X = I,\ Br)$$

The intermediate can be converted into (Z)-1-alkynyl-2-halo-1-alkenes by sequential reaction with a lithium acetylide and iodine (equation II).[3]

$$(II)\ \underset{X}{\overset{R^1}{\diagdown}}C{=}C\underset{B\bigcirc}{\overset{H}{\diagup}} \xrightarrow{LiC{\equiv}CR^2} \left[\underset{X}{\overset{R}{\diagdown}}C{=}C\underset{B^-\!-C{\equiv}CR^2}{\overset{H}{\diagup}} Li^+ \right] \xrightarrow[60-75\%]{I_2} \underset{X}{\overset{R^1}{\diagdown}}C{=}C\underset{C{\equiv}CR^2}{\overset{H}{\diagup}}$$

[1] H. C. Brown and S. U. Kulkarni, *J. Organometal. Chem.*, **168**, 281 (1977).
[2] S. Hara, H. Dojo, S. Takinami, and A. Suzuki, *Tetrahedron Letters*, **24**, 731 (1983).
[3] S. Hara, Y. Satoh, H. Ishiguro, and A. Suzuki, *ibid.*, **24**, 735 (1983).

1,3,3a,4,7,7a-Hexahydro-4,7-methanobenzo[*c*]thiophene 2,2-dioxide (1).
Preparation:[1]

1

(E)-1,3-Dienes;[2] (E,E)-conjugated dienes.[3]   Metallation and alkylation of **1** results in a single product formed by *exo*-attack. Thermolysis of the product results in a terminal (E)-1,3-diene. An example is the synthesis of **2**, a sex pheromone of the red bollworm moth.

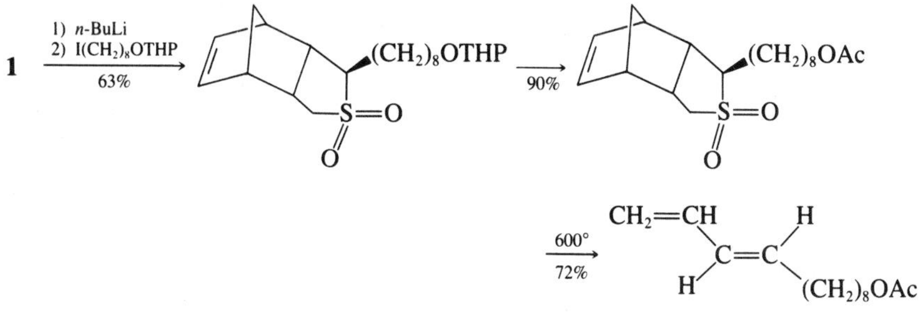

A second alkylation of **1** also results in a single product, also formed by *exo*-attack. Thermolysis of products of this type results in essentially pure (E,E)-conjugated dienes,

[1] Method of: J. M. Photis and L. A. Paquette, *Am. Soc.*, **96**, 4715 (1974).
[2] R. Bloch and J. Abecassis, *Tetrahedron Letters*, **23**, 3277 (1982).
[3] *Idem, ibid.*, **24**, 1247 (1983).

Hexahydro-4,4,7-trimethyl-4H-1,3-benzothiin (1). The optically pure **1** can be prepared from (+)-pulegone in about 30% yield. The isomer **2**, in which the oxathiane group is enantiomeric to that in **1**, is obtained as a by-product.

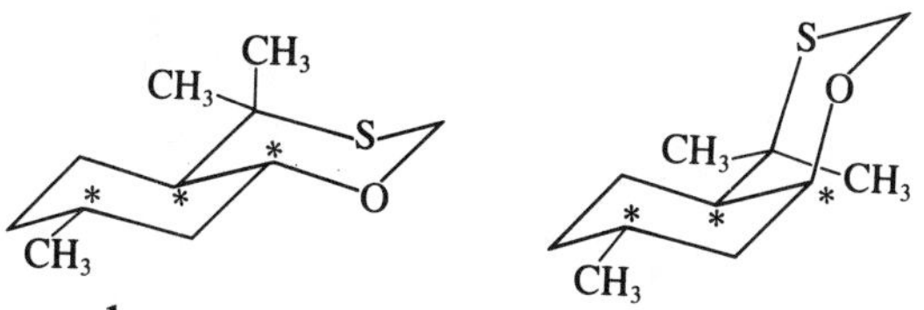

Chiral tertiary α-hydroxy aldehydes.[2]   This chiral 1,3-oxathiane (**1**) undergoes the highly selective reactions shown by 4,4,6-trimethyl-1,3-oxathiane (this volume) to give the optically active oxathiane carbinols **3** in greater than 90% diastereomeric purity. These are cleaved by NCS and silver nitrate (**4**, 216) to α-hydroxy aldehydes (**4**) and a sultine, which can be converted back to **1** (95% yield). The α-hydroxy aldehydes (**4**) are unstable

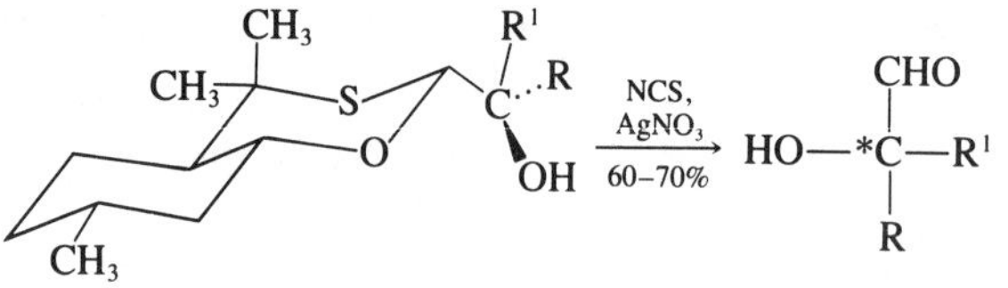

3   **4**

and are usually converted without isolation into α-hydroxy acids, glycols, and tertiary methylcarbinols (Scheme I).

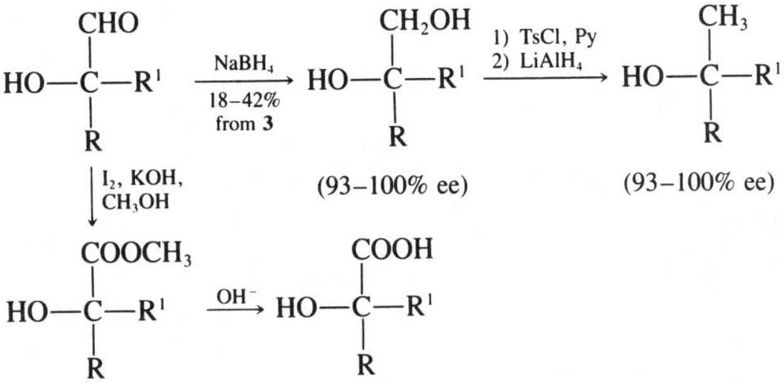

Scheme (I)

In theory, the epimeric α-hydroxy aldehydes can be obtained by reversing the order of introduction of R and R^1. When this procedure is impracticable or results in lower stereoselectivity, the oxathiane **2** can be used in place of **1**.

The reaction of a chiral Grignard reagent derived from ephedrine with the oxathiane **2** was used to prepare the δ-hydroxy acid **5** in 34% overall yield. On standing, **5** lactonizes to (−)-malyngolide (**6**).[3]

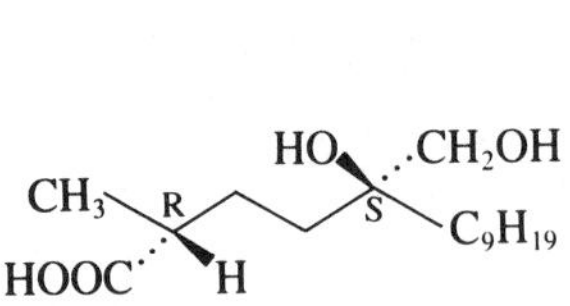

5 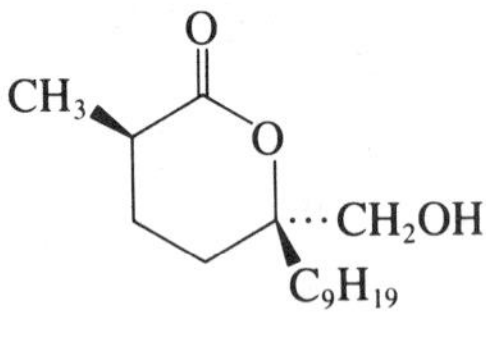   **6** (α_D − 13.4°)

Chiral secondary α-hydroxy aldehydes, R$\overset{*}{C}$HOHCHO.   The chiral acyl derivatives **3**, obtained from **1** by lithiation and reaction with an aldehyde followed by Swern oxidation, can be reduced stereoselectively before cleavage to the secondary α-hydroxy aldehydes.

Of the large number of reducing agents, the most useful are DIBAH and lithium tri-*sec*-butylborohydride. Regardless of the nature of R, DIBAH reduces **3** mainly to the alcohol **4** (the *anti*-Cram product) in 60–80% de. Reduction with L-Selectride usually proceeds in the opposite sense and in accordance with Cram's chelate rule, but high selectivity is observed only when R is a primary or tertiary alkyl group.

3 (R = CH$_3$)	Li(*sec*-Bu)$_3$BH	21:79
	DIBAH	78:22

[1] E. L. Eliel, J. E. Lynch, F. Kume, and S. V. Frye, *Org. Syn.*, submitted (1984).
[2] J. E. Lynch and E. L. Eliel, *Am. Soc.*, **106**, 2943 (1984).
[3] T. Kogure and E. L. Eliel, *J. Org.*, 49, 576 (1984).
[4] K.-Y. Ko, W. J. Frazee, and E. L. Eliel, *Tetrahedron*, **40**, 1333 (1984)

Hexamethyldisilizane, (CH$_3$)$_3$SiNHSi(CH$_3$)$_3$ (**1**), **1**, 427; **2**, 207–208; **5**, 323; **6**, 273; **7**, 167; **9**, 234.

Silylation–amination of hydroxy N-heterocycles.[1]   Silylation of a hydroxy N-heterocycle such as the pyridinone **2** gives a fully aromatic system (**a**), which undergoes ready amination catalyzed by various Lewis acids with formation of trimethylsilanol as the leaving group. The reactivity of the silylated heterocyclic system increases with an increasing number of nitrogen atoms and with an increasing number of annelated rings. In fact, 2-methoxy-1,2,4-triazines can be aminated directly.

[1] H. Vorbrüggen and K. Krolikiewicz, *Ber.*, **117**, 1523 (1984).

Hexamethylphosphoric triamide (HMPT), **1**, 425; **2**, 207; **3**, 148–149; **6**, 279–280; **9**, 235–236; **10**, 199; **11**, 253–254.

Stereoselective Michael reactions.[1]   Enolates of monosubstituted acetates (**1**) add to β-substituted (E)-α,β-unsaturated esters (**2**) in THF with slight *syn*-selectivity. The *syn*-selectivity can be dramatically increased by addition of HMPT.

Example:

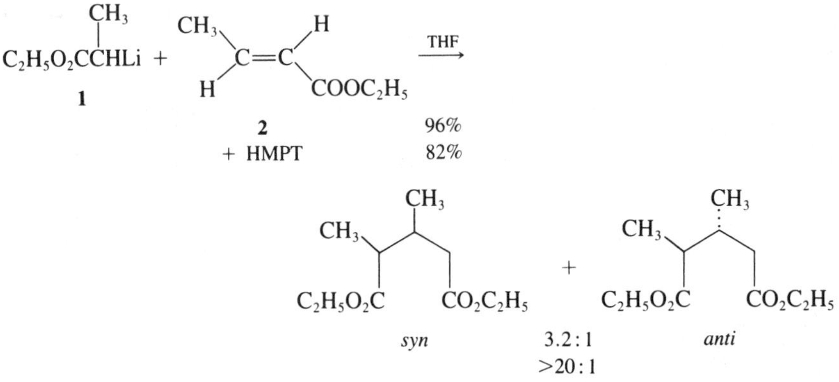

syn-Selectivity is favored to a lesser extent ($\sim 2:1$) in reactions of **1** with (Z)-isomers of **2** in THF/HMPT. However, *anti*-selectivity is obtained in reaction of the enolate of t-butyl propionate with **2** in the absence of HMPT ($\sim > 20:1$).

Asymmetric alkylation of esters.[2]   HMPT has a marked influence on the asymmetric alkylation of enolates of chiral esters, such as the propionates **1**, where R* is a derivative of (+)-camphor. A typical example is shown in equation (I). The esters are deprotonated with lithium cyclohexylisopropylamide (LICA) either in THF (A) or in THF/HMPT (B). In the first case, alkylation with benzyl bromide results mainly in (2S)-**2**; in the second case, formation of (2R)-**2** is markedly favored. The effect of HMPT is considered to result, at least in part, from preferential formation of the (E)-enolate of **1**.

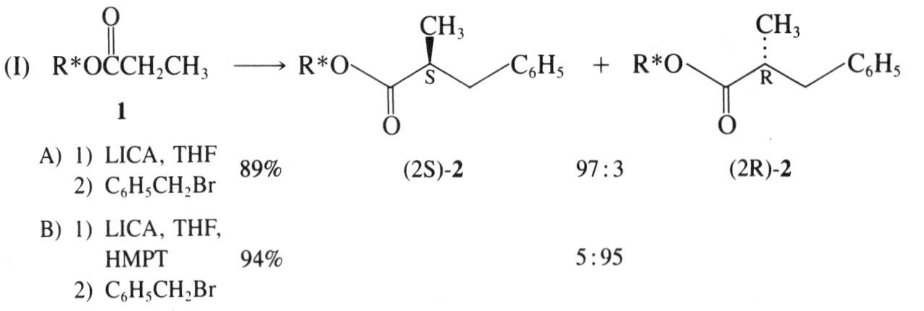

Arsonium ylides.[3]   Semistabilized allylic arsonium ylides are generally not useful in synthesis because they react with carbonyl compounds to form mixtures of epoxides and alkenes. Unexpectedly, the reaction of the semistabilized ylide triphenylarsonium cinnamylide (**1**) can be completely controlled to give either epoxides or alkenes. Epoxides are formed exclusively by reaction of **1** with the carbonyl compound in THF at $-78°$; alkenes are obtained mainly in reactions at $-30°$ in THF/HMPT in the ratio $70:30$ to $50:50$.

Example:

$$C_6H_5 \diagdown \diagup CH\!=\!As(C_6H_5)_3 + CH_3(CH_2)_2CHO$$

1

THF, $-78°$
85%
$$\longrightarrow CH_3(CH_2)_2\overset{\displaystyle O}{\overset{\diagup \diagdown}{CH}}\!-\!CHCH\!=\!CHC_6H_5$$

THF, HMPT
$-30°$
71%
$$\longrightarrow CH_3(CH_2)_2CH\!=\!CHCH\!=\!CHC_6H_5$$

[1] M. Yamaguchi, M. Tsukamoto, S. Tanaka, and I. Hirao, *Tetrahedron Letters*, **25**, 5661 (1984).
[2] G. Helmchen, A. Selim, D. Dorsch, and I. Taufer, *ibid.*, **24**, 3213 (1983).
[3] J. B. Ousset, C. Mioskowski, and G. Solladié, *Syn. Comm.*, **13**, 1193 (1983); *idem, Tetrahedron Letters*, **24**, 4419 (1983).

Hydrazine, 1, 434–445; **2**, 211; **3**, 153; **4**, 248; **5**, 327–329; **6**, 280–281; **7**, 170–171; **8**, 245; **9**, 236–237; **1**, 255.

Vinyl iodides[1] or selenides.[2] Some time ago Barton and Sternhall reported the conversion of ketone hydrazones to a mixture of vinyl iodides and *gem*-diiodides by reaction with iodine in the presence of triethylamine (**4**, 260). Formation of vinyl iodides is significantly enhanced if a strongly basic pentaalkylguanidine (**11**, 249–250) is used in place of triethylamine. Formation of diiodides is also suppressed by addition of the hydrazone to a solution of the base and an electrophile.

Example:

Use of benzeneselenenyl bromide as the electrophile in this reaction provides phenyl vinyl selenides in 70–90% yield.

[1] D. H. R. Barton, G. Bashiardes, and J.-L. Fourrey, *Tetrahedron Letters*, **24**, 1605 (1983).
[2] *Idem, ibid.*, **25**, 1287 (1984).

Hydrazoic acid, 1, 446–448; **2,** 211; **5,** 329.

Addition to alkenes.[1] Hydrazoic acid adds readily to enol ethers to give α-azido ethers. Addition to styrenes and 1,1-disubstituted or trisubstituted alkenes requires a Lewis acid catalyst ($TiCl_4$ or $AlCl_3$). Benzylic, allylic, and tertiary alcohols react with HN_3–$TiCl_4$ to form azides.

[1] A. Hassner, R. Fibiger, and D. Andisik, *J. Org.*, **49**, 4237 (1984).

Hydrogen bromide–Acetic acid; 1, 452; **5,** 335; **7,** 171.

Epoxides from **vic-***diols.*[1] The reaction of secondary, primary *vic*-diols with a mixture of 6 *M* hydrogen bromide in HOAc results in a *vic*-2-acetoxy-1-bromoalkane, which is converted into an alkyloxirane on treatment with base. The method is suitable for preparation of chiral alkyloxiranes.

Example:

(>98% S)

[1] B. T. Golding, D. R. Hall, and S. Sakrikar, *J.C.S. Perkin I*, 1214 (1973); M. K. Ellis and B. T. Golding, *Org. Syn.*, submitted (1983).

Hydrogen peroxide, 1, 457–471; **2,** 216–217; **3,** 154–155; **4,** 253–255; **5,** 337–339; **6,** 286; **7,** 174; **8,** 247–248; **9,** 241–242; **10,** 201–203.

Two-carbon ring expansion.[1] Some years ago Criegee[2] reported that the benzoate (**1**) of 9-decalyl hydroperoxide rearranges on standing by way of 1-benzoyloxy-1,6-epoxycyclodecane (**2**) to the hydroxycyclodecanone **3**.

This type of rearrangement has now been used to effect two-carbon ring expansion of cycloalkenylethyl brosylates by reaction with THF/90% H_2O_2 (*caution*). For example, **4** is converted in this way into **5**, presumably via a cyclopropyl hydroperoxide (**a**).

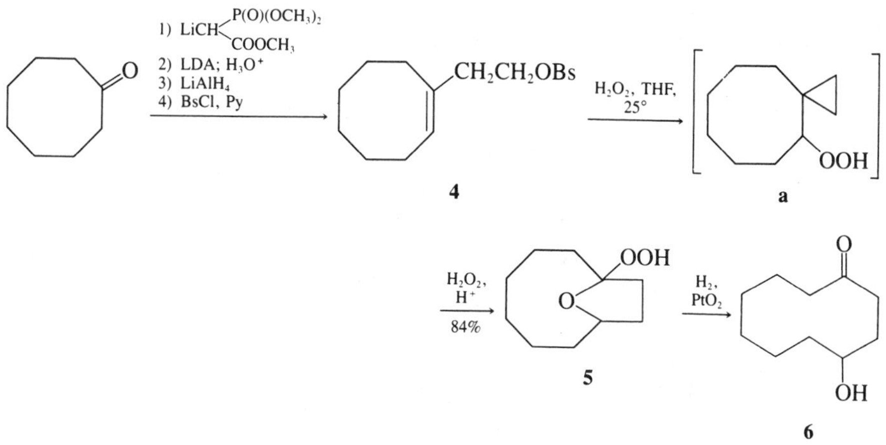

4

a

5

6

Another example:

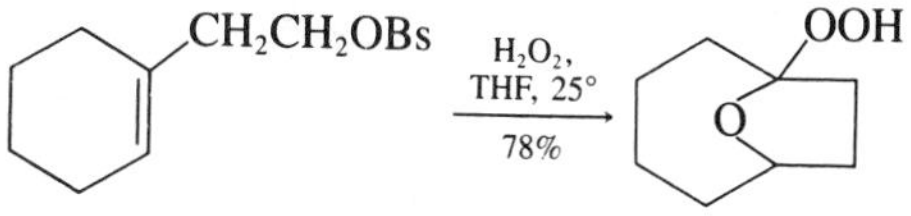

Oxidation of Si—C bonds. The carbon–silicon bond of organoalkoxysilanes is cleaved by 30% H_2O_2 or *m*-chloroperbenzoic acid in the presence of a base or under neutral conditions in the presence of KHF_2 (equation I). Only one alkoxy group on silicon

$$\text{(I) } RSiCH_3(OC_2H_5)_2 \xrightarrow[\sim 95\%]{\substack{H_2O_2,\ DMF,\\ KHF_2}} ROH$$

is essential, but yields are highest with substrates containing two such groups. The reaction proceeds with retention of configuration at the oxidized carbon atom.

This method can be used for a one-pot conversion of a terminal alkene into the *anti*-Markownikoff alcohol by a hydrosilylation–oxidation sequence (equation II).[3]

$$\text{(II) } R(CH_2)_nCH{=}CH_2 \xrightarrow[]{\substack{HSiCH_3(OC_2H_5)_2,\\ H_2PtCl_6}} [R(CH_2)_nCH_2CH_2SiCH_3(OC_2H_5)_2] \xrightarrow[70-80\%]{\substack{H_2O_2,\ KHF_2,\\ DMF}}$$

$$R(CH_2)_nCH_2CH_2OH$$

This oxidation effects selective cleavage of vinyl(alkoxy)silanes to give directly an aldehyde (or a carboxylic acid by further oxidation) or a ketone.[4]
Examples:

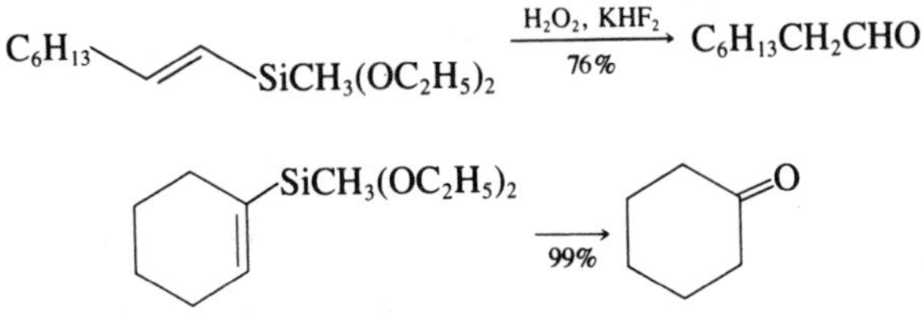

$$C_6H_{13} \diagdown \diagup \diagdown SiCH_3(OC_2H_5)_2 \xrightarrow[76\%]{H_2O_2, KHF_2} C_6H_{13}CH_2CHO$$

Hydroxymethylation of organic halides can be effected by metal-catalyzed cross-coupling with the Grignard reagent (diisopropoxymethylsilyl)methylmagnesium chloride (**1**) followed by oxidation. This two-step process is particularly useful for regio- and stereocontrolled conversion of vinyl halides into allylic alcohols and of allylic halides into homoallylic alcohols.[5]
Example:

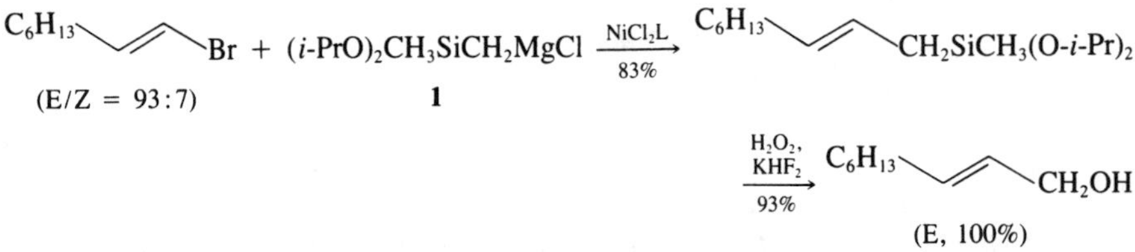

$$C_6H_{13} \diagdown \diagup \diagdown Br + (i\text{-PrO})_2CH_3SiCH_2MgCl \xrightarrow[83\%]{NiCl_2L} C_6H_{13} \diagdown \diagup \diagdown CH_2SiCH_3(O\text{-}i\text{-Pr})_2$$

(E/Z = 93:7)   **1**

$$\xrightarrow[93\%]{H_2O_2, KHF_2} C_6H_{13} \diagdown \diagup \diagdown CH_2OH$$

(E, 100%)

The Grignard reagent **1** does not undergo copper-catalyzed addition to α,β-enones, but the desired reaction can be effected with (allyldimethylsilyl)methylmagnesium chloride, $(CH_2=CHCH_2)(CH_3)_2SiCH_2MgCl$ (**2**), catalyzed by CuI. The allyl group of the adduct is converted into fluorine by reaction with KHF_2, and the resulting product is oxidized with H_2O_2.[6]
Example:

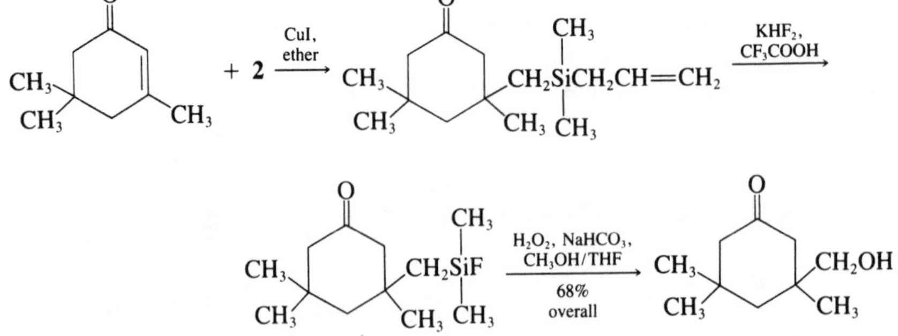

A general method for α-hydroxyalkylation[7] involves addition of a Grignard reagent or an alkyllithium to a vinyl(alkoxy)silane. Metal-catalyzed coupling of the resulting α-silycarbanion with an alkyl halide (or epoxide) followed by oxidation results in a secondary alcohol.

Example:

This method was used to prepare an optically active alcohol in 60% ee from the chiral vinylsilane **3**, derived from (−)-ephedrine, as outlined in equation (III).

Liberation of ligands from iron carbonyl complexes.[8]   This decomplexation can be conducted with 30% H_2O_2 and NaOH in CH_3OH at 0–20° in 75–95% yield. Partial oxidation of $Fe(CO)_5$ to $Fe_3(CO)_{12}$ can occur.

[1] R. C. Ronald and T. S. Lillie, *Am. Soc.*, **105**, 5709 (1983).

[2] R. Criegee, *Ber.*, **77**, 722 (1944); R. Criegee and W. Schnorrenberg, *Ann.*, **560**, 141 (1948).

[3] K. Tamao, N. Ishida, T. Tanaka, and M. Kumada, *Organometallics*, **2**, 1694 (1983).

[4] K. Tamao, M. Kumada, and K. Maeda, *Tetrahedron Letters*, **25**, 321 (1984).

[5] K. Tamao, N. Ishida, and M. Kumada, *J. Org.*, **48**, 2120 (1983).

[6] K. Tamao and N. Ishida, *Tetrahedron Letters*, **25**, 4249 (1984).

[7] K. Tamao, T. Iwahara, R. Kanetani, and M. Kumada, *ibid.*, **25**, 1909 (1984); K. Tamao, R. Kanetani, and M. Kumada, *ibid.*, **25**, 1913 (1984).

[8] M. Franck-Neumann, M. P. Heitz, and D. Martina, *ibid.*, **24**, 1615 (1983).

Hydrogen peroxide–Ammonium heptamolybdate, $(NH_4)_6Mo_7O_{24}·4H_2O$.

Oxidations.[1]   Secondary alcohols are oxidized to ketones in moderate to high yield by hydrogen peroxide (30%) in the presence of the ammonium molybdate, potassium carbonate, and tetra-*n*-butylammonium chloride in THF at 20°. Epoxidation is suppressed by the presence of potassium carbonate, which also accelerates the reaction. Primary alcohols are not oxidized under these conditions.

The paper also reports two examples of oxidation of aldehydes to acids under these conditions in satisfactory yield. In one case, addition of cerium chloride increased the yield.

[1] B. M. Trost and Y. Masuyama, *Tetrahedron Letters*, **25**, 173 (1984).

Hydrogen peroxide–Sodium tungstate.

Epoxidation of α,β-unsaturated acids (1,475).[1]   This reaction, first reported in 1959, has seen limited use because of moderate yields. Actually, yields are considerably improved by increasing the amount of catalyst and by increasing the pH to 5.8–6.8. The rate of epoxidation is markedly affected by substitution on the double bond, being enhanced by either an α-alkyl or a β-*cis*-alkyl substituent. The rate is also increased by addition of butanediol, lactic acid, or tartaric acid.

Examples:

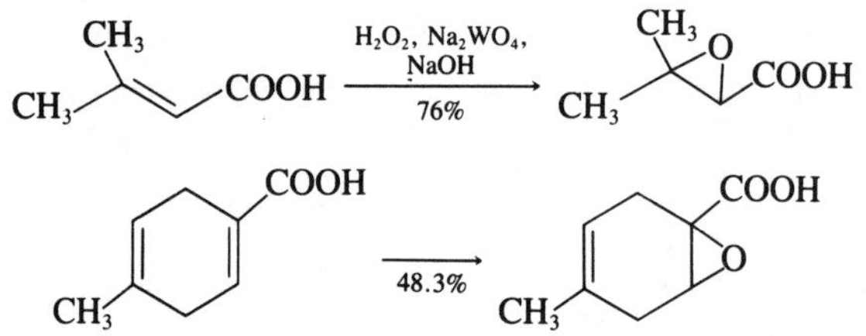

Nitrones.[2]   Secondary amines can be oxidized directly to nitrones with H_2O_2 (30%) catalyzed by $Na_2WO_4 \cdot 2H_2O$. Oxidation in the presence of an activated alkene gives the 1,3-dipolar adduct. The reaction can also be used to effect nucleophilic α-substitution of the amine.

Examples:

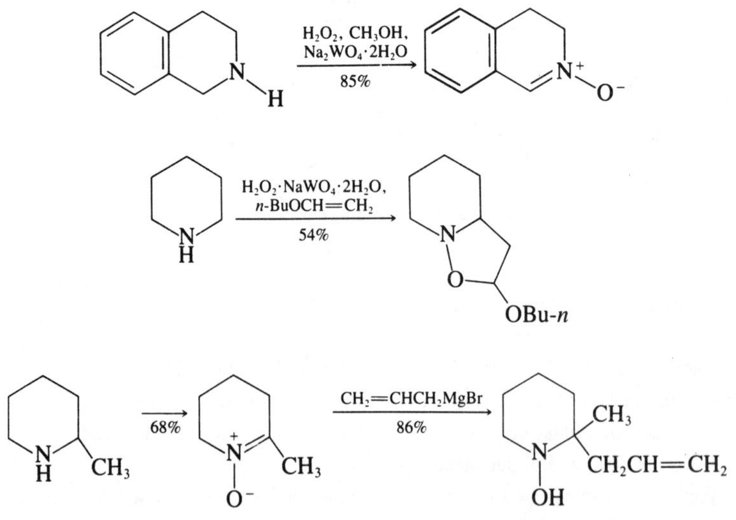

[1] K. Kirshenbaum and K. B. Sharpless, *J. Org.*, **50**, 1979 (1985).
[2] H. Mitsui, S. Zenki, T. Shiota, and S. Murahashi, *J.C.S. Chem. Comm.*, 874 (1984).

Hydrogen sulfide, H₂S, 4, 256.

Reduction of α,β-enones.[1]   The α,β-enone **1** was reduced by a new two-step process after usual methods failed. Reaction of **1** with H₂S and K₂CO₃ (1 equiv.) in DMSO at 23° results in 1,4-addition to form a β-mercapto ketone, which is desulfurized with tri-*n*-butylphosphine in benzene with UV irradiation; the overall yield is 65%.

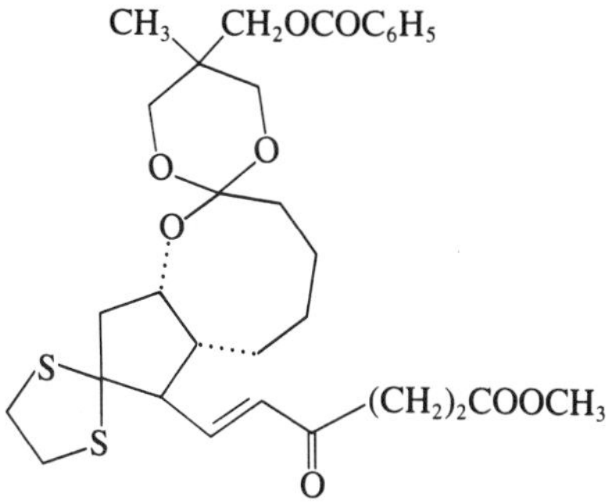

1

Thiono esters.[2]   Esters can be converted into the corresponding thiono esters by successive reaction in THF at −78° with LDA (1 equiv.) and ClSi(CH₃)₃ and then with H₂S at 0–25°. Typical thiono esters obtained in this way and the yield are formulated.

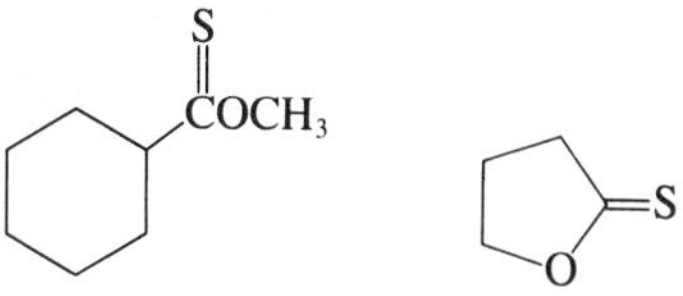

(73%)   (78%)

[1] E. J. Corey and K. Shimoji, *Am. Soc.*, **105**, 1662 (1983).
[2] E. J. Corey and S. W. Wright, *Tetrahedron Letters*, **25**, 2639 (1984).

1-Hydroxycyclopropanecarboxaldehyde, **(1).** This useful synthon cannot be isolated as such because it undergoes ready expansion to 2-hydroxycyclobutanone. However, the stable tetrahydropyranyl ether (**2**) is available (equation I).

(I)

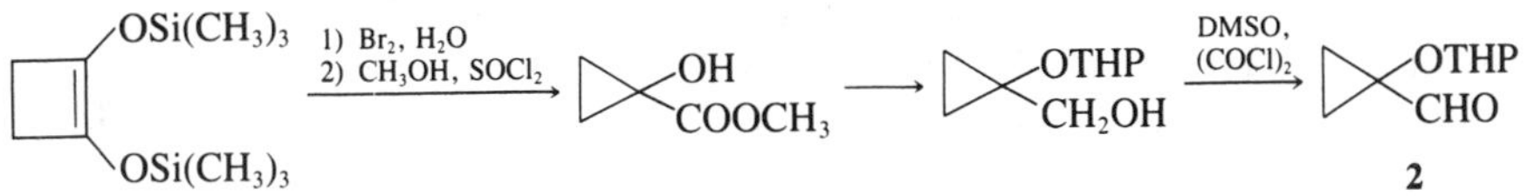

2

Cyclobutanones and cyclopentanones.[1] The 1-vinylcyclopropanols, readily obtainable from **2**, can rearrange either to cyclobutanones or to masked cyclopentanones. Thus the diol **3a** on treatment with BF_3 etherate undergoes dehydration and rearrangement to **4**. In contrast, the bissilyl ether **3b** on flash thermolysis undergoes ring enlargement to **5**, a useful precursor to 2,3-disubstituted cyclopentanones. This latter rearrangement

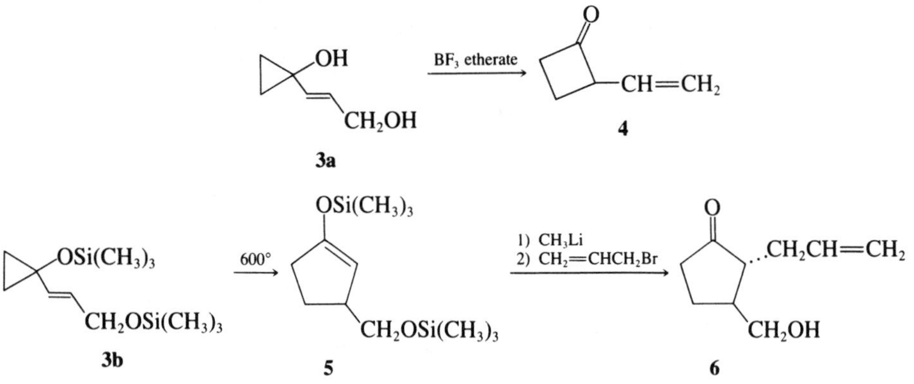

is a useful route to compounds containing the skeleton of spirovetivane sesquiterpenes (equation I).[2]

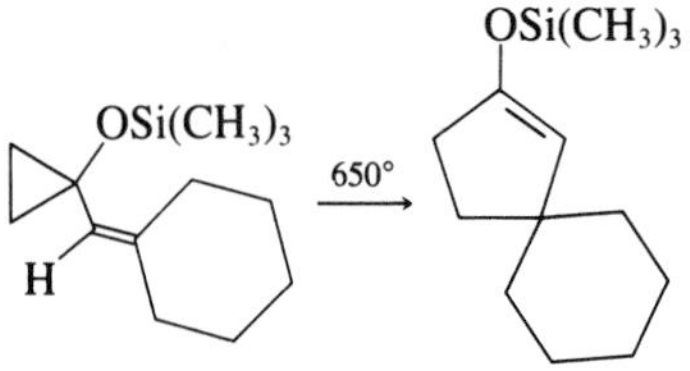

[1] J. Ollivier and J. Salaün, *Tetrahedron Letters*, **25**, 1269 (1984).
[2] J. P. Barnier and J. Salaün, *ibid.*, 1273 (1984).

(S)-(+)-α-Hydroxy-β,β-dimethylpropyl vinyl ketone,

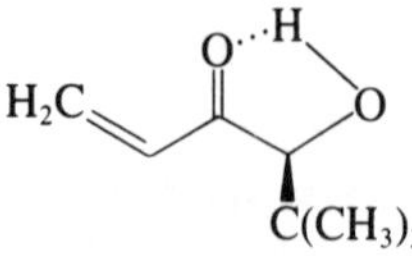

(1),

α_D +206°. The ketone is prepared by reaction of (S)-2-hydroxy-β,β-dimethylbutyric acid with vinyllithium in THF.

Asymmetric Diels-Alder reactions. The reaction of **1** with cyclopentadiene results in highly diastereoselective formation of one (**2a**) of the two possible *endo*-diastereomers. As expected, *endo*-adducts predominate over *exo*-adducts (*endo/exo* = 8:1). The high diastereoselectivity, even in the absence of a Lewis acid catalyst, is attributed, at least in part, to a rigid hydrogen-bonded cisoid conformation of **1**.[1]

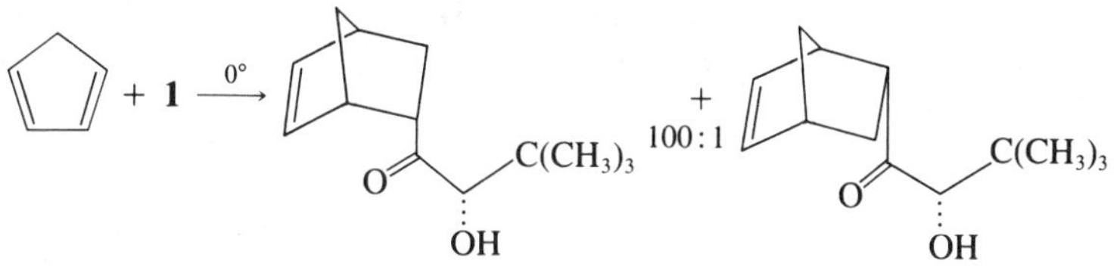

Similar diastereoselectivity is observed in reactions of **1** with acyclic dienes. Thus reaction with butadiene catalyzed by $ZnCl_2$ results in the single adduct **3**, which was converted into the optically pure alcohol **4** by oxidative removal of the chiral auxiliary group.[2]

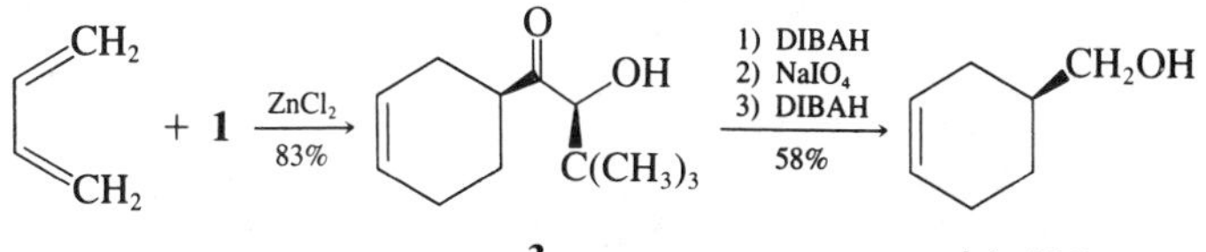

Substitution of a methyl group in the α-position of **1** has a deleterious effect on the stereoselectivity, but a β-methyl group increases the diastereoface selectivity.

[1] W. Choy, L. A. Reed III, and S. Masamune, *J. Org.*, **48**, 1137 (1983).
[2] S. Masamune, L. A. Reed III, J. T. Davis, and W. Choy, *ibid.*, 4441 (1983).

(1R)-(−)-*cis*-3-Hydroxyisobornyl neopentyl ether (1). This chiral alcohol and its antipode are available from (1R)- and (1S)-camphor, respectively.

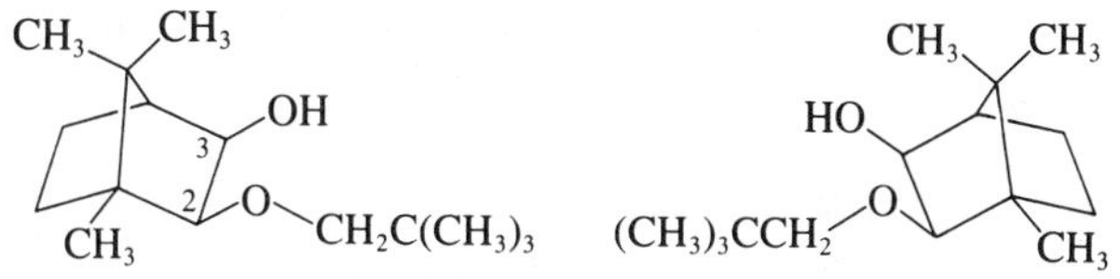

(1R)-(−)-**1**, α_D −42.55°
(R*OH)

(1S)-(+)-**1**, α_D +42.60°

Asymmetric Diels-Alder reactions.   The chiral acrylate (**2**) of (1R)-**1** undergoes cycloaddition with cyclopentadiene to give the adduct (2R)-**3** with high *endo*-selectivity and high asymmetric induction (99.3% de) in the presence of a mild Lewis acid catalyst, particularly TiCl$_2$(O-*i*-Pr)$_2$. Use of the antipode, (1S)-**1**, results in (2S)-**3** in equally high optical yield. The neopentyl group plays a significant role in directing the diastereoface

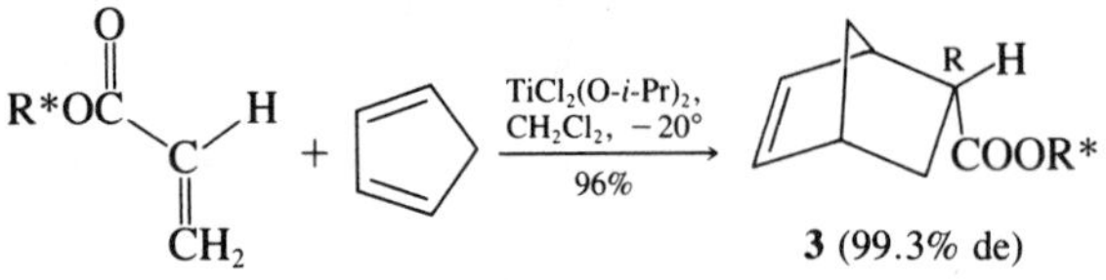

differentiation. Optical yields are considerably lower with the corresponding methylaryl ethers.[1]

The allenic ester **4** of (R)-**1** also undergoes a highly diastereoselective cycloaddition with cyclopentadiene to afford the *endo*-adduct **5** almost exclusively and in 99% de. The adduct provides a chiral intermediate to natural (−)-β-santalene (**6**).[2]

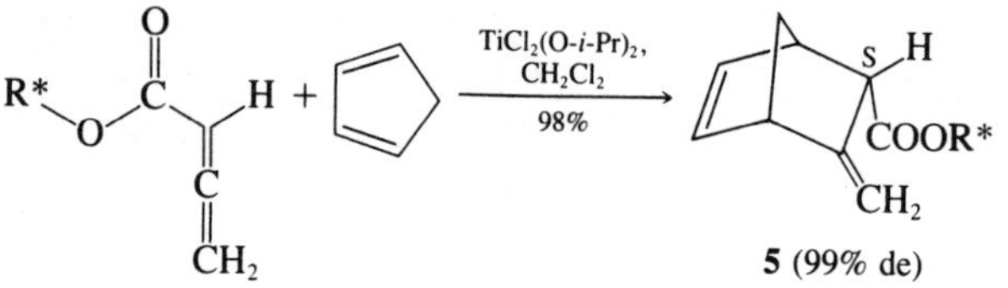

Asymmetric 1,4-addition of RCu·BF$_3$.[3]   Chiral α,β-unsaturated esters (**2**) derived from (1R)-**1** undergo a highly enantioselective Michael reaction with CH$_3$Cu·BF$_3$, particularly when stabilized with tri-*n*-butylphosphine or a cyano group (equation I).

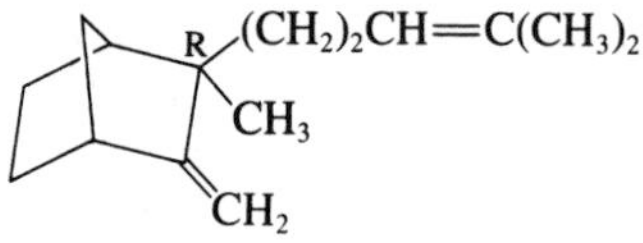

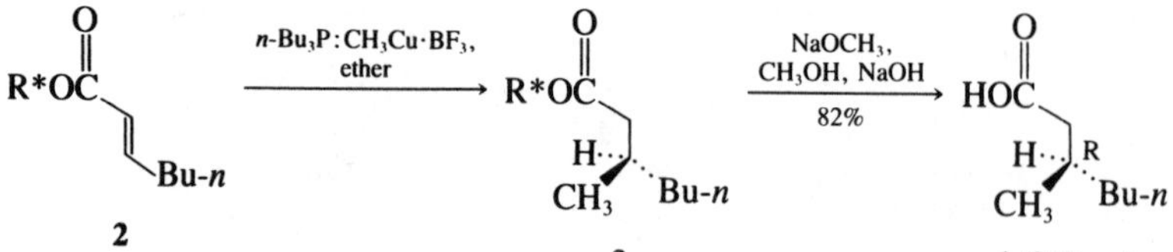

This asymmetric Michael addition was used for synthesis of (S)-citronellic acid (**6**) from the chiral crotonate **5** and a 4-methyl-3-pentenylcopper reagent.

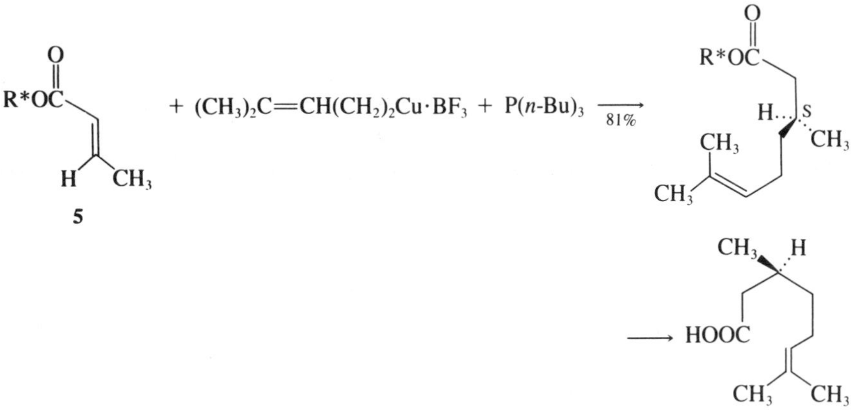

6 (98.5% ee)

[1] W. Oppolzer, C. Chapuis, G. M. Dao, D. Reichlin, and T. Godel, *Tetrahedron Letters*, **23**, 4781 (1982).

[2] W. Oppolzer and C. Chapuis, *ibid.*, **24**, 4665 (1983).

[3] W. Oppolzer, R. Moretti, T. Godel, A. Meunier, and H. Loher, *ibid.*, **24**, 4971 (1983).

Hydroxylamine, 1, 478–481; **2**, 217; **7**, 176–177; **9**, 245; **10**, 206–207; **11**, 257–258.

Arylamines.[1] A novel method for amination of organometallic reagents involves activation of an O-tosylhydroxylamine by attachment to an auxiliary group that can be recycled. Thus aryl Grignard or lithium reagents react with the O-tosylate oxime of tetraphenylcyclopentadienone (**1**) to form an imine (**2**) with elimination of the tosylate anion. The imine is then converted into an arylamine and the original oxime by treatment with excess hydroxylamine in aqueous pyridine.

Example:

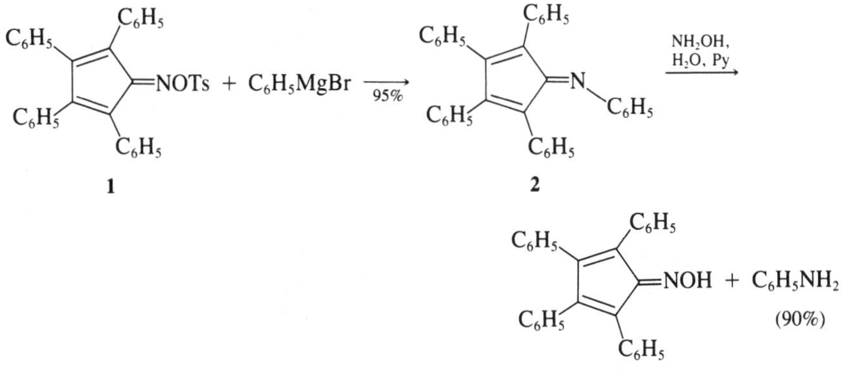

3 (95%)

[1] R. A. Hagopian, M. J. Therien, and J. R. Murdoch, *Am. Soc.*, **106**, 5753 (1984).

2-Hydroxymethyl-3-trimethylsilylpropene (1), 9, 454–455; **11**, 258–259.

Cyclocontraction–spiroannelation.[1]   The products (**3**) obtained by alkylation of cyclic β-keto sulfones with the mesylate (**2**) of **1** on treatment with ethylaluminum dichloride (2 equiv.) undergo spiroannelation to give **4** with contraction of the original ring. The expected product of cyclization (**a**) can be isolated from reactions conducted at low temperatures. Formation of **4** involves a pinacol-like rearrangement of **a** with the sulfone group functioning as a leaving group.

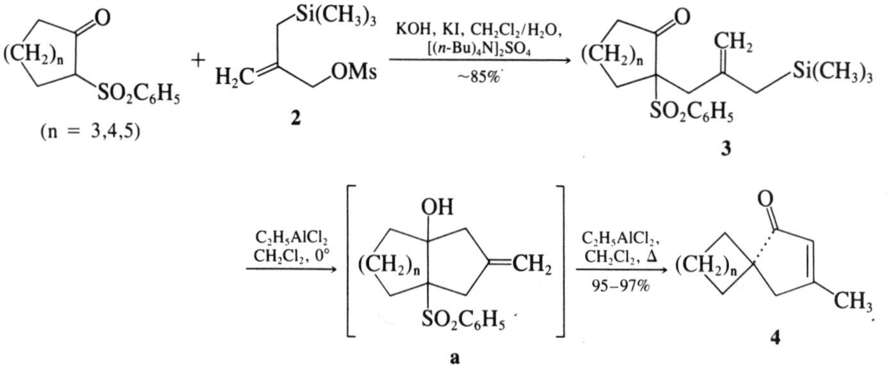

[1] B. M. Trost and B. R. Adams, *Am. Soc.*, **105**, 4849 (1983).

8-Hydroxyquinolinedihydroboronite,

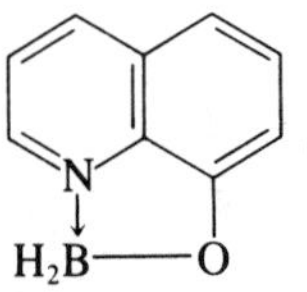

(**1**). The reagent is prepared by reaction of 8-hydroxyquinoline with $BH_3 \cdot S(CH_3)_2$ in THF.

Selective reduction of aldehydes.[1]   In the presence of BF_3 etherate at 25° this reagent reduces aldehydes more rapidly than it does ketones.

[1] S. Kim, H. J. Kang, and S. Yang, *Tetrahedron Letters*, **25**, 2985 (1984).

I

Iodine, 1, 495–500; 2, 220–222; 3, 159–160; 4, 258–260; 5, 346–347; 6, 293–295; 7, 179–181; 8, 256–260; 9, 248–249; 10, 210–211; 11, 261–267.

Iodolactonization of γ,δ-*unsaturated amides.*[1] Halolactonization (I$_2$ or NBS) of α-substituted γ,δ-unsaturated amides in DME/H$_2$O at room temperature gives predominantly *trans*-2,4-disubstituted γ-butyrolactones (equation I). This 1,3-stereoselectivity is in sharp contrast to the moderate 1,3-*cis*-selectivity observed with α-substituted γ,δ-unsaturated acids (8, 257; 9, 248). Both diastereomers of α,β-disubstituted γ,δ-unsaturated amides are converted into 2,4-*trans*-2,3,4-trisubstituted lactones (equation II).

90–99 : 10–1

97 : 3

Similar asymmetric induction is observed on halolactonization of γ,δ-unsaturated thioamides. These substrates, unlike the amides, undergo highly selective protiothiolactonization with *d*-(+)-10-camphorsulfonic acid.

Stereocontrolled iodolactonization (8, 257; 9, 248). Iodolactonization of 3-phenyl-4-pentenoic acid (1, I$_2$, CH$_3$CN, 0°, 12 hours) gives the more stable *trans*-iodolactone 2 as the major product. In the presence of NaHCO$_3$ (kinetic control), the *cis*-iodolactone 3 is the major product. The contrast between the thermodynamically controlled and the kinetically controlled reactions observed in this case is fairly general, but usually not so marked. Thus when the phenyl group of 1 is replaced by methyl the ratio of *trans*- to *cis*-iodolactonization obtained under thermodynamic control is only 11:1; however, the stereoselectivity observed under kinetically controlled iodolactonization is practically the same as with **1** (1:3).[2]

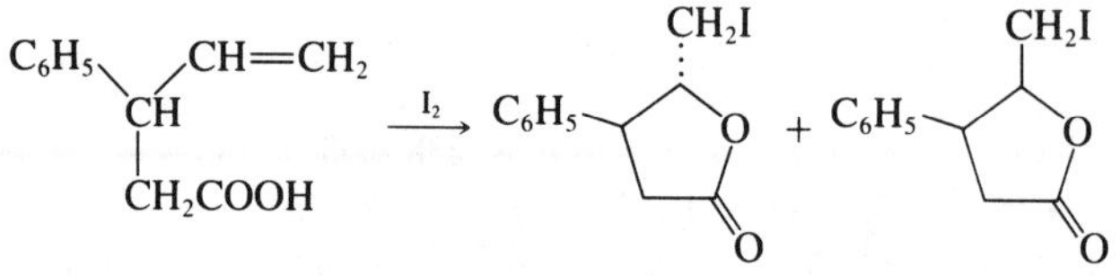

1	**2** **3**
CH$_3$CN	85–91%	>95:5
NaHCO$_3$, CHCl$_3$	68%	1:3.7

Iodocarbamation (**11**, 264–265).[3] High 1,2-asymmetric induction is possible in the iodocyclocarbamation of an acyclic chiral allylamine. Thus **1**, derived from Boc-D-phenylalanine, on treatment with iodine in CH$_2$Cl$_2$ at 0° is converted into the iodocar-

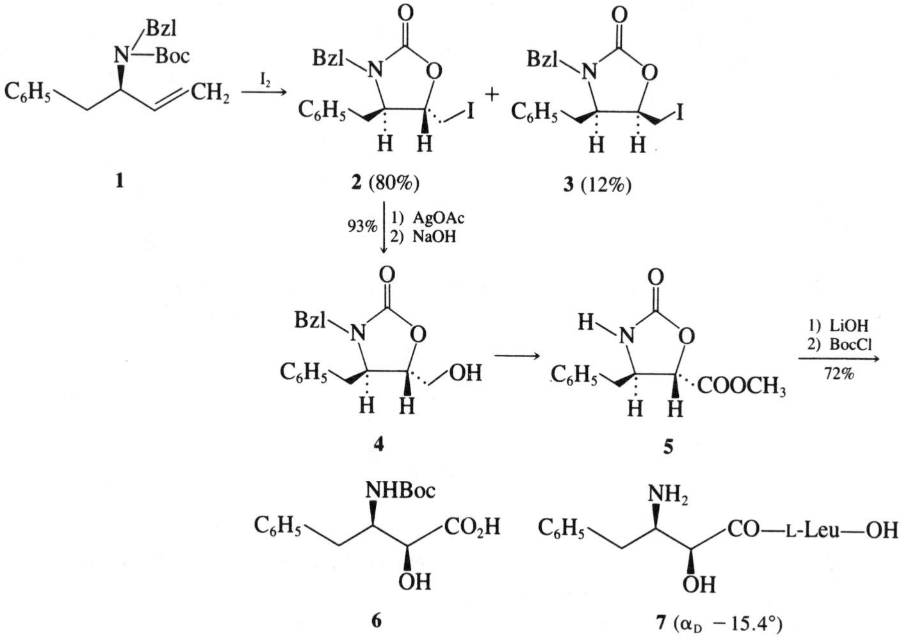

bamates **2** and **3** in the ratio 6.7:1. In the absence of the benzyl group on nitrogen, the reaction with iodine shows only a 1.5:1 preference in favor of the *trans*-isomer. The product **2** was converted as shown into **6**, (2S,3R)-3-benzyloxycarbonylamino-2-hydroxy-4-phenylbutanoic acid, a key component of (−)-bestatin (**7**).

Iodocyclization. γ,δ-Unsaturated alcohols or ethers cyclize to tetrahydrofuranes on treatment with iodine. With unsaturated alcohols, iodocyclization favors formation of *trans*-disubstituted tetrahydrofuranes. However, cyclization of olefinic benzyl ethers provides a stereoselective synthesis of *cis*-disubstituted tetrahydrofuranes (equation I).

(I)

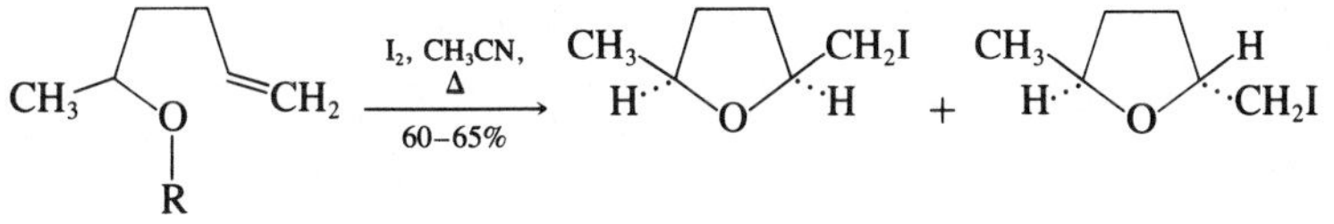

R	
R = H or CH$_3$	1 : 2
R = CH$_2$C$_6$H$_5$	2 : 1
R = CH$_2$C$_6$H$_3$-2,6-Cl$_2$	21 : 1

This cyclization can be used to prepare *cis*-linalyl oxide (**2**) in 70% overall yield from the benzyl ether **1**. Under the same conditions, the corresponding free alcohol is converted into *trans*-**2** in comparable yield.[4]

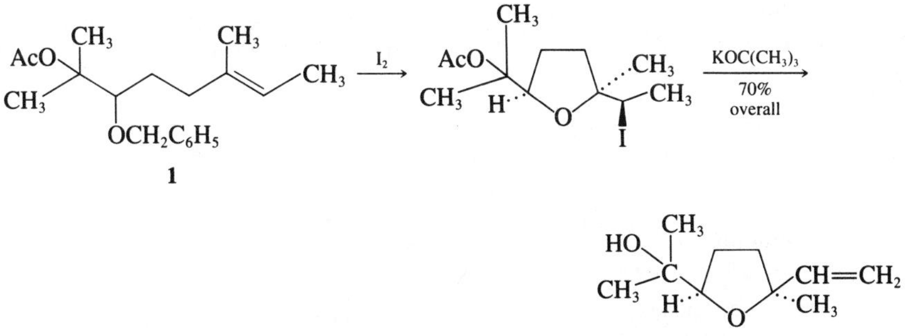

1

2

This cyclization provides a highly stereoselective synthesis of the sesquiterpene davanone (**5**).[5] Thus iodocyclization of the 4-bromobenzyl ether **3** followed by elimination and hydrolysis gives **4**, which consists mainly of the desired isomer with only 8% of two

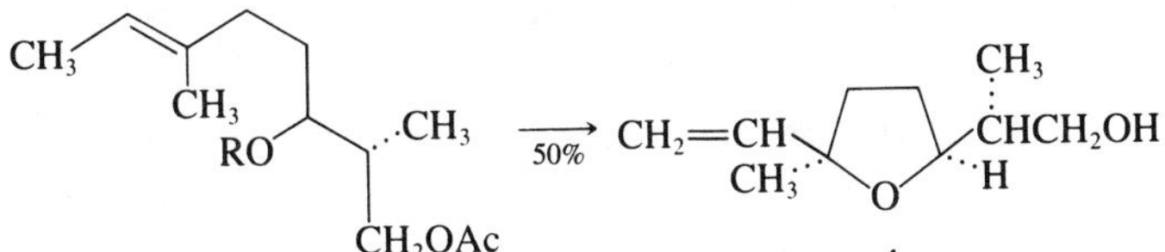

3, R = CH$_2$C$_6$H$_4$Br-*p*

4

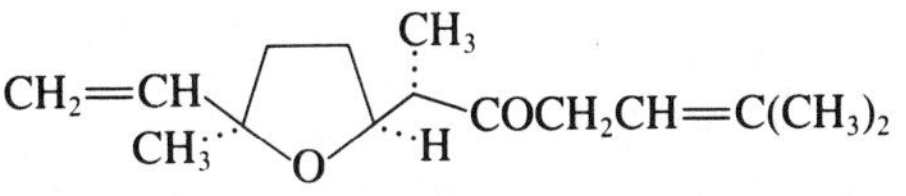

5

diastereomers. The synthesis of **5** is completed by Jones oxidation of the primary hydroxyl group to the acid followed by condensation with dimethallyllithium.

[1] Y. Tamaru, M. Mizutani, Y. Furukawa, S. Kawamura, Z. Yoshida, K. Yanagi, and M. Minobe, *Am. Soc.*, **106**, 1079 (1984).
[2] F. B. Gonzàlez and P. A. Bartlett, *Org. Syn.*, submitted (1983).
[3] S. Kobayashi, T. Isobe, and M. Ohno, *Tetrahedron Letters*, **25**, 5079 (1984).
[4] S. D. Rychnovsky and P. A. Bartlett, *Am. Soc.*, **103**, 3963 (1981).
[5] P. A. Bartlett and C. P. Holmes, *Tetrahedron Letters*, **24**, 1365 (1983).

Iodine–Copper(II) acetate, 8, 260; 10, 211; 11, 267.

α-Iodo carboxylic acids.[1]   α-Iodination of carboxylic acids can be effected with iodine in combination with a copper salt. The highest yields (85–90%) are obtained by reactions conducted at 120° with a slight excess of iodine and 0.5 equiv. each of CuCl and CuCl$_2$.

Aryl iodides.[2]   Activated arenes, mesitylene, aniline, or anisole, can be iodinated in high yield by I$_2$ and Cu(OAc)$_2$ in HOAc (2–10 hours). Although phenol is iodinated mainly at the *ortho*-position (64% yield), aniline and anisole are iodinated mainly at the *para*-position.

[1] C. A. Horiuchi and J. Y. Satoh, *Chem. Letters*, 1509 (1984).
[2] *Idem, Bull. Chem. Soc. Japan*, **57**, 2691 (1984).

Iodine–Pyridinium chlorochromate.

α-Iodo aldehydes and ketones.[1]   Enol silyl ethers are converted by this system into α-iodo carbonyl compounds in yields of >80%. The method fails with enamines, and is only marginally useful with enol ethers. This system converts cyclohexene into *trans*-1-chloro-2-iodocyclohexane (66% yield).

Examples:

$$C_2H_5CH=CCH_2C_2H_5 \xrightarrow[88\%]{\substack{I_2, \\ PCC}} C_2H_5CHCCH_2C_2H_5$$
$$| \qquad\qquad\qquad | \; ||$$
$$OSi(CH_3)_3 \qquad\qquad\quad I \;\; O$$

$$CH_3(CH_2)_4CH=CHOSi(CH_3)_3 \xrightarrow[82\%]{} CH_3(CH_2)_4CHCHO$$
$$|$$
$$I$$

[1] M. D'Auria, F. D'Onofrio, G. Piancatelli, and A. Scettri, *Syn. Comm.*, **12**, 1127 (1982).

Iodine–Silver acetate, 4, 261; 9, 249.

Prévost reaction.   A short synthesis of methyl shikimate (**5**) involves base-promoted cleavage of **1**, obtained by a Diels-Alder reaction of furane with methyl acrylate (**11**,

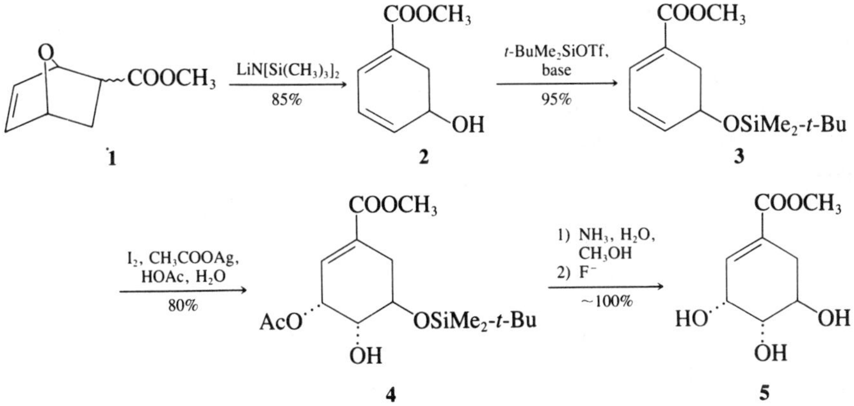

604–605), followed by *cis*-dihydroxylation (OsO₄) of the protected derivative **3**.[1] This *cis*-dihydroxylation is effected more conveniently by Prévost hydroxylation (**1**, 1007–1008) with iodine, silver acetate, and aqueous acetic acid, which results in **4** as the major product. A similar oxidation of **2** results in formation of seven products.[2]

[1] M. M. Campbell, A. D. Kaye, M. Sainsbury, and R. Yavarzadeh, *Tetrahedron*, **40**, 2461 (1984).
[2] M. M. Campbell, M. Sainsbury, and R. Yavarzadeh, *ibid.*, 5063 (1984).

Iodine–Silver nitrate, 11, 268.

ArCOCH₃ → ArCH₂COOCH₃.[1]   The rearrangement of the acetylpyrrole **1** to the pyrroleacetic acid **3** proceeds in higher yield if the intermediate iodoacetylpyrrole is trapped as the more stable chloroacetylpyrrole **2**, which is then converted into the ketal and

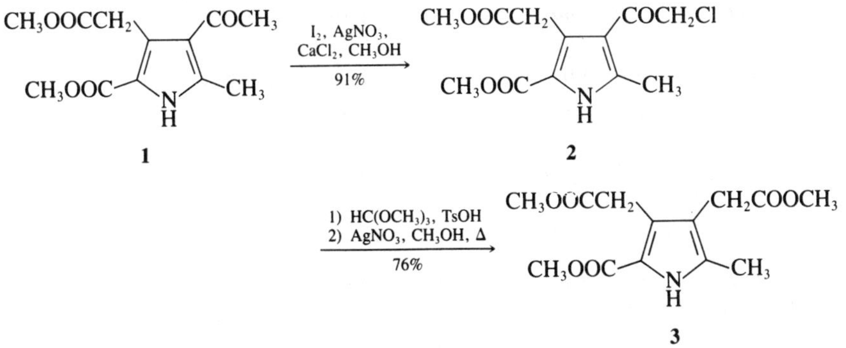

rearranged with AgNO₃ to **3**. The overall yield is comparable to that obtained with thallium(III) nitrate in trimethyl orthoformate (**5**, 656).

[1] T. Wollmann and B. Frank, *Angew. Chem., Int. Ed.*, **23**, 226 (1984).

Iodoacetoxy(tri-*n*-butyl)tin, $ICH_2COOSnBu_3$ (1). The reagent is prepared by reaction of iodoacetic acid with bis(tri-*n*-butyl)tin oxide. The stannyl ester $CH_3CHICOOSnBu_3$ (2) is prepared in the same way.

γ-Lactones.[1]   These esters react with electron-rich alkenes via a radical reaction to form γ-lactones. AIBN initiation is more effective and convenient than other initiation procedures.

Examples:

[1]G. A. Kraus and K. Landgrebe, *Tetrahedron Letters*, **25**, 3939 (1984).

N-Iodosuccinimide, 1, 510–511; **10**, 216. The reagent can be prepared by reaction of N-chlorosuccinimide with sodium iodide in acetone. Filtration from NaCl and evaporation of the solvent provides material comparable to that prepared according to the literature.[1]

Oxidative decarboxylation.[2]   *sec-* or *tert-*α-Hydroxy carboxylic acids undergo oxidative decarboxylation to aldehydes or ketones when treated with NIS (2 equiv). The rate of reaction is increased by irradiation. Yields are generally >80%.

[1] Y. D. Vankar and G. Kumaravel, *Tetrahedron Letters*, **25**, 233 (1984).
[2] T. R. Beebe, R. L. Adkins, A. I. Belcher, T. Choy, A. E. Fuller, V. L. Morgan, B. B. Sencherey, L. J. Russell, and S. W. Yates, *J. Org.*, **47**, 3006 (1982).

Iodosylbenzene, 1, 507–508; **10**, 213–214, **11**, 270.

RCONH$_2$ → RNH$_2$.[1]   Aliphatic or benzylic primary amides are converted to the amines by reaction with $C_6H_5I=O$ in the presence of formic acid (yields, 70–90%).

Oxidation of allylsilanes to α,β-enals.[2]   2-Substituted allylsilanes are oxidized to α,β-enals by reaction with $C_6H_5I=O$ (2 equiv.) activated with BF$_3$ etherate (1 equiv.) in dioxane.

Examples:

$$\underset{AcO(CH_2)_2\overset{\overset{\displaystyle CH_2}{\|}}{C}CH_2Si(CH_3)_3}{} \xrightarrow[63\%]{C_6H_5I=O,\ BF_3\cdot(C_2H_5)_2O,\ 25°} AcO(CH_2)_2\overset{\overset{\displaystyle CH_2}{\|}}{C}CHO$$

$$CH_2=CH(CH_2)_8\overset{\overset{\displaystyle OAc}{|}}{C}HCH_2\overset{\overset{\displaystyle CH_2}{\|}}{C}CH_2Si(CH_3)_3 \xrightarrow[63\%]{} CH_2=CH(CH_2)_8\overset{\overset{\displaystyle OAc}{|}}{C}HCH_2\overset{\overset{\displaystyle CH_2}{\|}}{C}CHO$$

[1] A. S. Radhakrishna, C. G. Rao, R. K. Varma, B. B. Singh, and S. P. Bhatnagar, *Synthesis*, 538 (1983).
[2] M. Ochiai, E. Fujita, M. Arimoto, and H. Yamaguchi, *Tetrahedron Letters*, **24**, 777 (1983).

o-Iodosylbenzoic acid,   This reagent is prepared by oxidation of *o*-iodobenzoic acid with fuming nitric acid or 40% peracetic acid. Suppliers: Pierce, Sigma.

α-*Hydroxylation of ketones* (**10**, 214).[1]   This reagent is preferred to iodosylbenzene, previously used for this reaction, because the *o*-iodobenzoic acid formed as a coproduct is easily removed by extraction with base. Overall yields of α-hydroxy ketones by this modified procedure are 65–85%.

Example:

$$\underset{C_6H_5\overset{\overset{\displaystyle O}{\|}}{C}CH_3}{} \xrightarrow[80\%]{KOH,\ o\text{-}HOOCC_6H_4I=O,\ CH_3OH} C_6H_5\overset{\overset{\displaystyle OCH_3}{|}}{\underset{\underset{\displaystyle OCH_3}{|}}{C}}-CH_2OH \xrightarrow[97\%]{H_2SO_4,\ CH_2Cl_2} C_6H_5\overset{\overset{\displaystyle O}{\|}}{C}CH_2OH$$

[1] R. M. Moriarty, K.-C. Hou, and S. K. Arora, *Org. Syn.*, submitted (1983); R. M. Moriarty and K.-C. Hou, *Tetrahedron Letters*, **25**, 691 (1984).

Iodotrimethylsilane, 8, 261–263; **9,** 251–252; **10,** 216–219; **11,** 271–275. ISi(CH_3)_3 can be prepared in 81–88% yield by reaction of ClSi(CH_3)_3 with LiI without a solvent at 25° for 4 hours. ISi(C_2H_5)_3 can be prepared in the same way in 65% yield[1]

Acetalization and allylation of carbonyl compounds.[2]   Iodotrimethylsilane is an effective catalyst for acetalization with an alkoxysilane. Addition of allyltrimethylsilane to this reaction results in allylation of the acetal to form a homoallyl ether.

Examples:

$$C_6H_5CHO + (CH_3O)_4Si \xrightarrow[87\%]{\substack{ISi(CH_3)_3, \\ CH_2Cl_2, -78°}} C_6H_5CH(OCH_3)_2$$

$$C_6H_5CHO + (CH_3O)_4Si + (CH_3)_3SiCH_2CH=CH_2 \xrightarrow[90\%]{} C_6H_5\underset{\underset{OCH_3}{|}}{C}HCH_2CH=CH_2$$

syn-*Selective aldol-type condensation*.[3]   Iodotrimethylsilane catalyzes the condensation of silyl enol ethers with acetals to afford β-alkoxy ketones with high *syn*-selectivity. Example:

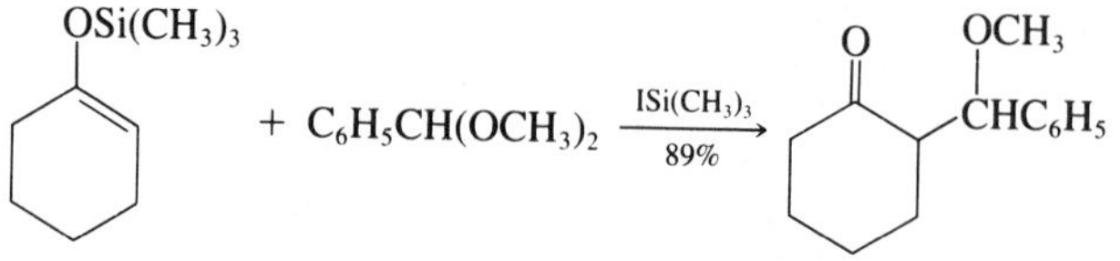

$$(syn/anti = 95:5)$$

α-*Methylene-γ-butyrolactones*.[4]   The last step in a synthesis of α-methylene-γ-butyrolactones (**4**) from *gem*-dibromocyclopropanes (**1**) involves reaction of a 1-(dimethylaminomethyl)propanecarboxylic ester (**3**) with iodotrimethylsilane (3 equiv.) followed by thermolysis. The reaction is considered to involve ester hydrolysis and quaternization to give **a**.

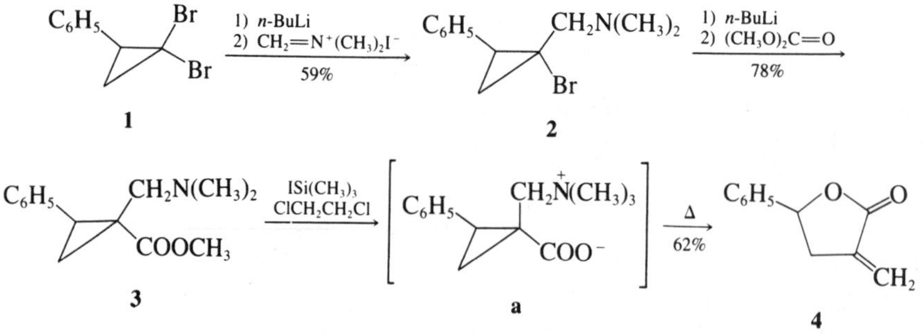

Lactonization of a δ-*hydroxy ester*.[5]   In the presence of 1 equiv. of ISi(CH_3)_3, the δ-hydroxy ester **1** undergoes lactonization at room temperature to afford **2**. Lactonization can also be effected by refluxing **1** in benzene containing TsOH for 20 hours.

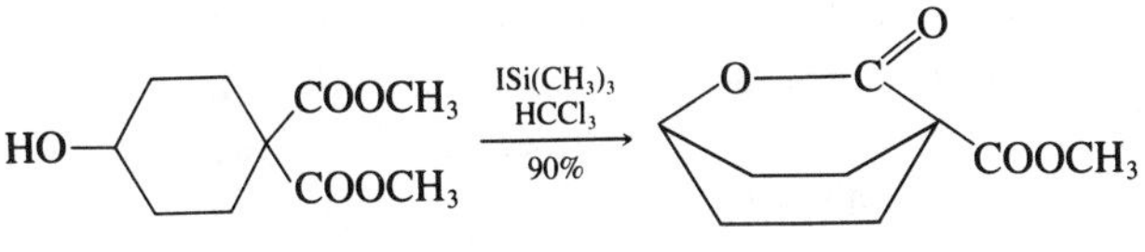

1   **2**

Conjugate addition to enones.   Iodotrimethylsilane can mediate conjugate addition of furanes to α,β-enones via the intermediate γ-iodo enol silyl ether (**9**, 252–253). The adduct can be isolated as the enol silyl ether or the corresponding ketone (equation I).[6]

(I)

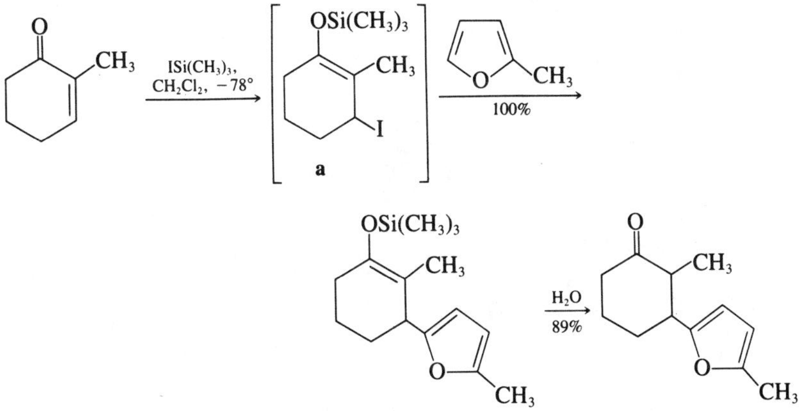

Conjugate addition of enol silyl ethers can also be effected in this way.[7] A similar reaction with dienol silyl ethers affords cyclized products. This variation of a Diels-Alder reaction can be useful because it can afford stereoisomers of the products formed by reactions catalyzed by conventional Lewis acids.

Example:

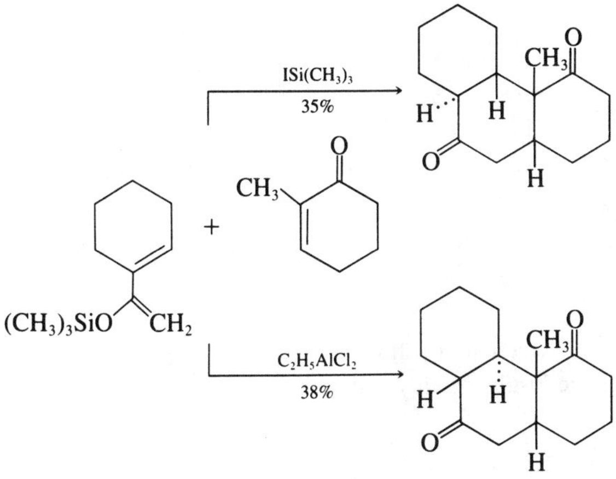

β-Alkoxy ketones.[8] In the presence of $ISi(CH_3)_3$ or $(CH_3)_3SiOTf$, silyl enol ethers react with α-chloro ethers to give β-alkoxy ketones selectively. Evidently $ISi(CH_3)_3$ activates only the C—Cl bond. Similarly, the reaction of silyl enol ethers with α-chloro sulfides gives rise only to β-alkylthio ketones.

Examples:

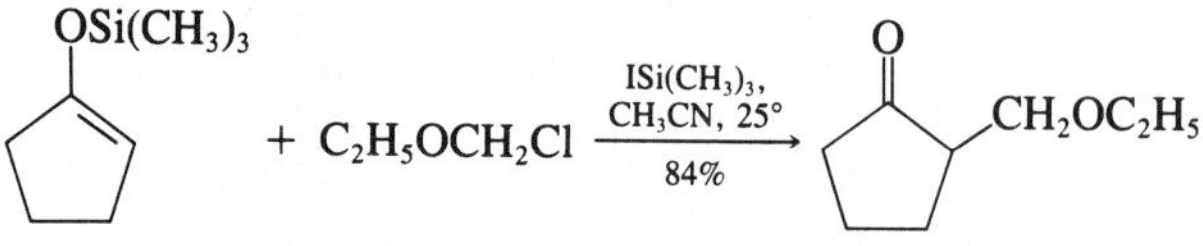

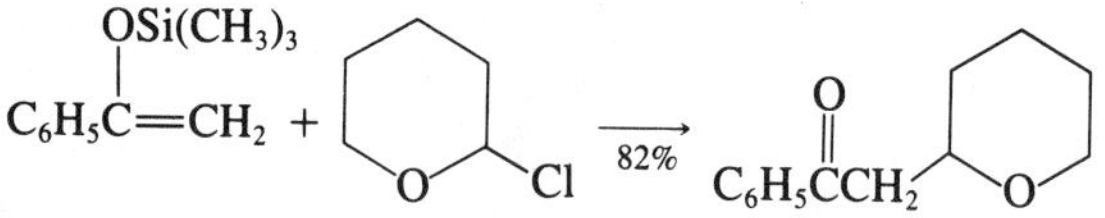

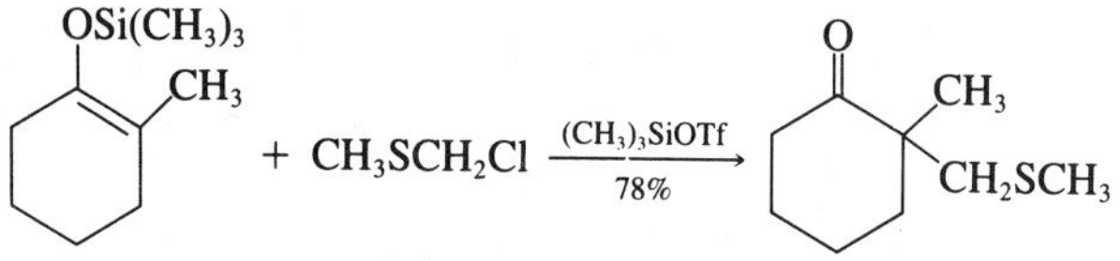

Deoxygenation.[9] The 17α-hydroxy group of 17α,21-dihydroxy-20-ketosteroids is selectively reduced by $ISi(CH_3)_3$ in CH_3CN at room temperature (equation I). A free 21-hydroxyl group is necessary for this dehydroxylation.

(I) 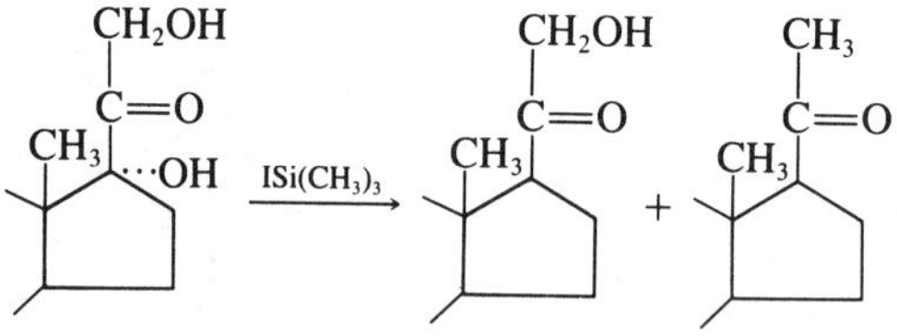

CH_3CN	72%	5%
$CHCl_3$	48%	38%

Vinyl sulfides. Reaction of sulfoxides with $ISi(CH_3)_3$ and a hindered base such as diisopropylethylamine results in vinyl sulfides, usually as mixtures of (E)- and (Z)-isomers.[10]

Examples:

$$CH_3(CH_2)_3\overset{\overset{\displaystyle O}{\|}}{S}(CH_2)_3CH_3 \xrightarrow[75\%]{\substack{ISi(CH_3)_3, \\ C_2H_5N(i\text{-}Pr)_2}} CH_3(CH_2)_3SCH=CHCH_2CH_3$$

$$(E/Z = 50:50)$$

$$C_6H_5\overset{\overset{\displaystyle O}{\|}}{S}CH_2CH=CHCH_3 \xrightarrow[85\%]{} C_6H_5S\diagdown\diagup\diagdown\diagup CH_2$$

$$(>95\% \ E)$$

In some cases, this rearrangement can also be effected with chlorotrimethylsilane in the absence of added base.[11]

Reaction with nitroalkanes.[12]   The reaction of primary and secondary nitroalkanes with $ISi(CH_3)_3$ (2 equiv.) results in formation of the corresponding oximes as the major product. However, benzylic nitroalkanes, $ArCH_2NO_2$, are converted into nitriles in high yield.

Tertiary nitroalkanes, on the other hand, are converted by $ISi(CH_3)_3$ (2 equiv.) into the t-alkyl iodides in high yield.

Examples:

$$CH_3(CH_2)_4CH_2NO_2 \xrightarrow{ISi(CH_3)_3} CH_3(CH_2)_4CH=NOH + CH_3(CH_2)_4C\equiv N$$

$$(81\%) \qquad\qquad\qquad (10\%)$$

$$(CH_3)_3CNO_2 \xrightarrow[94\%]{} (CH_3)_3I$$

[1] M. Lissel and K. Drecksler, *Synthesis*, 459 (1983).
[2] H. Sakurai, K. Sasaki, J. Hayashi, and A. Hosomi, *J. Org.*, **49**, 2808 (1984).
[3] H. Sakurai, K. Sasaki, and A. Hosomi, *Bull. Chem. Soc. Japan*, **56**, 3195 (1983).
[4] H. Saimoto, K. Nishio, H. Yamamoto, M. Shinoda, T. Hiyama, and H. Nozaki, *ibid.*, 3093 (1983).
[5] D. R. Walley and J. L. Belletire, *Syn. Comm.*, **14**, 401 (1984).
[6] G. A. Kraus and P. Gottschalk, *Tetrahedron Letters*, **24**, 2727 (1983).
[7] *Idem, J. Org.*, **49**, 1153 (1984).
[8] A. Hosomi, Y. Sakata, and H. Sakurai, *Chem. Letters*, 405 (1983).
[9] M. Numazawa, M. Nagaoka, and Y. Kunitama, *J.C.S. Chem. Comm.*, 31 (1984).
[10] R. D. Miller and D. R. McKean, *Tetrahedron Letters*, **24**, 2619 (1983).
[11] S. Lane, S. J. Quick, and R. K. Taylor, *ibid.*, **25**, 1039 (1984).
[12] G. A. Olah, S. C. Narang, L. D. Field, and A. P. Fung, *J. Org.*, **48**, 2766 (1983).

Iron, **1**, 519; **2**, 229; **3**, 167; **5**, 357; **9**, 257.

Deprotonation of aldehydes and ketones.[1]   Reaction of $FeCl_3$ with ethereal CH_3MgBr (3 equiv.) results in evolution of a gas (ethane/methane 2:1) and formation of an air-sensitive black powder that apparently consists of Fe(0) and magnesium chloride etherate. This iron powder deprotonates aldehydes and ketones, but not esters, as shown by trapping with chlorotrimethylsilane, to form silyl ethers in high yield (85–99%). Moreover, unsymmetrical ketones are converted into the more substituted silyl ether with high selectivity.

Example:

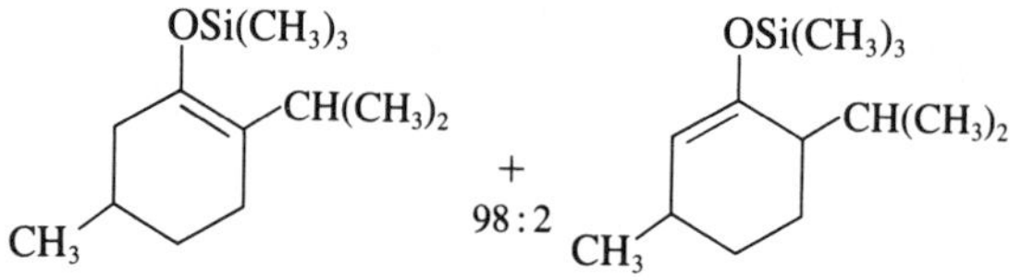

Trimethylsilyl dienol ethers.[2]   Reaction of cyclic enones with 1 equiv. of this activated Fe(0) and ClSi(CH$_3$)$_3$/N(C$_2$H$_5$)$_3$ results in predominant formation of the exocyclic through-conjugated dienol ether, rather than the cross-conjugated (kinetic) isomer, which is formed preferentially by deprotonation with LDA.

Examples:

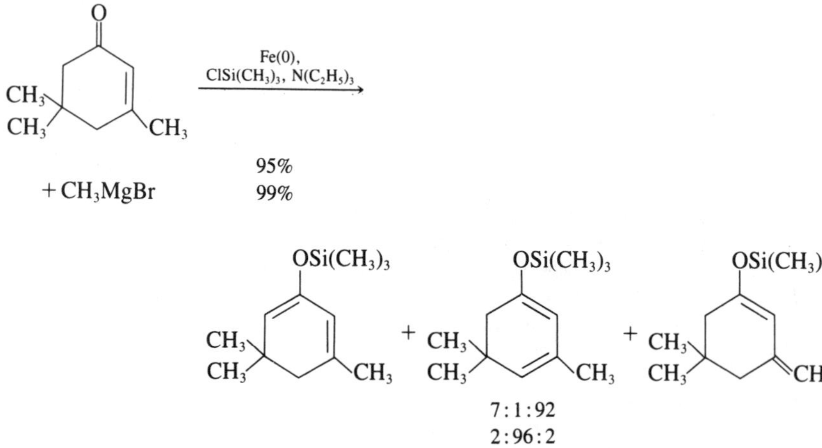

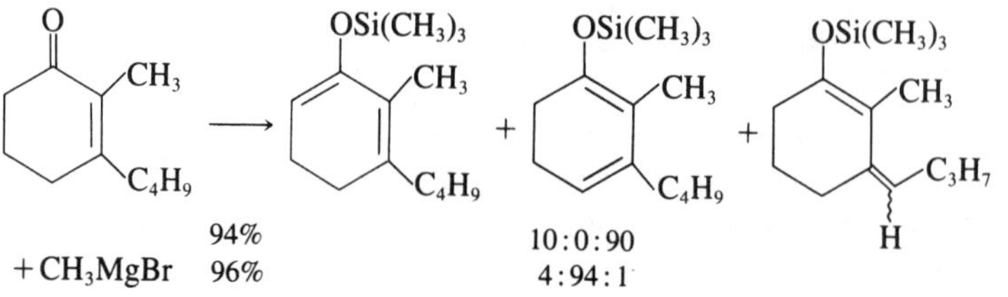

In contrast, use of 0.5–1 equiv. of Fe(0) and 1 equiv. of CH_3MgBr results in liberation of methane and predominant formation of the endocyclic through-conjugated dienol ether. This reaction requires exclusion of oxygen for reproducible results.

[1] M. E. Krafft and R. A. Holton, *J. Org.*, **49**, 3669 (1984).
[2] *Idem. Am. Soc.*, **106**, 7619 (1984).

Iron–carbene complexes, $(CO)_4Fe\!\!=\!\!C\overset{OC_2H_5}{\underset{R}{\diagdown}}$ The preparation of tetracarbonyl- (ethoxyphenylmethylidene)iron(0) (**1**, R = C_6H_5) is typical (equation I).[1]

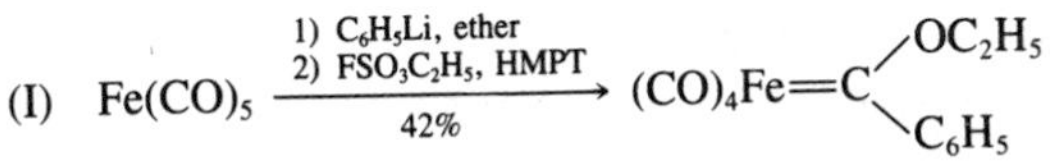

$$(I) \quad Fe(CO)_5 \xrightarrow[\text{42%}]{\substack{1)\ C_6H_5Li,\ \text{ether} \\ 2)\ FSO_3C_2H_5,\ \text{HMPT}}} (CO)_4Fe\!\!=\!\!C\overset{OC_2H_5}{\underset{C_6H_5}{\diagdown}}$$

$$\textbf{1},\ R\ =\ C_6H_5$$

Addition to alkynes; α-pyrones.   These complexes react with alkynes under carbon monoxide (55 psi) to form α-pyrone–Fe(CO)₃ complexes (**2**), which rearrange slowly to isomeric complexes (**3**) when heated. The reaction with unsymmetrical alkynes is regio-selective, with the less substituted carbon of the alkyne being coupled to the alkylidene carbon.

Examples:

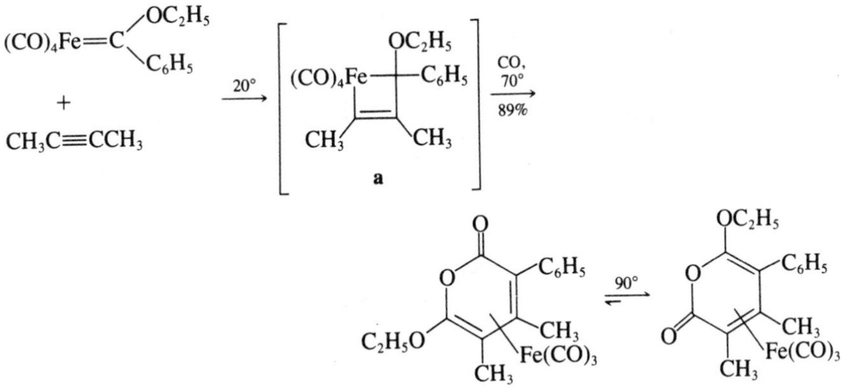

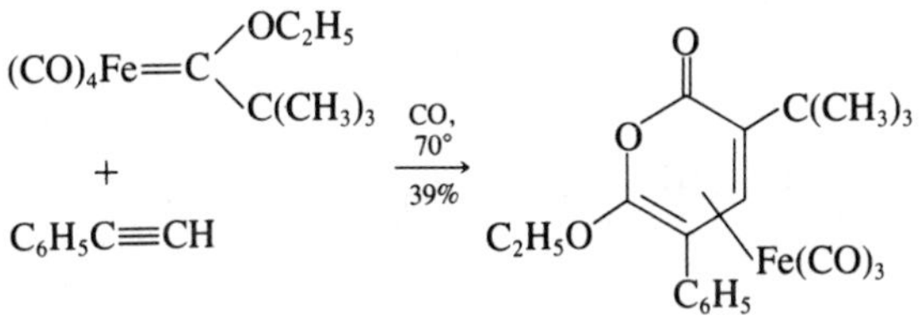

The unstable intermediate **a** has been obtained in a reaction conducted at 20° in the absence of carbon monoxide. The overall reaction involves incorporation of two molecules of CO and migration of the ethoxy group.[2]

[1] M. F. Semmelhack and R. Tamura, *Am. Soc.*, **105**, 4099 (1983).
[2] M. F. Semmelhack, R. Tamura, W. Schnatter, and J. Springer, *ibid.*, **106**, 5363 (1984).

Iron carbonyl, 1, 519; **2**, 229; **3**, 167; **4**, 268; **5**, 357–358; **6**, 304–305; **7**, 183; **8**, 265–267; **10**, 221–223.

Cyclohexadiene–Fe(CO)₃ complexes. Pearson[1] has reviewed the formation and reactions of these complexes. One particularly interesting property is the ability of a methoxy group to direct attack of nucleophiles to the *para*-position, even when this position is substituted. This selectivity provides a useful route to 4,4-disubstituted cyclohexenones (**3**), including spirocyclic compounds.

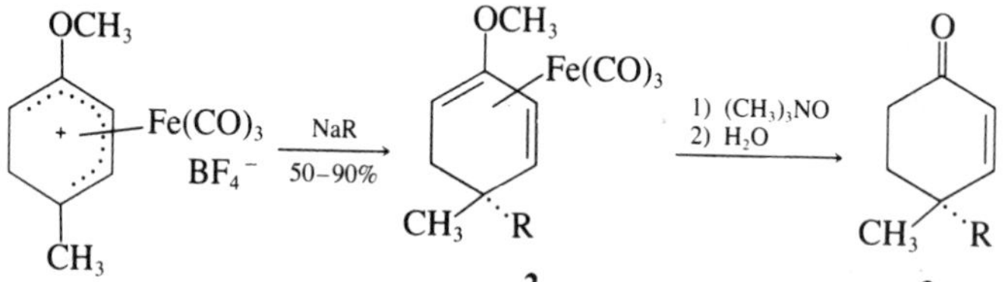

These iron complexes can be used to effect a stereocontrolled synthesis of 4,4,5-trisubstituted cyclohexenones.[2] An example is shown in equation (I).

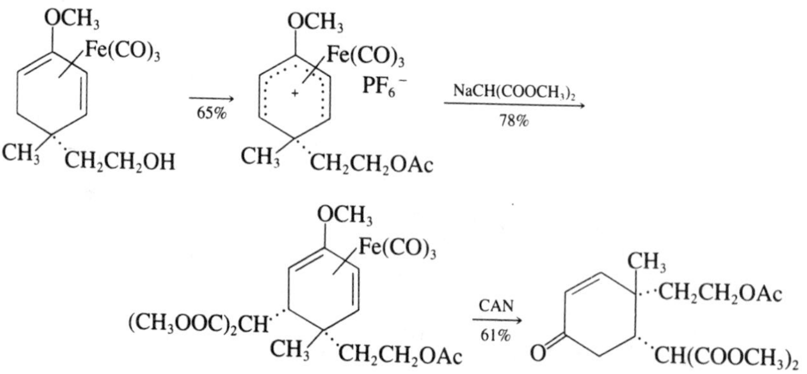

Cycloheptadienyl–iron complexes.[3]   Cycloheptadienyl–Fe(CO)$_3$ reacts with nucleophiles to give mixtures of products, usually in low yield. However, complexes **1**, in which one CO ligand is replaced by triphenylphosphine or triphenyl phosphite, react with most nucleophiles at the dienyl terminus and in high yield. Hydride abstraction followed by reaction with a second nucleophile results in a regio- and stereocontrolled substitution at the opposite terminus and again *anti* to the metal. Decomplexation affords *cis*-5,7-disubstituted cycloheptadienes.

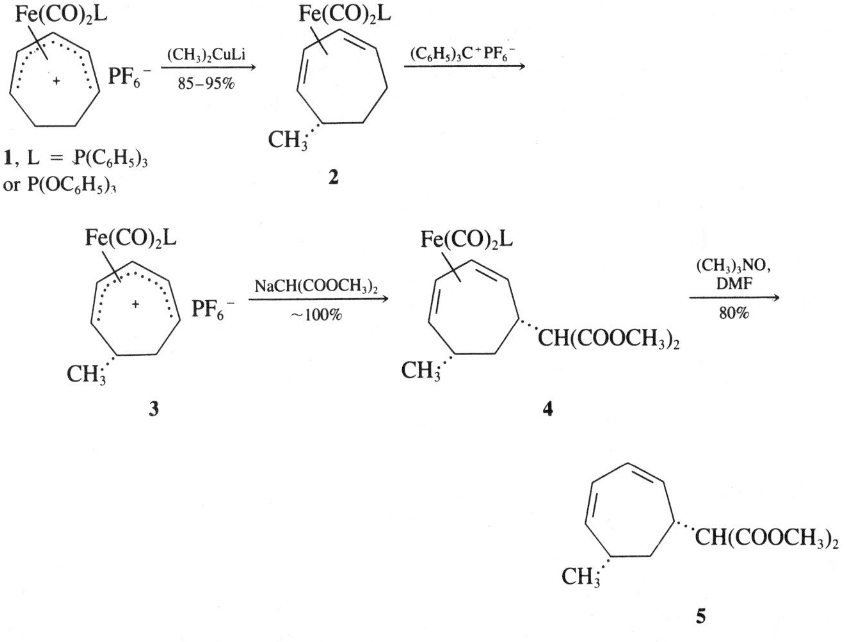

1, L = P(C$_6$H$_5$)$_3$
or P(OC$_6$H$_5$)$_3$

[1] A. J. Pearson, *Accts. Chem. Res.,* **13**, 463 (1980).
[2] A. J. Pearson and C. W. Ong, *J. Org.,* **47**, 3780 (1982).
[3] A. J. Pearson, S. L. Kole, and B. Chen, *Am. Soc.,* **105**, 4483 (1983).

K

Ketene bis(trimethylsilyl) ketals, $\overset{R^1}{\underset{R^2}{\diagup}}C=C[OSi(CH_3)_3]_2$ **(1)**. The ketals are obtained

by reaction of chlorotrimethylsilane with the anion of a trimethylsilyl ester or with the dianion of a carboxylic acid.[1]

Aldol-type reactions.[2] In the presence of $TiCl_4$, these ketals condense with aldehydes to form β-hydroxy carboxylic acids (equation I) or with imines to form β-lactams (equation II).

$$\text{(I) } RCHO + 1 \xrightarrow[\substack{75-90°}]{\substack{TiCl_4, \\ CH_2Cl_2, 25°}} RCH\overset{\overset{\displaystyle OH}{|}}{\underset{\underset{\displaystyle R^2}{|}}{-}}C\overset{\overset{\displaystyle R^1}{|}}{-}COOH$$

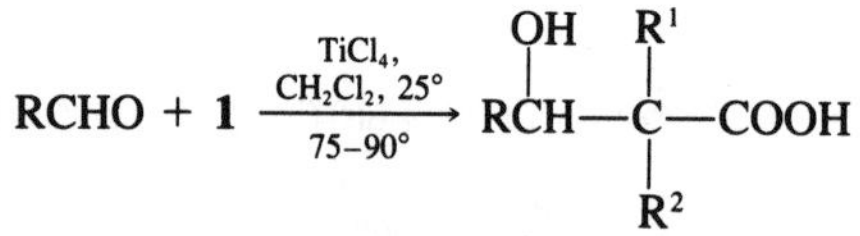

$$\text{(II)} \quad R^3CH=NR^4 + 1 \xrightarrow[60-75\%]{}$$

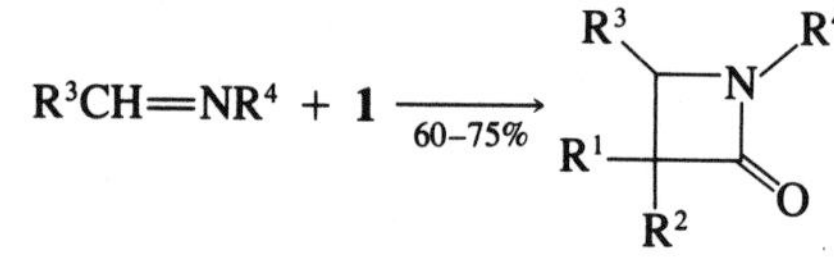

[1] C. Ainsworth and Y.-N. Kuo, *J. Organometal. Chem.*, **46**, 73 (1972).
[2] J.-E. Dubois and G. Axiotis, *Tetrahedron Letters*, **25**, 2143 (1984).

Ketenylidenetriphenylphosphorane, $(C_6H_5)_3P=C=C=O$ **(1)**, **10**, 450–451; **11**, 281.

Macrocyclic lactones.[1] The ylides **3**, formed by reaction of acetals (**2**) of long-chain hydroxy aldehydes with **1**, can be converted into the ylides **4**, which undergo an

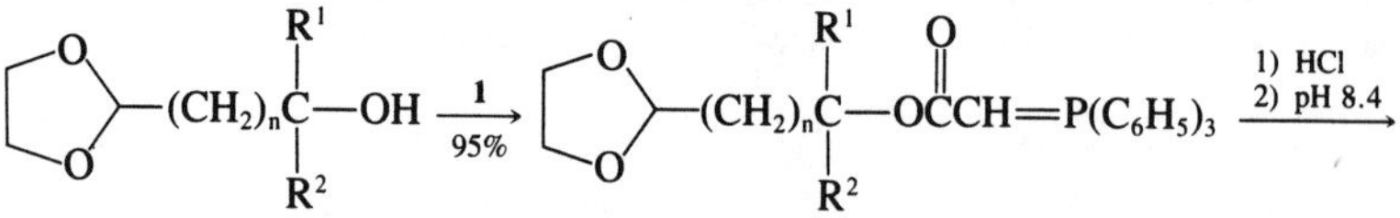

2, n = 7,9,11   **3**

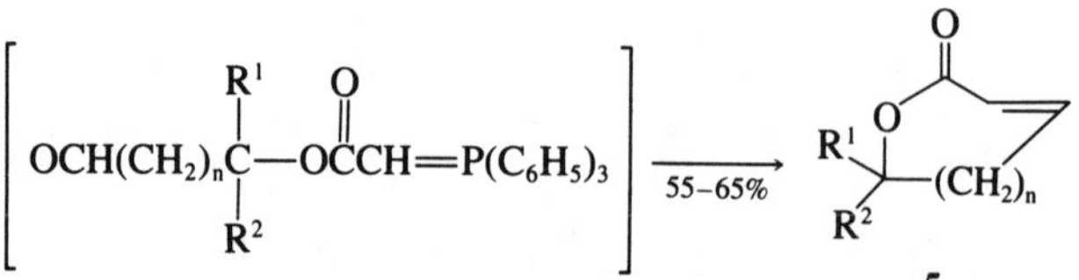

4   **5**

intramolecular Wittig reaction to provide (E)-α,β-unsaturated lactones (**5**). The conjugated double bond can be reduced by hydrogenation, or in the case of a diunsaturated lactone, it can be reduced selectively with sodium bis(2-methoxyethoxy)aluminum hydride and CuBr (**8**,120).

[1] H. J. Bestmann and R. Schobert, *Angew. Chem., Int. Ed.*, **22**, 780 (1983).

L

Lead tetraacetate, 1, 537–563; **2**, 234–238; **3**, 168–171; **4**, 278–282; **5**, 365–370; **6**, 313–317; **7**, 185–188; **8**, 269–272; **9**, 265–269; **10**, 228.

Tetrahydrofuranes.[1] Čeković, Bošnjak, and Mihailović have reviewed their work on the oxidative cyclization of aliphatic alcohols to tetrahydrofuranes by oxidation with lead tetraacetate. The reagent should be free of acetic acid. Anhydrous calcium carbonate can be added to neutralize the acetic acid in the oxidant and that formed during the reaction. The reaction involves selective oxygenation of a δ-carbon; the actual mechanism is uncertain, but is probably a radical reaction. Direct oxidation of the alcohol is generally of minor importance, but β-fragmentation can be a problem, particularly when a stable benzyl or allyl radical is formed.

Examples:

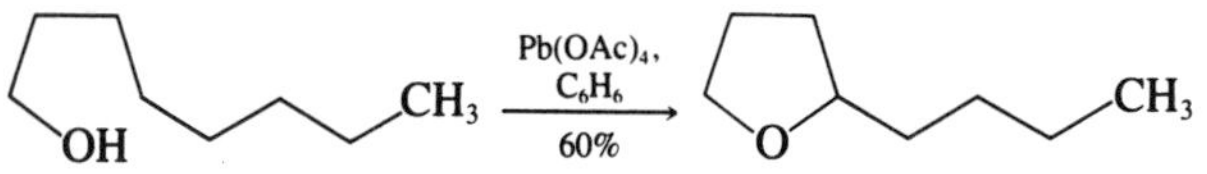

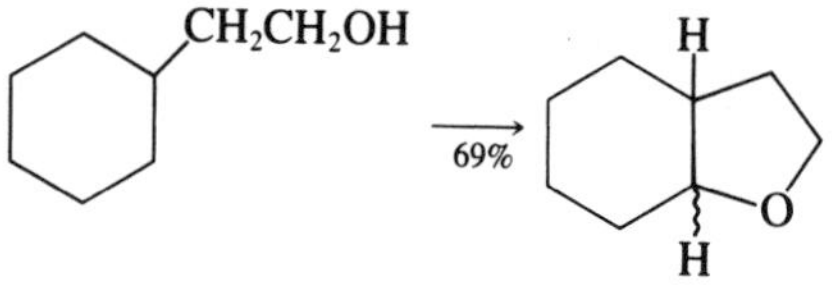

(*cis/trans* = 6:1)

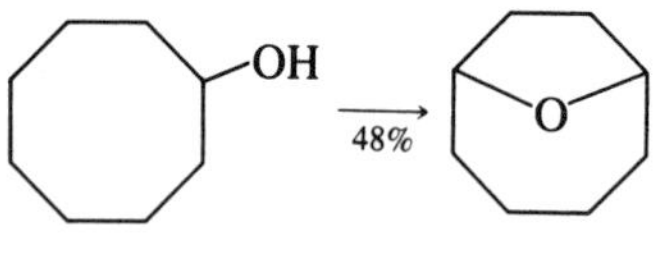

Cyclonucleosides.[2] A typical protected cyclopurine nucleoside (**2**) can be obtained by oxidation of a protected purine nucleoside (**1**) with LTA in refluxing benzene.

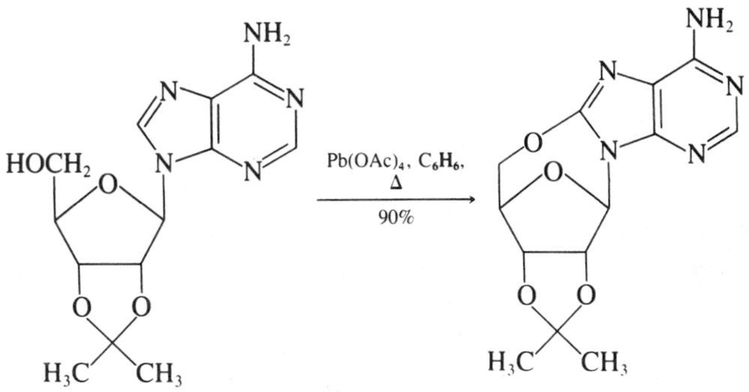

A novel type of cyclonucleoside (**4**) is obtained from a similar oxidation of a protected pyrimidine nucleoside (**3**).

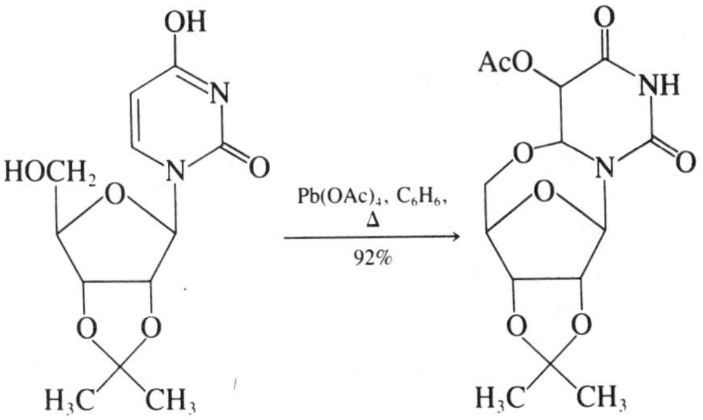

α-Acetoxy esters or lactones.[3]   Alkyl trimethylsilyl ketene acetals **1** and **2**, derived from esters and lactones, respectively, are oxidized by lead tetraacetate to α-acetoxy esters (**3**) and lactones (**4**), respectively.

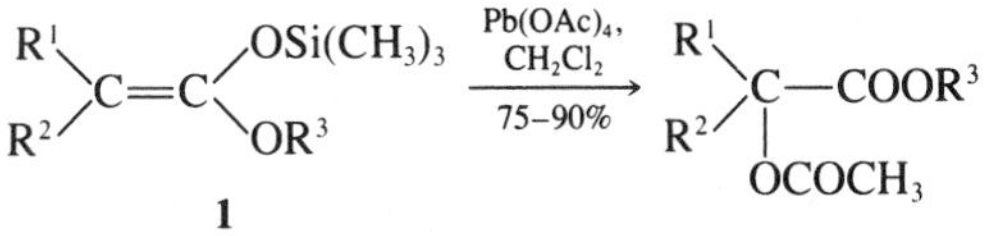

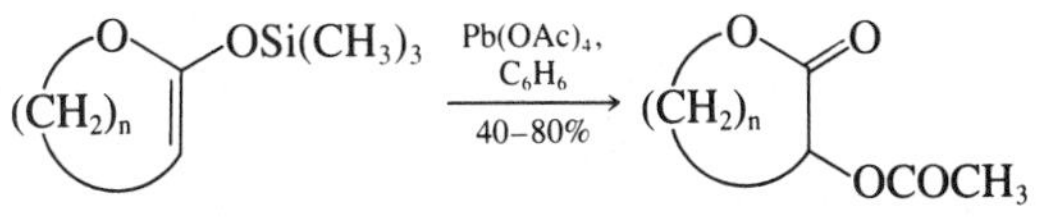

[1] Z. Čeković, J. Bošnjak, and M. L. Mihailović, *Org. Syn.*, submitted (1984).
[2] K. Kameyama, M. Sako, K. Hirota, and Y. Maki, *J.C.S. Chem. Comm.*, 1658 (1984).
[3] G. M. Rubottom, J. M. Gruber, R. Marrero, H. D. Juve, Jr., and C. W. Kim, *J. Org.*, **48**, 4940 (1983).

Lindlar catalyst, 1, 566–567; **3**, 171–172; **4**, 283; **6**, 319; **9**, 270–271.

MnCl$_2$-modified catalyst.[1]   The selectivity and reproducibility of Lindlar catalyst (Pd/CaCO$_3$/PbO) can be improved by doping with various salts, such as CdCl$_2$, SnCl$_2$, NiCl$_2$, and particularly MnCl$_2$. A secondary poison such as quinoline is still essential for satisfactory selectivity.

[1] J. Rajaram, A. P. S. Narula, H. P. S. Chawla, and S. Dev, *Tetrahedron*, **39**, 2315 (1983).

Lithium acetylide, 1, 573–574; **2**, 241; **6**, 324–325; **8**, 285–286.

Addition to cyclohexenones.[1]   Lithium acetylides undergo stereoselective axial addition to the carbonyl group of cyclohexenones that are conformationally rigid and free from 1,3-diaxial interactions. Thus the enone **1** reacts with lithium acetylide to give **2** as the only isolable adduct. Addition of ethyllithium is not stereoselective.

1 (R = H, CH$_3$)   2

[1] G. Stork and J. M. Stryker, *Tetrahedron Letters*, **24**, 4887 (1983).

Lithium aluminum hydride, 1, 581–595; **2**, 292; **3**, 176–177; **4**, 291–293; **5**, 382–389; **6**, 325–326; **7**, 196; **8**, 286–289; **9**, 274–277; **10**, 236–237; **11**, 289–292.

syn-1,3-Amino alcohols.[1]   A two-step stereoselective conversion of β-hydroxy ketones into *syn*-1,3-amino alcohols involves conversion into a mixture of the *syn*- and *anti*-benzyloximes, which are separable by chromatography. The *syn*-oximes are reduced with high asymmetric induction to *syn*-1,3-amino alcohols by lithium aluminum hydride. However, reduction of the *anti*-oximes shows only moderate selectivity for the *syn*-amino alcohol.

Example:

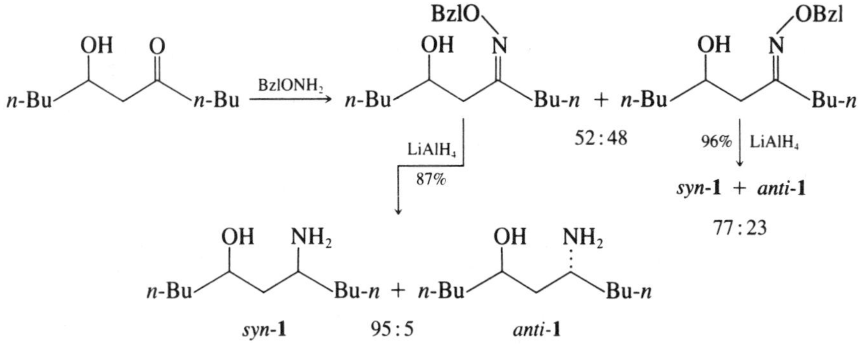

$$
n\text{-Bu} \xrightarrow[\text{BzlONH}_2]{}
$$

(OH O / n-Bu ... n-Bu) → BzlONH₂ → (BzlO, OH, N / n-Bu ... Bu-n) + (n-Bu, OH, N, OBzl / Bu-n)

52:48

96% | LiAlH₄

LiAlH₄
87%

syn-**1** + anti-**1**

77:23

(OH NH₂ / n-Bu ... Bu-n) + (n-Bu, OH, NH₂ / Bu-n)

syn-**1**   95:5   anti-**1**

α-Alkenylcarbinols.[2] Reduction of alkynyl chlorohydrins, readily available from reaction of α-chloro ketones with lithium acetylides, with lithium aluminum hydride and sodium methoxide results in partial reduction of the triple bond as well as a facile rearrangement to provide α-alkenylcarbinols. One useful feature is that only (E)-disubstituted alkenes are formed. The addition and reductive rearrangement can be conducted in a single operation.

Examples:

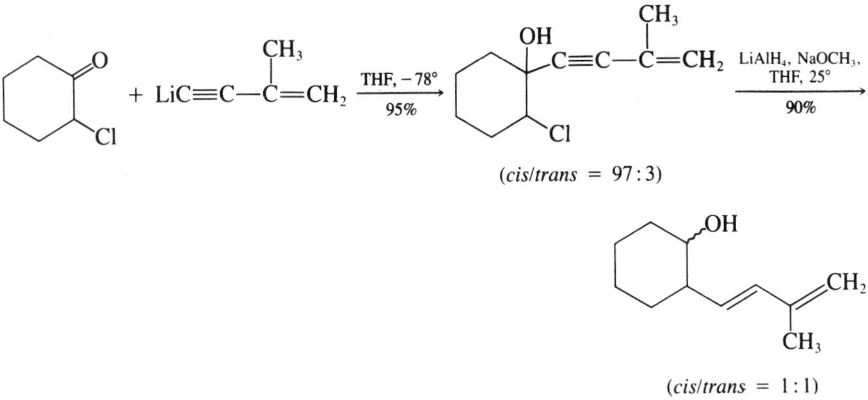

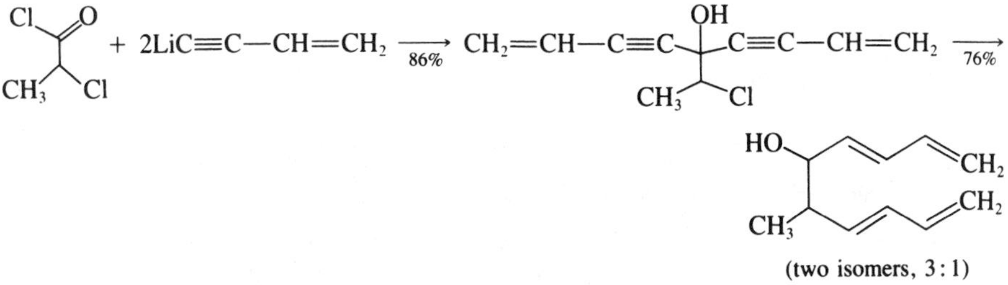

$$+ \text{LiC} \equiv \text{C} - \overset{\text{CH}_3}{\underset{}{\text{C}}} = \text{CH}_2 \xrightarrow[95\%]{\text{THF, }-78°}$$

(cis/trans = 97:3)

LiAlH₄, NaOCH₃,
THF, 25°
90%

(cis/trans = 1:1)

$$\underset{\text{CH}_3}{\overset{\text{Cl}}{\text{C}}}\overset{\text{O}}{\underset{\text{Cl}}{\text{C}}} + 2\text{LiC} \equiv \text{C} - \text{CH} = \text{CH}_2 \xrightarrow{86\%} \text{CH}_2 = \text{CH} - \text{C} \equiv \text{C} \underset{\underset{\text{CH}_3}{}\ \overset{}{\text{Cl}}}{\overset{\text{OH}}{\big|}} \text{C} \equiv \text{C} - \text{CH} = \text{CH}_2 \xrightarrow{76\%}$$

(two isomers, 3:1)

This method can be used to prepare muscone (**2**) from the chloro ketone **1**.

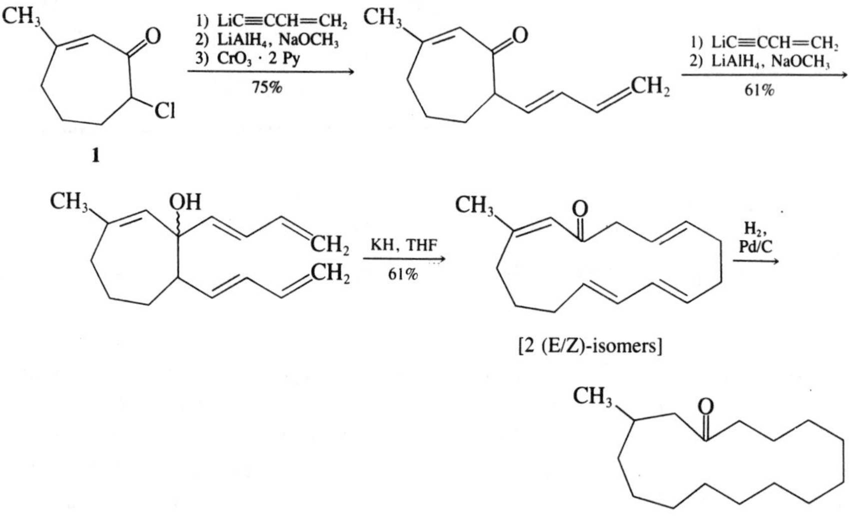

[2 (E/Z)-isomers]

2

Denitration of α-nitro ketones.[3]   These nitro ketones (**1**), readily prepared by the nitro-aldol reaction (equation I), can be converted into the corresponding ketones (**3**) by reduction of the tosylhydrazones (**2**).

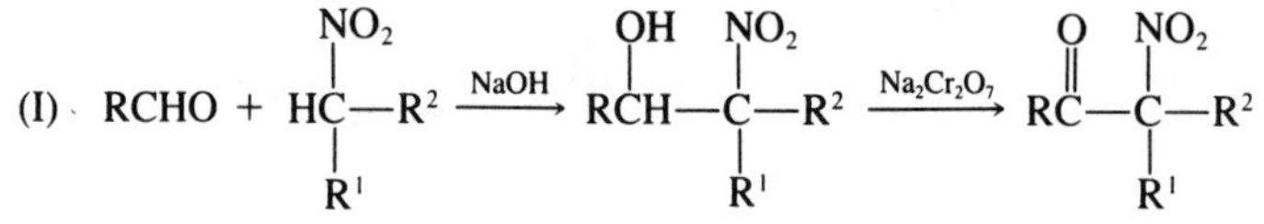

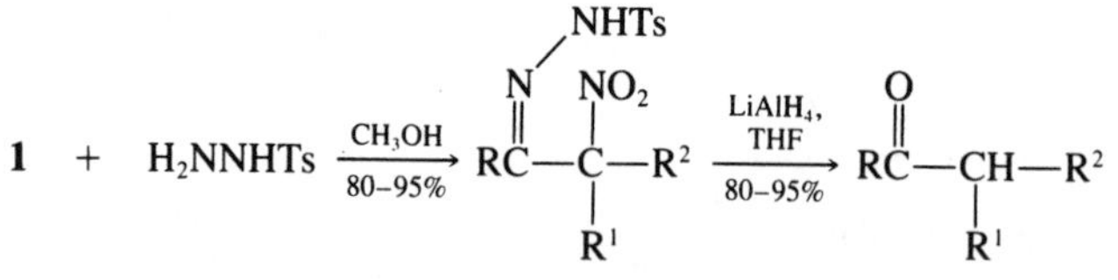

This reduction is applicable only to secondary or tertiary nitro compounds.

Regio- and stereospecific reduction of α-oxoketene thioacetals.[4]   These substrates undergo a novel two-step reduction with $LiAlH_4$. At low temperatures, only the carbonyl group is reduced, forming an allylic alcohol; this reduction is followed at higher temperatures by reduction of the double bond to give β-hydroxy thioacetals. If the double bond is substituted by an alkyl group, the final reduction results in an *anti*-alcohol.

Example:

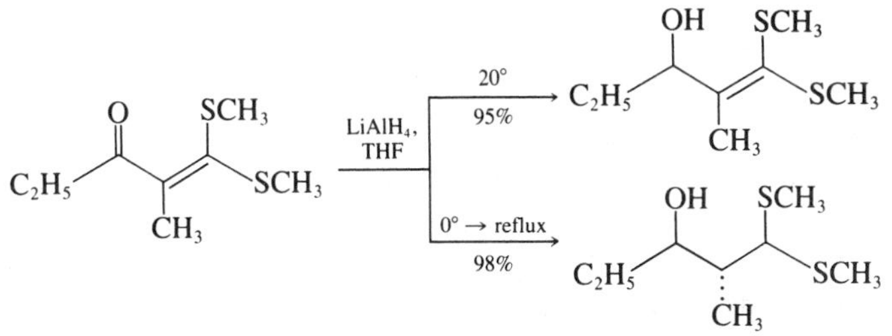

***Reduction of aryl and alkyl halides.*[5]** This reaction can be conducted in generally good yield with LiAlH$_4$ by a free-radical process initiated by irradiation of di-*t*-butyl peroxide. The order of reactivity of aryl halides is ArI > ArBr > ArCl > ArF. Alkyl and cycloalkyl halides are reduced efficiently, but vinyl bromides are reduced in only modest yield.

***Reductions in benzene or hexane.*[6]** Lithium aluminum hydride adsorbed on SiO$_2$ is useful in nonpolar solvents such as C$_6$H$_6$ or *n*-C$_6$H$_{14}$. Reductions of aldehydes, ketones, or acid chlorides proceed at room temperature; reduction of esters requires reflux temperatures.

[1] K. Narasaka and Y. Ukaji, *Chem. Letters*, 147 (1984).
[2] P. A. Wender, D. A. Holt, and S. M. Sieburth, *Am. Soc.*, **105**, 3348 (1983).
[3] G. Rossini, R. Ballini, and V. Zanotti, *Synthesis*, 137 (1983).
[4] R. B. Gammill, L. T. Bell, and S. A. Nash, *J. Org.*, **49**, 3039 (1984).
[5] A. L. J. Beckwith and S. H. Goh, *J.C.S. Chem. Comm.*, 907 (1983).
[6] Y. Kamitori, M. Hojo, R. Masuda, T. Izumi, and T. Inone, *Synthesis*, 387 (1983).

Lithium aluminum hydride–Diphosphorus tetraiodide.

***Reduction of alkyl aryl ketones.*[1]** Aromatic ketones can be reduced to the corresponding hydrocarbons by reaction with LiAlH$_4$ and P$_2$I$_4$ in refluxing benzene. The ketone and the two reagents are used in equimolar amounts. Yields range from 45 to ~100%. The method offers a mild alternative to Clemmensen and Wolff-Kishner reductions.

[1] H. Suzuki, R. Masuda, H. Kubota, and A. Osuka, *Chem. Letters*, 909 (1983).

Lithium 9-boratabicyclo[3.3.1]nonane, Li 9-BBNH (1). Standard solutions in THF are obtained by reaction of 9-BBN with LiH.

***Selective reductions.*[1]** As expected, this dialkylborohydride shows reducing properties intermediate between those of the mild lithium borohydride and of the powerful lithium trialkylborohydrides. As a result, it is particularly useful for selective reductions.

It reduces esters or lactones selectively in the presence of several other functional groups such as carboxylic acids, amides, nitriles, and halides. Simple amides are not reduced, but tertiary amides are reduced to the corresponding tertiary amines. $LiB(C_2H_5)_3H$ or 9-BBN reduces tertiary amides to the corresponding alcohols.

[1] H. C. Brown , C. P. Matthew, C. Pyun, J. C. Son, and N. M. Yoon, *J. Org.*, **49**, 3091 (1984).

Lithium borohydride, 1, 603; **4**, 296; **11**, 293.

Reductions in methanol-containing solvents.[1] Primary amides can be reduced selectively in the presence of carboxylic acid salts or of secondary amides by $LiBH_4$ in diglyme–CH_3OH. Esters and epoxides are reduced selectively in the presence of nitro, chloro, or amide groups by $LiBH_4$ in ether containing some CH_3OH.

Conversion of esters to secondary alcohols.[2] The reaction of esters with Grignard reagents results mainly in tertiary alcohols, which are formed by way of an intermediate ketone. Direct conversion of an ester to a secondary alcohol is possible by reaction with a Grignard reagent (2 equiv.) and lithium borohydride (0.5 equiv.), which reduces the intermediate ketone much more rapidly than it does the ester.

Examples:

$$C_6H_5(CH_2)_2COOC_2H_5 \xrightarrow[74\%]{\substack{LiBH_4,\ BuMgCl,\\ THF,\ -20^\circ,\ 24\ hr.}} C_6H_5(CH_2)_2\overset{\overset{\displaystyle OH}{|}}{C}HBu$$

$$C_6H_5COOC_2H_5 \xrightarrow[61\%]{THF,\ 0^\circ,\ 24\ hr.} C_6H_5\overset{\overset{\displaystyle OH}{|}}{C}HBu$$

[1] K. Soai, A. Ookawa, and H. Hayashi, *J.C.S. Chem. Comm.*, 668 (1983).
[2] D. L. Comins and J. J. Herrick, *Tetrahedron Letters*, **25**, 1321 (1984).

Lithium *n*-butyldiisobutylaluminum hydride (1). The reagent is prepared by reaction of DIBAH and *n*-butyllithium in mixtures of THF or toluene with hexane.

Reductions.[1] This hydride is a strong reducing agent comparable to other lithium trialkylhydrides. It is superior to DIBAH for selective 1,2-reduction of enones. Reduction of ketones, esters, acid chlorides, and anhydrides proceeds at -78°. However, ketones can be reduced selectively in the presence of an ester. Esters are reduced to a mixture of an alcohol and an aldehyde. Complete reduction to an alcohol can be effected by reduction at -78° with 2 equiv. of **1** and then with excess sodium borohydride. Tertiary amides are reduced by 1 equiv. of the reagent to aldehydes in generally high yield. Selective reduction of primary halides in the presence of secondary halides is possible.

[1] S. Kim and K. H. Ahn, *J. Org.*, **49**, 1717 (1984).

Lithium chloride–Diisopropylethylamine.

Wittig-Horner olefination.[1] This reaction can be effected with LiCl (1 equiv.) and either diisopropylethylamine or DBU (1 equiv.) in CH_3CN at room temperature. This variation is particularly useful in reactions with aldehydes or phosphonates that can undergo epimerization or aldol-type reactions under standard conditions (NaH or K_2CO_3). Yields are usually >80%. The reaction also shows a high (E)-selectivity. Presumably a chelated lithium enolate of the phosphonate is the reactive species.

[1] M. A. Blanchette, W. Choy, J. T. Davis, A. P. Essenfeld, S. Masamune, W. R. Roush, and T. Sakai, *Tetrahedron Letters*, **25**, 2183 (1984).

Lithium chloride–Hexamethylphosphoric triamide, 10, 240.

Dehydrobromination (**8**, 244). Dehydrobromination of 1,2-dibromocyclohexane with lithium chloride and lithium carbonate in HMPT at 160° provides 1,3-cyclohexadiene of >95% purity in yields of 75–90%. Even higher yields can be obtained by dehydro-bromination of 3-bromocyclohexene.[1]

Another example:

[1] A. Weisz and A. Mandelbaum, *J. Org.*, **49**, 2648 (1984).

Lithium diisopropylamide, 2, 249; **3,** 184–185; **4,** 298–302; **5,** 400–406; **6,** 334–339; **7,** 204–207; **8,** 292; **9,** 280–283; **10,** 241–243; **11,** 296–299. Molar quantities of the reagent can be prepared conveniently by reaction of lithium metal with the amine in the presence of styrene[1] or α-methylstyrene[2] (equation I).

(I) $\quad C_6H_5CH{=}CH_2 + 2Li + 2HN[CH(CH_3)_2]_2 \xrightarrow[95\%]{\text{ether}} C_6H_5CH_2CH_3 + 2LiN[CH(CH_3)_2]_2$

LDA prepared in this way is recommended for regioselective preparation of the less thermodynamically stable enol ether and for generation of dianions of carboxylic acids. The reaction of dianions prepared with this base with aldehydes provides a stereoselective route to γ,δ-unsaturated β-hydroxy carboxylic acids (equation II).

(*anti/syn* = 6:1)

Rearrangement of 2,4-alkadienoic esters.[3]   (2E,4Z)-Alkadienoates are rearranged by LDA/HMPT with 80–98% stereoselectivity to (3E,5E)-alkadienoates. In contrast, the (2E,4E)-isomers are rearranged mainly to (3E,5Z)-dienoates.

Example:

$$C_2H_5CH_2CH=CHCH=CHCOOCH_3 \xrightarrow[\text{HMPT}]{\text{LDA,}}$$

| (2E,4Z) | (77%) | (3%) |
| (2E,4E) | (13%) | (51%) |

α-Hydroxy-β-keto esters.[4]   A new route to these esters involves condensation of ethyl bromoacetate with a carboxylic acid in the presence of K_2CO_3 to form ethyl α-acyloxyacetates (**1**). These products when treated with LDA (2 equiv.) in THF at 0° rearrange to ethyl α-hydroxy-β-keto esters (**2**) in 50–65% yield. The intermediate anion can be silylated or acetylated.

Example:

$$C_2H_5COOH + BrCH_2COOC_2H_5 \xrightarrow{K_2CO_3} C_2H_5\overset{\displaystyle O}{\overset{\|}{C}}OCH_2COOC_2H_5 \xrightarrow[57\%]{\text{LDA,}\ \text{THF, 0°}}$$

1

$$C_2H_5\overset{\displaystyle O}{\overset{\|}{C}}CHCOOC_2H_5$$
$$\underset{\displaystyle OH}{|}$$

2

Succinic acids.[5]   The dianion salt of carboxylic acids, generated with LDA and freed from diisopropylamine, on reaction with I_2 couples to form derivatives of succinic acid. This coupling of 3-phenylpropionic acid dianion results largely in the *dl*-isomer (equation I).

(I) $$C_6H_5CH_2CH_2COOH \xrightarrow{\text{LDA}} \left[C_6H_5CH_2\overset{\displaystyle Li}{\overset{|}{C}}HCOOLi \right] \xrightarrow[77\%]{I_2} HOOC-\underset{\displaystyle CH_2C_6H_5}{\overset{\displaystyle C_6H_5CH_2}{\underset{|}{\overset{|}{C}H}}}-CHCOOH$$

Unsymmetrically substituted succinic acid derivatives can be prepared by reaction of carboxylic acid dianions with lithium α-iodocarboxylates (40–70% yield).

[1] M. T. Reetz and W. F. Maier, *Ann.*, 1471 (1980).
[2] J. Mulzer, P. de Lasalle, A. Chucholowski, W. Blaschek, G. Brüntrup, I. Jibril, and G. Hultner, *Tetrahedron*, **40**, 2211 (1984).

[3] S. Tsuboi, A. Kuroda, T. Masuda, and A. Takeda, *Chem. Letters*, 1541 (1984).
[4] S. D. Lee, T. H. Chan, and K. S. Kwon, *Tetrahedron Letters*, **25**, 3399 (1984).
[5] J. L. Belletire, E. G. Spletzer, and A. R. Pinhas, *ibid.*, 5969 (1984).

Lithium diisopropylamide–Chlorotrimethylsilane.

α-*Trimethylsilyl vinyl sulfides*.[1]   Alkyl sulfoxides in which the alkyl group is ethyl or a higher homolog are converted by reaction with excess LDA and $ClSi(CH_3)_3$ in THF into α-trimethylsilyl vinyl sulfides.

Examples:

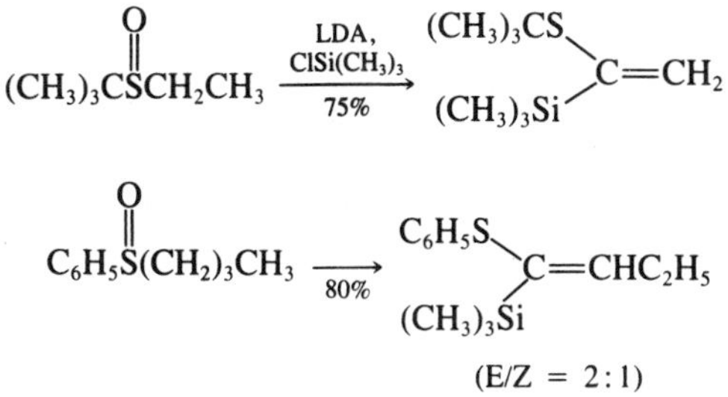

$$(E/Z = 2:1)$$

[1] R. D. Miller and R. Hassig, *Tetrahedron Letters*, **25**, 5351 (1984).

Lithium 1-(dimethylamino)naphthalenide (LDMAN), **10**, 244–245; **11**, 422.

α-*Lithio cyclic ethers*.[1]   The preparation of α-lithio ethers (**10**, 244) has been extended to the preparation of α-lithio cyclic ethers from γ- and δ-lactones. Thus the stereoisomeric α-phenylthio ethers **2**, obtained as shown from the δ-lactone **1**, undergo reductive lithiation with LDMAN to give exclusively the more stable *trans*-2-lithio-4-methyltetrahydropyrane (**3**). This product undergoes 1,2-addition to aldehydes and ketones.

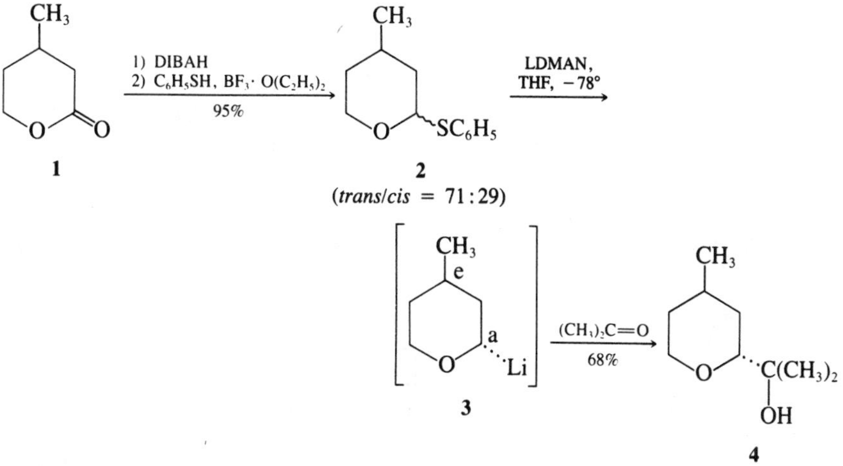

This process is also applicable to γ-lactones to obtain α-lithiotetrahydrofuranes, but in this series a mixture of epimers is obtained. The sequence is not applicable to seven-membered lactones.

α-Lithiosilanes; Peterson olefination.[2]   Secondary or tertiary α-lithiosilanes are conveniently prepared from diphenyl thioacetals or thioketals by reductive lithiation with LDMAN, silylation of the resulting anion, and a second reductive lithiation (equation I).

$$(I) \quad \begin{array}{c} R^1 \\ \diagdown \\ \quad C(SC_6H_5)_2 \\ \diagup \\ R^2 \end{array} \xrightarrow[\text{2) ClSi(CH}_3)_3]{\text{1) LDMAN}} \begin{array}{c} R^1 \quad SC_6H_5 \\ \diagdown \diagup \\ C \\ \diagup \diagdown \\ R^2 \quad Si(CH_3)_3 \end{array} \xrightarrow{\text{LDMAN}} \begin{array}{c} R^1 \quad Li \\ \diagdown \diagup \\ C \\ \diagup \diagdown \\ R^2 \quad Si(CH_3)_3 \end{array}$$

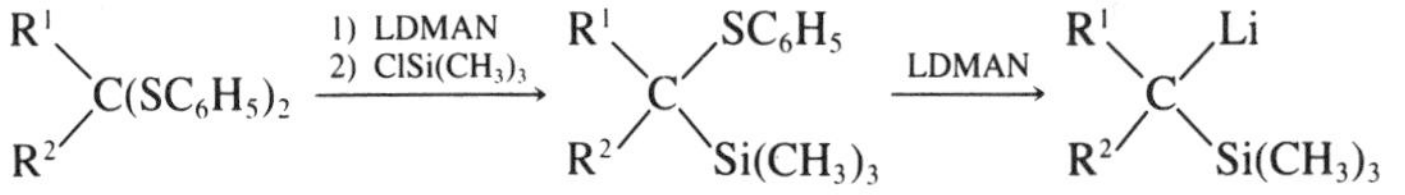

The same method can be used to obtain cyclopropyl and vinyl α-lithiosilanes. The α-lithiosilanes add to aldehydes and ketones to form β-silylcarbinols, which form olefins with loss of trimethylsilanol on treatment with potassium hydride (THF or diglyme) or in some cases with sulfuric acid.

Examples:

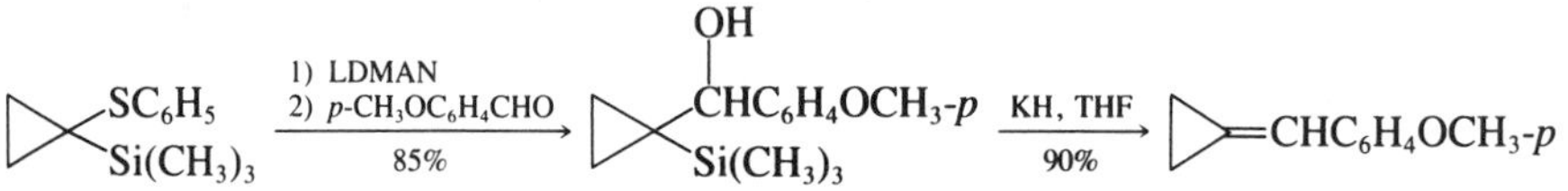

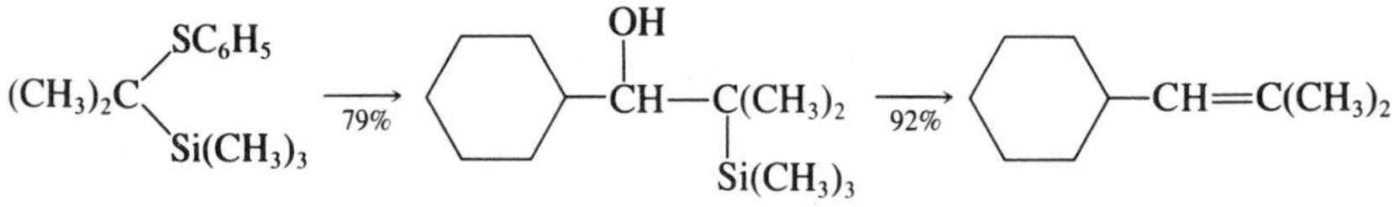

[1] T. Cohen and M.-T. Lin, *Am. Soc.*, **106**, 1130 (1984).
[2] T. Cohen, J. P. Sherbine, J. R. Matz, R. R. Hutchins, B. M. McHenry, and P. R. Willey, *ibid.*, **106**, 3245 (1984).

Lithium hexamethyldisilazide, 4, 296, 409; **5**, 393–394; **7**, 197; **11**, 300.
    Ester enolate Claisen rearrangement (4, 307–308; **6**, 276–277; **7**, 209–210).   Lithium hexamethyldisilazide is comparable to LDA in combination with HMPT for stereoselective Claisen rearrangement of ester enolates.[1]
    Ireland and Varney[2] have reported asymmetric induction in this rearrangement by use of the enantiomerically pure secondary allylic alcohol (**1**), obtained by resolution of the

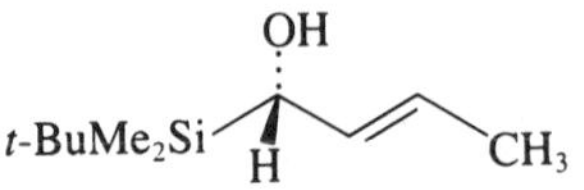

(S)-**1**

racemic alcohol with α-methoxy-α-trifluoromethylphenylacetic acid.[3] Thus rearrangement of the propionate ester (2) of 1 via the silylketene acetal obtained with lithium hexamethyldisilazide and *t*-butyldimethylchlorosilane results in 3 and 4 in the ratio 94:6. Rearrangement of 2 via the silylketene acetal obtained with LDA as base results in 4 and 3 in the ratio 93:7. These products are converted in several steps into the corresponding diastereomeric β,δ-unsaturated benzyl ethers 5 and 6. Desilylation of the intermediate vinylsilanes can be effected in high yield with aqueous tetrafluoroboric acid in acetonitrile. The α-silyl allylic alcohol 1 then can serve as the equivalent of a chiral primary alcohol.

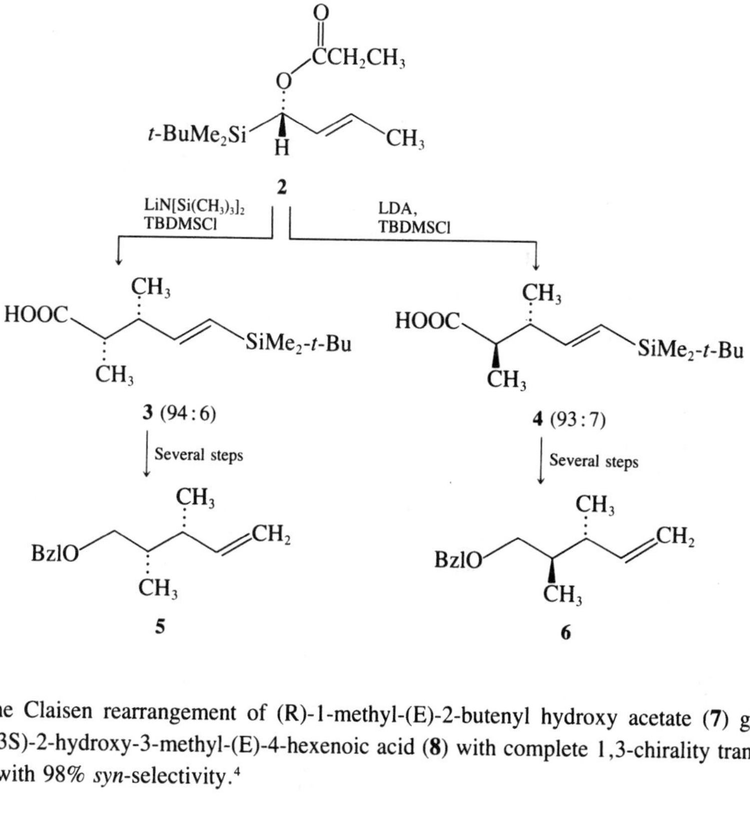

The Claisen rearrangement of (R)-1-methyl-(E)-2-butenyl hydroxy acetate (7) gives (2R,3S)-2-hydroxy-3-methyl-(E)-4-hexenoic acid (8) with complete 1,3-chirality transfer and with 98% *syn*-selectivity.[4]

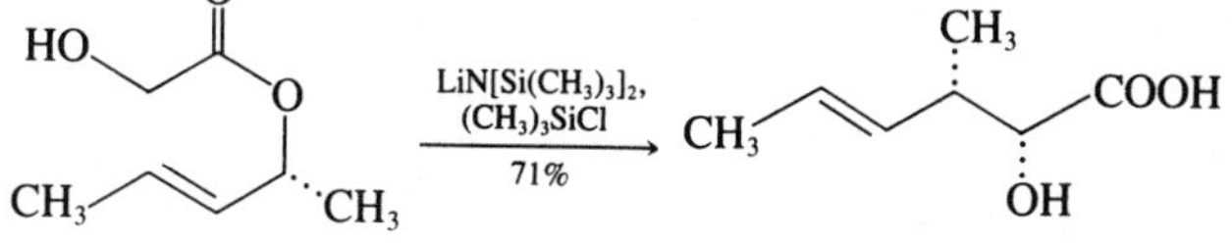

[1] R. E. Ireland and J. P. Daub, *J. Org.*, **46**, 479 (1981).
[2] R. E. Ireland and M. D. Varney, *Am. Soc.*, **106**, 3668 (1984).
[3] H. S. Mosher, J. A. Dale, and D. L. Dull, *J. Org.*, **34**, 2543 (1969).
[4] T. Fujisawa, K. Tajima, and T. Sato, *Chem. Letters*, 1669 (1984).

Lithium hydroxide, LiOH, 7, 208.

Hydrolysis of lactams.   The N-Boc derivatives of lactams are hydrolyzed to Boc derivatives of ω-amino acids by treatment with LiOH (3 equiv.) in aqueous THF at 25° in 85–95% yield. Treatment with $NaOCH_3$ in CH_3OH at 0° affords the esters of these products. The same reactions are useful for hydrolysis or methanolysis of secondary amides.[1]

Examples:

[1] D. L. Flynn, R. E. Zelle, and P. A. Grieco, *J. Org.*, **48**, 2424 (1983).

Lithium iodide, 5, 410–411; 7, 208; 9, 283; 10, 245–246; 11, 300–301.

Ketene thioacetals.[1]   The enolate (**1**) of 2-carbomethoxy-1,3-dithiane does not normally react with aldehydes or ketones, but a 1:1 mixture of trimethylacetyl chloride and an aldehyde reacts in THF to form the pivaloxy ester **2**. This product undergoes decarboalkoxylation (**9**, 283; **11**, 301) when heated in dry DMF with excess LiI to form ketene thioacetals (**3**).

Cleavage of glycosyl esters.[2]   Glycosyl esters of hindered carboxylic acids can be cleaved without decomposition of the resulting sugar by anhydrous lithium iodide in anhydrous methanol. This gives an anomeric mixture of a methyl glycoside and the aglycone in quantitative yield.

[1] J. L. Belletire, D. R. Walley, and S. L. Fremont, *Tetrahedron Letters*, **25**, 5729 (1984).
[2] K. Ohtani, K. Mizutani, R. Kasai, and O. Tanaka, *ibid.*, **25**, 4537 (1984).

Lithium o-lithiophenoxide,

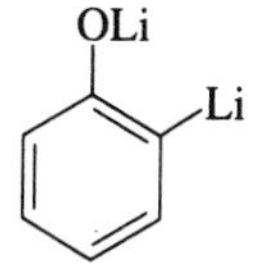

(**1**). The dianion is prepared in >90% yield by reaction of o-bromophenol with 2 equiv. of n-BuLi in hexane at 0°. The dianions of substituted phenols can be prepared in the same way.

The dianion reacts with enolizable and nonenolizable aldehydes and ketones to provide the adducts in ~70–90% yield (equation I).[1]

$$(I) \quad \begin{matrix} R^1 \\ R^2 \end{matrix} C{=}O + 1 \xrightarrow[70-90\%]{} $$

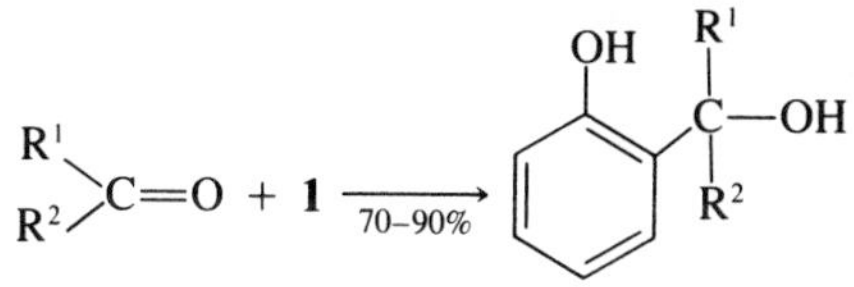

[1] J. J. Talley and I. A. Evans, *J. Org.*, **49**, 5267 (1984).

Lithium methylsulfinylmethylide (Lithium dimsylate), $CH_3\overset{\underset{\displaystyle |}{OLi}}{S}{=}CH_2$, **6**, 342.

Claisen rearrangement.   The Claisen rearrangement of highly substituted allyl vinyl ethers can proceed readily even when two vicinal quaternary centers are formed. Thus addition of an allylic alcohol (**1**) to an allenic sulfone (**2**) provides the allyl vinyl ether **3**, whose anion rearranges at 50° to **4**. Lithium dimsylate is the preferred base; other counterions promote an undesirable side reaction.[1]

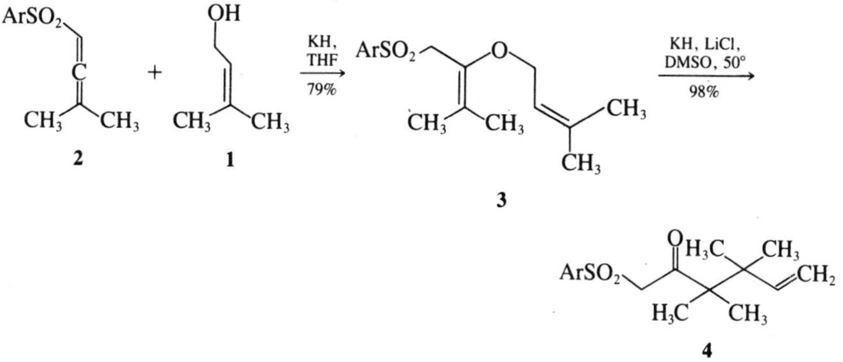

Another example:

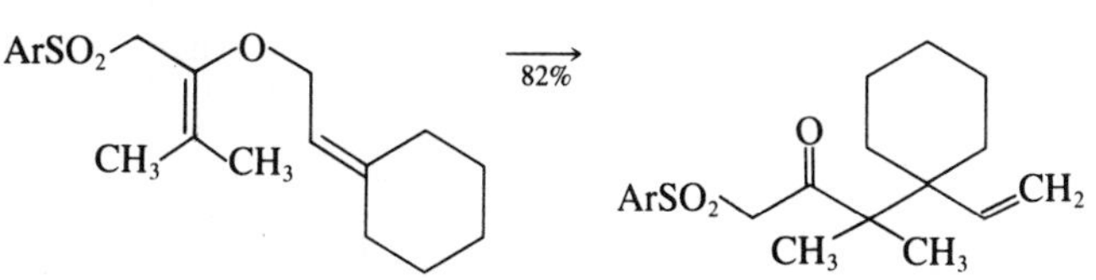

[1] S. E. Denmark and M. A. Harmata, *Tetrahedron Letters*, **25**, 1543 (1984).

Lithium morpholide,  (**1**). The lithium amide is prepared *in situ* by reaction of morpholine with *n*-BuLi in THF at −40°.

Protection of aldehydes.[1]   The reagent (**1**) or the related reagent lithium [N-methyl-N-(2-pyridyl)]amide (**2**), prepared from N-methyl-2-pyridylamine, converts aldehydes into lithium α-amino alkoxides, which are relatively stable to Grignard reagents. The derivatives on reaction with $ClSi(CH_3)_3$ are converted into silyl ethers, which are stable to $LiAlH_4$ and many organolithium reagents. Both types of protective groups are cleaved quantitatively by dilute acid.[1] This type of protection is useless for ketones, which are converted into enolate anions by **1**.

This strategy has been used to convert *o*-bromobenzaldehydes into phthalaldehydic acids (equation I).[2]

$$(I) \quad CH_3O\!-\!\!\langle o\text{-}C_6H_4\rangle\substack{CHO \\ Br} \xrightarrow[70–85\%]{\substack{1)\ \mathbf{1},\ THF \\ 2)\ n\text{-BuLi} \\ 3)\ CO_2}} CH_3O\!-\!\!\langle \ \rangle\!-\!O \ (lactone,\ OH)$$

[1] D. L. Comins and J. D. Brown, *Tetrahedron Letters*, **22**, 4213 (1981).
[2] A. K. Sinhababu and R. T. Borchardt, *J. Org.*, **48**, 2356 (1983).

Lithium naphthalenide, 2, 288–289; **3**, 208; **4**, 348–349; **5**, 468; **6**, 415; **8**, 305–306; **9**, 285; **10**, 247; **11**, 302–303.

Glycosyllithium reagent.[1]   The reactive glycopyranosyllithium **3** can be obtained

BzlO—(OBzl)—O **1** $\xrightarrow[100\%]{HCl}$ BzlO—(OBzl)—O, Cl **2** $\xrightarrow[-78°]{[C_{10}H_8]^{\cdot}Li^+,}$ BzlO—(OBzl)—O, Li **3**

$\Big\downarrow \substack{p\text{-}CH_3OC_6H_4CHO, \\ -78\%}$ 65%

BzlO—(OBzl)—O, $HOCHC_6H_4OCH_3\text{-}p$

4 (2 diastereomers)

by hydrochlorination of the D-glucal **1** followed by reductive lithiation with lithium naphthalenide. This derivative reacts with various electrophiles to provide C-glycosides.

[1] J.-M. Lancelin, L. Morin-Allory, and P. Sinaÿ, *J.C.S. Chem. Comm.*, 355 (1984).

Lithium *t*-octyl-*t*-butylamide   (LOBA), LiN$\overset{C(CH_3)_2CH_2C(CH_3)_3}{\underset{C(CH_3)_3}{\diagdown}}$ *t*-Octyl-*t*-butyl-

amine is prepared by a general method for preparation of di-*t*-alkylamines that involves radical *t*-butylation of a *t*-alkylnitroso compound to form a tri-*t*-alkylhydroxylamine, which is then reduced with sodium naphthalenide (equation I).[1]

(I) t-$C_8H_{17}NH_2 \xrightarrow[94\%]{CH_3CO_3H} t$-$C_8H_{17}N{=}O \xrightarrow[84\%]{\substack{t\text{-BuNHNH}_2, \\ PbO_2}} t$-$C_8H_{17}\overset{\overset{\text{O-}t\text{-Bu}}{|}}{N}$-$t$-Bu

$$\xrightarrow[91\%]{\substack{[C_{10}H_8]^+Na^+, \\ THF}} t\text{-}C_8H_{17}\overset{\overset{H}{|}}{N}\text{-}t\text{-Bu}$$

(b.p. 68–70°/1 mm)

Kinetic enolates.[2]   The kinetic enolate of a ketone or ester is generated with enhanced selectivity by a lithium dialkylamide in the presence of chlorotrimethylsilane. In addition, LOBA is superior to LDA for regioselective generation of enolates and for stereoselective formation of (E)-enolates.

Examples:

$$n\text{-}C_5H_{11}\overset{\overset{O}{\|}}{C}CH_3 \longrightarrow n\text{-}C_5H_{11}\overset{\overset{OSi(CH_3)_3}{|}}{C}{=}CH_2 \quad + \quad n\text{-}C_4H_9\overset{\overset{OSi(CH_3)_3}{|}}{C}{=}CCH_3$$

LDA	95:5	
LOBA	97.5:2.5	

$$CH_3CH_2\overset{\overset{O}{\|}}{C}CH_2CH_3 \xrightarrow{THF} \underset{CH_3}{\overset{H}{\diagdown}}C{=}C\underset{CH_2CH_3}{\overset{OSi(CH_3)_3}{\diagup}} \quad + \quad \underset{H}{\overset{CH_3}{\diagdown}}C{=}C\underset{CH_2CH_3}{\overset{OSi(CH_3)_3}{\diagup}}$$

LDA	77:23	
LOBA	98:2	

[1] E. J. Corey and A. W. Gross, *Tetrahedron Letters*, **25**, 491 (1984); *Org. Syn.*, submitted (1984).
[2] *Idem, Tetrahedron Letters*, **25**, 495 (1984).

Lithium tetramethylpiperidide, 4, 310–311; **5**, 417; **6**, 345–348; **7**, 213–215; **8**, 307–308; **9**, 285–286.

   o-Xylylene.[1]   This compound (**2**) can be generated as a reactive intermediate by treatment of methyl *o*-methylbenzyl ether with this base in the presence of a reactive dienophile (reflux). Yields of Diels-Alder adducts range from negligible (cyclohexene) to 70% (norbornene).

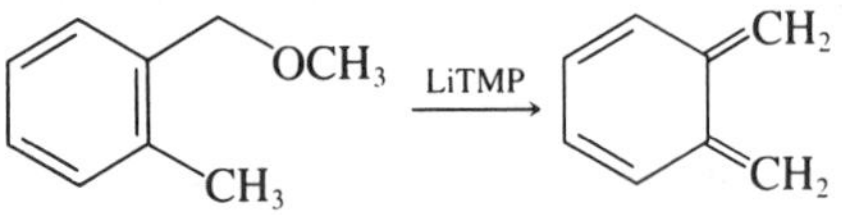

1   **2**

[1] T. Tuschka, K. Naito, and B. Rickborn, *J. Org.*, **48**, 70 (1983).

Lithium tri-*t*-butoxyaluminum hydride–Copper(I) bromide.

1,4-*Reduction of* 1-*acylpyridinium salts*.[1] These salts are reduced regiospecifically in moderate yield to 1-acyl-1,4-dihydropyridines by a copper hydride prepared from lithium tri-*t*-butoxyaluminum (3 equiv.) and CuBr (4.4 equiv.) (equation I).

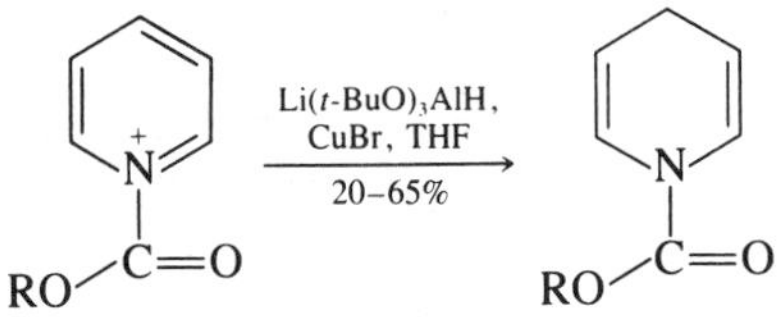

[1] D. L. Comins and A. H. Abdullah, *J. Org.*, **49**, 3392 (1984).

Lithium tri-*sec*-butylborohydride, 4, 312–313; 6, 348; 7, 307; 8, 308–309; 10, 248.

anti-*Selective reduction of* α-*methyl*-β,γ-*unsaturated ketones*.[1] The reduction of these ketones (**1**) with this borohydride in THF at −78° proceeds with high *anti*-selectivity regardless of the substitution on the alkenyl group. Exclusive formation of the *anti*-alcohol (**2**) is observed when R[1] is an alkyl or trimethylsilyl group. Reduction of **1** with DIBAH is nonstereoselective unless R[1] is CH$_3$ or (CH$_3$)$_3$Si, in which case the reduction is also *anti*-selective.

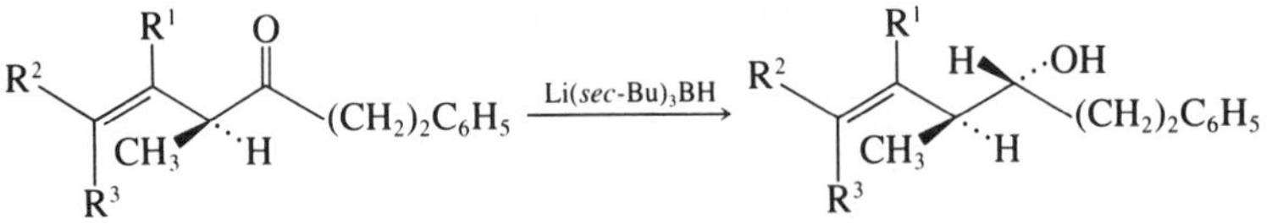

1   **2** (*anti/syn* > 99:1)

Reduction of α-*methylthio ketones*. α-Methylthio and α-phenylthio ketones (**1**) are reduced with high selectivity to *syn*-alcohols (**2**) by this complex hydride, except when

R^1 is a branched alkyl group. The products are converted exclusively into the *cis*-epoxides **3** by alkylation followed by internal displacement.[2]

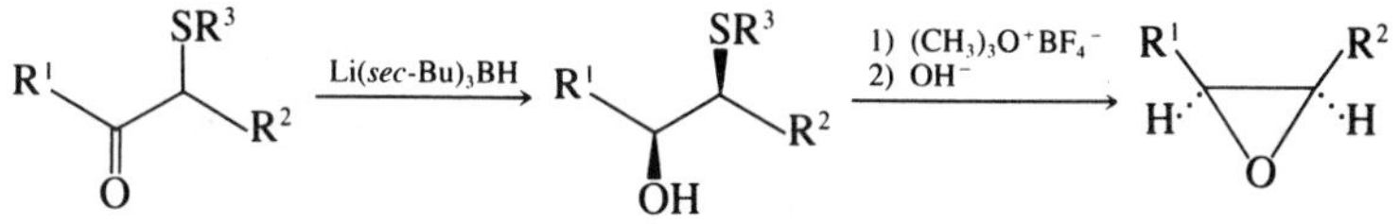

1, $R^3 = CH_3$, C_6H_5 **2** (*syn/anti* = 92–99 : 8–1)   **3**

Reduction of **1** ($R^3 = CH_3$) with $Zn(BH_4)_2$ results in the expected *anti*-alcohol only when R^1 is phenyl or an alkenyl group. However, the derived sulfonium salts (**4**) are reduced selectively to the *anti*-alcohols (**5**) by a variety of reducing agents, including $NaBH_4$, SMEAH, and $Zn(BH_4)_2$. The products are converted on treatment with base to *trans*-epoxides (**6**).[3]

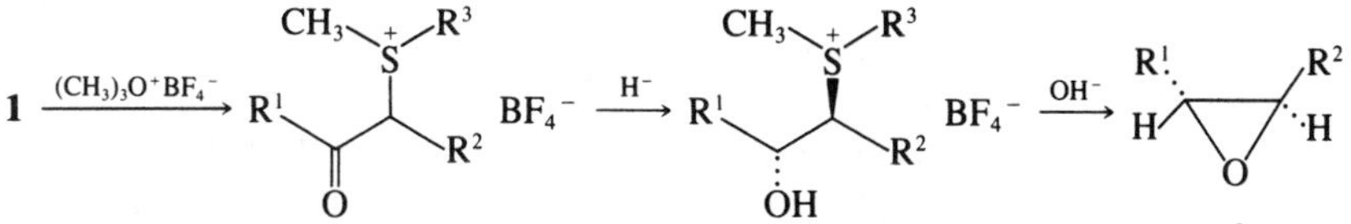

4   **5** (*anti/syn* = 95–99 : 5–1)   **6**

Stereoselective reduction of cyclohexylimines. Imines of alkyl-substituted cyclohexyl ketones are reduced by this borohydride stereoselectively (>90%) to axial secondary amines.[4] Axial primary cyclohexylamines[5] are prepared conveniently by reduction of the imine derived from *p,p'*-dimethoxybenzhydrylamine[6] and subsequent cleavage with formic acid (equation I).

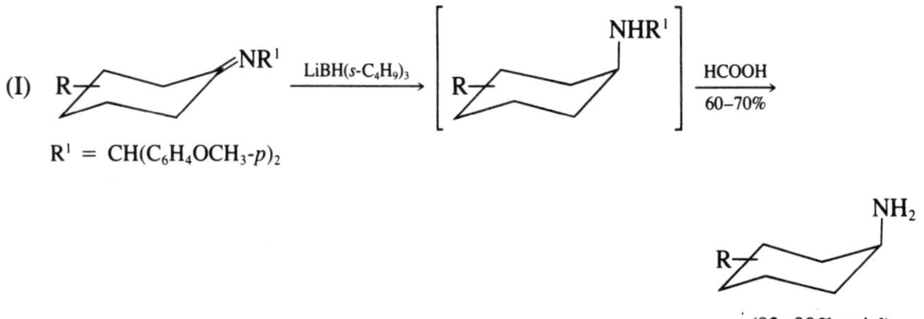

(I) $R^1 = CH(C_6H_4OCH_3\text{-}p)_2$

(82–99% axial)

Reduction of α,β-unsaturated nitroalkenes.[7]   These nitroalkenes are reduced by lithium tri-*sec*-butylborohydride or by lithium triethylborohydride to lithium nitronates, which are hydrolyzed by dilute acid or by silica gel to nitroalkanes (equation I).

$$
\text{(I)} \quad R^2CH{=}CR^1NO_2 \xrightarrow{\text{LiR}_3\text{BH}} \left[\underset{R^2CH_2}{\overset{R^1}{\underset{\quad}{C}}}{=}N\underset{OBR_3}{\overset{O}{\diagdown}} \right] \overset{+}{Li} \xrightarrow[65-80\%]{\overset{\text{H}_3\text{O}^+ \text{ or}}{\text{SiO}_2}} R^2CH_2\overset{R^1}{\underset{\quad}{C}}HNO_2
$$

$$(R^1 = H, CH_3)$$

[1] K. Suzuki, E. Katayama, and G. Tsuchihashi, *Tetrahedron Letters*, **25**, 2479 (1984).
[2] M. Shimagaki, T. Maeda, Y. Matsuzaki, I. Hori, T. Nakata, and T. Oishi, *ibid.*, 4775 (1984).
[3] M. Shimagaki, Y. Matsuzaki, I. Hori, T. Nakata, and T. Oishi, *ibid.*, 4779 (1984).
[4] R. O. Hutchins, W.-Y. Su, R. Sivakumar, F. Cistone, and Y. P. Stercho, *J. Org.*, **48**, 3412 (1983).
[5] R. O. Hutchins and W.-Y. Su, *Tetrahedron Letters*, **25**, 695 (1984).
[6] B. M. Trost and E. Kienan, *J. Org.*, **44**, 3451 (1979).
[7] R. S. Varma and G. W. Kabalka, *Syn. Comm.*, **14**, 1093 (1984).

Lithium trichloropalladate(II), LiPdCl₃.

Heteroannelation.[1]   A variety of oxygen- and nitrogen-containing heterocycles can be prepared by coupling of functionally substituted organomercurials with conjugated or nonconjugated dienes mediated by LiPdCl₃. The reaction involves transmetallation by LiPdCl₃ followed by addition to the diene to generate a (π-allyl)palladium intermediate. Addition of base liberates an oxygen or nitrogen nucleophile that displaces the palladium with formation of the carbon–heteroatom bond.

Examples:

[1] R. C. Larock, L. W. Harrison, and M. H. Hsu, *J. Org.*, **49**, 3662 (1984).

Lithium triethylborohydride, 4, 313–314; **6**, 348–349; **7**, 215–216; **8**, 309–310; **9**, 286; **10**, 249–250; **11**, 304–305.

Deoxygenation of **cis-***diol monotosylates.*[1]   Reduction of 2′-O-tosyladenosine (**1**) with an excess of this borohydride in THF or DMSO results in clean conversion to the inverted *threo*-alcohol **2** in high yield. A similar rearrangement occurs on reduction of 3′-O-tosyladenosine. The reduction involves a stereoselective 1,2-hydride shift with inversion at both C_2' and C_3'.

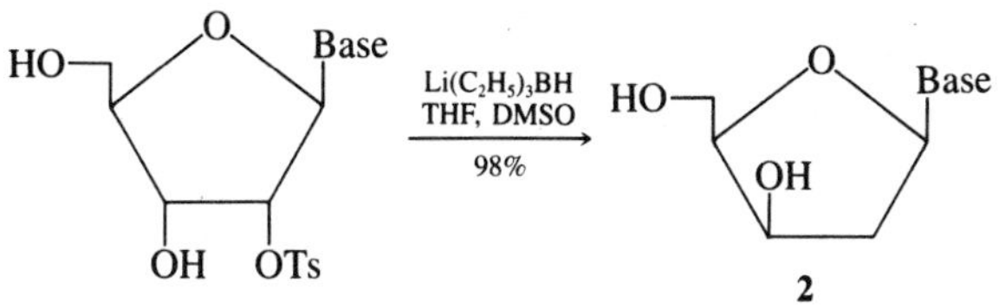

Stereoselective reduction of a γ-*hydroxy ketone.*[2]   Reaction of the γ-hydroxy ketone **1** with LiAlH$_4$, ZnBH$_4$, or NaBH$_4$ yields about equal amounts of the two possible diols. However, reduction with Li(C$_2$H$_5$)$_3$BH (2 equiv.) yields only one product (**2**) in 88% yield. The high stereoselectivity is attributed to formation of a chelated boron intermediate that favors attack by the hydride from the β-face. The diol (**2**) is converted by TsCl into ancistrofuran (**3**), a defensive secretion of certain termites.

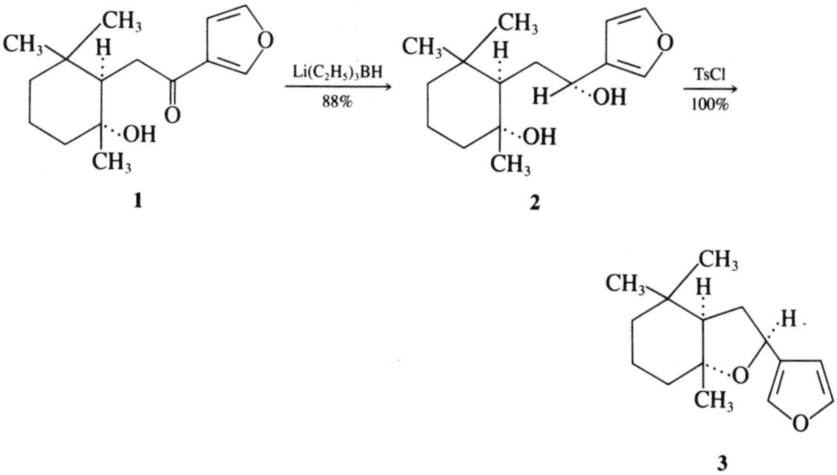

[1] F. Hansske and M. J. Robins, *Am. Soc.*, **105**, 6736 (1983).
[2] R. Baker, P. D. Ravenscroft, and C. J. Swain, *J.C.S. Chem. Comm.*, 74 (1984).

M

Magnesium, 1, 627–629; **2**, 259; **3**, 189; **4**, 315; **5**, 419; **6**, 351–352; **7**, 218; **10**, 251; **11**, 307–309.

Anthracene-activated magnesium. An activated form of Mg powder is obtained by equilibration with the anthracene adduct. The method involves sonication of a mixture of Mg with anthracene (0.02 equiv.) and CH$_3$I in THF at 20° for 14 hours.[1] This activated magnesium is particularly convenient for preparation of allylic Grignard reagents (**11**, 308).[2]

[1] H. Bönnemann, B. Bogdanović, R. Brinkmann, D.-W. He, B. Spliethoff, *Angew. Chem., Int. Ed.*, **22**, 728 (1983).
[2] W. Oppolzer and P. Schneider, *Tetrahedron Letters*, **25**, 3305 (1984).

Magnesium–Anthracene (1:1). This complex is obtained by reaction of magnesium bromide with sodium anthracenide or with anthracene, excess magnesium bromide, and Mg(0) in THF. It can be isolated as clear orange needles with the formula magnesium–anthracene complexed with 3THF. It is hydrolyzed by water to give dihydroanthracene and is therefore magnesium anthracene dianion.[1]

Grignard reagents.[2] The complex is particularly useful for conversion of benzylic chlorides (or bromides) into the Grignard reagents in yields generally of >92% by reaction in THF at 20°. Even benzylic di-Grignards can be prepared in high yield.

Examples:

[1] P. K. Freeman and L. L. Hutchinson, *J. Org.*, **48**, 879 (1983).
[2] C. L. Raston and G. Salem, *J.C.S. Chem. Comm.*, 1702 (1984).

Magnesium iodide etherate, MgI$_2$·O(C$_2$H$_5$)$_2$ (**1**). The reagent is prepared by reaction of Mg(0) and I$_2$ in benzene–ether.

Demethylation.[1] MgI$_2$ etherate can demethylate methoxy groups *ortho* to an acyl group under mild conditions that do not effect isomerization of an isopropenyl group.

Example:

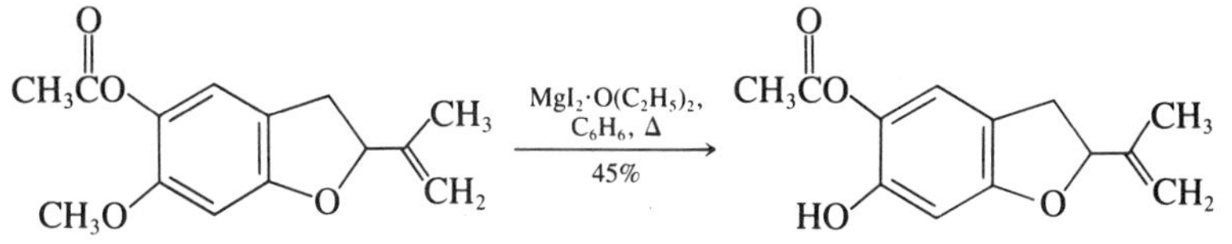

^{1} S. Yamaguchi, K. Sugiura, R. Fukuoka, K. Okazaki, M. Takeuchi, and Y. Kawase, *Bull. Chem. Soc. Japan*, **57**, 3607 (1984).

Malonyl dichloride, $ClCOCH_2COCl$ (**1**). Mol. wt. 140.95 b.p. 53–55°/19 mm. Suppliers: Aldrich, Fluka.

Phloroglucinols.[1] Malonyl dichloride reacts with keto enol ethers to form phloroglucinols (**2**) and/or 4-hydroxy-2H-pyran-2-ones (**3**), which are readily convertible into phloroglucinols.

Example:

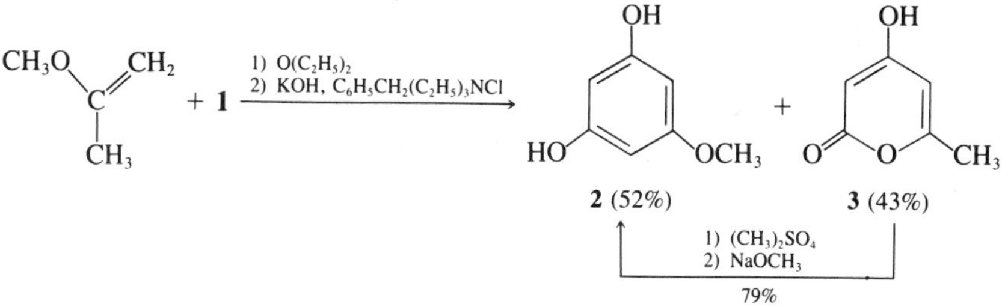

^{1} F. Effenberger and K.-H. Schönwälder, *Ber.*, **117**, 3270 (1984).

(S)-(+)- and (R)-(−)-Mandelic acid, $C_6H_5CH(OH)COOH$ (**1**). Suppliers: Aldrich, Fluka.

Enantioselec̣*ive aldol synthesis.*[1] The dioxolones formed from (S)- or (R)-**1** and aromatic aldehydes undergo a diastereoselective condensation with enol silyl ethers. Optically active aldols are obtained by removal of the chiral auxiliary by oxidative decarboxylation with Pb(OAc)$_4$. A typical example using the dioxolone (**2**) formed from (R)-**1** and benzaldehyde is shown in equation (I). However, only moderate diastereoselectivity

(83:17)

(>98% ee)

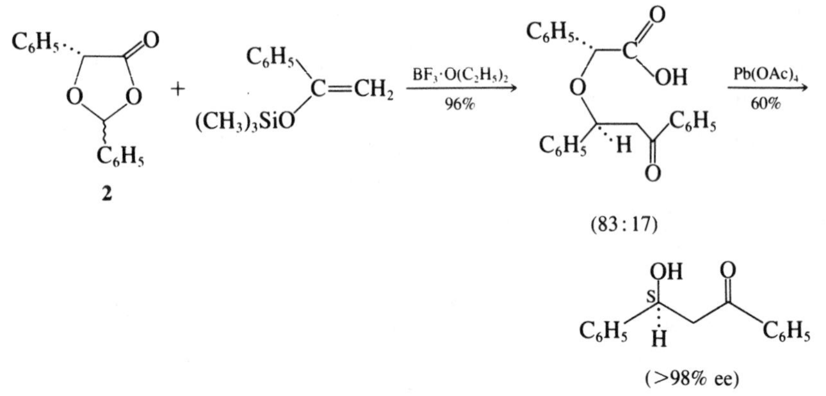

is observed in the reaction of a typical aromatic ketone (acetophenone). Dioxolones from aliphatic aldehydes do not react with enol silyl ethers under a variety of conditions.

[1] S. H. Mashraqui and R. M. Kellogg, *J. Org.*, **49**, 2513 (1984).

Manganese(III) acetate, 2, 263–264; **4**, 318; **6**, 355–357; **7**, 221.
    Structure: $Mn_3O(OCOCH_3)_7$.[1]
    Polycyclic γ-lactones.   An intramolecular version of the formation of γ-lactones by reaction of an alkene with acetic acid and manganese(III) acetate (**1**) (**6**, 355–356) provides a general route to polycyclic γ-lactones. Thus the unsaturated β-keto acid **2** cyclizes to the *cis*-bicyclic γ-lactone **3** when warmed in acetic acid with **1** (1.3 equiv.)[2]

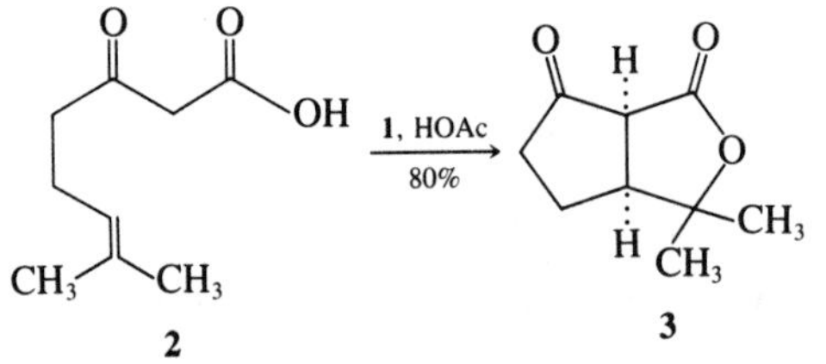

This reaction provides a short synthesis of the tetracyclic dilactone **5** from **4** (equation I).

(I)

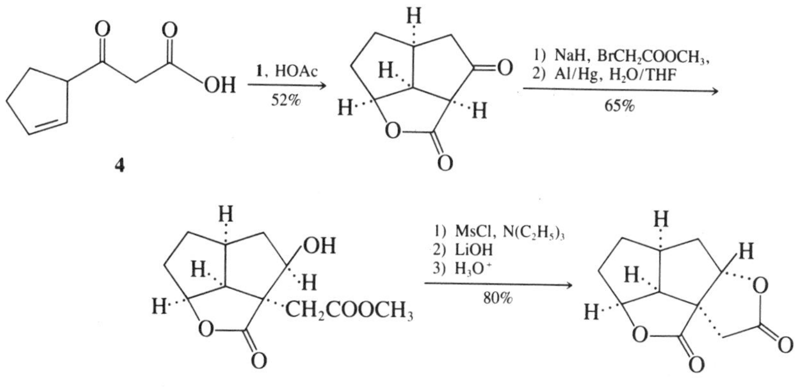

1) NaH, BrCH$_2$COOCH$_3$,
2) Al/Hg, H$_2$O/THF
65%

Reaction of the steroid unsaturated ketone **6** with this reagent results in formation of the γ-lactone **7** in addition to the acetate **8**. The lactone **6** is also obtained in the oxidation of **8**.[3]

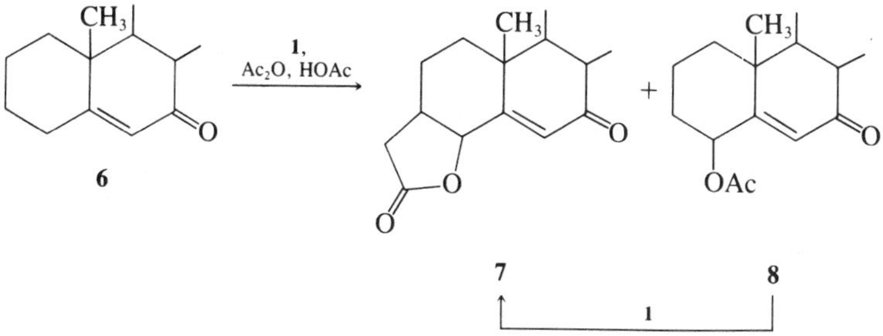

α′-Acetoxylation of enones[4] (7, 221).   This oxidation of an α′-methylene group can be effected in generally good yield in refluxing benzene with well-dried oxidant. Yields are lower in the oxidation of an α′-methyl group or of substrates with more than two γ-hydrogens.

Example:

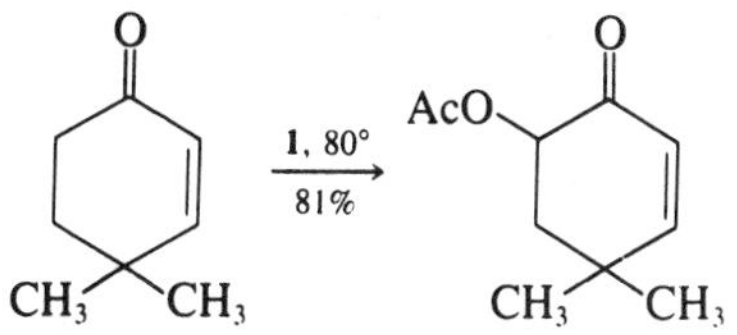

[1] L. W. Hessel and C. Romers, *Rec. Trav.*, **88**, 545 (1969).
[2] E. J. Corey and M. Kang, *Am. Soc.*, **106**, 5384 (1984).
[3] M. S. Ahmad, S. Z. Ahmad, and I. A. Ansari, *J. Chem. Res. (S)*, 374 (1984).
[4] N. K. Dunlop, M. R. Sabol, and D. S. Watt, *Tetrahedron Letters*, **25**, 5839 (1984).

d- and l-Menthol, 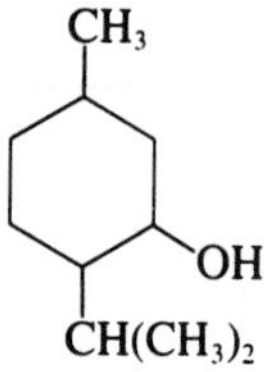   Mol. wt. 156.27. Both *d-* and *l*-**1** are available from Aldrich.

Enantioselective cyclopropanation.[1] The reaction of dimenthyl fumarates (**2**) with isopropylidene(triphenyl)phosphorane[2] proceeds in 74% de in THF at $-78 \rightarrow 20°$. (1S,3S)-**3** is obtained when the *l*-menthyl group is used as inductor; (1R,3R)-**3** is obtained in the same optical yield when the unnatural *d*-menthyl group is used. The concentration plays an important role; enantioselectivity is increased by dilution and decreased by concentration.

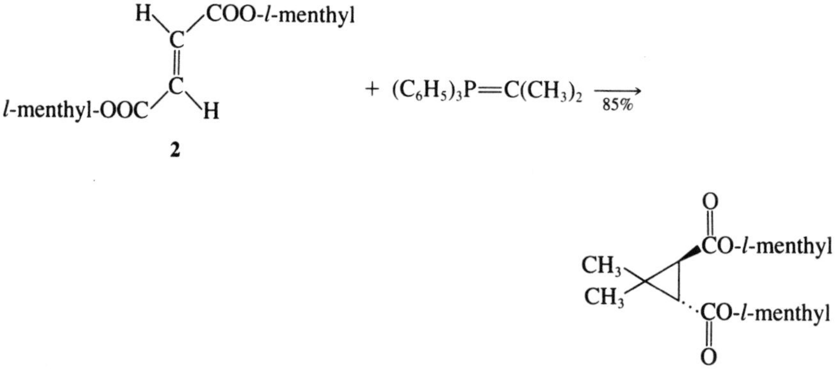

2

$+ (C_6H_5)_3P{=}C(CH_3)_2 \xrightarrow[85\%]{}$

(1S,3S)-**3** (74% de)

(1R,3R)-**3** has been converted into natural (1R)-*trans*-chrysanthemic acid (**4**), a constituent of natural pyrethrins.

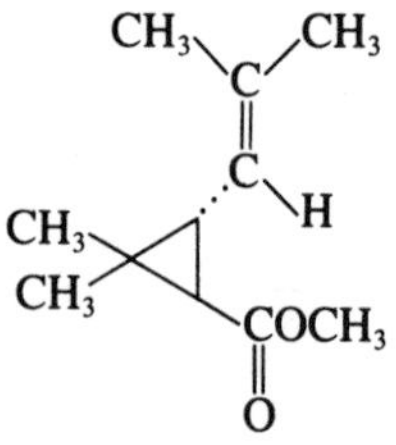

(1R)-**4**

[1] M. J. De Vos and A. Krief, *Tetrahedron Letters*, **24**, 103 (1983).
[2] P. A. Grieco and R. S. Findlehor, *Tetrahedron Letters*, 3781 (1972).

(S)- or (R)-Menthyl *p*-toluenesulfinate (1), 10, 406–4-7; **11**, 312–315.

Asymmetric synthesis of a chroman. Solladié and Moine[1] have effected an en-antiospecific synthesis of the chroman-2-carboxaldehyde **7**, a key intermediate in the synthesis of α-tocopherol, from (R)-(+)-**1**. The phosphonate **2**, derived from **1**, undergoes a Wittig-Horner reaction with the dimethyl ketal of pyruvaldehyde to afford the optically active vinyl sulfoxide **3**. Condensation of the aldehyde **4** with the lithio derivative of **3** affords, after silyl deprotection, the allylic alcohol **5** as the only diastereoisomer. This

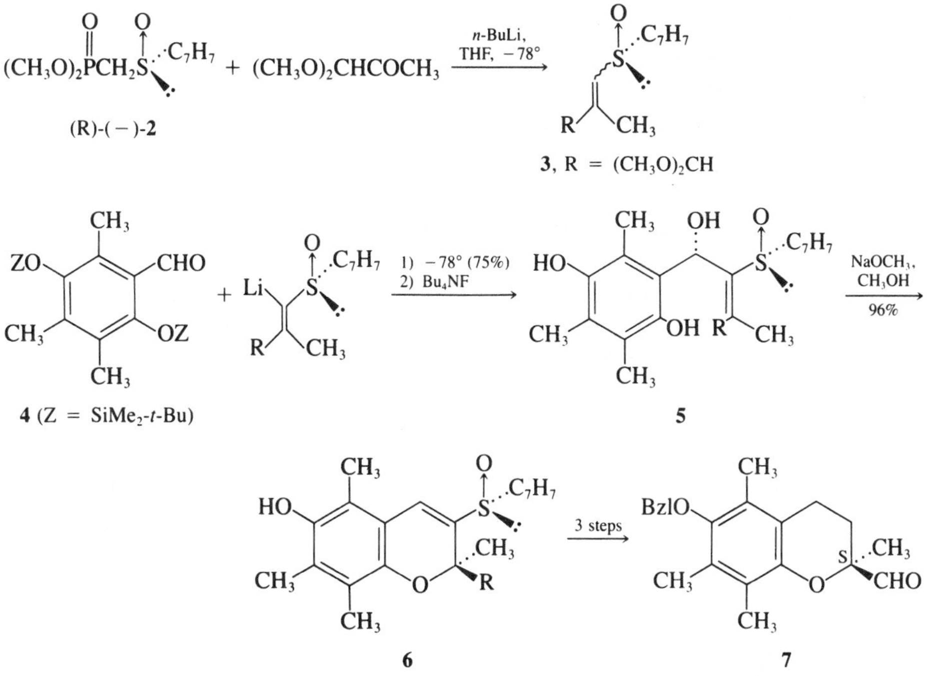

undergoes base-catalyzed cyclization to the chromene **6** in high yield. Remaining steps to **7** include reductive desulfuration, benzylation of the phenol, and liberation of the aldehyde group.

Chiral 3-substituted 4-butanolides.[2] (S)-(+)-2-(*p*-Tolylsulfinyl)-2-buten-4-olide (**2**), which can be prepared in several steps from propargyl alcohol (equation I), is an effective Michael acceptor for stereocontrolled conjugate addition of Grignard reagents.

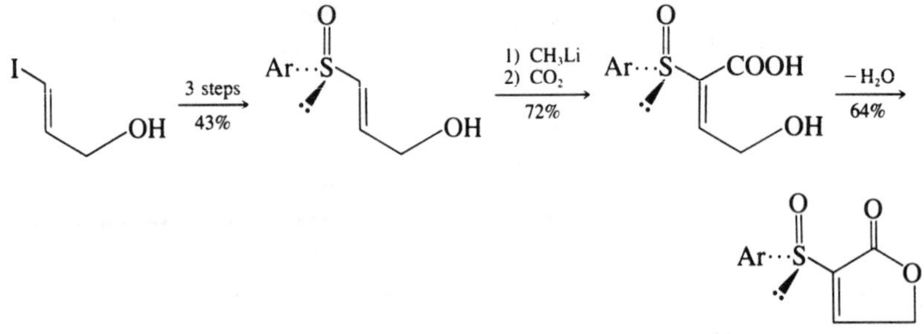

(+)-**2** (>98% ee)

Thus reaction of (+)-**2** complexed with $ZnBr_2$ (1 equiv.) with the Grignard reagent **3** in 2,5-dimethyltetrahydrofurane as solvent followed by elimination of the chiral auxiliary (Raney Ni) results in **4** in 70% yield. Acylation of **4** results in the *trans*-2,3-disubstituted lactone (−)-podorhizon (**5**) in 95% ee.

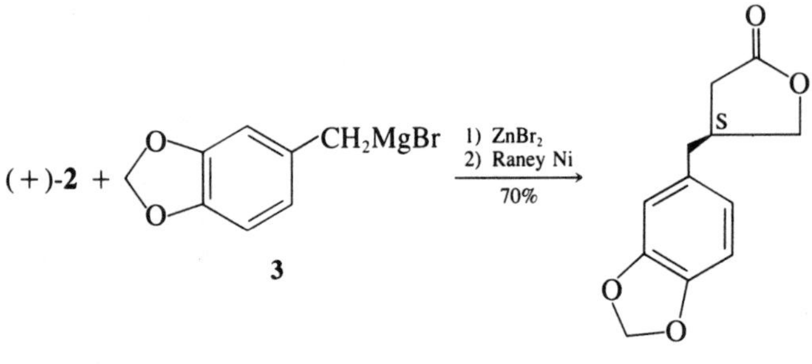

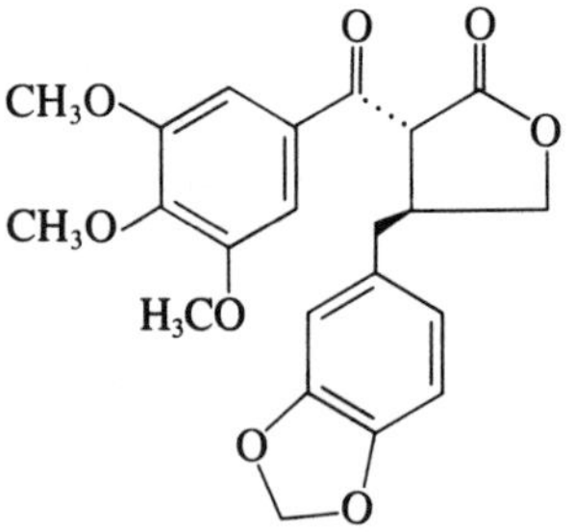

(−)-**5** (95% ee)

Chiral 3-substituted cycloalkanones (**11**, 314).[3]   The original route to these synthons is improved by use of (S)-(+)-2-(*p*-anisylsulfinyl)-2-cyclopentenones (**1**), which are pre-pared from (S)-(−)-menthyl *p*-methoxybenzenesulfinate, m.p. 110.5°, α_D − 188.9°. The conjugate addition of Grignard reagents to **1** proceeds mainly from the face of the enone opposite to that of the arylsulfinyl group to give, after removal of the sulfinyl group, (R)-(+)-3-substituted cyclopentanones (**2**). The enantioselectivity depends in part on the size of the organometallic reagent. Virtually complete asymmetric induction is observed in the addition of methyltitanium triisopropoxide, even in the absence of a chelating agent (equation I).

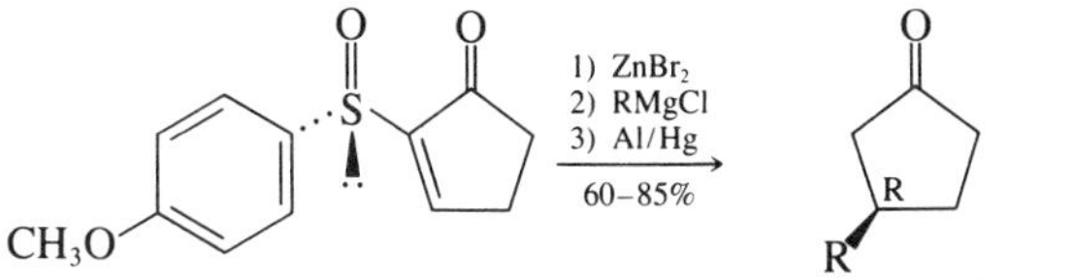

1   **2** (R/S = 5–14 : 1)

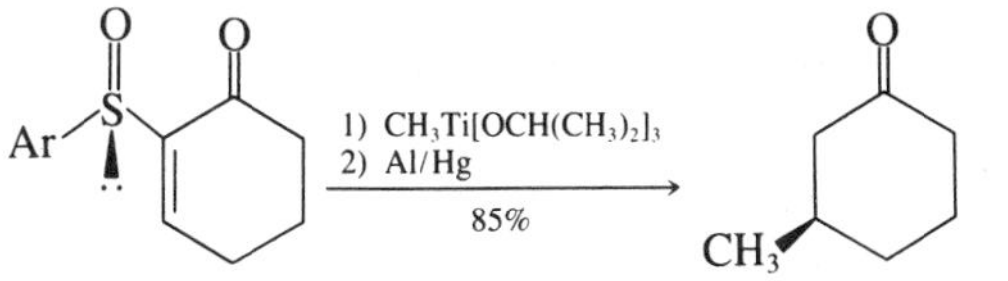

β-Hydroxy esters.[4] The lithium enolate of ethyl-N-methoxyacetimidate (**2**) reacts with (S)-(−)-**1** to provide (R)-(4-methylphenylsulfinyl)-ethyl-N-methoxyacetimidate (**3**). The enolate of **3** reacts with an aldehyde to give the adducts **4**, which are converted by desulfuration and hydrolysis into β-hydroxy acids (**5**). The stereochemical outcome depends on the experimental conditions. The reaction of the lithium enolate of **3** with benzaldehyde under thermodynamic control gives (S)-**5** in 75% ee. Use of the zinc enolate also gives (S)-**5**, in 86% ee, but use of zirconium enolate, obtained by addition of Cp_2ZrCl_2 to the lithium enolate, results in (R)-**5** in 88% ee.

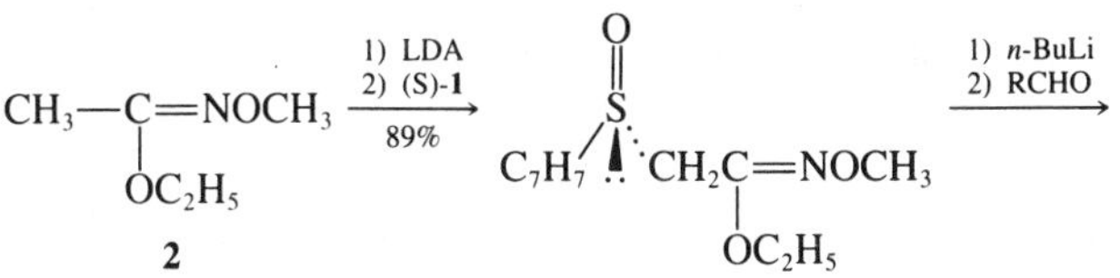

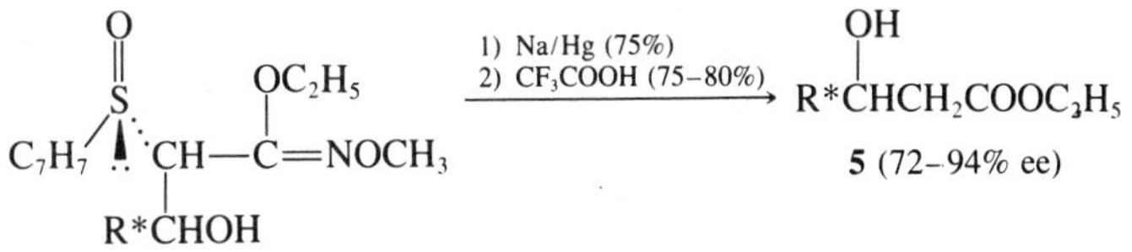

β-Hydroxy acetamides.[5]   The (R)-sulfinylacetamides (2) obtained by reaction of the enolate of N,N-dialkylacetamides with 1 undergo an aldol-type reaction with aldehydes in the presence of base. The adducts (3) are desulfurized to optically active β-hydroxy

acetamides (4). The enantioselectivity is highly dependent on the base. Optical yields of (−)-4 are only about 20–45% when the lithium enolate of 2 is used, but increase to 70–99% when *t*-butylmagnesium bromide is used as base and the reaction period is 60 minutes. The enantioselectivity is also dependent on the size of the R group, being highest when R is methyl and lowest with *t*-butyl.

[1] G. Solladié and G. Moine, *Am. Soc.*, **106**, 6097 (1984).
[2] G. H. Posner, T. P. Kogan, S. R. Haines, and L. L. Frye, *Tetrahedron Letters*, **25**, 2627 (1984).
[3] G. Posner, L. L. Frye, and M. Hulce, *Tetrahedron*, **40**, 1401 (1984).
[4] A. Bernardi, L. Colombo, C. Gennari, and L. Prati, *ibid.*, 3769 (1984).
[5] R. Annunziata, M. Cinquini, F. Cozzi, F. Montanari, and A. Restelli, *ibid.*, 3815 (1984).

Mercury(II) acetate, 1, 644–652; **2**, 264–267; **3**, 194–196; **4**, 319–323; **5**, 424–427; **6**, 358–359; **7**, 222–223; **8**, 315–316; **9**, 291; **10**, 252–253; **11**, 315–317.

Intramolecular solvomercuration of aryl alkynes[1] (*cf.* **11**, 320).   The triple bond of alkynylbenzenes substituted at the *ortho*-position with suitable groups undergoes intramolecular acetoxymercuration with Hg(OAc)₂ in HOAc to afford aromatic heterocycles such as benzofuranes, benzothiophenes, and chromones.

Examples:

The products undergo the usual replacements of the mercury group, such as reduction with alkaline sodium borohydride and iodination.[2] Carbonylation is particularly useful; an example is a new synthesis of coumestan (**1**, equation I).

(I)

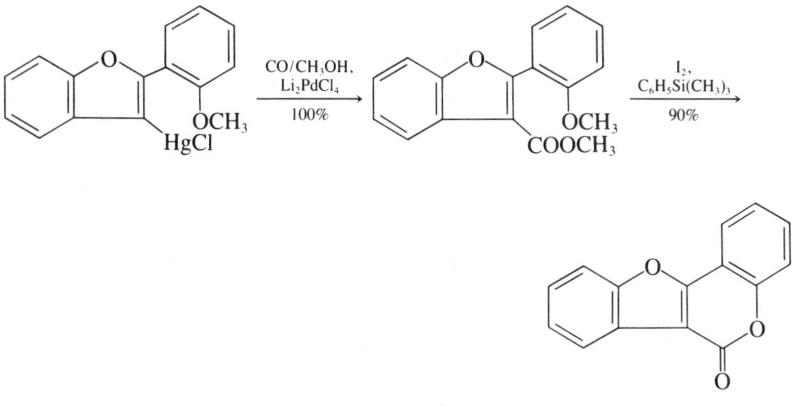

1

Mixed acetals.[3] A variety of functionalized mixed acetals can be prepared by acetoxymercuration of a mixture of a vinyl ether and an alcohol followed by demercuration with Na_2CS_3 (**10**, 369–370) or, in simple systems, with $NaBH_4$. Unsaturated alcohols can be used, and the vinyl ether can contain various functional groups, although not a cyano group.

Example:

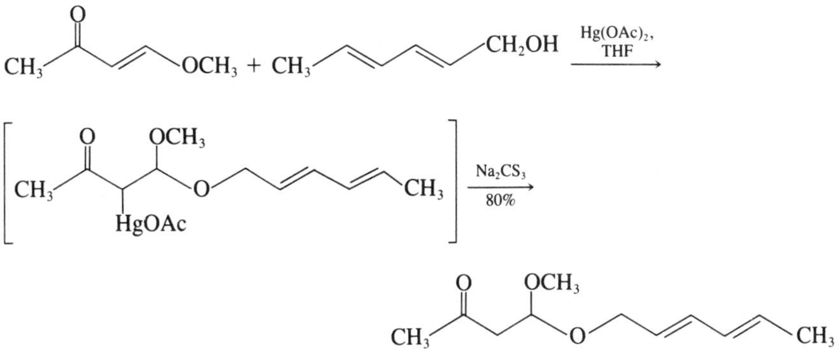

Diastereoselective alkoxymercuration.[4] Benzyloxymercuration–demercuration of protected chiral δ-hydroxy-α,β-unsaturated esters proceeds regioselectively to add the OR group in the β-position and with moderate diastereoselectivity. Higher diastereoselectivity obtains with substrates substituted at the γ-position by an alkyl group, which can override the inductive effect of the alkoxy group. In this case, the reaction results in 90–95% diastereoselectivity resulting in *anti*-1,2-induction.

Examples:

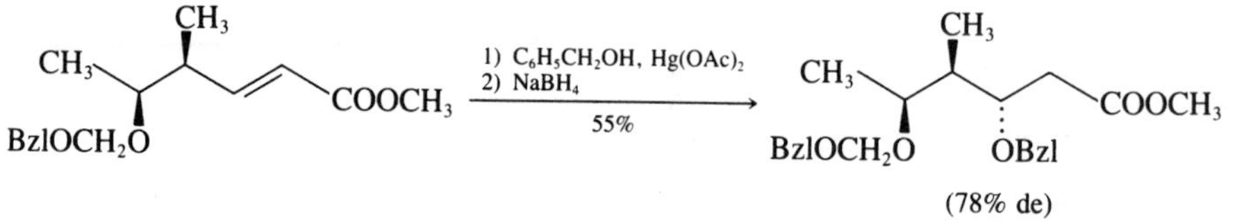

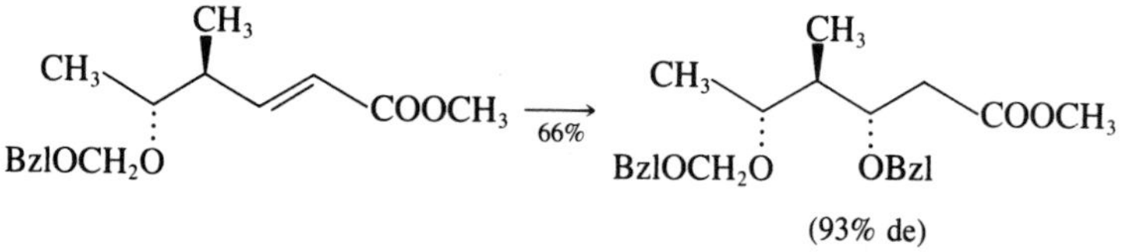

(78% de)

(93% de)

The substrates are available by diastereoselective alkylation of chiral β-hydroxy acids followed by conversion to the aldehyde and Wittig olefination. The overall process provides a diastereoselective synthesis of compounds with three consecutive asymmetric centers.

Tetrahydropyridines.[5]   Danishefsky has used an intramolecular version of the Giese

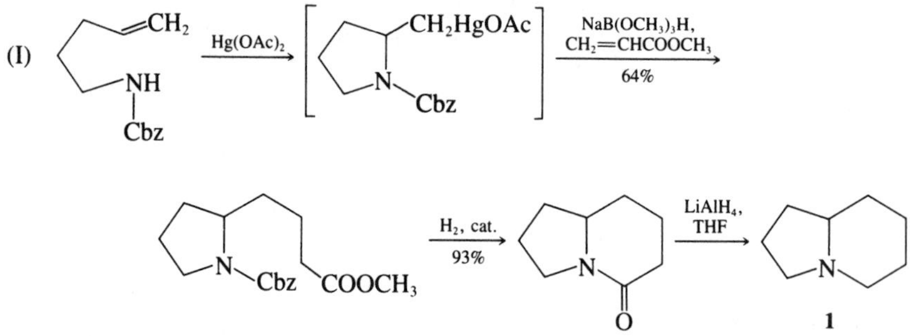

reaction for synthesis of alkaloids such as δ-coniceine (**1**) (equation I). A similar sequence was used to synthesize the benzo[*b*]indolizidinone **2**.

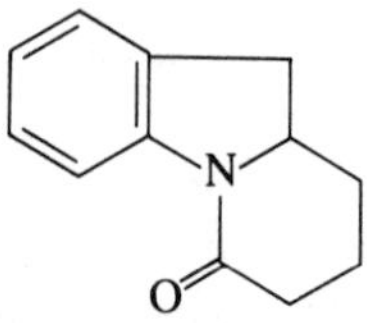

2

Intramolecular amidomercuration.[6] Kinetically controlled amidomercuration [Hg(OAc)$_2$, THF] of the δ-alkenylcarbamate **1** followed by reduction results in highly stereoselective cyclization to *trans*-2,5-dimethylpyrrolidines (equation I). Amidomercur-

(I)

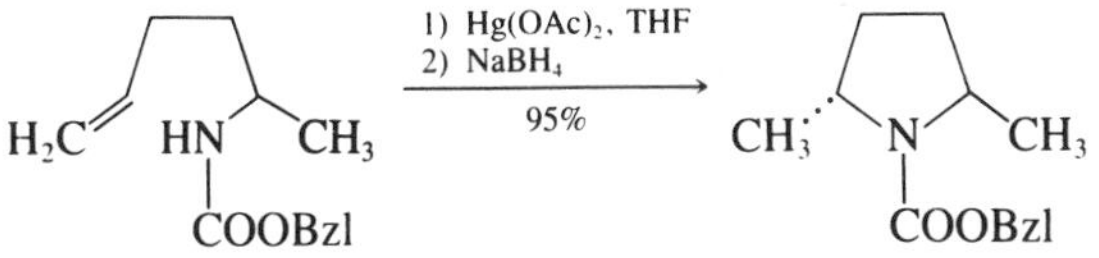

ation with Hg(OCOCF$_3$)$_2$ in CH$_3$NO$_2$ with equilibration of the organomercurial products results in a mixture of *cis*- and *trans*-products in which the former can even predominate.

The situation is reversed in the ε-alkenylcarbamate (**2**): Amidomercuration with Hg(OAc)$_2$ in THF results in a mixture of *cis*- and *trans*-products, whereas thermodynamically controlled reaction with Hg(OCOCF$_3$)$_2$ in CH$_3$NO$_2$ results in highly selective cyclization to a *cis*-2,6-dimethylpiperidine (equation II).

(II)

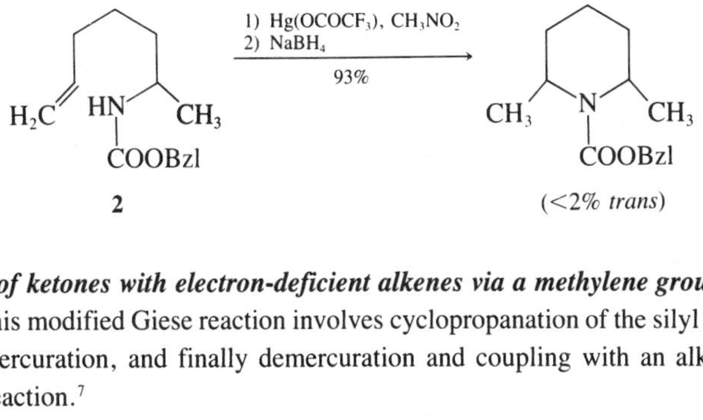

Coupling of ketones with electron-deficient alkenes via a methylene group (*cf.* **11**, 315–316). This modified Giese reaction involves cyclopropanation of the silyl enol ether of a ketone, mercuration, and finally demercuration and coupling with an alkene via a radical chain reaction.[7]

Example:

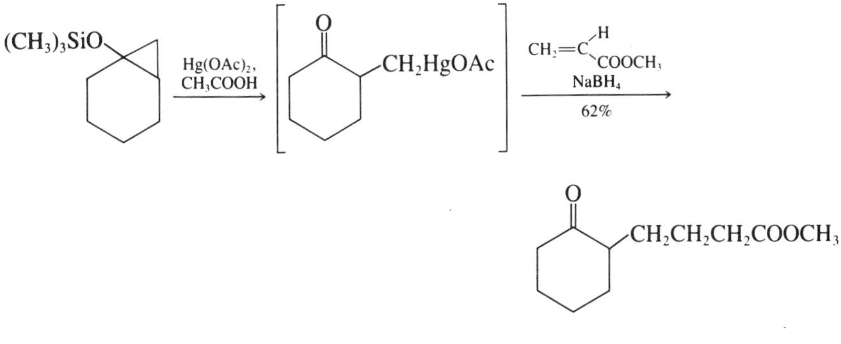

A similar sequence can be applied to aldehydes.[8]
Example:

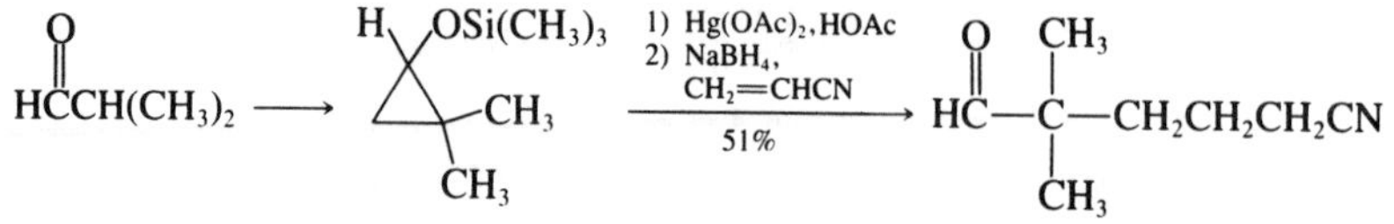

Acetoxymercuriation of cyclopropanes.[9]   Reaction of Hg(OAc)₂ with cyclopropanes results in attack at the least substituted carbon atom. Reduction of the product in the presence of an electron-deficient alkene results in coupling of the intermediate radical to the least substituted carbon atom of the alkene.
Example:

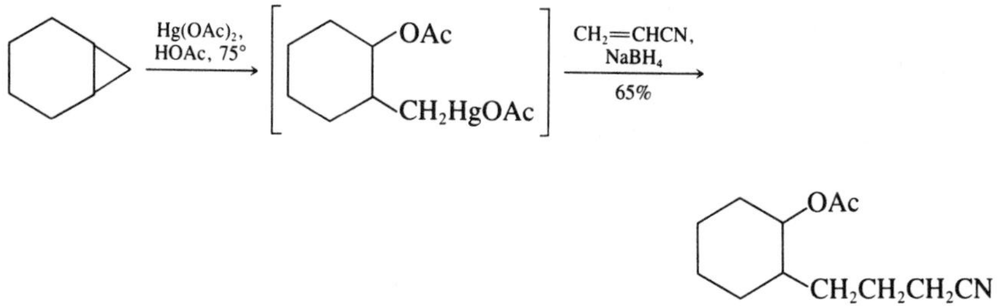

γ- and δ-Lactones.[10]   These lactones can be prepared by solvomercuration of alkenes followed by a radical reaction with acrylonitrile to give δ-acetoxy nitriles, which can be converted into either γ- or δ-lactones.
Example:

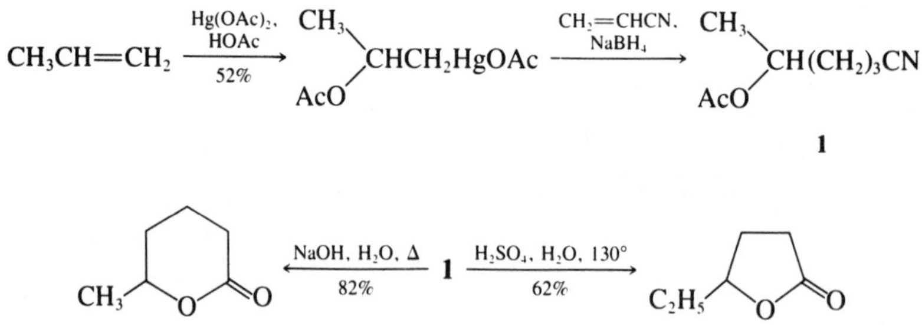

2-Deoxy sugars.[11]   Branched 2-deoxy sugars can be prepared by solvomercuration of a glycal followed by the Giese reaction with an electrophilic alkene, with marked preference for formation of an equatorial substituent.

Example:

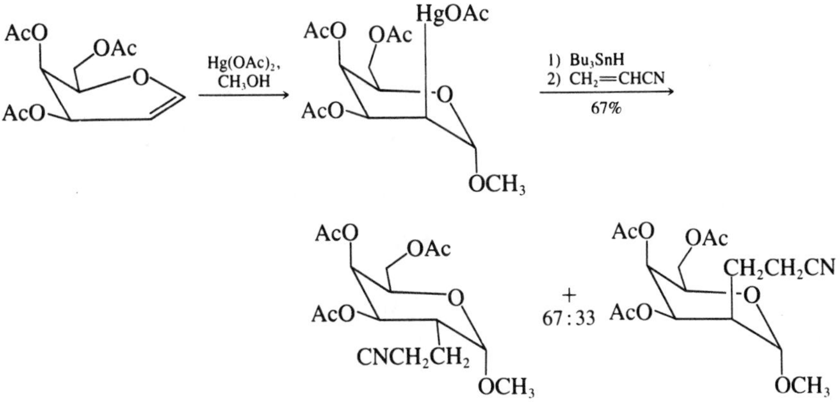

[1] R. C. Larock and L. W. Harrison, *Am. Soc.*, **106**, 4218 (1984).
[2] R. C. Larock, *Tetrahedron*, **38**, 1713 (1982).
[3] R. K. Boeckman, Jr., and C. J. Flann, *Tetrahedron Letters*, 4923 (1983).
[4] S. Thaisrivongs and D. Seebach, *Am. Soc.*, **105**, 7407 (1983).
[5] S. Danishefsky, E. Taniyama, and R. R. Webb II, *Tetrahedron Letters*, **24**, 11 (1983).
[6] K. E. Harding and S. R. Burks, *J. Org.*, 49, 40 (1984); K. E. Harding and T. H. Marman, *ibid.*, 2838 (1984).
[7] B. Giese, H. Horler, and W. Zwick, *Tetrahedron Letters*, **23**, 931 (1982).
[8] B. Giese and H. Horler, *ibid.*, **24**, 3221 (1983).
[9] B. Giese and W. Zwick, *Ber.*, **115**, 2526 (1982); *ibid.*, **116**, 1264 (1983).
[10] B. Giese, T. Hasskul, and U. Lüning, *ibid.*, **117**, 859 (1984).
[11] B. Giese and K. Gröninger, *Tetrahedron Letters*, **25**, 2743 (1984).

Mercury(II) acetate–Nafion-H.

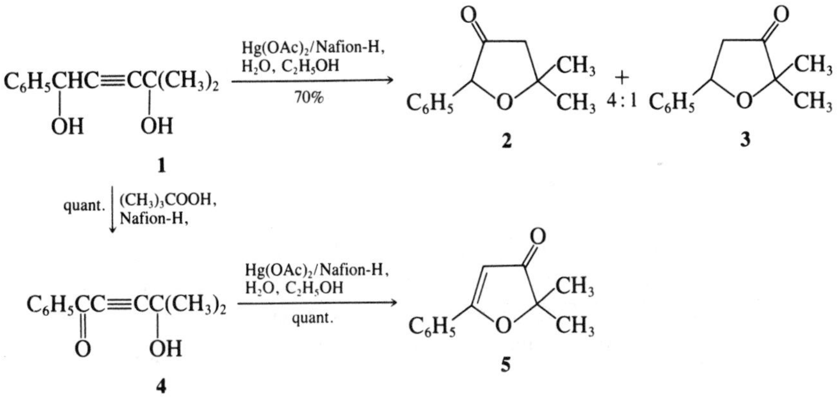

Hydration of alkynes. $Hg(OAc)_2$ adsorbed on Nafion-H is an effective catalyst for hydration of alkynes. It is more useful than the conventional catalyst, HgO, for regioselective hydration of 2-butyne-1,4-diols to form 4,5-dihydro-3(2H)-furanones. Thus use of this catalyst converts the diol **1** into **2** and **3** in the ratio 4:1.[1] Similar hydration of the keto alcohol **4** provides the 3(2H)-furanone **5** with complete regioselectivity.[2]

[1] H. Saimoto, T. Hiyama, and H. Nozaki, *Bull. Chem. Soc. Japan*, **56**, 3078 (1983).
[2] H. Saimoto, M. Shinoda, S. Matsubara, K. Oshima, T. Hiyama, and H. Nozaki, *ibid.*, 3088 (1983).

Mercury(II) chloride, 1, 652–654; **5**, 427–428; **6**, 359; **9**, 291–292.

Spiroketalization. The synthesis of talaromycin B (**3**) with four chiral centers by cyclization of an acyclic precursor presents stereochemical problems. A solution involves cyclization of a protected β-hydroxy ketone with only one chiral center.[1] Because of thermodynamic considerations (*i.e.*, all substituents being equatorial and the anomeric effect), cyclization of **1** with HgCl$_2$ in CH$_3$CN followed by acetonation results in the desired product (**2**, 65% yield) with a stereoselectivity of ~10:1. Final steps involve conversion of the hydroxymethyl group to ethyl by tosylation and displacement with lithium dimethylcuprate (80% yield) and hydrolysis of the acetonide group.

Protection of tertiary alcohols.[2] Methylthiomethyl (MTM) ethers have the advantage that they can be prepared from tertiary alcohols (**7**,135), but the disadvantage that they are prone to oxidation. They can be converted into 2-methoxyethoxymethyl (MEM) ethers, methoxymethyl (MOM) ethers, or ethoxymethyl (EOM) ethers by reaction with

$HgCl_2$ and 2-methoxyethanol, methanol, or ethanol, respectively, in 70–80% yield. All these ethers can be cleaved with trityl tetrafluoroborate in 10–15 minutes in 75–80% yield.

[1] S. L. Schreiber and T. J. Sommer, *Tetrahedron Letters*, **24**, 4781 (1983); A. P. Kozikowski and J. G. Scripko, *Am. Soc.*, **106**, 353 (1984).
[2] P. K. Chowdhury, D. N. Sarma, and R. P. Sharma, *Chem. Ind.*, 803 (1984).

Mercury(II) oxide, 1, 655–658; **2,** 267–268; **4,** 323–324; **5,** 428; **6,** 360; **7,** 224; **8,** 316; **9,** 293.

δ-*Methylene-δ-lactones.*[1]   In the presence of catalytic amounts of yellow HgO, δ-acetylenic acids cyclize in 95% yield to δ-methylene-δ-lactones (equation I). Under the same conditions, γ-allenic acids cyclize to γ,δ-unsaturated δ-lactones (equation II).

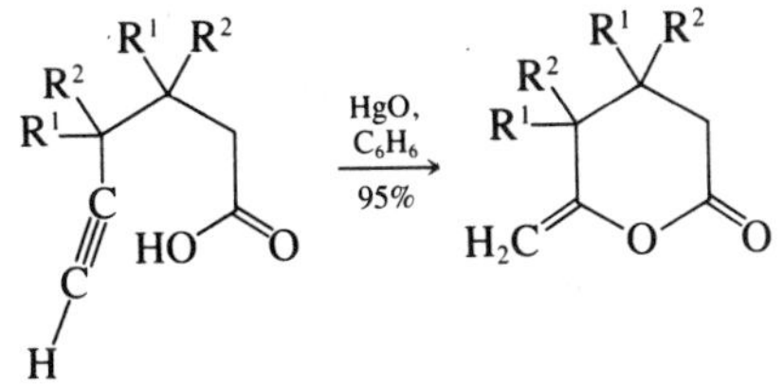

R^1, R^2 = H or CH_3

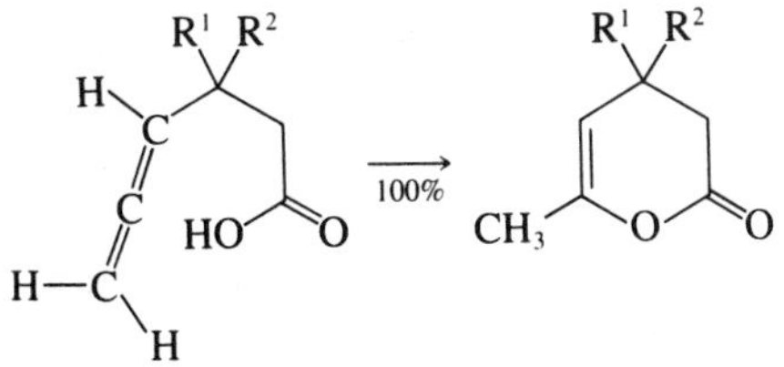

[1] A. Jellal, J. Grimaldi, and M. Santelli, *Tetrahedron Letters*, **25**, 3179 (1984).

Mercury(II) oxide–Iodine.

Cyclic ethers from cyclic ketones.[1]   The replacement of the carbonyl group of a cyclic ketone by oxygen can be effected in four steps: Baeyer-Villiger oxidation to a lactone, DIBAH reduction to a lactol, β-scission to an iodoformate, and finally cyclization.

The scission reaction is carried out by irradiation of the hypoiodite of the lactol. The process is formulated for conversion of 7-cholestanone (**1**) into 7-oxacholestane (**4**).

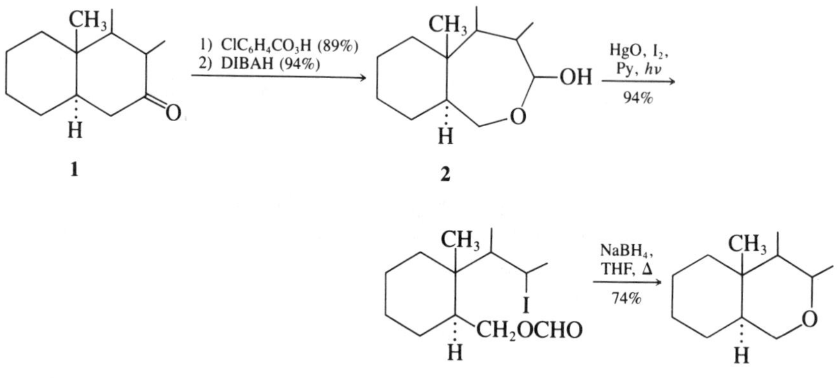

[1] H. Suginome and S. Yamada, *Tetrahedron Letters*, **25**, 3995 (1984).

Mercury(II) oxide–Tetrafluoroboric acid, 9, 293; 10, 254; 11, 318–319.

RBr → ROH.[1]   Alkyl bromides are converted into the corresponding alcohols by reaction with HgO (0.5 equiv.) and 35% aqueous HBF_4 in an organic solvent (THF, CH_2Cl_2, or dioxane) followed by successive treatment with $NaHCO_3$ and aqueous KOH. The HgO is recovered quantitatively. Ethers are obtained if stoichiometric amounts of alcohols are added to the above reaction. Yields of alcohols and ethers range from 55 to 95%.

Iodination of arenes.[2]   Iodination of arenes can be effected by reaction with $HgO·HBF_4$ in the presence of iodine. The orientation conforms to that observed in electrophilic aromatic substitution except that *ortho*-attack is favored over *para*-attack in activated arenes. The method is particularly useful for *meta*-iodination of deactivated arenes (99% selectivity).

[1] J. Barluenga, L. Alonso-Cires, P. J. Campos, and G. Asensio, *Synthesis*, 53 (1983).
[2] J. Barluenga, P. J. Campos, J. M. Gonzalez, and G. Asensio, *J.C.S. Perkin I*, 2623 (1984).

Mercury(II) trifluoroacetate, 3, 195; 4, 325; 6, 360; 8, 316–317; 9, 294–296; 11, 320–321.

Oxy-Cope rearrangement (**11**, 320–321).   Tertiary 1,5-hexadiene-3-ols undergo oxy-Cope rearrangement directly to δ,ε-unsaturated ketones in the presence of catalytic amounts of this Hg(II) salt and 1 equiv. of either lithium trifluoroacetate or lithium trifluoromethanesulfonate. The yields and the (E)-stereoselectivity are generally higher than those obtained by use of 1 equiv. of the mercury salt followed by demercuration.[1]

Example:

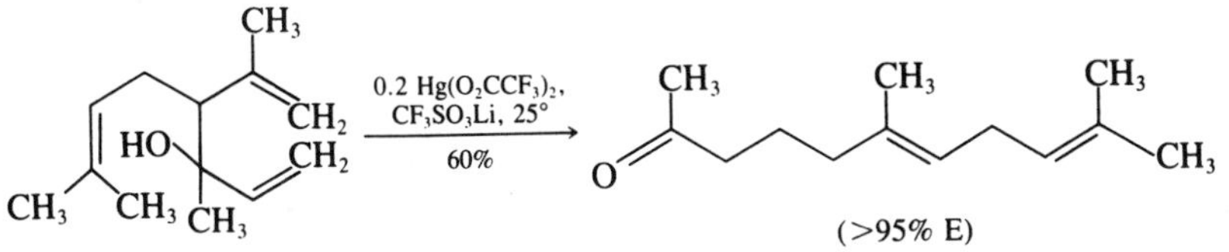

(>95% E)

[1] N. Bluthe, M. Malacria, and J. Gore, *Tetrahedron*, **40**, 3277 (1984).

Mercury(II) trifluoromethanesulfonate–N,N-Dimethylaniline (1). The reagent is prepared by reaction of triflic anhydride with yellow HgO in nitromethane with stirring for 18 hours. Then the amine is added, and a clear pale yellow solution is formed.

Cyclization of farnesol.[1] This complex (**1**) is superior to other Hg(II) reagents for regioselective cyclization of various farnesol derivatives. The product (**3**) obtained in this

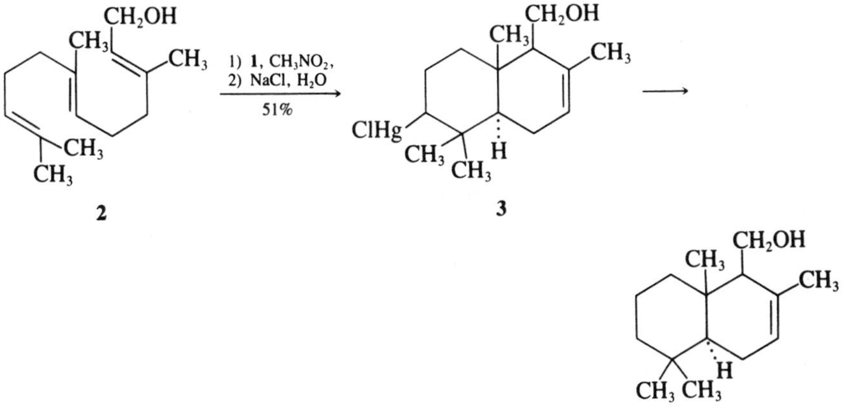

way from farnesol (**2**) was converted into (±)-drimenol (**4**) by reductive demercuration.

Onocerane triterpenes.[2] This reagent was used in the first synthesis of the unsymmetrical α,λ-onoceradienedione (**8**) for cyclization of two farnesol derivatives, (**2**) and (**6**), to A/B-*trans*, 9/10-*cis* decalin systems.

[1] M. Nishizawa, H. Takenaka, H. Nishide, and Y. Hayashi, *Tetrahedron Letters*, **24**, 2581 (1983); M. Nishizawa, H. Takenaka, and Y. Hayashi, *ibid.*, **25**, 437 (1984).
[2] M. Nishizawa, H. Nishide, and Y. Hayashi, *ibid.*, **25**, 5071 (1984).

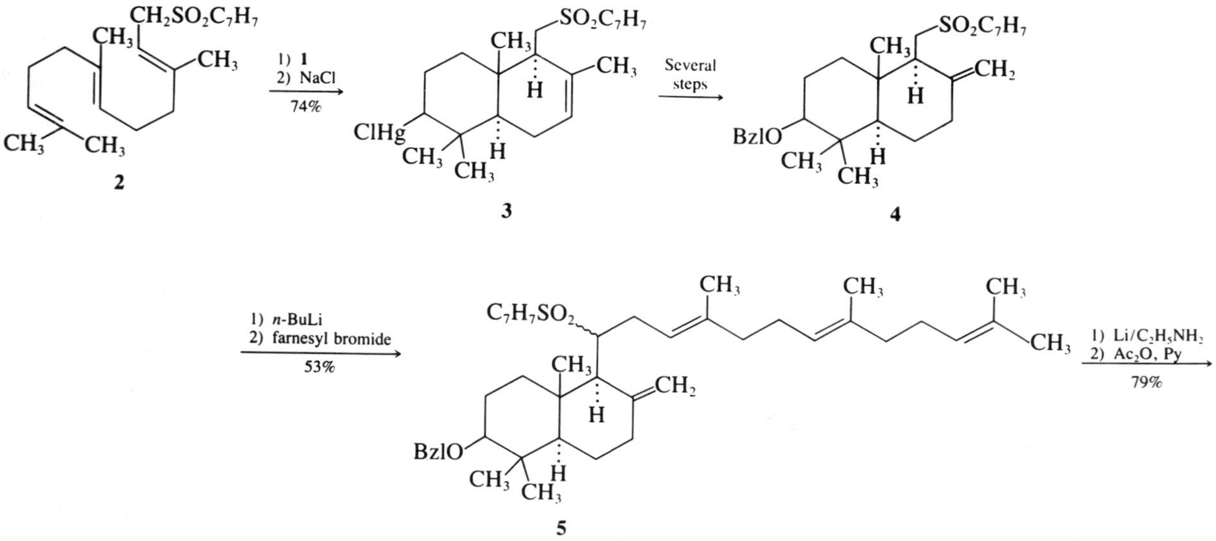

2

1) **1**
2) NaCl
74%

3

Several steps

4

1) n-BuLi
2) farnesyl bromide
53%

5

1) Li/C$_2$H$_5$NH$_2$
2) Ac$_2$O, Py
79%

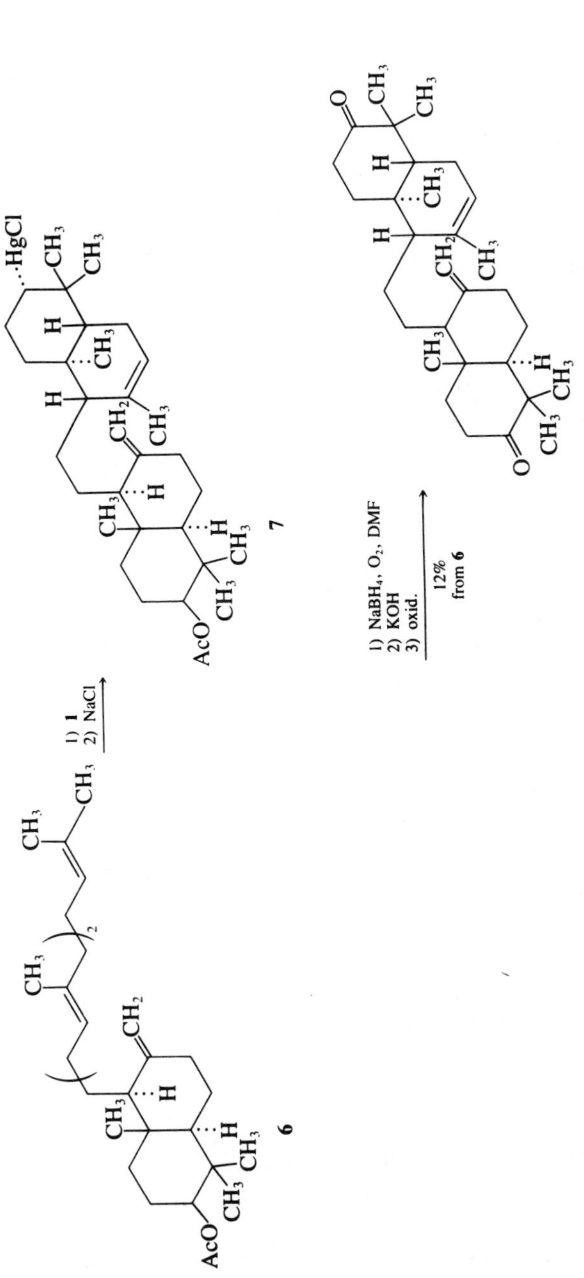

6

1) **1**
2) NaCl

7

...HgCl

1) NaBH₄, O₂, DMF
2) KOH
3) oxid.

12%
from **6**

8

Methoxyacetyl chloride, CH_3OCH_2COCl. Mol. wt. 108.52, b.p. 112–113°. Supplier: Aldrich.

Chloromethylation of arenes.[1]   This acid chloride can be used instead of the suspected carcinogen chloromethyl methyl ether (**9**, 102) for chloromethylation of some arenes. The reaction is conducted in CH_3NO_2 or CS_2 under standard Friedel-Crafts conditions ($AlCl_3$ catalysis). This chloromethylation is particularly useful for aryl ethers substituted in the *o-* or *p-*position by an electron-withdrawing group (yields ~50–90%).

[1] A. McKillop, F. A. Madjdabadi, and D. A. Long, *Tetrahedron Letters*, **24**, 1933 (1983).

α-Methoxyallene, $CH_3OCH=C=CH_2$ (**1**). The ether is obtained by rearrangement of methyl propargyl ether with potassium *t*-butoxide.[1]

Cyclopentannelation.   The adducts (**3**) of α-lithio-α-methoxyallene with the trimethylsilyl ether of a hydroxymethylene ketone undergo cyclization in the presence of BF_3 etherate to produce α-methylenecyclopentenones (**4**).[2]

[1] S. Hoff, L. Brandsma, and J. F. Arens, *Rec. Trav.*, **87**, 916 (1968).
[2] M. A. Tius and D. P. Astrab, *Tetrahedron Letters*, **25**, 1539 (1984).

(1S,2S)-(+)-1-Methoxy-2-amino-3-phenyl-3-hydroxypropane,

(1), **6**, 386, 388; **7**, 229–230.

Asymmetric addition to chiral oxazolines.[1]   The chiral 1-oxazolinylnaphthalene **2**, obtained by the reaction of (+)-**1** with α-naphthylamide, reacts with an organolithium to form an intermediate enolate that is trapped on the opposite face by an electrophile to give **3** as a mixture of diastereomers in the ratio 83:17. The major product results from attack of the organolithium at the β-face. The diastereomers are separable by flash chro-

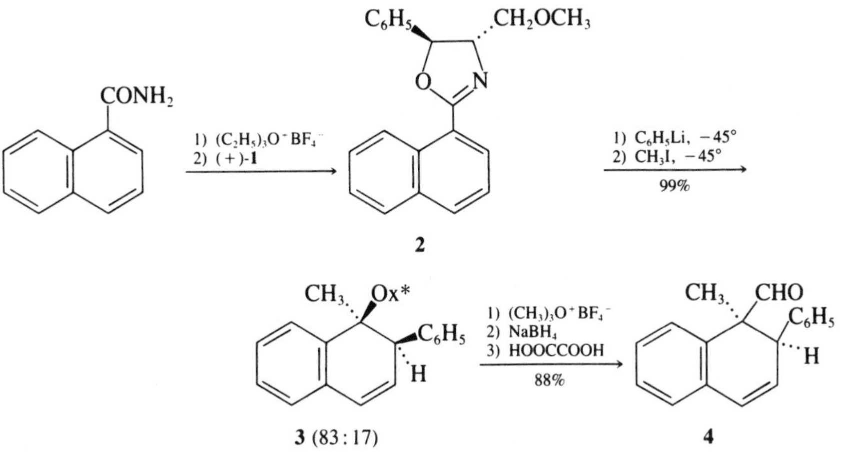

2

3 (83:17)   **4**

matography. The chiral auxiliary is removed by quaternization, reduction, and acid hydrolysis to provide the enantiomerically pure 1,2-dihydronaphthaldehydes **4**.

Chiral binaphthyls.   Chiral binaphthyls can be prepared by reaction of an α-naphthyl Grignard reagent with the chiral 1-methoxy-2-oxazolinyl naphthalene **2**. The adduct (**3**) is converted in two steps to a chiral binaphthyl alcohol (**4**).[2]

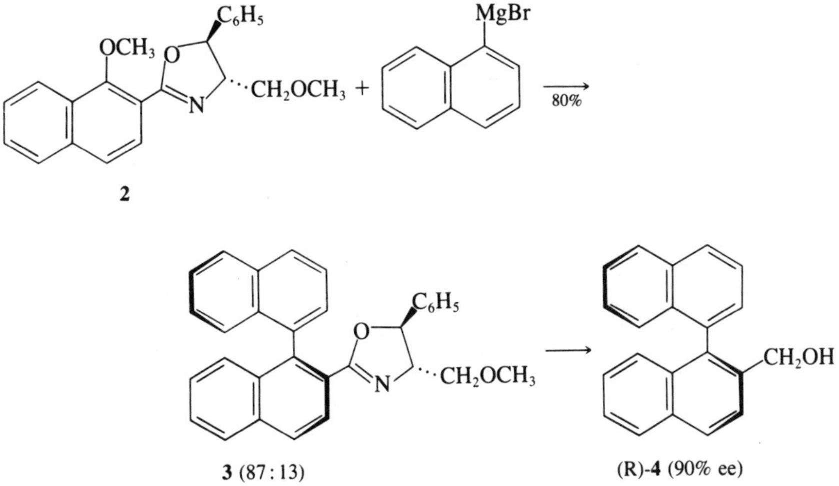

2

3 (87:13)   **(R)-4 (90% ee)**

An alternate route to chiral binaphthyls involves displacement of a chiral 1-alkoxy group from an achiral naphthyloxazoline with α-naphthyllithium or an α-naphthyl Grignard reagent.[3] The optical yields in this process depend on the choice of the leaving group.

The *l*-menthoxy group provides the highest optical yields ($\geq$95% ee); α-fenchoxy and bornoxy groups give the lowest optical yields.

[1] B. A. Barner and A. I. Meyers, *Am. Soc.*, **106**, 1865 (1984).
[2] A. I. Meyers and K. A. Lutomski, *ibid.*, 104, 879 (1982).
[3] J. M. Wilson and D. J. Cram, *ibid.*, **104**, 881 (1982).

1-Methoxy-2,4-dimethyl-3-trimethylsilyloxy-1,3-butadiene,   (1).

The diene is prepared by reaction of the corresponding ketone with ClSi(CH₃)₃ catalyzed with trimethylamine and zinc chloride.

Cyclocondensation with aldehydes (**11**, 332–333).   The cyclocondensation of aldehydes with trimethylsilyloxydienes to provide 2,3-dihydro-4-pyrones has been studied in the most detail with this diene. Two pathways have been identified. In reactions catalyzed by ZnCl₂,[1] MgBr₂, and a lanthanide shift reagent such as Eu(fod)₃,[2] a pericyclic pathway is suggested by isolation of the initial cycloadduct **2** with the expected *endo*-selectivity. In the presence of acid, this adduct is converted into a *cis*-2,3-dihydro-4-pyrone (**3**). Only traces of the *trans*-isomer (**4**) are formed. In contrast, the BF₃-catalyzed reaction results

in the *trans*-2,3-dihydro-4-pyrone (**4**) and the *cis*-isomer (**3**) in the ratio 3:1. In this case, the major product is believed to arise by an *anti*-selective aldol-type condensation followed by cyclization.[3]

The catalyst can affect the stereochemical outcome of the reaction. Thus the reaction of **1** with the aldehyde **5** catalyzed by MgBr₂ results in *exo*-addition to give a *trans*-

dihydropyrane **6**.[4] In contrast, catalysis of the same reaction with TiCl₄ results in a *cis*-dihydropyrane (**7**). In this case, chelation is coupled with a *syn*-aldol process.[4]

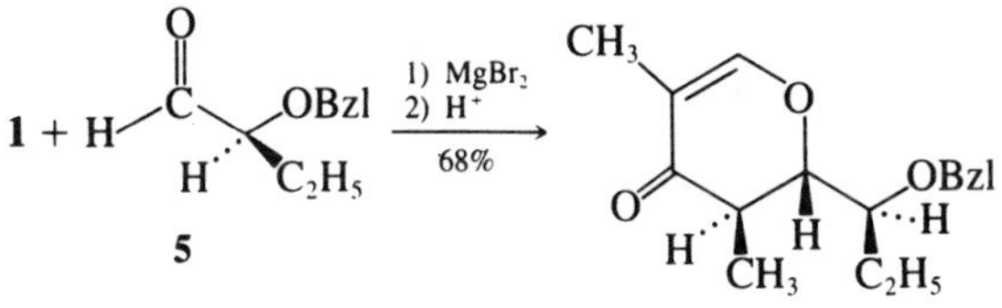

[1] E. R. Larson and S. Danishefsky, *Am. Soc.*, 104, 6458 (1982).
[2] M. Bednarski and S. Danishefsky, *ibid.*, **105**, 3716 (1983).
[3] S. Danishefsky, E. R. Larson, and D. Askin, *ibid.*, 104, 6457 (1982).
[4] S. J. Danishefsky, W. H. Pearson, and D. F. Harvey, *ibid.*, **106**, 2456 (1984).

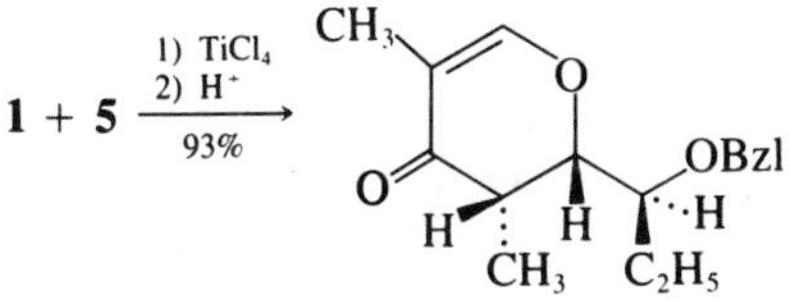

erythro-**2-Methoxy-1,2-diphenylethylamine**, (**1**). Resolution.[1]

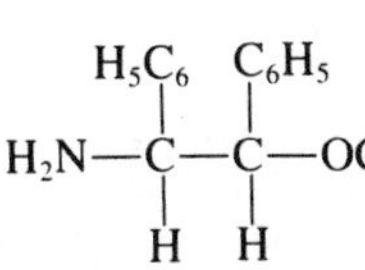

Enantioselective alkylation.[2]   The chiral imine of cyclohexanone derived from (−)- or (+)-**1** undergoes enantioselective alkylation to give chiral 2-alkylcyclohexanones. The highest optical and chemical yields are obtained by alkylation of the enaminozinc bromide (equation I). The products obtained from the imine formed from (−)-**1** have the (R)-configuration; those obtained from the imine formed from (+)-**1** have the (S)-configuration.

(I)

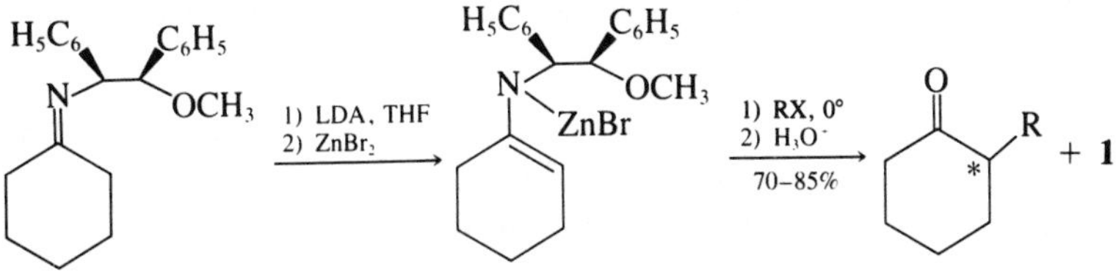

(79–92% ee)

[1] K. Saigo, S. Ogawa, S. Kikuchi, A. Kasahara, and H. Nohira, *Bull. Chem. Soc. Japan*, 55, 1568 (1982).
[2] K. Saigo, A. Kasahara, S. Ogawa, and H. Nohira, *Tetrahedron Letters*, 24, 511 (1983).

6-Methoxy-(E,E)-3,5-hexadienoic acid, CH_3O ⟍⟍═⟍═⟍ $COOH$ **(1).**

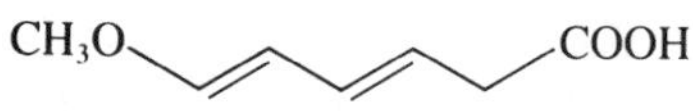

Preparation:

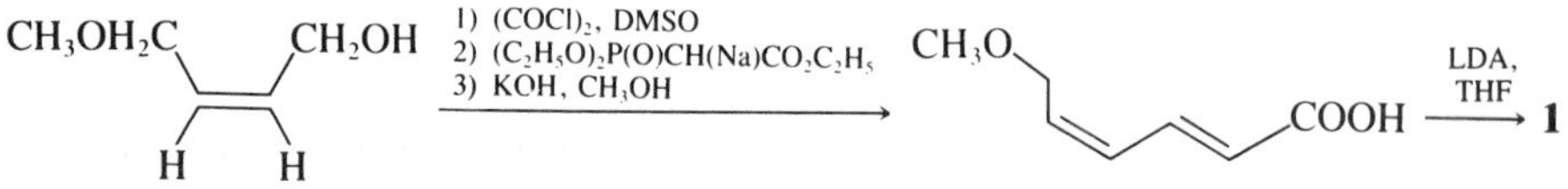

Diels-Alder reactions.[1] Investigation of use of the sodium salt of **1** in aqueous Diels-Alder reactions was prompted by reports from two laboratories[2] that water can function as a catalyst in this reaction. In fact, reaction of the sodium salt in water with representative dienophiles proceeds at lower temperatures than those required in reactions of the corresponding methyl ester in benzene. Moreover, the sensitive dienol ether system in **1** is stable to long reaction times (72 hours) and temperatures of 80°.

Example:

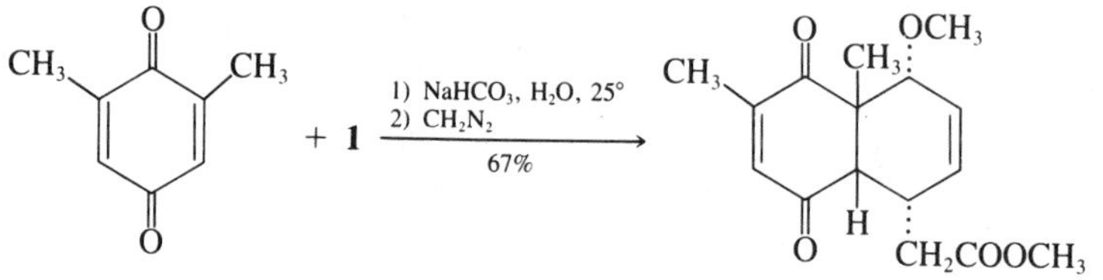

[1] P. A. Grieco, K. Yoshida, and Z. He, *Tetrahedron Letters,* **25,** 5715 (1984).
[2] R. Breslow, U. Maitra, and D. Rideout, *ibid.,* **24,** 1901 (1983); P. A. Grieco, K. Yoshida, and P. Garner, *J. Org.,* **48,** 3137 (1983).

N-Methoxy-N-methylacetamide, CH_3O⟍$NCCH_3$ (**1**), b.p. 40–44°/20 mm. The amide is prepared by acetylation of O,N-dimethylhydroxylamine with AcCl and pyridine in CH_2Cl_2 (65% yield).

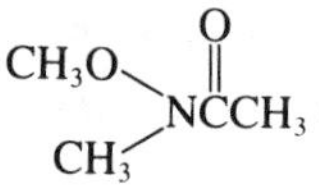

Acetylation. The reagent is useful for acetylation of lithium enolates of oligo-β-carbonyl compounds, which cannot be acetylated satisfactorily by ethyl acetate.[1]

Example:

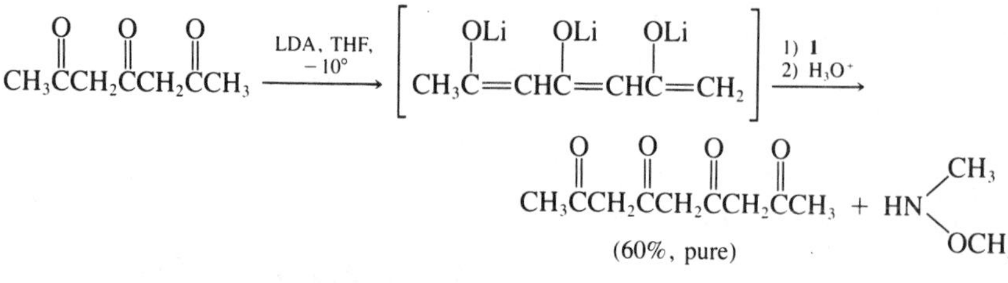

[1] T. A. Oster and T. M. Harris, *Tetrahedron Letters,* **24,** 1851 (1983).

(2S)-Methoxymethyl-(4S)-*t*-butylthio-N-pivaloylpyrroline,

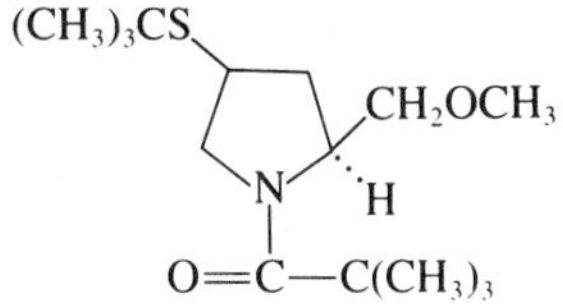

(1).

The pyrroline is prepared in several steps from 4-hydroxyproline.[1]

Asymmetric β-methylation of α,β-enones.[1]   Chiral bidendate ligands derived from L-prolinol can be used for asymmetric Michael additions to α,β-enones with cuprates of the type CH₃L*CuMgBr (**10**, 266). The highest optical yield in conjugate addition to chalcone is observed when L* is (S)-N-methylprolinol (88% ee).[2] The tridentate chiral ligand **1** is equally effective for asymmetric β-methylation of chalcone with CH₃L*CuLi and CuBr; the chemical yield is 95%. Reduction of the amide carbonyl group of **1** results in practically total loss of chiral induction.

[1] F. Leyendecker and D. Laucher, *Tetrahedron Letters*, **24**, 3517 (1983).
[2] F. Leyendecker, F. Jesser, and D. Laucher, *ibid.*, 3513 (1983).

3-Methoxymethylene-2,4-bistrimethylsilyloxy-1,4-pentadiene,

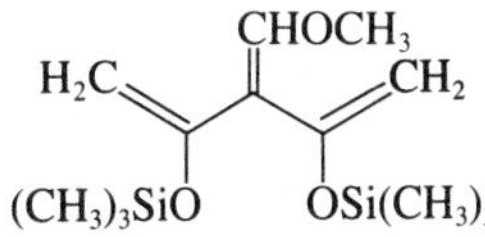

(1).

The triene is prepared by condensation of 2,4-pentanedione with trimethyl orthoformate (Ac₂O) followed by silylation [ZnO, N(C₂H₅)₃, C₆H₆].

Cross Diels-Alder reactions.[1]   The reaction of **1** with a cyclic olefinic dienophile (1 equiv.) results in an *endo*-monocycloadduct. A dienophile can react with the monoadduct to form an *endo*-bis-cycloadduct.

Example:

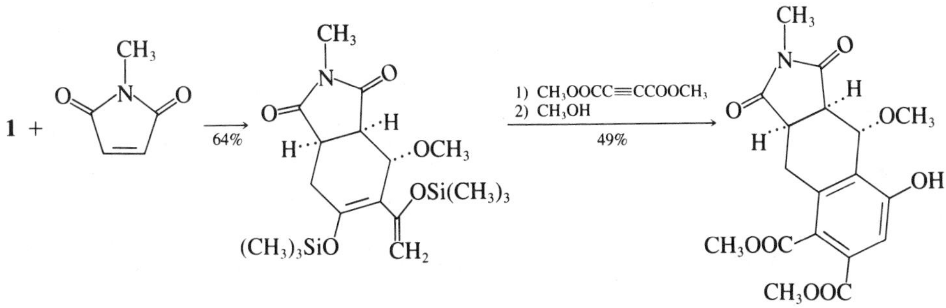

Only monoadducts are formed from **1** with a cyclic olefinic dienophiles, but the stereoselectivity is low. However, these monoadducts can also undergo slow cycloaddition with highly reactive dienophiles.

[1] O. Tsuge, E. Wada, S. Kanemasa, and H. Sakoh, *Chem. Letters*, 3221 (1984); O. Tsuge, S. Kanemasa, H. Sakoh, and E. Wada, *ibid.*, 3234 (1984).

Methoxymethyl phenyl thioether, $C_6H_5SCH_2OCH_3$ (**1**) **6,** 369   This ether can be prepared in 80% yield by reaction of thiophenol with the dimethyl acetal of formaldehyde (excess) in the presence of BF_3 etherate (0.5 equiv.).

Vinyl ethers.[1]   The anion (**2**) of **1** (*n*-BuLi, $-40°$) reacts with aldehydes to form adducts that are trapped with CS_2 and then CH_3I to give xanthates (**3**). Radical reduction (Bu_3SnH) of **3** results in elimination to form (Z)- and (E)-vinyl ethers (**4**).

Example:

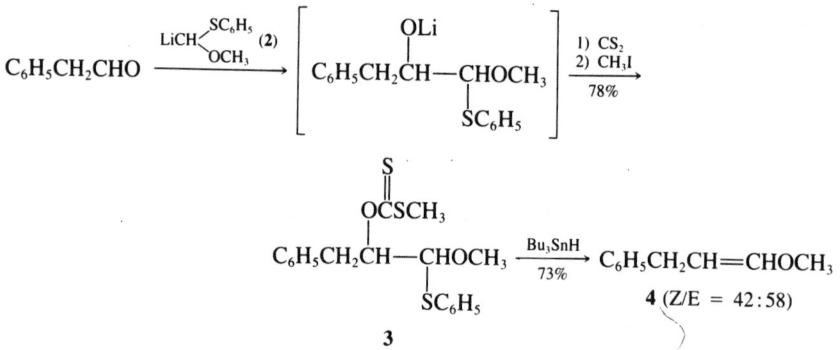

One-carbon homologation.[2]   The products (**3**) obtained by reaction of **2** with alkyl halides can be converted under mild conditions to one-carbon homologated aldehydes, carboxylic acids, or enol methyl ethers.

Example:

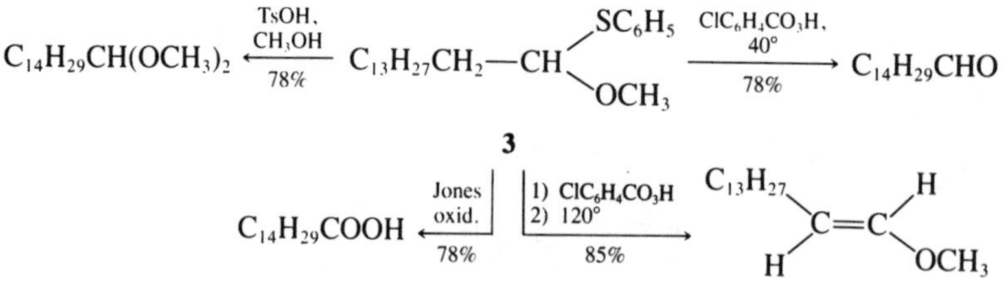

[1] J.-M. Vatele, *Tetrahedron Letters*, **25**, 5997 (1984).
[2] T. Mandai, K. Hara, T. Nakajima, M. Kawada, and J. Otera, *ibid.*, **24**, 4993 (1983).

Methoxy(phenylthio)trimethylsilylmethane,
$$
\underset{\underset{\displaystyle Si(CH_3)_3}{|}}{\overset{\overset{\displaystyle OCH_3}{|}}{HC}}-SC_6H_5 \ (\mathbf{1}).
$$
B.p. 120–122°/10

mm. The reagent is prepared by reaction of methoxy(phenylthio)methyllithium with chlorotrimethylsilane (88% yield).

Ester homologation.[1] The anion of **1** reacts with carbonyl compounds to form ketene-O,S-acetals (**2**), which are converted into homologated methyl esters (**3**) by reaction with methanol in the presence of $HgCl_2$.

[1] A. de Groot and B. J. M. Jansen, *Syn. Comm.*, **13**, 985 (1983).

Methoxy(phenylthio)trimethylsilymethyllithium,
$$
(CH_3)_3Si-\underset{\underset{\displaystyle CH_3O}{|}}{\overset{\overset{\displaystyle C_6H_5S}{|}}{C}}Li \ (\mathbf{1}).
$$
 The reagent is

prepared by sequential lithiation (*sec*-BuLi, TMEDA) and silylation of methoxymethyl phenyl sulfide.

Ketene-O,S-ketals.[1] The reagent converts carbonyl compounds into a mixture of the (Z)- and (E)-ketene-O,S-ketals (**2**) in 60–100% yield. The products can be converted

into the ketene-O-silyl,S-ketals **3** by reaction with iodotrimethylsilane and quenching with dimethyl carbonate. Alternatively, they can be cleaved directly to phenyl thiolesters by $ClSi(CH_3)_3$ and NaI followed by Al_2O_3. Conversion of **2** into amides can be effected by sequential treatment with lithium thiomethoxide and an amine (two examples, 95% yield).

[1] S. Hackett and T. Livinghouse, *Tetrahedron Letters*, **25**, 3539 (1984).

2-Methoxypropene, 2, 230–231; **5,** 360; **11,** 329.

Acetonation of pyranoid, **vic, trans-***glycols.* *vic, trans*-Glycols in a pyranoid ring do not undergo acetonation under usual conditions. However, such glycols can undergo kinetically controlled acetonation with 2-methoxypropene in DMF in the presence of TsOH.[1]

Example:

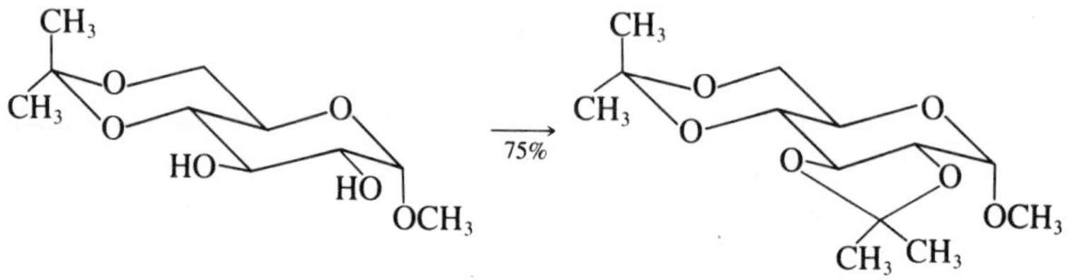

[1] J.-L. Debost, J. Gelas, D. Horton, and O. Mols, *Carbohydrate Res.*, **125**, 329 (1984).

(γ-**Methoxypropyl)-α-phenylethylamine,** $C_6H_5CHNH(CH_2)_3OCH_3$ (**1**). (R)- or (S)-**1** is prepared from (R)- or (S)-α-phenylethylamine:

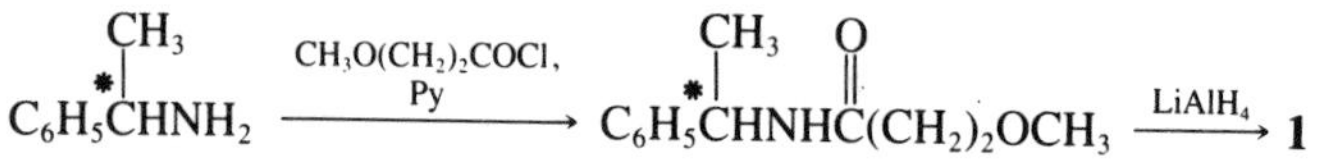

Bidentate base.[1]   The chiral lithium amide of (S)-**1** has been used to generate the anion of the ethyl *o*-toluate **2** and as a chiral complexing agent in reaction of the anion with acetaldehyde to give optically active mellein methyl ether (**3**) in 53% optical yield. Optical yields are markedly lower when a chiral base similar to **1** but lacking the OCH_3 group is used.

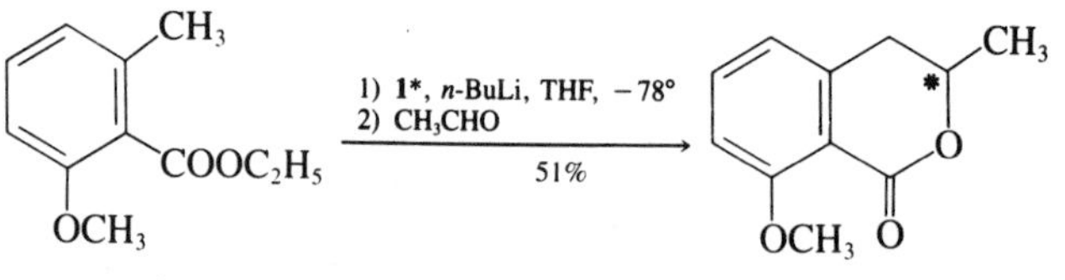

[1] A. C. Regan and J. Staunton, *J.C.S. Chem. Comm.*, 764 (1983).

trans-**1-Methoxy-3-trimethylsilyloxy-1,3-butadiene, 6**, 370–373; **7**, 233; **8**, 328–330; **9**, 303–304; **10**, 260; **11**, 332–334.

Enantiocontrolled Diels-Alder reaction.[1]   The β-D-glucopyranoside **1**, an analog of Danishefsky's diene, undergoes a diastereocontrolled Diels-Alder reaction with the

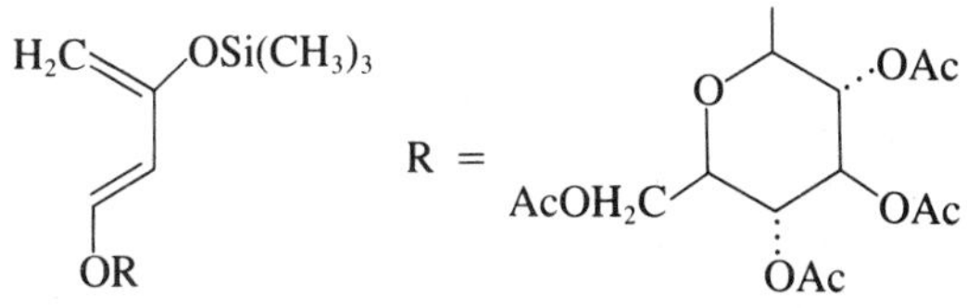

1

tricyclic dienophile **2** to give **3**, one of the two possible *endo*-adducts, as the major product. The adduct is converted in a few steps into (+)-4-demethoxydaunomycinone (**4**).

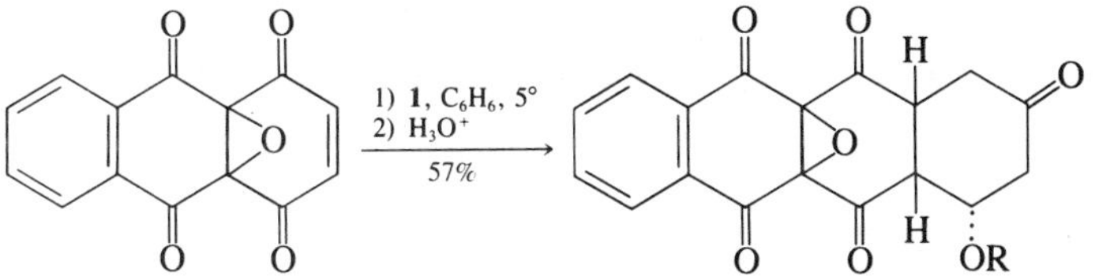

2   **3**

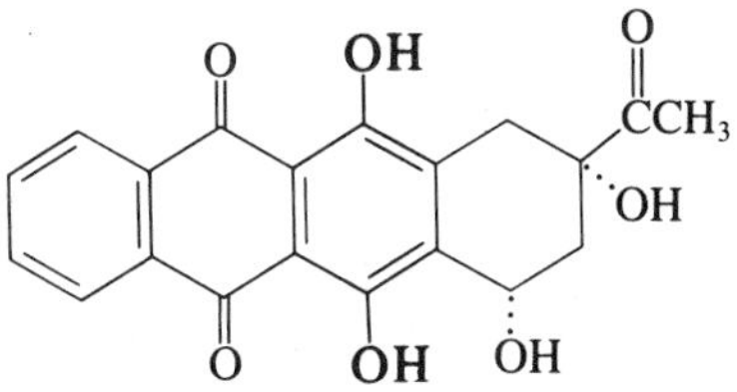

4

[1] R. C. Gupta, P. A. Harland, and R. T. Stoodley, *Tetrahedron*, **40**, 4657 (1984).

α-Methylbenzylamine (α-Phenylethylamine), **1**, 838; **2**, 272–273; **3**, 199–200; **6**, 457.

Optically active α-amino nitriles.[1]   Optically active α-amino nitriles can be obtained by reaction of α-silyloxy nitriles with (R)- or (S)-α-methylbenzylamine in methanol (equation I).

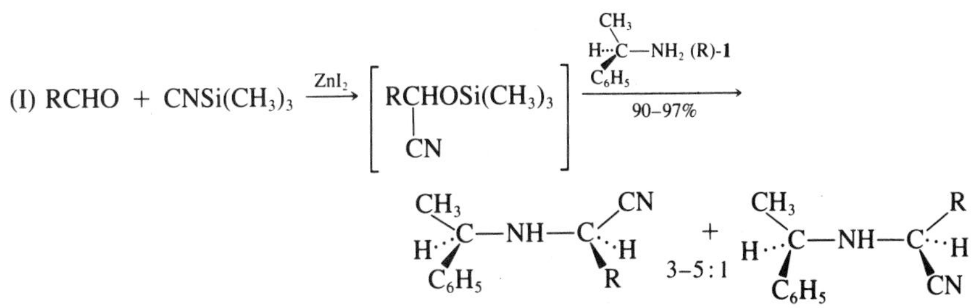

[1] K. Mai and G. Patil, *Syn. Comm.*, **14**, 1299 (1984).

Methyl bis(trifluoroethyl)phosphonoacetate, $(CF_3CH_2O)_2\overset{\overset{\displaystyle O}{\|}}{P}CH_2COOCH_3$ (1).
Preparation:

$$(CH_3O)_2P(O)CH_2COOCH_3 \xrightarrow{PCl_5} Cl_2P(O)CH_2COOCH_3 \xrightarrow[\substack{40\% \\ \text{overall}}]{\substack{CF_3CH_2OH, \\ C_2H_5N(i\text{-}Pr)_2}} 1$$

(Z)-α,β-Unsaturated esters.[1]   Wittig-Horner reactions generally show a preference for formation of (E)-alkenes. Thus (E)-α,β-unsaturated esters are obtained preferentially on reaction of aldehydes with trimethyl phosphonoacetate under usual conditions (potassium *t*-butoxide). Use of a highly dissociated base can favor (Z)-selectivity. The most effective base for this purpose is potassium hexamethyldisilazide, $KN[Si(CH_3)_3]_2$, in combination with 18-crown-6, although even potassium carbonate with the crown ether is fairly effective. The (Z)-selectivity can be further enhanced by use of **1** as the phosphonoester. Under these conditions, (Z)-unsaturated esters can be prepared from aliphatic and aromatic aldehydes with Z/E ratios as high as 50:1. The method is also useful for transformation of unsaturated aldehydes to (E,Z)-dienoates and (E,E,Z)-trienoates.

The related reagent $(CF_3CH_2O)_2P(O)CH(CH_3)COOCH_3$ shows higher (Z)-selectivity than **1** and provides a general route to (Z)-α-methyl-α,β-unsaturated esters.

[1] W. C. Still and C. Gennari, *Tetrahedron Letters*, **24**, 4405 (1983).

Methyl α-chloro-α-phenylthioacetate, $C_6H_5SCH(Cl)COOCH_3$ **(1).** The reagent is most conveniently prepared by NCS chlorination of methyl phenylthioacetate.

γ-Keto esters.   Two laboratories[1] have used the reaction of **1** with silyl enol ethers of ketones to obtain γ-keto esters.

Example:

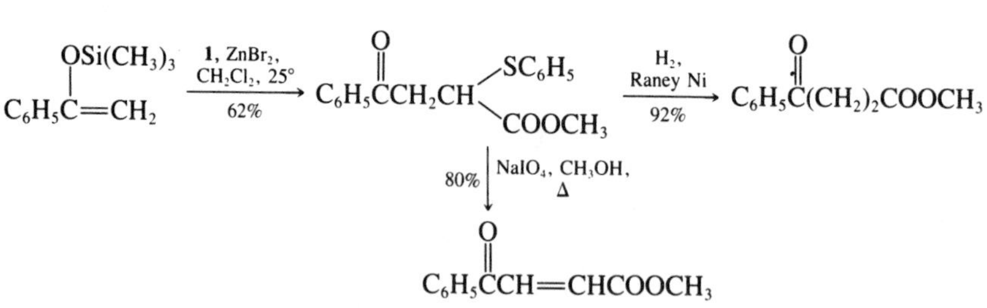

[1] T. V. Lee and J. O. Okonkwo, *Tetrahedron Letters*, **24**, 323 (1983); I. Fleming and J. Iqbal, *ibid.*, 327 (1983).

Methyl cyanoformate, $O=C\diagdown_{CN}^{OCH_3}$ **(1)**. The reagent is readily available by reaction

of KCN with methyl chloroformate in CH_2Cl_2 catalyzed by 18-crown-6 (76% yield).[1]

β-Keto esters.[2]   Methyl cyanoformate reacts regioselectively with preformed lithium enolates in the presence of HMPT at $-78°$ to give β-keto esters in generally high yield. Sodium and potassium enolates are unreactive.

Examples:

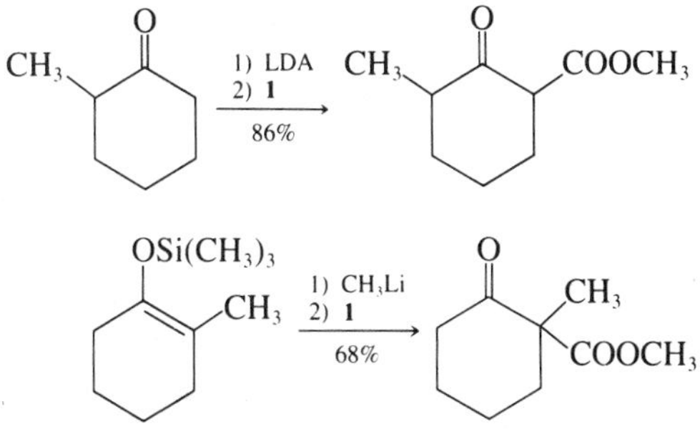

[1] M. E. Childs and W. P. Weber, *J. Org.*, **41**, 3486 (1976).
[2] L. N. Mander and S. P. Sethi, *Tetrahedron Letters*, **24**, 5425 (1983).

Methyldiphenylchlorosilane, 10, 91.

1-Alkenes.[1]   Esters can be converted into 1-alkenes by α-silylation with this silane followed by a reductive Peterson elimination.

Example:

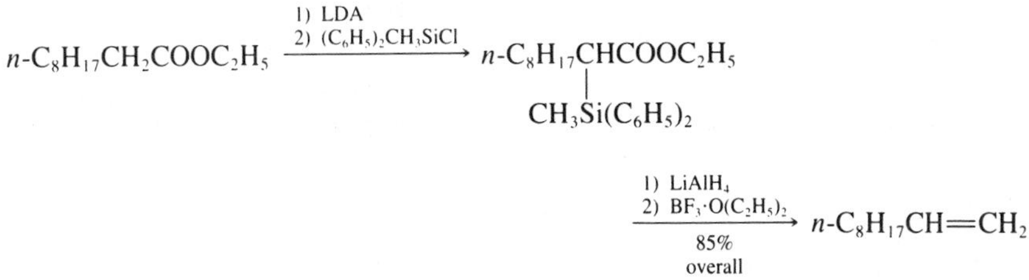

$$n\text{-}C_8H_{17}CH_2COOC_2H_5 \xrightarrow[\text{2) }(C_6H_5)_2CH_3SiCl]{\text{1) LDA}} n\text{-}C_8H_{17}\underset{\underset{CH_3Si(C_6H_5)_2}{|}}{C}HCOOC_2H_5$$

$$\xrightarrow[\substack{85\% \\ \text{overall}}]{\substack{\text{1) LiAlH}_4 \\ \text{2) BF}_3 \cdot O(C_2H_5)_2}} n\text{-}C_8H_{17}CH=CH_2$$

The method is useful for preparation of 1-alkenes labeled with deuterium at C_1 or C_2.

α-Alkyl-α,β-unsaturated esters.[2]   α-Silyl esters can be obtained in >80% yield by reaction of lithium ester enolates with this silane (**10,** 91; **11,** 247). Aldehydes and ketones react with the enolates of these α-silyl esters to give adducts that undergo a Peterson elimination to form α-alkyl-α,β-unsaturated esters in which the (Z)-isomer predominates.

Example:

$$\underset{\underset{CH_3Si(C_6H_5)_2}{|}}{CH_3CHCOOC_2H_5} \quad \xrightarrow[\underset{57\%}{}]{\begin{array}{l}1)\ LDA,\ THF,\ -78°\\2)\ CH_3(CH_2)_2CHO\end{array}}$$

(Z/E = 3 : 1)

[1] V. Cruz de Maldonado and G. L. Larson, *Syn. Comm.*, **13**, 1163 (1983).
[2] G. L. Larson, C. F. de Kaifer, R. Seda, L. E. Torres, and J. R. Ramirez, *J. Org.*, **49**, 3385 (1984).

Methylene bromide–Zinc–Titanium(IV) chloride, 8, 339; **11**, 357–358.

Methylenation.[1]   In contrast to the Wittig reagent, this reagent effects methylenation of ketones without epimerization of an adjacent chiral center. This problem is not encountered with aldehydes, which usually react readily at low temperatures. This reagent is also useful in the case of ketones containing other base-sensitive groups. A further advantage is that it can be stored at $-20°$ for an extended period of time. Reported yields are in the range 80–90%.

[1] L. Lombardo, *Org. Syn.*, submitted (1984).

(−)-N-Methylephedrine, 9, 308.

Asymmetric reduction of cyclic ketones.[1]   Prochiral cyclic ketones are reduced to (R)-alcohols in 75–96% ee by a chiral hydride obtained by refluxing a mixture of lithium aluminum hydride, (−)-N-methylephedrine (1 equiv.), and 2-ethylaminopyridine (2 equiv.) in ether for 3 hours. Reduction of prochiral acyclic ketones with this hydride also results in (R)-alcohols, but only in moderate yield.

[1] M. Kawasaki, Y. Suzuki, and S. Terashima, *Chem. Letters*, 239 (1984).

N-Methylhydroxylamine, 1, 478–481; **7**, 241.

Triquinanes.[1]   A new route to linearly fused triquinanes uses an intramolecular cycloaddition of a nitrone to a double bond (**1 → 2**). The product is a precursor to the natural product hirsutene (**4**).

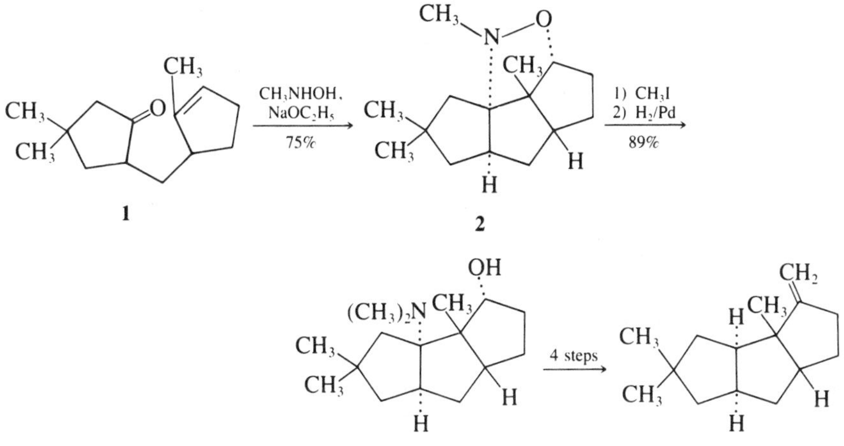

¹ R. L. Funk and G. L. Bolton, *J. Org.*, **49**, 5021 (1984).

3-Methyl-3-hydroxymethyloxetane (1). Mol. wt. 102.13, b.p. 80°/40 mm. Supplier: Aldrich.
Preparation:

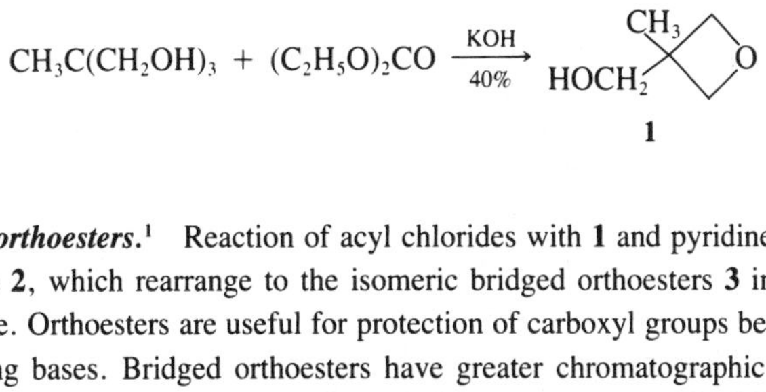

Bridged orthoesters.[1]   Reaction of acyl chlorides with **1** and pyridine results in the oxetane esters **2**, which rearrange to the isomeric bridged orthoesters **3** in the presence of BF₃ etherate. Orthoesters are useful for protection of carboxyl groups because they are stable to strong bases. Bridged orthoesters have greater chromatographic stability than acylic orthoesters. They are hydrolyzed by brief exposure to $NaHSO_4$ followed by saponification with LiOH.[2]

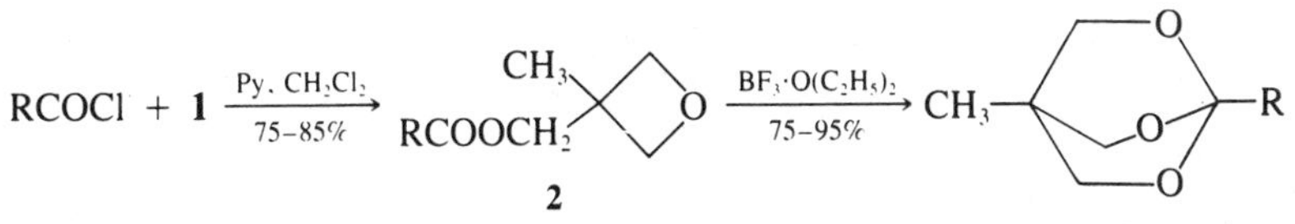

This protective group was useful in regio- and stereoselective syntheses of various unsaturated acids related to arachidonic acid.[3]

[1] E. J. Corey and N. Raju, *Tetrahedron Letters*, **24**, 5571 (1983).
[2] E. J. Corey and B. De, *Am. Soc.*, **106**, 2735 (1984).
[3] E. J. Corey and J. Kang, *ibid.*, **103**, 4618 (1981); *idem, Tetrahedron Letters*, **23**, 1651 (1982).

Methyllithium, 1, 686–689; **2**, 274–278; **3**, 202–204; **5**, 448–459; **6**, 384–385; **7**, 242–243; **8**, 342–344; **9**, 310–311.

Heteroconjugate addition (9, 311). The diastereoselective conjugate addition of methyllithium to a secondary allylic alcohol substituted with a $SO_2C_6H_5$ and a $Si(CH_3)_3$ group to give the *syn*-adduct as the only or major product (equation I)[1] has been extended to other alkyllithiums and used to prepare *syn-* and *anti*-diastereomers with different

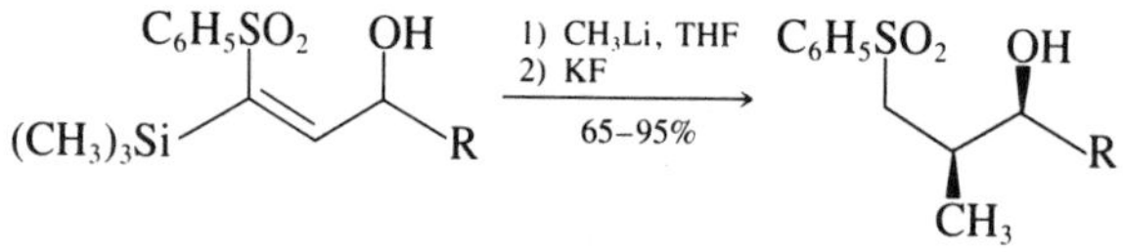

substituents on the asymmetric carbon. Thus the pyranosyl heteroalkene (**1**) can be converted into either the *anti*-product **2** or the *syn*-isomer **3**, depending on the order of addition of *n*-butyllithium or methyllithium (Scheme I).[2]

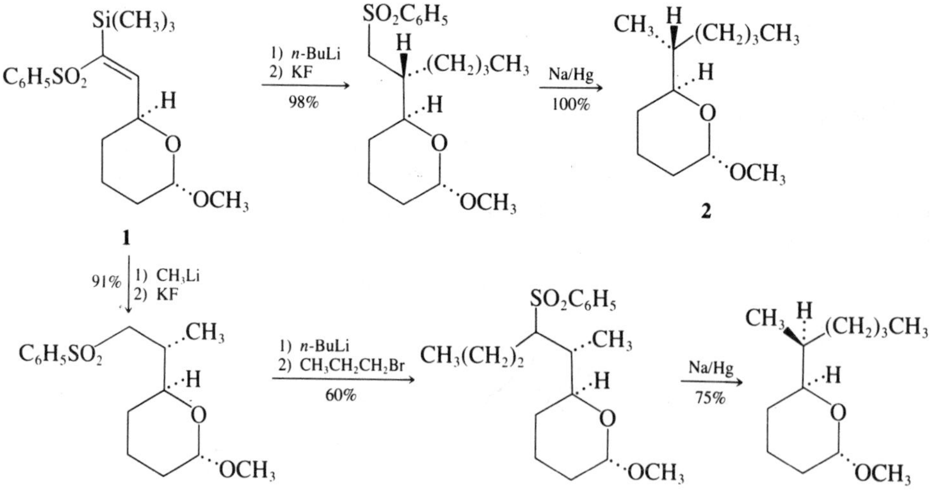

Scheme (I)

Ethoxyvinyllithium also adds stereoselectively to **1** to provide an adduct that is converted in several steps to the *anti*-carboxylic acid **4**. The *syn*-isomer **5** is available by addition of methyllithium, trapping with C_6H_5SeCl, and a sila-Pummerer rearrangement (Scheme II).

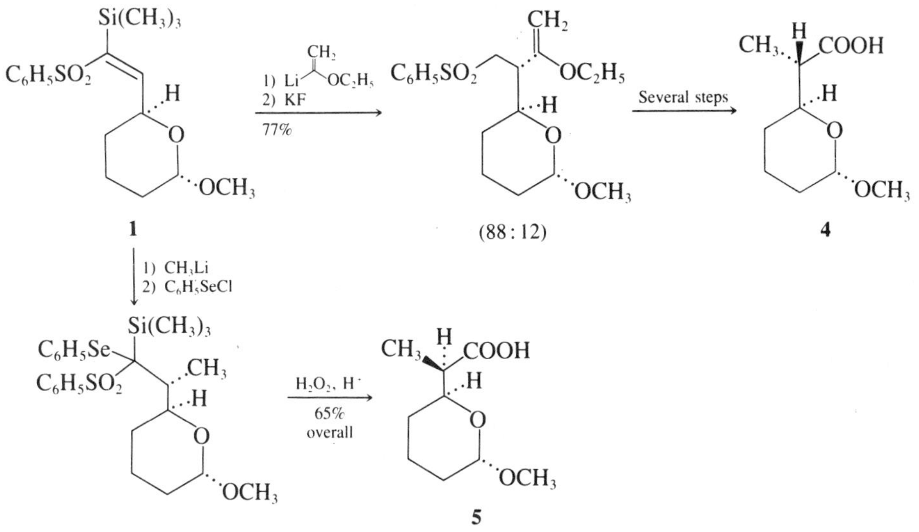

Scheme (II)

This addition to a pyranosyl heteroalkene provided an important diastereoselective step in a stereocontrolled synthesis of maytansinol, a 19-membered lactam with seven asymmetric centers.[3]

[1] M. Isobe, M. Kitamura, and T. Goto, *Tetrahedron Letters*, **21**, 4727 (1980).

[2] M. Isobe, Y. Funabashi, Y. Ichikawa, S. Mio, and T. Goto, *ibid.*, **25**, 2021 (1984).

[3] M. Kitamura, M. Isobe, Y. Ichikawa, and T. Goto, *Am. Soc.*, **106**, 3252 (1984).

Methyl methanethiosulfonate, 5, 454–455; **7**, 243–244. Supplier: Aldrich. The reagent can be prepared in 70–80% yield by reaction of excess DMSO with chlorotrimethylsilane and then with ethylene glycol (*cf.* **9**, 190).[1]

[1] P. Laszlo and A. Mathy, *J. Org.*, **49**, 2281 (1984); *Org. Syn.*, submitted (1984).

Methyl α-(methyldiphenylsilyl)acetate, $CH_3OCCH_2Si(C_6H_5)_2CH_3$ (**1**). This α-silylated ester is prepared by reaction of methyl bromoacetate with methyldiphenylchlorosilane and zinc wool (1 equiv.) catalyzed by iodine (76% yield).

α,β-Bifunctionalization of cyclopentenone.[1]   Although simple ester enolates do not undergo conjugate additions, the enolate of **1** undergoes conjugate addition to cyclopentenone in THF/HMPT (3.3:1) at $-20°$. This reaction coupled with *in situ* enolate alkylation provides a notably short synthesis of methyl jasmonate (**3**).

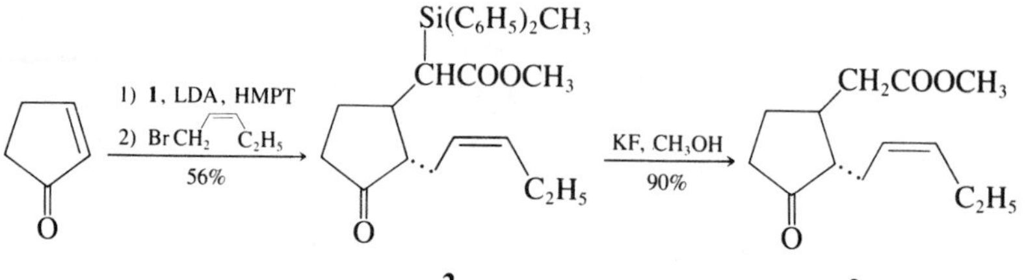

Lower yields are obtained when the anion of methyl α-(trimethylsilyl)acetate is used in this sequence.

[1] W. Oppolzer, M. Guo, and K. Baettig, *Helv.*, **66**, 2140 (1983).

N-Methylthiomethylpiperidine, 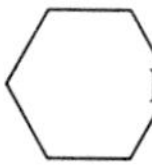 NCH_2SCH_3 (**1**). The reagent is prepared by reaction of piperidine with formalin and methanethiol.

Methylthiomethylenation.[1]   The reagent is useful for methylthiomethylenation of 1,3-dicarbonyl compounds. The products are useful precursors to 2-methylene-1,3-dicarbonyl compounds.

Example:

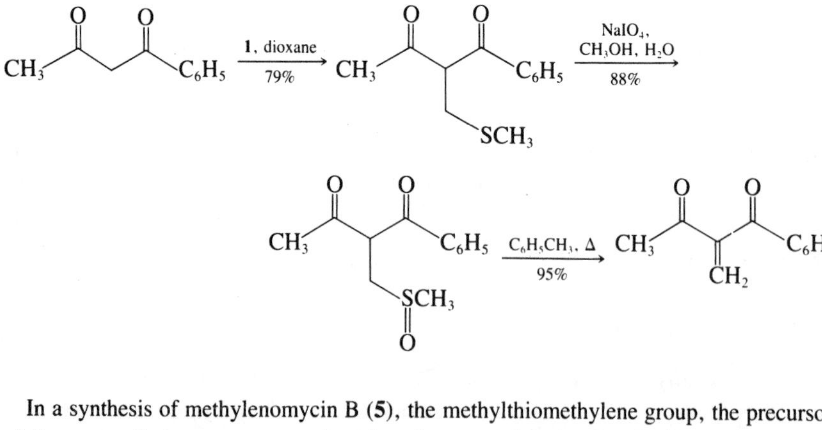

In a synthesis of methylenomycin B (**5**), the methylthiomethylene group, the precursor of the *exo*-methylene group, was introduced at an early stage by alkylation of a β-keto phosphonate (**2**). The last step in the synthesis involves sulfoxide elimination to provide **5**.[2]

$$(C_2H_5O)_2\overset{\overset{\displaystyle O}{\|}}{P}CH_2\overset{\overset{\displaystyle O}{\|}}{C}C_2H_5 \xrightarrow[74\%]{1} (C_2H_5O)_2\overset{\overset{\displaystyle O}{\|}}{P}\overset{|}{C}H\overset{\overset{\displaystyle O}{\|}}{C}C_2H_5 \xrightarrow{\text{Several steps}}$$

2

CH$_2$SCH$_3$

3

4   1) NaIO$_4$ 2) NaHCO$_3$ 63%   **5**

[1] M. Yamauchi, S. Katayama, and T. Watanabe, *Synthesis*, 935 (1982).
[2] M. Mikołajczyk and P. Bałaczewski, *ibid.*, 691 (1984).

Methylthiomethyl *p*-tolyl sulfone, CH$_3$SCH$_2$SO$_2$C$_6$H$_4$CH$_3$-*p* (**1**). The sulfone is prepared by reaction of CH$_3$SCH$_2$OAc with *p*-CH$_3$C$_6$H$_4$SO$_2$Na.

RCOOR^1 → *RCHO*.[1] This conversion can be affected by acylation of the anion of **1** with an ester, preferably a phenyl ester, to give **2**, which is reduced quantitatively by sodium borohydride to the alcohol **3**. In the presence of K$_2$CO$_3$, **3** dissociates into **1** and an aldehyde (**4**). Addition of 18-crown-6 markedly improves the yield in the last step in the case of aliphatic substrates.

RCHO + **1**

4

[1] K. Ogura, N. Yahata, K. Takahashi, and H. Iida, *Tetrahedron Letters*, **24**, 5761 (1983).

3-Methylthio-2-propenyl *p*-tolyl sulfone, CH$_3$S—⟍⟋—SO$_2$C$_6$H$_4$CH$_3$ (**1**). The reagent is obtained from 2-propenyl *p*-tolyl sulfone by chlorination followed by reaction with CH$_3$SNa.

α,β-*Enals*.[1]   This sulfone can be used for synthesis of β-substituted or β,β-disubstituted enals (equation I). The synthesis involves alkylation, hydrolysis of the vinyl

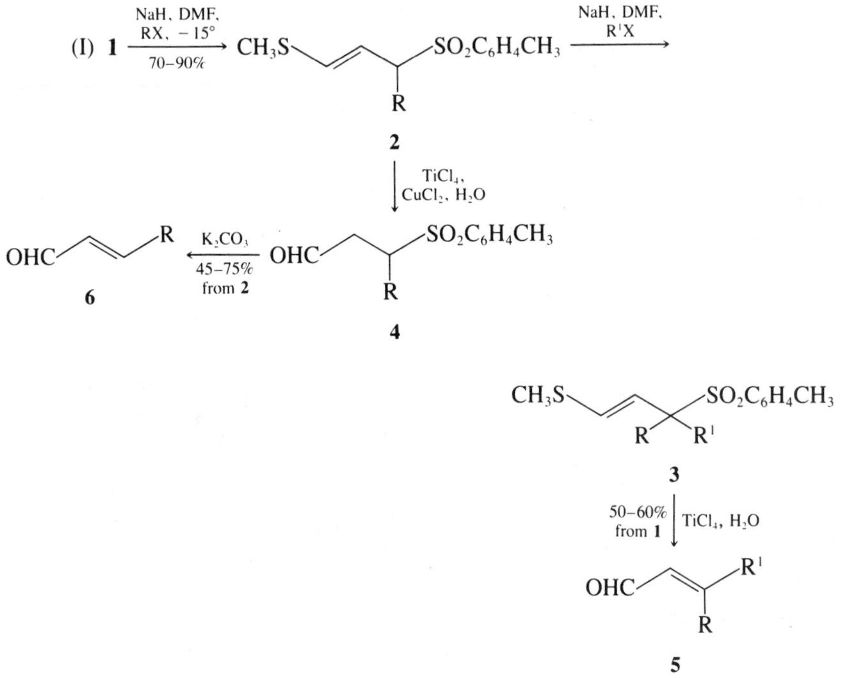

(I) 1 $\xrightarrow[\text{70–90\%}]{\substack{\text{NaH, DMF,}\\ \text{RX, } -15°}}$ CH₃S——=——SO₂C₆H₄CH₃ $\xrightarrow{\substack{\text{NaH, DMF,}\\ \text{R'X}}}$

with substituent R below the sulfone carbon

2

↓ TiCl₄, CuCl₂, H₂O

OHC——=——R $\xleftarrow[\substack{45–75\%\\ \text{from 2}}]{K_2CO_3}$ OHC——CH(R)——SO₂C₆H₄CH₃

6   **4**

CH₃S——=——C(R)(R')——SO₂C₆H₄CH₃

3

↓ $\substack{50–60\%\\ \text{from 1}}$ TiCl₄, H₂O

OHC——=——C(R)(R')

5

sulfide, and elimination of *p*-toluenesulfinic acid. The synthesis of dihydrocitral (**7**) is typical.

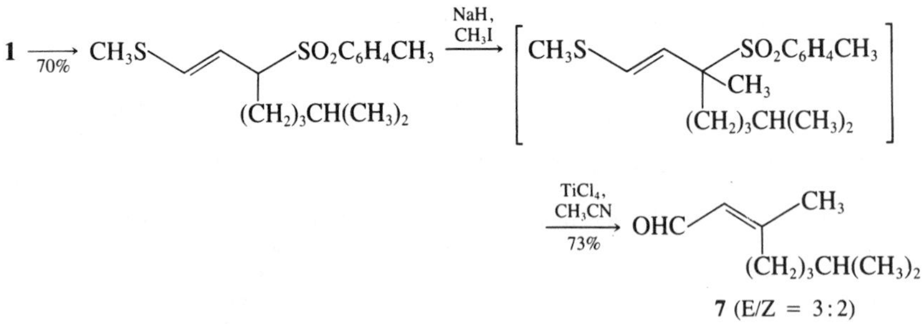

1 $\xrightarrow{70\%}$ CH₃S——=——CH((CH₂)₃CH(CH₃)₂)——SO₂C₆H₄CH₃ $\xrightarrow{\substack{\text{NaH,}\\ \text{CH}_3\text{I}}}$ [CH₃S——=——C((CH₂)₃CH(CH₃)₂)(CH₃)——SO₂C₆H₄CH₃]

$\xrightarrow[73\%]{\substack{\text{TiCl}_4,\\ \text{CH}_3\text{CN}}}$ OHC——=——C((CH₂)₃CH(CH₃)₂)(CH₃)

7 (E/Z = 3:2)

[1] K. Ogura, T. Iihama, K. Takahashi, and H. Iida, *Tetrahedron Letters*, **25**, 2671 (1984).

Methyl trichloroacetate, CCl_3COOCH_3 (**1**). Mol. wt. 149.48, b.p. 154°. Suppliers: Aldrich, Fluka.

Aryl ethers.[1] Phenols are converted into methyl ethers by reaction with **1** in the presence of potassium carbonate and 18-crown-6 at 150°. The reaction involves liberation of chloroform and carbon dioxide (equation I). If the reaction is conducted in the presence of an alkyl halide, alkyl aryl ethers are formed in generally good yield (equation II).

$$\text{(I)}\quad ArOH \ + \ \mathbf{1} \ \xrightarrow[-CHCl_3]{\substack{K_2CO_3, \\ 18\text{-crown-6}}} \ \left[\, ArOCO_2CH_3 \,\right] \ \xrightarrow[80-95\%]{-CO_2} \ ArOCH_3$$

$$\text{(II)}\quad ArOH \ + \ RX \ + \ \mathbf{1} \ \longrightarrow \ ArOR \ + \ CHCl_3 \ + \ CH_3X \ + \ CO_2$$
$$(65-90\%)$$

The reaction of carboxylic acids with **1** under similar conditions results in methyl esters in 70–100% yield. Primary alkyl and allyl trichloroacetates can also be employed in the same way.

[1] J. M. Renga and P.-C. Wang, *Syn. Comm.*, **14**, 69, 77 (1984).

Methyl trifluoromethanesulfonate, 6, 406.

N-Methylamino acids.[1] N-Methylation of amino acids can be effected with minimal racemization by alkylation of the amidine ester (**8**, 191) with methyl triflate. Methylation with dimethyl sulfate results in extensive racemization.

Example:

$$\underset{\overset{|}{NH_2}}{C_6H_5CHCOOCH_3} \ \xrightarrow[CH_2Cl_2]{(CH_3)_2NCH(OCH_3)_2,} \ \underset{\overset{|}{\underset{CHN(CH_3)_2}{N}}}{C_6H_5CHCOOCH_3} \ \xrightarrow[64\%]{\substack{1)\ CH_3OTf \\ 2)\ H_2O}} \ \underset{\overset{|}{NHCH_3}}{C_6H_5CHCOOH}$$

[1] M. J. O'Donnell, W. A. Bruder, B. W. Daugherty, D. Liu, and K. Wojciechowski, *Tetrahedron Letters*, **25**, 3651 (1984).

Methyl vinyl ketone, 1, 697–703; **2**, 283–285; **5**, 464; **6**, 407–409; **7**, 247; **10**, 272.

Spirocyclohexenones.[1] The piperidine enamine of cycloalkanecarboxaldehydes reacts with methyl vinyl ketone to afford, after treatment with NaOH, a spirocyclohexenone in

30–47% yield. The product can be dehydrogenated with DDQ in dioxane to a spiro-hexadienone (56–58% yield).

Example:

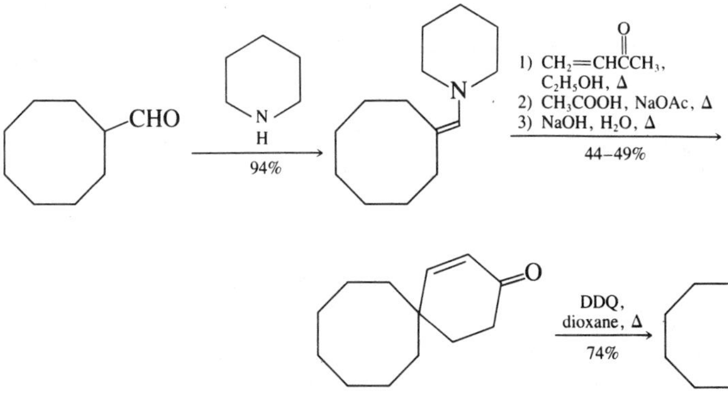

¹ V. V. Kane and M. Jones, Jr., *Org. Syn.*, **61**, 129 (1983).

Molybdenum carbonyl, 2, 287; **3**, 206–207; **4**, 346; **7**, 247–248; **9**, 317; **10**, 273–274; **11**, 350.

*Allylic alkylation.*¹ In general, allylic alkylation catalyzed by transition metals results from attack at the less substituted carbon atom of the π-allyl intermediate. Deviation from this pattern is observed with some nucleophiles when $Mo(CO)_6$ is used as catalyst. For example, the anion of dimethyl malonate generated with O,N-bis(trimethylsilyl)acetamide (BSA) reacts with the allylic acetate **1** mainly by attack at the tertiary center to give **2**.

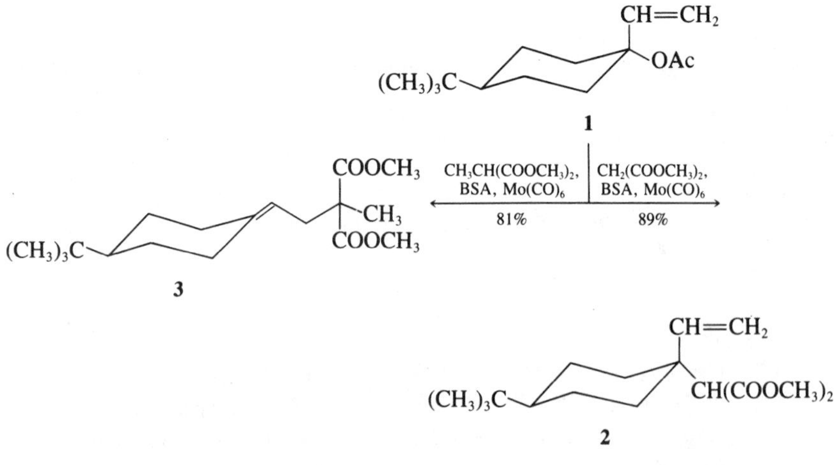

However, the anion of dimethyl methylmalonate preferentially attacks the terminal primary center to give **3**.

This regiocontrol permits a tandem alkylation–Diels-Alder addition (equation I).

(I)

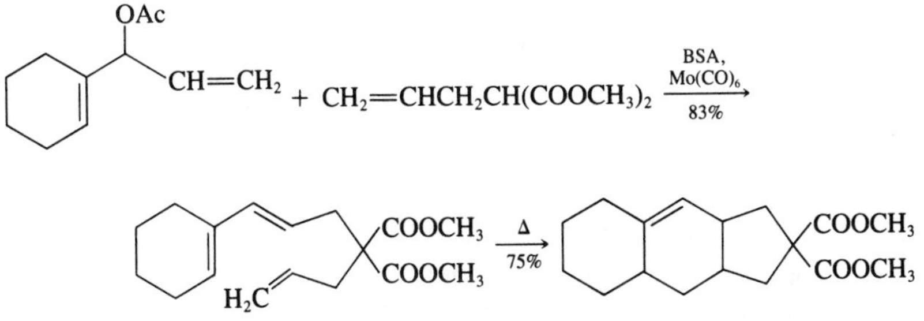

1,3-Dienes.[2] Allylic acetates when heated in toluene with $Mo(CO)_6$ and bis(trimethylsilyl)acetamide (BSA) are converted into 1,3-dienes by loss of acetic acid. The reaction is most useful when only one diene can be formed. It provides one step in a transformation of aldehydes into dienoates.

$$CH_3(CH_2)_4CHO \xrightarrow[61\%]{1)\ C_6H_5SOCH_2COOCH_3\quad 2)\ Ac_2O,\ DMAP} CH_3(CH_2)_3\underset{}{\overset{OAc}{|}}{C}HCH{=}CHCOOCH_3$$

$$\xrightarrow[68\%]{Mo(CO)_6,\ BSA} CH_3(CH_2)_2CH{=}CHCH{=}CHCOOCH_3$$

(E,E/E,Z = 5:1)

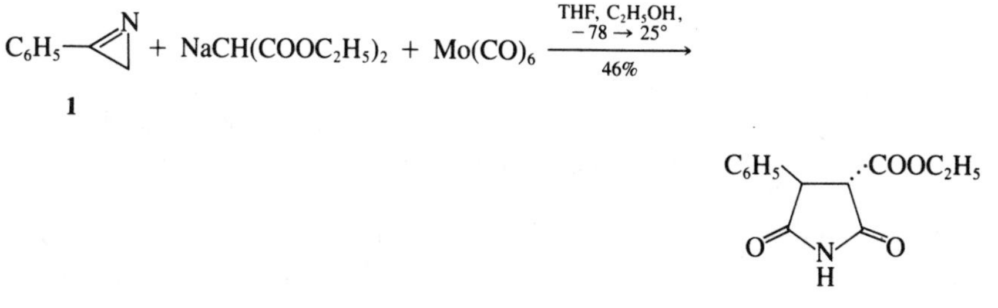

Succinimide synthesis.[3] Reaction of 2-phenylazirine (**1**) and $Mo(CO)_6$ (1 equiv.) with diethyl sodiomalonate furnishes the succinimide **2** in 46% yield. This reaction is applicable generally to 2-aryl-substituted azirines and carbanions of β-dicarbonyl com-

1 **2**

pounds (ethyl benzoylacetate, ethyl cyanoacetate), and results in *trans*-disubstituted suc-cinimides in 27–68% yield. Only one imide (**4**) is formed in the reaction of 2-phenyl-3-methylazirine (**3**).

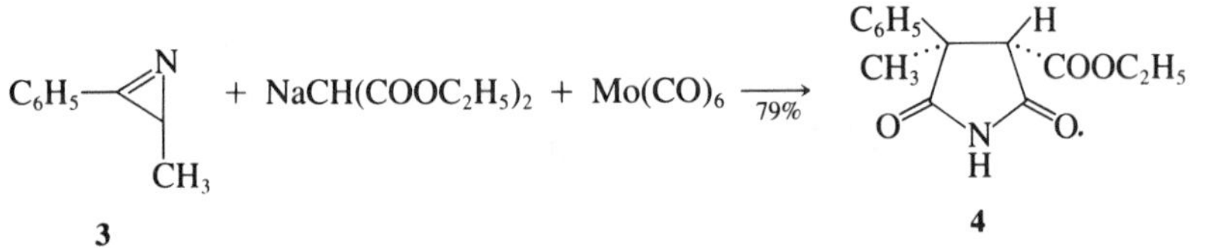

Cyclohexadiene–Mo(CO)₂Cp complexes.   The molybdenum complex **1** is prepared from 3-bromo-1-cyclohexene by reaction with molybdenum carbonyl and lithium cyclo-pentadienide followed by hydride abstraction. It undergoes a highly regio- and stereo-selective reaction with a Grignard reagent to give an adduct (**2**) with an axial substituent

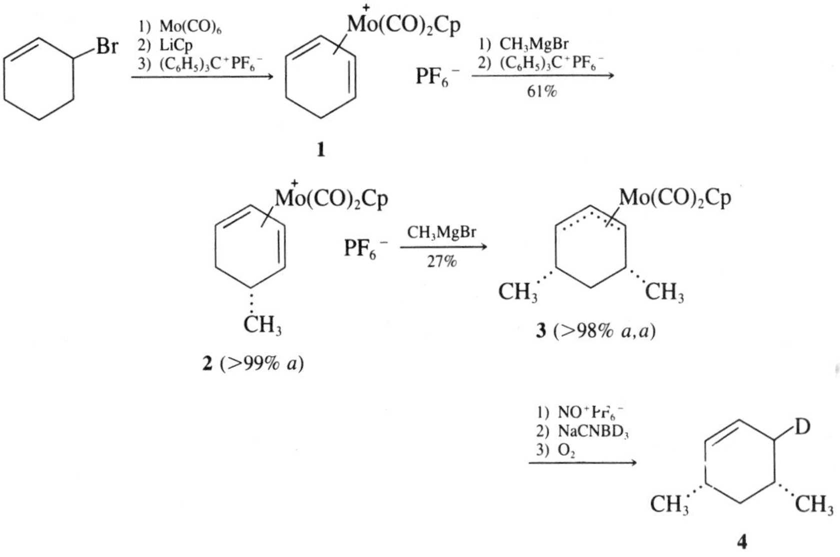

on the nonmetal side of the ring. This complex undergoes further reaction with a Grignard reagent to give a 4,6-dialkylated complex (**3**) in which both substituents have the *cis*-configuration. Similar stereo- and regioselective reactions are possible with other nucleo-philes such as enamines and sodium malonates.[4]

This method provides a useful route to *cis*-3,5-disubstituted cyclohexenes. However, the original method for decomplexation to give **4** proceeds in low yield, and the conditions are too drastic for sensitive functional groups. A novel demetallation of complexes such as **5** is possible by iodolactonization, which proceeds at room temperature in high yield to give an unsaturated bicyclic *cis*-lactone (**6**).[5]

5   **6**

[1] B. M. Trost and M. Lautens, *Am. Soc.*, **105**, 3343 (1983).
[2] B. M. Trost, M. Lautens, and B. Peterson, *Tetrahedron Letters*, **24**, 4525 (1983).
[3] H. Alper, C. P. Mahatantila, F. W. B. Einstein, and A. C. Willis, *Am. Soc.*, **106**, 2708 (1984).
[4] J. W. Faller, H. H. Murray, D. L. White, and K. H. Chao, *Organometallics*, **2**, 400 (1983).
[5] A. J. Pearson and M. N. I. Khan, *Am. Soc.*, **106**, 1872 (1984).

Monochloroalane, AlH_2Cl, **1**, 595–599.

Azetidines.[1]   N-Substituted 2-azetidinones (**1**) can be reduced in high yield by AlH_2Cl or $AlHCl_2$ to azetidines (**2**). The usual reducing agents reduce these substrates to derivatives of 3-aminopropanols by cleavage of the 1,2-bond.

1   **2**

[1] M. Yamashita and I. Ojima, *Am. Soc.*, **105**, 6339 (1983).

Monoisopinocampheylborane, $IpcBH_2$, **8**, 267; **9**, 317; **10**, 224; **11**, 350–351.

Asymmetric hydroboration.[1]   Hydroboration of 1-phenyl-1-cyclopentene with $IpcBH_2$ (100% ee) results in a dialkylborane (**1**) containing the *trans*-2-phenylcyclopentyl group of 100% ee. However, hydroboration of prochiral trisubstituted alkenes usually results in alkylisopinocampheylboranes of 50–85% ee. Most of these products are solids, and selective crystallization (usually from ether) can give the optically pure dialkyboranes. In some cases resolution can be achieved by allowing the impure borane to age for several

1

hours in THF at 0°. Treatment of these optically pure dialkylboranes with acetaldehyde liberates α-pinene and results in diethyl alkylboronates $R^*B(OC_2H_5)_2$ of high optical purity.

[1] H. C. Brown and B. Singaram, *Am. Soc.*, **106**, 1797 (1984).

N-Morpholinomethyldiphenylphosphine oxide (1), 9, 318–319.

Enamines.[1]   This Wittig-Horner reagent (**1**) is not useful for reactions with ketones because of its basicity. This difficulty is overcome by use of (N-methylanilino)methyldiphenylphosphine oxide (**2**), which undergoes Wittig-Horner reactions with a wide variety of ketones as well as aldehydes to give N-methylanilino enamines (**3**) as a mixture of (E)- and (Z)-isomers. The less stable (Z)-isomers can be isolated by crystallization at low temperature.

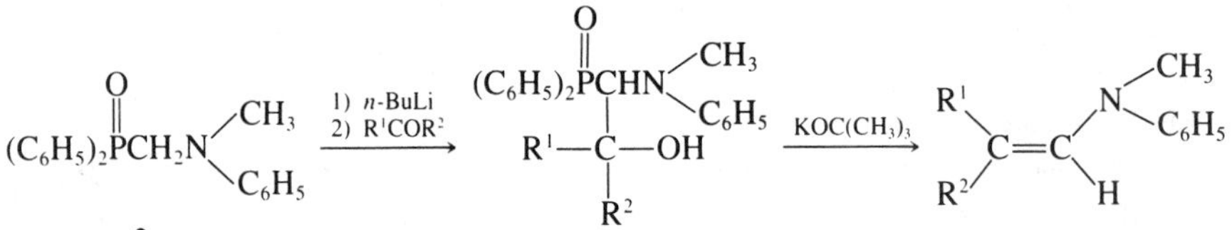

α-Substituted (aminomethyl)diphenylphosphine oxides, $(C_6H_5)_2P(O)—CH(R^3)—NR^1R^2$, are readily available and are useful for conversion of aldehydes into the enamines of homologous ketones.

[1] N. L. J. M. Broekhof and A. van der Gen, *Rec. Trav.*, **103**, 305 (1984); N. L. J. M. Broekhof, P. van Elburg, D. J. Hoff, and A. van der Gen, *ibid.*, **103**, 317 (1984).

N

Nafion-H, 9, 320; **10**, 275–276; **11**, 354–355.

Cleavage of α-keto acetals.[1]   Nafion-H is recommended as a catalyst for hydrolysis of these acetals to the unstable α-keto aldehydes. The reaction is carried out in two steps (equation I).

$$\text{(I)} \quad \underset{\displaystyle R\overset{\textstyle O}{\overset{\|}{C}}CH(OCH_3)_2}{} \quad \xrightarrow[\substack{94\%}]{\substack{\text{Nafion-H,} \\ Ac_2O,\ CHCl_3}} \quad R\overset{O}{\overset{\|}{C}}CH\underset{OCOCH_3}{\overset{OCH_3}{\big<}} \quad \xrightarrow[\substack{100\%}]{\substack{KHCO_3, \\ DME,\ H_2O}} \quad R\overset{O}{\overset{\|}{C}}CH(OH)_2$$

$$R = C_6H_5CH{=}CH{-}$$

[1] K. S. Petrakis and J. Fried, *Synthesis*, 891 (1983).

Nickel(0), activated. Activated nickel powder is obtained by reduction of NiX_2 with lithium (2.3 equiv.) and small amounts of naphthalene as the electron carrier in glyme.

Homocoupling of halobenzenes.[1]   Iodo- and bromobenzene couple to biphenyls in the presence of this activated Ni(0). Yields are definitely higher with the former substrate. *ortho*-Substituents, particularly nitro groups, inhibit coupling. The main by-products result from reduction. This coupling reaction is generally superior to the Ullmann reaction.

Benzyl ketones.   Ni(0) is a useful catalyst for coupling of benzyl halides with acyl halides to form benzyl ketones, $ArCH_2COR$, in 40–85% yield.[2]

[1] H. Matsumoto, S. Inaba, and R. D. Rieke, *J. Org.*, **48**, 840 (1983).
[2] S. Inaba and R. D. Rieke, *Tetrahedron Letters*, **24**, 2451 (1983).

Nickel boride, 1, 720; **2**, 289–290; **3**, 208–210; **4**, 351; **5**, 471–473.

Reductive deseleneniation.[1]   Aryl selenides and selenoketals can be reduced efficiently by nickel boride, generated *in situ* from $NiCl_2·6H_2O$ and $NaBH_4$. Selective deseleneniation is possible in the presence of aryl sulfides and carbon–carbon double bonds. This method is a useful alternative to reduction with triphenyltin hydride (**8**, 521–522; **10**, 451–452).

[1] T. G. Back, *J.C.S. Chem. Comm.*, 1417 (1984).

Nickel carbonyl, 1, 720–723; **2**, 290–293; **3**, 210–212; **4**, 353–355; **5**, 472–474; **6**, 417–419; **7**, 250; **10**, 276; **11**, 356–358.

Cyclopropanecarboxylates.   The reaction of *gem*-dibromocyclopropanes with an alcohol (or an amine) in the presence of $Ni(CO)_4$ (6 equiv.) results in cyclopropanecarboxylates (or carboxamides) in 50–75% yield. The reaction is not possible with monobromocyclopropanes or with *gem*-dichlorocyclopropanes.[1]

Example:

$$(trans/cis = 34:66)$$

Bicyclic lactones.   A related reaction of $Ni(CO)_4$ with 2,2-dibromo-3,3-dimethylcyclopropanemethanol (**1**) results in the bicyclolactone **2**.[1]

Dialkyl ketones (**4**, 353).[2]   Alkylmercuric bromides react with nickel carbonyl in DMF to give only modest yields of symmetrical ketones. Yields can be markedly improved by *in situ* conversion to alkylmercuric iodides by addition of KI to the reaction.

[1] T. Hirao, Y. Harano, Y. Yamana, Y. Ohshiro, and T. Agawa, *Tetrahedron Letters*, **24**, 1255 (1983).
[2] I. Ryu, M. Ryang, I. Rhee, H. Omura, S. Murai, and N. Sonoda, *Syn. Comm.*, **14**, 1175 (1984).

Nickel(II) chloride, Mol. wt. 129.62.

Rearrangement of dienols.[1]   Dienols of the type **2** bearing methylthio or phenylthio substituents are readily available by sequential condensation of 1-(methylthio)-1-(trimethylsilyl)-2-propene (**1**)[2] with two ketones (equation I). On treatment with $NiCl_2$ [$Ni(OAc)_2$, $Ni(acac)_2$, and $PdCl_2$ are less useful] in aqueous *t*-BuOH, **2** rearranges to the dienol **3**. This rearrangement was used in a synthesis of (20R)-25-hydroxycholesterol (**4**) from pregnenolone to transpose a C_{20}-hydroxy group to the C_{25}-position (equation II).

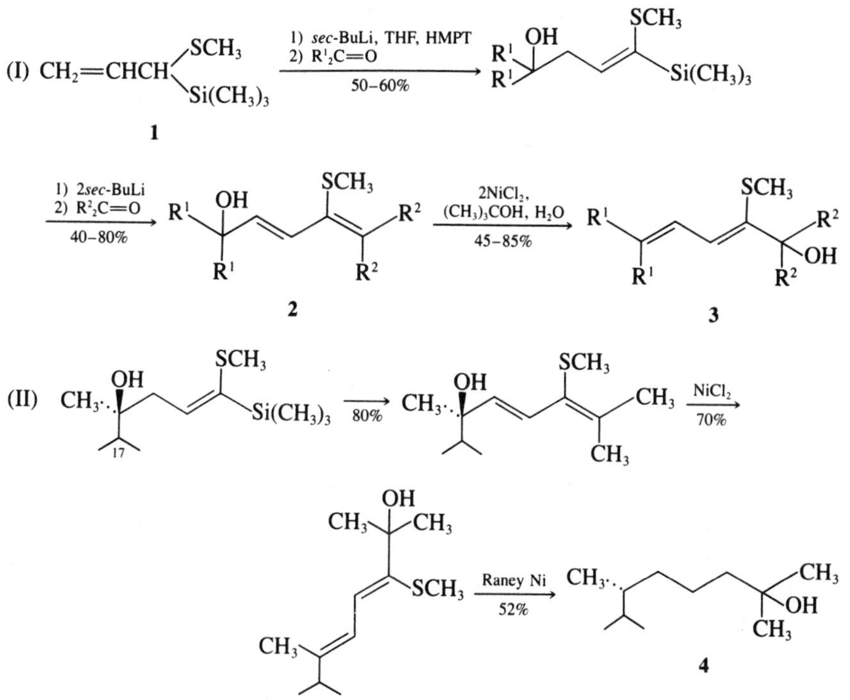

(I) $CH_2=CHCH \overset{SCH_3}{\underset{Si(CH_3)_3}{\big\langle}}$ $\xrightarrow[\text{50-60\%}]{\begin{array}{l}\text{1) } sec\text{-BuLi, THF, HMPT}\\ \text{2) } R^1{}_2C=O\end{array}}$ $R^1{}_2 \overset{OH}{\underset{}{\big\langle}}$ ⟶ $\overset{SCH_3}{\big\langle}Si(CH_3)_3$

1

$\xrightarrow[\text{40-80\%}]{\begin{array}{l}\text{1) 2} sec\text{-BuLi}\\ \text{2) } R^2{}_2C=O\end{array}}$ $R^1 \overset{OH}{\underset{R^1}{\big|}} \quad \overset{SCH_3}{\big|} \quad R^2 \atop R^2$

2

$\xrightarrow[\text{45-85\%}]{\begin{array}{l}\text{2NiCl}_2,\\ \text{(CH}_3)_3\text{COH, H}_2\text{O}\end{array}}$ $R^1 \quad \overset{SCH_3}{\big|} \quad R^2 \atop R^2 \atop OH$

3

(II) $CH_3 \overset{OH}{\underset{}{\cdot\cdot}} \quad \overset{SCH_3}{\big\langle} Si(CH_3)_3$ $\xrightarrow{80\%}$ $CH_3 \overset{OH}{\underset{}{\cdot\cdot}} \quad \overset{SCH_3}{\big|} CH_3 \atop CH_3$ $\xrightarrow[\text{70\%}]{NiCl_2}$

17

$CH_3 \overset{OH}{\underset{}{\diagup}} CH_3$ ... SCH_3 CH_3 $\xrightarrow[\text{52\%}]{\text{Raney Ni}}$ $CH_3 \cdots \overset{}{\diagdown} \overset{CH_3}{\underset{OH}{\big\langle}} CH_3$

4

Homocoupling of alkenyl halides; 1,3-dienes.[3] Alkenyl bromides or iodides couple to 1,3-dienes in the presence of NiCl$_2$, zinc (excess), and KI in HMPT, N-methyl-2-pyrrolidone (NMP), or tetramethylurea. The actual catalyst is Ni(0), generated *in situ* by zinc and KI. Unfortunately, the reaction is not stereospecific.

Example:

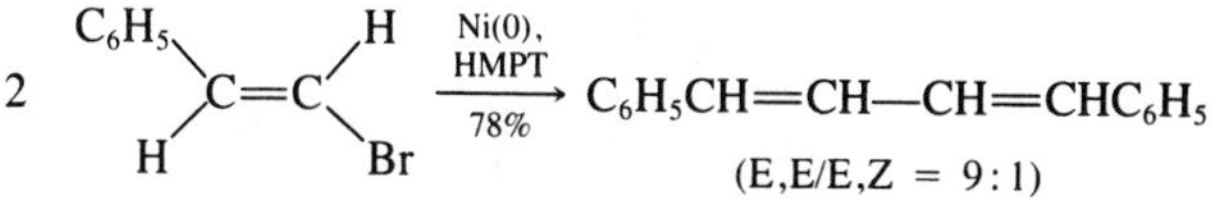

2 $\overset{C_6H_5}{\underset{H}{\big\rangle}}C=C\overset{H}{\underset{Br}{\big\langle}}$ $\xrightarrow[\text{78\%}]{\begin{array}{l}\text{Ni(0),}\\ \text{HMPT}\end{array}}$ $C_6H_5CH=CH-CH=CHC_6H_5$
$(E,E/E,Z = 9:1)$

[1] K. S. Kyler and D. S. Watt, *Am. Soc.*, **105**, 619 (1983).
[2] *Idem, J. Org.*, **46**, 5182 (1981).
[3] K. Takagi and N. Hayama, *Chem. Letters*, 637 (1983); K. Takagi, H. Mimura, and S. Inokawa, *Bull. Chem. Soc. Japan*, **57**, 3517 (1984).

Nitromethane, 1, 739; **4**, 437; **9**, 323–324.

Nitroalkenes.[1] Preparation of nitroalkenes by reaction of CH_3NO_2 with aldehydes and ketones is not generally useful, even when catalyzed by primary amines. The reaction of 17-ketosteroids is further complicated by steric hindrance. This reaction can be effected in useful yields (70–100%), however, by catalysis with a bifunctional amine such as ethylenediamine.

Example:

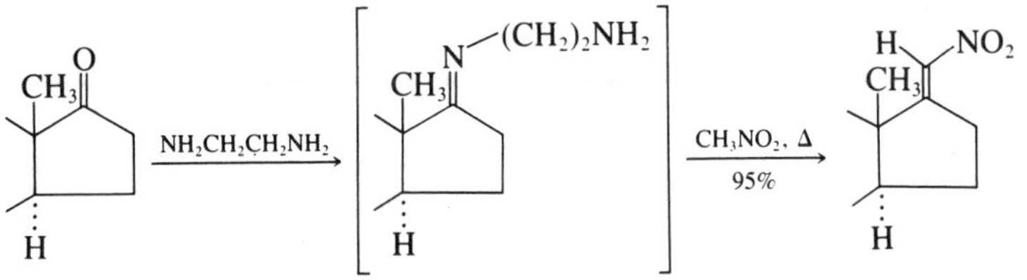

[1] D. H. R. Barton, W. B. Motherwell, and S. Z. Zard, *Bull. Soc.*, II-61 (1983).

Nonafluorobutanesulfonyl fluoride, 5, 479–480.

Deoxygenation of phenols.[1]   Aryl nonaflates, readily available by reaction of phenols with this reagent and triethylamine, undergo hydrogenolysis (H_2, Pd/C) more rapidly and in higher yield (80–90%) than do aryl mesylates or tosylates.

Similar hydrogenolysis (PtO_2 or Raney Ni) of enol triflates provides a general method for conversion of a carbonyl group to a methylene group (60–90% yield).

[1] L. R. Subramanian, A. G. Martinez, A. H. Fernandez, and R. M. Alvarez, *Synthesis*, 481 (1984).

(−)-Norephedrine, $C_6H_5CH(OH)CH(CH_3)NH_2$. Suppliers: Aldrich, Fluka.

Aldol reactions of methyl ketones.[1]   The optically active 1,3-oxazolidine (1) formed from a methyl ketone and (−)-norephedrine after conversion to the tin(II) enolate undergoes enantioselective aldol condensation with aldehydes. The enantioselectivity is partic-

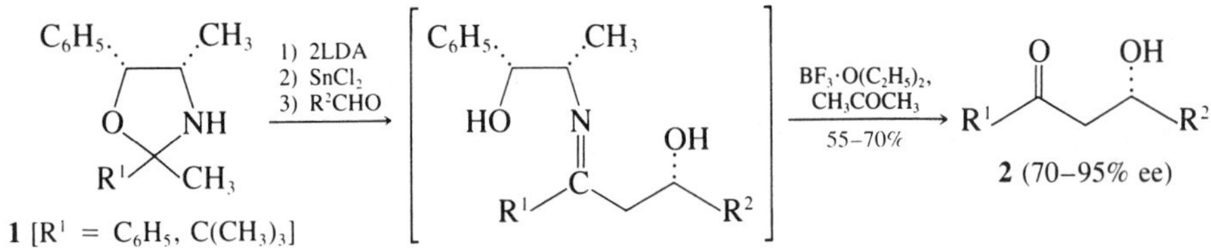

ularly high when both R^1 and R^2 are *t*-butyl groups, but is significant (58–86%) even when R^1 is methyl.

[1] K. Narasaka, T. Miwa, H. Hayashi, and M. Ohta, *Chem. Letters*, 1399 (1984).

O

[(2S)-(2α, 3aα, 4α, 7α, 7aα)]-2,3,3a,4,5,6,7,7a-Octahydro-7,8,8-trimethyl-4,7-methanobenzofurane-2-ol (MBF-OH, 1). Mol. wt. 196.3, b.p. 120°/0.005 mm, α$_D$ + 100°.

1

The lactol is prepared[1] in several steps from *d*-camphor. In the presence of a trace of acid **1** forms a crystalline dimer [MBF-O-MBF (**2**)], m.p. 151–152°, α$_D$ + 193° (available from Aldrich).

Resolution.[2]   The reagent **1** reacts with racemic alcohols, thiols, amines, and cyanohydrins to give diastereomeric derivatives that can be separated by column chromatography or crystallization. Subsequent hydrolysis gives the enantiomers and **1** is recovered. With racemic alcohols, preferential acetalization of the (R)-enantiomers is observed.

Enantioselective alkylation of HSCH₂COOH.[3]   S-MBF-mercaptoacetic acid (**3**), colorless oil, α$_D$ + 270°, obtained by reaction of the acid with **2** in the presence of a trace of HCl, is stereoselectively methylated or ethylated in favor of the (R)-diastereomer (54% ee and 60% ee, respectively). The (S)-diastereomer is removed by esterification of the mixture and subsequent chromatography.

[1] C. R. Noe, *Ber.*, **115**, 1576 (1982).
[2] *Idem, ibid.*, 1591 (1982).
[3] *Idem, ibid.*, 1607 (1982).

Organoaluminum compounds.

n-Bu₃SnAl(C₂H₅)₂.   This Sn–Al reagent (**1**) is prepared from (C₂H₅)₂AlCl and *n*-Bu₃SnLi in THF. It can be used to effect a Reformatsky-type reaction of α-bromo carbonyl compounds with aldehydes or ketones to afford β-hydroxy carbonyl compounds. Yields are generally improved by addition of catalytic amounts of Pd[P(C₆H₅)₃]₄.

Examples:

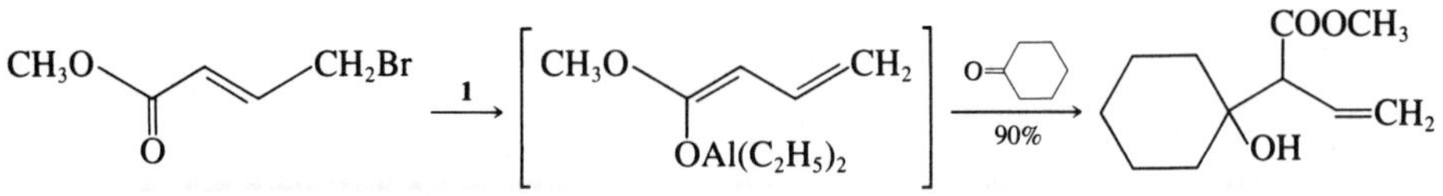

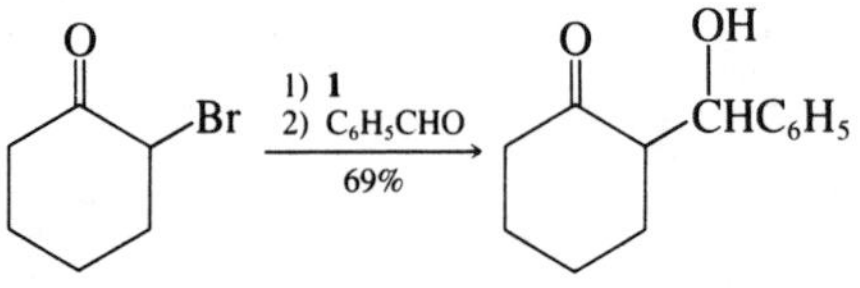

CH₃O‑...‑CH₂Br $\xrightarrow{1}$ [CH₃O‑...‑CH₂, OAl(C₂H₅)₂] $\xrightarrow[90\%]{}$ (cyclohexane ring with COOCH₃, =CH₂, OH)

(cyclohexanone with Br) $\xrightarrow[69\%]{\text{1) }\mathbf{1}\text{, 2) }C_6H_5CHO}$ (cyclohexanone with CH(OH)C₆H₅)

(syn/anti = 25:75)

$Bu_3PbAl(C_2H_5)_2$ can be used in the same way, but $(CH_3)_2C_6H_5SiAl(C_2H_5)_2$ or $(C_6H_5)_3GeAl(C_2H_5)_2$ is ineffective.[1]

Addition of $Bu_3SnAl(C_2H_5)_2$ or $Bu_3SnMgBr$ to 1-alkynes catalyzed by CuCN provides vinylstannanes in good yield (equation I). The intermediate adducts undergo electrophilic substitution to provide alkenylstannanes.[2]

(I) 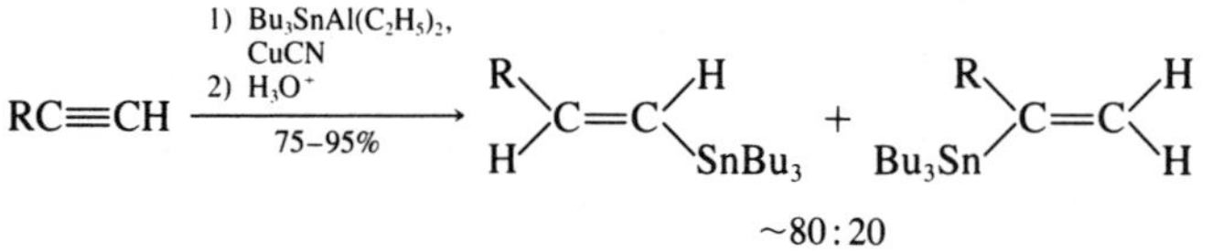 $RC{\equiv}CH$ $\xrightarrow[75-95\%]{\substack{\text{1) }Bu_3SnAl(C_2H_5)_2,\\ \text{CuCN}\\ \text{2) }H_3O^-}}$ (R,H)C=C(H,SnBu₃) + (R,Bu₃Sn)C=C(H,H)

~80:20

In the presence of $Pd[P(C_6H_5)_3]_4$ this reagent reacts with the less substituted carbon of allyl acetates to form allylstannanes with inversion of configuration.

Examples:

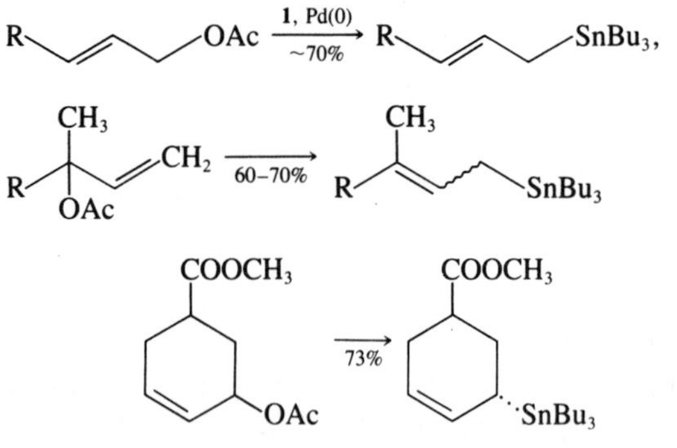

R‑...‑OAc $\xrightarrow[\sim70\%]{\text{1, Pd(0)}}$ R‑...‑SnBu₃,

R‑C(CH₃)=...‑CH₂ $\xrightarrow[60-70\%]{}$ R‑C(CH₃)=...‑SnBu₃

(cyclohexene with COOCH₃ and OAc) $\xrightarrow{73\%}$ (cyclohexene with COOCH₃ and ‑‑SnBu₃)

(trans/cis = 7:3)

Reaction of the crude products with aldehydes catalyzed by BF_3 etherate provides homoallylic alcohols (**9**, **8**).[3]

Cleavage of epoxides.[4]   The ready cleavage of epoxides by $(C_2H_5)_2AlC\equiv CR$ (**4**, 145–146) has been shown to proceed with high regioselectivity in reactions with *cis*-2,3-epoxycyclopentanol (**1**). In this case the usual *trans*-diaxial cleavage is not applicable, but nevertheless the cleavage proceeds with high regioselectivity at the β-position (equation

I). Actually, the highest regioselectivity is observed with organoalane derivatives, possibly because aluminum coordinates with both oxygens in the substrate. This reaction was useful in syntheses of prostaglandin analogs.

Alkylation of arylhydroxylamines.[5]   The trimethylsilyl ether of phenylhydroxylamines undergo nucleophilic *o*- and *p*-alkylation with trialkylaluminums. Only N-alkylation occurs in the absence of the trimethylsilyl group. This reaction is particularly useful for preparation of *o*-alkynyl derivatives of aromatic amines, which provide a convenient access to indoles.

Example:

2-Alkyl-3-chloro-1,4-naphthoquinones.[6]   The reaction of 2,3-dichloro-1,4-naphthoquinone (**1**) with $(n\text{-}C_{12}H_{25})_3Al$ and $ZnCl_2$ in THF results in the 2-alkylnaphthoquinone **2** in 82% yield. Yields are only 25–40% with aluminum reagents in which the R group is propyl or ethyl because of formation of 2,2-dialkyl-3-chlorodihydronaphthoquinones. A similar reaction occurs with n-$C_{12}H_{25}ZnCl$ to give **2** in 61% yield. Tetraallyltin reacts with **1** to give 2,2-diallyl-3-chlorodihydronaphthoquinone in 65% yield.

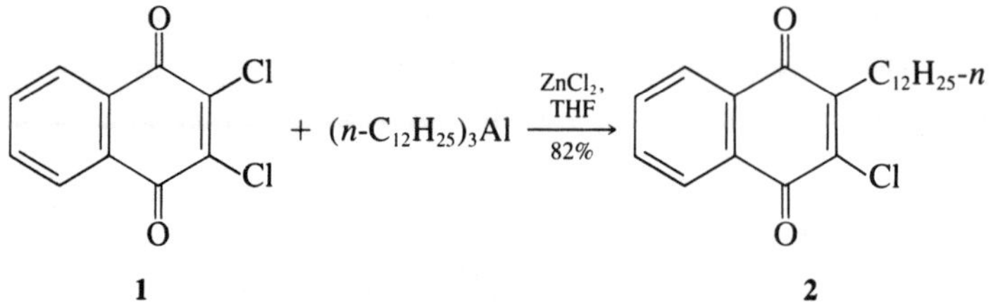

1   **2**

Diethyl[dimethyl(phenyl)silyl]aluminum. $C_6H_5(CH_3)_2SiAl(C_2H_5)_2$ (**1**).[7] The aluminum reagent is prepared *in situ* by addition of $(C_2H_5)_2AlCl$ to $C_6H_5(CH_3)_2SiLi$ in THF at 0°. The reagent converts allylic phosphates into allylsilanes. Rearrangement products can be formed from unsymmetrical allylic phosphates.
Example:

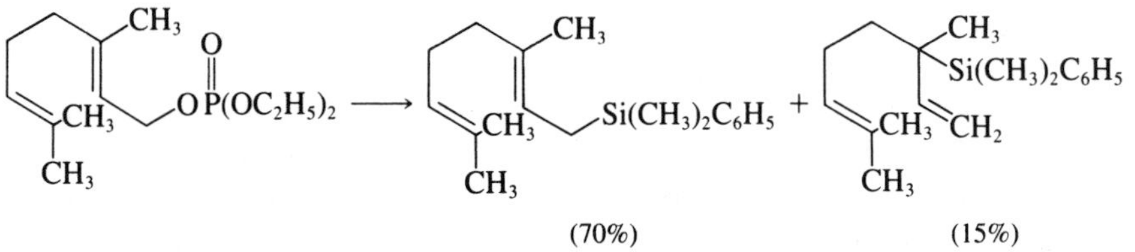

(70%)   (15%)

In the presence of a Pd catalyst prepared from $Pd(OAc)_2$ and $(o\text{-}CH_3C_6H_4)_3P$ (4 equiv.), **1** converts enol phosphates into vinylsilanes. Higher yields are usually obtained for this coupling with a related organomagnesium reagent, $C_6H_5(CH_3)_2SiMgCH_3$ (**2**), prepared *in situ* from $C_6H_5(CH_3)_2SiLi$ and CH_3MgI.
Examples:

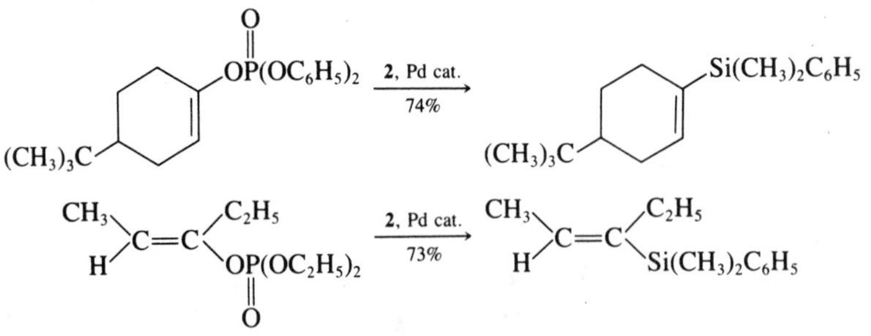

$[(CH_3)_2CHCH_2]_2AlS(CH_2)_2SAl[CH_2CH(CH_3)_2]_2$ (**1**).[8]   The reagent is prepared by addition of $HS(CH_2)_2SH$ to DIBAL (2 equiv.) in C_6H_6 at 0°. The reagent does not react with ketones, but it does convert 1,3-dioxolanes into 1,3-dithiolanes in modest to good yield.

Example:

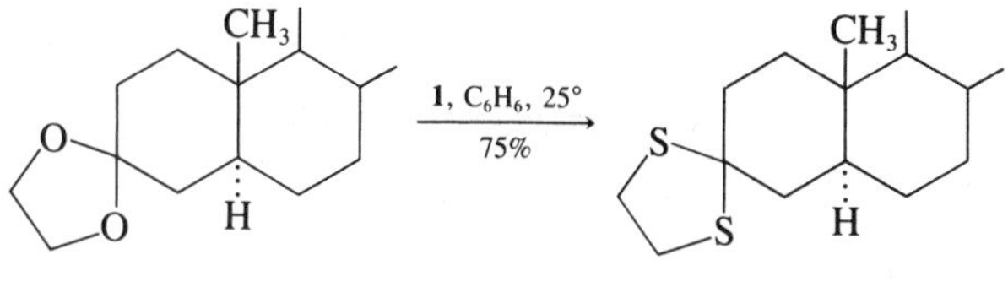

Rearrangement/alkylation of cycloalkylhydroxylamine carbonates.[9] These hydroxylamines, which can be prepared from cycloalkyl amines or cyclic ketones, on reaction with a trialkylaluminum rearrange to an α-alkylated nitrogen-containing heterocycle. The complete sequence is formulated for synthesis of an α-alkylated piperidine (equation I).

(I)

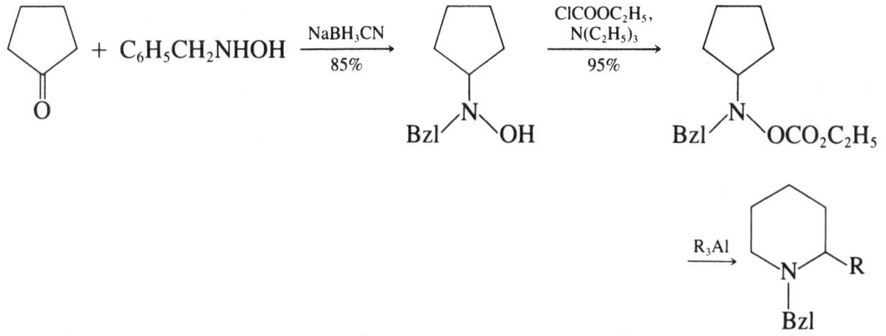

$$R = CH_3 \qquad 74\%$$
$$R = C{\equiv}CSi(CH_3)_3 \qquad 71\%$$

Claisen rearrangement of allyl vinyl ethers.[10] Allyl vinyl ethers undergo Claisen rearrangement at 25° in $ClCH_2CH_2Cl$ in the presence of 2 equiv. of $(C_2H_5)_2AlSC_6H_5$ (prepared from triethylaluminum and thiophenol) or $(C_2H_5)_2AlCl/P(C_6H_5)_3$.

Example:

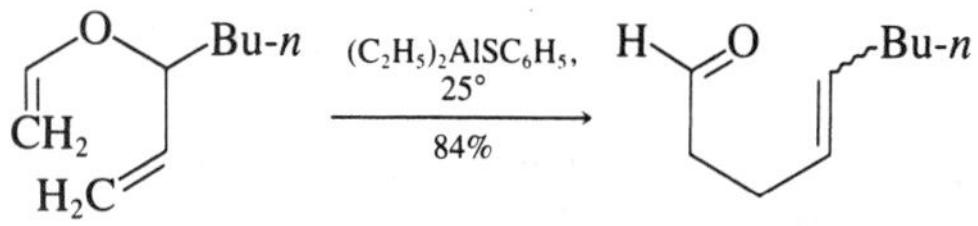

$$(E/Z = 39:61)$$

Tris(trimethylsilyl)aluminum, $Al[Si(CH_3)_3]_3$ (1). The reagent is obtained as the THF complex by reaction of $Hg[Si(CH_3)_3]_2$ with aluminum powder in THF (90% yield). Nonsolvated reagent is prepared by reaction of sodium tetrakis(trimethylsilyl)aluminate with $AlCl_3$ in pentane. It ignites spontaneously in air; dec. at 50°.[11]

The reagent undergoes 1,2-addition to α,β-enones in ether at room temperature. In contrast, 1,4-addition occurs at $-78°$ in THF or DME.[12]

Example:

$$(CH_3)_2C\!\!=\!\!CHCOCH_3 \xrightarrow[\substack{2)\ CH_3OH \\ 85\%}]{1)\ \mathbf{1},\ 25°} (CH_3)_2C\!\!=\!\!CHCCH_3 \underset{\overset{|}{Si(CH_3)_3}}{\overset{\overset{OH}{|}}{}}$$

$$85\% \Big\downarrow \substack{1)\ \mathbf{1},\ -78° \\ 2)\ CH_3OH}$$

$$(CH_3)_2\overset{\overset{\displaystyle Si(CH_3)_3}{|}}{C}\!\!-\!\!CH_2COCH_3$$

In the presence of Pd(0) or Mo(CO)$_6$, allyl acetates couple with **1** to give allyltrimethylsilanes.[13] The regioselectivity depends on the catalyst. The silyl group becomes attached to the less hindered end of the allyl group when Mo(CO)$_6$ is the catalyst. The regioselectivity of the Pd(0)-catalyzed reaction depends on the solvent and, particularly, on the ligands. The two catalysts also differ with respect to the stereochemistry. The Mo catalyst gives net retention; the Pd catalyst, net inversion. The reagent does not react with acetals, esters, enones, and unsaturated centers.

Example:

Pd[P(C$_6$H$_5$)$_3$]$_4$, THF	72:28	42%
Pd[P(C$_6$H$_5$)$_3$]$_4$, C$_6$H$_6$	75:25	56%
Mo(CO)$_6$, C$_6$H$_5$CH$_3$	30:70	42%

[1] N. Tsuboniwa, S. Matsubara, Y. Morizawa, K. Oshima, and H. Nozaki, *Tetrahedron Letters*, **25**, 2569 (1984).

[2] J. Hibino, S. Matsubara, Y. Morizawa, K. Oshima, and H. Nozaki, *ibid.*, 2151 (1984).

[3] B. M. Trost and J. W. Herndon, *Am. Soc.*, **106**, 6835 (1984).

[4] R. S. Matthews, E. D. Mihelich, L. S. McGowan, and K. Daniels, *J. Org.*, **48**, 409 (1983).

[5] J. Fujiwara, Y. Fukutani, H. Sano, K. Maruoka, and H. Yamamoto, *Am. Soc.*, **105**, 7177 (1983).

[6] W. G. Peet and W. Tam, *J.C.S. Chem. Comm.*, 853 (1983).

[7] Y. Okuda, M. Sato, K. Oshima, and H. Nozaki, *Tetrahedron Letters*, **24**, 2015 (1983).

[8] T. Satoh, S. Uevaya, and K. Yamakawa, *Chem. Letters*, 667 (1983).

[9] J. Fujiwara, H. Sano, K. Maruoka, and H. Yamamoto, *Tetrahedron Letters*, **25**, 2367 (1984).

[10] K. Takai, I. Mori, K. Oshima, and H. Nozaki, *Bull. Chem. Soc. Japan*, **57**, 446 (1984).

[11] L. Rösch, *Angew. Chem., Int. Ed.*, **16**, 480 (1977); L. Rösch and G. Altnau, *J. Organometal. Chem.*, **195**, 47 (1980).

[12] G. Altnau, L. Rösch, and G. Jas, *Tetrahedron Letters*, **24**, 45 (1983).

[13] B. M. Trost, J. Yoshida, and M. Lautens, *Am. Soc.*, **105**, 4494 (1983).

Organocerium reagents. Reaction of RLi with $CeCl_3$ in THF at $-78°$ results in an organocerium reagent formulated as $RCeCl_2$, although an ate complex may be present. Alkyl-, alkenyl-, and alkynylcerium reagents can be prepared. These reagents form adducts with ketones, even hindered and easily enolizable ones, in high yield. A further advantage is that they are less basic than Grignard or lithium reagents.[1]

Example:

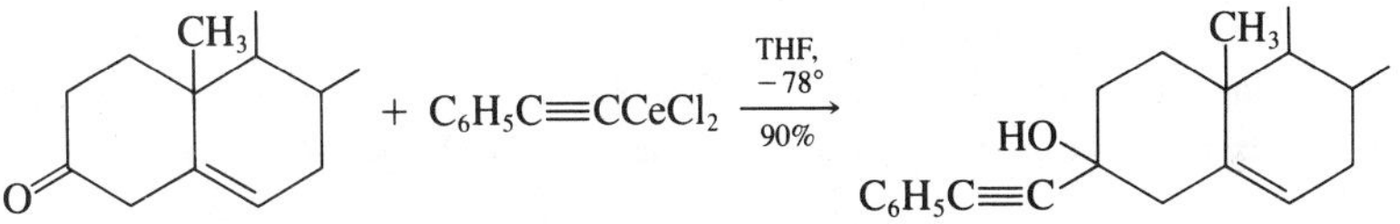

[1] T. Imamato, Y. Sugiura, and N. Takiyama, *Tetrahedron Letters*, **25**, 4233 (1984).

Organocopper reagents, **9**, 328–334; **10**, 282–290; **11**, 365–373.

Allénylsilanes. A general method for preparation of these silanes involves reaction of (trimethylsilylpropynyl)methanesulfonates or -sulfinates with organoheterocuprates

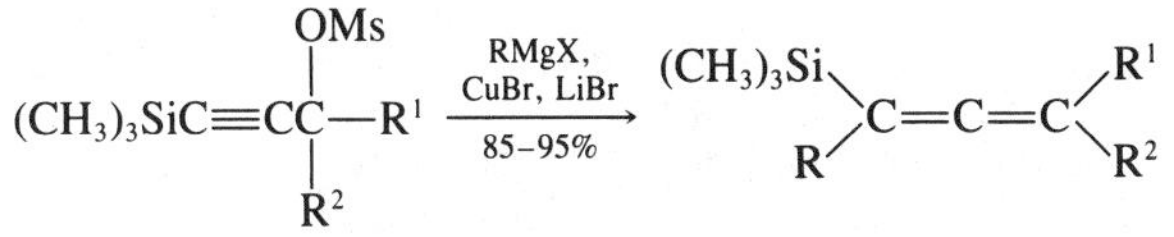

(equation I).[1,2] Traces of a substituted alkyne are also formed, but this by-product can be removed by complexation with $AgNO_3$.

These silanes undergo regiospecific [3 + 2] cycloaddition with α,β-enones in the presence of $TiCl_4$ to form cyclopentenes stereoselectively (**10**, 429).[3]

Example:

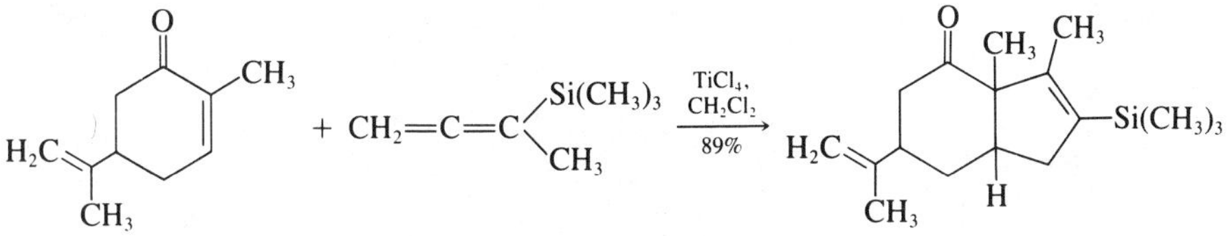

Ketone synthesis.[4] Organocuprates do not ordinarily react with esters at low temperatures, but they react satisfactorily with 2-pyridyl esters to provide ketones. These esters are readily available by reaction of 2-pyridyl chloroformate, an acid, and $N(C_2H_5)_3$ catalyzed by DMAP.

Example:

One advantage of this synthesis is that both alkyl groups of R_2CuLi are utilized.

Lithium divinyl cuprates.[5] The addition of vinylic cuprates to chiral γ-alkoxy-α,β-unsaturated ketones and esters proceeds with high diastereoselectivity; the major product is that in which the vinyl group is *anti* to the allylic alkoxyl group. The geometry of the unsaturated system does not affect the stereochemistry of the addition.

Example:

Addition to β-alkylthio-α,β-enones.[6] The stereochemistry of conjugate addition–elimination of organocuprates with (E)-β-alkylthio-α,β-enones is dependent on the temperature and on the solvent. The reaction proceeds with predominant inversion in ether and with predominant retention in THF. The degree of stereoselectivity is dependent on the reaction temperature; it is highest at −78°. Since (Z)-alkylthio-α,β-enones react with R_2CuLi with retention in either ether or THF, a mixture of (E)- and (Z)-alkylthio-α,β-enones can be converted into the same stereoisomer by reaction with R_2CuLi in ether.

Reaction with allylic nitro compounds.[7] Allylic nitro compounds such as **1** undergo S_N2' substitution with lithium dialkyl cuprates to form (E)-trisubstituted alkenes selectively.

Reaction with amino acid derivatives.[8]   N-Protected amino acid esters substituted by I, Br, or OTs at the γ-position undergo substitution reactions with lithium dialkyl cuprates without detectable racemization.

Example:

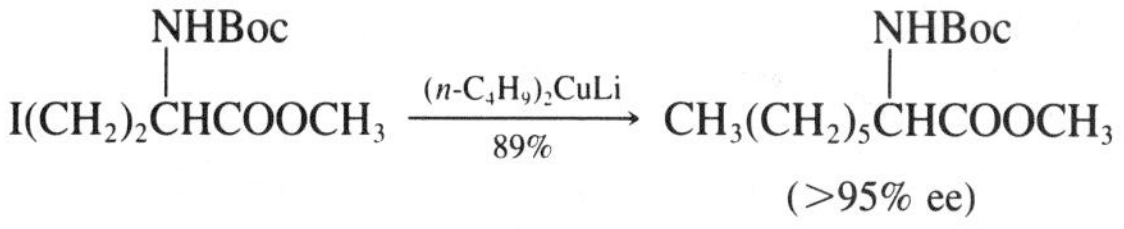

$$I(CH_2)_2\overset{\overset{\displaystyle NHBoc}{|}}{C}HCOOCH_3 \xrightarrow[89\%]{(n\text{-}C_4H_9)_2CuLi} CH_3(CH_2)_3\overset{\overset{\displaystyle NHBoc}{|}}{C}HCOOCH_3$$

(>95% ee)

Reductive decarboxylation.   The reaction of γ-carbamoyloxy-α,β-unsaturated esters with a lithium dialkyl cuprate (10 equiv.) in ether–HMPT results in loss of CO_2 and formation of a β,γ-unsaturated ester.[9] Zinc–acetic acid has been used for this reaction,[10] but yields are lower.

Examples:

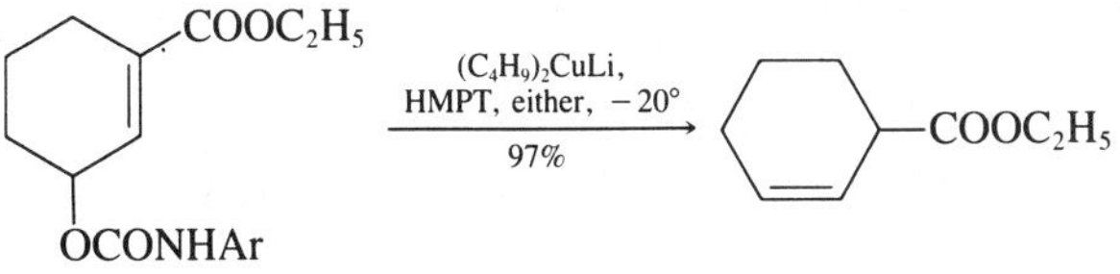

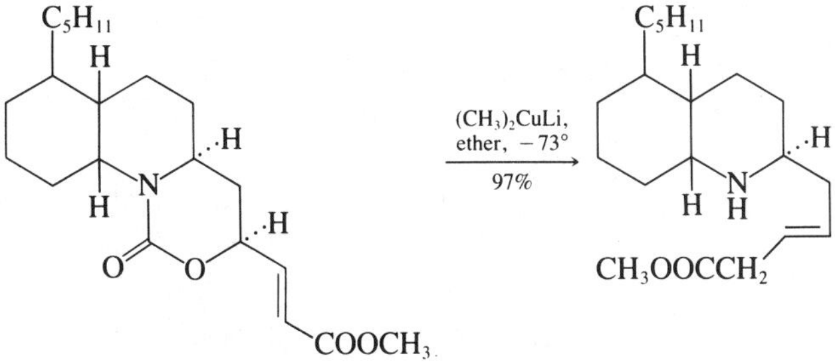

Cleavage of acetals by RCu or R₂CuLi.[11]   Acetals are cleaved by organocopper and cuprate reagents complexed with BF_3 with substitution of one alkoxy group. The reaction in ether is rapid, even at −30°.

Example:

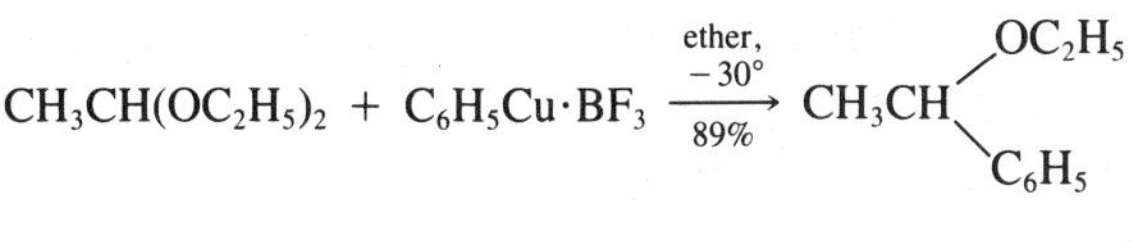

$$CH_3CH(OC_2H_5)_2 + C_6H_5Cu\cdot BF_3 \xrightarrow[89\%]{\overset{\text{ether,}}{-30°}} CH_3CH\overset{\nearrow OC_2H_5}{\searrow_{C_6H_5}}$$

This reaction can be used to cleave chiral acetals with moderate to high diastereoselectivity.

Example:

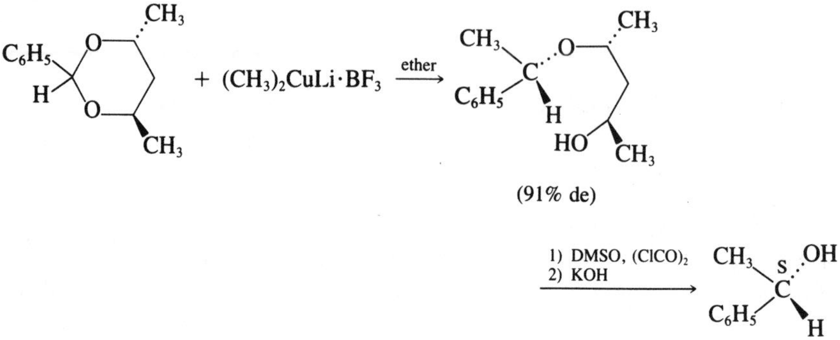

(91% de)

$RCu(CN)_2N(Bu)_4Li.$[12]   The complex n-Bu$_4$NCu(CN)$_2$ (1) is prepared from CuCN and n-Bu$_4$NCN (1 equiv.). This new class of cuprate reagents was introduced to effect coupling of the vinyllithium 2 with the iodide 3, which proceeds in low yield with a variety of organocopper reagents. However, the cuprate prepared from 2 and 1 equiv. of 1 reacts with 3 to form the desired coupling product in 69% yield.

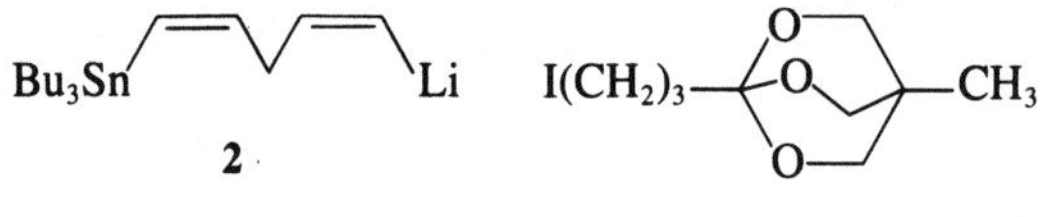

Cuprates of this type also undergo addition to enones in high yield (equation I). They should be particularly useful for effecting complete utilization of an organolithium reagent.

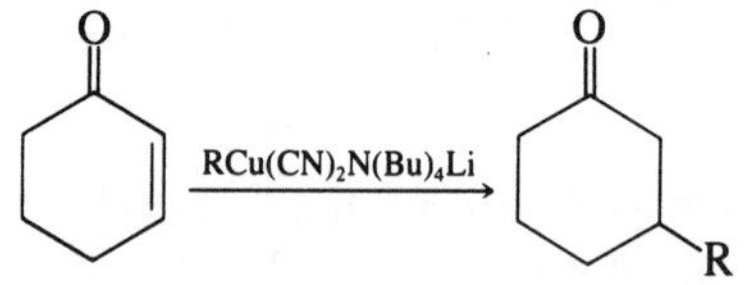

R = n-Bu   97%
R = C$_6$H$_5$ 92%
R = CH$_2$=CH   66%

Conjugate addition–cyclization.   3-Substituted 2-carbomethoxycyclopentanones are prepared conveniently in one step from dimethyl (2E)-hexenedioate by conjugate addition of lithium dialkyl cuprates or higher-order cyanocuprates (11, 366–367) followed by Dieckmann cyclization.[13]

Example:

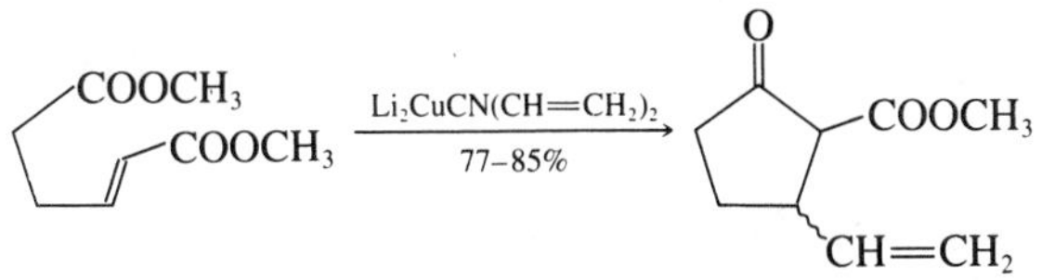

A similar reaction using a dimethyl 2-hexynedioate as the Michael acceptor results in the corresponding cyclopentenones.[14]

Example:

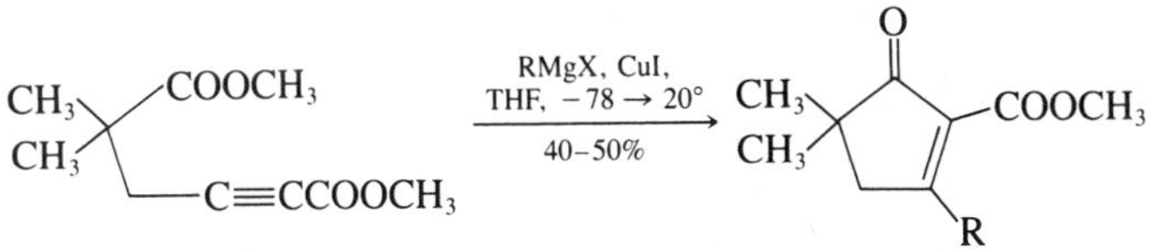

***Chiral all*-syn-1,3-*polyols*.**[15]   A reiterative route to these polyols from an optically active epoxide (**1**) involves ring opening with a cuprate derived from vinyllithium and copper(I) cyanide (**11**, 366–367) to give an optically active homoallylic alcohol (**2**). This is converted into the epoxide (**4**) via a cyclic iodocarbonate (**3**) by a known procedure (**11**, 263). Repetition of the cuprate cleavage results in a homoallylic 1,3-diol (**5**). The ratio of desired *syn*- to *anti*-diols is 10–15:1. This two-step sequence can be repeated, with each 1,3-diol unit formed being protected as the acetonide. The strategy is outlined in Scheme (I).

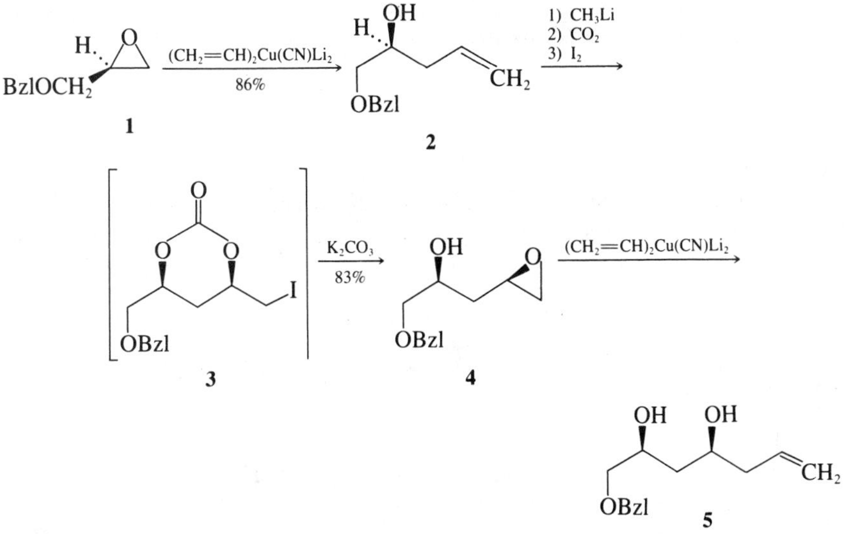

Scheme (I)

R₂Cu(CN)Li₂; arene synthesis.[16]   Aryl triflates or nonaflates do not couple with R_2CuLi, but they do couple with the mixed cuprates $R_2Cu(CN)Li_2$ to give alkylarenes in 40–95% yield. Substituents on the aromatic ring have little effect in the reaction.
   Example:

R = CH₃	85%
= *t*-Bu	96%
= C₆H₅	45%

[1] H. Westmijze and P. Vermeer, *Synthesis*, 390 (1979).
[2] R. L. Danheiser, Y.-M. Tsai, and D. M. Fink, *Org. Syn.*, submitted (1984).
[3] R. L. Danheiser, D. J. Carini, D. M. Fink, and A. Basak, *Tetrahedron*, **39**, 935 (1983).
[4] S. Kim and J. I. Lee, *J. Org.*, **48**, 2608 (1983).
[5] W. R. Roush and B. M. Lesur, *Tetrahedron Letters*, **24**, 2231 (1983).
[6] R. K. Dieter and L. A. Silks, *J. Org.*, **48**, 2786 (1983).
[7] N. Ono, I. Hamamoto, and A. Kaji, *J.C.S. Chem. Comm*, 274 (1984).
[8] A. Bernardini, A. El Hallaoui, R. Jacquier, C. Pigière, and P. Viallefont, *Tetrahedron Letters*, **24**, 3717 (1983); J. A. Bajgrowicz, A. El Hallaoui, R. Pigière, and P. Viallefont, *ibid.*, **25**, 2231 (1984).
[9] T. Ibuka, G.-N. Chu, and F. Yoneda, *ibid.*, **25**, 3247 (1984).
[10] L. E. Overman and C. Fukaya, *Am. Soc.*, **102**, 1454 (1980).
[11] A. Ghribi, A. Alexakis, and J. F. Normant, *Tetrahedron Letters*, **25**, 3075, 3079 (1984).
[12] E. J. Corey, K. Kyler, and N. Raju, *ibid.*, 5115 (1984).
[13] W. A. Nugent and F. W. Hobbs, Jr., *J. Org.*, **48**, 5364 (1983); *idem, Org. Syn.*, submitted (1984).
[14] M. T. Crimmins, S. W. Mascarella, and J. A. DeLoach, *J. Org.*, **49**, 3033 (1984).
[15] B. H. Lipshutz and J. A. Kozlowski, *ibid.*, 1147 (1984).
[16] J. E. McMurry and S. Mohanraj, *Tetrahedron Letters*, **24**, 2723 (1983).

Organolithium reagents.

t-Butoxymethyllithium (1).   *t*-Butyl methyl ether can be metallated at $-78°$ by *sec*-butyllithium activated by potassium *t*-butoxide (**5**, 552) to give *t*-butoxymethyllithium after addition of LiBr in THF to remove K^+ as KBr.[1]

$$(CH_3)_3COCH_3 \xrightarrow[\text{2) LiBr, THF}]{\text{1) } sec\text{-BuLi, KOC(CH}_3)_3} (CH_3)_3COCH_2Li$$

1

   The reagent forms the expected adducts with aldehydes and ketones in high yield. It can be used for preparation of *t*-butoxymethyl ketones from acid chlorides (equation I). In combination with CuBr, it undergoes conjugate addition to α,β-enones (equation II).

(I) $C_6H_5(CH_2)_4COCl$ + $(CH_3)_3COCH_2Cu[OC(CH_3)_3]Li$ $\xrightarrow[90\%]{}$ $C_6H_5(CH_2)_4\overset{\text{O}}{\overset{\|}{C}}CH_2OC(CH_3)_3$

(II)

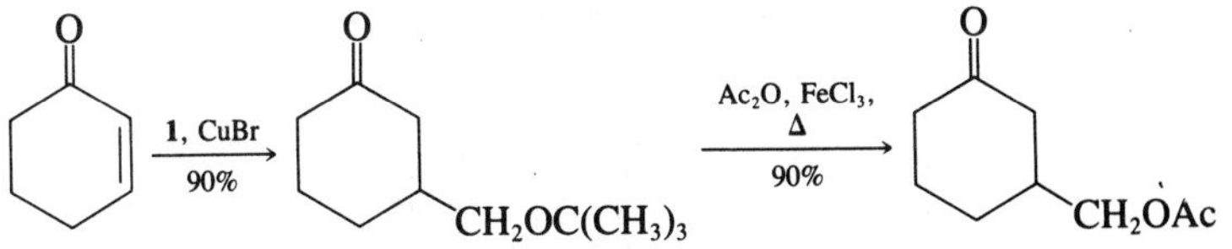

A more general route to alkoxymethyllithium reagents involves sequential treatment of alkyl or aryl chloromethyl ethers with $SnCl_2$–LiBr in THF and n-butyllithium (equation III).[2]

(III) $ROCH_2Cl$ $\xrightarrow[\text{THF}]{SnCl_2,\ LiBr,}$ $[ROCH_2SnBrCl_2]$ $\xrightarrow{4BuLi}$ $ROCH_2Li$ + $SnBu_4$ + $LiBr$ + $2LiCl$

α,β-Enones.[3] Pyrrolidino-enaminoketones (1) can be converted into α,β-enones (2) by reaction with organolithium compounds. Satisfactory yields are obtained only in hydrocarbon solvents, and are highest with vinyl- and ethynyllithium.

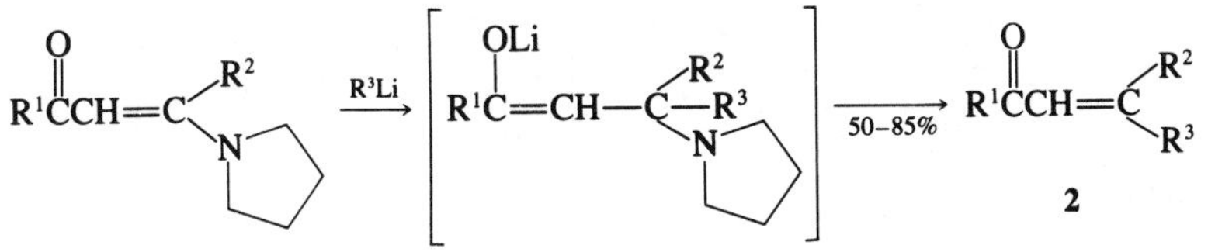

Related reactions:

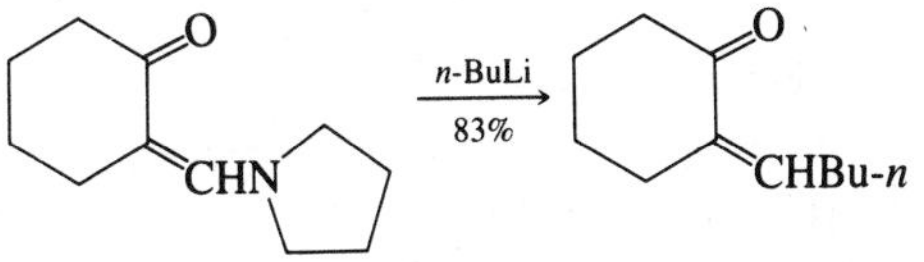

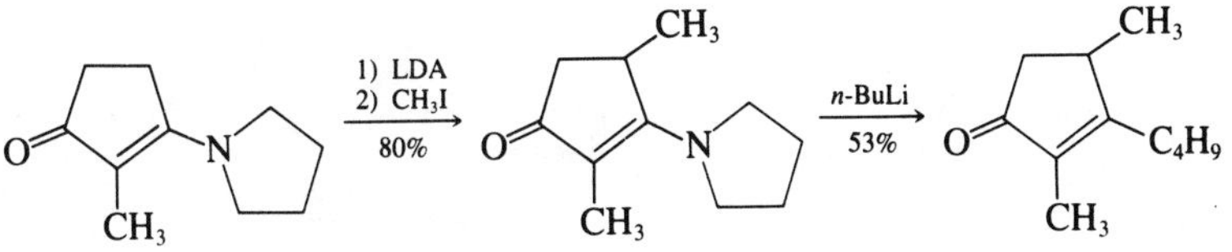

[1] E. J. Corey and T. M. Eckrich, *Tetrahedron Letters*, **24**, 3165 (1983).
[2] *Idem, ibid.*, 3163 (1983).
[3] T. Mukaiyama and T. Ohsumi, *Chem. Letters*, 875 (1983).

Organomagnesium reagents.

Silylmagnesiation of alkynes.   The reagent (**1**) prepared from $C_6H_5(CH_3)_2SiLi$ and CH_3MgI reacts with terminal alkynes in the presence of CuI or Pt catalysts to give exclusively (E)-1-silyl-1-alkenes (equation I).[1]

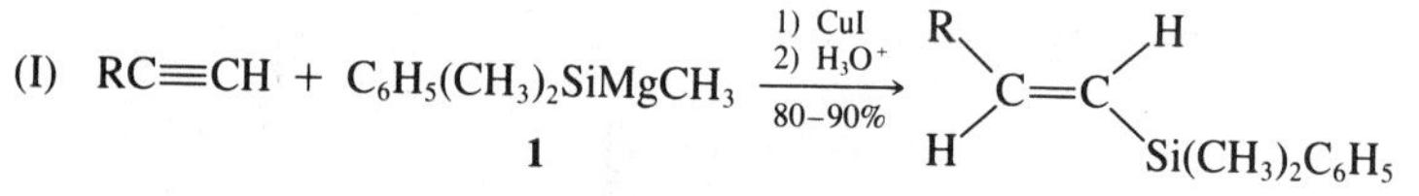

$$\text{(I)   } RC{\equiv}CH + C_6H_5(CH_3)_2SiMgCH_3 \xrightarrow[\substack{80-90\%}]{\substack{1)\ CuI \\ 2)\ H_3O^+}} \underset{1}{} \begin{array}{c} R \\ \diagdown \\ H \end{array} C{=}C \begin{array}{c} H \\ \diagup \\ Si(CH_3)_2C_6H_5 \end{array}$$

The initial adducts can undergo an intramolecular cyclization when the R group is substituted at the terminal position by a leaving group to provide cycloalkenes with a silylmethylene substituent.[2]

Examples:

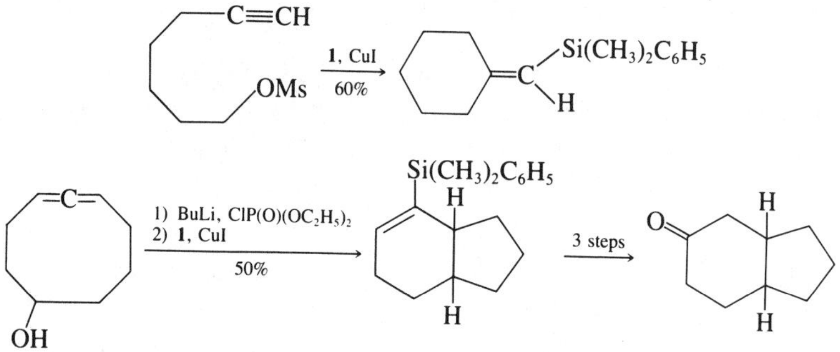

[1] H. Hayami, M. Sato, S. Kanemoto, Y. Morizawa, K. Oshima, and H. Nozaki, *Am. Soc.*, **105**, 4491 (1983).

[2] Y. Okuda, Y. Morizawa, K. Oshima, and H. Nozaki, *Tetrahedron Letters*, **25**, 2483 (1984).

Organomolybdenum reagents.

Methylenemolybdenum reagents.[1]   The novel methylenation reagents **1** and **2** are prepared *in situ* from molybdenum(V) chloride by reaction with methyllithium (equations I and II).

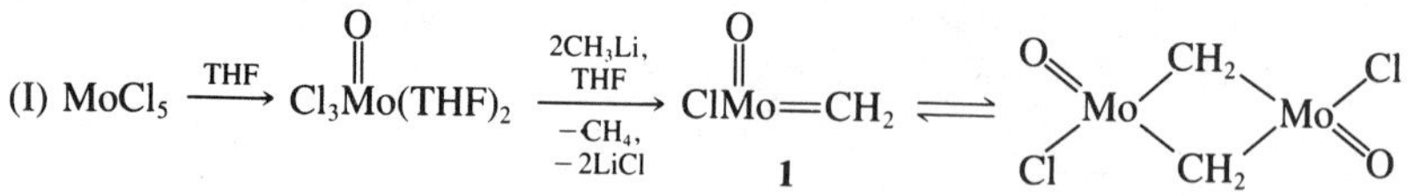

2

Aldehydes and ketones undergo methylenation in 50–85% yield on reaction with either **1** or **2** (2 equiv.) in THF; aldehydes react much more rapidly than ketones. Reactions with **1** can also be conducted in ethanol or aqueous THF at $-70°$, but reagent **2** is much less effective in the presence of ethanol or water. Both reagents are useful for methylenation of substrates containing hydroxyl groups. Methylenation of *o*-hydroxybenzaldehyde with **1** in anhydrous THF proceeds in 76% yield.

Reaction of $MoCl_5$ with 3–4 equiv. of CH_3Li results in a methylating reagent, possibly $Cl_2Mo(CH_3)_2$.

[1] T. Kauffmann, B. Ennen, J. Sander, and R. Wieschollek, *Angew. Chem., Int. Ed.*, **22**, 244 (1983); T. Kauffmann, P. Fiegenbaum, and R. Wieschollek, *ibid.*, **23**, 531 (1984); T. Kauffmann and G. Kieper, *ibid.*, **23**, 532 (1984).

Organotitanium reagents, 10, 138, 271, 422; 11, 374–378.

β-Acetylenic alcohols (11, 378).[1] The propargyltitanium reagent derived from 1-trimethylsilyl-3-(tetrahydropyranyloxy)propyne (**1**) reacts with crotonaldehydes to afford the β-acetylenic alcohol **2** with high stereoselectivity. This alcohol was used in a stereocontrolled synthesis of the natural antibiotic (±)-asperlin (**6**).

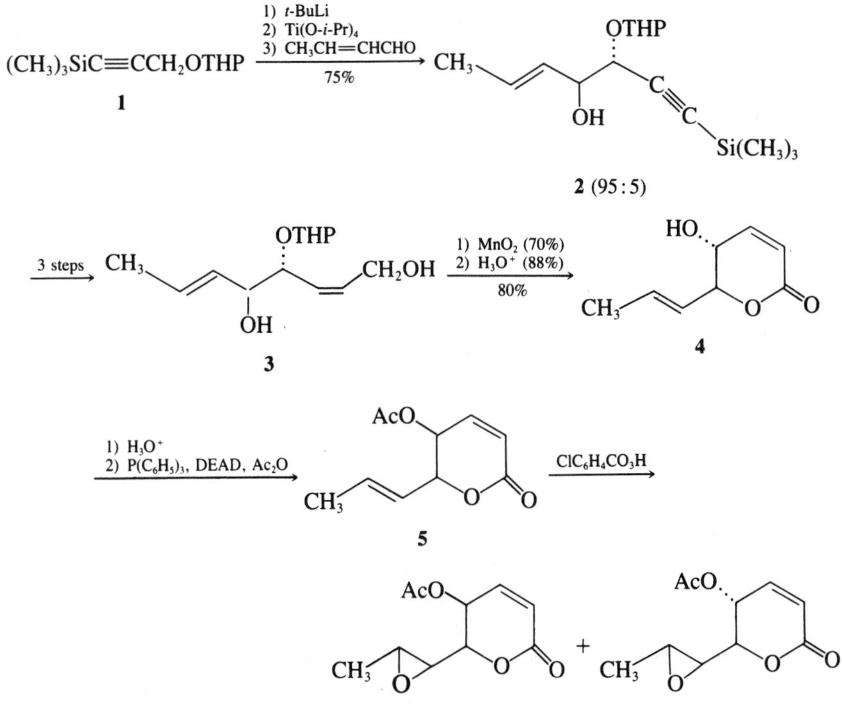

2 (95:5)

3

4

5

6 (39% from **4**)   **7** (8% from **4**)

N-Methyl-C-(trichlorotitanio)formimidoyl chloride, $\quad CH_3N{=}C\underset{TiCl_3}{\overset{Cl}{\diagdown}}$ (**1**). The

reagent is prepared *in situ* by reaction of methyl isocyanide with $TiCl_4$ in CH_2Cl_2. The reagent reacts with aldehydes or ketones to give an adduct that is converted into an α-hydroxy amide (equation I). Yields as high as 90% can be obtained with unhindered aldehydes or ketones, but are rather poor with α,β-unsaturated or hindered carbonyl compounds.[2]

$$
(I)\quad \underset{R^2}{\overset{R^1}{\diagdown}}C{=}O + \mathbf{1} \xrightarrow{CH_2Cl_2} \left[\underset{R^2}{\overset{R^1}{\diagdown}}C\underset{\underset{Cl}{\overset{|}{C}}{=}NCH_3}{\overset{OTiCl_3}{\diagup}} \right] \xrightarrow{H_3O^+} \underset{R^2}{\overset{R^1}{\diagdown}}C\underset{\underset{O}{\overset{\|}{C}}NHCH_3}{\overset{OH}{\diagup}}
$$

Crotyltitanium compounds. The diastereoselectivity of the reaction of (E)-halo-bis(cyclopentadienyl)crotyltitanium reagents (**1**) with aldehydes is dependent in part on the halide ligand. When X = Br or I, *anti*-selectivity is highly favored, but when X = Cl, only slight *anti*-selectivity obtains.[3] However, in the presence of BF_3, the reaction is highly *syn*-selective regardless of the halogen ligand.[4] The selectivity thus corresponds to that observed in the BF_3-catalyzed reaction of tri-*n*-butylcrotyltin (**10**, 411). Surprisingly, BF_3 does not reverse the *anti*-selectivity of crotyltris(isopropoxy)titanium.

$$(CH_3)_2CHCHO + \underset{CH_3}{\diagdown}\diagup\diagdown\diagup TiCp_2X \longrightarrow$$

1, X = Br	87%
+ BF_3	75%

$$(CH_3)_2CH\diagup\overset{OH}{\underset{\overset{|}{CH_3}}{\diagdown}}\diagup\diagdown CH_2 + (CH_3)_2CH\diagup\overset{OH}{\underset{\overset{|}{CH_3}}{\diagdown}}\diagup\diagdown CH_2$$

$$99:1$$
$$9:91$$

Ti(NR$_2$)$_4$ (**1**).[5] $Ti(NR_2)_4$, R = CH_3 or C_2H_5, is an excellent reagent for *in situ* protection of aldehydes and ketones by formation of adducts that revert to the original carbonyl compounds on aqueous work-up. Both **1a**, R = CH_3, and **1b**, R = C_2H_5, react more readily with aldehydes than with ketones, but **1a** is more reactive in general than **1b**. Thus selective reactions can be conducted on a ketone group in the presence of an aldehyde group (equation I). The method can also be used to carry out selective reaction at the more hindered of two carbonyl groups by selective protection of the less hindered carbonyl group with **1b**.

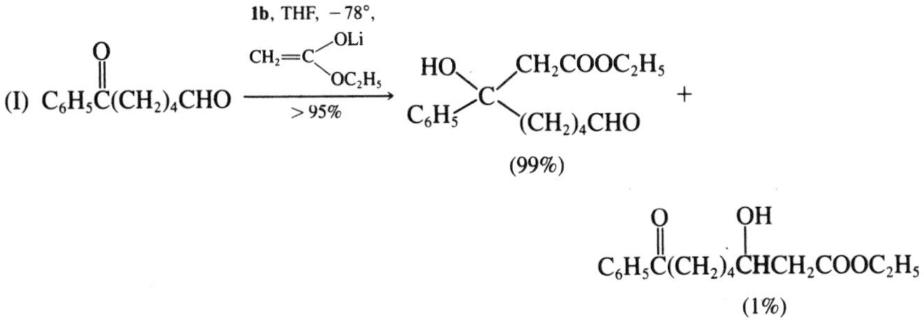

$$\text{(I)}\quad C_6H_5\overset{O}{\overset{\|}{C}}(CH_2)_4CHO \quad\xrightarrow[>95\%]{\substack{\textbf{1b, THF, } -78°,\\ CH_2=C\overset{\displaystyle OLi}{\underset{\displaystyle OC_2H_5}{}}}}\quad \underset{C_6H_5}{\overset{HO}{>}}C\overset{CH_2COOC_2H_5}{\underset{(CH_2)_4CHO}{<}} \quad +$$

(99%)

$$C_6H_5\overset{O}{\overset{\|}{C}}(CH_2)_4\overset{OH}{\underset{|}{C}}HCH_2COOC_2H_5$$

(1%)

Chiral 1,3-diols.[6] Chiral β-alkoxy aldehydes react with compounds of the type RTiCl$_3$ at $-78°$ to form derivatives of 1,3-diols with high 1,3-asymmetric induction (equation I). High 1,3-asymmetric induction is also observed in aldol condensations of these aldehydes catalyzed by TiCl$_4$ (equations II and III).

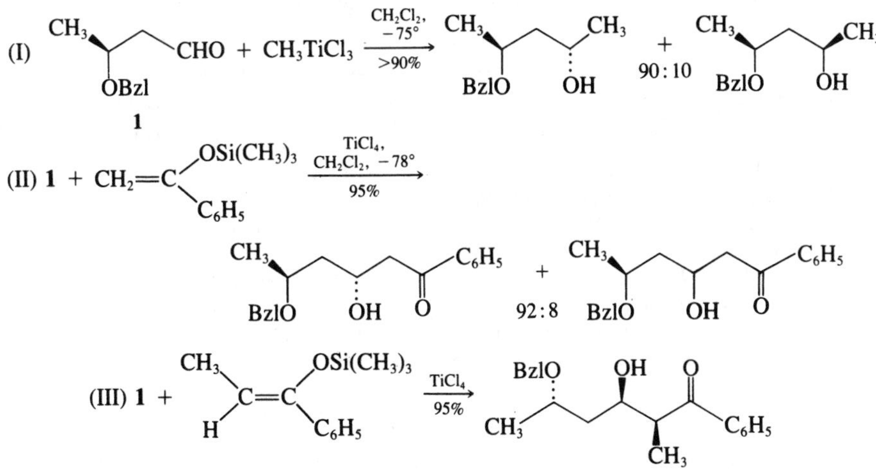

Cleavage of acetals.[7] The acetals derived from (2R,4R)-2,4-pentanediol (this volume) are cleaved by organotitanium reagents of the type RTiCl$_3$ or R$_2$TiCl$_2$ with high chemo- and stereoselectivity. Removal of the chiral auxiliary gives chiral secondary alcohols in high purity. Acetals complexed with TiCl$_4$ are also cleaved by treatment with an alkyllithium.

Example:

(96:4)

(89:11)

Review.[8]   A particularly interesting property of reagents of the type R^{1}Ti(OR^2)$_3$ is their ability to react at a much faster rate with aldehydes than with ketones. The selectivity can be further increased by an increase in the bulk of the R^2 group. Ordinarily titanium(IV) n-alkoxides exist as tetramers. However, aggregate formation is decreased by steric hindrance at the metal center, and t-butyl titanates exist as monomers, which are more reactive. Another advantage of these reagents over RMgBr is that chlorinated solvents, acetonitrile, or pyridine can be used. The very useful diastereoselective addition of allylic organotitanium reagents to ketones and of chiral RTi(OR*)$_3$ to aldehydes is discussed.

[1] H. Hiraoka, K. Furuta, N. Ikeda, and H. Yamamoto, *Bull. Chem. Soc. Japan*, **57**, 2777 (1984).
[2] M. Schiess and D. Seebach, *Helv.*, **66**, 1618 (1983).
[3] F. Sato, K. Iida, S. Iijima, H. Moriya, and M. Sato, *J.C.S. Chem. Comm.*, 1140 (1981).
[4] M. T. Reetz and M. Sauerwald, *J. Org.*, **49**, 2292 (1984).
[5] M. T. Reetz, B. Wenderoth, and R. Peter, *J.C.S. Chem. Comm.*, 406 (1983).
[6] M. T. Reetz and A. Jung, *Am. Soc.*, **105**, 4833 (1983).
[7] A. Mori, K. Maruoka, and H. Yamamoto, *Tetrahedron Letters*, **25**, 4421 (1984).
[8] B. Weidmann and D. Seebach, *Angew. Chem., Int. Ed.*, **22**, 31 (1983).

Organozinc reagents.

Allenylzinc reagents.[1]   These reagents (**1**) are obtained by lithiation (t-BuLi) of monosubstituted allenes followed by transmetallation with ZnCl$_2$. They react with alde-

2 (*anti/syn* = 96–99:4–1)

hydes regio- and stereoselectively to form *anti*-homopropargylic alcohols. The related zinc reagent **3** also undergoes a similar reaction with aldehydes to form the *anti*-alcohols **4**.

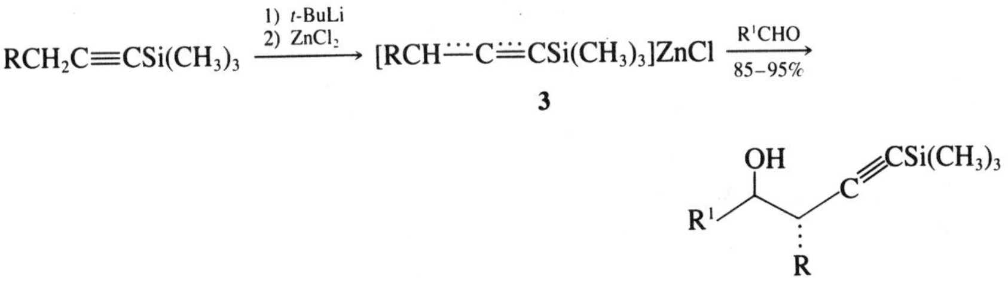

4 (*anti/syn* = 92–99 : 8–1)

R_2Zn. Dialkylzinc reagents can be prepared from an organic halide, lithium wire, and zinc bromide at 0° in toluene–THF by sonication in a cell-disruptor generator. Yields are essentially quantitative.[2] Diarylzinc reagents[3] and dimethylzinc[4] can be prepared by sonication in an ultrasonic laboratory cleaner.

These organozinc reagents undergo facile conjugate addition to α,β-enones in the presence of $Ni(acac)_2$ in generally good yield. Methylation occurs very easily; secondary and even tertiary groups also can be introduced.

Examples:

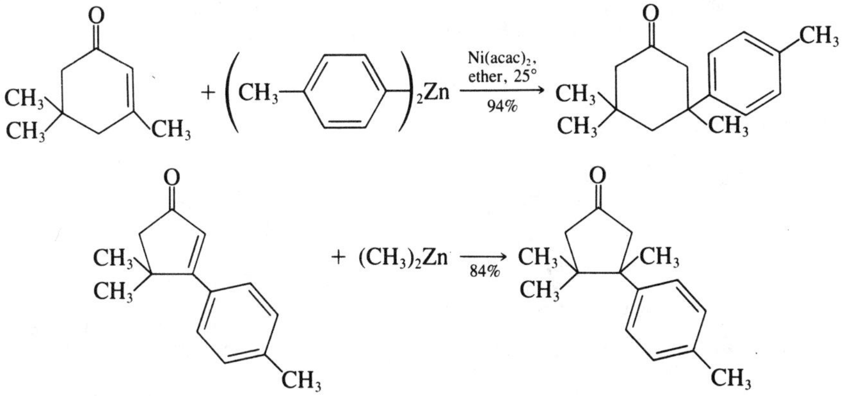

[1] G. Zweifel and G. Hahn, *J. Org.*, **49**, 4565 (1984).
[2] C. Petrier, J.-L. Luche, and C. Dupuy, *Tetrahedron Letters*, **25**, 3463 (1984).
[3] J.-L. Luche, C. Petrier, J.-P. Lansard, and A. E. Greene, *J. Org.*, **48**, 3837 (1983).
[4] A. E. Greene, J. P. Lansard, J.-L. Luche, and C. Petrier, *ibid.*, **49**, 931 (1984).

Organozirconium reagents, 11, 378–379.

Review.[1] In general, these reagents are less basic than the corresponding organo-titanium compounds. Thus they add satisfactorily to easily enolizable ketones such as α- and β-tetralones. They can transfer even the *t*-butyl group to carbonyl compounds. Vinylzirconium compounds are sufficiently stable to add to carbonyl groups (equation I).

$$(I) \quad C_6H_5CHO + CH_3(CH_2)_3CH{=}CHZr(OBu)_3 \xrightarrow[78\%]{} C_6H_5\overset{\displaystyle OH}{\overset{|}{C}}H{-}CH{=}CH(CH_2)_3CH_3$$

[1] B. Weidmann and D. Seebach, *Angew. Chem., Int. Ed.*, **22**, 31 (1983).

Osmium tetroxide, 1, 759–767; **10**, 290.

Oxidation of chiral allylic alcohols. Based on a comprehensive examination of the osmium tetroxide oxidation of a variety of allylic alcohols, Kishi *et al.*[1] have suggested an empirical guide for predicting the favored stereochemistry. The relative configuration between the original hydroxyl or alkoxyl group and the adjacent introduced hydroxyl group of the major product is *syn*. The stereoselectivity ranges from about 3:1 to 11:1 and is higher for *cis*-olefins than for the corresponding *trans*-olefins. If the allylic alcohol exists in conformation **A**, then osmium tetroxide selectively attacks the side of the double bond opposite to the hydroxyl group.

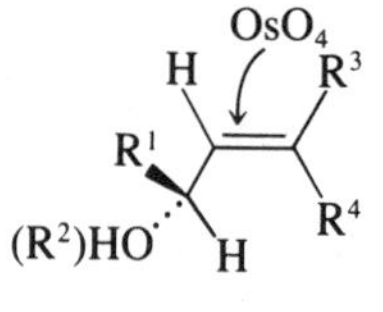

A

Examples:

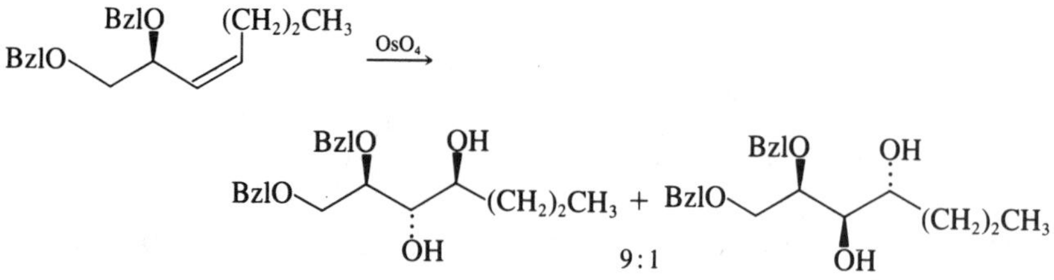

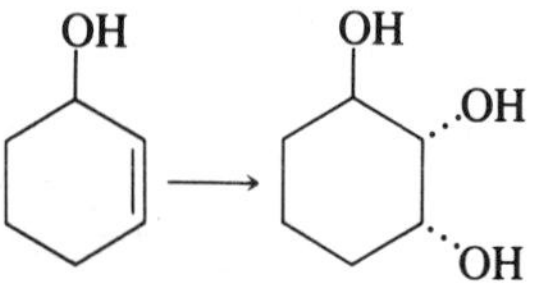

Stork and Kahn[2] have reported similar highly stereoselective OsO_4-catalyzed hydroxylation of the γ-hydroxy α,β-unsaturated esters **1** and **3**. The corresponding acids exhibit lower stereoselectivity.

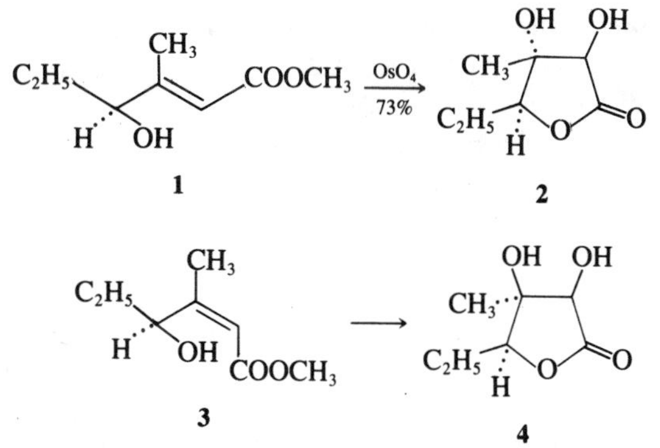

[1] J. K. Cha, W. J. Christ, and Y. Kishi, *Tetrahedron Letters*, **24**, 3943 (1983); W. J. Christ, J. K. Cha, and Y. Kishi, *ibid.*, 3947 (1983).
[2] G. Stork and M. Kahn, *ibid.*, 3951 (1983).

2-Oxazolidones, chiral, **11**, 379–381.

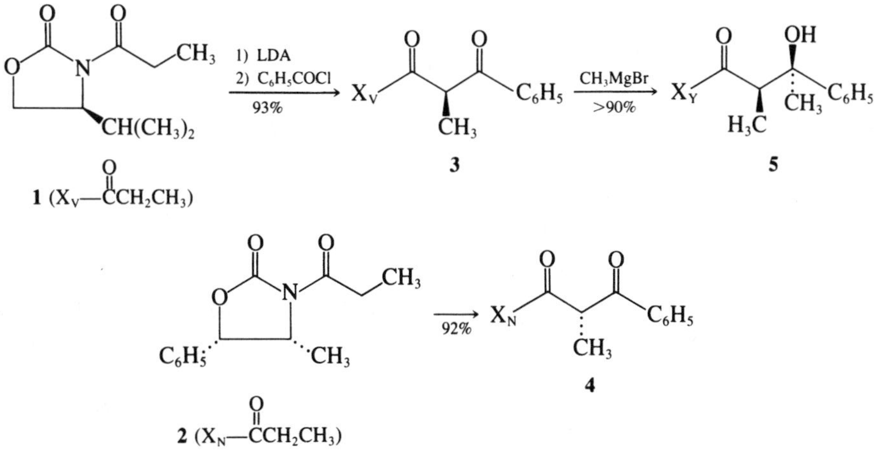

Chiral β-keto imides.[1]   The (Z)-enolates of chiral propionimides **1** and **2**, derived from valine (X_V) and norephedrine (X_N), respectively, undergo highly diastereoselective acylation (usually >95% de) to give chiral β-keto imides (**3** and **4**) as the major products. Surprisingly, the newly generated asymmetric center in **3** and **4** is relatively stable to acids and mild bases because of steric effects. These β-keto imides undergo highly diastereoselective additions to the carbonyl group. Thus **3** reacts with methylmagnesium bromide to give **5** in 90% yield. It is reduced by $Zn(BH_4)_2$ in the same sense to a single β-hydroxy imide in >95% yield. In contrast, reduction with $NaBH_4$ is nonstereoselective. Evidently chelation by a metal ion is crucial for diastereoface selection.

Asymmetric Diels-Alder reactions.   Chiral α,β-unsaturated N-acyl oxazolidones exhibit high diastereoface selection in Diels-Alder reactions, particularly those conducted in the presence of diethylaluminum chloride (1.2 equiv.).

In reactions of the chiral acrylate and crotonate imides with cyclopentadiene, *endo*-products are formed almost exclusively, with diastereoselection of about 95%.[2]

Examples:

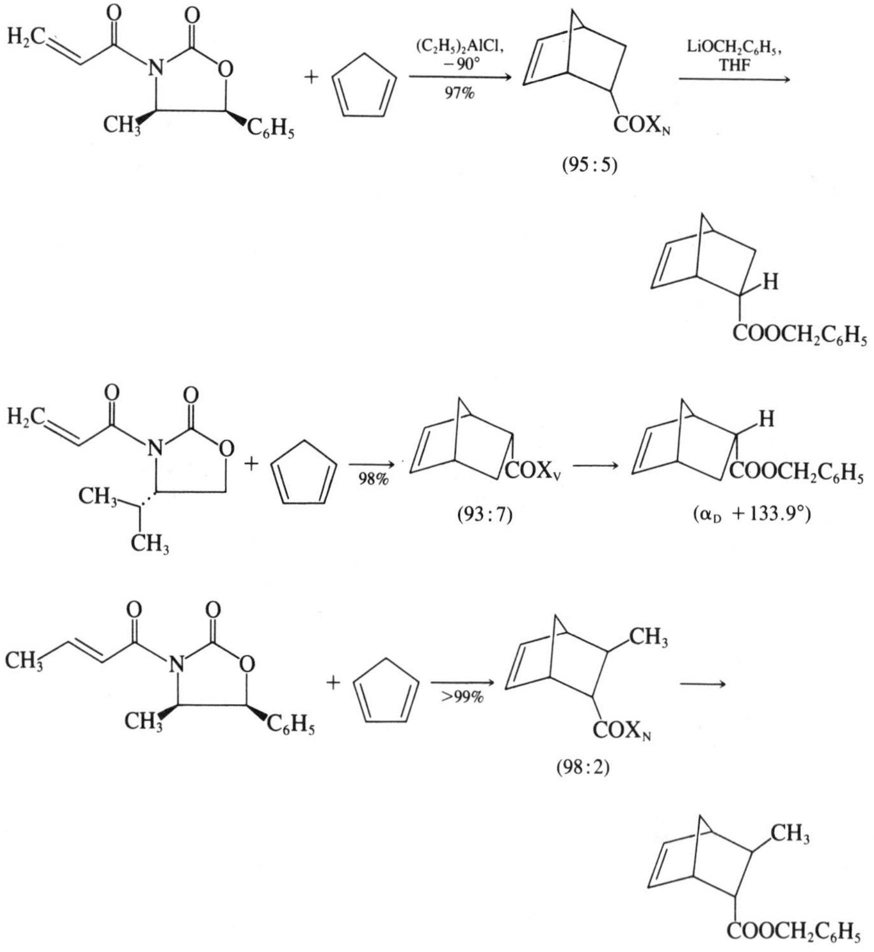

In reactions with acyclic dienes, the stereoselection with the (S)-valinol-derived crotonate imides is only about 3:1, but diastereoselection is high with N-acyl oxazolidines
derived from phenylalanine. The chiral auxiliary is cleaved by transesterification with
lithium benzyloxide in 85–95% yield.[3]

Examples:

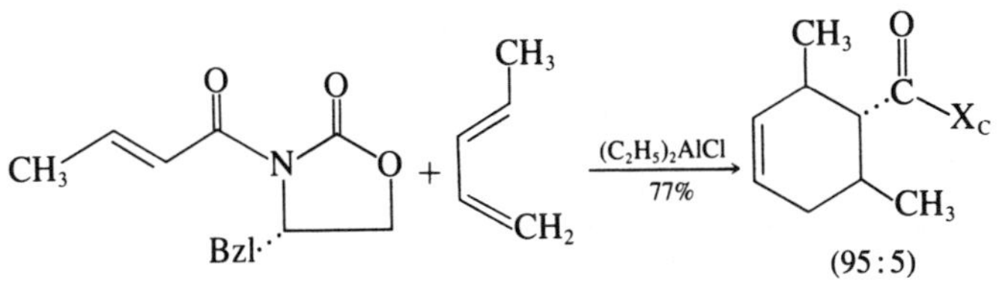

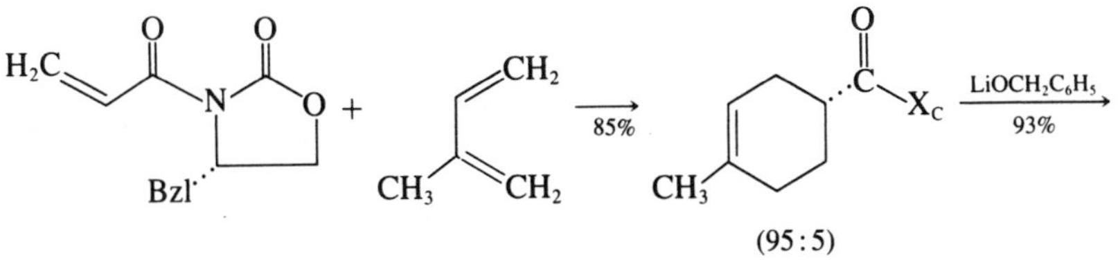

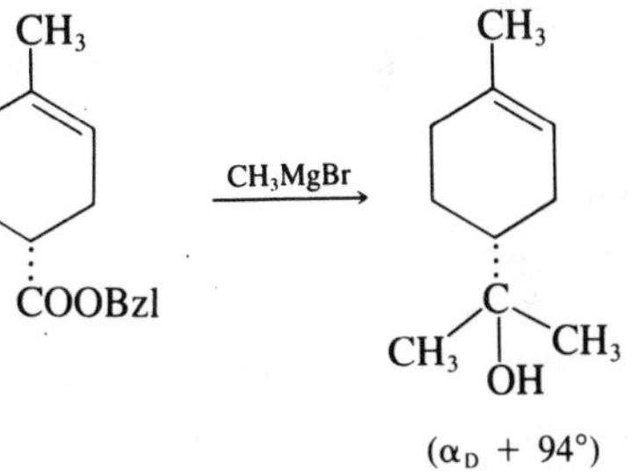

The absolute stereochemistry of the cycloaddition was confirmed by a synthesis of α-
terpinol (second example).

Diastereofacial-selective intramolecular Diels-Alder reactions.[4] Trienecarbox-
imides derived from chiral oxazolidones undergo Lewis-acid-catalyzed intramolecular
Diels-Alder reactions with high *endo*- and diastereofacial selectivity (*endo/exo* ≈100:1).

The diastereofacial selectivity is controlled by the asymmetric center at C_4 of the chiral auxiliary (Scheme I).

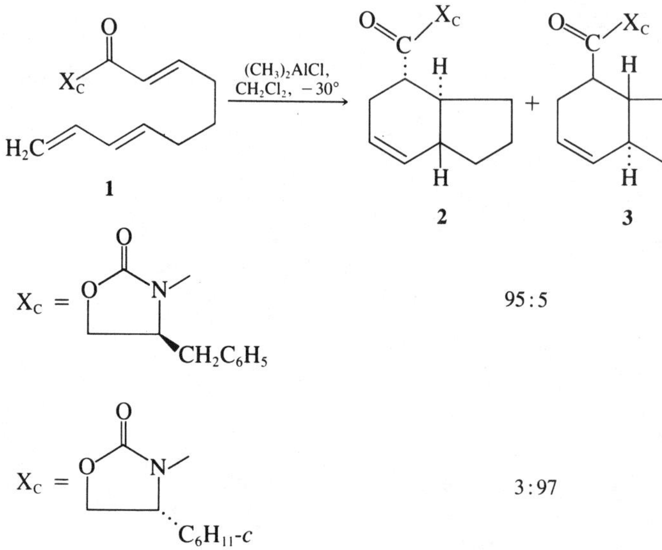

Scheme (I)

[1] D. A. Evans, M. D. Ennis, T. Le, N. Mandel, and G. Mandel, *Am. Soc.*, **106**, 1154 (1984).
[2] D. A. Evans, K. T. Chapman, and J. Bisaha, *ibid.*, **106**, 4261 (1984).
[3] D. A. Evans, M. D. Ennis, and D. J. Mathre, *ibid.*, **104**, 1737 (1982).
[4] D. A. Evans, K. T. Chapman, and J. Bisaha, *Tetrahedron Letters*, **25**, 4071 (1984).

Oxygen, 4, 362; **5**, 482–486; **6**, 426–430; **7**, 258–260; **8**, 366–367; **9**, 335–337; **10**, 293–294; **11**, 384–385.

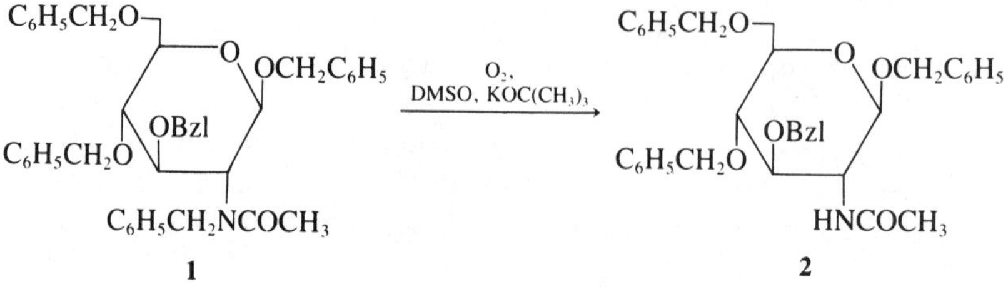

N-Debenzylation.[1] The β-D-glucopyranoside **1**, when dissolved in DMSO containing KOC(CH$_3$)$_3$, is rapidly converted into the N-acetamide derivative **2** when air is passed into the stirred solution. This conversion of an N-benzylacetamido group into an acetamido group is a general reaction. It can also be used to prepare acetylated amines. Thus alkylation of N-benzylacetamide with dodecyl bromide (NaH, DMF) results in N-acetyl-N-benzyldodecylamine, which can be debenzylated by this reaction to N-acetyldodecylamine.

Isoalloxazines.[2]   The pyrimidinediones **1**, available by condensation of anilines with 5-bromo-6-alkylaminopyrimidinediones, cyclize to isoalloxazines (**2**) when heated under oxygen at 120° in DMF, HMPT, or lutidine. The intermediate **a** has been isolated.

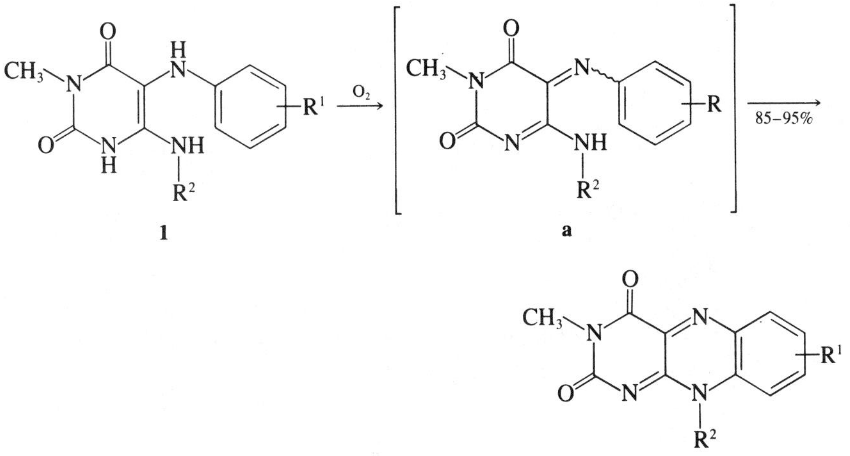

[1] R. Gigg and R. Conant, *J.C.S. Chem. Comm.*, 465 (1983).
[2] M. Sako, Y. Kojima, K. Hirota, and Y. Maki, *ibid.*, 1691 (1984).

Oxygen, singlet, **4**, 362–363; **5**, 486–491; **6**, 431–436; **7**, 261–269; **8**, 367–374; **9**, 338–341; **10**, 294–295; **11**, 385–387.

α,β-Enones.[1]   Photooxygenation (*meso*-tetraphenylporphine) of cycloalkenes followed by *in situ* acetylation of the resulting hydroperoxide produces cycloalkenones directly. The method is also applicable to acyclic alkenes.

Examples:

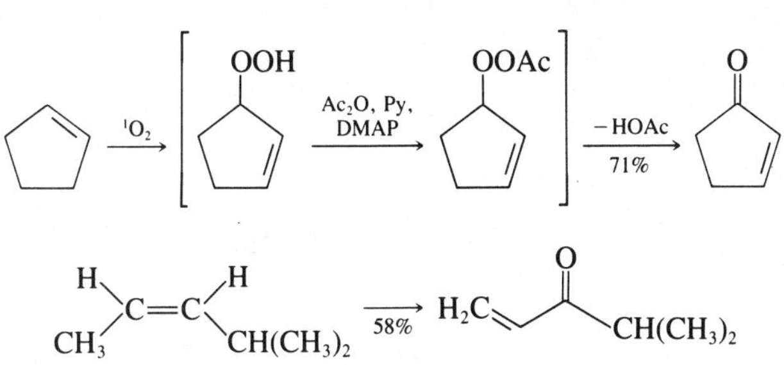

cis-*Enediones*.[2]   The *endo*-peroxides obtained by photosensitized oxidation of furanes **1** are usually unstable and difficult to isolate in pure form. However, they can be

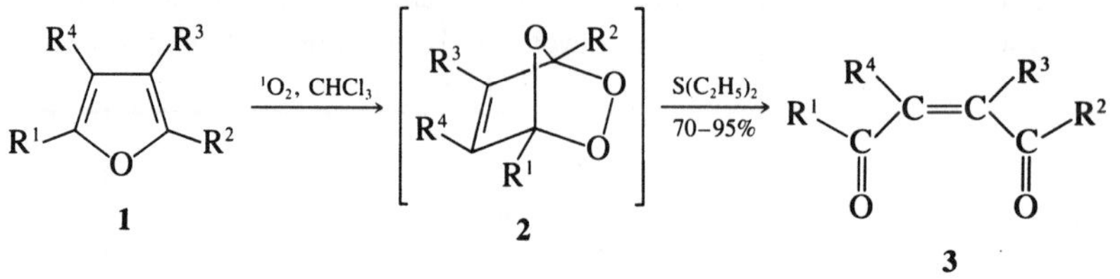

reduced *in situ* by diethyl sulfide (which is oxidized to diethyl sulfoxide) to *cis*-enediones in generally high yield.

Cleavage of imidazoles; dehydroamino acids.[3]   2,4-Disubstituted imidazoles[4] undergo a Diels-Alder-like reaction with singlet oxygen in the presence of DBU (1–2 equiv.) to provide imine diamides, which isomerize to dehydroamino acid derivatives in the presence of base. Hydrogenation in the presence of catalysts with chiral phosphine ligands results in optically active amino acid diamides. The overall process is illustrated for the synthesis of the N-benzylamide of N-acetylleucine (equation I).

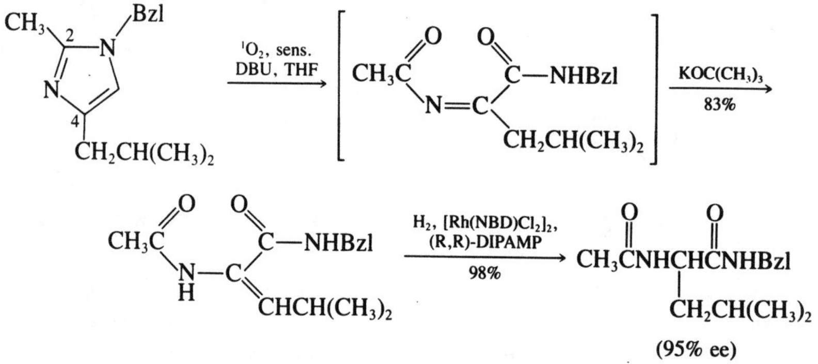

vic-*Tricarbonyl compounds*.[5]   Singlet oxygen is more efficient than ozone for cleavage of the enamines **2**, obtained by reaction of a β-dicarbonyl compound (**1**) with dimethylformamide dimethyl acetal.

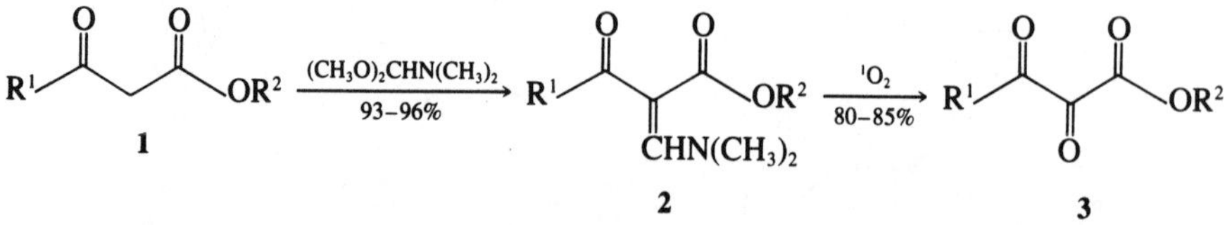

This reaction provides a novel access to carbacephams (equation I).

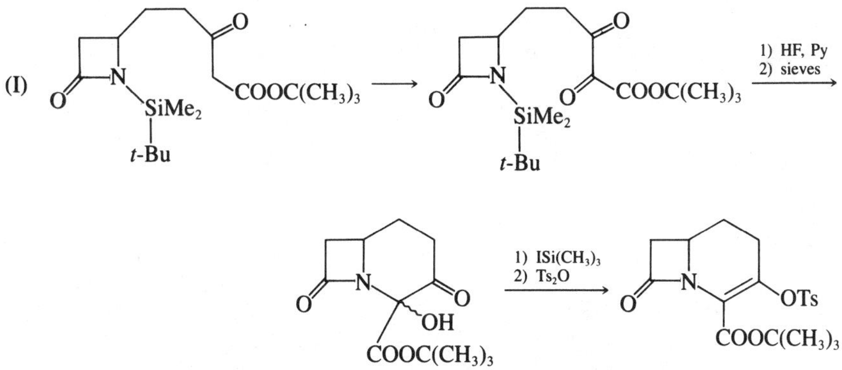

δ-*Lactones.*[6] Photosensitized (methylene blue) oxidation of 3-alkyl-1,2-cyclohex-anediones (**1**) results in δ-keto carboxylic acids (**2**) in high yield. They are converted by reduction and lactonization into δ-lactones (**3**). The method is attractive for substrates bearing labile, unsaturated side chains.

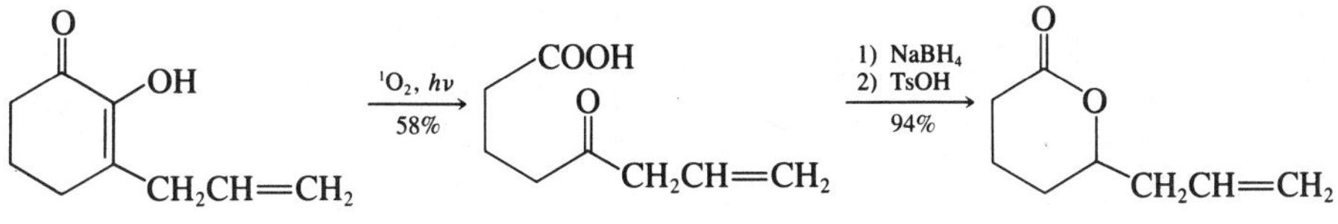

[1] E. D. Mihealich and D. J. Eickhoff, *J. Org.*, **48**, 4135 (1983).
[2] M. L. Graziano, M. R. Iesce, B. Carli, and R. Scarpati, *Synthesis*, 125 (1983).
[3] B. H. Lipshutz and M. C. Morey, *Am. Soc.*, **106**, 457 (1984).
[4] *Idem, J. Org.*, **48**, 3745 (1983).
[5] H. H. Wasserman and W. T. Han, *Tetrahedron Letters*, **25**, 3743, 3747 (1984).
[6] M. Utaka, H. Kuriki, T. Sakai, and A. Takeda, *Chem. Letters*, 911 (1983).

Ozone, 1, 773–777; **4**, 363–364; **5**, 491–495; **6**, 436–441; **7**, 269–271; **8**, 374–377; **9**, 341–343; **10**, 295–296; **11**, 387–388.

Primary and secondary nitroalkanes.[1] A new route to nitro compounds involves conversion of azides to phosphine imines by reaction with triphenyl- or tri-*n*-butylphosphine at 25–35° followed by ozonolysis in CH_2Cl_2 at $-78°$. Presumably the initial ozonide is converted into a nitroso compound, which is then oxidized to the nitro compound (equation I). The method is limited to substrates that are stable to ozone. Benzylic azides, however, are converted mainly to benzaldehydes under these conditions.

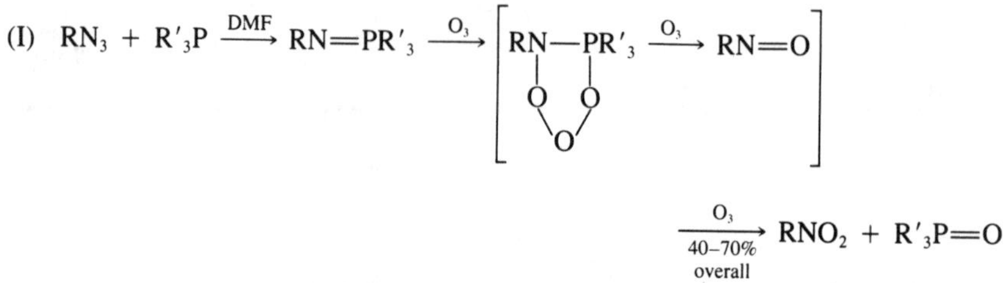

$$\text{(I)} \quad RN_3 + R'_3P \xrightarrow{\text{DMF}} RN{=}PR'_3 \xrightarrow{O_3} \left[\underset{\underset{O}{\overset{|}{\underset{\diagdown \diagup}{O \qquad O}}}}{RN{-}PR'_3} \xrightarrow{O_3} RN{=}O \right]$$

$$\xrightarrow[\substack{40-70\% \\ \text{overall}}]{O_3} RNO_2 + R'_3P{=}O$$

The method is particularly useful for preparation of 6-nitrogalactopyranosides, which cannot be prepared by the usual Kornblum route (S_N2 displacement of I by $AgNO_2$).[2]

Criegee rearrangement[3] of α-alkoxy hydroperoxides.[4]   The α-alkoxy hydroperoxides formed by ozonolysis in CH_3OH rearrange on treatment with Ac_2O to esters or lactones.

Examples:

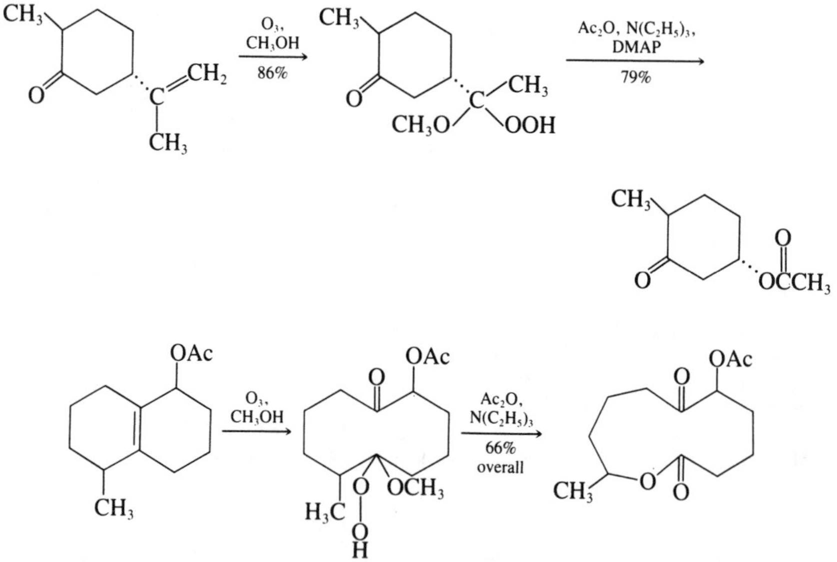

[1] E. J. Corey, B. Samuelsson, and F. A. Luzzio, *Am. Soc.*, **106**, 3682 (1984).
[2] N. Kornblum, *Org. React.*, **12**, 101 (1962).
[3] R. Criegee, *Ber.*, **77**, 722 (1944).
[4] S. L. Schreiber and W.-F. Liew, *Tetrahedron Letters*, **24**, 2363 (1983).

P

Palladium(II) acetate, 1, 778; **2**, 203; **4**, 365; **5**, 496–497; **6**, 442–443; **7**, 274–277; **8**, 378–382; **9**, 297–298; **11**, 389–392.

1,4-*Diacetoxylation of* 1,3-*dienes*.[1] Palladium-catalyzed oxidation of 1,3-cyclohexadiene with benzoquinone (used in catalytic amounts with MnO_2 as the external oxidant) in acetic acid gives a 1 : 1 mixture of *cis*- and *trans*-1,4-diacetoxy-2-cyclohexene. Addition of LiCl or LiOAc has a profound effect on the stereochemistry. Oxidation in the presence of lithium acetate results in selective *trans*-diacetoxylation, whereas addition of lithium chloride results in selective *cis*-diacetoxylation (equation I).[2]

(I)

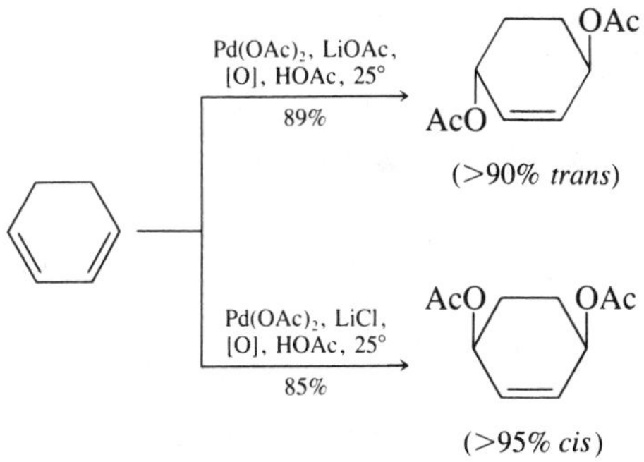
Pd(OAc)₂, LiOAc,
[O], HOAc, 25°
89%
OAc
AcO
(>90% trans)
AcO
OAc
Pd(OAc)₂, LiCl,
[O], HOAc, 25°
85%
(>95% cis)

This diacetoxylation reaction is general for cyclic and acyclic dienes. In the latter reaction, the new double bond has the (E)-configuration.[3]

Examples:

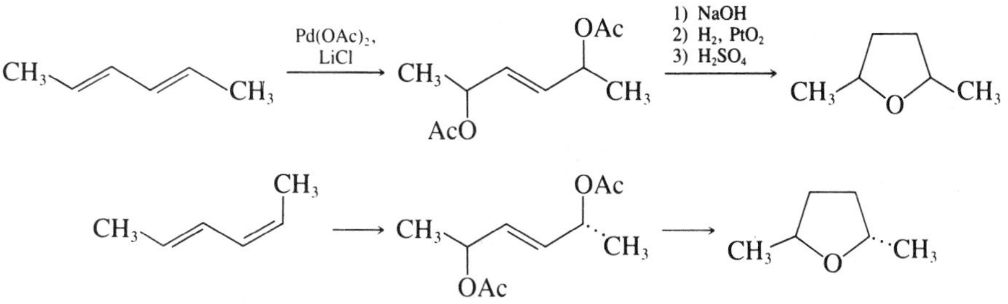
Pd(OAc)₂,
LiCl
CH₃
CH₃
OAc
AcO
1) NaOH
2) H₂, PtO₂
3) H₂SO₄
O
OAc

If the reaction is carried out in acetic acid containing excess lithium chloride, *cis*-1,4-chloro acetates become the only products.[3] This reaction is useful because the chloro and acetoxy groups can undergo selective consecutive substitution reactions.

Examples:

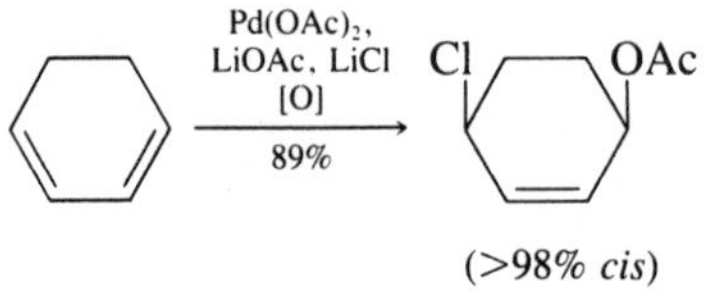

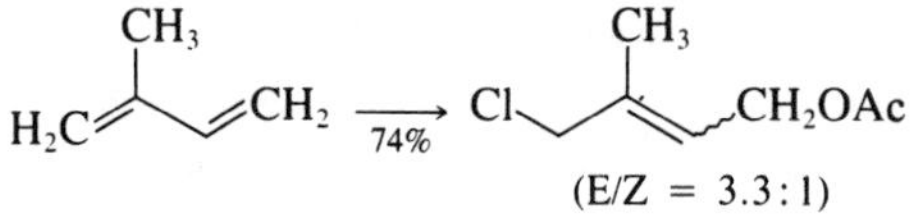

(>98% cis)

(E/Z = 3.3 : 1)

1,4-Acetoxytrifluoroacetoxylation of 1,3-dienes.[4] Oxidation of a 1,3-diene with MnO_2/p-benzoquinone catalyzed by Pd(OAc)$_2$ in acetic acid containing CF_3COOH/CF_3COOLi results in 1,4-addition of acetoxy and trifluoroacetoxy groups. The unsymmetrical product is obtained selectively from (E,Z)-2,4-hexadiene and cyclic dienes. The overall addition is *trans* with cyclohexadiene and cyclooctadiene, but *cis* with cycloheptadiene. The trifluoroacetoxy group is selectively hydrolyzed by Na_2CO_3 in aqueous methanol.

Examples:

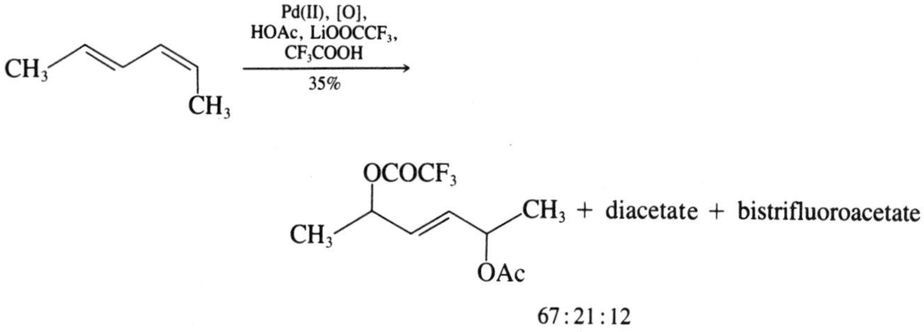

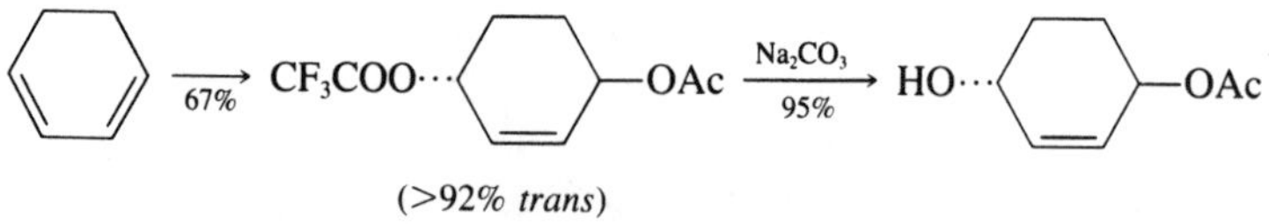

+ diacetate + bistrifluoroacetate

67 : 21 : 12

(>92% trans)

Vinylation of alkyl or aryl halides.[5] The Pd-catalyzed vinylation of organic halides[6] can proceed at room temperature when conducted under solid–liquid phase-transfer conditions with Bu_4NCl as catalyst and $NaHCO_3$ as base.

Examples:

$$(E)\text{-}C_4H_9CH\!=\!CHI \ + \ CH_2\!=\!CHCO_2CH_3 \xrightarrow[85\%]{\substack{Pd(OAc)_2,\ cat.,\\ NaHCO_3,\ DMF}} C_4H_9CH\!=\!CHCH\!=\!CHCO_2CH_3$$

$$(E,E/E,Z \ = \ 80:5)$$

$$C_6H_5I \ + \ CH_2\!=\!CHCH_2OH \xrightarrow[91\%]{} C_6H_5CH_2CH_2CHO$$

β-Aryl aldehydes (**7**, 274).[7]   These aldehydes can be obtained by reaction of allylic alcohols with aryl halides and a base (usually triethylamine) in the presence of Pd(OAc)₂ (equation I).

$$(\text{I}) \ C_6H_5I \ + \ CH_2\!=\!\overset{\overset{\displaystyle CH_3}{|}}{C}CH_2OH \ + \ N(C_2H_5)_3 \xrightarrow[82\%]{\substack{Pd(OAc)_2,\\ CH_3CN}} C_6H_5CH_2\overset{\overset{\displaystyle CH_3}{|}}{C}HCHO \ + \ (C_2H_5)_3\overset{+}{N}HI^-$$

$$(90\% \ \text{pure})$$

Hydrogenolysis of allylic acetates.[8]   Allylic acetates (or carbonates or chlorides) are converted into terminal alkenes by ammonium formate using a Pd(II) catalyst complexed with a trialkylphosphine, particularly P(n-Bu)₃.
   Example:

$$CH_2\!=\!CH(CH_2)_3CH\!=\!CHCH_2OAc$$

or

$$CH_2\!=\!CH(CH_2)_3CH(OAc)CH\!=\!CH_2 \xrightarrow[\hspace{2cm}]{\substack{HCO_2NH_4,\\ Pd(II),\ P(n\text{-}Bu)_3}} CH_2\!=\!CH(CH_2)_4CH\!=\!CH_2$$

Enones (cf. **11**, 391–392).[9]   Pd(OAc)₂ is the preferred catalyst for dehydrogenation of alkenyl allyl carbonates to give α,β-enals or -enones. Use of a phosphine ligand can catalyze allylation/protonation in addition to β-elimination.
   Example:

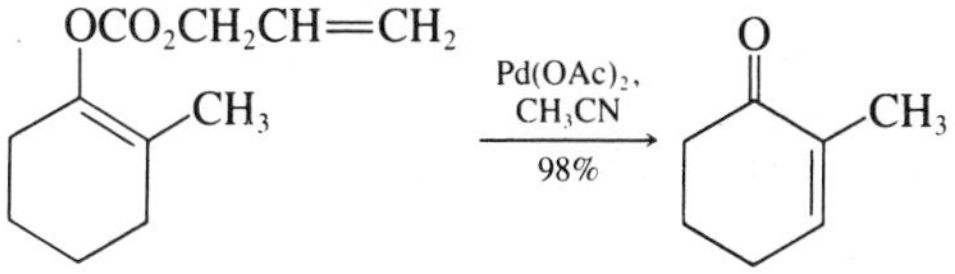

A related reaction can be used to convert silyl enol ethers or enol acetates into enones. Phosphine-free Pd(OAc)₂ is also the preferred catalyst for these dehydrogenations.

Examples:

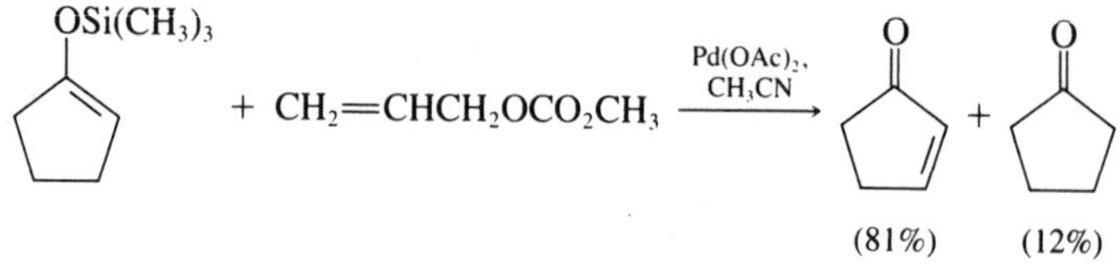

(81%)   (12%)

73%

ortho-*Alkylation of acetanilides*.[10]   Reaction of acetanilide with a slight excess of Pd(OAc)$_2$ in HOAc (3 hours) followed by addition of an alkyl halide results in exclusive *ortho*-alkylation. Reaction with an excess of Pd(OAc)$_2$ and the alkyl halide results in exclusive di-*ortho*-substitution. The reaction involves formation of an *ortho*-metallated acetanilide that can be isolated and that undergoes facile alkylation.

Example:

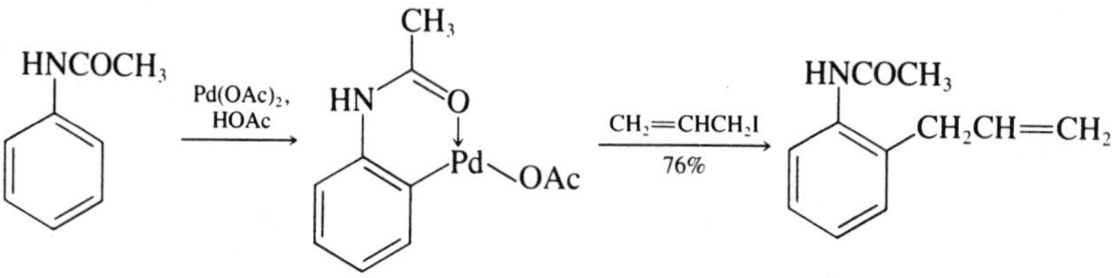

[1] J.-E. Bäckvall, *Pure Appl. Chem.*, **55**, 1669 (1983).
[2] J.-E. Bäckvall and R. E. Nordberg, *Am. Soc.*, **103**, 4959 (1981).
[3] J.-E. Bäckvall, R. E. Nordberg, and S. E. Byström, *Tetrahedron Letters*, **23**, 1617 (1982); *idem*, *J. Org.*, **49**, 4619 (1984).
[4] J.-E. Bäckvall, J. Vågberg, and R. E. Nordberg, *Tetrahedron Letters*, **25**, 2717 (1984).
[5] T. Jeffery, *J.C.S. Chem. Comm.*, 1287 (1984).
[6] R. F. Heck, *Org. React.*, **27**, 345 (1982).
[7] S. A. Buntin and R. F. Heck, *Org. Syn.*, **61**, 82 (1983).
[8] J. Tsuji, I. Shimizu, and I. Minami, *Chem. Letters*, 1017 (1984).
[9] J. Tsuji, I. Minami, I. Shimizu, and H. Kataoka, *ibid.*, 1133 (1984).
[10] S. J. Tremont and H. U. Rahman, *Am. Soc.*, **106**, 5759 (1984).

Palladium(II) acetate–Sodium hydride–*t*-Amyl alkoxide.

Hydrogenation catalyst.[1]   A catalyst (Pdc) prepared from Pd(OAc)$_2$ (10 equiv.), *t*-AmOH (20 equiv.), and NaH (60 equiv.) in combination with quinoline effects highly selective semihydrogenation of alkynes to *cis*-alkenes. Pdc is superior to the Lindlar catalyst and comparable to a related nickel catalyst (Nic, **10**, 365) with respect to selectivity and stereoselectivity. A catalyst prepared from only Pd(OAc)$_2$ and NaH effects complete hydrogenation of alkynes to alkanes.

[1] J.-J. Brunet and P. Caubere, *J. Org.*, **49**, 4058 (1984).

Palladium(II) chloride, 1, 782; **3**, 303–305; **4**, 367–370; **5**, 500–503; **6**, 447–450; **7**, 277; **8**, 384–385; **9**, 352; **10**, 300–302; **11**, 393–395.

Allyl–allyl coupling.[1]   Coupling of a (π-allyl)palladium(II) complex with an alkenylzirconium species to give a 1,4-diene (**11**, 394) has been extended to a general synthesis of 1,5-dienes. Thus crotylpalladium chloride couples with crotylmagnesium chloride to give a bis(allylic)palladium(II) species from which the 1,5-diene is eliminated by addition of a π-acid ligand (maleic anhydride or benzoquinone) to give a mixture of 1,5-dienes in which the two products of head-to-head coupling predominate over that of head-to-tail coupling (equation I).

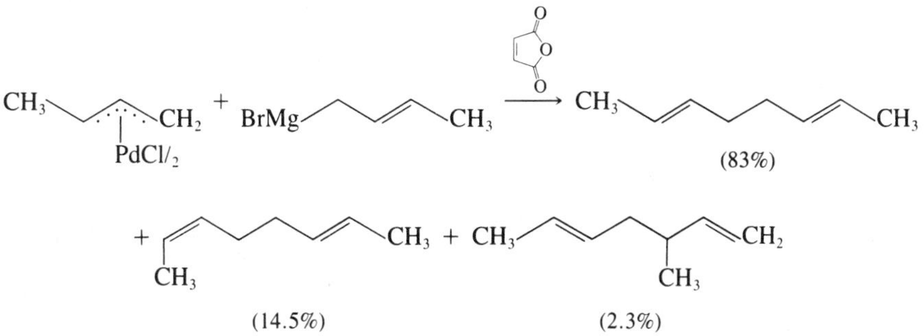

Cross-coupling of two unsymmetrical allylic complexes is possible by this transmetallation, particularly in the presence of maleic anhydride.

Example:

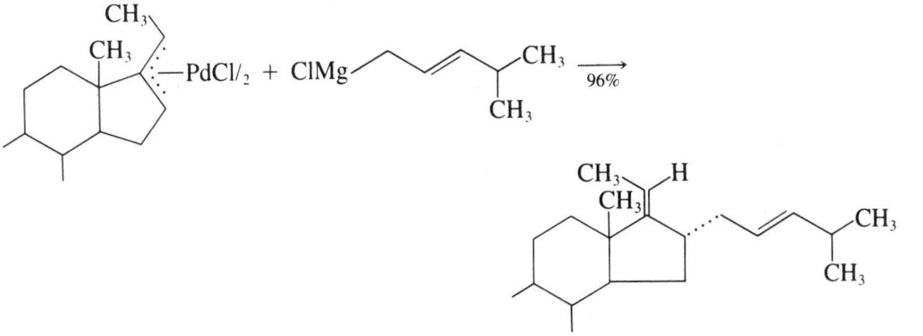

Cyclization of hydroxyalkenes to furanes and pyranes.[2]   Alkenes suitably substituted by a hydroxyl group undergo cyclization and carbonylation in the presence of $PdCl_2$/$CuCl_2$ to pyranes and/or furanes. The preference for five- or six-membered rings depends mainly on the geometry of the alkene; (E)-alkenes cyclize mainly to pyranes and (Z)-alkenes cyclize mainly to furanes.

Example:

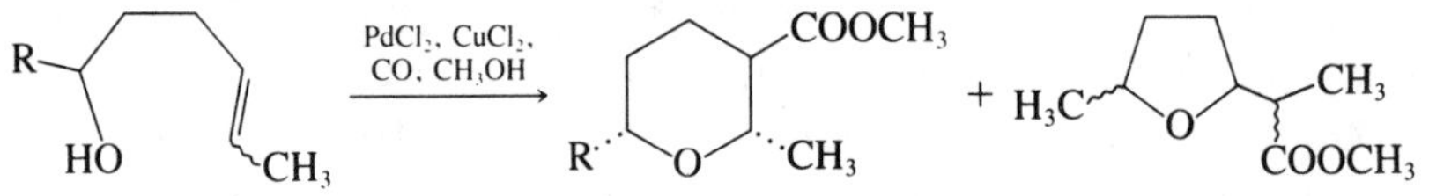

(E)	84%	70:30
(Z)	65%	15:85

This intramolecular alkoxypalladation/carbonylation has been used to synthesize the ester (**2**) of optically pure civet cat acid from **1**.

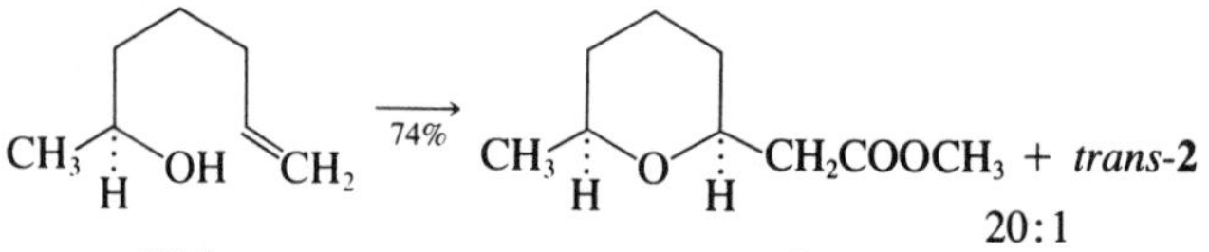

[1] A. Goliaszewski and J. Schwartz, *Am. Soc.*, **106**, 5028 (1984).
[2] M. F. Semmelhack and C. Bodurow, *ibid.*, 1496 (1984).

Palladium(II) chloride–Copper(I) chloride, 7, 278; 11, 396.

Acetalization of 1-alkenes.[1]   The Wacker conversion of 1-alkenes to methyl ketones by oxidation catalyzed by $PdCl_2$–$CuCl$ takes a different course when applied to vinyl ketones (**1**). Thus oxygenation of mixtures of **1** and 1,3- or 1,2-diols catalyzed by $PdCl_2$–$CuCl$ results in cyclic acetals formed by exclusive attack at the terminal carbon atom. A similar reaction occurs with 1-alkenes substituted with $COOCH_3$.

Examples:

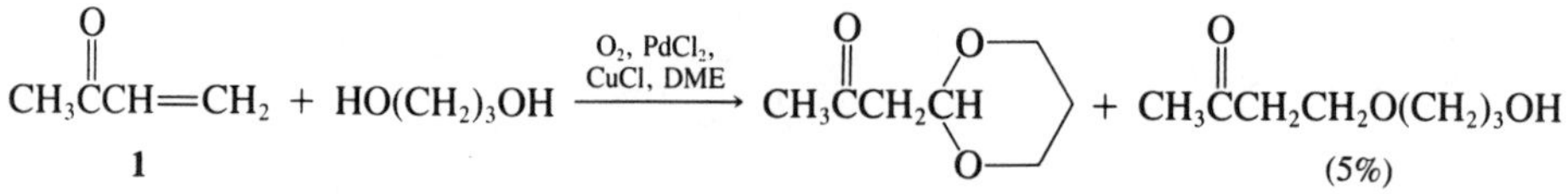

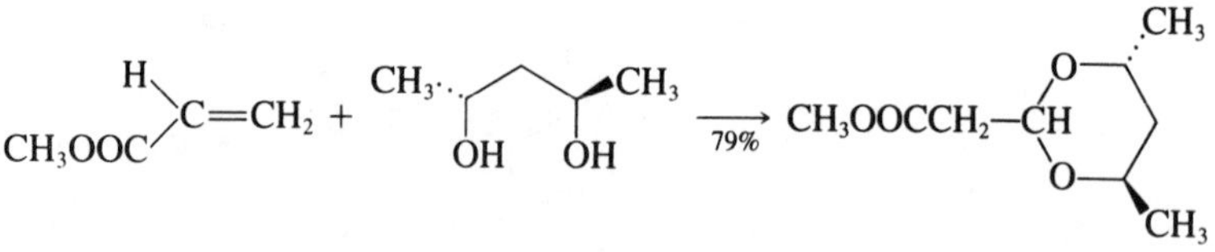

Monoesterification of HO(CH₂)ₙOH.[2]   Alkenes react with diols under oxidative carbonylation conditions catalyzed by $PdCl_2$ and $CuCl_2$ to give hydroxy esters (equation I). Internal alkenes show similar reactivity.

(I) $RCH{=}CH_2 + HO(CH_2)_nOH + CO \xrightarrow[\text{THF, HCl}]{PdCl_2,\ CuCl_2,\ O_2,} \underset{\text{major product}}{R\overset{\displaystyle CH_3}{\underset{|}{C}}HCO_2(CH_2)_nOH}$ + isomers

[1] T. Hosokawa, T. Ohta, and S.-I. Murahashi, *J.C.S. Chem. Comm.*, 848 (1983).
[2] S. B. Fergusson and H. Alper, *ibid.*, 1349 (1984).

Palladium(II) trifluoroacetate, 10, 302–303.

Allylic acetoxylation.[1]   A new method for regioselective allylic oxidation of geranylacetone (1) is based on oxidation of the π-allylpalladium complex known to be formed selectively with the terminal allylic methyl group. The most satisfactory results are obtained

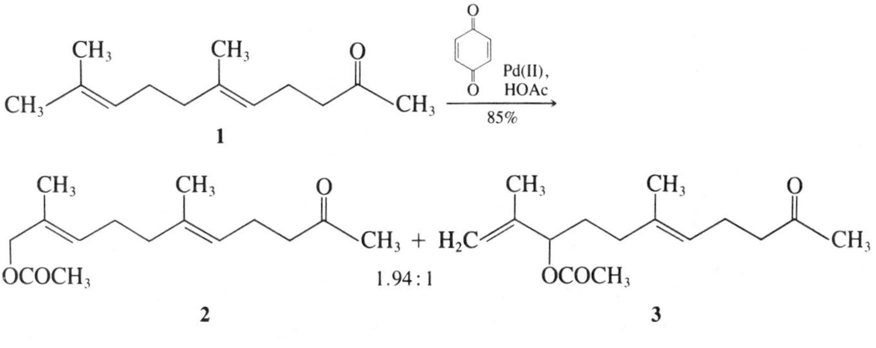

using a catalytic amount of $Pd(OCOCF_3)_2$, 1 equiv. of benzoquinone, an added ligand, *o*-methoxyacetophenone, and acetic acid as solvent. The usual oxidants are not useful, and the regioselectivity is markedly dependent on the choice of the quinone and the ligand. The method can be used with simple alkenes.

Example:

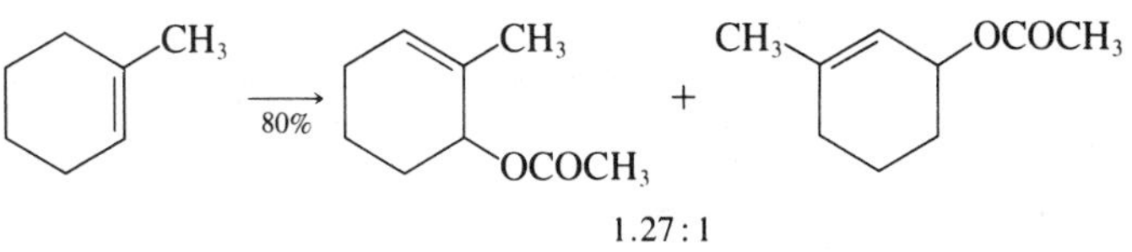

1.27 : 1

[1] J. E. McMurry and P. Kočovský, *Tetrahedron Letters*, **25**, 4187 (1984).

Pentadienyl-1-pyrrolidinecarbodithioate, H_2C ⎓⎓⎓SCN ⟩ (1). The reagent is prepared by reaction of pentadienyl bromide with sodium 1-pyrrolidinecarbo-
dithioate.

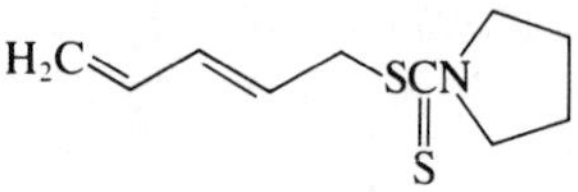

***All*-trans-*polyenes*.**[1] The anion of **1** is alkylated by a primary alkyl iodide mainly
at the α-position to give **2** with higher regioselectivity than that shown in alkylation of
pentadienyl anion. When heated, **2** undergoes a double [3.3] sigmatropic rearrangement
with 100% *trans*-selectivity to afford **3**. The newly formed α-methylene group can undergo

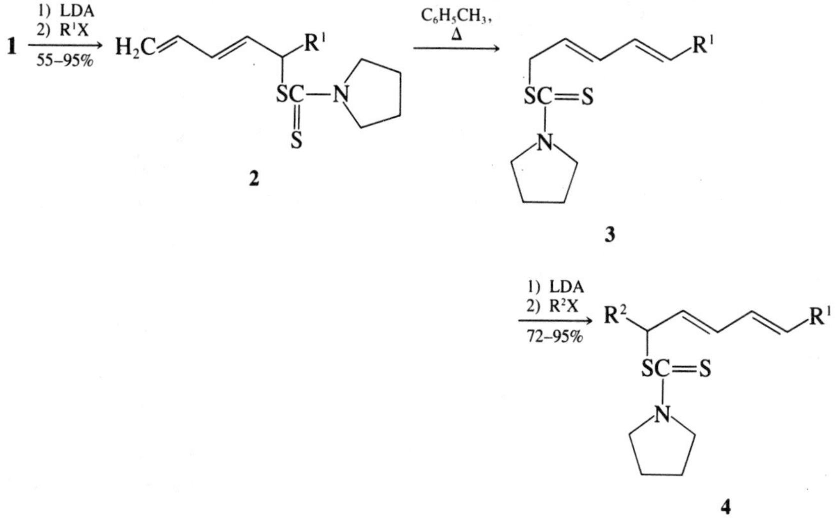

further alkylation and rearrangement. The α-selectivity decreases with increasing length
or branching of the chain. The method can be used to obtain all-*trans*-dienes, -trienes,
and -tetraenes. The thiocarbamoylthio group can be removed by desulfuration or by
reaction with CH_3I, LiF, and Li_2CO_3 in DMF.

Example:

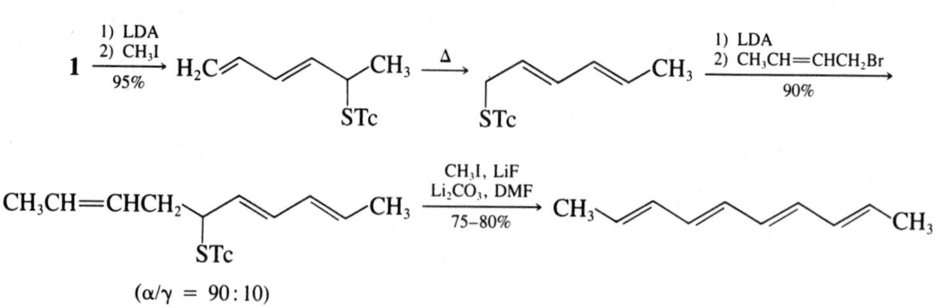

$(\alpha/\gamma = 90:10)$

[1] T. Hayashi, I. Hori, and T. Oishi, *Am. Soc.,* **105**, 2909 (1983).

(2R,4R)-Pentanediol, 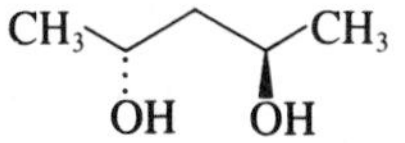 (1). Supplier: Aldrich.

Chiral homoallylic alcohols.   The chiral acetals **2** formed from an aldehyde and **1**, undergo titanium-catalyzed coupling with allyltrimethylsilane with marked stereoselectivity. Highest stereoselectivity is usually obtained with the mixed catalyst TiCl$_4$–Ti(O-*i*-Pr)$_4$ (6:5). Cleavage of the chiral auxiliary, effected by oxidation to the ketone followed by β-elimination, provides optically active alcohols (**4**) with ~95% ee (equation I).[1]

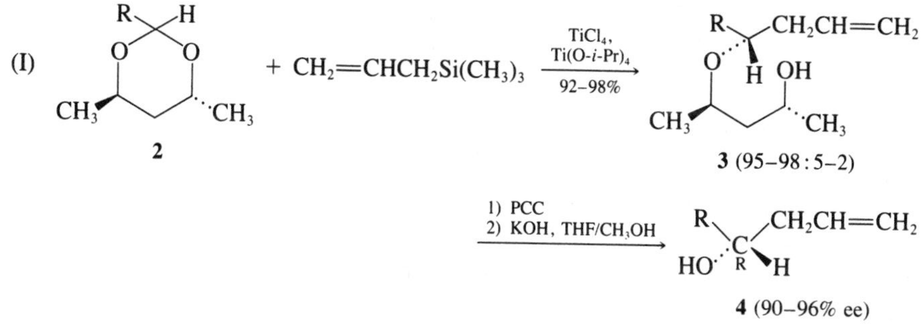

Reactions of **2** with methallyltrimethylsilane are even more selective (equation II).

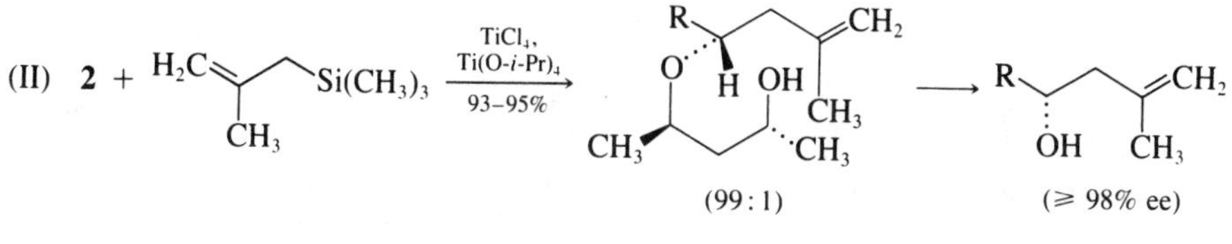

The acetals **2** undergo a similar coupling with 1-trimethylsilylalkynes, also with high diastereoselectivity, to furnish secondary propargylic alcohols after elimination of the chiral auxiliary (equation III).[2]

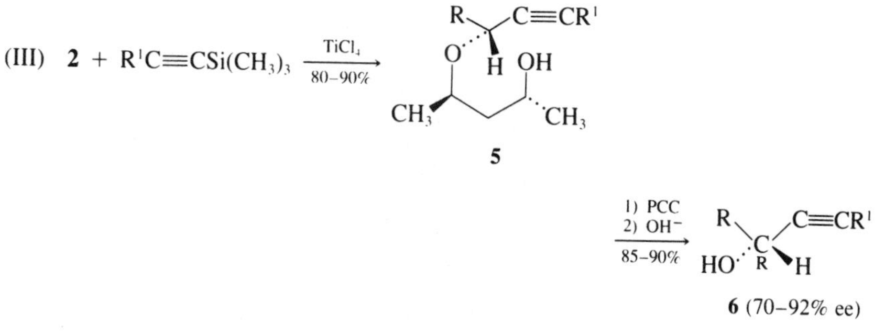

Chiral secondary alcohols.[3] The reaction of the chiral acetals **2** with Grignard reagents (or RLi) catalyzed by $TiCl_4$ proceeds stereoselectively to give the adducts **3**, which are cleaved by the usual procedure into the chiral secondary alcohols **4**.
Example:

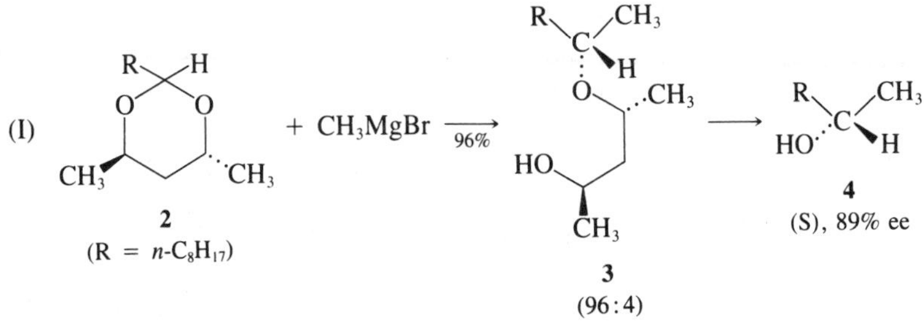

(I)

$+ CH_3MgBr \xrightarrow{96\%}$

2
$(R = n\text{-}C_8H_{17})$

3
(96:4)

4
(S), 89% ee

This reaction was extended to a synthesis of a chiral tertiary alcohol in 72% ee from the corresponding chiral ketal.

Chiral cyanohydrines.[4] The $TiCl_4$-catalyzed reaction of $(CH_3)_3SiCN$ with (2R,4R)-pentanediol acetals (**2**) gives cyanohydrin ethers (**3**) with high diastereoselectivity (~95% de).

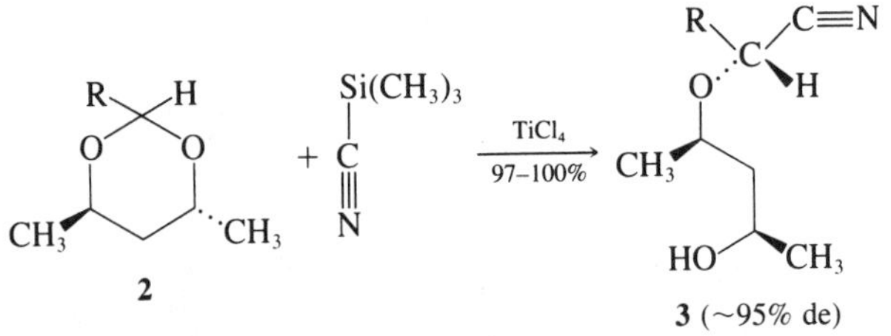

2

$+ \underset{N}{\overset{Si(CH_3)_3}{\underset{\|}{C}}}$ $\xrightarrow[97-100\%]{TiCl_4}$

3 (~95% de)

The products can be converted into chiral cyanohydrins or α-hydroxy esters by oxidation and β-elimination of the chiral side chain (equation I). Alternatively, the cyano group can be reduced ($BH_3 \cdot THF$) prior to oxidation and β-elimination for a synthesis of chiral β-amino secondary alcohols.

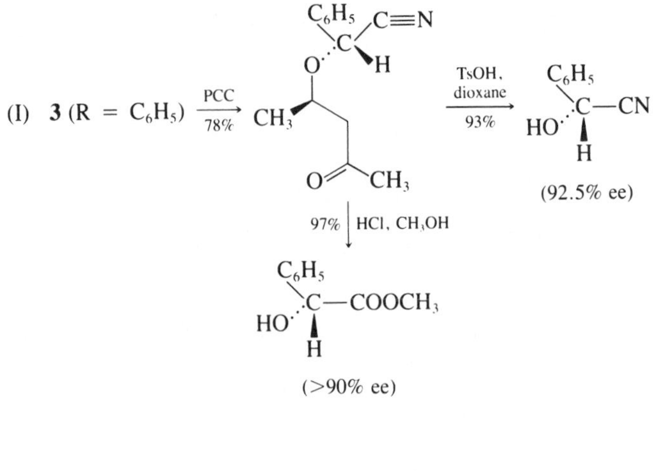

(I) **3** (R = C_6H_5) $\xrightarrow[78\%]{PCC}$

$\xrightarrow[93\%]{\substack{TsOH, \\ dioxane}}$ (92.5% ee)

$97\% \downarrow HCl, CH_3OH$

(>90% ee)

Asymmetric reduction of ketones.[5]   Chiral ketals **2**, obtained by reaction of **1** with prochiral ketones, are reduced diastereoselectively to **3** by several aluminum hydride reagents, the most selective of which is dibromoalane ($LiAlH_4$–$AlBr_3$ 1:3). Oxidation and cleavage of the chiral auxiliary furnishes optically active alcohols (**4**) in optical yields of 78–96% ee (equation I).

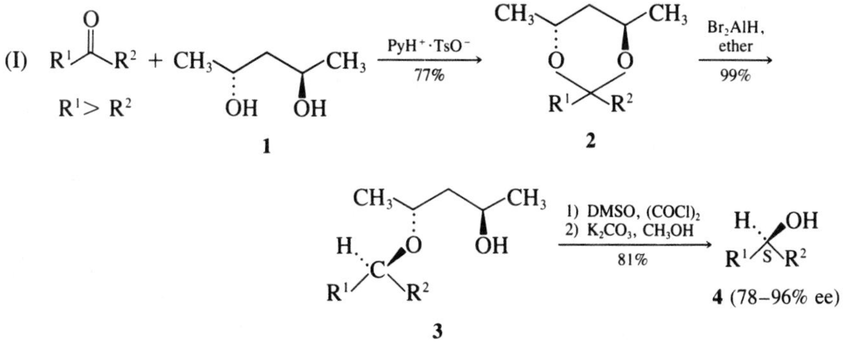

Aldol coupling of chiral acetals.[6]   The acetals (**2**) prepared from an aldehyde and (2R,4R)-pentanediol react with α-silyl ketones or enol silyl ethers in the presence of $TiCl_4$ to form aldol ethers **3** and **4** with high diastereoselectivity (>95:5). Removal of the chiral auxiliary usually results in decomposition of the aldol, but can be effected after reduction

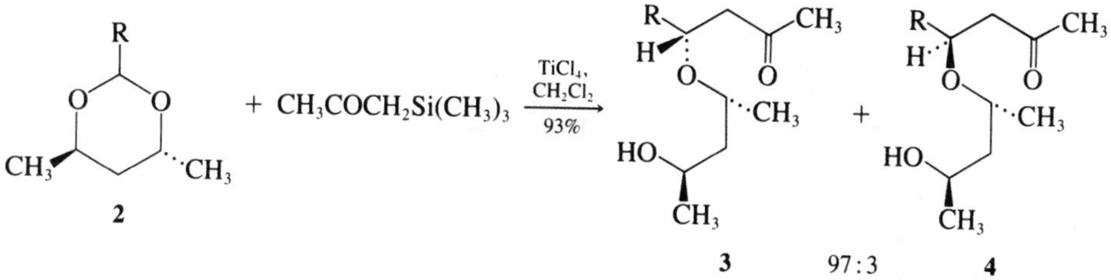

of the carbonyl group (lithium tri-*sec*-butylborohydride) to provide chiral 1,3-diols. Thus the aldol ether **3** ($R = CH_2$=$CHCH_2CH_2$) can be converted in several steps into the (2S,4R)-diol (**5**), one of the intermediates to nonactic acid.

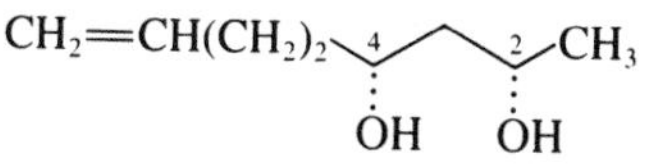

[1] P. A. Bartlett, W. S. Johnson, and J. D. Elliott, *Am. Soc.*, **105**, 2088 (1983); S. D. Lindell, J. D. Elliott, and W. S. Johnson, *Tetrahedron Letters*, **25**, 3951 (1984).

[2] W. S. Johnson, R. Elliott, and J. D. Elliott, *Am. Soc.*, **105**, 2904 (1983).

[3] S. D. Lindell, J. D. Elliott, and W. S. Johnson, *Tetrahedron Letters*, **25**, 3947 (1984).

[4] J. D. Elliott, V. M. F. Choi, and W. S. Johnson, *J. Org.*, **48**, 2294 (1983).
[5] A. Mori, J. Fujiwara, K. Maruoka, and H. Yamamoto, *Tetrahedron Letters*, **24**, 4581 (1983).
[6] W. S. Johnson, C. Edington, J. D. Elliott, and I. R. Silverman, *Am. Soc.*, **106**, 7588 (1984).

Peracetic acid, 1, 785–791; **2**, 307–309; **3**, 219; **4**, 372; **5**, 505–506; **6**, 452–453; **8**, 386; **11**, 402.

Butenolides. Δ^3-Butenolides can be prepared regioselectively by alkylation of 2-trimethylsilylfurane followed by oxidation with 40% peracetic acid (equation I).[1]

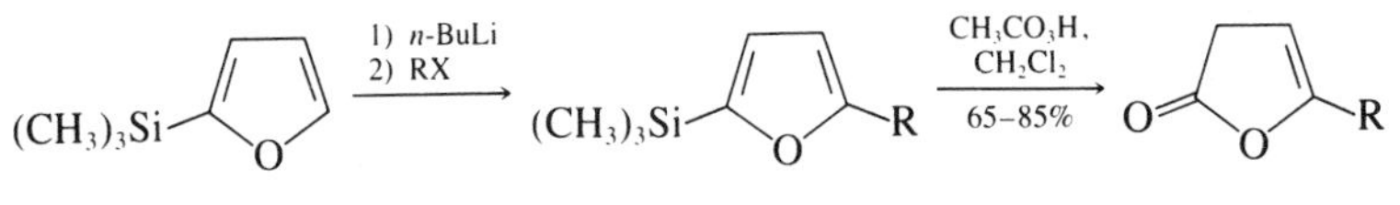

This peracid oxidation is also useful for preparation of Δ^2-butenolides (equation II).[2]

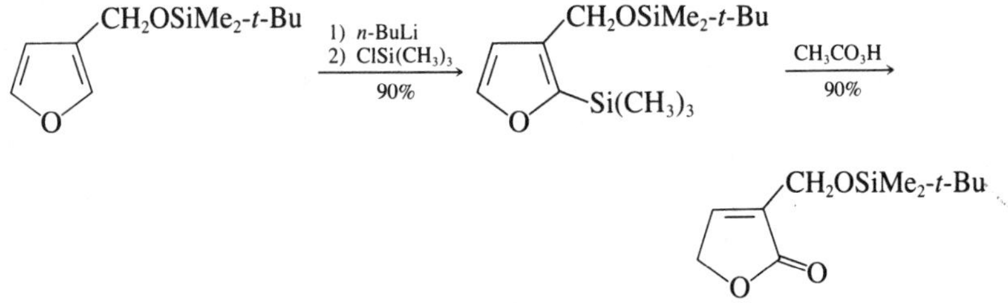

[1] I. Kuwajima and H. Urabe, *Tetrahedron Letters*, **22**, 5191 (1981).
[2] D. Goldsmith, D. Liotta, M. Saindane, L. Waykole, and P. Bowen, *ibid.*, **24**, 5835 (1983).

Periodinane (1). Mol. wt. 424.13, m.p. 124–126° (dec.), fairly stable to water. Preparation:

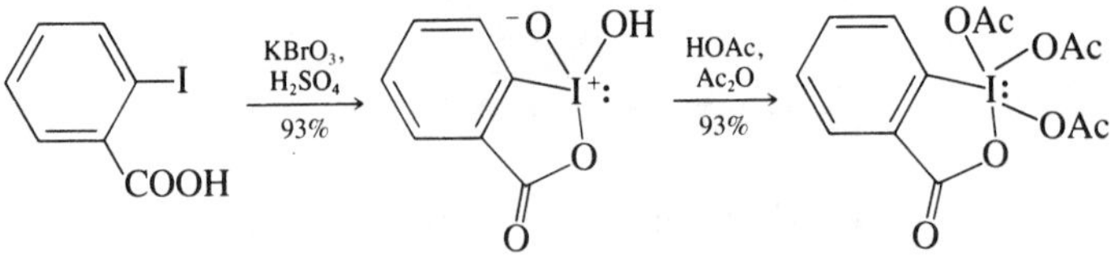

1

Oxidation.[1] This pentavalent iodine compound oxidizes primary or secondary alcohols to aldehydes or ketones in generally high yield (85–95%). Benzylic alcohols are oxidized more readily and also in high yield. Work-up of the reactions involves conversion of the trivalent iodine coproduct **2** either to 2-iodosylbenzoic acid by alkaline hydrolysis or to 2-iodobenzoic acid by reduction with sodium thiosulfate.

2

[1] D. B. Dess and J. C. Martin, *J. Org.*, **48**, 4155 (1983).

Permaleic acid, 1, 819.

Baeyer-Villiger reaction.[1]   This peracid is recommended for large-scale Baeyer-Villiger oxidation of unreactive ketones. It is almost as reactive as trifluoroperacetic acid, but is less expensive. The 90% H_2O_2 originally used for *in situ* preparation of the peracid can be replaced by the less hazardous 30% H_2O_2. By suitable modifications, dodecanolide (**2**) can be prepared in 77% yield from cyclododecanone.

[1] I. Bidd, D. J. Kelly, P. M. Ottley, O. I. Paynter, D. J. Simmonds, and M. C. Whiting, *J.C.S. Perkin I*, 1369 (1983).

Peroxytrichloroacetimidic acid, $Cl_3C\overset{NH}{\underset{OOH}{}}$ (1).

Peroxytrichloroacetimidic acid, (**1**).   The reagent is prepared *in situ* from $Cl_3CC{\equiv}N$ and aqueous 30% H_2O_2 in CH_2Cl_2.

Epoxidation.[1]   This reagent is more reactive than the similar reagents of Payne derived from aceto- and benzonitrile (**1**, 469–470). Its reactivity is comparable to that of *m*-chloroperbenzoic acid. It epoxidizes α,β-enones; it does not effect Baeyer-Villiger oxidation or oxidation of nitrile groups.

[1] L. A. Arias, S. Adkins, C. J. Nagel, and R. D. Bach, *J. Org.*, **48**, 888 (1983).

Phase-transfer catalysts, 8, 387–391; 9, 356–361; 10, 305–306; 11, 403–407.

N-p-Trifluoromethylbenzylcinchoninium bromide (**1**).[1]   Merck chemists have reported a highly efficient enantioselective alkylation with a chiral phase-transfer catalyst

1

(**1**) derived from a cinchona alkaloid. Initial observations that N-benzylcinchoninium chloride effected alkylation of **2a** with modest enantioselectivity (20–30% ee) eventually led to use of **1**, which can effect alkylation of **2a** to give **3a** in 92% ee and 95% yield

2a, R = CH_3
2b, R = CH_2COOH

(S)-(+)-**3**

under suitable conditions. Enantioselectivity is favored by use of low temperatures, non-polar solvents, high dilution, and, particularly, efficient agitation. CH_3Cl is markedly superior to CH_3Br or CH_3I as the electrophile. Use of iodide as the counterion is deleterious.

The (S)-(+)-2-methylindanone **3b** (MK-0197) stimulates excretion of uric acid, whereas the (R)-(−)-enantiomer stimulates sodium excretion.

Gomberg-Bachmann biphenyl synthesis.[2]   Reaction of stable arenediazonium tetrafluoroborates or hexafluorophosphates in an aromatic solvent with potassium acetate (2 equiv.) and a phase-transfer catalyst results in biaryls in high yield. Crown ethers, Aliquat 336, and tetrabutylammonium hydrogen sulfate are all effective catalysts. The reaction is useful for synthesis of unsymmetrical biaryls. The *ortho*-isomer predominates in reactions with a monosubstituted benzene. The most selective method is to couple a substituted arenediazonium salt with a symmetrical arene.

The same conditions are useful in the intramolecular version of the reaction (Pschorr cyclization).

Wacker oxidation.[3]   The oxidation of 1-alkenes to methyl ketones by oxygen catalyzed by $PdCl_2$ and $CuCl_2$ can be carried out under phase-transfer conditions with cetyltrimethylammonium bromide or a closely related salt as the phase-transfer catalyst. Yields are in the range 50–75%. Several rhodium and ruthenium complexes can be used as the metal catalyst, but the yields are lower.

[1] U.-H. Dolling, P. Davis, E. J. J. Grabowski, *Am. Soc.*, **106**, 446 (1984).
[2] J. R. Beadle, S. H. Korzeniowski, D. E. Rosenberg, B. J. Garcia-Slanga, and G. W. Gokel, *J. Org.*, **49**, 1594 (1984).
[3] K. Januszkiewicz and H. Alper, *Tetrahedron Letters*, **24**, 5159, 5163 (1983).

Phenacyl bromide (α-Bromoacetophenone), $C_6H_5COCH_2Br$. Suppliers: Aldrich, Fluka.

Protection of carboxyl groups.[1]   Phenacyl esters can be cleaved in ~90% yield with Zn–HOAc/DMF. For peptides that are poorly soluble in DMF, cleavage can be effected with Zn–anthranilic acid–pyridine in a mixture of N-methylpyrrolidone and DMSO (86–92% yield).

[1] D. Hagiwara, M. Neya, Y. Miyazaki, K. Hemmi, and M. Hashimoto, *J.C.S. Chem. Comm.*, 1676 (1984).

(1S, 2S)-(+)-1-Phenyl-2-amino-3-methoxypropanol-1,

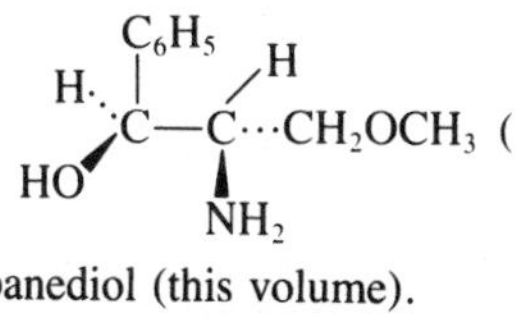

Preparation.[1] See also (1S, 2S)-1-Phenyl-2-amino-1,3-propanediol (this volume).

Chiral binaphthyls.[2]   The chiral oxazoline **3**, prepared by reaction of **1** with the α-methoxynaphthalene derivative **2**, undergoes displacement of the o-methoxy group by naphthyl Grignard reagents to give the chiral binaphthyls **4**. These are converted into the chiral binaphthyls **5** by hydrolysis and hydride reduction. The absolute configuration (R)

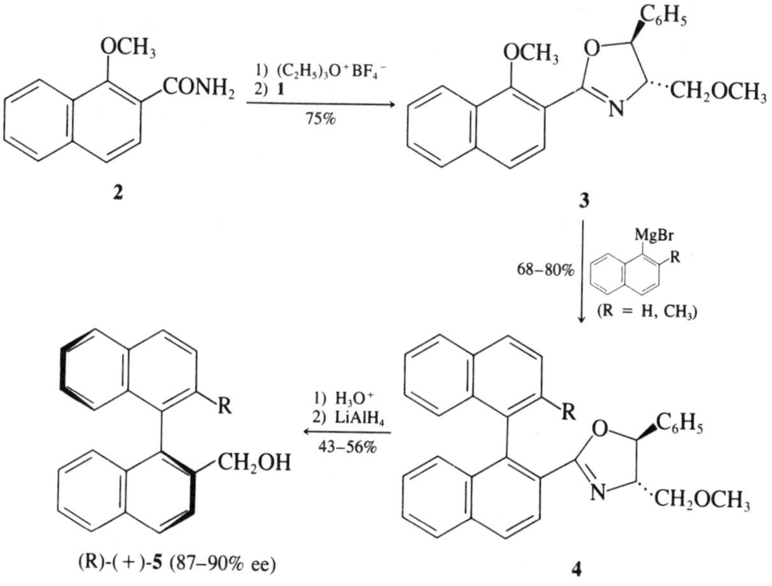

and the enantiomeric excess of **5** were determined by conversion to binaphthyls of known configuration. An alternative route to chiral binaphthyls involves a similar displacement of a 1-menthoxy group from an achiral aryloxazoline.[3]

Chiral 4-naphthylquinolines.[4] The quinoline **2**, substituted at C_3 by the chiral oxazoline group, undergoes diastereoselective addition with naphthyllithium to give **3** in 88% de. On oxidation with DDQ it is converted into (S)-**4** (84% ee). This oxidation proceeds with almost complete conservation of chirality.

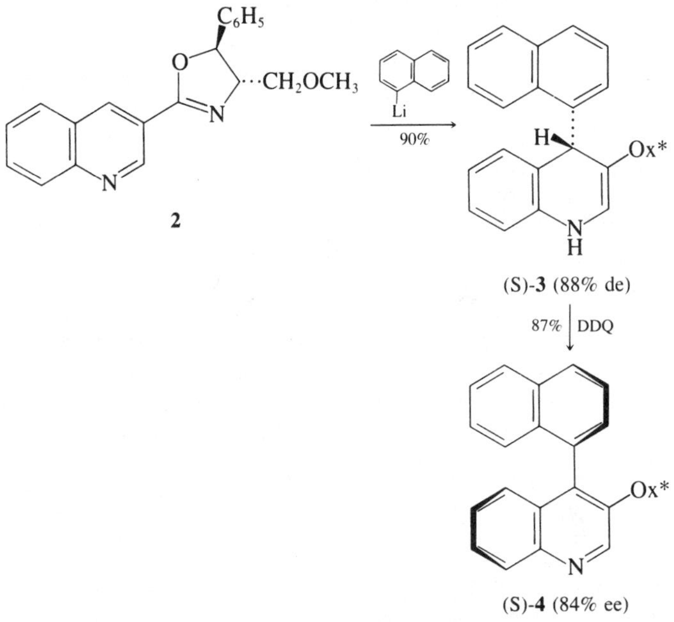

2

(S)-3 (88% de)

87% | DDQ

(S)-4 (84% ee)

This transfer of chirality can occur in the absence of the chiral oxazoline group. Thus **5**, obtained by cleavage of the oxazoline group of **3**, was oxidized by DDQ to the biaryl aldehyde (S)-(−)-**6** (90% ee).

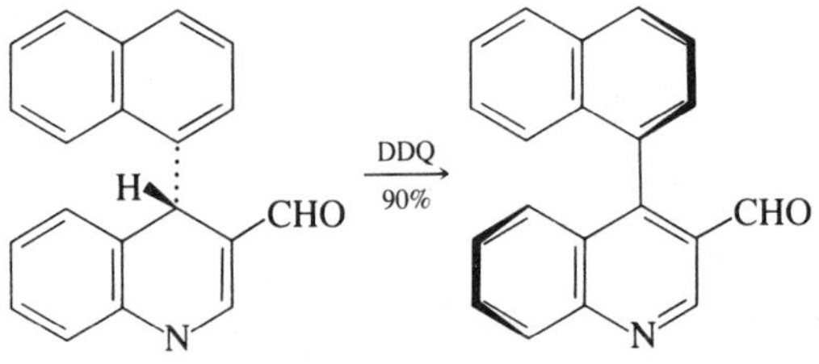

(S)-5

(S)-(−)-6 (90% ee)

¹ A. I. Meyers, G. Knaus, K. Kamata, M. E. Ford, *Am. Soc.*, **98**, 567 (1976).
² A. I. Meyers and K. A. Lutomski, *ibid.*, **104**, 879 (1982).
³ J. M. Wilson and D. J. Cram, *ibid.*, **104**, 881 (1982).
⁴ A. I. Meyers and D. G. Wettlaufer, *ibid.*, **106**, 1135 (1984).

$$\underset{HO}{\overset{C_6H_5}{C}}\!-\!\underset{NH_2}{\overset{H}{C}}\cdots CH_2OH$$

(1S, 2S)-1-Phenyl-2-amino-1,3-propanediol, H$\cdots$C—C$\cdots$CH$_2$OH   **(1)**, **6**, 386–387.

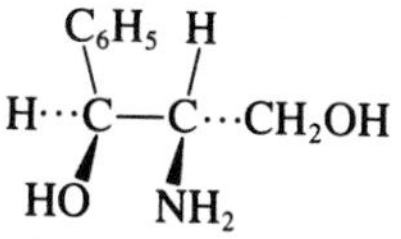

α-Alkylation of N-heterocycles. Formamidine derivatives of N-heterocycles are deprotonated selectively at the α-position. After alkylation the activating group can be removed by reduction (LiAlH₄), alkaline hydrolysis, or hydrazinolysis. An example is the use of the tetramethyleneformamidine, $(CH_2)_4NCH{=}NC(CH_3)_3$, of *t*-butylamine (equation I).[1]

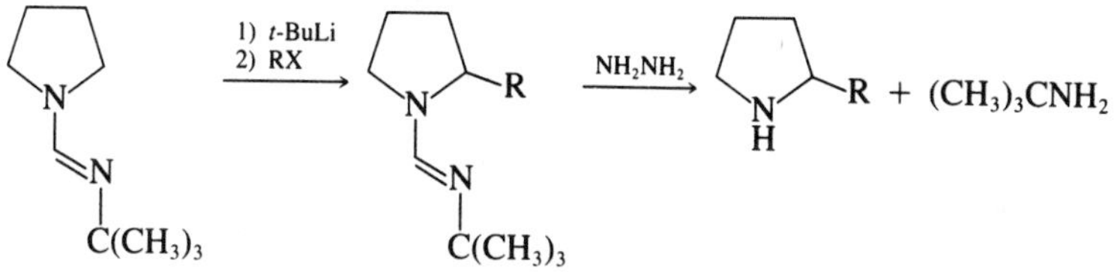

Metallation–alkylation of chiral formamidine derivatives of 1,2,3,4-tetrahydroisoquin-oline provides optically active 1-alkyl-1,2,3,4-tetrahydroisoquinolines.[2] The formam-idines of 10 optically active amino alcohols have been examined as the chiral auxiliaries and of these, the bistrimethylsilyl ether **2** (S,S-BISPAD) of **1** proved to be the most efficient as well as consistent (equation II). The configuration (S) was established by synthesis of the benzoquinolizine (S)-**5**, a degradation product of an alkaloid.

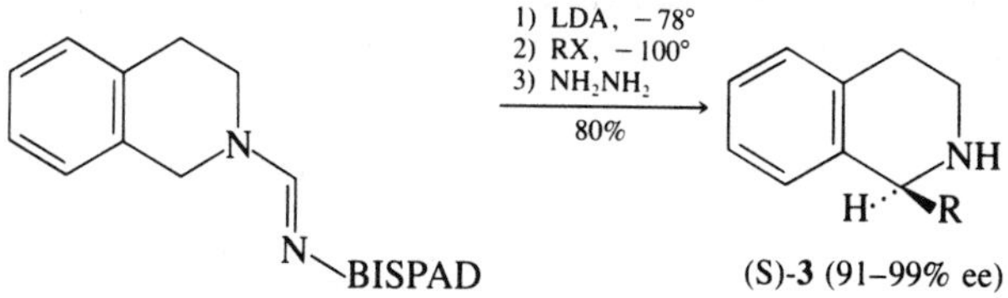

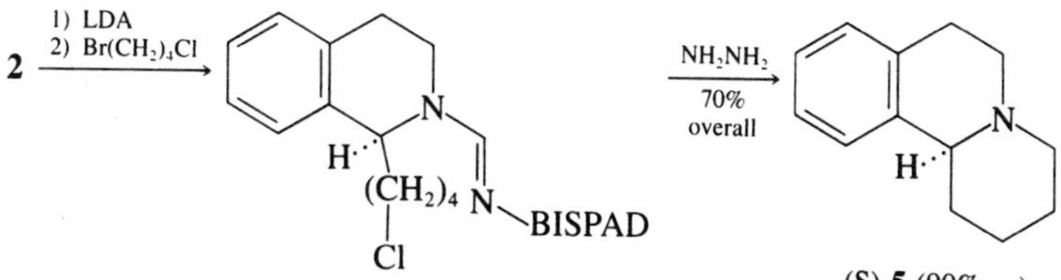

[1] A. I. Meyers, P. D. Edwards, W. F. Rieker, and T. R. Bailey, *Am. Soc.*, **106**, 3270 (1984).
[2] A. I. Meyers and L. M. Fuentes, *ibid.*, **105**, 117 (1983); A. I. Meyers, L. M. Fuentes, and Y. Kubota, *Tetrahedron*, **40**, 1361 (1984).

2-Phenylbenzothiazoline, (1). The reagent is prepared in nearly quantitative yield by reaction of *o*-aminothiophenol with benzaldehyde (C_2H_5OH, 25°).

Conjugate reduction of enones. In the presence of 1 equiv. of aluminum chloride, 1 can effect conjugate reduction of open-chain α,β-enones, usually in 75–100% yield. Cyclic enones also are reduced, but in lower yield (50–65%). This reagent is not useful for conjugate reduction of α,β-unsaturated esters or aldehydes.[1]

The reagent is effective for reduction of the C=C bond of α,β-unsaturated dinitriles.[2]

[1] H. Chickashita, M. Miyazaki, and K. Itoh, *Synthesis,* 308 (1984).
[2] H. Chickashita, S. Nishida, M. Miyazaki, and K. Itoh, *Syn. Comm.*, **13**, 1033 (1983).

Phenyl dichlorophosphate (1), 1, 847.

Acyl azides.[1] These azides are prepared in almost quantitative yield by reaction of a carboxylic acid or the potassium salt with phenyl dichlorophosphate, pyridine, sodium azide, and a catalytic amount of tetrabutylammonium bromide in CH_2Cl_2 at 25°.

[1] J. M. Lago, A. Arrieta, and C. Palomo, *Syn. Comm.*, **13**, 289 (1983).

Phenyliodine(III) diacetate (Iodosobenzene diacetate), 1, 508–509; **3,** 166; **4,** 266; **10,** 214–215; **11,** 411–412.

Oxidation of enones.[1] α,β-Enones **1** and **3** are oxidized by $C_6H_5I(OAc)_2$ and KOH in CH_3OH to α-hydroxy-β-methoxy dimethyl ketals in 50–65% yield.

Coumarane-3-ones.[2]   Oxidation of *o*-hydroxyphenyl alkyl ketones with $C_6H_5I(OAc)_2$ results in 2,2-dimethoxycoumarane-3-ones (**2**).
Example:

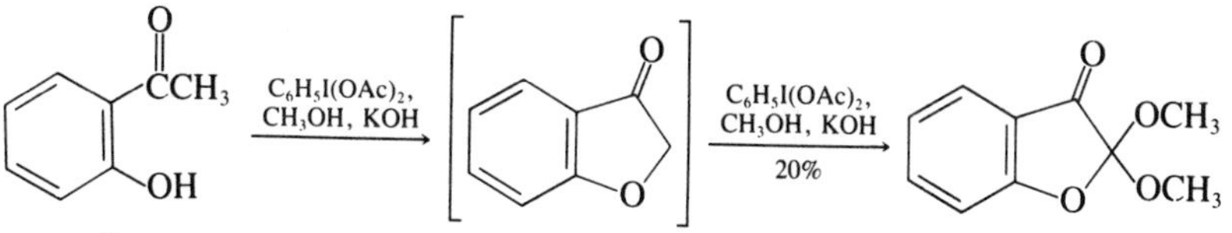

The reaction of **3** provides a route to aurone (**5**) and isoaurone (**6**).

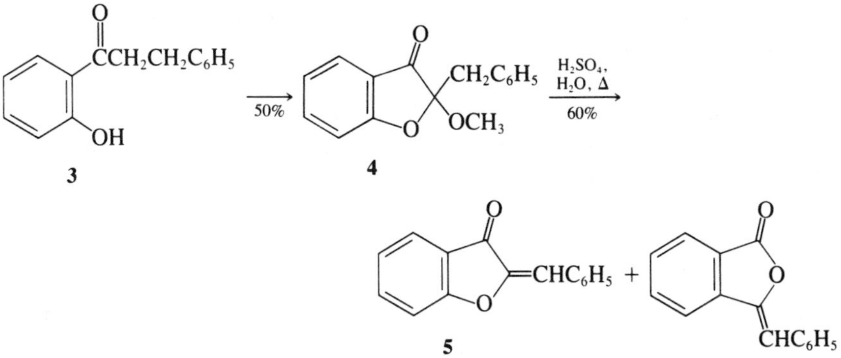

vic-*Triketones*.[3]   These triketones are usually isolated as the hydrate. Unsolvated triketones (**3**) can be obtained by ozonation of phenyliodonium β-diketonates (**2**), prepared by reaction of β-diketones with iodosobenzene diacetate.
Example:

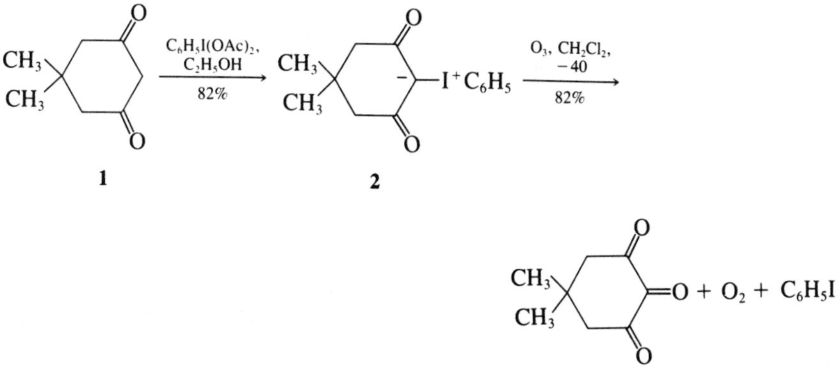

Review.[4]   Varvoglis has reviewed the chemistry of tri- and pentavalent iodine (217 references).

[1] R. M. Moriarty, O. Prakash, and W. A. Freeman, *J.C.S. Chem. Comm.*, 927 (1984).
[2] R. M. Moriarty, O. Prakash, I. Prakash, and H. A. Musallam, *ibid.*, 1342 (1984).
[3] K. Schank and C. Lick, *Synthesis*, 392 (1983).
[4] A. Varvoglis, *ibid.*, 709 (1984).

Phenyl isocyanate, 1, 843–844; **2,** 322–323; **4,** 378; **7,** 284–285; **8,** 396; **10,** 309.

Macrocyclization of an unsaturated nitrile oxide (**10,** 309).[1]   Reaction of the ω-nitroalkene **1** with *p*-chlorophenyl isocyanate results in a nitrile oxide intermediate that undergoes an intramolecular [3 + 2] cycloaddition to give **2.** The yield is considerably higher than that obtained by oxidation of the corresponding unsaturated oxime. The product is converted in several steps to the maytansinoid **3.**

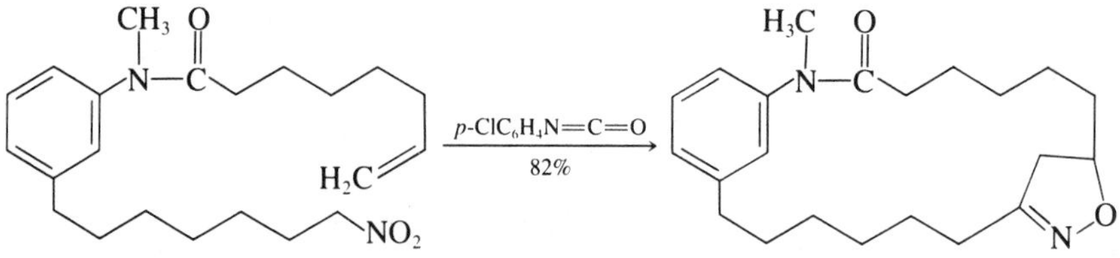

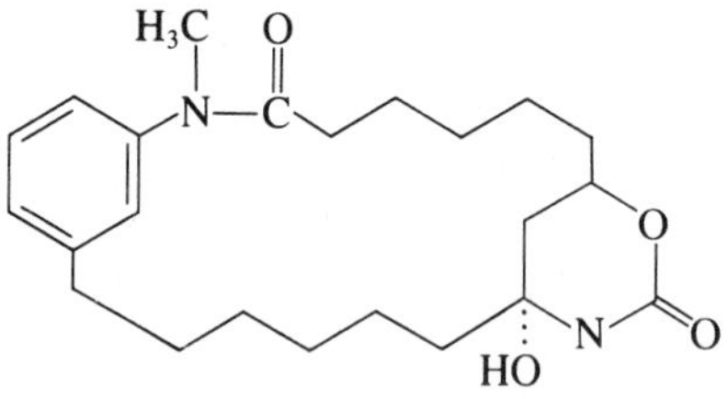

[(4 + 2) + (3 + 2)] *Triannelation.*[2]   A Diels-Alder reaction with a nitro diene such as **1** can be followed by an intramolecular nitrile oxide cycloaddition to afford multiply fused ring systems.

Example:

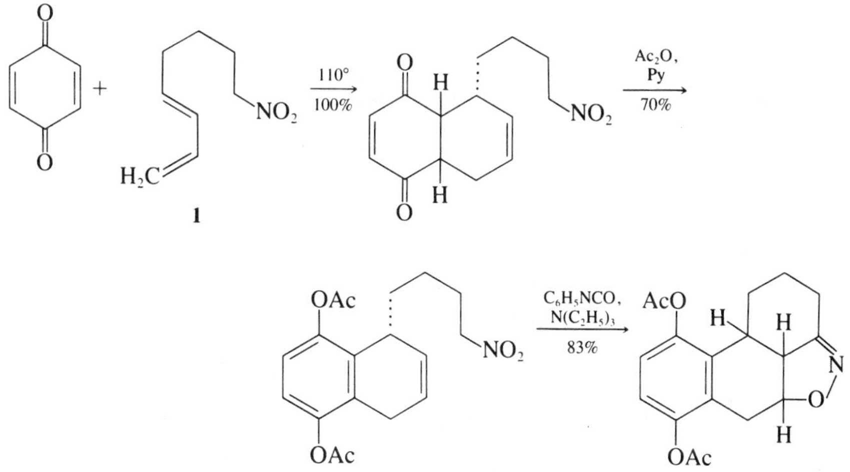

[1] P. N. Confalone and S. S. Ko, *Tetrahedron Letters*, **25**, 947 (1984).
[2] A. P. Kozikowski, K. Hiraga, J. P. Springer, B. C. Wang, and Z.-B. Xu, *Am. Soc.*, **106**, 1845 (1984).

N-Phenylketeniminyl(triphenyl)phosphorane (1).
Preparation:[1]

$$3(C_6H_5)_3P{=}CH_2 + Cl_2C{=}NC_6H_5 \xrightarrow[80\%]{} (C_6H_5)_3P{=}C{=}C{=}NC_6H_5 + 2[(C_6H_5)_3PCH_3]^+Cl^-$$

1, m.p. 151–152°

Acyl ylides. The reagent reacts with a free carboxylic acid to form the phosphorane **2**. When heated, **2** is converted into the acyl ylide **3** with loss of phenyl isocyanate. These

$$RCOOH + \mathbf{1} \longrightarrow \underset{\mathbf{2}}{RCOC({=}P(C_6H_5)_3){-}CONHC_6H_5} \xrightarrow[-C_6H_5NCO]{} \underset{\mathbf{3}}{RCOCH{=}P(C_6H_5)_3}$$

acyl ylides, when generated from keto carboxylic acids, undergo intramolecular Wittig reactions to provide cycloalkenones.[2]

Examples:

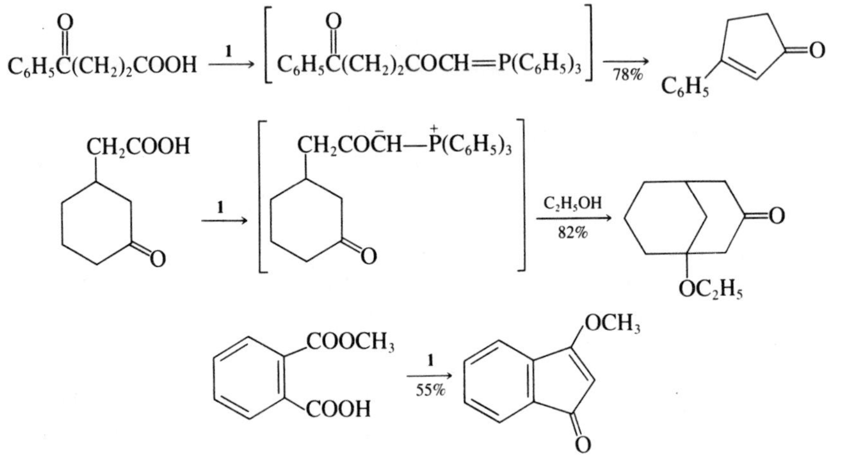

This method provides an attractive route to unsaturated macrocyclic ketones.[3]

Example:

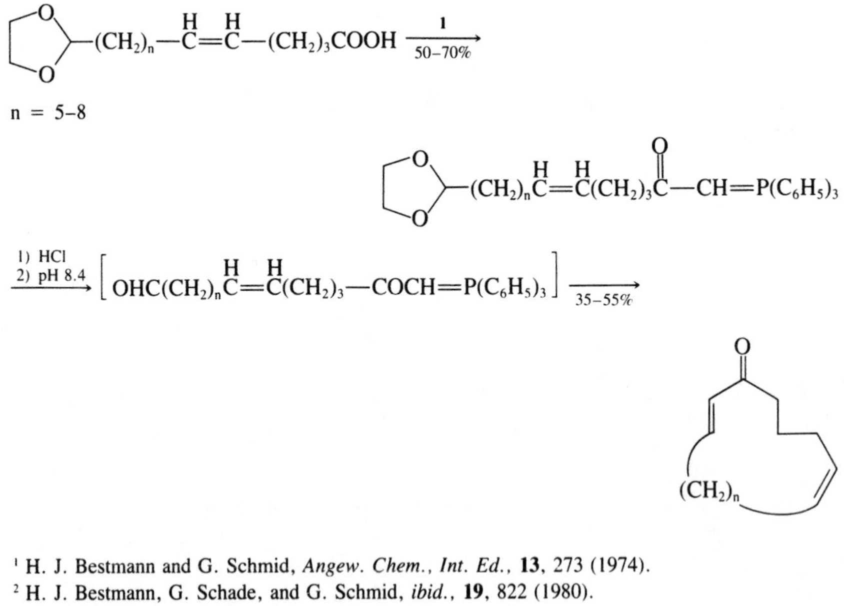

n = 5-8

[1] H. J. Bestmann and G. Schmid, *Angew. Chem., Int. Ed.*, **13**, 273 (1974).
[2] H. J. Bestmann, G. Schade, and G. Schmid, *ibid.*, **19**, 822 (1980).
[3] H. J. Bestmann and H. Lütke, *Tetrahedron Letters*, **25**, 1707 (1984).

8-Phenylmenthol (1), **11**, 412–415. The material obtained from commercial (+)-pule-gone by the original procedure of Corey and Ensley consists of (−)-8-phenylmenthol and the C_1,C_6-diastereomer in the ratio 88:12. Separation of the pure reagent can be accomplished by column chromatography or, more conveniently, by crystallization of the 3,5-dinitrobenzoate ester, m.p. 119°, α_D −217°.[1]

Asymmetric reactions of α-keto esters of (−)-1.[2]   Reduction of the α-keto ester (**2**) of (−)-8-phenylmenthol (**1**) with potassium triisopropoxyborohydride proceeds in 90% de to give the lactate ester (R)-**3** with chirality opposite to that obtained by reaction of

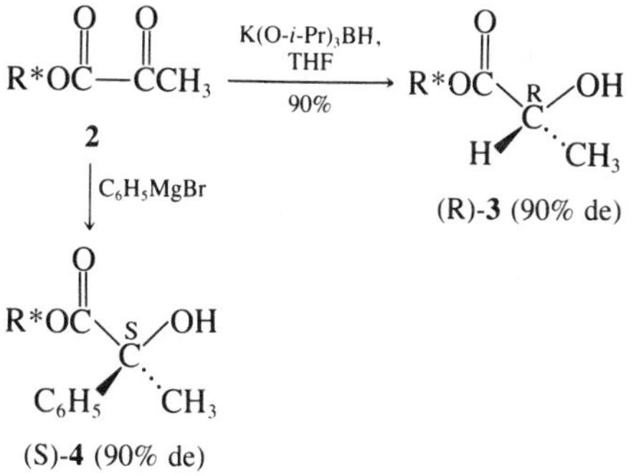

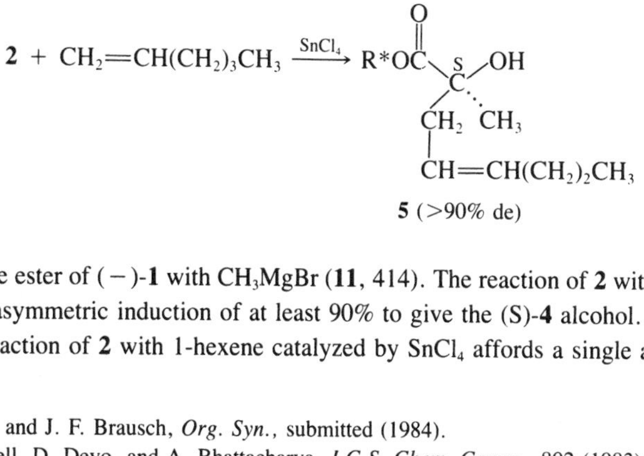

the glyoxylate ester of (−)-**1** with CH_3MgBr (**11**, 414). The reaction of **2** with C_6H_5MgBr occurs with asymmetric induction of at least 90% to give the (S)-**4** alcohol.

The ene reaction of **2** with 1-hexene catalyzed by $SnCl_4$ affords a single adduct (**5**) in >90% de.

[1] H. E. Ensley and J. F. Brausch, *Org. Syn.*, submitted (1984).
[2] J. K. Whitesell, D. Deyo, and A. Bhattacharya, *J.C.S. Chem. Comm.*, 802 (1983).

4-Phenyloxazole, (1). This reagent is obtained as a pale yellow oil in one step by reaction of phenacyl bromide and ammonium formate in formic acid (~50% yield).[1]

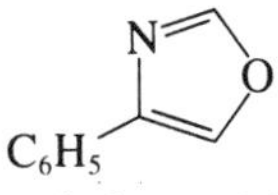

3-Substituted furanes.[1,2]   These rather inaccessible furanes can be readily prepared by a Diels-Alder addition of **1** with alkylacetylenes followed by a retro-Diels-Alder reaction with elimination of $C_6H_5C\equiv N$. The one-step reaction occurs at temperatures of 180–210°.

Example:

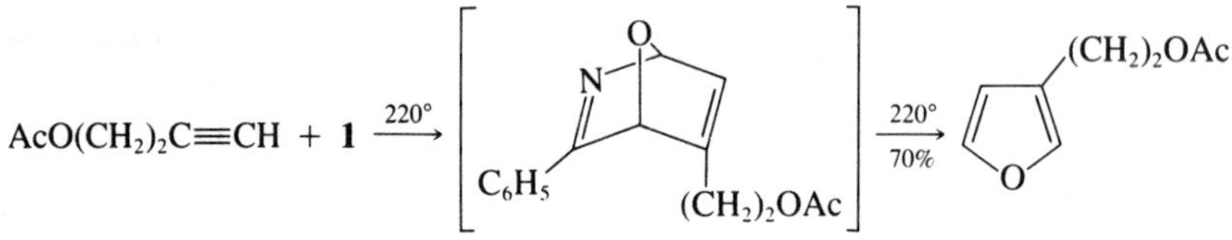

[1] J. Hutton, B. Potts, and P. F. Southern, *Syn. Comm.*, **9**, 789 (1979).
[2] D. Liotta, M. Saindane, and W. Ott, *Tetrahedron Letters*, **24**, 2473 (1983).

Phenylselenenyl benzenesulfonate, $C_6H_5SO_2SeC_6H_5$ **(1), 10,** 315; **11,** 407.

Diels-Alder reactions with unreactive alkenes.[1]   The adducts of **1** with alkenes are oxidized by H_2O_2 to vinylsulfones, which are usually satisfactory dienophiles. After the cycloaddition desulfonylation can be effected with zinc in HOAc. The sequence can be used to effect cycloadditions of unreactive alkenes.

Example:

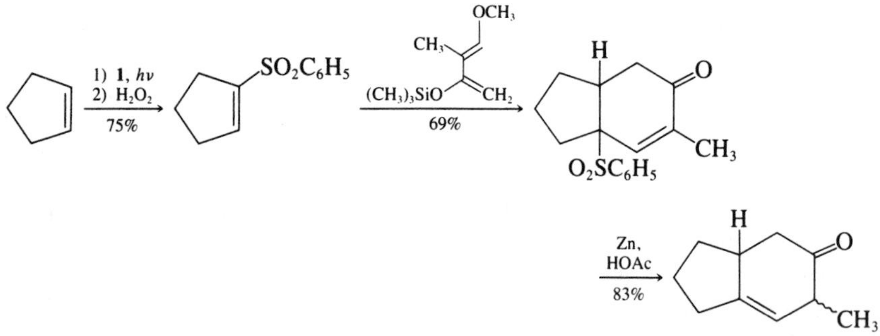

Selenosulfonation can also be used to achieve regiochemical reversal in Diels-Alder reactions.

[1] L. A. Paquette and G. D. Crouse, *J. Org.*, **48**, 141 (1983).

Phenylselenophthalimide (1), 9, 366–367; **10,** 312–313; **11,** 417–418.

Intramolecular ureidoselenenylation.[1]   Intramolecular cyclization of unsaturated urethanes with **1** followed by allylative deselenenylation results in net ureidoallylation.

Examples:

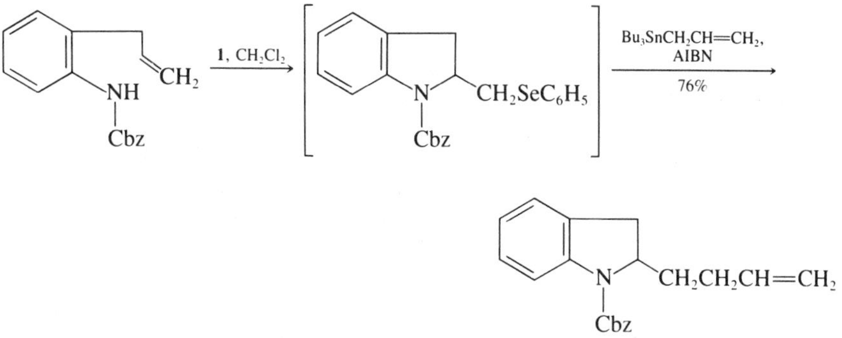

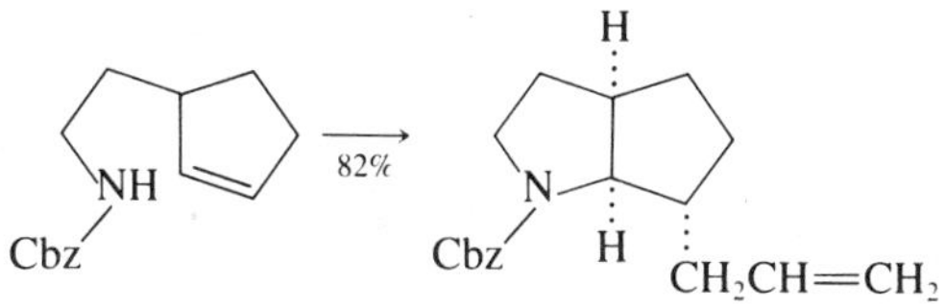

[1] R. R. Webb II, and S. Danishefsky, *Tetrahedron Letters*, **24**, 1357 (1983).

α-(Phenylsulfinyl)acetonitrile, $C_6H_5\overset{\overset{O}{\|}}{S}CH_2CN$ **(1)**, **11**, 418–419. The reagent is prepared by reaction of α-chloroacetonitrile with thiophenol followed by oxidation ($NaIO_4$).

γ-Hydroxy-α,β-unsaturated nitriles.[1]   The reaction of hexanal with **1** and piperidine affords directly the (E)-γ-hydroxy-α,β-unsaturated nitrile **2**. The reaction of a methyl

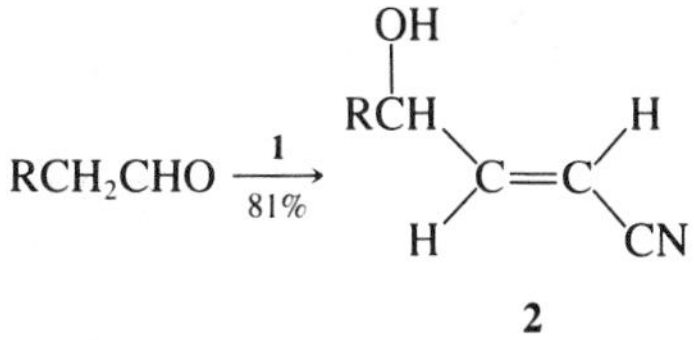

2

ketone (2-octanone) with **1** can be controlled to effect hydroxylation of either the methyl or the methylene group adjacent to the carbonyl group (equation I).

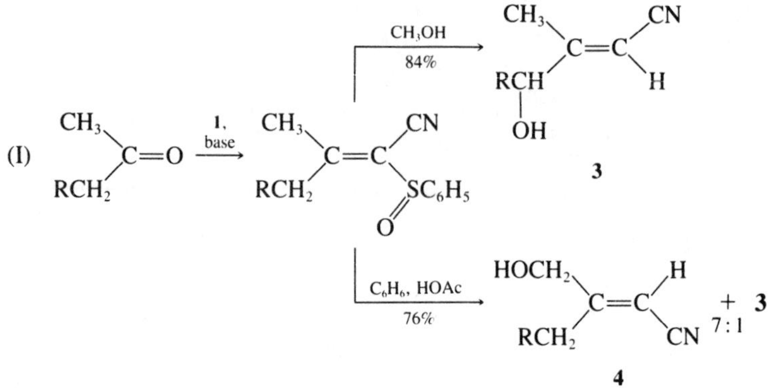

This reaction was used to prepare a conjugated trienoic acid related to vitamin A (equation II).

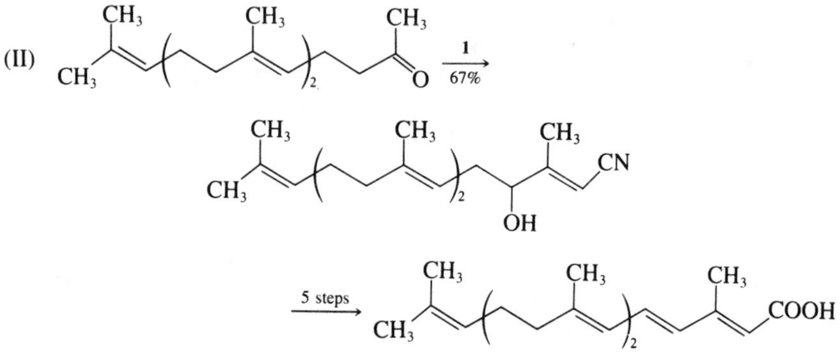

[1] T. Ono, T. Tamaoka, Y. Yuasa, T. Matsuda, J. Nokami, and S. Wakabayashi, *Am. Soc.*, **106**, 7890 (1984).

2-(Phenylsulfonyl)-3-phenyloxaziridine, $C_6H_5SO_2N\overset{O}{\underset{}{\triangle}}CHC_6H_5$ (**1**).
    Preparation:[1]

$$C_6H_5SO_2NH_2 + C_6H_5CH(OC_2H_5)_2 \xrightarrow{150°} C_6H_5SO_2N=CHC_6H_5 \xrightarrow[92\%]{\substack{ClC_6H_4CO_3H, \\ CHCl_3, H_2O, R_4NCl}} \mathbf{1}$$

Enolate hydroxylation (*cf.* **11**, 108).[2] Enolates of ketones or esters are oxidized by this oxaziridine to α-hydroxy carbonyl compounds. Yields are highly dependent on the base; they are highest with potassium hexamethyldisilazide. Yields are generally higher than those obtained with the Vedejs reagent (MoOPH, **8**, 207).

Examples:

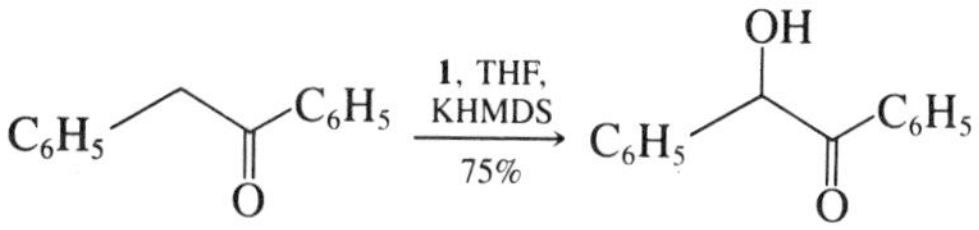

$$C_6H_5CH_2COOCH_3 \xrightarrow[87\%]{} C_6H_5\overset{OH}{\overset{|}{C}}HCOOCH_3$$

This reagent is particularly useful for stereoselective hydroxylation. Thus the lactone **2** is oxidized by **1** to a single product (**3**), whereas oxidation with MoOPH results in a mixture of **3** and the epimer **4**.

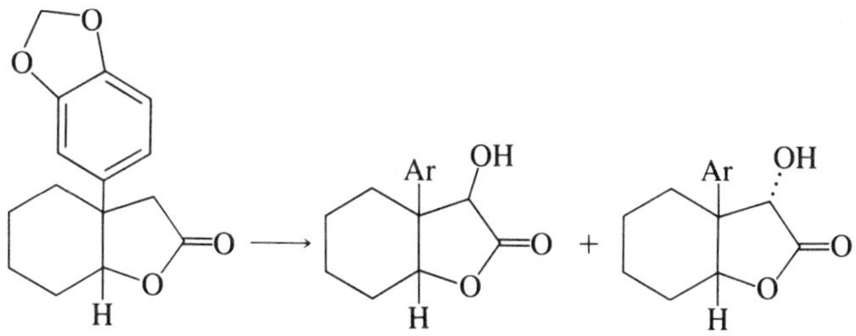

	2	**3**	**4**
LDA, MoOPH	15%	3:1	
KHMDS, **1**	91%	1:0	

[1] F. A. Davis and O. D. Stringer, *J. Org.*, **47**, 1774 (1982).
[2] F. A. Davis, L. C. Vishwakarma, J. M. Billmers, and J. Finn, *ibid.*, **49**, 3241 (1984).

Phenylsulfonyl(trimethylsilyl)methane, $(CH_3)_3SiCH_2SO_2C_6H_5$ (**1**). B.p. 160°/6 mm.
Preparation:[1]

$$(CH_3)_3SiCH_2Cl + C_6H_5SNa \xrightarrow[67\%]{} (CH_3)_3SiCH_2SC_6H_5 \xrightarrow[65\%]{\text{oxid.}} \mathbf{1}$$

Vinyl sulfones.[2] The anion of **1** (*n*-BuLi) undergoes a Peterson reaction with aldehydes or ketones to afford vinyl sulfones directly in 50–85% yield. Use of dimethoxyethane as solvent is essential for satisfactory results. The reaction is not stereoselective.

$$\begin{matrix}R^1\\ \quad\diagdown\\ \quad C{=}O\\ \quad\diagup\\ R^2\end{matrix} + (CH_3)_3Si\overset{Li}{\overset{|}{C}}HSO_2C_6H_5 \xrightarrow[50\text{–}85\%]{\text{DME, } -78°} \begin{matrix}R^1\\ \quad\diagdown\\ \quad C{=}CHSO_2C_6H_5\\ \quad\diagup\\ R^2\end{matrix} + (CH_3)_3SiOH$$

[1] G. D. Cooper, *Am. Soc.*, **76**, 3713 (1954).
[2] S. V. Ley and N. S. Simpkins, *J.C.S. Chem. Comm.*, 1281 (1983).

1-Phenylthio-1-trimethylsilylethylene, $CH_2{=}C{\overset{\displaystyle SC_6H_5}{\underset{\displaystyle Si(CH_3)_3}{\Big\langle}}}$ (**1**). Preparation.[1]

Aldehyde synthesis.[2]   An aldehyde synthesis involves addition of an alkyllithium to **1** in the presence of TMEDA to give 1-phenylthio-1-trimethylsilylalkanes (**2**), which are converted to aldehydes by oxidation, rearrangement, and hydrolysis (**10**, 314).

$$\mathbf{1} \xrightarrow[\substack{75-90\%}]{\substack{\text{RLi, ether,}\\ \text{TMEDA}}} \underset{\mathbf{2}}{C_6H_5SCHSi(CH_3)_3^{\overset{\displaystyle CH_2R}{|}}} \xrightarrow[\substack{2)\ \Delta}]{\substack{1)\ ClC_6H_4CO_3H}} \underset{\mathbf{3}}{RCH_2CHO}$$

Ketone synthesis.[3]   The intermediate **a** in the synthesis of aldehydes can be alkylated in the presence of TMEDA to give the ketone equivalent **2**. The silanes are converted into ketones (**3**) by a sila-Pummerer rearrangement (**10**, 314). Unfortunately, only primary alkyl halides give satisfactory yields.

$$\underset{\mathbf{a}}{R^1CH_2\overset{\overset{\displaystyle SC_6H_5}{|}}{\underset{\underset{\displaystyle Si(CH_3)_3}{|}}{C}}{-}Li} \xrightarrow[\substack{70-80\%}]{\substack{R^2CH_2I,\\ \text{TMEDA}}} \underset{\mathbf{2}}{R^1CH_2\overset{\overset{\displaystyle SC_6H_5}{|}}{\underset{\underset{\displaystyle Si(CH_3)_3}{|}}{C}}{-}CH_2R^2} \xrightarrow[\substack{50-70\%}]{\substack{1)\ ClC_6H_4CO_3H\\ 2)\ \Delta\\ 3)\ H_3O^-}} \underset{\mathbf{3}}{R^1CH_2\overset{\overset{\displaystyle O}{\|}}{C}CH_2R^2}$$

[1] B. Harirchian and P. Magnus, *J.C.S. Chem. Comm.*, 522 (1977).
[2] D. J. Ager, *Tetrahedron Letters*, **22**, 587 (1981).
[3] *Idem, ibid.*, **24**, 95 (1983).

Se-Phenyl *p*-tolueneselenosulfonate, $TsSeC_6H_5$, **10**, 315; **11**, 407–408.

Vinyl sulfones.[1]   The adducts (**2**) of internal or terminal alkynes with this reagent (**11**, 408) undergo stereoselective substitution of the SeC_6H_5 group by the alkyl group of organocuprates to give vinyl sulfones. Cuprates of the type $R^1Cu(SeC_6H_5)Li$, obtained by addition of R^1Li to $CuSeC_6H_5$, are more useful than lithium dialkyl cuprates, which also give products of conjugate addition. This sequence effects regioselective alkylation of the alkyne.

Example:

$$\underset{\mathbf{2}}{\overset{\displaystyle n\text{-Bu}}{\underset{\displaystyle C_6H_5Se}{\Big\rangle}}C{=}C\overset{\displaystyle Ts}{\underset{\displaystyle Bu\text{-}n}{\Big\langle}}} + CH_3Cu(SeC_6H_5)Li \xrightarrow[\substack{92\%}]{\substack{25^\circ,\ \text{THF}}} \underset{\mathbf{3}}{\overset{\displaystyle n\text{-Bu}}{\underset{\displaystyle CH_3}{\Big\rangle}}C{=}C\overset{\displaystyle Ts}{\underset{\displaystyle Bu\text{-}n}{\Big\langle}}}$$

[1] T. G. Back, S. Collins, and K.-W. Law, *Tetrahedron Letters*, **25**, 1689 (1984).

N-Phenyl-1,2,4-triazoline-3,5-dione (1), **1**, 849–850; **2**, 324–326; **3**, 223–224; **4**, 381–383; **5**, 528–530; **6**, 467; **7**, 287–288; **9**, 372.

Improved *in situ* preparation from 4-phenylurazole:

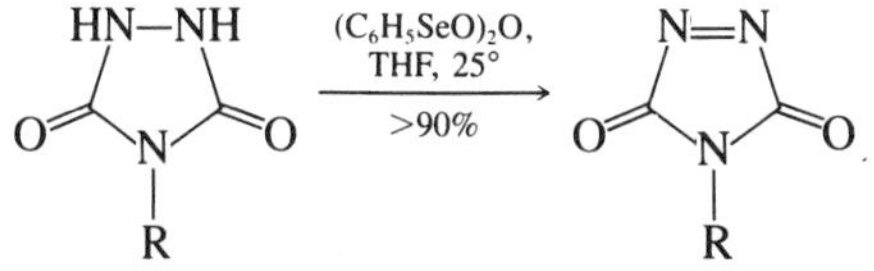

1, R = C_6H_5
2, R = $C_6H_4NO_2$-*p*

Protection of steroidal 5,7-dienes (**3**, 223–224; **6**, 467; **9**, 372).   The adducts of **1** with an ester of ergosterol can be obtained in >90% yield by reaction with **1** prepared *in situ*. The diene system can be regenerated by treatment with 2 *N* KOH in 95% ethanol. The derivative prepared from **2** is particularly readily hydrolyzed (98% yield), and consequently this reagent is generally preferable to **1**.[1]

[1] D. H. R. Barton, X. Lusinchi, and J. S. Ramirez, *Tetrahedron Letters*, **24**, 2995 (1983).

N-Phenyltrifluoromethanesulfonimide (N-Phenyltriflimide), $(CF_3SO_2)_2NC_6H_5$ (**1**). Supplier: PCR.

Enol triflates.[1]   Lithium enolates are converted into enol triflates in satisfactory yield by reaction with **1** in DME or THF without rearrangement.

Examples:

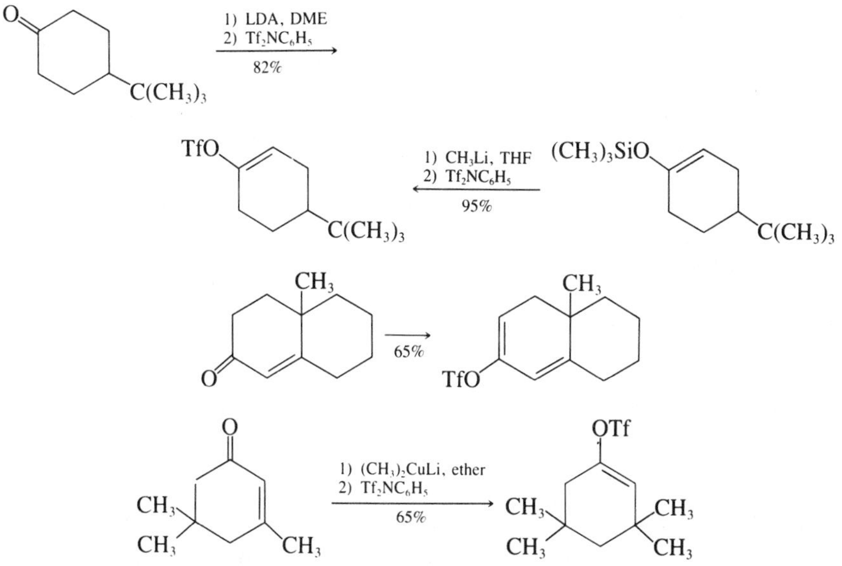

[1] J. E. McMurry and W. J. Scott, *Tetrahedron Letters*, **24**, 979 (1983).

Phosphonamides, chiral. Hanessian *et al.*[1] have prepared the enantiomeric bicyclic phosphonamides (R,R)-**1** and (S,S)-**1** from (R,R)- and (S,S)-1,2-diaminocyclohexane, respectively.

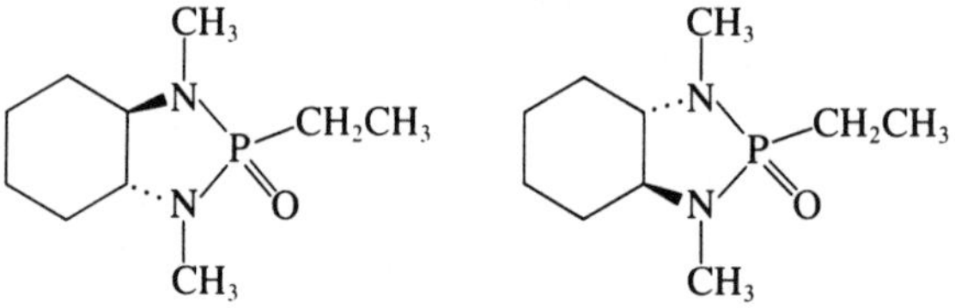

(R,R)-**1**, α_D $-92°$ (S,S)-**1**, α_D $+90°$

Asymmetric olefination.[1] The anion (KDA) of the enantiomers of **1** react with substituted cyclohexanones with the opposite sense of induction to give optically active olefins.

Examples:

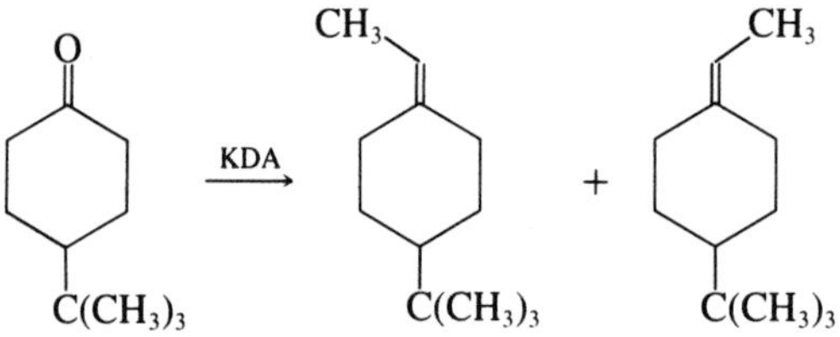

	(R)	(S)	
(R,R)-**1** 82%	95:5	90% ee	
(S,S)-**1**	5:95	90% ee	

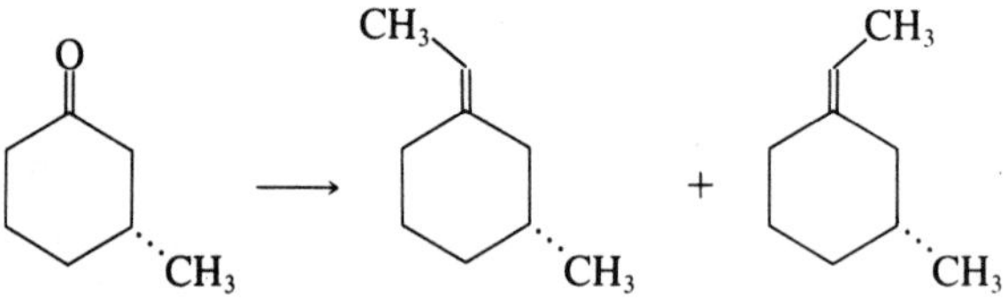

(R)-(+)   (E,R)   (Z,R)

	(E,R)	(Z,R)
(R,R)-**1**	93:7	86% de
(S,S)-**1**	15:85	69% de

Alkylation of (S,S)- and (R,R)-**1** also affords about 80% of one diastereomer.

[1] S. Hanessian, D. Delorme, S. Beaudoin, and Y. Leblanc, *Am. Soc.*, **106**, 5754 (1984).

Pinacol chloromethaneboronate,

BCH_2Cl (1). Mol. wt. 176.46, b.p.

103°/20 mm. The reagent is prepared by reaction of dichloromethaneboronic acid, $Cl_2CHB(OH)_2$, with pinacol hydrate with removal of water (89% yield).[1]

Allylboronates; homoallylic alcohols.[2]   Vinyllithium reagents react regio- and stereospecifically with this α-chloroboronic ester (1) to form an ate complex (a) that rearranges to an allylboronate (2). Allylboronates are known from the work of Schlosser[3] and Hoff-

mann,[4] as well as of others, to react stereoselectively with aldehydes to form homoallylic alcohols (3). Thus the *anti/syn* selectivity corresponds to the E/Z ratio of the allylboronate. The overall process can be carried out in two steps or as a one-pot operation. This route to allylboronates is attractive, even though the yield is only moderate, because it is compatible with various functional groups, such as orthoester and alkoxy groups.

[1] P. G. M. Wuts and P. A. Thompson, *J. Organometal. Chem.*, **234**, 137 (1982).
[2] P. G. M. Wuts, P. A. Thompson, and G. R. Callen, *J. Org.*, **48**, 5398 (1983).
[3] M. Schlosser and K. Fujita, *Angew. Chem., Int. Ed.*, **21**, 309 (1982).
[4] R. W. Hoffmann, *ibid.*, 555 (1982).

B-3-Pinanyl-9-borabicyclo[3.3.1]nonane (1), 8, 403; 9, 320–321; 10, 429–430; 11, 429–430.

Asymmetric reduction of ketones (10, 429).   Elevated pressures (6000 atm.) not only increase the rate of reduction of ketones, but also improve the enantioselectivity, which is largely dependent on differences in the size of the two alkyl groups. Although a highly hindered ketone (*t*-butyl methyl ketone) is not reduced even under pressure, ketones with one moderately bulky group are reduced under pressure with enantioselectivities of >92%. Even 2-octanone is reduced under pressure with about 60% ee.[1]

Asymmetric reduction of α-halo ketones.[2]   An α-halo substituent increases the rate and enantioselectivity of reduction of aryl ketones by neat **1**. Chemical yields of chloro- and bromohydrins thus obtained are usually >90%; optical yields are usually also >90%. Chemical yields in reduction of α-iodo ketones are lower because of deiodination. The absolute configuration is (R) when **1** is derived from (+)-α-pinene. As expected, optical yields are lower when the reduction is conducted with aliphatic α-halo ketones.

Asymmetric reduction of α-keto esters.[3]   Alkyl pyruvates are rapidly reduced by this borane to alkyl (S)-lactates (equation I). The extent of asymmetric induction depends to some extent on the temperature and the size of the ester group. Quantitative optical

induction is realized with the *t*-butyl ester in reductions conducted at 0°. Similar results are obtained when the methyl group of pyruvic esters is replaced by an alkyl group of moderate size. However, asymmetric induction is slight in the case of substrates with branching α to the carbonyl group.

Similar results are obtained on reduction of alkyl benzoylformates to alkyl (R)-mandelates (equation II). Again, the *t*-butyl ester is reduced with quantitative optical induction.

Stereoselective reduction of a chiral α-methylpropargyl ketone.[4]   (R)-**1**, prepared from (+)-α-pinene, reduces the steroidal propargyl ketone **2** to **3** and **4** with higher than expected diastereoselectivity (24:1). In contrast, (S)-**1** reduces **2** slowly and incompletely, with low enantioselectivity.

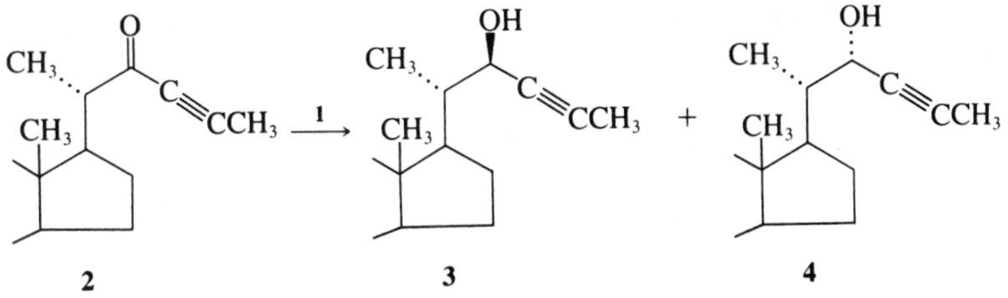

	2	**3**	**4**
	(R)-1 (92% ee)	95%	125 : 1
	(S)-1 (92% ee)	39%	1 : 2.7

[1] M. M. Midland and J. I. McLoughlin, *J. Org.*, **49**, 1316 (1984).
[2] H. C. Brown and G. G. Pai, *ibid.*, **48**, 1784 (1983).
[3] H. C. Brown, G. G. Pai, and P. K. Jadhav, *Am. Soc.*, **106**, 1531 (1984).
[4] M. M. Midland and Y. C Kwon, *Tetrahedron Letters*, **25**, 5981 (1984).

Polyethylene glycol, 9, 366, 376.

Hydrogenation of alkynes to cis-alkenes.[1] This reduction can be carried out with sodium borohydride and a catalytic amount of $PdCl_2$, which is more effective than Pd/C, in polyethylene glycol and CH_2Cl_2. This system is superior to $PdCl_2$ and H_2 in polyethylene glycol.

[1] N. Suzuki, T. Tsukanaka, T. Nomoto, Y. Ayaguchi, and Y. Izawa, *J.C.S. Chem. Comm.*, 515 (1982).

Polyphosphoric acid (PPA), 1, 894–895; **2**, 334–336; **3**, 231–233; **4**, 395–397; **5**, 540–542; **6**, 474–475; **7**, 294–295.

Cyclopentenones.[1] γ,δ-Unsaturated acids when treated with PPA cyclize to cyclopentenones. Yields are dependent on the substitution pattern; they are very high when

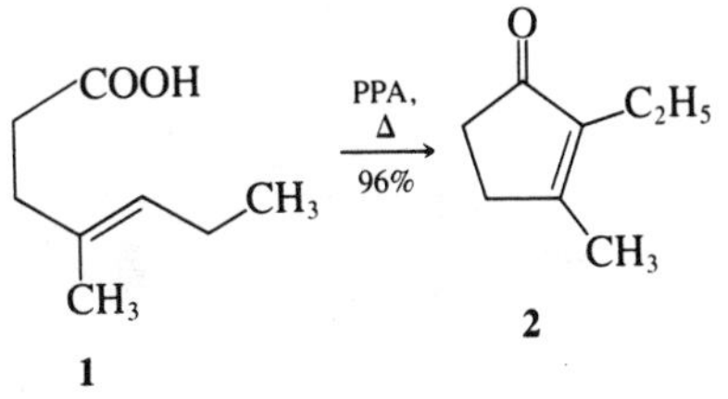

2,3-disubstituted cyclopentenones are formed (equation I). This reaction was used in an iterative cyclopentane annelation to provide the tricyclic cyclopentenone **3** as outlined

briefly in equation (II). Note that PPA can induce isomerization of *endo-* to *exo*-double bonds.

[1] M. Dorsch, V. Jager, and W. Sponlein, *Angew. Chem., Int. Ed.*, **23**, 798 (1984).

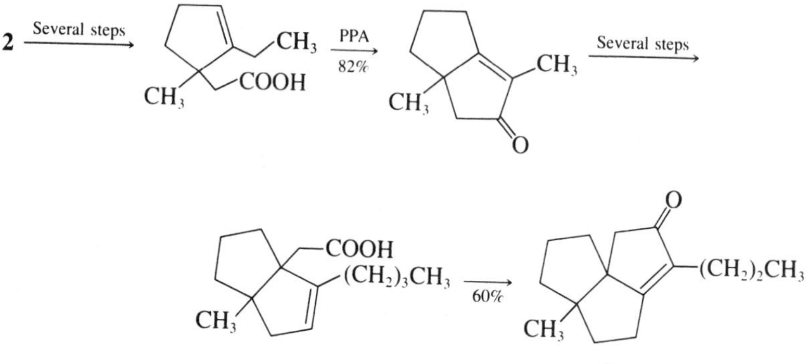

Potassium, 1, 905–906.

    Colloidal metal.[1]   A fine suspension of colloidal potassium is obtained by sonication in toluene or xylene under argon at 10°. This material is highly effective for Dieckmann condensation of dicarboxylic esters.

[1] J. L. Luche, C. Petrier, and C. Dupuy, *Tetrahedron Letters*, **25**, 753 (1984).

Potassium–Graphite, 4, 397; **7**, 396; **8**, 405–406; **9**, 378; **10**, 326.

    Phenanthrenequinones.[1]   Reaction of benzil with C_8K in THF results in cyclization to phenanthrenequinone (equation I). A similar reaction of *p,p'*-dimethylbenzil gives 3,6-dimethylphenanthrenequinone (72% yield).

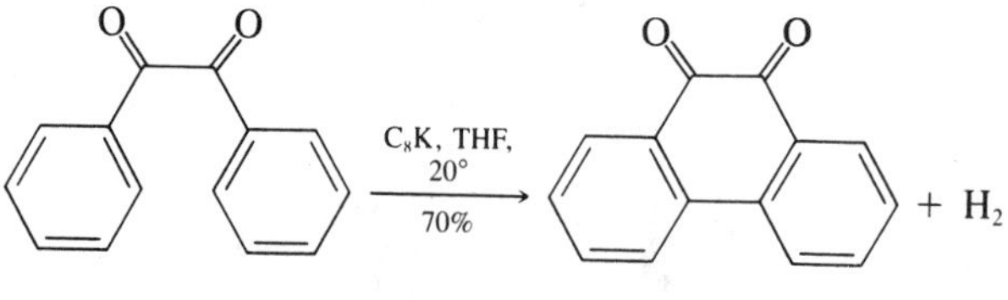

[1] D. Tamarkin, D. Benny, and M. Rabinovitz, *Angew. Chem., Int. Ed.*, **23**, 642 (1984).

Potassium 3-aminopropylamide, 6, 476; **7**, 296; **8**, 406–407; **9**, 378–379; **11**, 411. The base is obtained conveniently by reaction of potassium and 1,3-propanediamine under sonication. Addition of $Fe(NO_3)_3$ increases the rate, but is not essential. The base can also be prepared by bubbling ammonia gas into a mixture of potassium and the amine. The ammonia may function as a solvent or may react to form potassium amide.[1]

***Isomerization of* exo-*cyclic alkenes*.**[2]   This base is recommended for isomerization of (−)-β-pinene (**1**) to the rather inaccessible (−)-α-pinene (**2**) in high yield without racemization. The product is purified by distillation from lithium aluminum hydride, which removes traces of water and the amine.

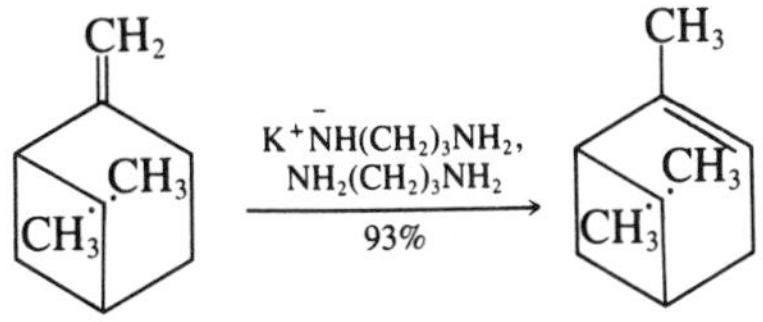

1 (92.1% ee)   **2** (92% ee)

[1] T. Kimmel and D. Becker, *J. Org.*, **49**, 2494 (1984).
[2] C. A. Brown and P. K. Jadhav, *Org. Syn.*, submitted (1984).

Potassium *t*-butoxide, 1, 911–927; **2**, 338–339; **3**, 233–234; **4**, 399–405; **5**, 544–553; **6**, 477–479; **7**, 296–298; **8**, 407–408; **9**, 380–381; **10**, 323; **11**, 432.

Dehydrosulfonylation.   Reaction of α-methoxyalkyl phenyl sulfones with potassium *t*-butoxide effects 1,2-elimination of benzenesulfinic acid to give enol ethers (equation I).[1] A related elimination (equation II) has been reported.[2]

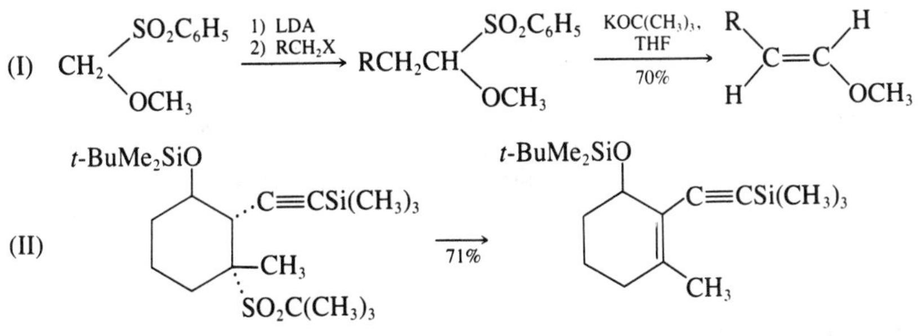

Oxidative desulfonylation of β-acetoxy or β-tetrahydropyranyloxy sulfones provides a novel route to diynes, enynes, and polyenes.

Examples:

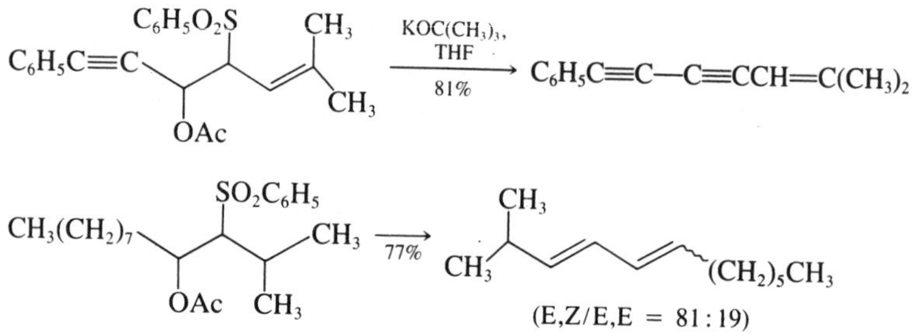

$$C_6H_5C\equiv C-C\equiv CCH=C(CH_3)_2$$

This reaction is a key step in a route to methyl retinoate (**1**).[3]

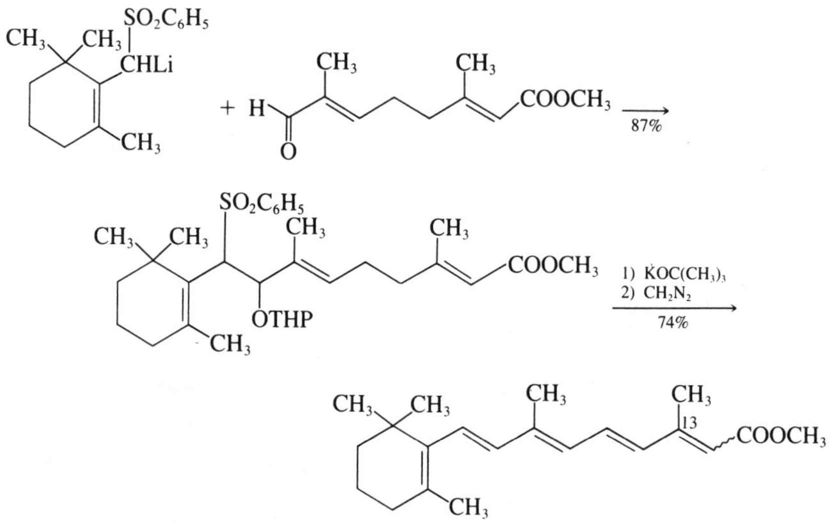

1 (13-*cis* and -*trans*, 1:1)

[1] T. Mandai, K. Hara, T. Nakajima, M. Kawada, and J. Otera, *Tetrahedron Letters*, **24**, 4993 (1983).
[2] P. L. Fuchs and P. R. Hamann, *J. Org.*, **48**, 914 (1983).
[3] T. Mandai, T. Yanagi, K. Araki, Y. Morisaki, M. Kawada, and J. Otera, *Am. Soc.*, **106**, 3670 (1984).

Potassium *t*-butoxide–*t*-Butyl alcohol complex. Review.[1]

Baker-Venkataraman rearrangement.[2] This rearrangement of *o*-aroyloxyaceto-phenones (equation I) is generally carried out with sodium or sodium ethoxide under

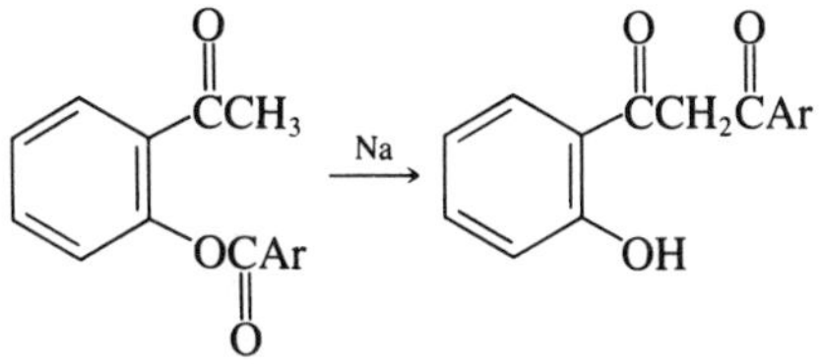

vigorous conditions. The KOC(CH₃)₃–(CH₃)₃COH complex effects this transfer of both
aromatic and aliphatic acyl groups at 0°. Formation of the substrate and the transfer
reaction are conveniently carried out in a one-pot reaction.[3]

Example:

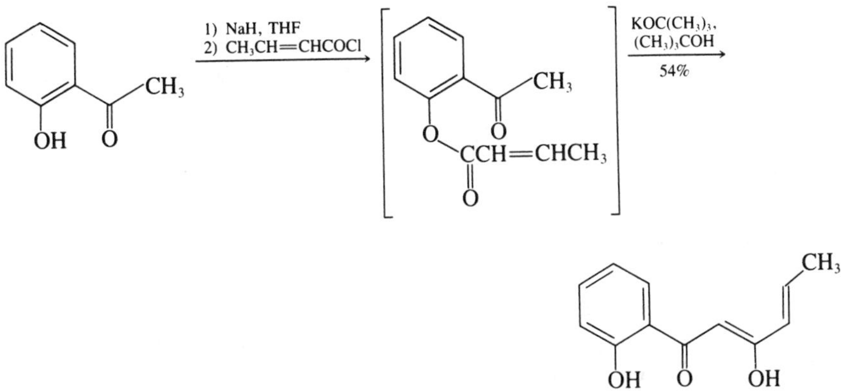

Yields are low, however, when the ester group bears acidic α-protons.

[1] D. E. Pearson and C. A. Buehler, *Chem. Rev.*, **74**, 45 (1974).
[2] C. R. Hauser, F. W. Swamer, and J. T. Adams, *Org. React.*, **8**, 59 (1954).
[3] G. A. Kraus, B. S. Fulton, and S. H. Woo, *J. Org.*, **49**, 3212 (1984).

Potassium carbonate–18-Crown-6.

Intramolecular Wittig-Horner reactions. Aristoff[1] effected intramolecular cyclization of **1** to the prostacyclin analog **2** in high yield by treatment with K₂CO₃ (1 equiv.)
and 18-crown-6 (2 equiv.) in toluene. Usual conditions for this reaction resulted in de-

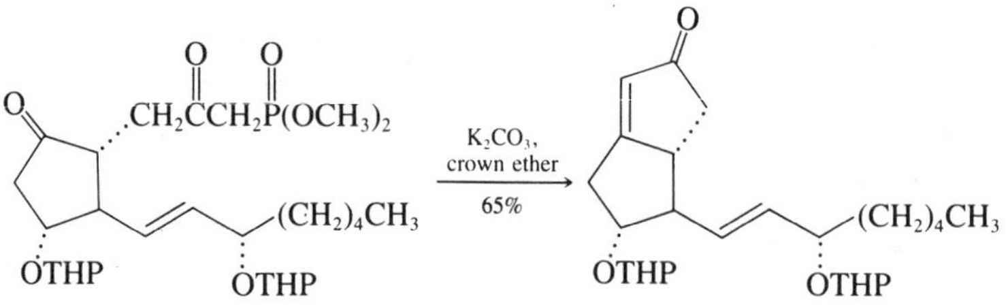

composition. Similar conditions effect cyclization of **3** to **4**. Use of NaH in DMF results only in tars.[2]

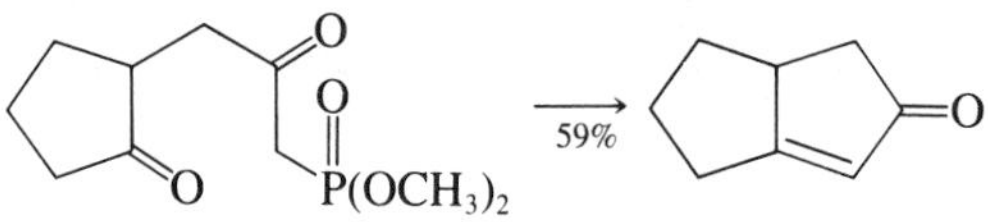

[1] P. A. Aristoff, *J. Org.*, **46**, 1954 (1981).
[2] *Idem, Syn. Comm.*, **13**, 145 (1983).

Potassium cryptate[2.2.2], 5, 156; **6,** 137–138; **8,** 130.

Bicyclocyclization.[1]   The key step in a short, stereocontrolled synthesis of the antitumor agent aklavinone (**5**) is a biomimetic cyclization of the tricarbonylnaphthalene derivative **1** to the tetracyclic product **3**. The cyclization involves a Michael addition followed by aldol condensation. The first step to provide **2** can be effected in high yield

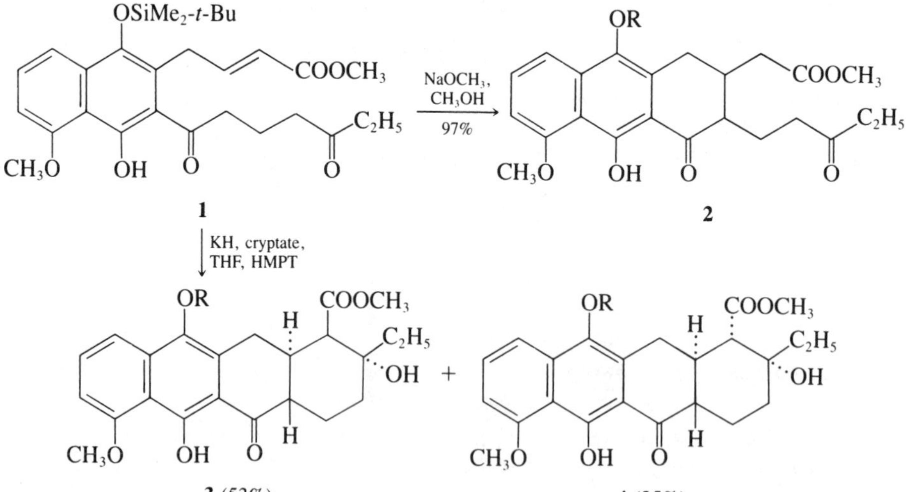

O   COOCH₃

(structure 5 diagram)

$$\text{5}$$

by K_2CO_3 or $NaOCH_3$ in methanol. Stereoselective bicyclization to the desired *trans-syn-trans*-product **3** can be effected with KH and cryptate[2.2.2] in THF/HMPT. The isomeric *trans-anti-cis*-product **4** is formed as a minor product, but is obtained almost exclusively by use of several bases: LiH and K211 (91%), KH and 18-crown-6 (100%), and NaH and DMF (89%).

[1] H. Uno, Y. Naruta, and K. Maruyama, *Tetrahedron,* **40**, 4725 (1984).

Potassium cyanide, 5, 553; **7**, 299; **8**, 409; **10**, 324; **11**; 433–434.

Cyanoesterification.[1] A typical plant type-IV cyanolipid (**1**) can be prepared by reaction of methacrolein with oleoyl chloride and KCN in the presence of 18-crown-6 in toluene at room temperature (equation I).

(I) $\begin{array}{c}CH_3 \\ {}^{\diagdown}C{-}CHO \\ H_2C^{\diagup}\end{array}$ + $CH_3(CH_2)_7CH{=}CH(CH_2)_7COCl$ $\xrightarrow[90\%]{\substack{KCN, \\ crown\ ether}}$ $\begin{array}{c}CH_3 \\ {}^{\diagdown}C{-}CH{\diagdown}^{O_2CR} \\ H_2C^{\diagup} \qquad {}_{CN}\end{array}$
(RCOCl)

1

[1] M. Nishizawa, K. Adachi, and Y. Hayashi, *J.C.S. Chem. Comm.,* 1637 (1984).

Potassium dichromate, 8, 410; **9**, 383.

α-Nitro ketones.[1] α-Nitro alcohols are readily oxidized to α-nitro ketones by potassium dichromate under phase-transfer conditions (tetrabutylammonium sulfate, methylene chloride, 30% H_2SO_4). A one-pot synthesis is possible by reaction of an aldehyde and a nitroalkane on alumina followed by *in situ* oxidation (equation I).

(I) R^1CHO + $R^2CH_2NO_2$ $\xrightarrow{Al_2O_3}$ $\begin{array}{c}OH \\ | \\ R^1CHCHR^2 \\ | \\ NO_2\end{array}$ $\xrightarrow[\substack{72-80\% \\ overall}]{\substack{K_2Cr_2O_7,\ H_2SO_4, \\ cat.,\ CH_2Cl_2}}$ $\begin{array}{c}O \\ \| \\ R^1CCHR^2 \\ | \\ NO_2\end{array}$

[1] G. Rosini, R. Ballini, P. Sorrenti, and M. Petrini, *Synthesis,* 607 (1984).

Potassium 9-(2,3-dimethyl-2-butoxy)-9-boratabicyclo[3.3.1]nonane (1).

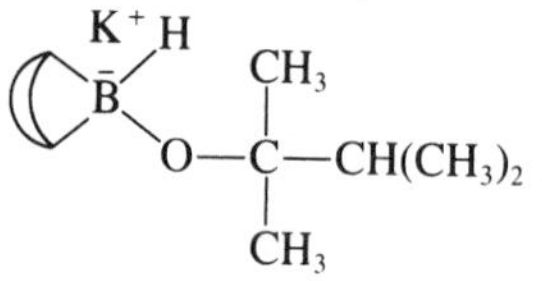

1

The reagent is prepared by reaction of 2,3-dimethyl-2-butanol with 9-BBN and then with potassium hydride. The borohydride is stable for more than a year in THF solution.

Stereoselective reduction of ketones.[1]   This borohydride (1) is comparable to lithium tri-*sec*-butylborohydride (**4**, 312–313) for stereoselective reduction of cyclic ketones to the less stable alcohols, but less stereoselective than lithium trisiamylborohydride (**7**, 216–217). The by-product formed in reductions with **1** can be removed as an insoluble ate complex formed by addition of water, simplifying isolation of the reduction product.

[1] H. C. Brown, J. S. Cha, and B. Nazer, *J. Org.*, **49**, 2073 (1984).

Potassium fluoride, 1, 933–935; **2**, 346; **5**, 555–556; **6**, 481–482; **8**, 410–412; **10**, 325–326; **11**, 434–435.

1,4-Dihydroxyxanthones.[1]   KF in DMF is a more efficient catalyst than K_2CO_3 for condensation of methyl salicylates with chlorobenzoquinones to provide phenoxybenzoquinones (**1**). The products on reduction ($Na_2S_2O_4$) and cyclization (H_2SO_4) afford 1,4-dihydroxyxanthones (**2**) with high regioselectivity (equation I).

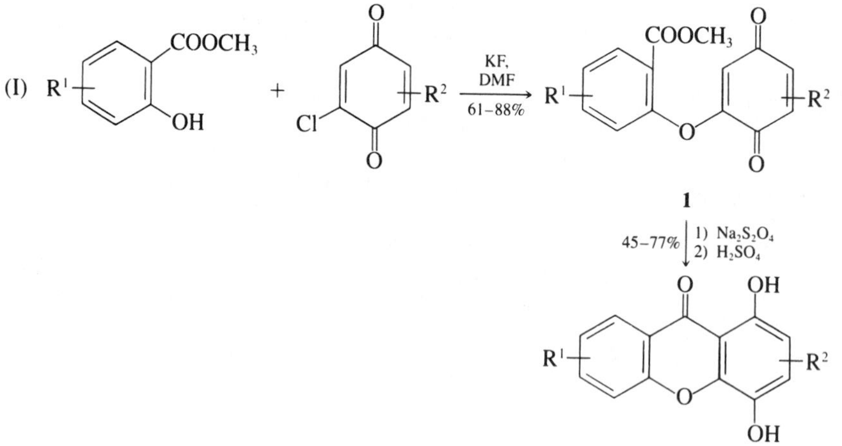

(I)

1

2

Alkylation catalyst.   KF or CsF impregnated on Al_2O_3 is an effective catalyst for O-alkylation of phenols and alcohols in CH_3CN or DME. Both are preferable to Bu_4NF/Al_2O_3, because they are more easily handled as well as cheaper.[2]

The reagent is also useful for dehydrohalogenation of *sec*-alkyl bromides and of *vic*-dibromides, for some Michael additions, and for aldol condensation.[3]

[1] B. Simoneau and P. Brassard, *J.C.S. Perkin I,* 1507 (1984).
[2] T. Ando, J. Yamawaki, T. Kawate, S. Sumi, and T. Hanafusa, *Bull. Chem. Soc. Japan,* **55**, 2504 (1982).
[3] J. Yamawaki, T. Kawate, T. Ando, and T. Hanafusa, *ibid.,* **56**, 1885 (1983).

Potassium hexamethyldisilazide, 10, 326.

Deconjugation of α,β-*unsaturated esters* (**11**, 297).[1]   This base is more effective than LDA/HMPT for isomerization of α,β-unsaturated esters to (Z)-β,γ-unsaturated esters. The (Z)-selectivity is increased somewhat with increasing size of the ester group. Isobutyl esters are isomerized to (Z)-3-alkenoates almost exclusively (Z/E = 97:3).

[1] Y. Ikeda and H. Yamamoto, *Tetrahedron Letters,* **25**, 5181 (1984).

Potassium hydride, 1, 935; 2, 346; 4, 409; 5, 557; 6, 482–483; 7, 302–303; 8, 412–415; 9, 386–387; 10, 327–328; 11, 435–436.

Potassium enoxyborates (**9**, 482).   Reaction of 2-methylcyclohexanone (**1**) with potassium hydride at 25° and then with triethylborane generates the most stable enolate (**2**) with ≥90% regioselectivity, as judged by conversion to **4**. Use of potassium

bis(trimethylsilyl)amide as base generates the less stable enolate (**3**) with comparable regioselectivity, as judged by conversion to **5**. The regioselectivity obtained by the former method is considerably higher than that reported previously in generation of thermodynamic enolates. Comparable regioselectivity is observed in two other cases, one being the acyclic ketone 2-heptanone.[1]

Dehydration of a γ-hydroxy ketone.[2]   The synthesis of the antibiotic nonactin (**1**) presents an unusual problem in that the subunit, nonactic acid, occurs as both optical antipodes. In a recent synthesis, this problem was simplifed by synthesis of both antipodes

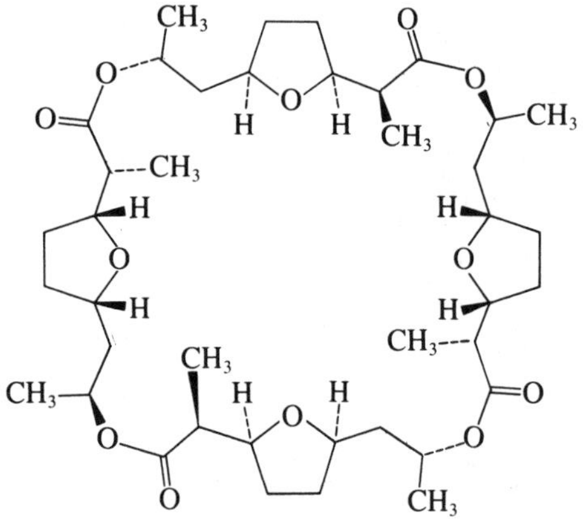

1

from a common intermediate (**2**) prepared from (S)-(−)-malic acid. Thus oxalic acid-catalyzed dehydration of the hydroxy ketone **2** provides (+)-**3**, which on hydrogenation provides the ester of (−)-8-epinonactic acid (**4**).

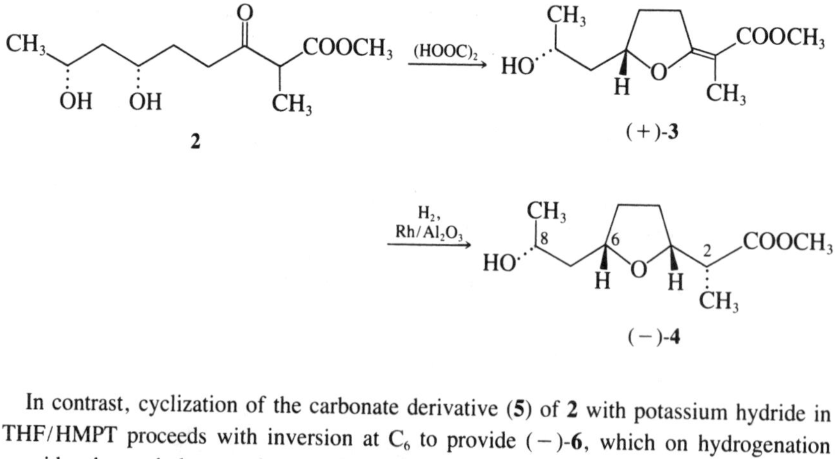

In contrast, cyclization of the carbonate derivative (**5**) of **2** with potassium hydride in THF/HMPT proceeds with inversion at C_6 to provide (−)-**6**, which on hydrogenation provides the methyl ester of (+)-nonactic acid (**7**).

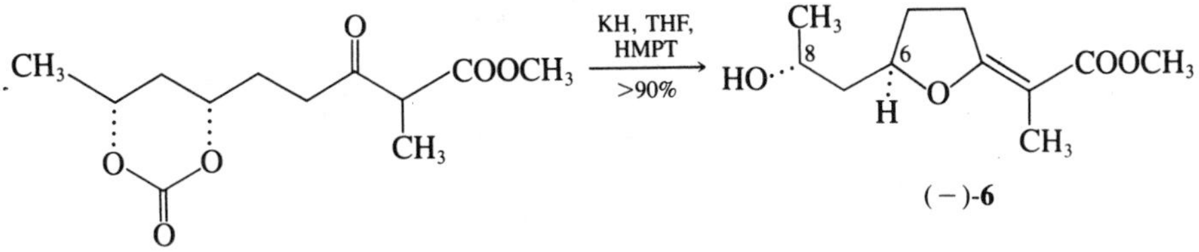

$$(-)\text{-}6$$

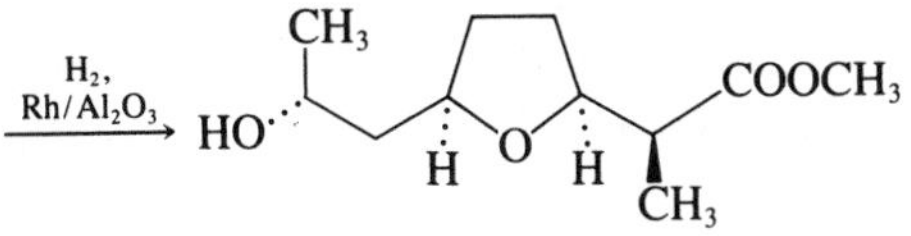

$$(+)\text{-}7$$

Assembly of nonactin involves coupling the potassium salt of (+)-nonactic acid with the mesylate of (−)-**4** with inversion of C_8 to give methyl (+)-nonactyl (−)-nonactate (**8**) in 86% yield. The corresponding free acid was simultaneously dimerized and cyclized via the mixed anhydride formed from diphenyl phosphorochloridate (**11**, 223) to give nonactin (**1**) in 15–20% yield.

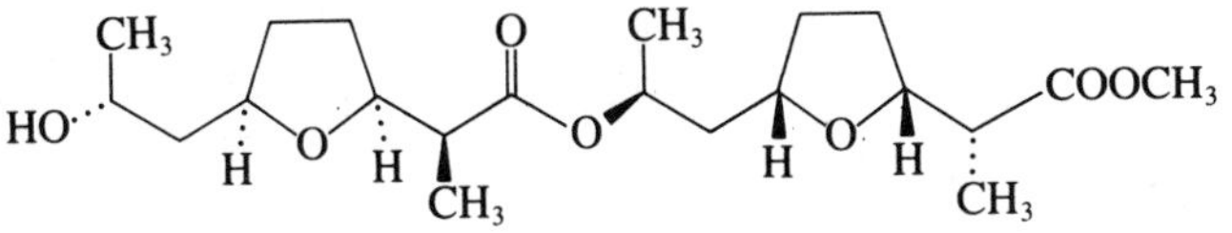

8

α-Alkylidenetetrahydrofuranes.[3]   Reaction of epoxysilanes of structure **1** with KH in THF provides α-alkylidenetetrahydrofuranes (**2**) with ≥95% retention of stereochemistry. The precursor vinylsilanes can be prepared stereoselectively from an alkyne by

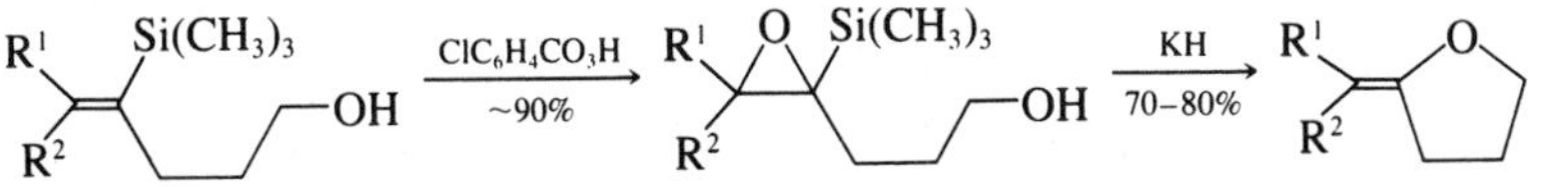

hydroalumination or carbocupration. This method thus allows preparation of either (E)- or (Z)-**2**.

A related stereoselective synthesis of γ-alkylidenebutyrolactones (**4**) from the same starting materials is outlined in equation (I). In this case epoxidation is accompanied by cyclization to give **3**. Elimination of $(CH_3)_3SiOH$ to give **4** results in 99% inversion of stereochemistry.

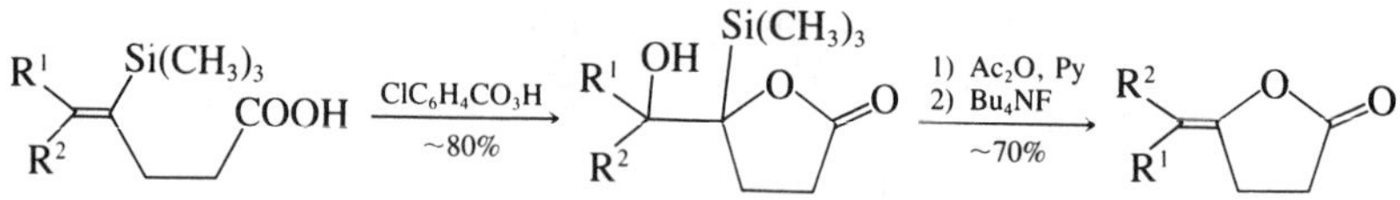

Cyclooctenones.[4]   Neutral or anionic oxy-Cope rearrangement of 1,2-dialkenylcyclobutanols provides a useful route to cyclooctenones.
Examples:

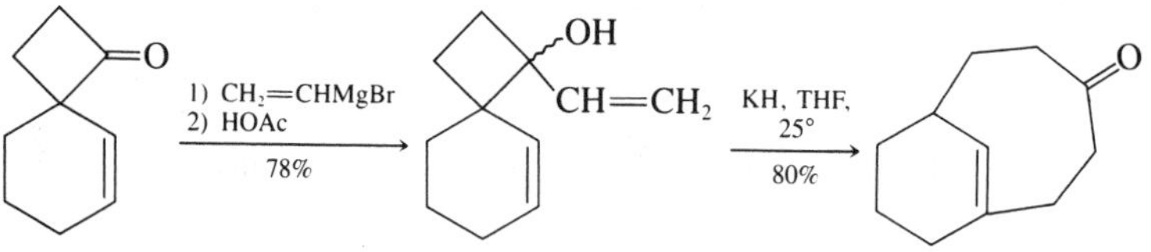

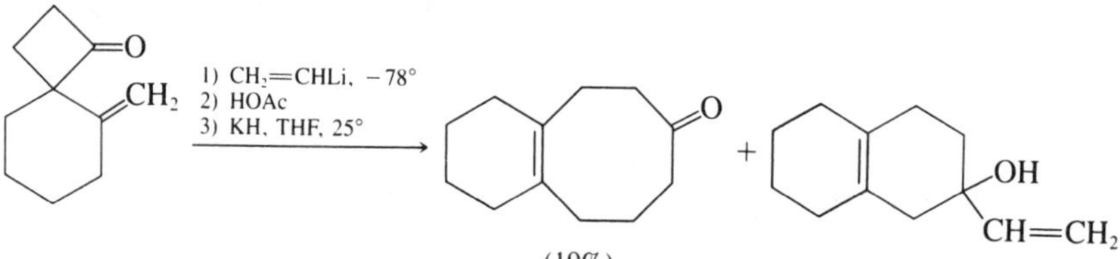

This ring expansion when applied to **1** in a synthesis of the sesquiterpene poitediol (**3**) proceeds in low yield. However, the 1-alkynyl-2-alkenylcyclobutanol obtained by addition of lithium acetylide to **1** rearranges to the desired cyclooctadienone **2** in satisfactory yield at 50°.[5]

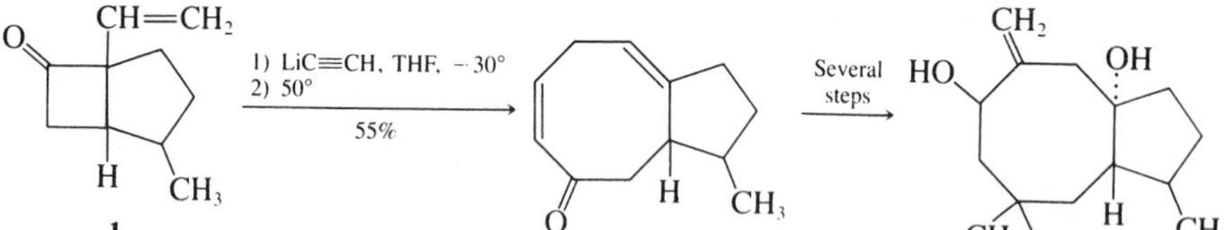

Spiroketal cyclization.[6]   Intramolecular spiroketal cyclization via Michael addition of an alcohol to a chiral α,β-unsaturated sulfoxide can proceed with high stereoselectivity. Reactions with KH are more stereoselective than those with NaH or n-BuLi. Thus cyclization of the sulfoxide **1** results in exclusive formation of **2**, which is desulfurized to (E)-2-methyl-1,6-dioxaspiro[4.5]decane (**3**). The isomeric sulfoxide **4** under the same conditions is converted in comparable yield to (Z)-**3**. These spiroketals are pheromones of wasps.

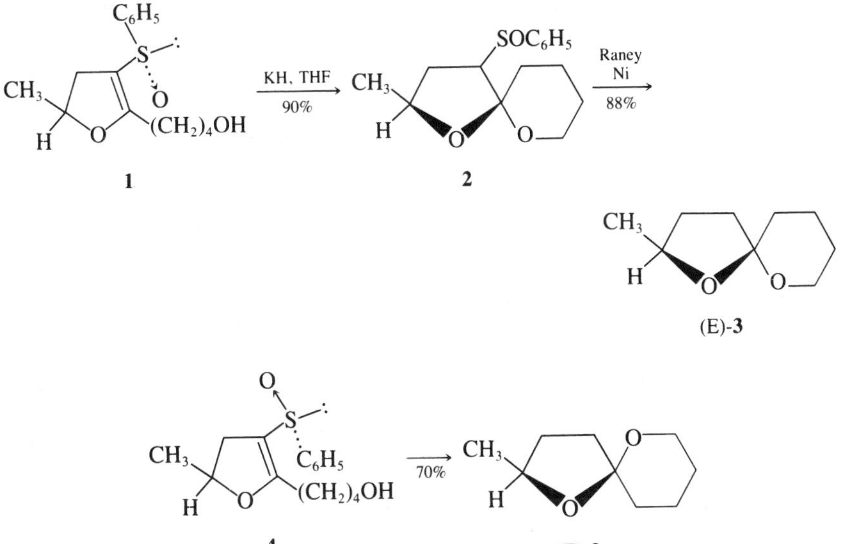

Review. Pinnick[7] has reviewed use of KH as a base and as a nucleophile in organic synthesis. He also covers various rearrangements induced by KH.

[1] E. Negishi and S. Chatterjee, *Tetrahedron Letters*, **24**, 1341 (1983).
[2] P. A. Bartlett, J. D. Meadows, and E. Ottow, *Am. Soc.*, **106**, 5304 (1984).
[3] F.-T. Luo and E. Negishi, *J. Org.*, **48**, 5144 (1983).
[4] R. C. Gadwood and R. M. Lett, *ibid.*, **47**, 2268 (1982).
[5] R. C. Gadwood, R. M. Lett, and J. E. Wissinger, *Am. Soc.*, **106**, 3869 (1984).
[6] C. Iwata, K. Hattori, S. Uchida, and T. Imanishi, *Tetrahedron Letters*, **25**, 2995 (1984).
[7] H. W. Pinnick, *Org. Prep. Proc. Int.*, **15**, 199 (1983).

Potassium hydroxide, 5, 557–560; **6,** 486; **7,** 303–304; **8,** 415–416; **11,** 439.

α-Hydroxy aldehydes.[1]   A new method for one-carbon homologation of ketones to α-hydroxy aldehydes involves Darzens condensation with chloromethyl phenyl sulfone to give an α,β-epoxy sulfone followed by ring opening by hydroxide ion. The anhydrous potassium hydroxide obtained by controlled addition of water to potassium *t*-butoxide (**8,**

415) is uniquely suited to this purpose. The overall transformation proceeds with high stereoselectivity in the case of 17- and 20-ketosteroids.

Examples:

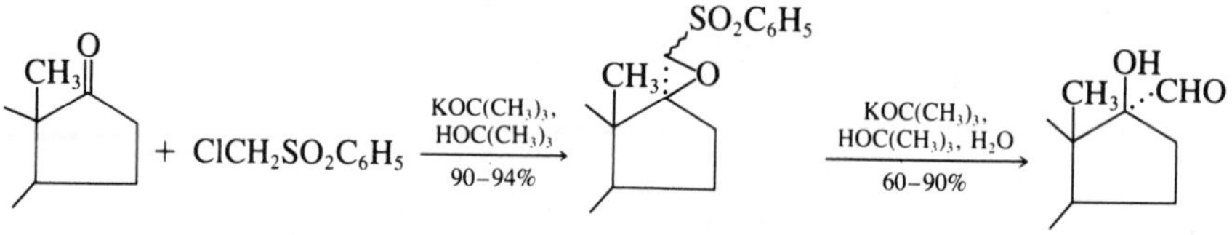

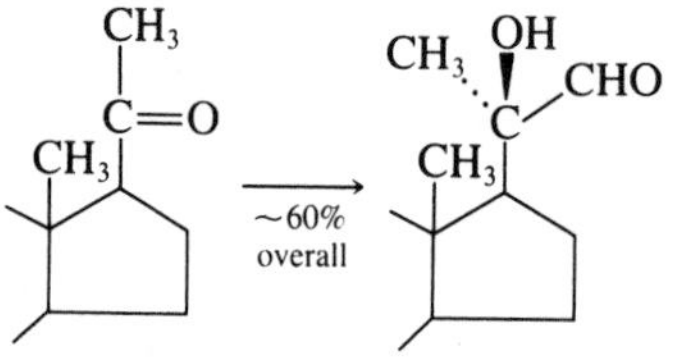

[1] M. Adamczyk, E. K. Dolence, D. S. Watt, M. R. Christy, J. H. Reibenspies, and O. P. Anderson, *J. Org.*, **49**, 1378 (1984).

Potassium *o*-nitrobenzeneperoxysulfonate, $o\text{-NO}_2\text{C}_6\text{H}_4\overset{\text{O}}{\underset{\text{O}}{\overset{\|}{\underset{\|}{\text{S}}}}}\text{OOK}$ **(1).**   This peroxysulfur compound, considered to have the structure indicated, is generated *in situ* by reaction of KO_2 (2 equiv.) with *o*-nitrobenzenesulfonyl chloride at $-35°$. Similar reagents are formed from *m*- and *p*-nitrobenzenesulfonyl chlorides, but **1** is more useful because of greater stability.

Epoxidation of alkenes.[1]   The reagent effects epoxidation of alkenes in yields generally higher than 70%. Epoxidation of alkenes with two double bonds shows regioselectivity as high or higher than that obtained with hydrogen peroxide or monoperphthalic acid.

Examples:

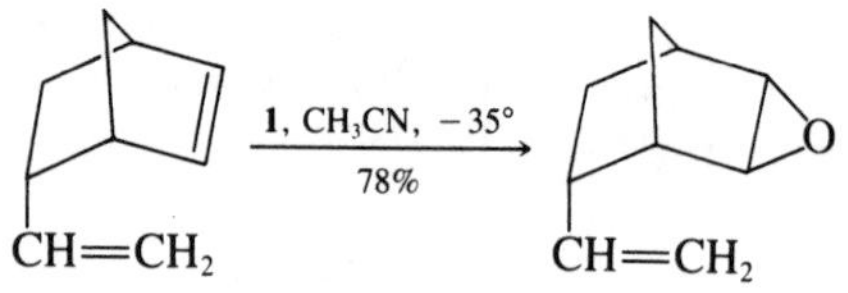

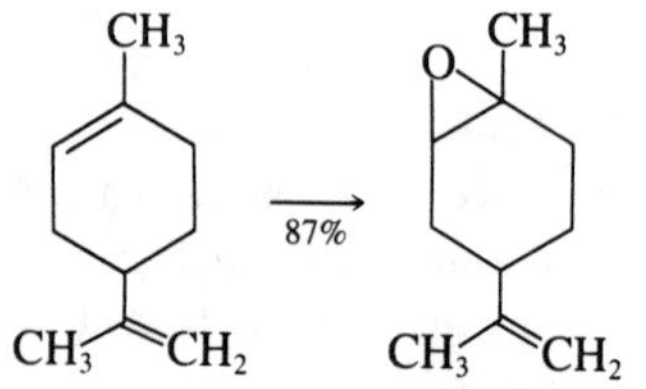

[1] Y. H. Kim and B. C. Chung, *J. Org.*, **48**, 1562 (1983).

Potassium permanganate, 1, 942–952; **2,** 348; **4,** 412–413; **5,** 562–563; **8,** 416–417; **9,** 388–391; **10,** 330; **11,** 440–441.

2,3-Dihydroxy-γ-butyrolactones.[1] γ-Butenolides (1) are oxidized by $KMnO_4$ in the presence of a crown ether (dicyclohexyl-18-crown-6) to mixtures of the cis-diols 2 and 3 with some preference for the former. The stereoselectivity is influenced by the substituent at the γ-position. It is highest with bulky γ-alkoxymethyl groups and lowest with alkyl substituents.

RCH₂ [structure 1] → KMnO₄, CH₂Cl₂, cat., −42°, 25–65% → RCH₂ [structure 2] + RCH₂ [structure 3]   3–50:1

1   2 (HO OH)   3

[1] T. Mukaiyama, F. Tabusa, and K. Suzuki, *Chem. Letters,* 173 (1983).

Potassium peroxomonosulfate (Oxone), **10,** 328; **11,** 442.

Arene oxides.[1] The acetone–oxone system (**11,** 442) converts several polycyclic aromatic hydrocarbons into the K-region oxide, often in synthetically useful yield. Thus the 9,10-oxide of phenanthrene is formed with this system in 60% yield.

[1] R. Jeyaraman and R. W. Murray, *Am. Soc.,* **107,** 2462 (1984).

Potassium superoxide, 6, 488–490; **7,** 304–307; **8,** 417–419; **9,** 391–393; **11,** 442–445.

Oxidation of pyrocatechol.[1] Pyrocatechol (1) is oxidized in the presence of oxygen by KO_2 in DMSO to the very unstable product 2, characterized by transformation into

[structure 1: benzene with two OH] → KO₂, O₂, DMSO → [structure 2: CHO, COOH, OH] → NH₃ → [structure 3: pyridine N, COOH]

1   2   3

picolinic acid (3). This cleavage reaction is a known enzymatic process, but has not been realized previously with chemical systems, which usually cleave the bond between the hydroxyl groups.

Epoxidation.[2] In the presence of catalytic amounts of 18-crown-6 and an acyl chloride, KO_2 can effect epoxidation of alkenes and of some polycyclic arenes derived from phenanthrene. The most effective acyl chlorides are phosgene and benzoyl chloride. The highest yields of epoxides are formed from *trans*-stilbene (89%) and cyclohexene (80%). The 9,10-epoxide of phenanthrene is obtained in 38% yield.

Heterogeneous oxidations. KO_2 can effect oxidation of either 1,2- or 1,3-dihydroxynaphthalene to 2-hydroxy-1,4-naphthoquinone in an aprotic medium in 60% yield. Addition of a crown ether is not essential, but results in a somewhat higher yield (80%). The yield is somewhat lower (50%) in the absence of oxygen.[3]

The α- and β-tetralones are also oxidized to 2-hydroxy-1,4-naphthoquinone by KO_2 and a crown ether in 75% yield. In the absence of a crown ether, the oxidation proceeds slowly and in low yield (~10%).[4] The α- and β-naphthols are intermediates.[5]

[1] R. Müller and F. Lingens, *Angew. Chem., Int. Ed.*, **23**, 79 (1984).
[2] T. Nagano, K. Yokoohji, and M. Hirobe, *Tetrahedron Letters*, **24**, 3481 (1983).
[3] D. Vidril-Robert, M.-T. Maurette, and E. Oliveros, *ibid.*, **25**, 529 (1984).
[4] M. Hocquaux, B. Jacquet, D. Vidril-Robert, M.-T. Maurette, and E. Oliveros, *ibid.*, **25**, 533 (1984).
[5] M. Lissel, *ibid.*, **25**, 2213 (1984).

Potassium triisopropoxyborohydride, 5, 565; **9,** 393. As originally prepared, $K(O\text{-}i\text{-}Pr)_3BH$ contains an impurity, probably $K(O\text{-}i\text{-}Pr)_4B$. Essentially pure reagent can be prepared by reaction of $(O\text{-}i\text{-}Pr)_3B$ with excess KH in refluxing THF.[1] It should be stored over a small amount of KH to prevent disproportionation. This material provides essentially quantitative yields in the hydridation of thexylBR[1]Cl in the presence of a second olefin to form thexylBR[1]R[2].

[1] H. C. Brown, B. Nazer, and J. A. Sikorski, *Organometallics*, **2**, 634 (1983).

Potassium trimethylsilanolate, $KOSi(CH_3)_3$. Supplier: Petrarch.

Potassium carboxylates.[1] Anhydrous potassium salts of carboxylic acids can be prepared in high yield by reaction of methyl or trimethylsilyl esters or acid chlorides or fluorides with $KOSi(CH_3)_3$ in an anhydrous solvent (equations I and II). Sodium or lithium carboxylates can be obtained from $NaOSi(CH_3)_3$ or $LiOSi(CH_3)_3$.

$$\text{(I)} \quad RCOOCH_3 + KOSi(CH_3)_3 \xrightarrow[65-95\%]{\text{ether}} RCOOK + (CH_3)_3SiOCH_3$$

$$\text{(II)} \quad RCOCl + KOSi(CH_3)_3 \longrightarrow [RCOOSi(CH_3)_3] \xrightarrow[75-85\%]{KOSi(CH_3)_3} RCOOK + [(CH_3)_3Si]_2O$$

[1] E. D. Laganis and B. L. Chenard, *Tetrahedron Letters*, **25**, 5831 (1984).

(S)-(−)-Proline, 6, 492–493; **7,** 307; **8,** 421–424; **10,** 331–332; **11,** 446–447.

Asymmetric aldol cyclization. Complete details are available for the cyclization of the triketone **1** to the optically active bicyclic aldol product **2**, first reported in 1974 (**6,** 411). The high asymmetric induction is attributed to formation of the rigid intermediate **A** by virtue of two hydrogen bonds. Other amino acids are considerably less effective for this asymmetric cyclization.[1]

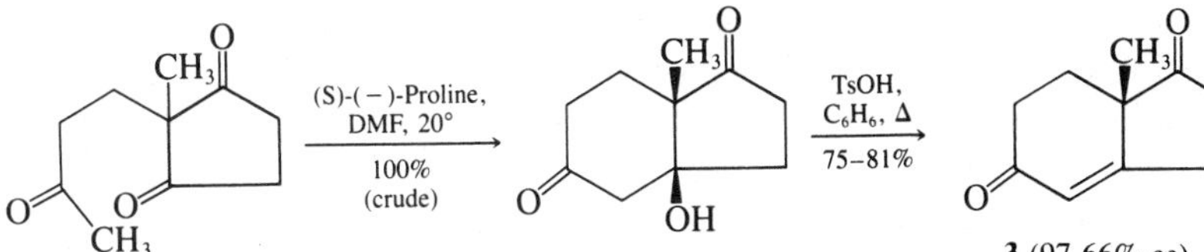

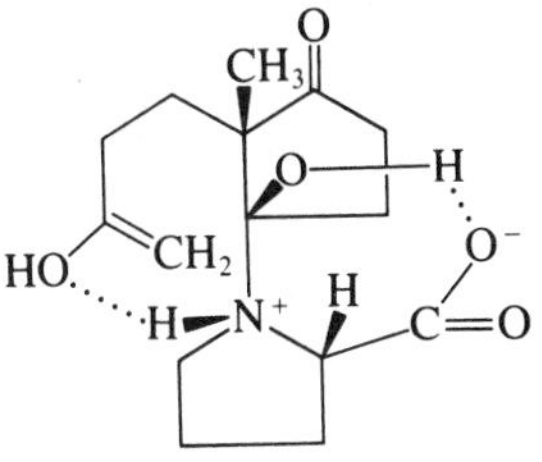

A

The cyclization of the acyclic 1,5-diketone (**4**) catalyzed by (S)-proline gives the enone (R)-(−)-**5** in 43% optical purity. In this case (S)-phenylalanine is much less effective than (S)-proline.[2]

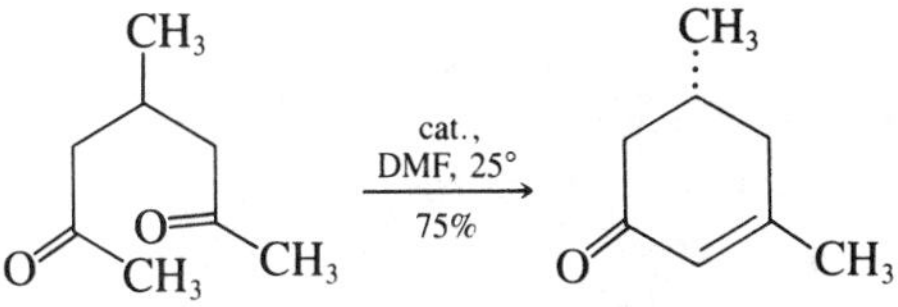

[1] Z. G. Hajos and D. R. Parrish, *Org. Syn.*, **63**, 26 (1985).
[2] C. Agami and H. Sevestre, *J.C.S. Chem. Comm.*, 1385 (1984).

Propionic acid, C_2H_5COOH (**1**), b.p. 141°.

Dealkoxycarbonylation.[1] β-Keto esters and malonic esters with an enolizable proton α to the ester group undergo dealkoxycarbonylation in 80–95% yield when heated in propionic acid for 24–72 hours. β-Keto esters can also be cleaved by refluxing in acetic acid. The requirement for an α-proton suggests that the reaction involves loss of ROH to give a ketene intermediate. The propionic acid is converted into the anhydride.

[1] R. T. Brown and M. F. Jones, *J. Chem. Res. (S)*, 332 (1984).

2-Propynyltriphenyltin, $HC\equiv CCH_2Sn(C_6H_5)_3$ (**1**). The reagent is prepared by reaction of $HC\equiv CCH_2MgBr$ with $ClSn(C_6H_5)_3$ in ether (70% yield).[1]

Terminal allenes.[2] In the presence of a radical initiator, alkyl bromides or iodides react with **1** to provide allenes in moderate to good yield. The synthesis of a natural amino acid (**3**) from **2** by this reaction proceeds without racemization.

[1] M. Le Quan and P. Cadiot, *Compt. Rend. C,* **254**, 133 (1962).
[2] J. E. Baldwin, R. M. Adlington, and A. Basak, *J.C.S. Chem. Comm.,* 1284 (1984).

Pyridine, 1, 958–963; **2**, 349–351; **4**, 914–915; **6**, 497; **11**, 448.

Deconjugation of α,β-enones.[1] The photochemical deconjugation of α,β-unsaturated acids and esters via a dienol is often a useful preparative route to the β,γ-unsaturated isomers. The corresponding reaction with α,β-enones is erratic; it can result in reversion to the original enone or the *trans*-isomer. Irradiation in the presence of a mild base such as pyridine (or imidazole) can promote conversion of α,β-enones to β,γ-enones. A stronger base such as triethylamine catalyzes reconjugation of the intermediate dienol. Base-catalyzed deconjugation is more efficient in polar solvents (DMF, CH_3CN, CH_3OH) than in ether or hydrocarbon solvents. Thus under proper conditions several α,β-enones previously considered inert to irradiation can be isomerized to the β,γ-isomers.

[1] S. L. Eng, R. Ricard, C. S. K. Wan, and A. C. Weedon, *J.C.S. Chem. Comm.,* 236 (1983).

2-Pyridineethanol, CH_2CH_2OH (**1**). Mol. wt. 123.16, b.p. 114–116°/9 mm. Supplier: Aldrich.

Protection of carboxylic acids.[1] The 2-(2-pyridyl)ethyl group is useful for protection of amino acids. The esters are obtained by reaction with **1** and DCC/1-hydroxybenzotriazole. The group provides higher solubility in protic solvents, and is stable to both acids and bases. It is removed by treatment with methyl iodide and a weak base.

[1] H. Kessler, G. Becker, H. Kogler, and M. Wolff, *Tetrahedron Letters,* **25**, 3971 (1984).

2-Pyridinethiol-1-oxide,   [structure] —SH   **(1)**.   Mol. wt. 127.17, m.p. 69–72°. Supplier:
Aldrich.

    Decarboxylation.   Acid chlorides react with the sodium salt of **1** (DMAP catalysis) to form esters (**2**) derived from N-hydroxypyridine-2-thione. These esters undergo radical chain decarboxylation to the noralkane on heating with tri-*n*-butyltin hydride (equation I).[1]

$$(I) \quad RCO-N \xrightarrow[70-95\%]{\substack{Bu_3SnH, \\ C_6H_6, \ 40-80°}} RH + CO_2 + \ N{-}SSnBu_3$$

    2

    The esters are converted by a radical chain reaction into alkyl chlorides in refluxing carbon tetrachloride (equation II). Reaction with bromotrichloromethane or iodoform results in alkyl bromides or iodides in similar yields.[2]

$$(II) \quad 2 + CCl_4 \xrightarrow[70-95\%]{} RCl + CO_2 + \ N{-}SCCl_3$$

    The N-hydroxypyridine-2-thione esters of Boc-protected amino acids undergo decarboxylation on photolysis in the presence of a hydrogen-atom-transfer reagent such as *t*-butyl mercaptan in 80–95% yield. Decarboxylation of the side-chain carboxy groups of suitably protected aspartic and glutamic acids can also be effected, but yields are lower (50–80%).[3]

[1] D. H. R. Barton, D. Crich, and W. B. Motherwell, *J.C.S. Chem. Comm.*, 939 (1983).
[2] *Idem, Tetrahedron Letters*, **24**, 4979 (1983).
[3] D. H. R. Barton, Y. Hervé, P. Potier, and J. Thierry, *J.C.S. Chem. Comm.*, 1298 (1984).

Pyridinium chlorochromate, 6, 498–499; **7**, 308–309; **8**, 425–427; **9**, 397–399; **10**, 334–335; **11**, 450–452.

    Oxidation of 5,6-dihydropyranes.[1]   These cyclic allylic ethers are oxidized by PCC directly to α,β-unsaturated δ-lactones.

Examples:

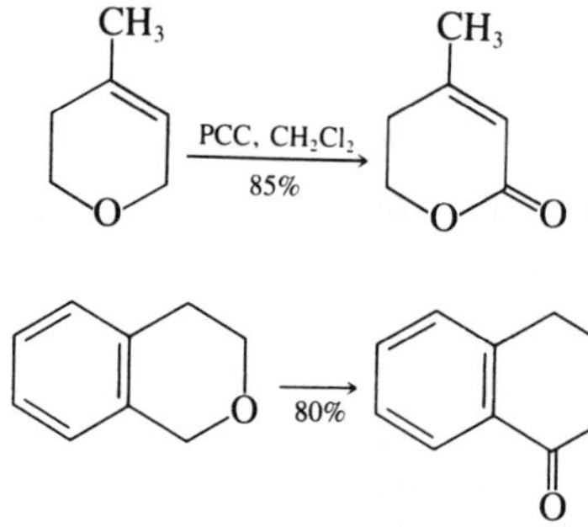

$$R^1CH(OH)CH(NO_2)R^2 \longrightarrow R^1\overset{\overset{O}{\|}}{C}CH(NO_2)R^2.^2$$ 2-Nitroalkanols, available by a nitro-aldol reaction,[3] undergo a retroaldol reaction under acidic conditions, but can be oxidized to α-nitro ketones with PCC in CH_2Cl_2 in 60–85% yield.

Lactonization of a δ,ε-unsaturated ester.[4] In a projected synthesis of quassinoid diterpenes, the final step involves lactonization of the unsaturated ester **1**. Acid-catalyzed solvolysis (HCl in THF) or treatment with an arylsulfonic acid in toluene results mainly

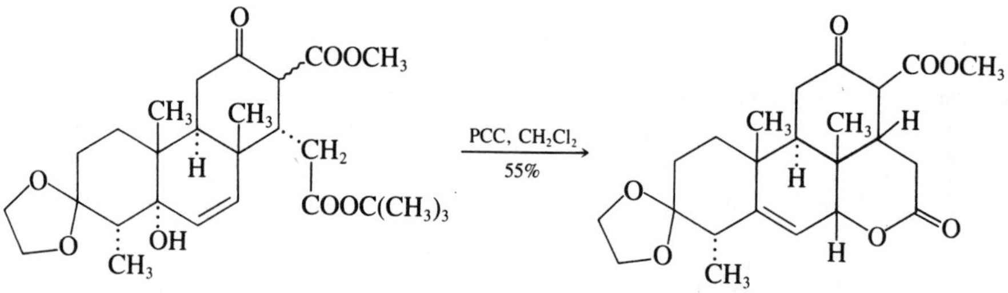

in hydrolysis of the ester and/or cleavage of the ketal group. The desired lactonization to **2** unexpectedly is effected in 55% yield by reaction with PCC in methylene chloride.

[1] F. Bonadies, R. Di Fabio, and C. Bonini, *J. Org.*, **49**, 1647 (1984).
[2] G. Rosini and R. Ballini, *Synthesis*, 543 (1983).
[3] D. Seebach, A. K. Beck, T. Mukhopadhyay, and E. Thomas, *Helv.*, **65**, 1101 (1982).
[4] C. H. Heathcock, C. Mahaim, M. F. Schlecht, and T. Utawanit, *J. Org.*, **49**, 3264 (1984).

Pyridinium chlorochromate–Pyrazole.

Selective oxidation of allylic alcohols. The complex of PCC with pyrazole[1] or 3,5-dimethylpyrazole[2] is recommended for selective oxidation of allylic alcohols in the presence of nonallylic hydroxyl groups. The complex of chromic anhydride and dimethylpyrazole (**8**, 110; **9**, 116) has been used for this oxidation.

[1] E. J. Parish, S. Chitrakorn, and S. Lowery, *Lipids,* **19**, 550 (1984).
[2] E. J. Parish and A. D. Scott, *J. Org.,* **48**, 4766 (1983).

Pyridinium dichromate–Acetic anhydride.

Oxidation of alcohols.[1] The reactivity of PDC is increased by acetic anhydride. Primary and secondary alcohols are oxidized efficiently with 0.6–0.7 equiv. of PDC and 3 equiv. of Ac_2O in CH_2Cl_2 at 40°. Addition of DMF in the oxidation of primary alcohols retards further oxidation to the carboxylic acid. Acid-sensitive groups are stable to the conditions.

[1] F. Andersson and B. Samuelsson, *Carbohydrate Res.,* **129**, Cl (1984).

Pyridinium fluorochromate (PFC), **11**, 453.

Selective oxidation of secondary alcohols.[1] This oxidant is almost as effective as PCC, but is less acidic and does not attack *t*-butyldimethylsilyl ethers. This property allows selective oxidation of a secondary hydroxyl group in the presence of the silyl ether of a primary hydroxyl group.

[1] T. Nonaka, S. Kanemoto, K. Oshima, and H. Nozaki, *Bull. Chem. Soc. Japan,* **57**, 2019 (1984).

Pyridinium *m*-nitrobenzenesulfonate, 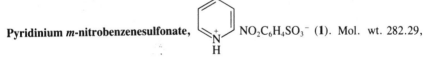 $NO_2C_6H_4SO_3^-$ (**1**). Mol. wt. 282.29,

m.p. 120–123°. In contrast to pyridinium *p*-toluenesulfonate, **1** is not hygroscopic and is highly soluble in CH_2Cl_2. It is useful as a catalyst for ester exchange reactions.[1]

[1] M. Sekine and T. Hata, *Am. Soc.,* **105**, 2044 (1983).

Pyridinium poly(hydrogen fluoride), 5, 528–529; **6**, 473–474; **7**, 294; **9**, 399–400; **11**, 453–454.

Glycosyl fluorides.[1,2] The anomeric hydroxyl or acetoxy group of monosaccharides is converted into a fluoro group by reaction with Olah's reagent in 70–90% yield. This reaction is useful for preparation of the thermodynamically more stable β-glycosyl fluoride ($\beta/\alpha \approx >95:5$).

α-Halocarboxylic acids.[3] Diazotization ($NaNO_2$) of α-amino acids in HF/pyridine (48:52 w/w) in the presence of KCl or KBr results in α-chloro- or α-bromocarboxylic acids, respectively. Yields are ~50–85%. Use of KI in this reaction leads to complex mixtures.

[1] M. Hayashi, S. Hashimoto, and R. Noyori, *Chem. Letters,* 1747 (1984).
[2] W. A. Szarek, G. Grynkiewicz, B. Doboszewski, and G. W. Hay, *ibid.,* 1751 (1984).
[3] G. A. Olah, J. Shih, and G. K. S. Prakash, *Helv.,* **66**, 1028 (1983).

Pyridinium *p*-toluenesulfonate, 8, 427–428; **9,** 400.

2-*Methoxybutadienes*.[1]   These dienes can be obtained by alkylation of the anion of methoxyallene followed by isomerization to the diene with pyridinium tosylate. Other acid catalysts are not useful. The (E)-isomer of the diene is formed predominantly or exclusively.

$$CH_2\!=\!C\!=\!CHOCH_3 \xrightarrow[\text{2) RCH}_2\text{X}]{\text{1) }n\text{-BuLi, THF}} CH_2\!=\!C\!=\!C\!\begin{array}{l} \nearrow CH_2R \\ \searrow OCH_3 \end{array} \xrightarrow[\text{40–50\%}]{\substack{\text{PyH·Ts,} \\ \text{CH}_2\text{Cl}_2}}$$

$$(E/Z = 5:1\text{--}100:1)$$

Deprotection of MEM and MOM ethers.[2]   Methoxyethoxymethyl and methoxymethyl ethers are cleaved to the alcohol by this reagent in high yields. Either 2-butanone or 2-methyl-2-propanol is recommended as solvent. The procedure is particularly useful for cleavage of ethers of allylic alcohols, for which $ZnBr_2$ and $TiCl_4$ are not useful.

[1] A. Kucerovy, K. Neuenschwander, and S. M. Weinreb, *Syn. Comm.,* **13**, 875 (1983).
[2] H. Monti, G. Léandri, M. Klos-Ringuet, and C. Corriol, *ibid.,* 1021 (1983).

2-(2-Pyridyl)ethyl-*p*-nitrophenyl carbonate, The reagent is prepared by reaction of (2-pyridyl)ethanol and *p*-nitrophenyl chloroformate.

Protection of amines.[1]   The 2-(2-pyridyl)ethoxycarbonyl group (Pyoc) is useful for protection of the terminal amino group in peptide synthesis. The group provides solubility in water and stability to both acids and bases. Peptide synthesis can be effected with a water-soluble carbodiimide in the presence of 1-hydroxybenzotriazole. The group is removed by conversion to a pyridinium salt with methyl iodide followed by elimination with dimethylamine.

[1] H. Kunz and R. Barthels, *Angew. Chem., Int. Ed.*, **22**, 783 (1983).

(S)-(+)-2-(1-Pyrrolidinyl)methylpyrrolidine, **(1)**. Preparation.[1]

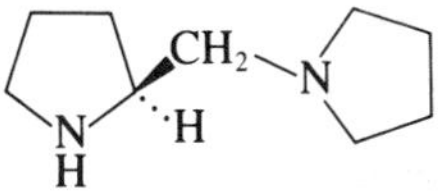

Asymmetric rearrangement of cyclohexene oxide.[2] Cyclohexene oxide is rearranged to (S)-2-cyclohexene-1-ol in 92% ee by the chiral lithium amide (**2**) prepared from *n*-butyllithium and **1**. Several related (S)-2-(disubstituted aminomethyl)pyrrolidines prepared from (S)-proline are almost as stereoselective.[3]

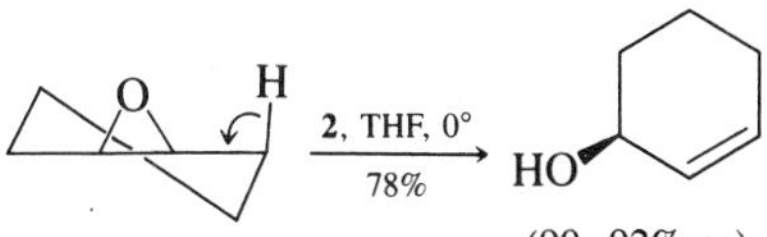

(90–92% ee)

[1] T. Sone, K. Hiroi, and S. Yamada, *Chem. Pharm. Bull. Japan*, **21**, 2331 (1973).
[2] M. Asami, *Chem. Letters*, 829 (1984).
[3] R. P. Thummel and B. Rickborn, *Am. Soc.*, **92**, 2064 (1970).

R

Raney nickel, 1, 723–731; **2**, 293–294; **5**, 570–571; **6**, 502; **7**, 312; **8**, 433; **9**, 405–406; **10**, 339–340; **11**, 457–458.

Reduction of nitroalkenes.[1] Vinyl nitro compounds are reduced to saturated ketones or aldehydes by Raney nickel and an aqueous solution of sodium hypophosphite. Ester groups, C=C bonds, and aryl NO_2 and halo groups are not reduced. Under these conditions nitroalkanes are reduced to amines, and oximes are converted into carbonyl compounds in high yield.

Examples:

$$o\text{-}HOC_6H_4CH=C \underset{NO_2}{\overset{CH_3}{<}} \xrightarrow[70\%]{\text{Raney Ni,}\ Na_2H_2PO_2} o\text{-}HOC_6H_4CH_2COCH_3$$

$$p\text{-}CH_3OC_6H_4CH=CHNO_2 \xrightarrow[53\%]{} p\text{-}CH_3OC_6H_4CH_2CHO$$

Reduction of oximes to ketones.[2] Ketoximes can be converted to ketones by hydrogenation catalyzed by Raney nickel (deactivated with acetone) in $THF/CH_3OH/H_2O$ or in CH_3OH/H_2O containing boric acid to facilitate hydrolysis of the intermediate imine (equation I).

$$(I) \quad \overset{R^1}{\underset{R^2}{>}}C=NOH \xrightarrow{H_2} \left[\overset{R^1}{\underset{R^2}{>}}C=NH \right] \xrightarrow[70-100\%]{H_2O} \overset{R^1}{\underset{R^2}{>}}C=O$$

[1] D. Monti, P. Gramatica, G. Speranza, and P. Manitto, *Tetrahedron Letters*, **24**, 417 (1983).
[2] D. P. Curran, J. F. Brill, and D. M. Rakiewicz, *J. Org.*, **49**, 1654 (1984).

Reformatsky reagent, 1, 1285–1286.

γ, δ-Unsaturated esters.[1] These esters can be prepared stereo- and regioselectively by coupling of the Reformatsky reagent with allylic bromides or chlorides in the presence of $Cu(acac)_2$.

Examples:

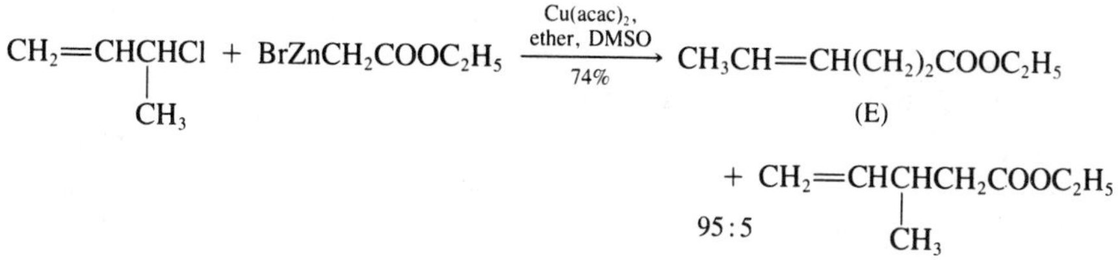

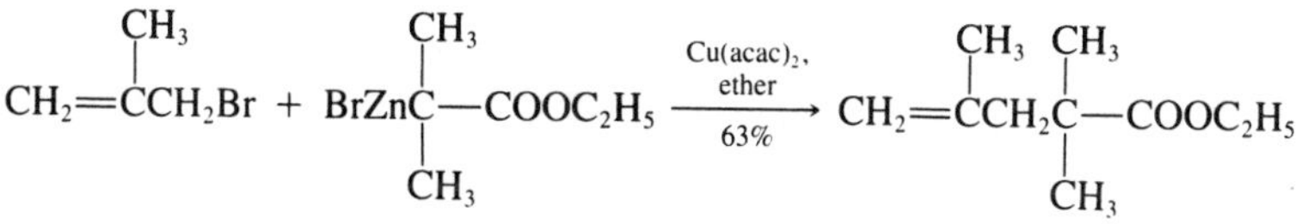

[1] M. Gaudemar, *Tetrahedron Letters*, **24**, 2749 (1983).

Rhodium(II) carboxylates, 5, 571–572; **7**, 313; **8**, 434–435; **9**, 406–408; **10**, 340–342; **11**, 458–460.

δ-Lactones.[1] In a synthesis of pentalenolactone E (**4**), the lactone ring was constructed by acylation of the alcohol **1** with the tosylhydrazone of glyoxalyl chloride followed by elimination of TsOH to give the diazoacetate **2**. In the presence of $Rh_2(OAc)_4$ in an inert Freon solvent, **2** undergoes an intramolecular carbene insertion at the adjacent bridgehead to generate the δ-lactone ring of **3** in satisfactory yield. Two factors favor the

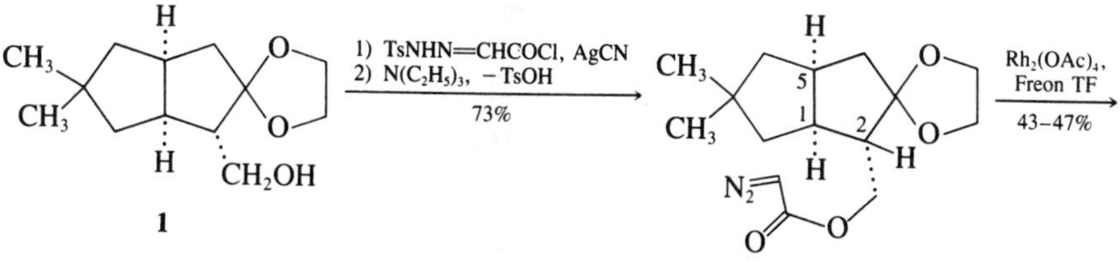

1 2

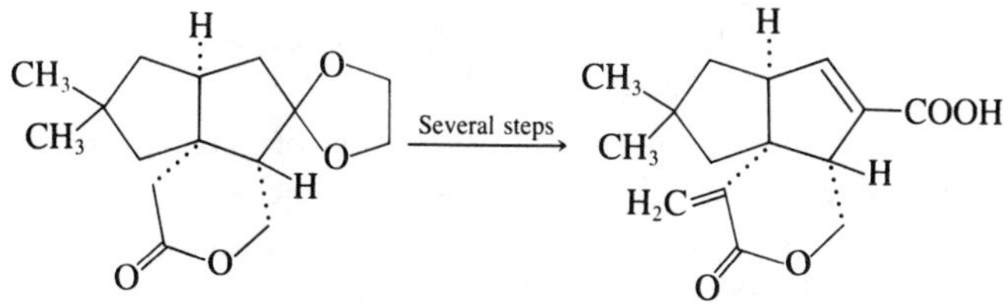

3 4

desired insertion reaction: It involves a tertiary center and results in a six-membered lactone. Evidently insertion at C_2 is disfavored by steric factors, and insertion at C_5 leads to a seven-membered ring. Further studies suggest that intramolecular carbene insertions particularly at secondary centers, are unusually sensitive to steric interactions.

Chiral cyclopentanones.[2]   The regioselective cyclization of α-diazo-β-keto esters to cyclopentanones (**11**, 459) is also enantioselective with substrates derived from chiral alcohols. Preliminary studies show that steric factors affect the diastereoselectivity; the highest diastereoselectivity is obtained with esters of the alcohol **1**, which is available from camphor,[3] and in which both the bornane and the naphthalene rings can exert steric effects on the diastereoselectivity.

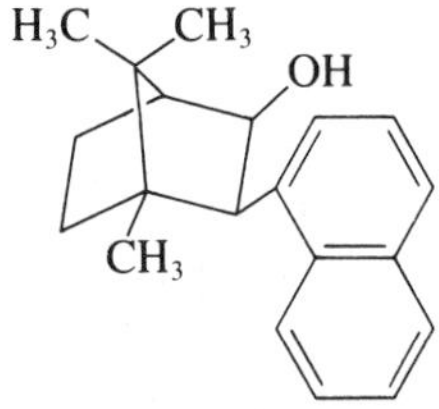

1 = R*OH

Example:

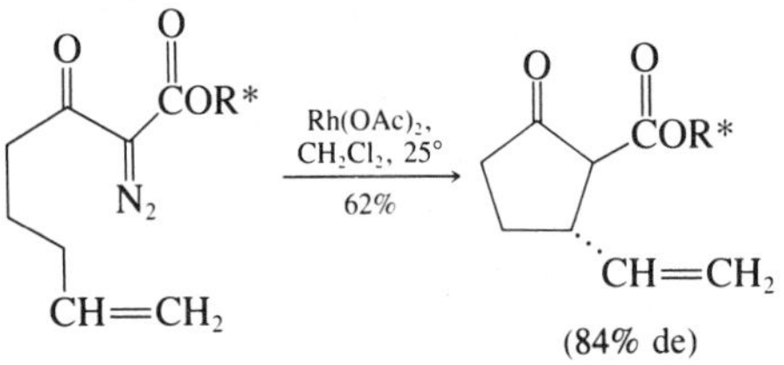

***Intramolecular Buchner reaction.*[4]**   Rh(OAc)$_2$ is more efficient than CuCl[5] as the catalyst for cyclization of the α-diazo ketone **1**, derived from 3-phenylpropionic acid, to bicyclo[5.3.0]decatrienone (**2**). This product isomerizes in the presence of triethylamine to the more stable trienone **4**. A useful isomerization to β-tetralone (**3**) occurs in the presence of trifluoroacetic acid.

The substitution pattern in the benzene ring of **1** controls the regioselectivity of this β-tetralone synthesis. 7-Substituted-2-tetralones are obtained from precursors substituted in the *p*-position (CH$_3$, OCH$_3$, OAc). An *o*-methyl substituent in **1** results in 5-methyl-2-tetralone (86%), whereas a *m*-methoxyl substituent in **1** results in a one-step direct conversion to 6-methoxy-2-tetralone (86%).

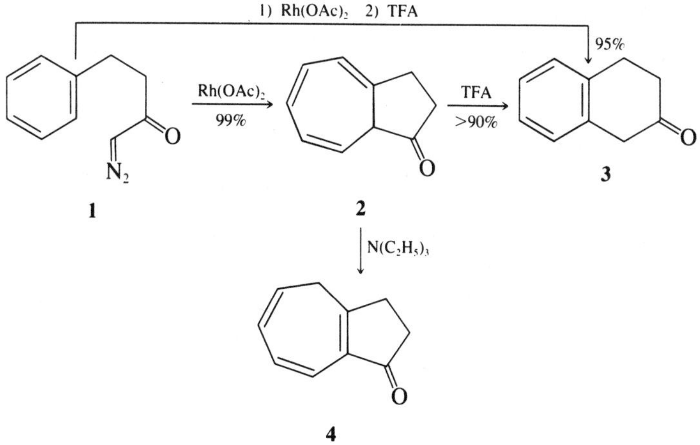

Hydroformylation.   Rhodium catalysts are generally preferred to cobalt catalysts by industry for hydroformylation of alkenes to give the homologous aldehydes because of the former's greater activity and higher regioselectivity.[6]

Hydroformylation of a secondary homoallylic alcohol such as **1** followed by PCC oxidation provides a useful route to the δ-lactone **2**. The product was converted by known reactions into the Prelog-Djerassi lactone (**3**).[7]

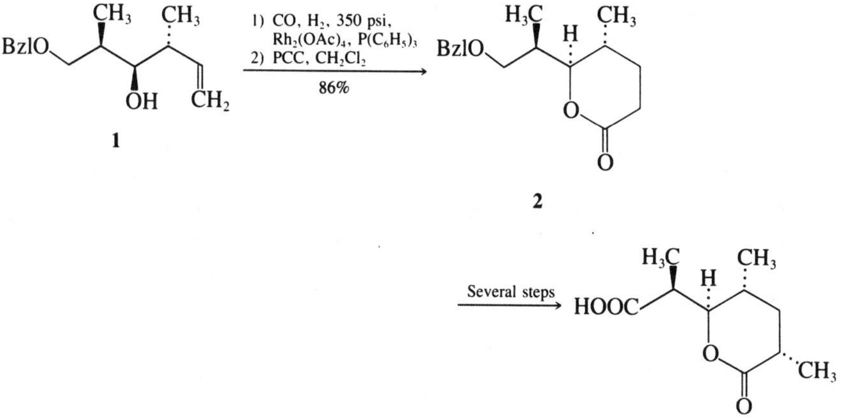

[1] D. E. Cane and P. J. Thomas, *Am. Soc.*, **106**, 5295 (1984).
[2] D. F. Taber and K. Raman, *ibid.*, **105**, 5935 (1983).
[3] J. M. Coxon, M. P. Hartshorn, and A. J. Lewis, *Aust. J. Chem.*, **24**, 1017 (1971).
[4] M. A. McKervey, S. M. Tuladhar, and M. F. Twohig, *J.C.S. Chem. Comm.*, 129 (1984).

[5] L. T. Scott, M. A. Minton, and M. A. Kirms, *Am. Soc.*, **102**, 6311 (1980).
[6] H. Siegel and W. Himmele, *Angew. Chem., Int. Ed.*, **19**, 178 (1980).
[7] P. G. M. Wuts, M. L. Obrzut, and P. A. Thompson, *Tetrahedron Letters*, **25**, 4051 (1984).

Rhodium catalysts, 1, 982–983; **4**, 418–419; **6**, 503; **8**, 433–434; **11**, 460.

Alkoxycarbonylation of amines.[1] Carbamates can be prepared by reaction of a primary amine with carbon monoxide, oxygen, and an alcohol catalyzed by either 5% rhodium on activated carbon or palladium black and an alkali metal halide, particularly an iodide such as CsI or KI (equation I). Essentially no reaction occurs in the absence of the salt. Dialkylureas are intermediates in the reaction, and can be isolated as the major product when less active catalysts such as $IrCl_3$ are used.

$$(I)\ R^1NH_2 + CO + R^2OH + \tfrac{1}{2}O_2 \xrightarrow[-H_2O]{\substack{Rh/C,\\ CsI}} R^1NHCOOR^2 + R^1NHC\!\!\!\underset{\substack{\|\\O}}{}\!\!\!NHR^1$$

[1,4-*Bis(diphenylphosphine)butane*]*norbornadienerhodium(I) tetrafluoroborate* (1). Highly stereoselective hydrogenation with the cationic Rh(I) catalyst **1** of the acyclic

allylic alcohol **2** and a related homoallylic alcohol, but not of the corresponding acetates, indicates that a proximal hydroxyl group at a chiral center can direct the course of reduction of a double bond, presumably by chelate complexation with the catalyst.[2]

Highly diastereoselective hydrogenation of the chiral allylic alcohols **3** and **4** is obtained with this Rh(I) catalyst at elevated hydrogen pressures (640 psi). At low pressures, the catalyst effects interconversion of **3** and **4**. Hydrogenation of **3** and **4** catalyzed by Crabtree's cationic iridium catalyst, $Ir(COD)(Py)(PCy_3)PF_6$ (**8**, this volume), which is known to be very effective for stereoselective hydrogenation of cyclic allylic and homoallylic alcohols, shows slight diastereoselectivity.[3] Surprisingly, in this case, a decrease in the catalyst/substrate ratio results in an increase in the diastereoselection. Even so, the Rh catalyst **1** is superior to the Ir catalyst **8** for stereoselective reduction of **3** and **4**. The catalyst $Ir(1,4\text{-bisdiphos})(COD)BF_4$ is somewhat more effective than the Ir catalyst **8**, but less effective than **1**.

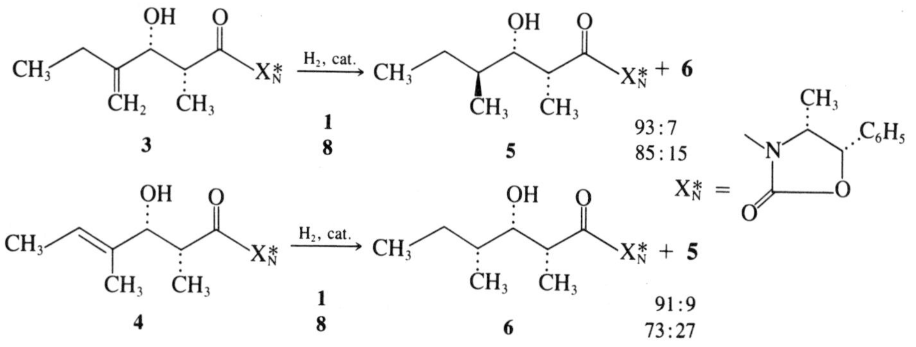

The Rh(I) catalyst **1** is also superior to the Ir(I) catalyst **8** for diastereoselective reduction of cyclic allylic and homoallylic alcohols (equations I and II).[4]

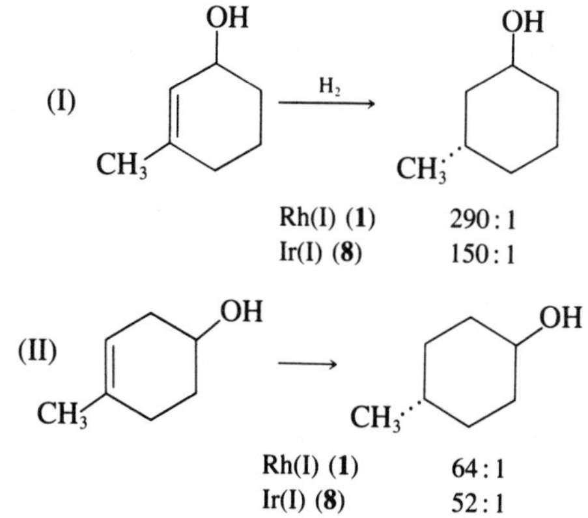

The Rh(I) catalyst can effect hydrogenation of highly hindered allylic alcohols that resist reduction with the Ir(I) catalyst or Pd and Pt catalysts.

Example:

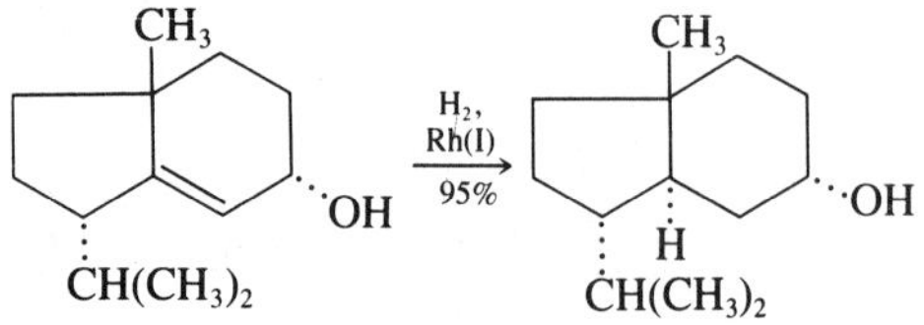

[1] S. Fukuoka, M. Chono, and M. Kohno, *J. Org.*, **49**, 1458 (1984).
[2] J. M. Brown and R. G. Naik, *J.C.S. Chem. Comm.*, 348 (1982).
[3] D. A. Evans and M. M. Morrissey, *Am. Soc.*, **106**, 3866 (1984).
[4] *Idem, Tetrahedron Letters,* **25**, 4637 (1984).

Ruthenium(IV) oxide, 1, 986–989; **2**, 357–359; **3**, 293–294; **4**, 420–421; **6**, 504–506; **7**, 315; **8**, 438; **10**, 343; **11**, 462–463.

Oxidation of benzyl ethers.[1]   Benzyl ethers are oxidized to benzoate esters in yields of 54–96% by ruthenium tetroxide using the Sharpless conditions (**11**, 462). This reaction provides a useful method for cleavage of benzyl ethers, which is usually effected by hydrogenation.

Oxidation of allylic alcohols.[2]   Ruthenium(IV) oxide, particularly the hydrate, is more efficient than MnO_2 for oxidation of allylic alcohols to the corresponding aldehydes. Only catalytic amounts are required if the oxidation is conducted under oxygen. An antioxidant is also required to prevent further oxidation. Either system oxidizes primary allylic alcohols in high yield (76–98%); yields are lower in oxidations of secondary allylic alcohols.

[1] P. F. Schuda, M. B. Cichowicz, and M. R. Heimann, *Tetrahedron Letters,* **24**, 3829 (1983).
[2] M. Matsumoto and N. Watanabe, *J. Org.*, **49**, 3435 (1984).

S

Salcomine, 2, 360; **3**, 245; **6**, 507.

Naphthoquinones.[1] 1,5-Dihydroxynaphthalene (**2a**) and the monomethyl ether (**2b**) are conveniently oxidized to the corresponding 1,4-naphthoquinones by oxygen in the presence of this catalyst (**1**).

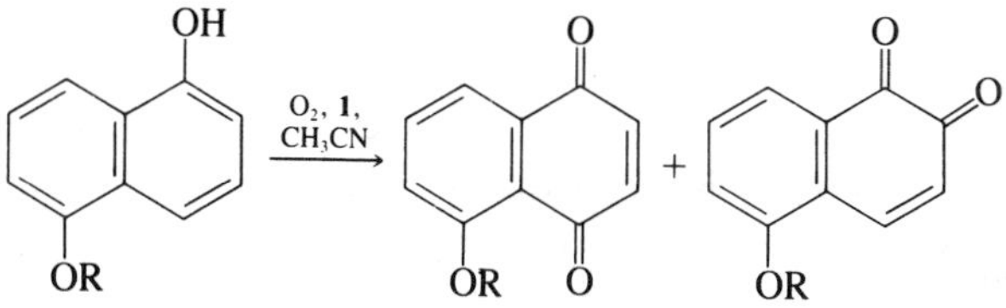

2a, R = H   3 (71%)   4 (14%)
2b, R = CH₃   86%   4 : 1

[1] T. Wakamatsu, T. Nishi, T. Ohnuma, and Y. Ban, *Syn. Comm.*, **14**, 1167 (1984).

Samarium(II) iodide, 8, 439; **10**, 344; **11**, 464–465.

α-Hydroxy ketones.[1] In the presence of SmI_2 (2 equiv.), acid chlorides react with aldehydes or ketones to give α-hydroxy ketones (30–85% yield). The reaction is believed

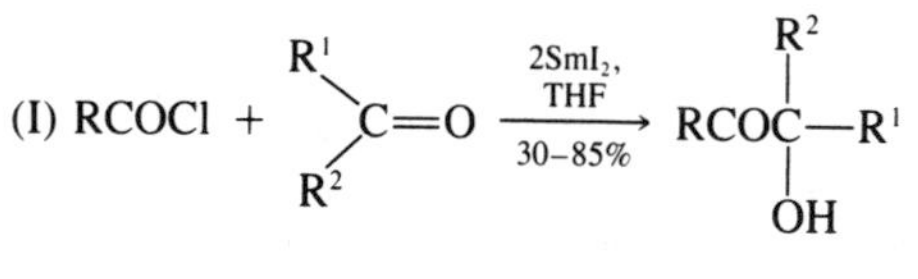

to involve reduction of the acid chloride by SmI_2 to an acyl anion, which is then trapped by the aldehyde or ketone.

Barbier cyclization (*cf.* **11**, 307–308).[2] SmI_2 is effective for intramolecular cyclization of α-(ω-iodoalkyl)cycloalkanones to bicyclic alcohols. Yields can be improved by addition of Fe(III) catalysts such as tris(dibenzoylmethane)iron (**6**, 259). Cyclizations induced with ytterbium(II) iodide are also stereoselective.

Example:

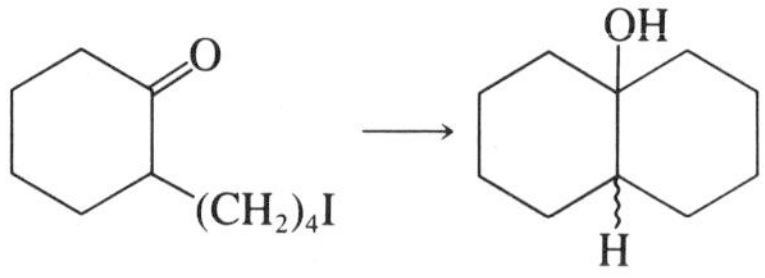

		cis/trans
SmI_2	71%	1:3
SmI_2, Fe(III)	95%	1:1.5
YbI_2	68%	1:5.6

Hydroxymethylation of carbonyl compounds.[3] Alkyl chloromethyl ethers react in the presence of SmI_2 (2 equiv.) with ketones in THF to give adducts in 50–80% yield. The reaction with aldehydes is conducted in tetraethylene glycol dimethyl ether, which suppresses pinacol reduction. Hydroxymethylation of carbonyl compounds can be effected by use of benzyl chloromethyl ether followed by hydrogenolysis of the adduct.

Example:

$$CH_3CO(CH_2)_3COCH_3 + ClCH_2OCH_2C_6H_5 \xrightarrow[56\%]{2SmI_2,\ THF} CH_3CO(CH_2)_3C(OH)(CH_3)CH_2OCH_2C_6H_5$$

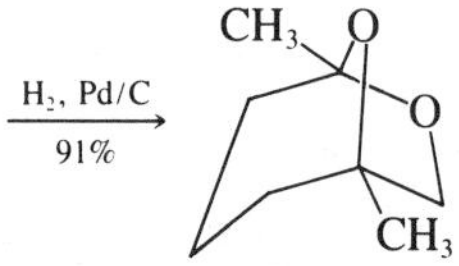

Pinacol coupling.[4] Pinacols are formed in 80–95% yield by reduction of aldehydes and ketones with SmI_2 (1 equiv.) in THF at 25°. The reaction is slower with ketones than with aldehydes.

[1] J. Souppe, J.-L. Namy, and H. B. Kagan, *Tetrahedron Letters*, **25**, 2869 (1984).
[2] G. A. Molander and J. B. Etter, *ibid.*, **25**, 3281 (1984).
[3] T. Imamoto, T. Takeyama, and M. Yokoyama, *ibid.*, **25**, 3225 (1984).
[4] J.-L. Namy, J. Souppe, and H. B. Kagan, *ibid.*, **24**, 765 (1983).

L(−)-Serine, $H_2N{-}CH(COOH){-}CH_2OH$. Mol. wt. 105.09, m.p. 222°, α_D +14.7°.

D-α-*Amino acids.* Rapoport *et al.*[1] have developed a general synthesis of unnatural D-α-amino acids with L-serine as the chiral educt. The method involves aminoacylation of an appropriate Grignard reagent with the dilithium salt of N-phenylsulfonyl-L-serine (**1**) to give N-protected α-amino ketones (**2**). Reduction of the carbonyl group followed by oxidation of the primary hydroxyl group results in an N-protected D-amino acid (**4**). The method is formulated for the synthesis of D-norleucine (**5**).

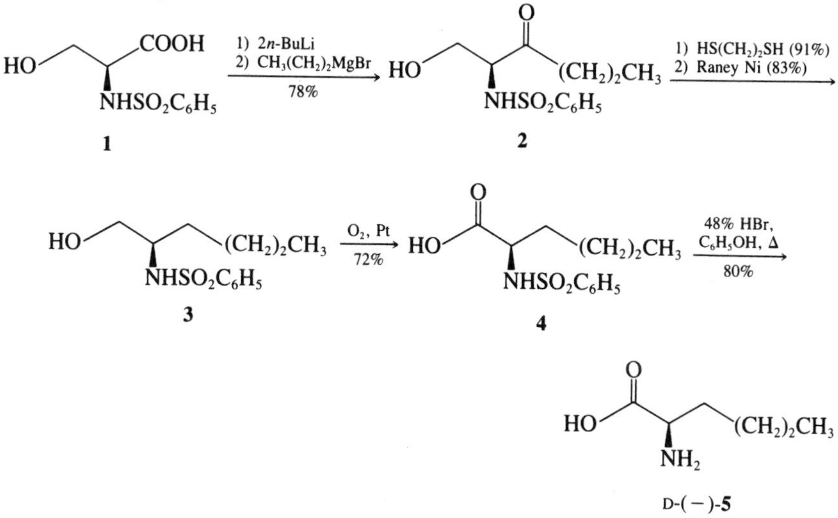

[1] P. J. Maurer, H. Takahata, and H. Rapoport, *Am. Soc.*, **106**, 1095 (1984).

Silica, **6**, 510; **9**, 910; **10**, 346–347; **11**, 466.

Lactonization of γ-halo esters.[1]   Silica gel (2–6 equiv.) promotes lactonization of these esters in refluxing xylene (3–15 hours).

Examples:

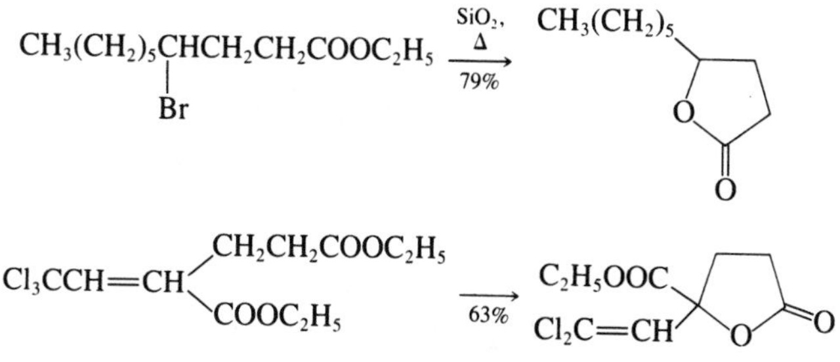

Bis-γ-butyrolactones can also be obtained in this way. An example is the synthesis of canadensolide (1).

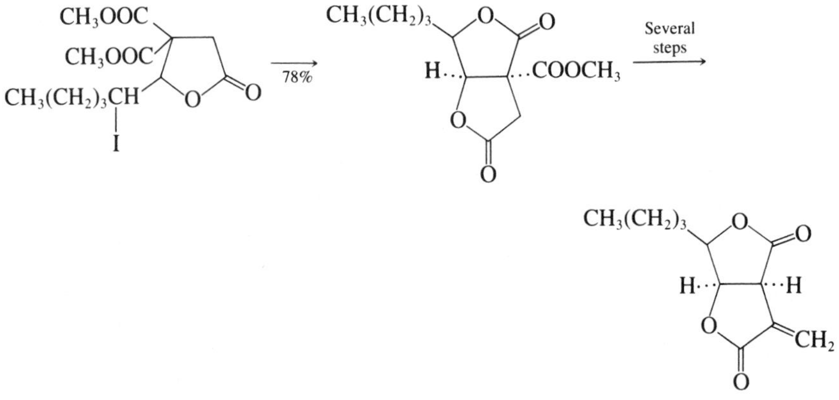

[1] S. Tsuboi, H. Fujita, K. Muranaka, K. Seko, and A. Takeda, *Chem. Letters,* 1909 (1982).

Silver carbonate, 1, 1005; **2,** 363; **4,** 425; **6,** 511; **8,** 441.

Aromatic methoxyamidation.[1]   The reaction of silver salts with N-chloro-N-methoxyamides in an acid medium generates a nitrenium ion that is sufficiently stable to permit aromatic substitution (equation I). The products can be converted into anilines by hydrogenation and hydrolysis.

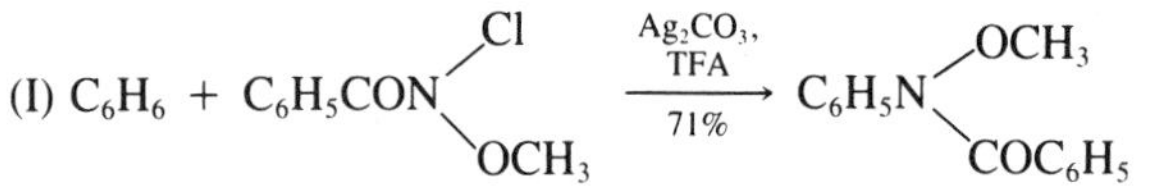

The intramolecular version of this aromatic substitution provides a route to nitrogen heterocycles, in particular 1-methoxy-2-oxindoles.

Example:

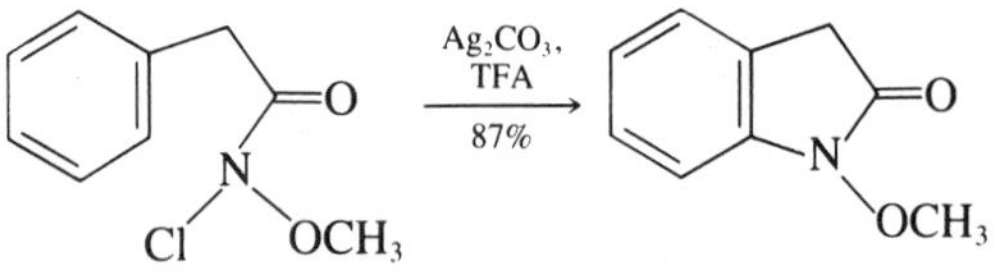

[1] Y. Kikugawa and M. Kawase, *Am. Soc.,* **106,** 5728 (1984).

Silver carbonate–Celite, 2, 363; **3**, 247–249; **4**, 425–428; **5**, 577–580; **6**, 511–514; **7**, 319–320; **8**, 441–442.

Butenolides.[1]   Oxidation of (Z)-2-butene-1, 4-diols with this oxidant results in butenolides in good yield (equation I). Furanes are formed on oxidation with usual Cr and Mn reagents.

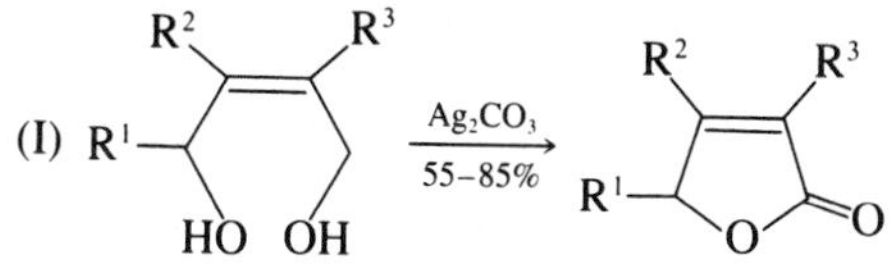

(I)

[1] T. K. Chakraborty and S. Chandrasekaran, *Tetrahedron Letters*, **25**, 2891 (1984).

Silver(I) nitrate, 1, 1008–1011; **2**, 366–368; **3**, 252; **4**, 429–430; **5**, 582; **7**, 321; **9**, 411; **10**, 350.

Cyclization of allenic alcohols.[1]   Silver nitrate (1–2 equiv.) promotes cyclization of the secondary allenic alcohols **1** to give mainly *cis*-2,6-disubstituted tetrahydropyranes (**2**). Catalytic amounts of the salt can be used, but then the reaction is very slow.

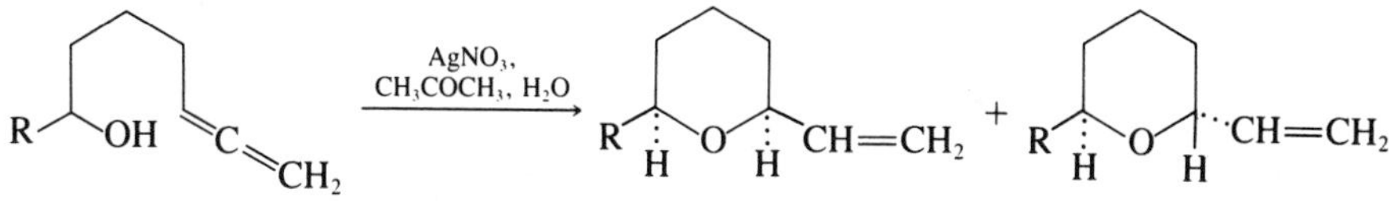

The product **2** (R = CH₃) obtained in this way is converted by hydroboration and Jones oxidation into **4**, a minor constituent of civet.

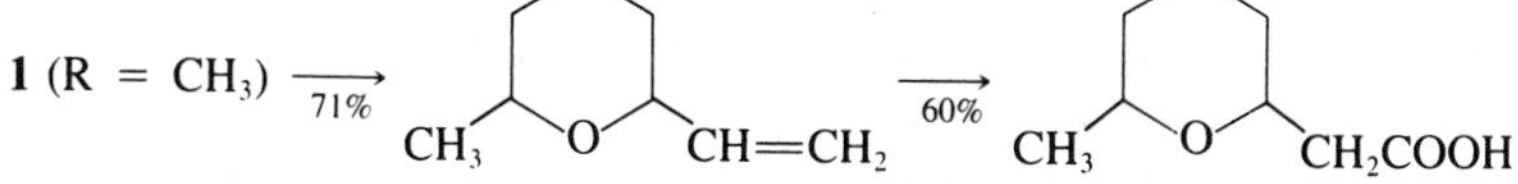

Enol lactonization.[2]   The last step in a synthesis of (±)-cyanobacterin (**2**), an antibiotic isolated from a cyanobacterium, is the enol lactonization of the γ,δ-acetylenic carboxylic acid **1**. Cyclization catalyzed by AgNO₃ results in the desired exocyclic γ-methylene-γ-butyrolactone **2** in >90% yield. In contrast, cyclization catalyzed by HgO or Hg(OAc)₂ results in the unsaturated δ-valerolactone **3** as the major product.

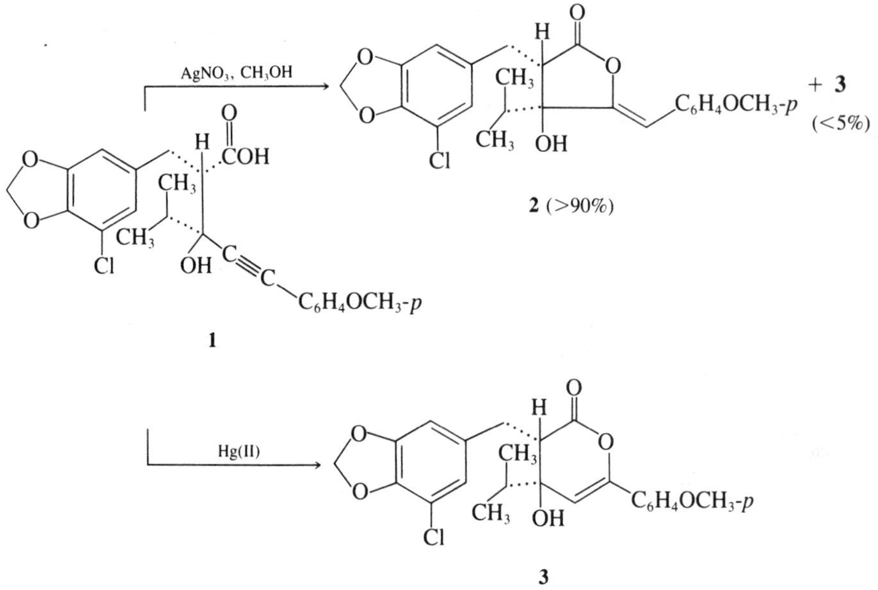

¹ T. Gallagher, *J.C.S. Chem. Comm.*, 1554 (1984).
² T.-T. Jong, P. G. Williard, and J. P. Porwoll, *J. Org.*, **49**, 735 (1984).

Silver tetrafluoroborate, 1, 1015–1018; **2,** 365–366; **3,** 250–251; **4,** 428–429; **5,** 587–588; **6,** 519–520; **8,** 443–444; **9,** 414; **11,** 471.

*Allylation of cyclic β-bromo ethers.*¹ In the presence of AgBF₄ (2 equiv.), cyclic β-bromo ethers can be allylated with allyltrimethylsilane or crotyltrimethylsilane. In the case of α-(bromomethyl)tetrahydrofuranes, two isomeric ethers are obtained. However, α-(bromomethyl)tetrahydropyranes form only one ether with high stereoselectivity. BF₃ etherate or TiCl₄ are ineffective.

Examples:

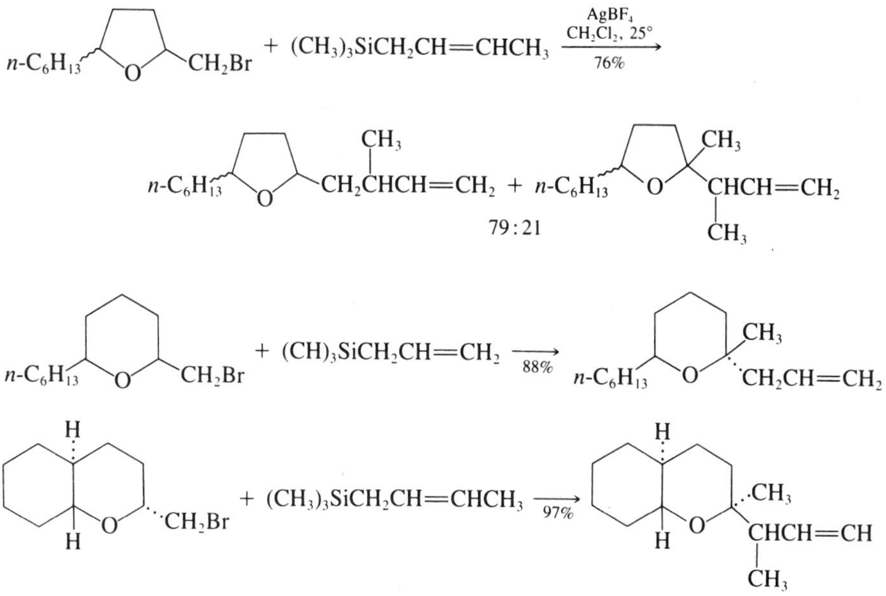

Acetylenic ketones.[2]   Reaction of thiol esters with trimethylsilylalkynes catalyzed by AgBF$_4$ in CH$_2$Cl$_2$ provides acetylenic ketones in moderate to high yield (equation I).

$$(I) \quad R^1COSC_2H_5 + R^2C{\equiv}CSi(CH_3)_3 \xrightarrow[\substack{55-83\% \\ (\text{isolated})}]{\substack{AgBF_4, \\ CH_2Cl_2}} R^1COC{\equiv}CR^2$$

[1] H. Nishiyama, T. Naritomi, K. Sakuta, and K. Itoh, *J. Org.*, **48**, 1557 (1983).
[2] Y. Kawanami, T. Katsuki, and M. Yamaguchi, *Tetrahedron Letters*, **24**, 5131 (1983).

Silver(I) trifluoromethanesulfonate, 6, 520–521; **7**, 324; **8**, 444–445; **9**, 414.

Electrophilic substitutions.   The key step in a synthesis[1] of the ring system (6) of the antibiotic bicyclomycin (7) is the reaction of the silver triflate complex of the glycine anhydride derivative 1 with 2 to give the *syn*-lactone 3 as the major product, with retention of configuration. No bis-coupled products are observed. The diol **4** obtained by reduction of **3** undergoes cyclization in the presence of AgOTf to give **5**. The transformation of **5**

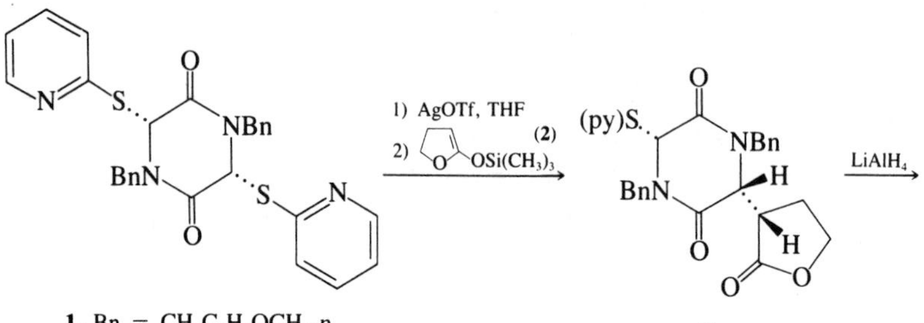

1, Bn = $CH_2C_6H_4OCH_3\text{-}p$

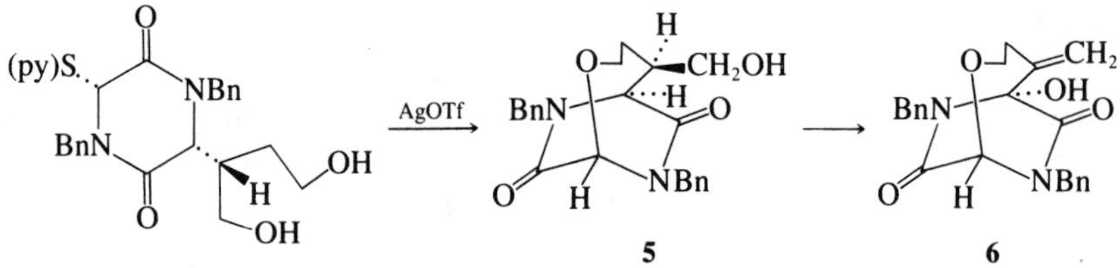

to **6** includes dehydration of the primary alcohol group and bridgehead hydroxylation. The final steps to the antibiotic involve an aldol condensation to introduce the side chain and removal of the *p*-methoxybenzyl groups as *p*-methoxybenzaldehyde by oxidation with CAN in CH_3CN.[2]

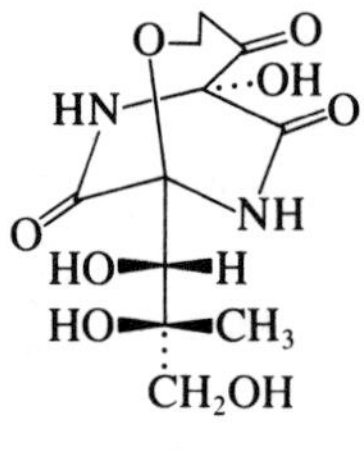

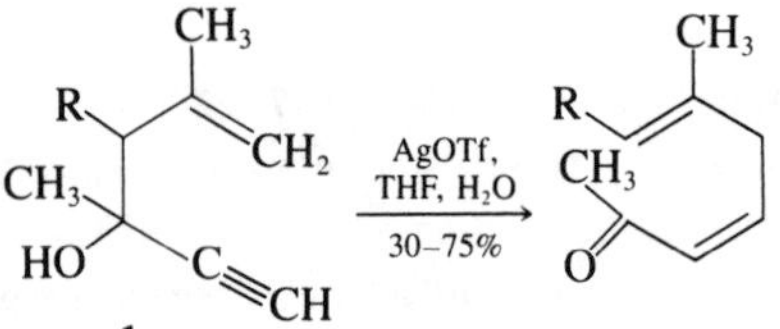

Oxy-Cope rearrangement.[3]   Hexene-5-yne-1-ols-3 (**1**) rearrange at 20–60° to α,δ-diunsaturated ketones (**2**) when treated with silver triflate (1 equiv.) in aqueous THF. Other silver salts (nitrate, tetrafluoroborate, fluoride) are ineffective.

trans-*Cycloheptenes*.[4]   Deamination of a norcarane nitrosamide (**1**) in the presence of silver triflate generates a complex (**2**) of a *trans*-cycloheptene that can be stable enough for isolation. Such complexes undergo Diels-Alder reactions with simple dienes at room temperature to give selectively the adducts to the *trans*-double bond.

Example:

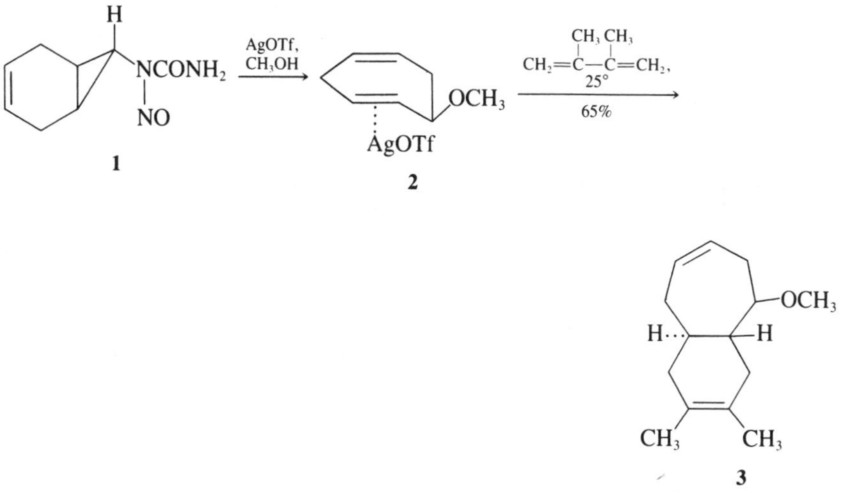

[1] R. M. Williams, R. W. Armstrong and J.-S. Dung, *Am. Soc.*, **106**, 5748 (1984).
[2] J. Yoshimura, M. Yamaura, T. Suzuki, and H. Hashimoto, *Chem. Letters*, 1001 (1983).
[3] N. Bluthe, M. Malacria, and J. Gore, *Tetrahedron Letters*, **25**, 2873 (1984).
[4] H. Jendralla, *Angew. Chem., Int. Ed.*, **19**, 1032 (1980); H. Jendralla and B. Spur, *J.C.S. Chem. Comm.*, 887 (1984).

Simmons-Smith reagent, 1, 1019–1022, **2,** 371–372; **3,** 255–258; **4,** 436–437; **5,** 588–589; **6,** 521–523; **8,** 445; **9,** 415; **11,** 472.

Acylation of alkenes.[1]   Friedel-Crafts acylation ($AlCl_3$) of alkenes suffers from lack of selectivity and low yields. The reaction is markedly improved by use of CH_2I_2–Zn/Cu (**3,** 255) as catalyst. No cyclopropanation is observed.

Examples:

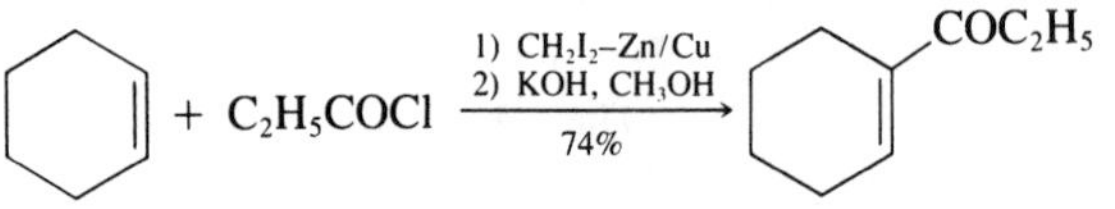

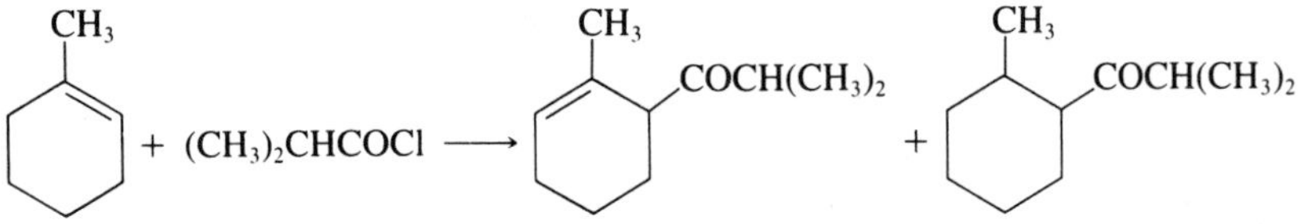

(70%)   (6%)

Angular methylation.[2]   A recent synthesis of 11-ketoprogesterone (**6**) requires introduction of the C_{19}-methyl group into the β,γ-unsaturated ketone **1**, obtained by Birch
reduction of a precursor with an aromatic ring A. Reaction of **1** with CH_2I_2/C_2H_5ZnI, the
Sawada modification[3] of the Simmons-Smith reagent, is not successful, and the diethyl

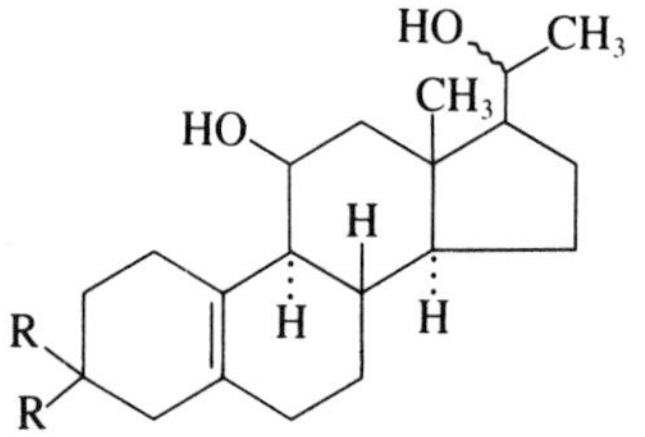

1, R = O
2, R = OC_2H_5
3, R = $OCH_2CH(CH_3)_2$

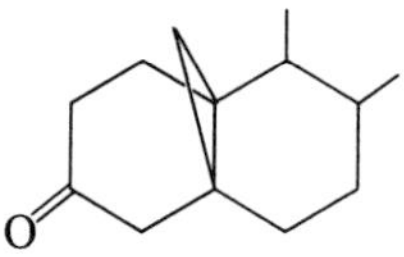

4

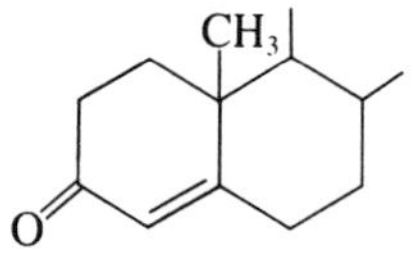

5

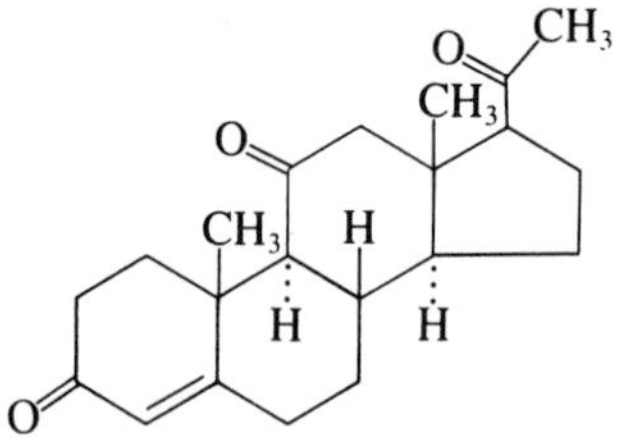

6

ketal **2** undergoes cyclopropanation in low yield. The desired product, **4**, can be obtained
in 70% yield by cyclopropanation of **3** followed by mild acid hydrolysis. This product is

cleaved to the enone **5** in 52% yield by potassium *t*-butoxide in DMSO. Finally, oxidation (PDC, CH_2Cl_2, 95% yield) of **5** provides 11-ketoprogesterone (**6**).

[1] T. Shono, I. Nishiguchi, M. Sasaki, H. Ikeda, and M. Kurita, *J. Org.*, **48**, 2503 (1983).
[2] F. E. Ziegler and T.-F. Wang, *Am. Soc.*, **106**, 718 (1984).
[3] S. Sawada and Y. Inouye, *Bull. Chem. Soc. Japan*, **42**, 2669 (1969).

Sodium amalgam, 1, 1030–1033; **2**, 373; **3**, 259; **7**, 326; **8**, 171, 296; **9**, 416–417; **10**, 355–356; **11**, 473–475.

Alkyne synthesis.[1] A synthesis of alkynes from phosphorus ylides (**1**) involves acylation to give **2**. Thermal cleavage of **2** gives pure alkynes only when both R^1 and R^2 are aryl groups. However, reductive cleavage of the corresponding triflates (**3**) with 2% sodium amalgam results in the pure alkynes **4** in which the R groups can be alkyl or aryl. The highest yields are obtained with one R group is aryl.

$$(C_6H_5)_3P{=}CHR^1 \xrightarrow[75-95\%]{R^2COCl} (C_6H_5)_3P{=}CR^1 \xrightarrow[81-93\%]{\underset{C_6H_6,\ -5\ \to\ 25°}{(CF_3SO_2)_2O,}}$$

$$\underset{\textbf{1}}{} \qquad\qquad \underset{\textbf{2}}{\underset{R^2C{=}O}{}}$$

$$\underset{\textbf{3}}{(C_6H_5)_3\overset{+}{P}{-}CR^1\ \underset{R^2COSO_2CF_3}{\|}}\ \overset{-OSO_2CF_3}{}\ \xrightarrow[45-80\%]{\underset{THF,\ 5°}{Na/Hg,}}\ \underset{\textbf{4}}{R^1C{\equiv}CR^2}$$

[1] H.-J. Bestmann, K. Kumar, and W. Schaper, *Angew. Chem., Int. Ed.*, **22**, 167 (1983).

Sodium benzeneselenoate, 5, 273; **6**, 548–549; **8**, 447–448; **10**, 356; **11**, 475.

Ketone synthesis.[1] The reaction of $NaSeC_6H_5$ (excess) with α,β-epoxy sulfoxides, available by reaction of aldehydes or ketones with the anion of chloromethyl phenyl sulfoxide (**8**, 94–95), results in α-substituted ketones and diphenyl diselenide. Examples:

A similar reaction with sodium benzenethiolate results in α-phenylthio ketones. Example:

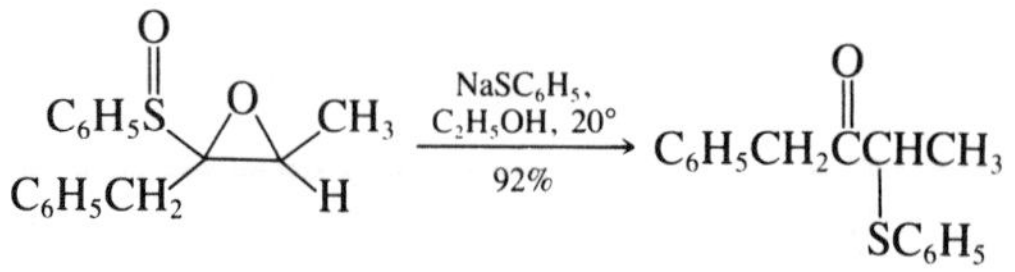

[1] T. Satoh, Y. Kaneko, T. Kumagawa, T. Izawa, K. Sakata, and K. Yamakawa, *Chem. Letters,* 1957 (1984).

Sodium bis(2-methoxyethoxy)aluminum hydride (SMEAH), **3**, 260–261; **4**, 441–442; **5**, 596; **6**, 528–529; **7**, 327–329; **8**, 998–999; **9**, 418–420; **10**, 357; **11**, 476–477.

Reduction of propargyl alcohols.[1] 3-Trimethylsilyl-2-propyne-1-ol (**1**), prepared as shown from propargyl alcohol, is reduced stereospecifically by SMEAH to (E)-3-trimethylsilyl-2-propene-1-ol (**2**). Reductions of propargyl alcohols with LiAlH$_4$ are less *trans*-selective.

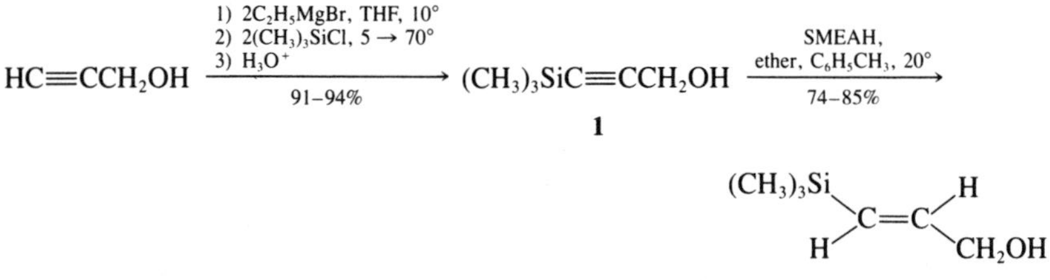

Reduction of α-silyloxy ketones.[2] α-Hydroxy ketones are reduced by zinc borohydride with the expected *anti*-selectivity, the extent of which varies somewhat with the substitution pattern. Preparation of the isomeric *syn*-diols can be effected by reduction of the α-*t*-butyldiphenylsilyloxy ketones with SMEAH in toluene at −78° followed by desilylation (equation I). Again, the selectivity varies with the nature of R^1 and R^2, and is low when R^1 is a bulky alkyl group.

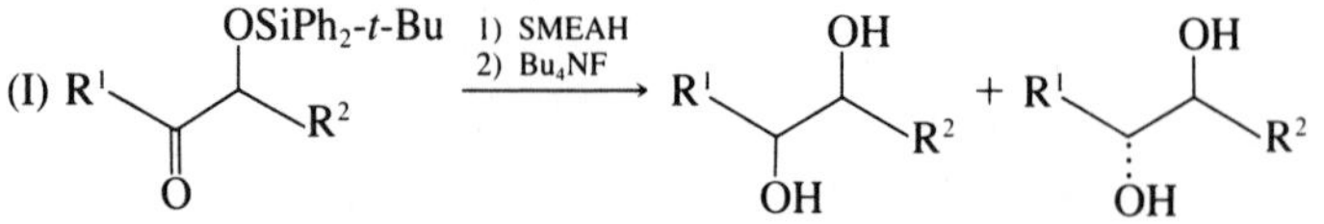

[1] T. K. Jones and S. E. Denmark, *Org. Syn.*, submitted (1983).
[2] T. Nakata, T. Tanaka, and T. Oishi, *Tetrahedron Letters*, **24**, 2653 (1983).

Sodium bis(trimethylsilyl)amide, 1, 1046–1047; **3**, 261–262; **4**, 442–443; **6**, 529–530; **7**, 329.

Primary amines.[1]   This amide reacts with a primary bromide, iodide, or tosylate in hexamethyldisilazane or HMPT to give a N,N-bis(trimethylsilyl)amine, which is hydrolyzed by dilute hydrochloric acid to the primary amine (equation I).

$$\text{(I)} \quad RX + NaN[Si(CH_3)_3]_2 \xrightarrow[50-80\%]{-NaX} RN[Si(CH_3)_3]_2 \xrightarrow[96-100\%]{H_3O^+} RNH_3Cl + [(CH_3)_3Si]_2O$$

[1] H. J. Bestmann and G. Wolfel, *Ber.*, **117**, 1250 (1984).

Sodium borohydride, 1, 1049–1055; **2**, 377–378; **3**, 262–264; **4**, 443–444; **5**, 597–601; **6**, 530–539; **7**, 329–331; **8**, 449–451; **9**, 420–421; **10**, 357–359; **11**, 477–479.

Stereocontrolled reduction of alkylidene malonates.[1]   The stereochemistry of conjugate hydride reduction of alkylidene malonates such as **1** can be controlled by a homoallylic substituent. Thus hydride reduction of **1a** results in predominant *syn*-reduction.

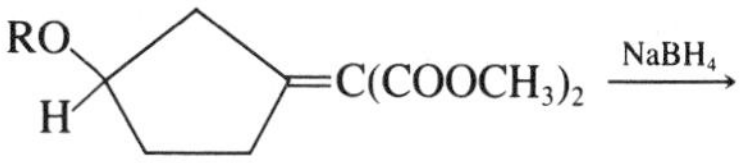

1a, R = H
1b, R = MEM

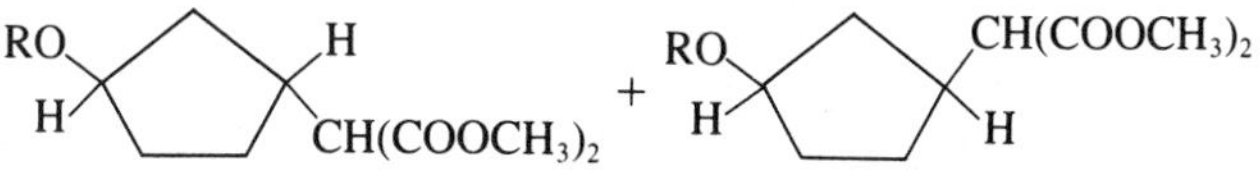

trans-**2**   1:0.37   cis-**2**
                  1:5

In contrast, the corresponding MEM ether is reduced mainly by *anti*-approach of the hydride because of steric hindrance.

This stereocontrol is useful in a synthesis of a spatane diterpene (**5**). Thus hydride reduction of **3** affords only **4**. In this case the *syn*-influence of the hydroxyl group is

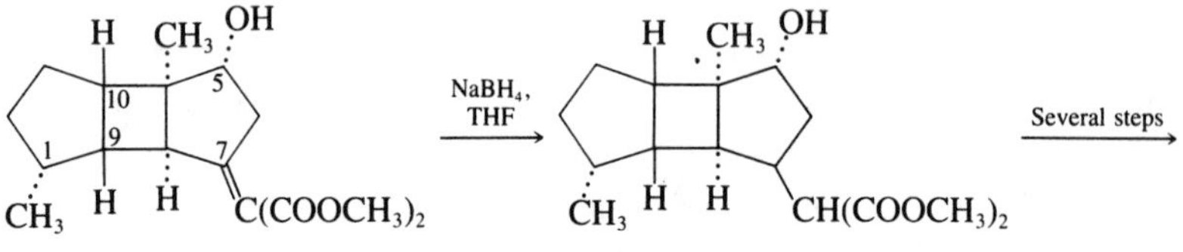

reinforced by the hydrogens at C_9 and C_{10}. In contrast, reduction of the corresponding MEM ether followed by removal of the MEM group resulted in **4** and the corresponding C_7-epimer in the ratio 1:2.

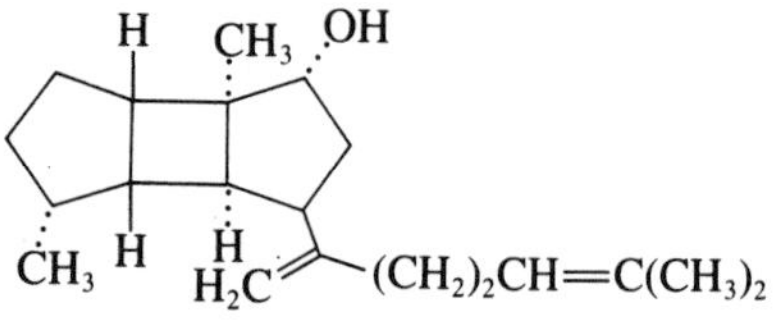

Deprotection of phthalimides.[2]   Phthalimides are useful for complete protection of primary amino groups, including amino acids, even though hydrazinolysis is usually necessary for deprotection. A mild, two-step but one-flask cleavage involves reduction with $NaBH_4$ in aqueous 2-propanol[3] followed by cyclization to a lactone with release of the amine (equation I). Yields are generally 80–97%. Racemization does not occur in deprotection of amino acids (four examples).

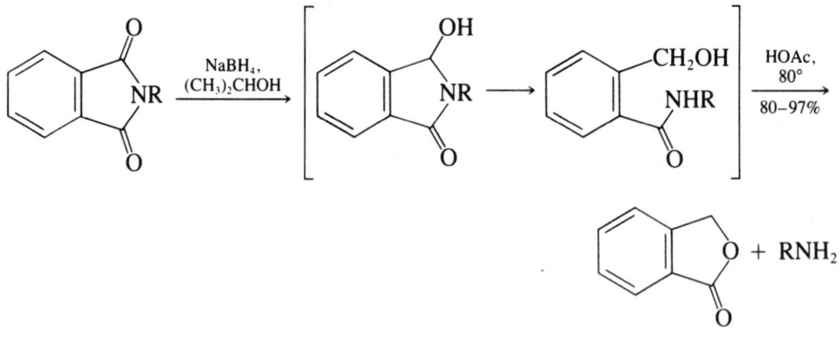

Reduction of β-keto esters to 1,3-diols.[4]   β-Keto esters can be reduced selectively to 1,3-diols by $NaBH_4$ in a mixed solvent system, either methanol–THF or methanol–*t*-butyl alcohol.

Deoxygenation of alcohols.   Primary mesylates or tosylates of carbohydrates in the presence of secondary ones can be selectively reduced with $NaBH_4$ in DMSO at 85° (60–65% yield).[5] If the temperature is increased to 140°, secondary mesylates also can be reduced in high yield.[6] In fact, this method can even be more useful than reductions with lithium triethylborohydride, which has been the reagent of choice even though it is expensive.

syn-α-Amino-β-hydroxy carboxylic acids.[7]   The α-dibenzylamino-β-keto esters (**2**) obtained by acylation of the anion of methyl N,N-dibenzylglycinate (**1**) are reduced by

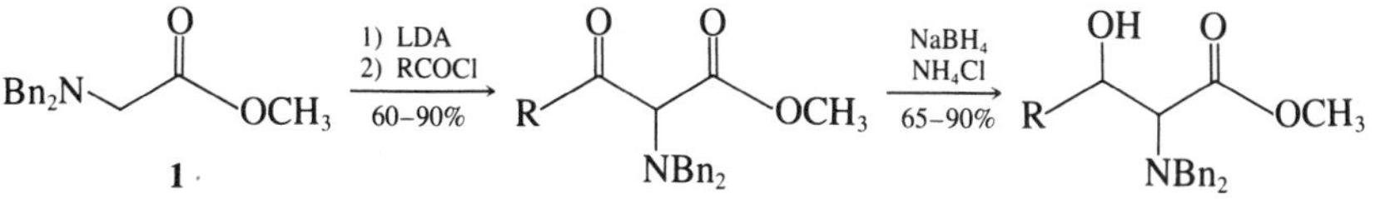

sodium borohydride in aqueous ethanol buffered with ammonium chloride with high diastereoselectivity to *syn*-2-dibenzylamino-β-hydroxy esters (**3**). The high stereoselectivity is attributed to the bulky dibenzylamino group. The esters (**3**) can be obtained directly by aldol condensations with the anion of **1**, but the diastereoselectivity of this reaction is variable and only moderate.

Stereoselective reduction of β-hydroxy ketones.[8]   Boron chelates of β-hydroxy ketones are reduced by sodium borohydride stereoselectively to *syn*-1,3-diols (equation I). Even higher *syn*-selectivity obtains on similar reduction of *syn*-α-substituted-β-hydroxy

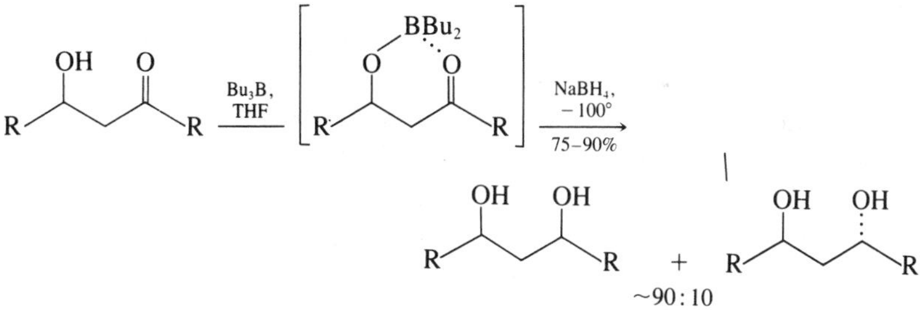

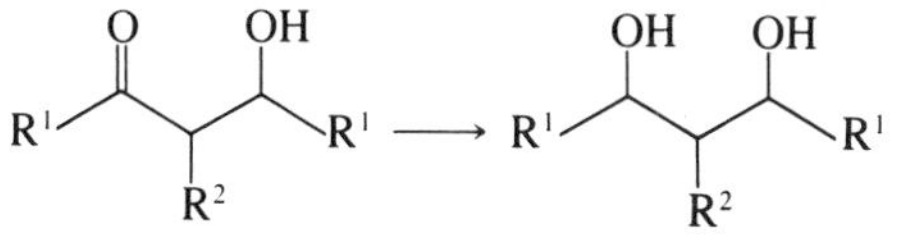

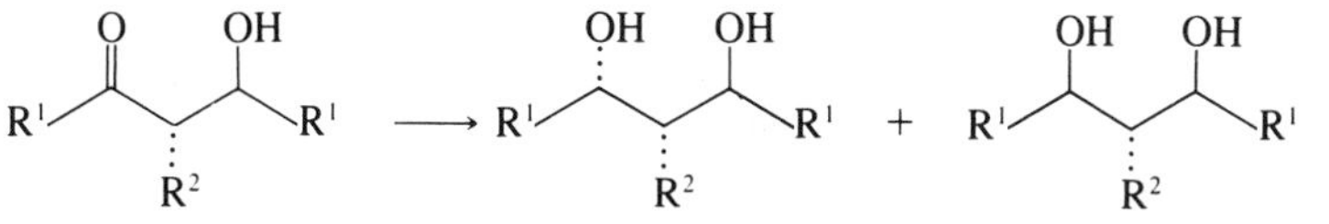

R[1] = *i*-Pr, R[2] = CH₃   27:73
R[1] = *n*-Pr, R[2] = C₂H₅   100:0

ketones (equation II). However, the stereoselectivity of reduction of the *anti*-isomers of α-substituted-β-hydroxy ketones depends on the alkyl substituents, R^1 and R^2 (equation III).

Lithium and magnesium chelates have no effect on the stereoselection of this reduction. Triisobutylborane is somewhat more effective than *n*-butylborane; no reduction occurs if tri-*sec*-butylborane is used.

Stereoselective reduction of β-*keto sulfoxides.*[9]   The reduction of the β-keto sulfoxide **1** with $NaBH_4$ in CH_3OH/aq. NH_3 (9:1) proceeds with efficient 1,3-asymmetric induction and epimerization at C_2 (equation I). Only slight stereoselectivity obtains in a neutral medium. The chirality of the sulfinyl group has no effect.

(99% ee)

NaBH₄ in polyethylene glycols.[10] $NaBH_4$ in PEG, particularly PEG 400, is more reactive than $NaBH_4$ in the usual hydroxylic solvents, possibly because a sodium dialkoxy-borohydride is generated. This system reduces esters and acyl chlorides to primary alcohols, and halides or tosylates to hydrocarbons.

[1] R. G. Salomon, N. D. Sachinvala, S. R. Raychaudhuri, and D. B. Miller, *Am. Soc.*, **106**, 2211 (1984).
[2] J. O. Osby, M. G. Martin, and B. Ganem, *Tetrahedron Letters*, **25**, 2093 (1984).
[3] F. C. Uhle, *J. Org.*, **26**, 2998 (1961).
[4] K. Soai and H. Oyamada, *Synthesis*, 605 (1984).
[5] H. Weidmann, N. Wolf, and W. Timpe, *Carbohydrate Res.*, **24**, 184 (1972).
[6] P. Kocienski and S. D. A. Street, *Syn. Comm.*, **14**, 1087 (1984).
[7] G. Guanti, L. Banfi, E. Narisano, and C. Scolastico, *Tetrahedron Letters*, **25**, 4693 (1984).
[8] K. Narasaka and F.-C. Pai, *Tetrahedron*, **40**, 2233 (1984).
[9] K. Ogura, M. Fujita, T. Inaba, K. Takahashi, and H. Iida, *Tetrahedron Letters*, **24**, 503 (1983).
[10] E. Santaniello, A. Fiecchi, A. Manzocchi, and P. Ferraboschi, *J. Org.*, **48**, 3074 (1983).

Sodium borohydride–Praseodymium chloride.

Stereoselective reduction of a ketone.[1]   Reduction of **1** with $NaBH_4$ gives a 1:1 mixture of **2** and the α-isomer as well as products of conjugate addition. Reduction with $NaBH_4$–$PrCl_3$ results in the desired alcohol **2** exclusively.

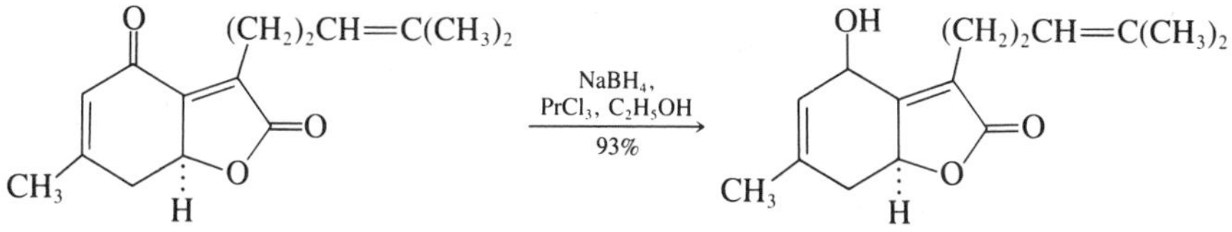

1 → 2

[1] P. A. Jacobi, C. S. R. Kaczmarek, and U. E. Udodong, *Tetrahedron Letters*, **25**, 4859 (1984).

Sodium bromide.

RCH₂Cl → RCH₂Br. This conversion can be effected in good yield with NaBr in DMF at 100° in the presence of methylene bromide as a chloride ion scavenger.[1] The transformation can also be effected with NaBr and C_2H_5Br in N-methyl-2-pyrrolidinone at 60–70°,[2] but this reaction requires several days.

[1] J. H. Babler and K. P. Spina, *Syn. Comm.*, **14**, 1313 (1984).
[2] W. E. Willy, D. R. McKean, and B. A. Garcia, *Bull. Chem. Soc. Japan*, **49**, 1989 (1976).

Sodium bromite, $NaBrO_2$.

Lactonization of α,ω-diols.[1] Diols of this type (**1**) are converted into lactones (**2**) when treated with $NaBrO_2$ (3 equiv.). A primary, secondary diol is oxidized under these conditions to a hydroxy ketone; a primary alcohol, RCH_2OH, is converted to an ester, $RCOOCH_2R$, in 70–95% yield.

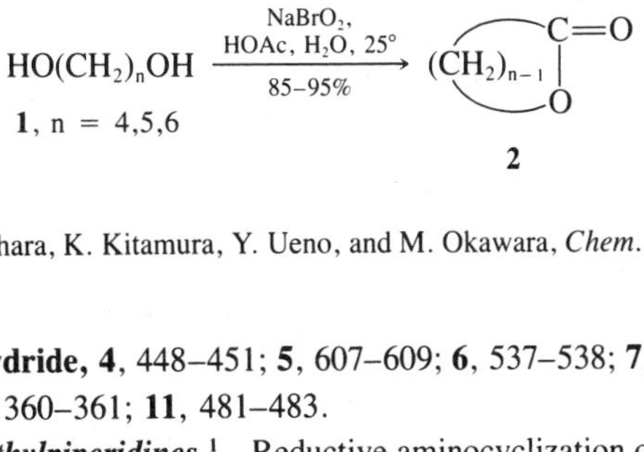

$$HO(CH_2)_nOH$$
1, n = 4,5,6

[1] T. Kageyama, S. Kawahara, K. Kitamura, Y. Ueno, and M. Okawara, *Chem. Letters*, 1097 (1983).

Sodium cyanoborohydride, **4**, 448–451; **5**, 607–609; **6**, 537–538; **7**, 334–335; **8**, 454–455; **9**, 424–426; **10**, 360–361; **11**, 481–483.

cis-6-*Alkyl-2-methylpiperidines.*[1] Reductive aminocyclization of 2,6-diones (**1**) affords cis-2,6-dialkylpiperidines (**2**) exclusively.

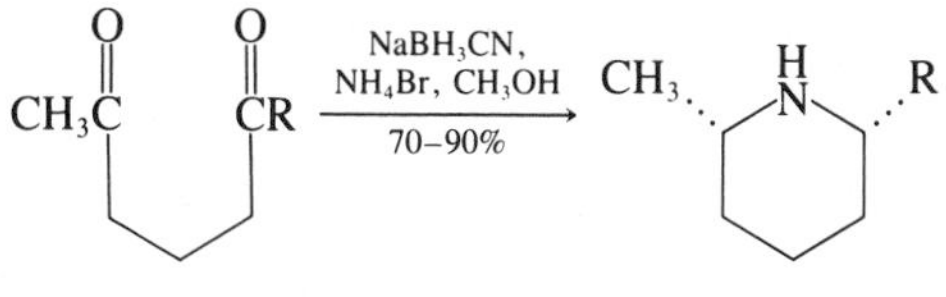

This reaction has been used to prepare *dl*-pumiliotoxin C (**2**) and an isomer (**3**).[2]

[1] K. Abe, H. Okumura, T. Tsugoshi, and N. Nakamura, *Synthesis*, 597 (1984).
[2] K. Abe, T. Tsugoshi, and N. Nakamura, *Bull. Chem. Soc. Japan*, **57**, 3351 (1984).

Sodium cyanoborohydride–Zinc chloride.

Reduction of tertiary, allylic, and benzylic halides.[1]   NaCNBH$_3$ and ZnCl$_2$ in a 2:1 molar ratio in ether reduce these halides selectively in ~70–90% yield in the presence of primary, secondary, vinyl, and aryl halides. The selectivity is comparable to that of lithium 9,9-di-*n*-butyl-9-borabicyclo[3.3.1]nonate, the ate complex obtained by reaction of *n*-butyllithium with 9-BBN.[2]

[1] S. Kim, Y. J. Kim, and K. H. Ahn, *Tetrahedron Letters*, **24**, 3369 (1983).
[2] H. Toi, Y. Yamamato, A. Sonoda, and S.-I. Murahashi, *Tetrahedron*, **37**, 2261 (1981).

Sodium (dimethylamino)borohydride, Na(CH$_3$)$_2$NBH$_3$; Sodium (*t*-butylamino)borohydride, Na(CH$_3$)$_3$CHNBH$_3$.

These reagents are prepared by reaction of NaH with the aminoboranes (Callery Chem. Co.). They are more active reducing agents than NaBH$_4$. Thus they reduce esters in moderate to high yield. They do not reduce secondary amides, and reduce primary amides only sluggishly to primary amines. However, they reduce tertiary amides to primary alcohols and/or amines, depending on the substituents on nitrogen. N,N-Dimethylamides are reduced selectively to alcohols, whereas N,N-diisopropylamides are reduced selectively to amines.[1]

[1] R. O. Hutchins, K. Learn, F. El-Telbany, and Y. P. Stercho, *J. Org.*, **49**, 2438 (1984).

Sodium hexamethyldisilazide, NaN[Si(CH$_3$)$_3$]$_2$, 4, 407–409.

Macrocyclic ketones.   A recent method for synthesis of macrocyclic ketones involves intramolecular alkylation of protected cyanohydrins. Sodium hexamethyldisilazide is the preferred base, particularly for the synthesis of α,β-enones. An example is the synthesis of *trans*-2-cyclopentadecenone (**1**),[1] as outlined in Chart (I).

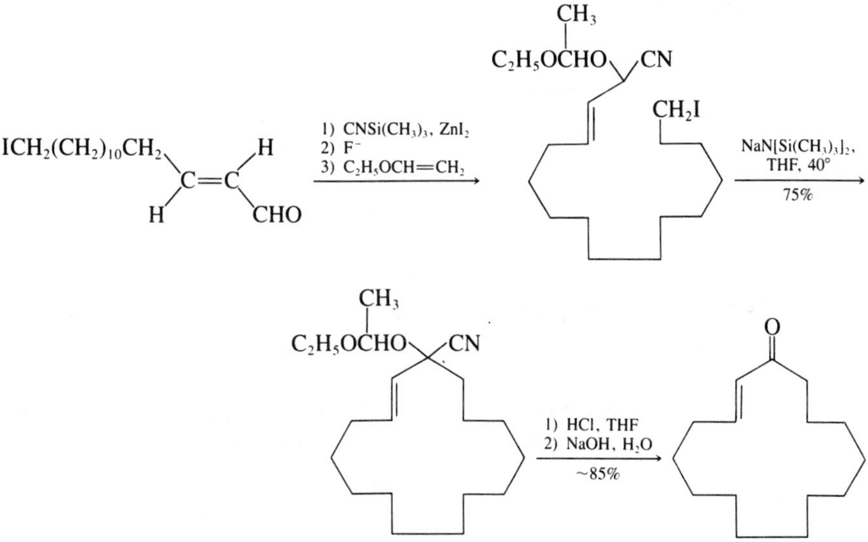

Chart (I)

The same methodology has been used to synthesize various dimethyl-2,6-cyclodecadienones, a ring system encountered in various 10-membered sesquiterpenes. An example is outlined in equation (I).[2]

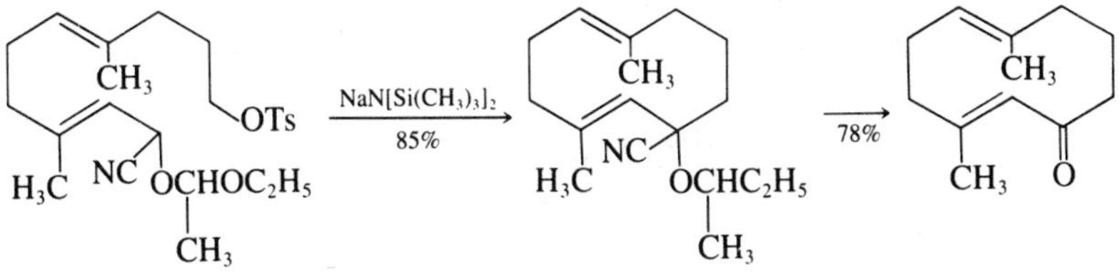

[1] T. Takahashi, T. Nagashima, and J. Tsuji, *Tetrahedron Letters*, **22**, 1359 (1981).
[2] T. Takahashi, H. Nemoto, and J. Tsuji, *ibid.*, **24**, 2005 (1983).

Sodium hydride, 1, 1075–1081; **2**, 380–383; **4**, 452–455; **5**, 610–614; **6**, 591–592; **8**, 458–459; **9**, 427–428.

Stereoselective intramolecular Michael cyclizations.[1] The cyclization of **1** with sodium hydride (catalytic) in benzene at 23° results only in the *trans*-disubstituted cyclo-pentanone **2**. In contrast, use of sodium methoxide in CH_3OH results in **2** and the *cis*-isomer in the ratio 1:3. Similar results obtain in related cyclizations leading to *vic*-disubstituted cyclohexanes.

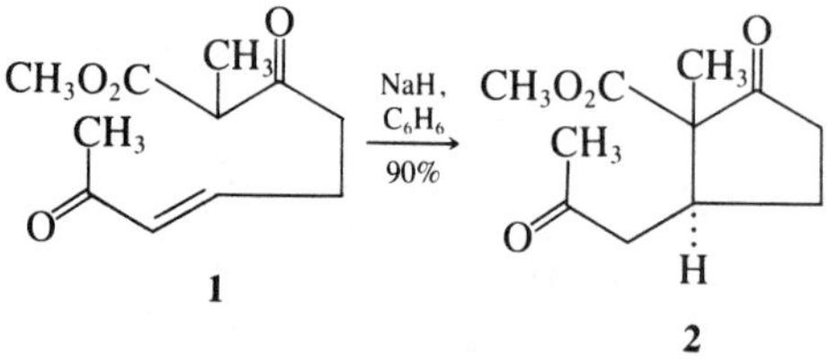

Cycloadditions with homophthalic anhydride.[2]   Homophthalic anhydride can undergo thermal cycloaddition with dienophiles with loss of carbon dioxide, but the yields are generally low. The reaction can proceed at room temperature with dramatic improvement of the yield if the anhydride is first converted into the sodium salt (NaH) or lithium salt (LDA). The cycloaddition provides a regioselective route to *peri*-hydroxy polycyclic compounds.

Example:

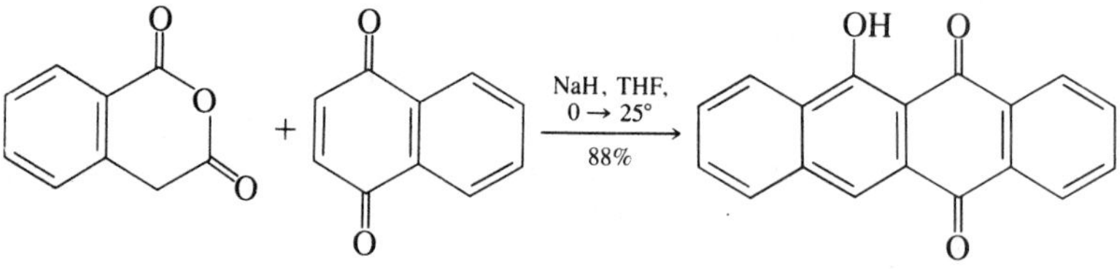

Allyl silyl ethers.[3]   β-Hydroxyvinylsilanes (**1**), readily available by hydromagne-siation of silylalkynes (**10**, 130), are stable to sulfuric acid or to sodium hydride in THF,

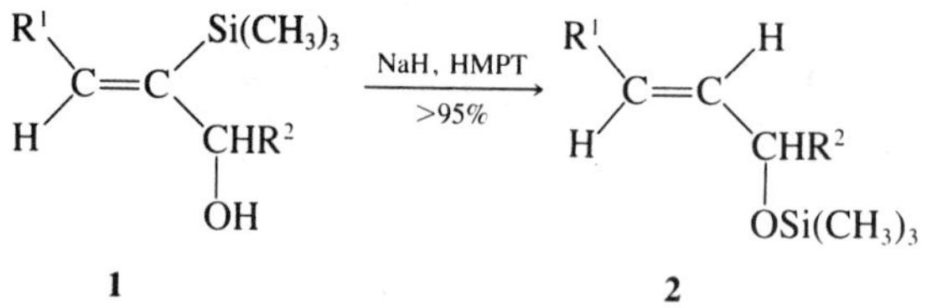

but rearrange to allyl silyl ethers (**2**) in the presence of a catalytic amount of sodium hydride in HMPT. The reaction must be conducted under strictly anhydrous conditions, because the products are very sensitive to aqueous acid or base.

Darzens reaction.[4]   Under the usual conditions (sodium alkoxides as base and the corresponding alcohol as solvent), the condensation of aliphatic aldehydes with α-halo esters proceeds in generally low yield, particularly with α-chloro esters (**4**, 296). Sodium hydride in acetonitrile is recommended for Darzens condensations using ethyl chloroacetate. Yields of 45–75% can be obtained with aliphatic aldehydes. Reactions with ketones and aromatic aldehydes proceed in yields of about 75%.

[1] G. Stork, J. D. Winkler, and N. A. Saccomano, *Tetrahedron Letters,* **24**, 465 (1983).
[2] Y. Tamura, M. Sasho, K. Nakagawa, T. Tsugoshi, and Y. Kita, *J. Org.,* **49**, 473 (1984).
[3] F. Sato, Y. Tanaka, and M. Sato, *J.C.S. Chem. Comm.,* 165 (1983).
[4] R. Della Pergola and P. DiBattista, *Syn. Comm.,* **14**, 121 (1984).

Sodium hydrogen telluride, 8, 459; **10**, 365.

Reduction of α,β-epoxy ketones.[1]   α,β-Epoxy ketones are reduced by NaTeH regioselectively to β-hydroxy ketones. Unactivated epoxides, such as stilbene oxide and α,β-epoxy alcohols, are not reduced.

Examples:

Reductive amination of carbonyl compounds.[2]   This reagent effects reductive amination of aldehydes or ketones with amines (equation I). Yields tend to be only moderate with anilines, but are high with aliphatic amines.

[1] A. Osuka, K. Taka-Oka, and H. Suzuki, *Chem. Letters,* 271 (1984).
[2] M. Yamashita, M. Kadokura, and R. Suemitsu, *Bull. Chem. Soc. Japan,* **57**, 3359 (1984).

Sodium iodide, 1, 1087–1090; **2**, 384; **3**, 267; **4**, 456–457; **6**, 543; **7**, 338; **10**, 365–366.

Pschorr phenanthrene synthesis.[1]   Although use of sodium iodide rather than copper powder generally improves the yield in the synthesis of phenanthrenes from diazonium 2-amino-α-phenylcinnamic acids (**3**, 267), yields are poor when the acceptor ring is substituted by electron-donating alkoxy groups, which are often present in natural phenanthrenes. However, yields are considerably improved by replacing an alkoxy group by an electron-withdrawing phenylsulfonyloxy group.[2]

Example:

$$R = CH_3 \qquad 45\%$$
$$R = SO_2C_6H_5 \qquad 71\%$$

[1] D. F. Detar, *Org. React.*, **9**, 409 (1957).
[2] R. I. Duclos, Jr., J. S. Tung, and H. Rapoport, *J. Org.*, **49**, 5243 (1984).

Sodium iodide–Acetonitrile.

Acyl iodides.[1]   Aliphatic and aromatic acyl iodides and diacyl iodides can be prepared from the corresponding chlorides by reaction with anhydrous NaI in absolute acetonitrile, and isolated by liquid–liquid extraction with pentane, which is almost immiscible with acetonitrile. The products can be stored under N_2 over copper powder as stabilizer. The same technique can be used to obtain iodoglyoxalic esters, ROOCCOI,[2] and iodoformic esters, ROCOI.[3]

[1] H. M. R. Hoffmann and K. Haase, *Synthesis*, 715 (1981); H. M. R. Hoffmann, K. Haase, and P. M. Geschwinder, *ibid.*, 237 (1982).
[2] P. M. Geschwinder, S. Preftitsi, and H. M. R. Hoffmann, *Ber.*, **117**, 408 (1984).
[3] H. M. R. Hoffmann and L. Iranshahi, *J. Org.*, **49**, 1174 (1984).

Sodium methaneselenolate; sodium ethanethiolate.

O- and Se-Demethylation.   Sodium methaneselenolate cleaves methyl aryl ethers and sulfides. HMPT is preferred to DMF as solvent. Sodium ethanethiolate cleaves various alkyl aryl selenides in refluxing HMPT or DMF.

Aryl diselenides can be prepared directly by treatment of aryl chlorides with $NaSeCH_3$ in refluxing HMPT.[1]

Examples:

$$p\text{-CH}_3\text{SC}_6\text{H}_4\text{OCH}_3 \xrightarrow[90\%]{\substack{\text{NaSeCH}_3, \\ \text{HMPT, } \Delta}} p\text{-CH}_3\text{SC}_6\text{H}_4\text{OH}$$

$$2\text{C}_6\text{H}_5\text{Cl} \xrightarrow[73\%]{\substack{\text{1) 3NaSeCH}_3 \\ \text{2) I}_2}} (\text{C}_6\text{H}_5\text{Se})_2$$

[1] M. Evers and L. Christiaens, *Tetrahedron Letters*, **24**, 377 (1983).

Sodium methylsulfinylmethylide (dimsylsodium), **1**, 310–313; **2**, 166–169; **3**, 123–124; **4**, 195–196; **5**, 621; **6**, 546–547; **7**, 338–339; **9**, 431; **11**, 489–490.

Anionic Claisen rearrangement.   Acceleration of the Claisen rearrangement of allyl vinyl ethers was originally observed with potassium hydride in HMPT,[1] but even milder conditions are possible using sodium or lithium dimsylate.[2] The catalyzed rearrangement is as stereoselective as the thermal counterpart. Thus vicinal asymmetric centers are formed selectively on rearrangement of (E)- and (Z)-crotyl ethers (equations I and II).

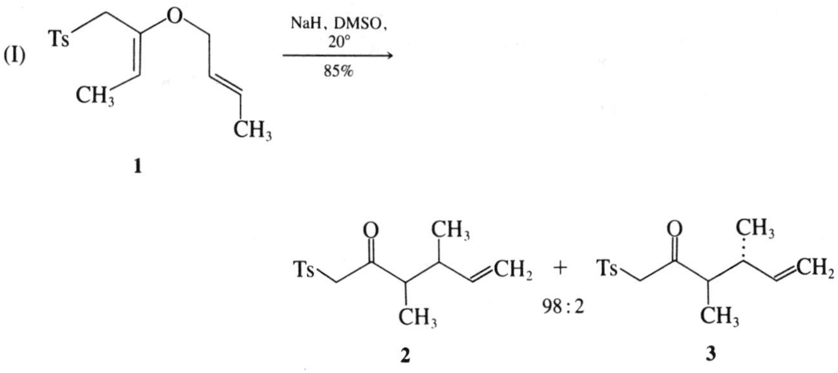

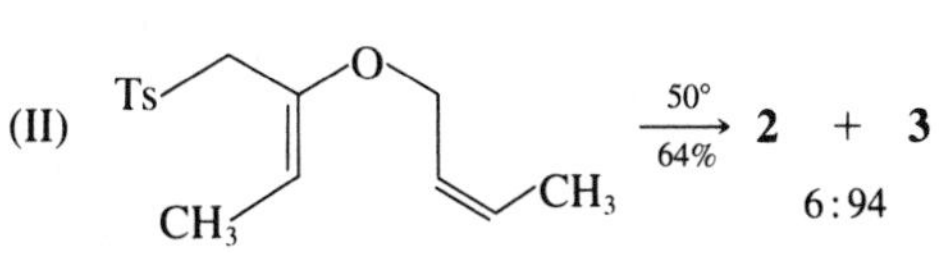

[1] S. E. Denmark and M. A. Harmata, *Am. Soc.*, **104**, 4972 (1982).
[2] *Idem, J. Org.*, **48**, 3369 (1983).

Sodium perborate (1), 1, 1102.

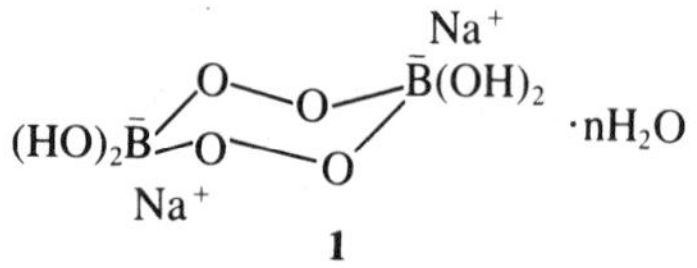

Oxidations.[1]   Anilines substituted by halo, nitro, cyano, and alkylalkoxy groups are oxidized by **1** in HOAc to the corresponding nitro compounds in 75–90% yield. Over-oxidation is usual when this oxidation is applied to anilines substituted by electron-donating groups. In successful reactions, yields are comparable to those obtained with CF_3CO_3H (**1,** 821–822).

Sodium perborate is also an excellent reagent for oxidation of sulfides to either sulfoxides or sulfones, particularly the latter compounds (90–99% yield).

[1] A. McKillop and J. A. Tarbin, *Tetrahedron Letters,* **24,** 1505 (1983).

Sodium selenophenolate, 5, 272–273; **6,** 548–549; **7,** 341; **9,** 432–434.

2-Arylpropanoic acids.[1]   The ethylene ketals (**2**) of aryl α-phenylselenoethyl ketones, prepared by reaction of aryl α-bromo ketals (**1**) with diphenyl diselenide and sodium wire, are converted into the hydroxyethyl esters (**3**) of 2-arylpropanoic acids (**4**) on oxidation with excess *m*-chloroperbenzoic acid. The reaction probably involves a selenone intermediate, in which the rearrangement of the aryl group occurs. The acetal group of **2** is essential for this rearrangement.

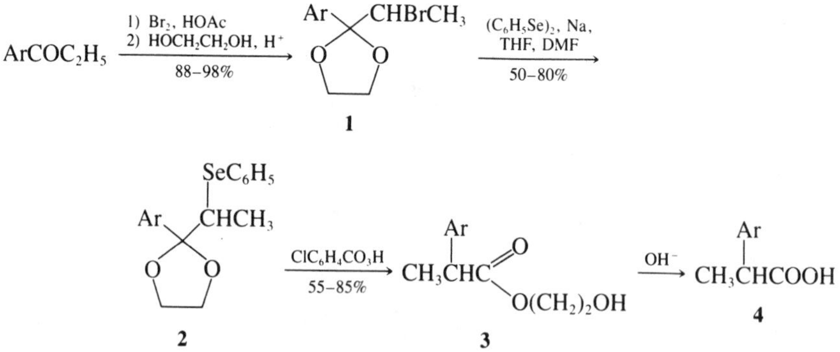

[1] S. Uemura, S. Fukuzawa, T. Yamauchi, K. Hattori, S. Mizutaki, and K. Tamaki, *J.C.S. Chem. Comm.,* 426 (1984).

Sodium sulfide, 3, 269–272; **5**, 623–624; **11**, 492.

Debromination of **vic-***dibromides.*[1]   *vic*-Dibromides are reduced to alkenes by treatment with Na_2S or NaSH in benzene/H_2O in the presence of catalytic amounts of methyltrioctylammonium chloride (Adogen 464). The debromination occurs by stereoselective *anti*-elimination. The phase-transfer catalyst is not required with water-soluble substrates.

Example:

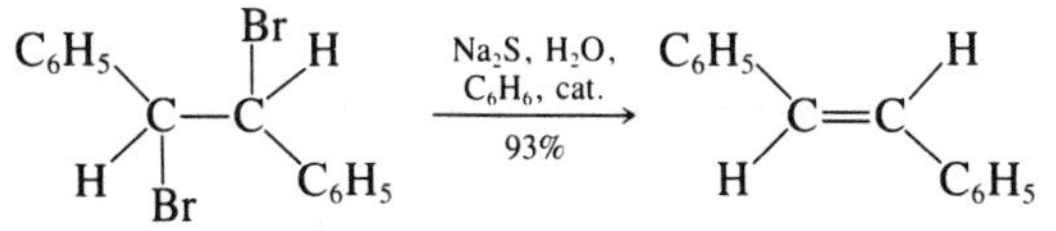

[1] J. Nakayama, H. Machida, and M. Hoshino, *Tetrahedron Letters*, **24**, 3001 (1983).

Sodium triacetoxyborohydride, 6, 553.

Stereo- and regioselective reduction of ketones.[1]   The germine derivative **1** is reduced by $NaBH_4$ in *t*-butyl alcohol or ethanol mainly to the corresponding 7α-alcohol. In contrast, use of $NaBH(OAc)_3$ (1.7 equiv.) in HOAc gives the 7β-alcohol in 92% yield.

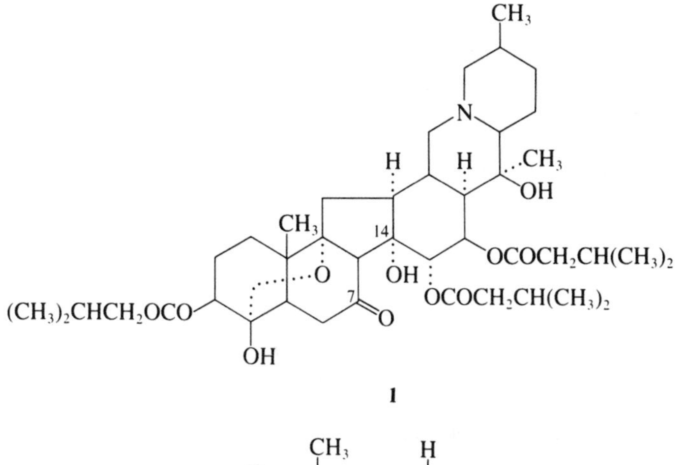

1

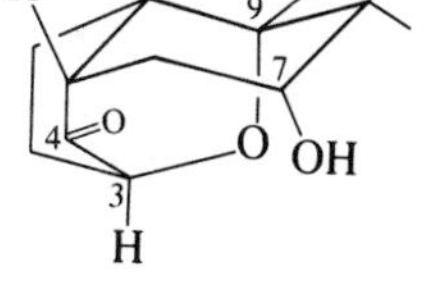

2

Similarly, the bicyclic ketone **2** (a rearrangement product of germine) is reduced by $NaBH_4$ exclusively to the *endo*-alcohol, and by $NaBH(OAc)_3$ to the *exo*-alcohol in 85% yield. The stereoselectivity in the case of **1** is attributed to the presence of a nearby hydroxyl group at C_{14}, which exchanges with one of the acetoxy groups of $NaBH(OAc)_3$. The selectivity in the reduction of **2** is attributed to a similar complexation of the C_7-hydroxyl group.

[1] A. K. Saksena and P. Mangiaracina, *Tetrahedron Letters*, **24**, 273 (1983).

Sodium tris[(S)-N-benzyloxycarbonylprolyoxy]hydroborate,

Mol. wt. 779.55, m.p. 55–65° (dec.). The borohydride is obtained in 94% yield by reaction of 3 equiv. of (S)-N-benzyloxycarbonylproline with 1 equiv. of $NaBH_4$ in THF at 5–10°. It can be stored at 5–10° for at least 1 month.

Asymmetric reduction of cyclic imines.[1] Of a series of chiral sodium acyloxyborohydrides obtained from various chiral N-acyl-α-amino acids, **1** affords the highest optical yields (70–86% ee) in the reduction of cyclic imines.

Examples:

[1] K. Yamada, M. Takeda, and T. Iwakuma, *J.C.S. Perkin I*, 265 (1983).

Sulfene, 11, 494.

Carboxylic anhydrides.[1] Carboxylic acids can be converted into the anhydrides by reaction with sulfene (generated from methanesulfonyl chloride and triethylamine). Yields are in the range 82–95%.

[1] A. Nangia and S. Chandrasekaran, *J. Chem. Res. (S),* 100 (1984).

N-Sulfinyltoluenesulfonamide, TsN=S=O (**1**). Mol. wt. 217.26, m.p. 53°. Preparation (**8**, 43).

Diastereoselective synthesis of homoallylic amines.[1] The Diels-Alder adducts (**2**) formed from 1,3-dienes and **1** undergo basic hydrolysis to a single unsaturated sulfonamide (**3**) via a concerted retro-ene reaction.

Examples:

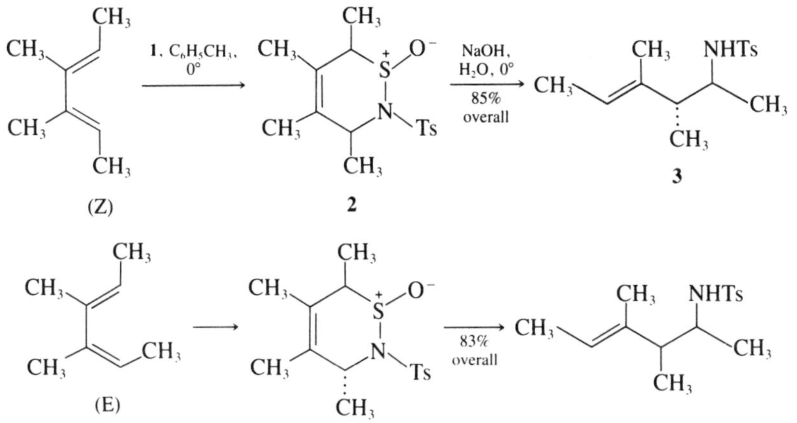

[1] R. S. Garigipati, J. A. Morton, and S. M. Weinreb, *Tetrahedron Letters*, **24**, 987 (1983).

3-Sulfolene, 2, 389–391; **3**, 272–273; **4**, 468.

(E,Z)- or (E,E)-1,3-Dienes. Dialkylation of 3-sulfolene results selectively in *trans*-2,5-dialkyl derivatives (**1**) in 50–70% yield. The products are converted into (E,Z)-1,3-dienes (**2**) on thermal or reductive desulfonylation. In the presence of base, *trans*-**1** isomerizes to *cis*-**1**, which undergoes thermal desulfonylation to (E,E)-1,3-dienes (**2**).[1]

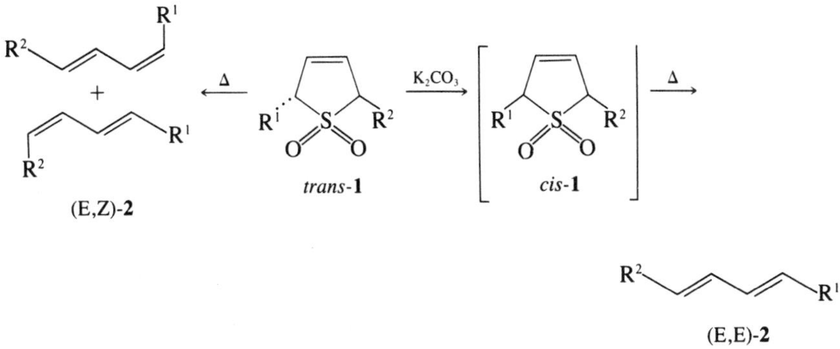

A convenient route to (E,E)-1,3-dienes involves dialkylation of the sulfone **3** to give exclusively *cis*-2,5-derivatives (**4**). On thermolysis of **4**, *cis*-**1** is generated and is converted into (E,E)-**2**.[2]

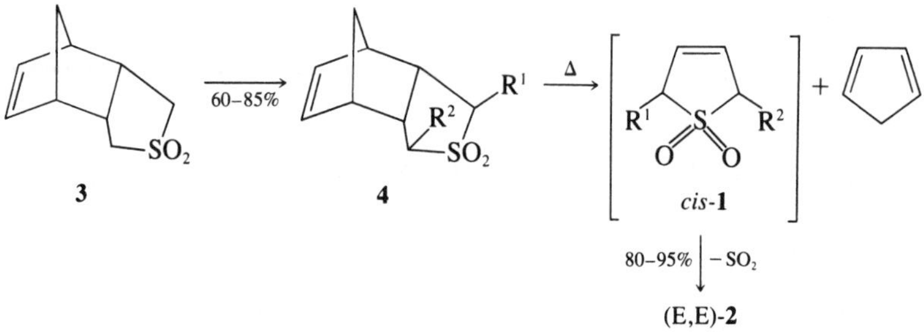

[1] S. Yamada, H. Ohsawa, T. Suzuki, and H. Takayama, *Chem. Letters*, 1003 (1983).
[2] R. Bloch and J. Abecassis, *Tetrahedron Letters*, **24**, 1247 (1983).

T

Tellurium(IV) chloride, TeCl$_4$ **4**, 476; **5**, 643; **9**, 443; **10**, 377. Supplier: Alfa.

Addition to double bonds.[1] Alkenes react with TeCl$_4$ to give mainly products of *syn*-addition of Cl and TeCl$_3$. Only Markownikoff addition occurs with terminal alkenes. The reaction with allylic esters is unusual; it involves *syn*-1,3-addition with a 1,2-rear-

$$(I) \quad CH_2 \diagup\diagdown\diagup OCC_6H_5 + TeCl_4 \xrightarrow[94\%]{CHCl_3, \; 25°} Cl_3Te\diagdown\diagup\diagdown Cl$$

rangement of the acyloxy group (equation I). Allylic ethers and homoallylic esters react normally to give 1,2-adducts. The adducts can be used for synthesis of epoxides, since hydrodetelluration with Raney nickel results in chlorohydrin esters. The sequence provides a synthesis of a chiral epoxide from a chiral allylic ester.

Example:

[1] J. E. Bäckvall and L. Engman, *Tetrahedron Letters*, **22**, 1919 (1981); J. E. Bäckvall, J. Bergman, and L. Engman, *J. Org.*, **48**, 3918 (1983).

2,4,4,6-Tetrabromo-2,5-cyclohexadienone (1), 4, 476–477; **5**, 643–644; **6**, 563; **7**, 351–352; **10**, 377; **11**, 498.

trans-2,5-Disubstituted tetrahydrofuranes. Bromine- (or iodine-) induced cyclization of γ,δ-unsaturated alcohols results in a mixture of *trans*- and *cis*-2,5-disubstituted tetrahydrofuranes, with some preference for the *trans*-isomer. Steric effects can be used to effect selective formation of the *cis*-isomer (**11**, 264). Thermodynamic control can be used indirectly to obtain the *trans*-isomer selectively. Thus the unsaturated alcohol **2** on treatment with **1** cyclizes to the tetrahydropyrane **3** and the tetrahydrofurane regioisomers **4** in the ratio 3:1. Ring contraction of **3** with silver tetrafluoroborate in acetone results in the *trans*-disubstituted tetrahydrofurane **5**.

Preparation of 2,3-*cis*-2,5-*trans*-trisubstituted tetrahydrofuranes by this methodology is possible, but presents some problems. Cyclization is not stereospecific or results only in tetrahydropyranes because of unfavorable 1,3-interactions. Moreover, ring contraction can result in the undesired 2,5-*cis*-diastereomer, even as the major product.

[1] P. C. Ting and P. A. Bartlett, *Am. Soc.*, **106**, 2668 (1984).

Tetrabutylammonium chlorochromate, $Bu_4NCrO_3Cl^-$ (**1**). Mol. wt. 377.91, m.p. 185°. The reagent is prepared by reaction of CrO_3 with Bu_4NHSO_4 in hydrochloric acid (80% yield). Supplier: Fluka.

Oxidation.[1] This mild oxidizing reagent can be used for selective oxidation of benzylic and allylic alcohols. Complete conversion requires 3 equiv. of oxidant. Primary and secondary alcohols are oxidized slowly in refluxing chloroform, but require a large excess of oxidant. An example is the selective oxidation of 1-phenyl-1,3-propanediol to 3-hydroxy-1-phenyl-1-propanone (52% yield).

[1] E. Santaniello, F. Milano, and R. Casati, *Synthesis*, 749 (1983).

Tetra-*n*-butylammonium fluoride (TBAF), **4**, 477–478; **5**, 645; **7**, 353–354; **8**, 467–468; **9**, 444–446; **10**, 378–381; **11**, 499–500.

"Anhydrous reagent." Two laboratories[1,2] have reported that the commercially available trihydrate is converted into almost anhydrous reagent when heated at 40–45° under high vacuum. This material is very effective for various S_N2 fluoride ion displacements, but elimination and hydrolysis to an alcohol can be significant side reactions.

Aldol condensation (**8**, 467).[3]   Complete details have been published concerning the fluoride-ion-catalyzed aldol reaction of enol silyl ethers with aldehydes. The primary product is the silyl ether of the aldol, and yields can be markedly improved by addition of FSi(CH$_3$)$_3$ to the reaction. The reaction exhibits only slight *anti–syn* selectivity, but does show high axial selectivity when the substituent is added to the α-position of a cyclohexanone (equation I).

(I)

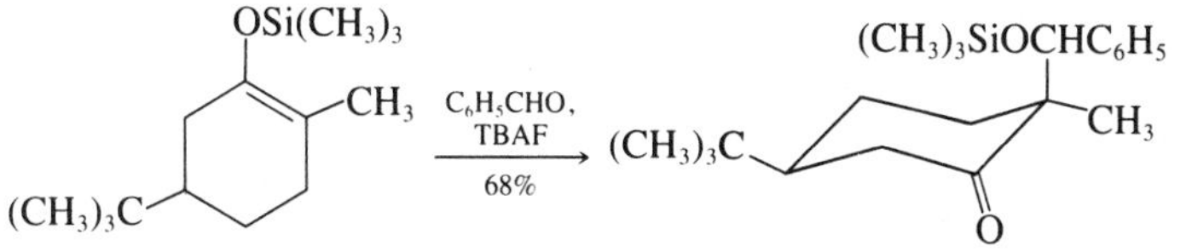

(*anti*/*syn* = 6:4)

The aldol reaction can also be effected by reaction of the ketone and the aldehyde in the presence of excess ethyl trimethylsilylacetate (**7**, 150–151) and TBAF in THF (equation II).

(II)

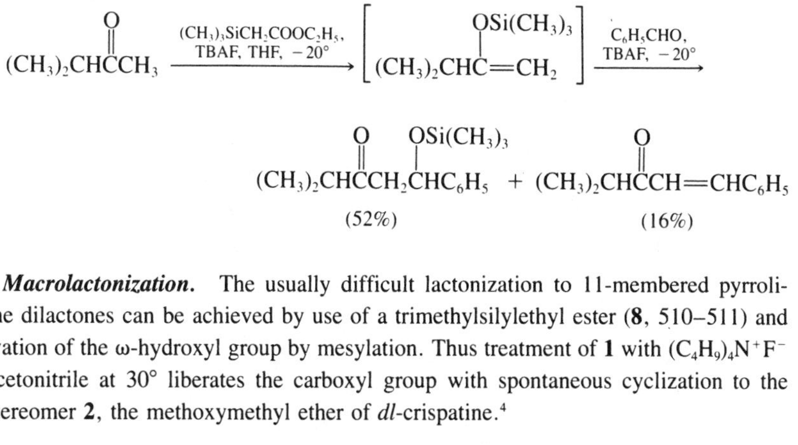

(52%)   (16%)

Macrolactonization.   The usually difficult lactonization to 11-membered pyrrolizidine dilactones can be achieved by use of a trimethylsilylethyl ester (**8**, 510–511) and activation of the ω-hydroxyl group by mesylation. Thus treatment of **1** with (C$_4$H$_9$)$_4$N$^+$F$^-$ in acetonitrile at 30° liberates the carboxyl group with spontaneous cyclization to the diastereomer **2**, the methoxymethyl ether of *dl*-crispatine.[4]

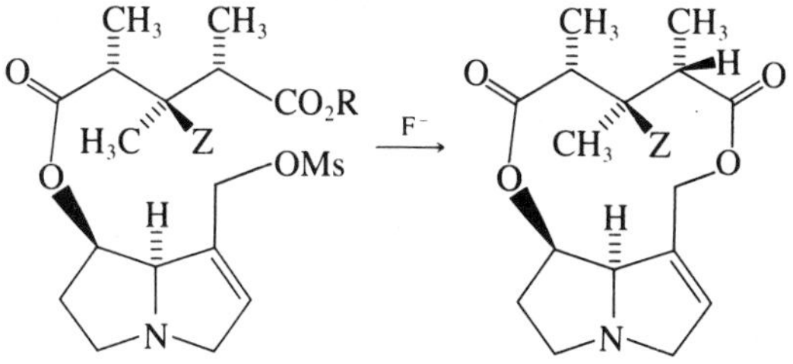

1, R = (CH$_2$)$_2$Si(CH$_3$)$_3$,   **2**
Z = OCH$_2$OCH$_3$

A general method for conversion of a cyclic ketone into a four-membered ring enlarged lactone is illustrated by the synthesis of phoracantolide I (**6**) from α-nitrocyclohexanone.[5] The Michael adduct (**3**) of the nitro ketone to acrylaldehyde is converted on selective methylation with $(CH_3)_2Ti(O$-i-$Pr)_2$ into a single hemiacetal (**4**). This product is transformed into the ring enlarged nitrodecanolide (**5**) in the presence of catalytic amounts of Bu_4NF. Conversion of **5** to **6** is effected by conversion of the nitro group to a keto group followed by desulfuration of the dithioethylene ketal.

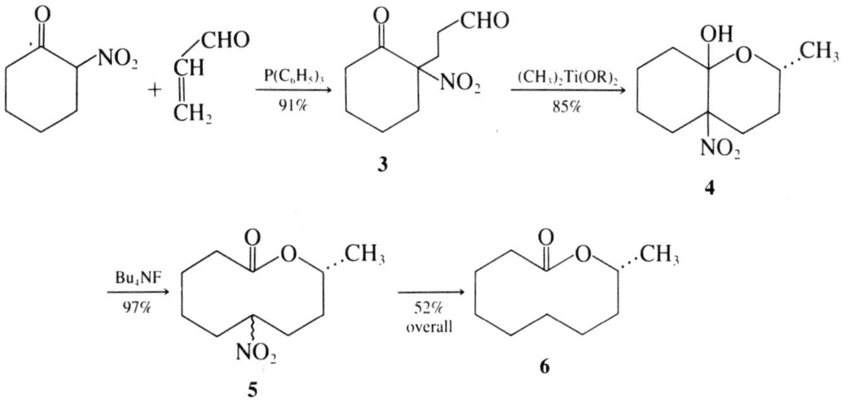

3

4

5

6

α-*Hydroxy*-β-*diketone system*.[6]   The α-hydroxy-β-diketone system of **3**, a common feature of tetracycline antibiotics, is generally obtained as a mixture of *cis*- and *trans*-isomers by hydroxylation of a β-diketone. The desired *cis*-isomer can be obtained selec-

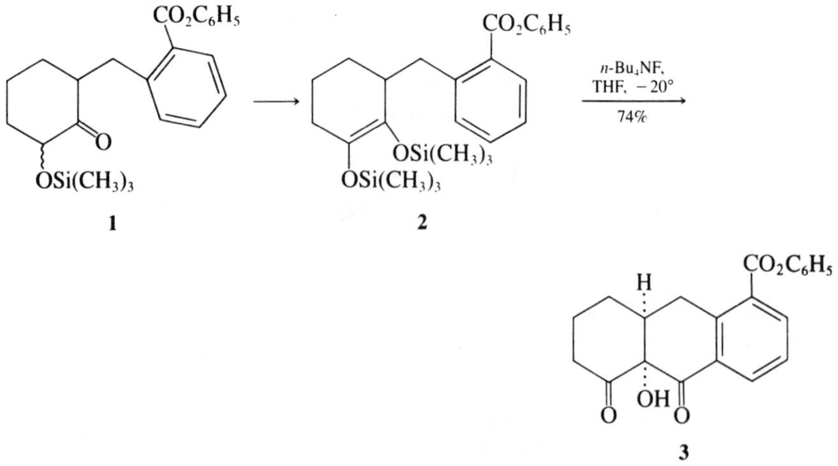

1

2

3

tively by reaction of fluoride ion with the enediol disilyl ether **2** formed from the keto ether **1**. Attempts at cyclization of the lithium enolate of **1** fail, as does cyclization of the methyl ester corresponding to **2**.

Silyl anions.[7] A symmetrical disilane such as hexamethyldisilane is cleaved by TBAF (or CsF, but not by KF) to trimethylsilyl fluoride and the trimethylsilyl anion **1**. An unsymmetrical silane is cleaved selectively at the less hindered Si atom.

$$(CH_3)_3Si\text{---}Si(CH_3)_3 \xrightarrow{Bu_4N^-F^-} Bu_4N^+Si(CH_3)_3^- + (CH_3)_3SiF$$

1

Aliphatic aldehydes react with **1** in the presence of HMPT to form 1-(trimethylsilyl)-1-alkanols, $RCH(OH)Si(CH_3)_3$, in 30–70% yield. Aryl aldehydes are converted into pinacols by reaction with **1**, generally in high yield. Ketones and epoxides are inert to **1**.

Reaction of **1** and related silyl anions with 1,3-dienes gives 1,4-disilyl-2-butenes. The (E)-isomer is the exclusive or major product. HMPT or 1,3-dimethyl-2-imidazolidinone (DMI) is essential. 1,3-Dienes substituted at C_1 and/or C_4 do not undergo this reductive disilylation.

Example:

$(E/Z >99:1)$

Cyclopropyl carbanions.[8] Attempts to deprotonate cyclopropanes substituted by a electronegative group result in dimerization or trimerization. The desired carbanions can be generated *in situ* by desilylation of an α-(trimethylsilyl)cyclopropane derivative with Bu_4NF or $C_6H_5CH_2N(CH_3)_3F$ and used to effect aldol condensations.

Examples:

Nitro-aldol reaction (**9**, 444; **10**, 381).[9]   2-Methylcyclohexanone does not react with nitromethane under usual basic conditions, but does react under high pressure and with fluoride ion catalysis to give the nitro alcohol (**1**) in moderate yield.

1

[1] R. K. Sharma and J. L. Fry, *J. Org.*, **48**, 2112 (1983).
[2] D. P. Cox, J. Terpinski, and W. Lawrynowicz, *ibid.*, **49**, 3216 (1984).
[3] E. Nakamura, M. Shimizu, I. Kuwajima, J. Sakata, K. Yokoyama, and R. Noyori, *ibid.*, **48**, 932 (1983).
[4] E. Vedejs and S. D. Larsen, *Am. Soc.*, **106**, 3030 (1984).
[5] K. Kostova and M. Hesse, *Helv.*, **67**, 1713 (1984).
[6] G. Stork and Y. K. Yee, *Can. J. Chem.*, **62**, 2627 (1984).
[7] T. Hiyama, M. Obayashi, I. Mori, and H. Nozaki, *J. Org.*, **48**, 912 (1983).
[8] L. A. Paquette, C. Blankenship, and G. J. Wells, *Am. Soc.*, **106**, 6442 (1984).
[9] K. Matsumoto, *Angew. Chem., Int. Ed.*, **23**, 617 (1984).

1,1,6,6-Tetra-*n*-butyl-1,6-distanna-2,5,7,10-tetraoxacyclodecane (1). Mol. wt. 585.93, m.p. 223–226°. The stannoxane is obtained in 91% yield by reaction of dibutyltin oxide and ethylene glycol (Dean-Stark trap).[1]

1

Macrocyclic lactones.   This stannoxane serves as a template to catalyze the condensation of propiolactone to a series of macrocyclic lactones (**2**). The trimer, n = 1, is the major product (22%). The related linear stannoxane $Bu_2Sn(OCH_2CH_3)_2$ is ineffective for this macrocyclization of lactones.[2]

2, n = 1–5

This cyclization has been used for a synthesis of enterobactin (**5**), a natural iron carrier (previous synthesis, **8**, 215–216) by cyclization of a linear trimer of serine. Thus the lactone **3**, derived from L-serine, cyclizes in the presence of **1** to the tricyclic lactone **4** in 23% yield. This product was converted into **5** by detritylation and acylation.[3]

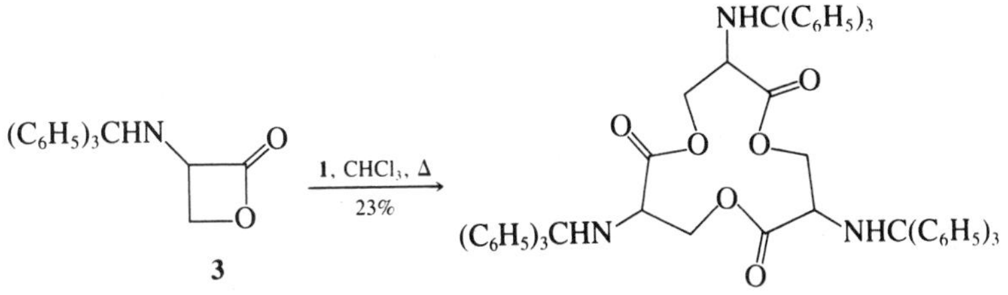

3

4

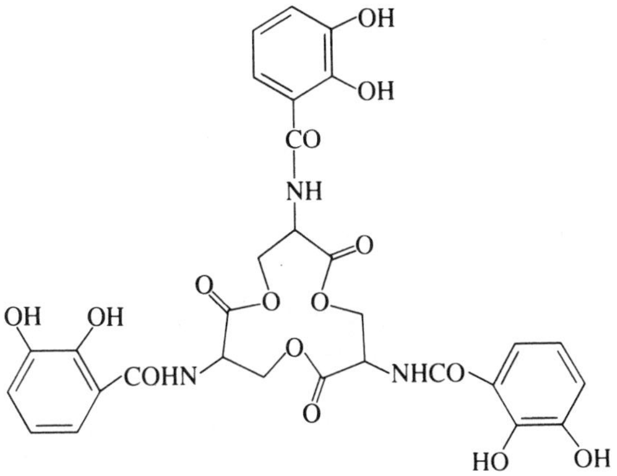

5

[1] W. J. Considine, *J. Organometal. Chem.*, **5**, 263 (1966).
[2] A. Shanzer, J. Libman, and F. Frolow, *Am. Soc.*, **103**, 7339 (1981).
[3] A. Shanzer and J. Libman, *J.C.S. Chem. Comm.*, 846 (1983).

Tetrachlorosilane, SiCl₄, 3, 277; **4**, 277; **10**, 347.

β-Halo acetals.[1] Reaction of an α,β-enal or -enone with a halotrimethylsilane in the presence of ethylene glycol results in a β-halo acetal in generally good yield.

Example:

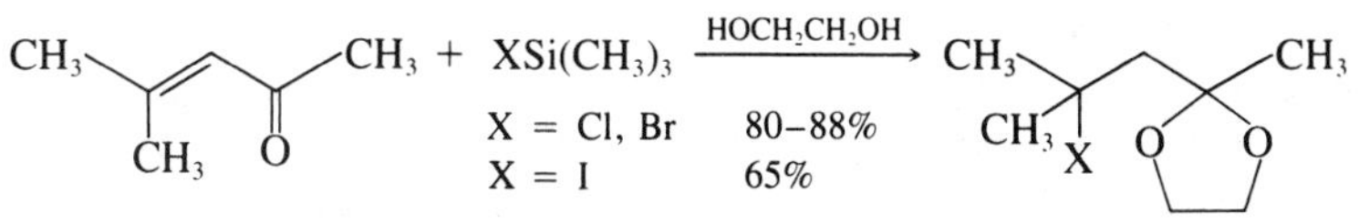

SiCl$_4$ is generally preferred for preparation of β-chloro acetals; the reagent is cheap and yields are often higher than those obtained with ClSi(CH$_3$)$_3$. It can also serve as a catalyst for a similar reaction of azidotrimethylsilane with enals or enones.

Example:

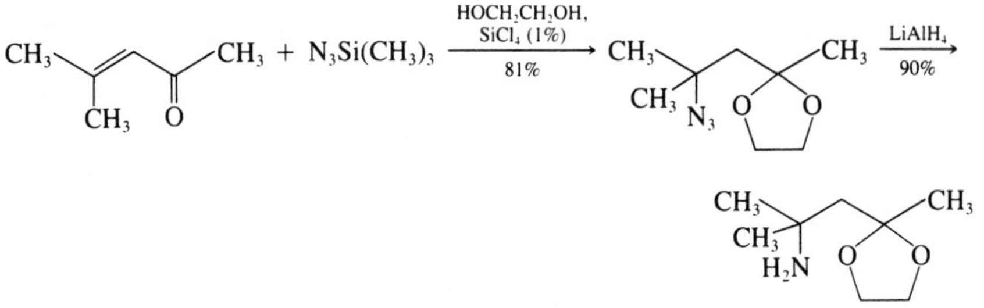

[1] G. Gil, *Tetrahedron Letters*, **25**, 3805 (1984).

7,7,8,8-Tetracyanoquinodimethane (1), **1**, 1136.
Improved synthesis:[1]

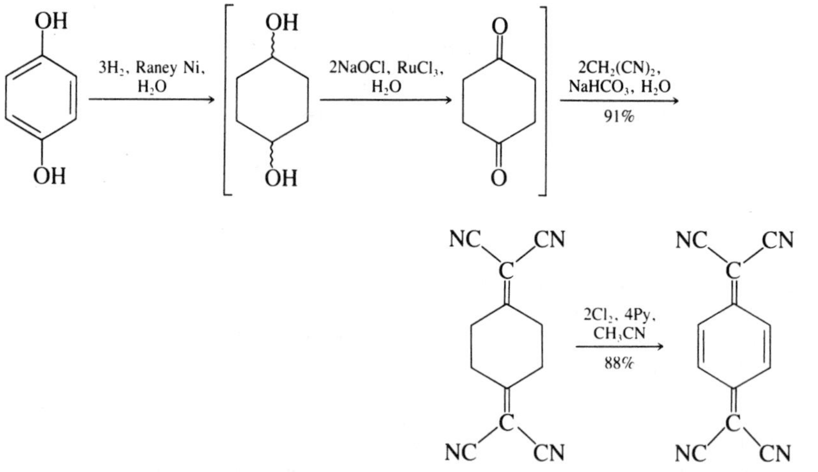

1, m.p. 287–289°

α-Chlorination of carboxylic acids.[2]   A Hell-Volhard-Zelinsky type of α-chlorination of carboxylic acids with chlorosulfonic acid is possible if a competing free radical chlorination is suppressed. For this purpose, **1** is superior to oxygen;[3] in addition, the reaction is applicable to both short- and long-chain acids (equation I).

(1) $CH_3(CH_2)_nCH_2COOH + Cl_2 \xrightarrow[80-90\%]{\overset{\text{ClSO}_3\text{H. 1.}}{150°}} CH_3(CH_2)_nCHClCOOH + HCl$

(n = 3–15)

[1] R. J. Crawford, *J. Org.*, **48**, 1366 (1983).
[2] *Idem, ibid.*, **48**, 1364 (1983).
[3] Y. Ogata, T. Harada, K. Matsuyama, and T. Ikejiri, *ibid.*, **40**, 2960 (1975).

Tetraethylammonium iodide, $(C_2H_5)_4NI$. Suppliers: Aldrich, Fluka.

β-*Iodo ketones*.[1]   This salt reacts with cyclic or acyclic α,β-enones in anhydrous TFA to form β-iodo ketones in yields generally of 85–95%. The method can be used for synthesis of β-bromo and β-chloro ketones, but the rate of these reactions is much slower. The reaction fails with β-substituted cyclohexenones and cyclopentenones, and with α,β-unsaturated esters and nitriles.

[1] J. N. Marx, *Tetrahedron*, **39**, 1529 (1983).

Tetrafluoroboric acid, 1, 1139; **2,** 397; **10,** 382–383.

1,3-*Carbonyl transposition*.[1]   Tetrafluoroboric acid is the most effective acid for rearrangement of α-hydroxy ketene dithioketals to α,β-unsaturated thiol esters. This rearrangement is particularly useful for rearrangement of the tertiary allylic alcohols formed by addition of organometallic reagents to α-keto ketene dithioketals, which are readily available by reaction of ketone enolates with carbon disulfide followed by alkylation with methyl iodide.

Application of this rearrangement provides a short synthesis of the furanosesquiterpene myodesmone (**1**).

1

[1] R. K. Dieter, Y. J. Lin, and J. W. Dieter, *J. Org.*, **49**, 3183 (1984).

Tetrafluorosilane, SiF_4. This gas is prepared by thermal decomposition of $BaSiF_6$.[1]

Glycosylation.[2] This silane (or trimethylsilyl triflate) is an effective catalyst for glycosylation of benzyl-protected glycopyranosyl fluorides with trimethylsilyl ethers or even alcohols. Of particular interest is the fact that the stereochemistry is independent of the anomeric configuration, but markedly dependent on the solvent. Glycosylation in acetonitrile gives β-glycosides with moderate to high stereoselectivity, whereas the same reaction in ether affords α-glycosides selectively.

Example:

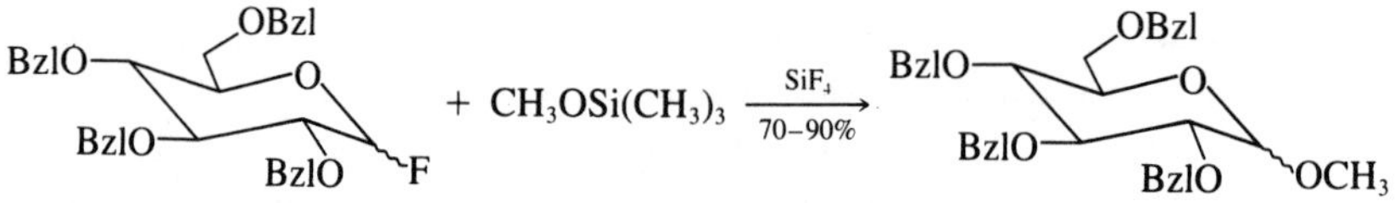

1 (α or β) **2**

CH_3CN	(α/β = 15:85)
ether	(α/β = 75–85:25–15)

[1] C. J. Hoffman and H. S. Gutowsky, *Inorg. Syn.*, **4**, 145 (1953).
[2] S. Hashimoto, M. Hayashi, and R. Noyori, *Tetrahedron Letters*, **25**, 1379 (1984).

Tetraisopropylthiuram disulfide, $(i\text{-Pr})_2N\overset{S}{\overset{\|}{C}}\text{—S—S—}\overset{S}{\overset{\|}{C}}N(i\text{-Pr})_2$ (**1**).

Preparation:[1]

$$[(CH_3)_2CH]_2NH \xrightarrow[CS_2]{NaOH,\ H_2O,} [(i\text{-Pr})_2N\overset{S}{\overset{\|}{C}}\text{—SNa}] \xrightarrow[81-83\%]{K_3Fe(CN)_6} \mathbf{1}$$

Aryl thiols.[2] Aryllithiums react with this disulfide to form crystalline dithiocarbamates, which are cleaved to aryl thiols on alkaline hydrolysis (equation I).

$$\text{(I)}\quad ArLi \xrightarrow{\mathbf{1}} ArS\underset{\underset{S}{\|}}{C}N(i\text{-Pr})_2 \xrightarrow{KOH,\ H_2O} ArSH$$

[1] R. Rothstein and K. Binovic, *Rec. Trav.* **73**, 561 (1954).
[2] K.-Y. Jen and M. P. Cava, *Tetrahedron Letters*, **23**, 2001 (1982); *idem, Org. Syn.*, submitted (1984).

Tetrakis(acetonitrile)palladium tetrafluoroborate, $(CH_3CN)_4Pd(BF_4)_2$ (**1**). Supplier: Strem. The cationic complex is prepared by reaction of Pd sponge and nitrosonium tetrafluoroborate in acetonitrile.[1]

Dimerization of acrylates.[2]   Acrylates and methacrylates undergo tail-to-tail dimerization at moderate temperatures in the presence of this catalyst. Added lithium tetrafluoroborate increases the rate and prolongs catalyst life. This additive can be omitted in dimerizations conducted in nitromethane.

Example:

$$CH_2 = CHCOOCH_3 \xrightarrow[91-93\%]{1,\ LiBF_4,\ 40°} CH_3O_2C(CH_2)_2 \diagup\diagup COOCH_3$$

$$(\Delta^2,\ 93-96\%)$$

[1] B. B. Wayland and R. F. Schramm, *Inorg. Chem.*, **8**, 971 (1969).
[2] W. A. Nugent and F. W. Hobbs, Jr., *J. Org.*, **48**, 5364 (1983); *idem, Org. Syn.*, submitted (1984).

Tetrakis(triphenylphosphine)nickel, **6**, 570; **7**, 357; **9**, 450; **11**, 503.

Coupling of vinyl chlorides with Grignard reagents.[1]   (E)- or (Z)-Vinyl chlorides, available by Ni(0)-catalyzed coupling of (E)- or (Z)-dichloroethylene with a Grignard reagent, undergo cross-coupling with Grignard reagents in the presence of tetrakis(triphenylphosphine)nickel with high selectivity. Tetrakis(triphenylphosphine)palladium is somewhat less efficient. The reaction provides a useful preparation of unsaturated acetates typical of insect pheromones.

Example:

[1] V. Ratovelomanana and G. Linstrumelle, *Syn. Comm.*, **14**, 179 (1984).

Tetrakis(triphenylphosphine)palladium(0), 6, 571–573; **7**, 357–358; **8**, 472–476; **9**, 451–458; **10**, 384–391; **11**, 503–514.

1,3-Dienes.[1]   (E)-1-Alkenylalanes, readily obtained by hydroalumination of 1-alkynes,[2] couple with vinyl iodides in the presence of this Pd(0) catalyst and zinc chloride to give 1,3-dienes with retention of the configuration. Alkenylzinc derivatives may be the actual reactants. Nickel catalysts are less stereoselective than palladium–phosphine catalysts.

Example:

$$n\text{-}C_8H_{17}C\equiv CH \xrightarrow[\text{hexane}]{\text{DIBAH,}} \underset{\textbf{1}}{\overset{n\text{-}C_8H_{17}}{\underset{H}{}}C=C\overset{H}{\underset{Al(C_4H_9\text{-}i)_2}{}}}$$

$$\textbf{1} + \overset{I}{\underset{H}{}}C=C\overset{C_4H_9\text{-}n}{\underset{H}{}} \xrightarrow[72\%]{\text{Pd(0), ZnCl}_2,\ \text{THF}} \overset{n\text{-}C_8H_{17}}{\underset{H}{}}C=C\overset{H}{\underset{}{}}\ \overset{}{\underset{H}{}}C=C\overset{C_4H_9\text{-}n}{\underset{H}{}}$$

Heterosubstituted 1,3-dienes.[3]   Some alkenylmetals substituted by OR, SR, or SiR₃ can be coupled with alkenyl halides (or aryl halides) in the presence of this Pd(0) catalyst. Alkenylzinc reagents, prepared by reaction of alkenyllithiums with ZnCl₂, and alkenylalanes are the most useful. Yields from reactions with organoboronates are low.

$$CH_2=CH\overset{OC_2H_5}{\underset{ZnCl}{}} + I\diagup\diagdown\diagup C_5H_{11}\text{-}n \xrightarrow[74\%]{\substack{\text{Pd(0),}\\ \text{THF, 25°}}} \underset{C_2H_5O}{\overset{C_5H_{11}\text{-}n}{}}CH_2$$

$$\overset{n\text{-}C_6H_{13}}{\underset{H}{}}C=C\overset{Si(CH_3)}{\underset{Al(i\text{-}Bu)_2}{}} + BrCH=CH_2 \xrightarrow[74\%]{\substack{\text{Pd(0),}\\ \text{ZnCl}_2}} \underset{(CH_3)_3Si}{\overset{CH_2}{}}\underset{C_6H_{13}\text{-}n}{}$$

1,3-Diynes.[4]   A general method for preparation of 1,3-diynes, particularly terminal ones, involves palladium-catalyzed coupling of alkynylzinc derivatives with (E)-1-iodo-2-chloroethylene. This alkene is obtained in 83% yield by reaction of acetylene with iodine monochloride in 6 N HCl. Coupling results in a 1-chloro-1,3-enyne (**1**), which is converted into a 1-sodio-1,3-diyne (**2**), which in turn can be reduced or alkylated to give a 1,3-diyne.

Example:

$$CH_3(CH_2)_5C\equiv CH \xrightarrow[\text{2) ZnCl}_2]{\text{1) } n\text{-BuLi}} [RC\equiv CZnCl] + \underset{H}{\overset{I}{}}C=C\overset{H}{\underset{Cl}{}} \xrightarrow[73\%]{\text{Pd(0)}}$$

$$(= RC\equiv CH)$$

$$\underset{H}{\overset{RC\equiv C}{}}C=C\overset{H}{\underset{Cl}{}} \xrightarrow[\text{NH}_3]{\text{NaNH}_2,} [RC\equiv CC\equiv CNa] \xrightarrow[68\%]{\text{NH}_4Cl} RC\equiv CC\equiv CH$$

$$\mathbf{1} \qquad\qquad\qquad \mathbf{2}$$

$$67\% \big\downarrow CH_3I$$

$$RC\equiv CC\equiv CCH_3$$

An alternative method involves coupling the alkynylzinc derivative with vinyl bromide to give a 1-en-3-yne, which is converted into a terminal 1,3-diyne selectively by bromination of the double bond followed by debromination (equation I).

(I) $RC\equiv CZnCl + CH_2=CHBr \xrightarrow[78\%]{\text{Pd(0)}} RC\equiv CCH=CH_2 \xrightarrow[\text{2) Na, NH}_3 (82\%)]{\text{1) Br}_2 (76\%)}$

$$RC\equiv CC\equiv CH$$

Allenynes.[5]   An attractive route to allenynes involves the reaction of allenic bromides with 1-alkynes in the presence of catalytic amounts of this Pd(0) complex and CuI in diethylamine at 25° (equation I). The synthesis presumably involves cross-coupling between an allenic palladium σ-compound and a copper acetylide.

(I) $\underset{R^2}{\overset{R^1}{}}C=C=C\overset{H}{\underset{Br}{}} + HC\equiv CR^3 \xrightarrow[70-94\%]{\substack{\text{Pd(0), CuI,}\\ (C_2H_5)_2NH}} \underset{R^2}{\overset{R^1}{}}C=C=C\overset{H}{\underset{\underset{CR^3}{\|\|}}{}}$

Coupling of vinyl triflates with organotins.   Vinyl triflates couple with organotin reagents in the presence of catalytic amounts of Pd(0) and lithium chloride (2 equiv.). The reaction is general, but is particularly useful for stereoselective synthesis of 1,3-dienes.[6]

Example:

This coupling in conjunction with the regioselective formation of vinyl triflates from enolates (see N-phenyltrifluoromethanesulfonimide, this volume) provides a short convergent synthesis of pleraplysillin-1 (**1**, Chart I).

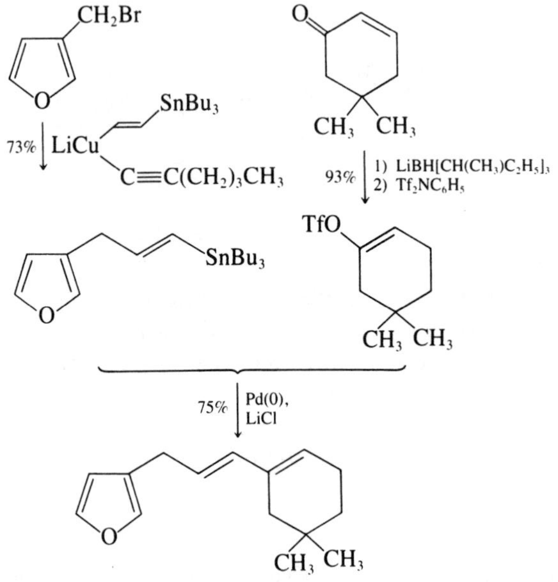

Chart (I)

This coupling when carried out in the presence of carbon monoxide (15–50 psi) results in cross-coupled ketones in generally good yield.[7] This reaction is a particularly attractive route to divinyl ketones, which are substrates for Nazarov cyclization. The geometry of the vinyl triflate is retained.

Example:

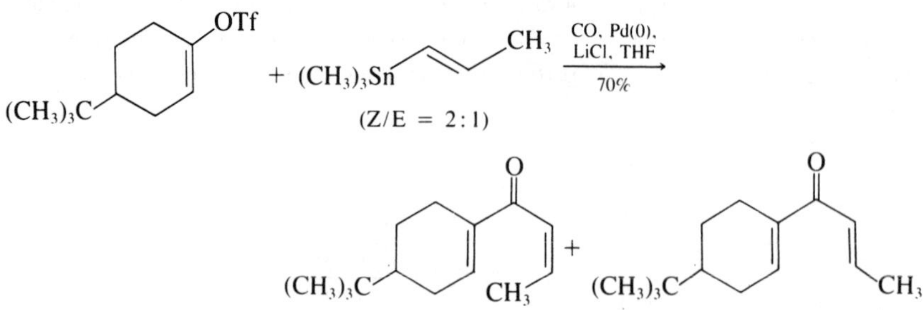

This reaction was used for an iterative cyclopentenone annelation in a synthesis of $\Delta^{9(12)}$-capnellene (**2**), formulated in part in Scheme (I).

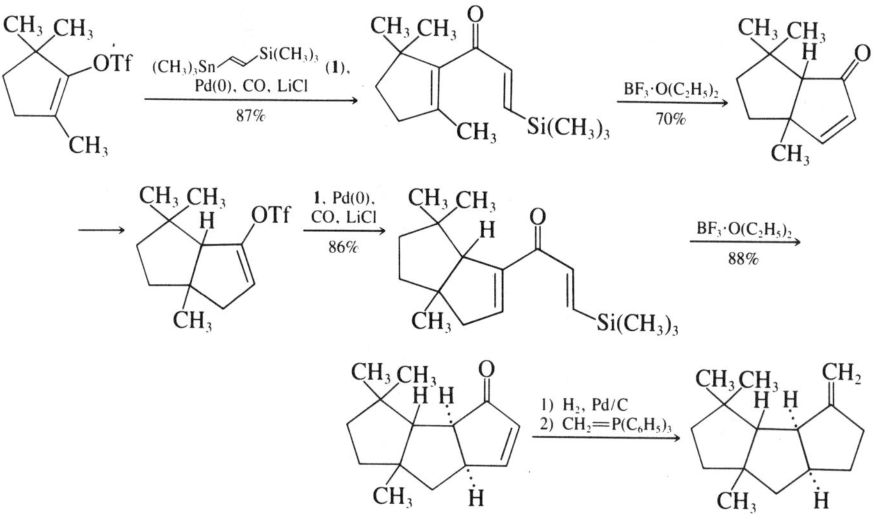

Scheme (I)

Acylation of organozinc reagents.[8]   Reaction of acyl chlorides with organozinc compounds catalyzed by palladium–phosphine complexes provides a general synthesis of ketones. The organozincs are readily available by treatment of the corresponding organo-lithium with $ZnCl_2$. Alkenylzinc compounds and α,β-unsaturated acyl chlorides react with retention of the stereochemistry. Isolated yields of ketones are 55–90%.

Example:

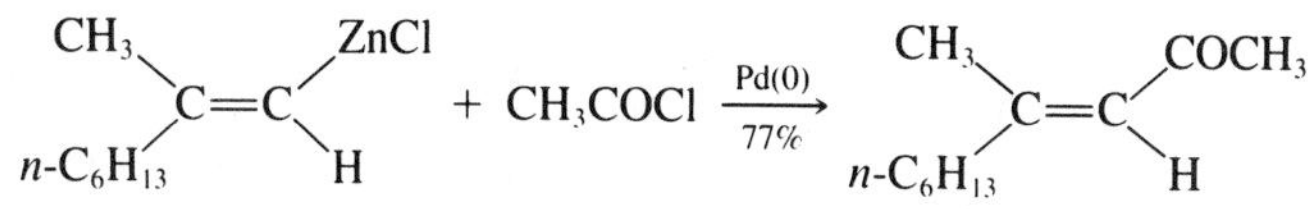

(E, 99%)

β-Carboline synthesis.[9]   The 3-amino-4-(*o*-bromophenyl)pyridine **1** cyclizes to the β-carboline **2** in the presence of tetrakis(triphenylphosphine)palladium(0) (1.2–1.5 equiv.,

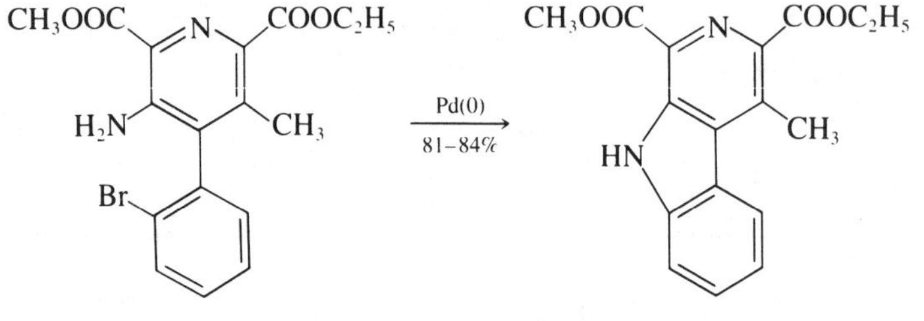

1   **2**

THF, 80°). The precursor (**1**) was prepared in several steps from the Diels-Alder adduct (**5**) of the triazine **3** with the α-phenyl enamine **4**.

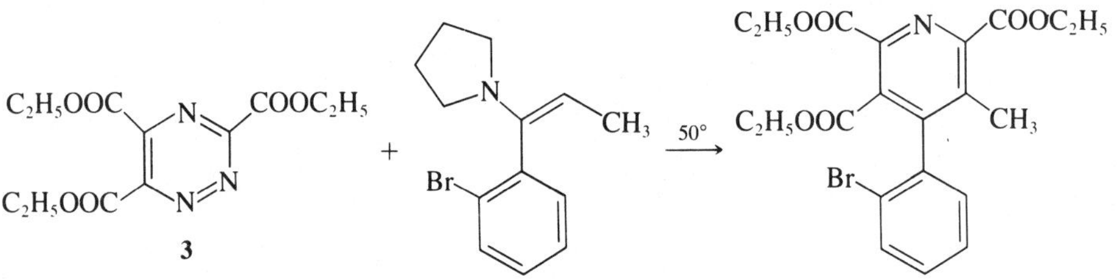

3   **4**   **5** (major product)

Allylation of potassium enoxyborates (**11**, 506).[10]   This reaction can be used to prepare 1,4- and 1,5-diketones by allylation with 2,3-dichloropropene or 1,3-dichloro-2-butene followed by hydrolysis of the vinyl chloride group (**9**, 294).

Example:

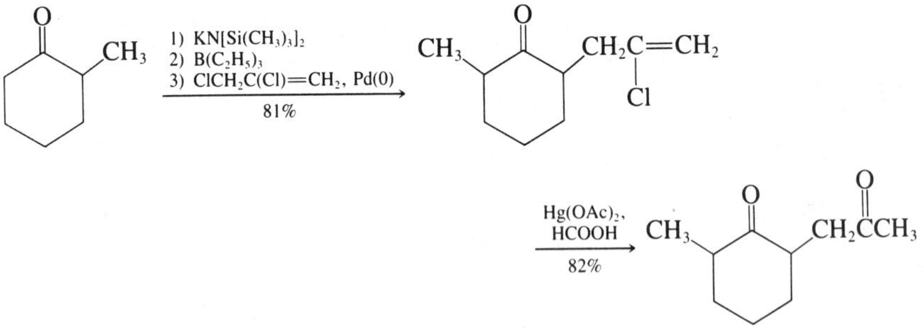

Allylic amination.[11]   In the presence of Pd(0) catalysts, primary and secondary amines react with (E)- or (Z)-γ-acetoxy or γ-chloro allylic alcohols or acetates to form (E)-γ-amino allylic alcohols or acetates.

Examples:

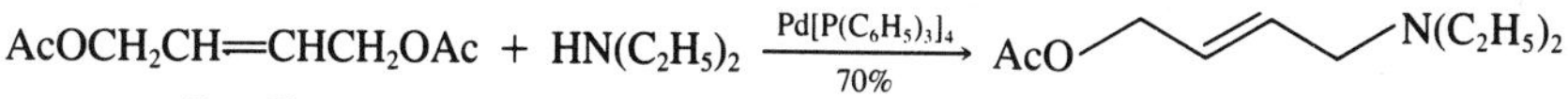

$$\text{AcOCH}_2\text{CH}\!=\!\text{CHCH}_2\text{OAc} + \text{HN}(C_2H_5)_2 \xrightarrow[\;70\%\;]{\text{Pd[P(C}_6\text{H}_5)_3]_4}$$

(E or Z)

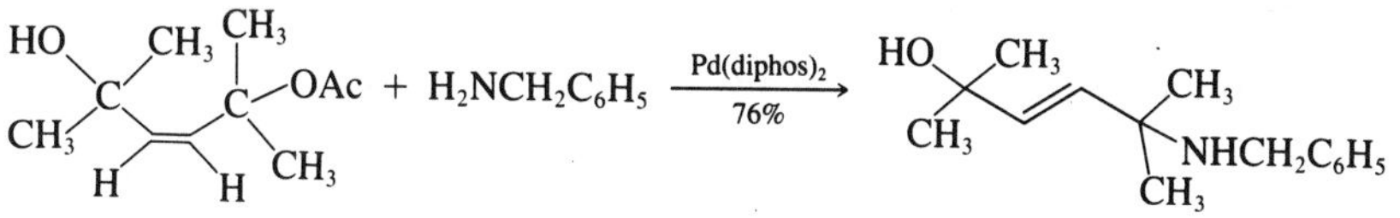

AcO⌇⌇⌇N(C$_2$H$_5$)$_2$

(Z)

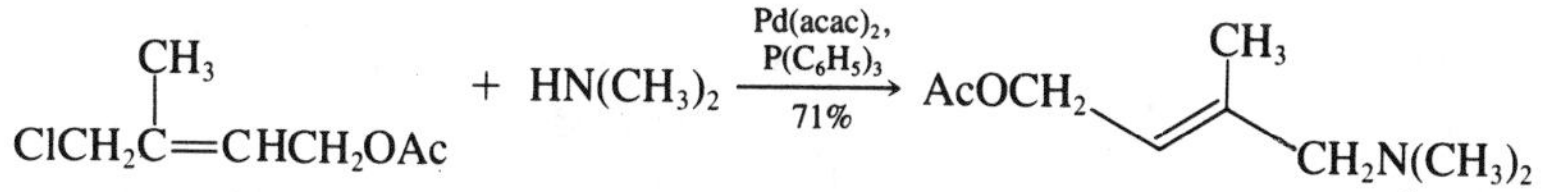

$$+ \text{H}_2\text{NCH}_2\text{C}_6\text{H}_5 \xrightarrow[76\%]{\text{Pd(diphos)}_2}$$

(E/Z = 3.6)

$$+ \text{HN}(CH_3)_2 \xrightarrow[71\%]{\substack{\text{Pd(acac)}_2, \\ \text{P(C}_6\text{H}_5)_3}}$$

Macrocyclization by allylation–alkylation.[12]   The key step in a synthesis of the antibiotic A26771B (**3**) is cyclization of the substrate **1** using O,N-bis(trimethylsilyl)acetamide (**1**, 61; **2**, 30; **3**, 23–24) as base and Pd[P(C$_6$H$_5$)$_3$]$_4$ as catalyst. In addition a bidentate phosphorus ligand is essential. The highest yields were obtained with 1,4-bis(diphenylphosphine)butane (dppb).

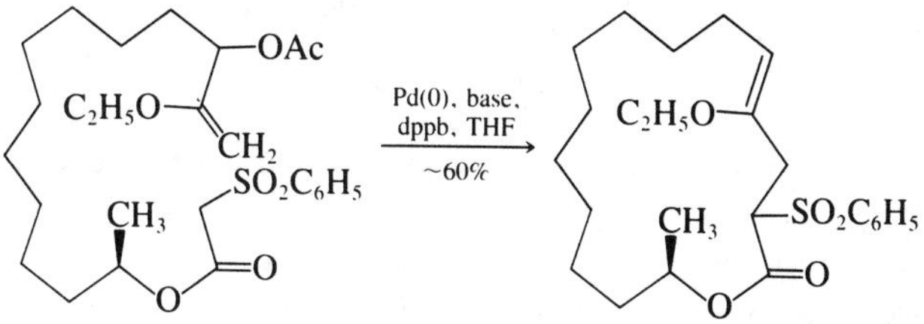

1   **2** (major product)

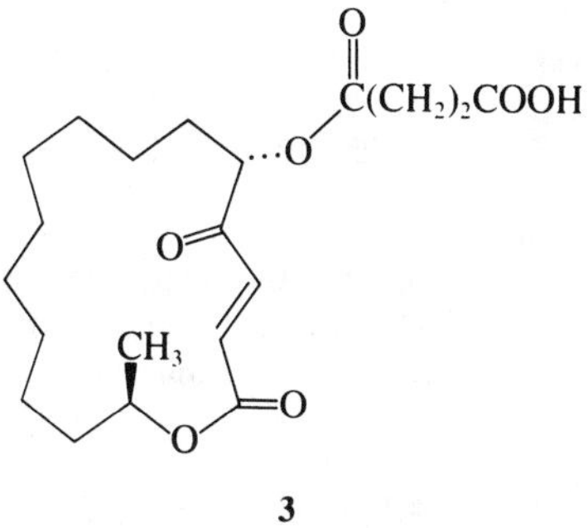

3

Steroidal 6β-hydroxy-2,4-diene-1-ones.[13]   This system is present in a number of natural steroids such as the withanolides. A novel approach to this system is typified by conversion of **1** into **2** by reaction with Pd(0) (1 equiv.) in THF at 25°.

trans- → cis-*Cinnamic acids.*[14]   This conversion is effected in 10–40% yield by irradiation. A more efficient, although longer, method is outlined in equation (I). The key step, carbonation of a vinyl bromide, is regiospecific when catalyzed with $Pd[P(C_6H_5)_3]_4$ or with bis[bis(1,2-diphenylphosphine)ethane]palladium(0) [Pd(diphos)$_2$] under phase-transfer conditions.

RX → RCHO.[15]   Alkyl halides can be converted directly into aldehydes in moderate to high yield by reaction with carbon monoxide (1–3 atm.) and tri-*n*-butyltin hydride catalyzed by the palladium(0) complex. The reaction involves insertion of carbon monoxide to form an acyl halide, which is known to be reduced to an aldehyde under these conditions (**10**, 411). Direct reduction of the halide can be minimized by slow addition of the tin hydride to the reaction and by an increase in the carbon monoxide pressure.

[1] E. Negishi, T. Takahashi, and S. Baba, *Org. Syn.*, submitted (1984).
[2] G. Zweifel and J. A. Miller, *Org. React.*, **32**, 375 (1984).
[3] E. Negishi and F.-T. Luo, *J. Org.*, **48**, 1560 (1983).
[4] E. Negishi, N. Okukado, S. F. Lovich, and F.-T. Luo, *ibid.*, **49**, 2629 (1984).
[5] T. Jeffery-Luong and G. Linstrumelle, *Synthesis*, 31 (1983).
[6] W. J. Scott, G. T. Crisp, and J. K. Stille, *Am. Soc.*, **106**, 4630 (1984).
[7] G. T. Crisp, W. J. Scott, and J. K. Stille, *ibid.*, **106**, 7500 (1984).
[8] E. Negishi, V. Bagheri, S. Chatterjee, F.-T. Luo, J. A. Miller, and A. T. Stoll, *Tetrahedron Letters*, **24**, 5181 (1983).
[9] D. L. Boger and J. S. Panek, *ibid.*, **25**, 3175 (1984).
[10] E. Negishi, F.-T. Luo, A. J. Pecord, and A. Silveira, Jr., *J. Org.*, **48**, 2427 (1983).

[11] J. P. Genêt, M. Balabane, J. E. Bäckvall, and J. E. Nyström, *Tetrahedron Letters*, **24**, 2745 (1983).
[12] B. M. Trost and S. J. Brickner, *Am. Soc.*, **105**, 568 (1983).
[13] E. Keinan, M. Sahai, and I. Kirson, *J. Org.*, **48**, 2550 (1983).
[14] V. Galamb and H. Alper, *Tetrahedron Letters*, **24**, 2965 (1983).
[15] V. P. Baillargeon and J. K. Stille, *Am. Soc.*, **105**, 7175 (1983).

Tetrakis(triphenylphosphine)palladium(0)–Copper(I) iodide.

Coupling of 1-alkynes with vinyl halides.   Two laboratories[1] reported a few years ago that vinyl halides couple with retention of configuration with terminal alkynes to form conjugated enynes in diethylamine or *n*-propylamine in the presence of CuI and either Pd[P(C$_6$H$_5$)$_3$]$_4$ or Cl$_2$Pd[P(C$_6$H$_5$)$_3$]$_2$. Nicolaou *et al.*[2] have used this reaction to advantage in syntheses of some of the linear eicosanoid metabolites of arachidonic acid. Thus the C$_1$–C$_{10}$ fragment (**3**) of leukotriene B$_4$ (**4**) is obtained in high yield by coupling of the vinyl bromide **1** with the optically active propargylic alcohol derivative **2**. The final step

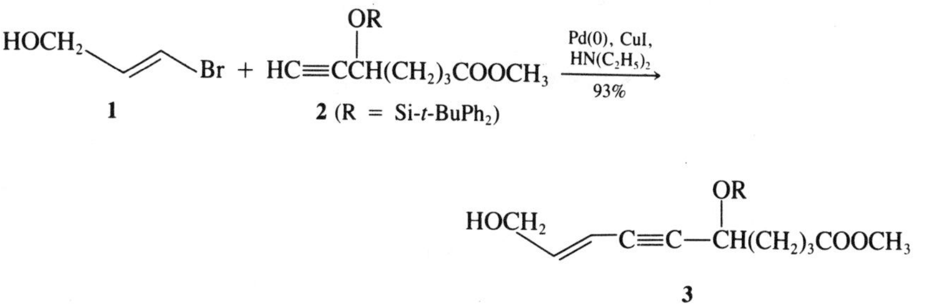

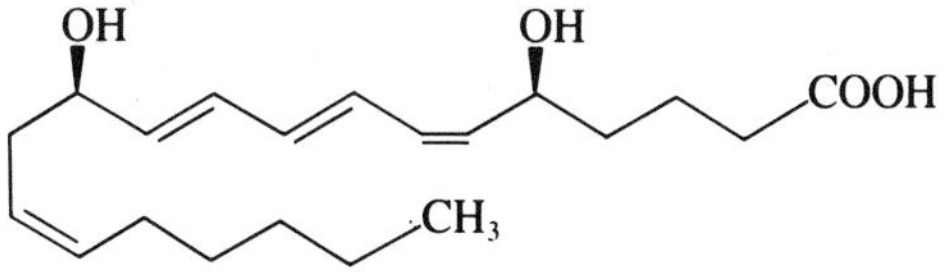

in the synthesis of **4** involves *cis*-hydrogenation (Lindlar, 60% yield) of the triple bond and deprotection.

This Pd/Cu coupling was also used to provide the *cis,trans*-diene systems of the related 5,15- and 8,15-DiHETES. The synthesis of the 5,15-isomer (**8**) involves coupling of the

1,4-dyne **5** with **6** and then, after liberation of the terminal acetylene group, with **7**, to give a product that on *cis*-hydrogenation and deprotection is converted into **8**.

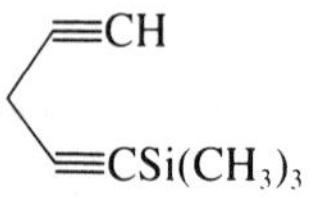

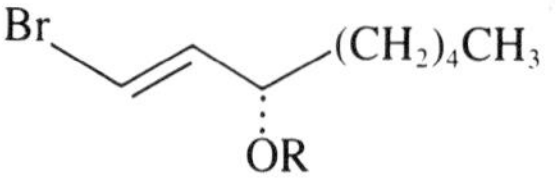

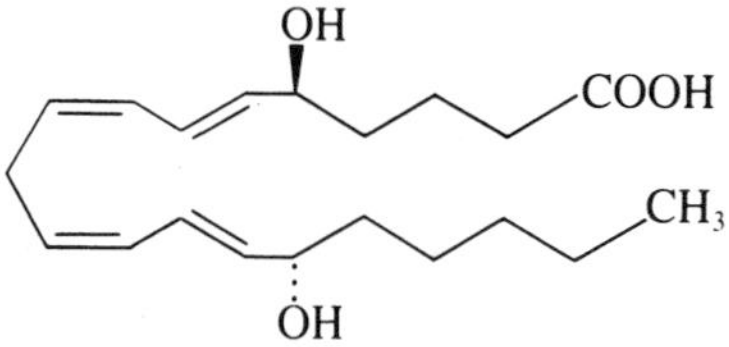

[1] K. Sonogashira, Y. Tohda, and N. Hagihara, *Tetrahedron Letters*, 4467 (1975); V. Ratovelomana and G. Linstrummelle, *Syn. Comm.*, **11**, 917 (1981).
[2] K. C. Nicolaou, R. E. Zipkin, R. E. Dolle, and B. D. Harris, *Am. Soc.*, **106**, 3548 (1984); K. C. Nicolaou and S. E. Webber, *ibid.*, 5734 (1984).

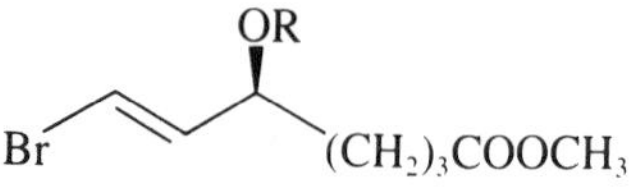

1,1,4,4-Tetramethyl-1,4-bis(N,N-dimethylamino)disilethylene,

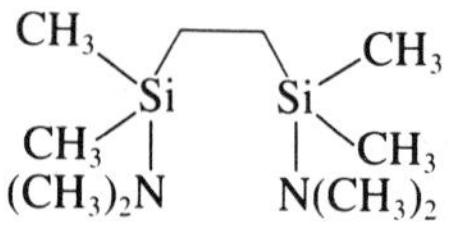

Supplier: Petrarch.

Protection of arylamines.[1]   Anilines react with **1** in the presence of zinc iodide as catalyst at 140° to form the adducts **2**. The protecting group can be removed quantitatively with methanol containing a trace of TsOH.

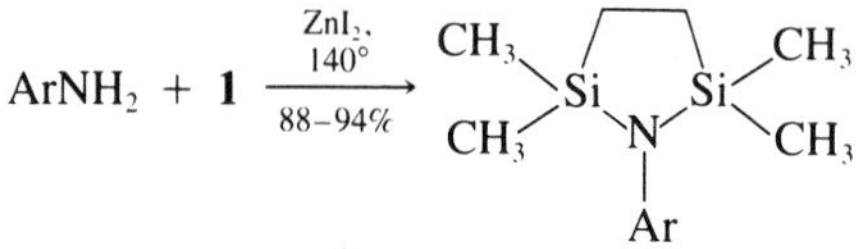

[1] T. L. Guggenheim, *Tetrahedron Letters*, **25**, 1253 (1984).

N,N,N',N'-Tetramethylchloroformamidinium   chloride, $(CH_3)_2NC(Cl)={}^+\!\!=N\!-$
$(CH_3)_2Cl^-$ **(1)**. The salt is prepared by reaction of tetramethylurea with oxalyl chloride (0.5 equiv.) in CH_2Cl_2 at 65° (2 hours).

Esterification.[1] The related salt dimethylchloroforminium chloride has been used for esterification (**8**, 186), but has the disadvantage that alkyl chlorides are also formed as by-products. The salt **1** does not activate an alcohol group and, in the presence of pyridine as base, it effects esterification of acids (even hindered ones) with primary and secondary alcohols in high yield. Esterification with hindered tertiary alcohols is possible, but slow.

Lactonization.[1] The salt also effects lactonization of ω-hydroxy carboxylic acids in the presence of collidine in a dilute solution of acetonitrile and ether. The yield of the lactone varies with the number of carbon atoms in the ring, but medium-sized lactones can be prepared in ~50% yield. 15-Pentadecanolide is obtained in 90% yield.

[1] T. Fujisawa, T. Mori, K. Fukumoto, and T. Sato, *Chem. Letters*, 1891 (1982).

Tetramethylethylenediamine (TMEDA), **2**, 403; **3**, 284–285; **4**, 485–489; **5**, 652; **6**, 576–577; **6**, 576–577; **7**, 358–359.

Wittig-Horner reaction.[1] High *syn*-selectivity is possible in the Wittig-Horner reaction of aldehydes with alkyl diphenylphosphonates (**10**, 2), formulated in equation (I).

Little selectivity is observed in hydrocarbon solvents, but use of DME or THF markedly improves *syn*-selectivity, particularly when TMEDA or 1,3-dimethyl-2-imidazolidinone (DMI) is added. Low temperatures are also important. Even so, little stereoselectivity obtains if R^1 and R^2 are of similar size. For highest selectivity, R^2 should be larger than R^1.

[1] A. D. Buss and S. Warren, *Tetrahedron Letters*, **24**, 3931 (1983).

1,1,3,3-Tetramethylguanidine, 1, 1145; **4**, 489–490. Suppliers: Fluka, Sigma.

Vinyl iodides.[1] The reaction of iodine in the presence of triethylamine with ketone hydrazones can result in *gem*-diiodides or vinyl iodides (**4**, 260). Hindered hydrazones are converted mainly into the latter products; however, yields are generally only moderate.

Yields of vinyl iodides are improved by inverse addition of the hydrazone to iodine and the base and by strictly anhydrous conditions, which prevent hydrolysis to the ketone. Use of a strong guanidine base markedly improves the yield. Tetramethylguanidine is almost as satisfactory as a pentasubstituted guanidine.

3α,5-Cyclo-5α-6-ketosteroids (2). 3β-Tosyloxy- or 3β-chloro-6-ketosteroids (1) have been converted into these *i*-steroids (2), but only under drastic conditions (alcoholic

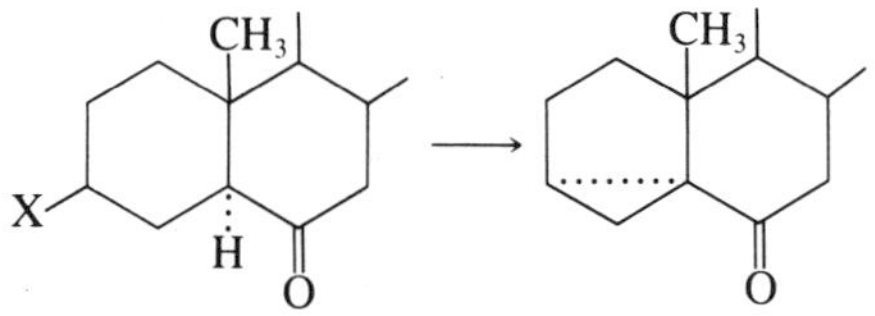

1 (X = TsO, Cl)   2

NaOH or KOH, reflux). The conversion can be conducted in 90–98% yield by treatment of **1** with tetramethylguanidine at 60° for 5 minutes. Use of benzyltrimethylammonium hydroxide in pyridine is also satisfactory.[2]

This conversion can be used for protection of **1**, since it can be reversed by mild acid treatment (HCl in HOAc for regeneration of 3β-chlorosteroids, aqueous HCl for 3β-hydroxysteroids).

Michael additions (4, 489–490).[3] A recent synthesis of sarkomycin (5), an antibiotic and antitumor agent, involves conjugate addition of nitromethane to the ketal ester **1** in the presence of this base.

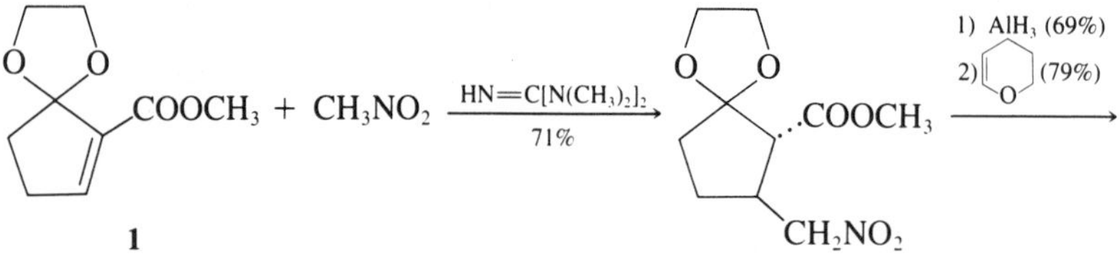

1

2

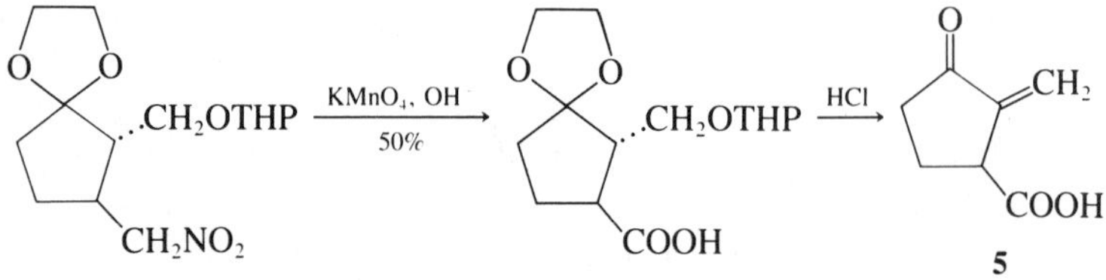

3   4

5

Solid-phase peptide synthesis.[4]   A major problem in solid-phase peptide synthesis is the final step, cleavage of the protected peptide from the solid support. Recently, a multidetachable Pop resin has been developed for the stepwise synthesis of a peptide–Pop resin (**1**) from which the peptide was originally cleaved slowly by photolysis. The cleavage can be effected rapidly and in high yield by treatment with the hindered bases tetramethylguanidine or DBU in N-methylpyrrolidinone without effect on the commonly used protective groups.

1

Silyl ethers.[5]   This base is an effective catalyst for *t*-butyldimethylsilylation of alcohols with the silyl chloride, particularly in CH_3CN (primary alcohols) or DMF (secondary alcohols).

[1] D. H. R. Barton, G. Bashiardes and J.-L. Fourrey, *Tetrahedron Letters,* **24**, 1605 (1983).
[2] M. Anastasia, P. Allevi, P. Cuiffreda, and A. Fiecchi, *Synthesis,* 123 (1983).
[3] A. T. Hewson and D. T. MacPherson, *Tetrahedron Letters,* **24**, 647 (1983).
[4] D. B. Whitney, J. P. Tam, and R. B. Merrifield, *Tetrahedron,* **40**, 4237 (1984).
[5] S. Kim and H. Chang, *Syn. Comm.,* **14**, 899 (1984).

2,2,6,6-Tetramethylpiperidinyl-1-oxyl (1), 6, 110–111; 11, 160–161.

Oxidation of alcohols and amines.   Allylic and benzylic alcohols can be oxidized to the aldehydes in high yield by oxygen in the presence of catalytic amounts of **1** and

CuCl. The actual oxidant is the nitrosonium ion (**2**), formed on oxidation of **1** by $CuCl_2$. The oxygen is required for conversion of Cu(I) to Cu(II). Oxidation of primary alcohols catalyzed by **1** requires a stoichiometric amount of $CuCl_2$ and an added base (CaH_2) to neutralize the HCl that is generated. This system is useful for selective oxidation of primary hydroxyl groups in the presence of secondary ones.[1]

Electrooxidation of alcohols at a very low potential ($\sim +0.4$ V) is possible with catalytic amounts of **1**. A weak base, usually 2,6-lutidine, is added to avoid a high acid concentration. This method is particularly successful for oxidation of primary alcohols to aldehydes with no overoxidation to acids. Benzylic and allylic primary alcohols are oxidized particularly readily. Secondary alcohols are oxidized slowly and incompletely.[2]

The oxidation of primary amines under similar conditions results in aldehydes (in an aqueous medium) or in nitriles (anhydrous acetonitrile).[3]

[1] M. F. Semmelhack, C. R. Schmid, D. A. Cortés, and C. S. Chou, *Am. Soc.*, **106**, 3374 (1984).
[2] M. F. Semmelhack, C. S. Chou, and D. A. Cortés, *ibid.*, **105**, 4492 (1983).
[3] M. F. Semmelhack and C. R. Schmid, *ibid.*, **105**, 6732 (1983).

(R,R)- and (S,S)-Tetramethyltartaric acid diamide (1).
Preparation:[1]

(R,R)   1 (m.p. 190°, α_D +43°)

Chiral β-substituted ketones.[2]   Conjugate addition of $(CH_3)_3Al$ to the ketal (**2**) of cyclohexenone derived from (S,S)-**1** followed by acetylation results in the adduct **3**, which is hydrolyzed by acid to (S)-3-methylcyclohexanone (**4**) in 77% ee. The ketal derived from (2R,3R)-2,3-butanediol undergoes a similar conjugate addition with low asymmetric

induction ($\sim$23% ee). This reaction is applicable to both acyclic and cyclic ketals. Surprisingly, the optical yields appear to decrease with decreasing reaction temperatures.

[1] D. Seebach, H.-O. Kalinowski, W. Langer, G. Crass, and E.-M. Wilka, *Org. Syn.*, **61**, 24 (1983).
[2] Y. Fukutani, K. Maruoka, and H. Yamamoto, *Tetrahedron Letters*, **25**, 5911 (1984).

Thallium(III) nitrate (TTN), **4**, 492–497; **5**, 656–657; **6**, 578–579; **7**, 362–365; **8**, 476–478; **9**, 460–462; **10**, 395–396.

Oxidation of α-methylpyrroles.[1] Treatment of the pyrrole **1** with TTN supported on montmorillonite clay effects the expected oxidative rearrangement of the β-acetyl group

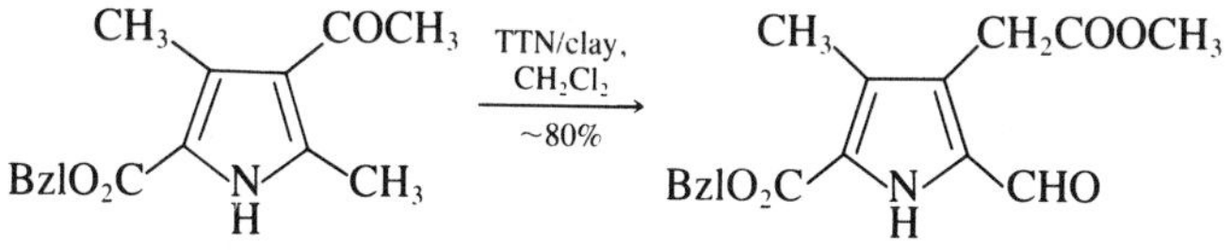

1   **2**

(**5**, 656) and, in addition, oxidation of the α-methyl group to a formyl group to give **2**. This oxidation is a general reaction of pyrroles (equation I).

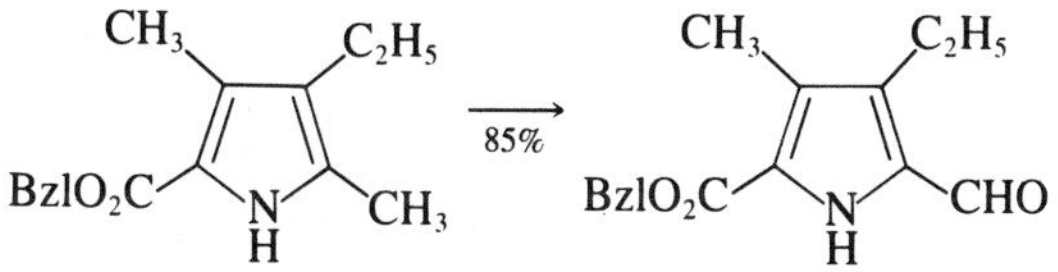

[1] A. H. Jackson, K. R. N. Rao, N. Sim Ooi, and E. Adelakun, *Tetrahedron Letters*, **25**, 6049 (1984).

Thallium(III) trifluoroacetate, 3, 286–289; **4**, 496–501; **5**, 658–659; **7**, 365; **8**, 478–481; **9**, 462–464; **10**, 397; **11**, 515–516.

Thallation–carbonylation of aromatics.[1] *ortho*-Thallation of benzoic acids, benzylic and β-phenethyl alcohols, and benzamides followed by Pd(II)-catalyzed carbonylation provides a route to aromatic carbonyl compounds such as phthalides, anhydrides, and imides.

Examples:

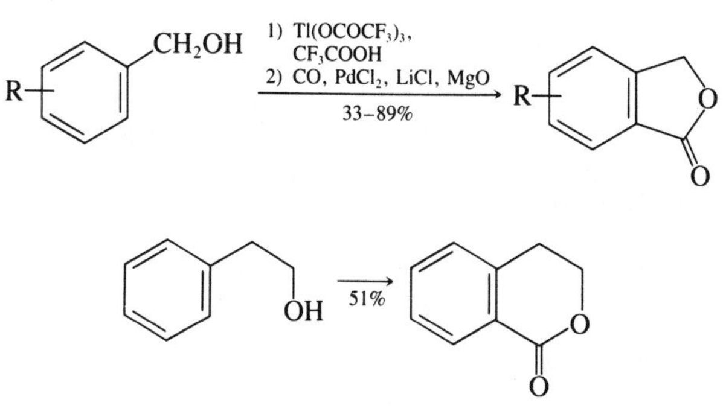

Isocoumarins.[2]   Isocoumarins can be prepared from benzoic acids by *ortho*-thallation followed by olefination promoted by PdCl$_2$. Reaction with simple alkenes requires 1 equiv. of PdCl$_2$ and a base-catalyzed reaction to effect the final cyclization. The reaction with vinyl halides or acetates generates isocoumarins directly and requires only catalytic amounts of PdCl$_2$.

Examples:

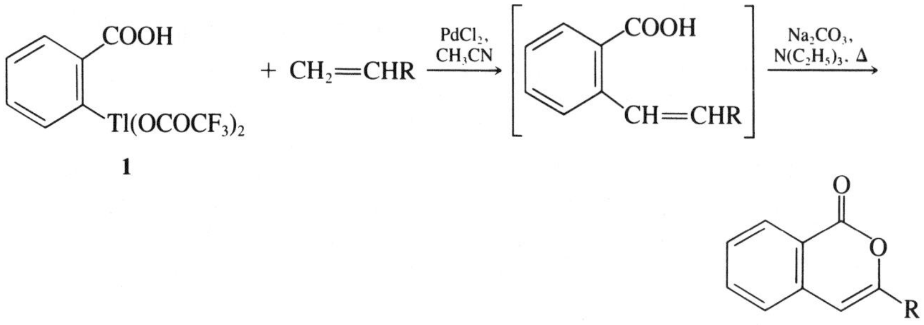

R = H,   37%
    n-Bu, 40%
    C$_6$H$_5$, 79%

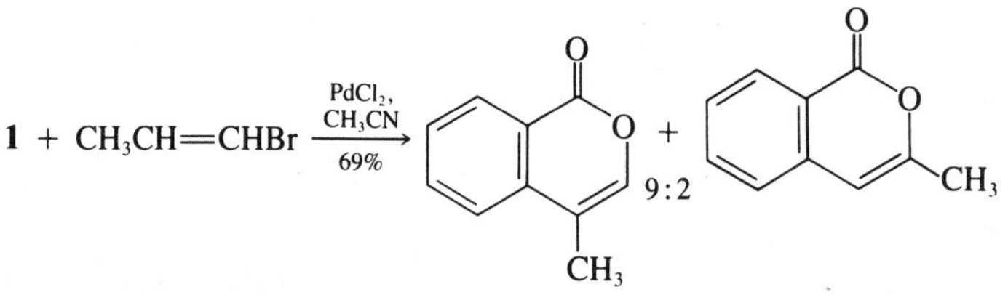

Heterocycles.[3]   This thallation–olefination sequence is widely applicable to synthesis of oxygen and nitrogen heterocycles. Thus it is applicable to tolylacetic acids, benzamides, and acetanilides.

Examples:

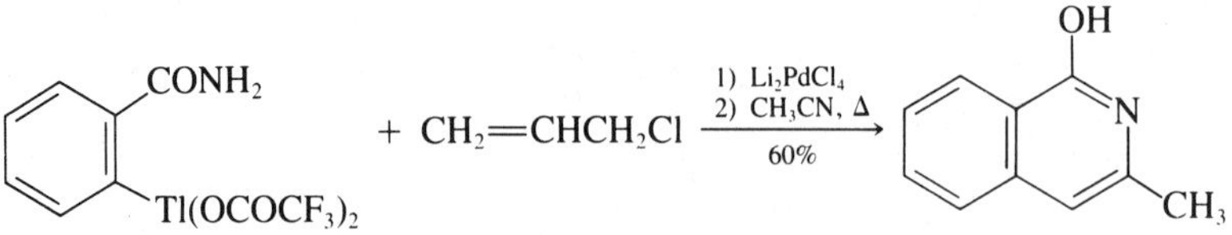

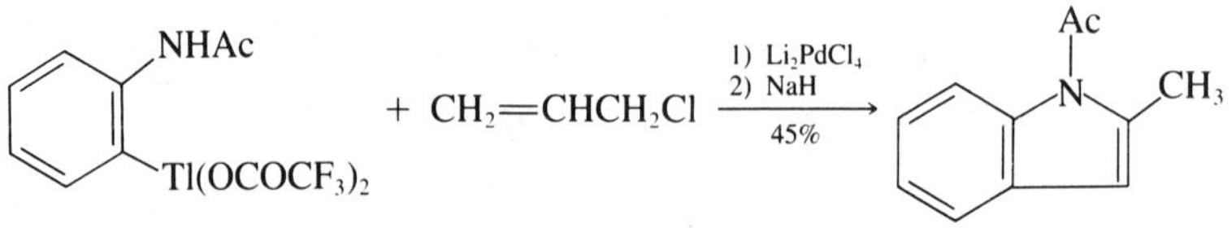

Oxidative biaryl coupling.[4]   A recent approach to the ring system of steganone (**3**) involves an oxidative biaryl coupling of the (E)-α,β-unsaturated ester **1** to give **2**, which is effected with thallium(III) trifluoroacetate in unusually high yield. Remaining steps involve enlargement of the seven-membered ring by cyclopropanation, acid-catalyzed solvolysis, and adjustment of the biaryl twist.

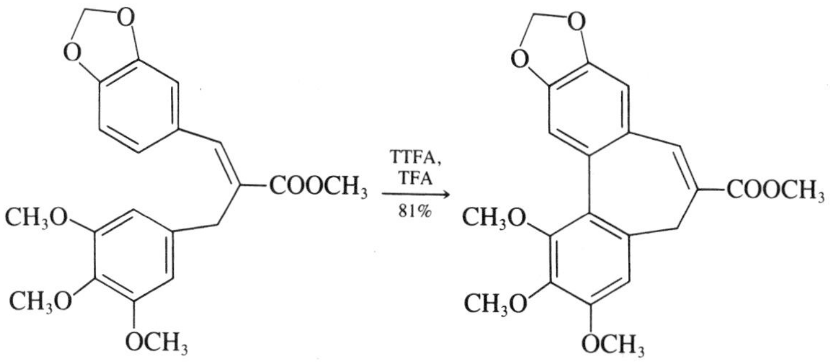

Aryl nitriles.[5]   Arylthallium bis(trifluoroacetates) react with CuCN in refluxing CH_3CN to give aryl nitriles in 60–85% yield. A one-pot synthesis is possible by thallation of the arene in CH_3CN.

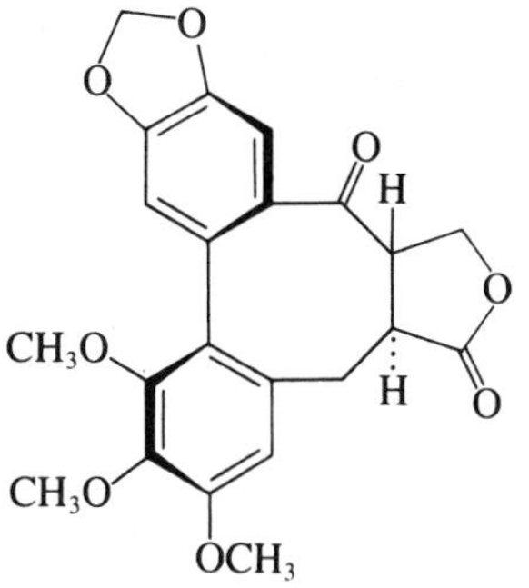

[1] R. C. Larock and C. A. Fellows, *Am. Soc.*, **104**, 1900 (1982).
[2] R. C. Larock, S. Varaprath, H. H. Lau, and C. A. Fellows, *ibid.*, **106**, 5274 (1984).
[3] R. C. Larock, C.-L. Liu, H. H. Lau, and S. Varaprath, *Tetrahedron Letters*, **25**, 4459 (1984).
[4] P. Magnus, J. Schultz, and T. Gallagher, *J.C.S. Chem. Comm.*, 1179 (1984).
[5] E. C. Taylor, A. H. Katz, and A. McKillop, *Tetrahedron Letters*, **25**, 5473 (1984).

Thallium(III) trifluoroacetate–Palladium(II) acetate, 11, 516.

Biaryls.[1]   Biaryls can be prepared by oxidative coupling of arenes with palladium(II) compounds, but the coupling is not regioselective. Regioselectivity is considerably improved by use of TTFA as oxidant and only catalytic amounts of Pd(OAc)$_2$. Formation of 4,4′-biaryls is favored from arenes substituted with either electron-donating or moderately electron-withdrawing substituents. The first step is thallation to form ArTl(OCOCF$_3$)$_2$.

Example:

[1] A. K. Yatsimirsky, S. A. Deiko, and A. D. Ryabov, *Tetrahedron,* **39,** 2381 (1983).

Thexylborane, 1, 276; **2,** 148–149; **4,** 175–176; **5,** 232–233; **6,** 207–208; **10,** 397–398; **11,** 516–517.

Stereoselective hydroboration of allylic alcohols.[1]   Acyclic secondary allylic alcohols of the types **1** and (E)- and (Z)-**2** undergo hydroboration stereoselectively to yield *threo*-1,3-diols (**3**). Highest *threo*-selectivity in hydroboration of **1** is obtained using 9-

BBN (~11:1 diastereoselection). Hydroboration of **2** with 9-BBN is too sluggish to be useful; but use of thexylborane can result in >15:1 diastereoselection. Somewhat higher stereoselectivity is obtained with (E)-**2** than with (Z)-**2**. Different protecting groups of the alcohol can affect the stereoselectivity. For example, tritylation decreases the *threo*-stereoselectivity, whereas trimethylsilylation shows little effect.

This stereoselective hydroboration can be used to obtain the nine chiral centers of the polyol **4**, which is present in the ansa bridge of rifamycin.

[1] W. C. Still and J. C. Barrish, *Am. Soc.,* **105,** 2487 (1983).

Thexylchloroborane–Dimethyl sulfide (1). This borane can be prepared by treating thexylborane–dimethyl sulfide with hydrogen chloride or by hydroboration of 2,3-dimethyl-2-butene with monochloroborane–dimethyl sulfide.

RCOOH → RCHO.[1] This borane reduces aliphatic carboxylic acids to aldehydes in >90% yield within minutes at 25°. Aromatic acids are reduced slowly and in significantly lower yields.

[1] H. C. Brown, J. S. Cha, B. Nazer, and N. M. Yoon, *Am. Soc.*, **106**, 8001 (1984).

1,3-Thiazolidine-2-thione (1), 11, 518–519.

Peptide synthesis.[1] The 1,3-thiazolidine-2-thione amides (**2**), readily obtained by reaction of **1** with a N-protected amino acid, react chemoselectively with multifunctional

2 (yellow)

3 (colorless)

1

amino acids to form dipeptides in high yield without significant racemization. Thus the amides (**2**) react selectively with the α-amino group of lysine, arginine, and histidine; protection of SH and OH groups is not necessary. A further advantage is that the reaction can be monitored by loss of the yellow color of **2**. The aminolysis reaction is readily extended to synthesis of polypeptides.

[1] Y. Nagao, T. Miyasaka, K. Seno, D. Shibata, E. Doi, and E. Fujita, *J.C.S. Perkin I*, 2439 (1984).

Thiophenol, 4, 505; 5, 585–586; 6, 458–459, 585; 7, 367–368; 9, 465; 10, 399.

Rearrangement of α-ethynylcarbinols.[1] Thiophenol adds regioselectively to α-ethynylcarbinols (**1**) to form phenylthiovinylcarbinols (**2**), which undergo biphasic aqueous

1

2

3

acid hydrolysis to α,β-enals. Secondary carbinols are converted into a mixture of (E)-
and (Z)-**3** in which the former predominates; tertiary carbinols are also converted mainly
into (E)-**3**.

[1] M. Julia and C. Lefebvre, *Tetrahedron Letters*, **25**, 189 (1984).

Tin–Aluminum.

Homoallylic alcohols. Aldehydes or ketones react with an allylic halide in the
presence of tin and aluminum in an aqueous medium to give homoallylic alcohols.[1]
Example:

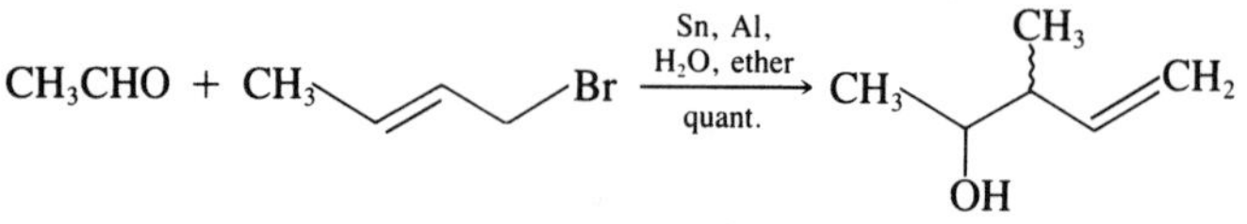

(*syn/anti* = 61 : 39)

An intramolecular version of this reaction to obtain cyclic homoallylic alcohols has
been reported.[2]
Example:

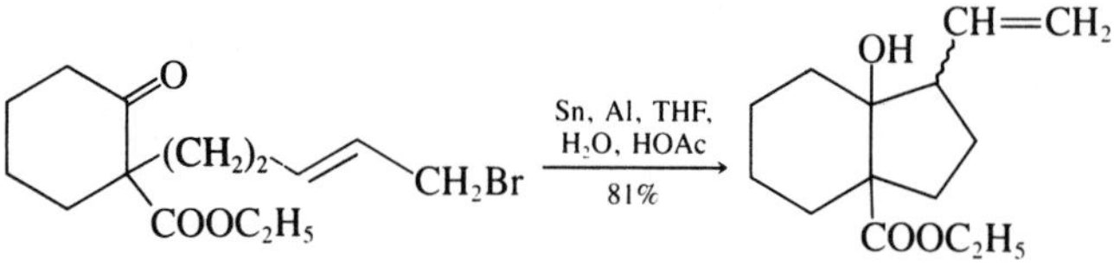

[1] J. Nokami, J. Otera, T. Sudo, and R. Okawara, *Organometallics*, **2**, 191 (1983).
[2] J. Nokami, S. Wakabayashi, and R. Okawara, *Chem. Letters*, 869 (1984).

Tin(IV) chloride, 1, 1111–1113; **3**, 269; **5**, 627–631; **6**, 553–554; **7**, 342–345; **9**, 436–
438; **10**, 370–373; **11**, 522–524.

Intramolecular acylation of vinylsilanes.[1] This reaction can be used for synthesis
of cyclopentenones. A typical sequence is formulated in equation (I). Either TiCl₄ or

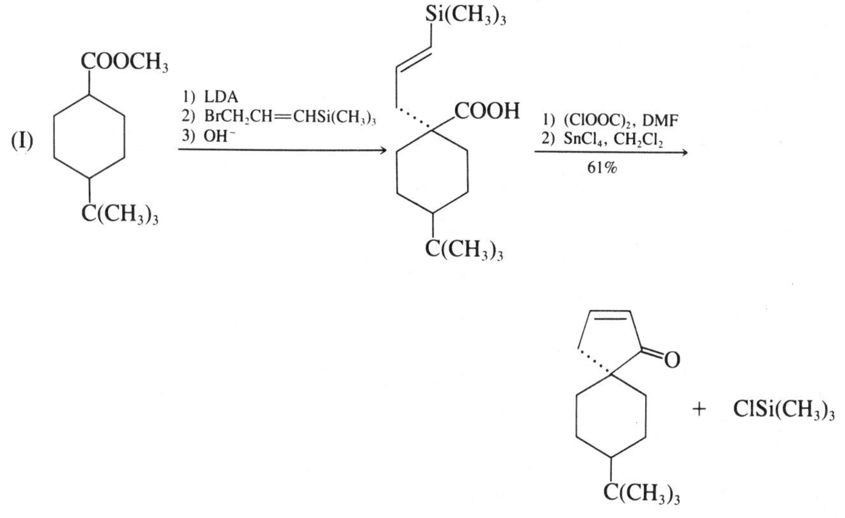

Si(CH$_3$)$_3$

COOCH$_3$

(I)

1) LDA
2) BrCH$_2$CH=CHSi(CH$_3$)$_3$
3) OH⁻

C(CH$_3$)$_3$

COOH

C(CH$_3$)$_3$

1) (ClOOC)$_2$, DMF
2) SnCl$_4$, CH$_2$Cl$_2$

61%

O

C(CH$_3$)$_3$

+ ClSi(CH$_3$)$_3$

SnCl$_4$ can be used to effect the cyclization. The method has been used for synthesis of the bicyclic dienone **1** (equation II).

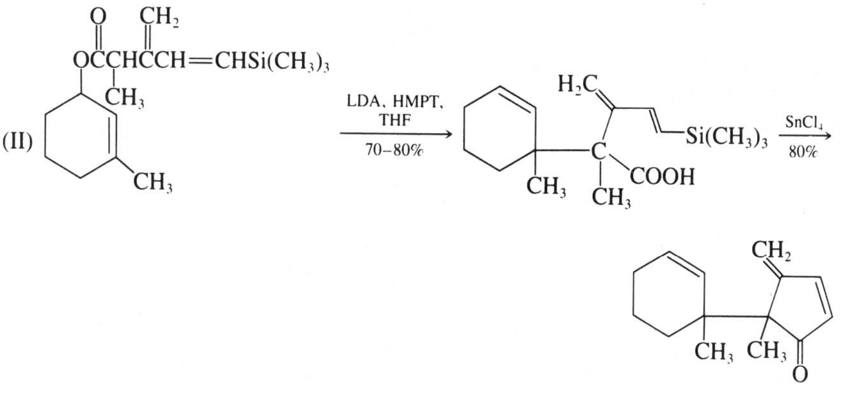

(II)

O   CH$_2$
OCCHCCH=CHSi(CH$_3$)$_3$
CH$_3$

CH$_3$

LDA, HMPT,
THF

70–80%

H$_2$C

Si(CH$_3$)$_3$

CH$_3$ CH$_3$

COOH

SnCl$_4$

80%

CH$_2$

CH$_3$ CH$_3$

O

1

Diastereoselective aldol reactions.[2] The diastereoselectivity in the Lewis acid-catalyzed aldol reaction of chiral α-hydroxy aldehydes is independent of the geometry of the enol silyl ether. Also, the reaction does not involve prior Si–Ti or Si–Sn exchange.

Apparently *syn*-chelation is involved. In fact, diastereoselectivity is notable in aldol addition to an achiral aldehyde (equation I).

(I) BzlOCH₂CHO +

90:10
>95:5

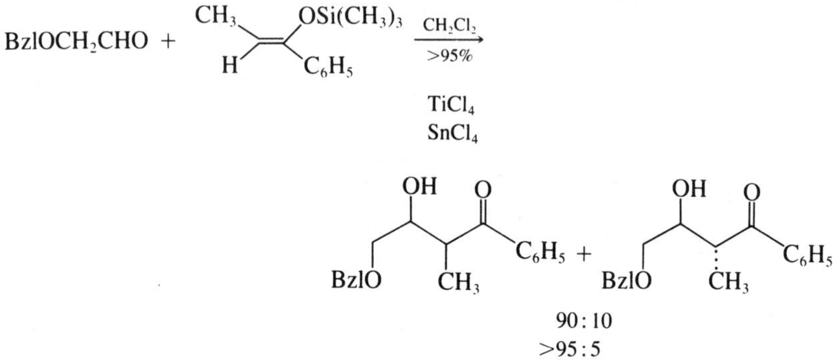

Reaction of a chiral α-alkoxyaldehyde with a prochiral enol silyl ether catalyzed by SnCl₄ results in a single diastereomer with additional *syn*-selectivity (equation II).

(II)

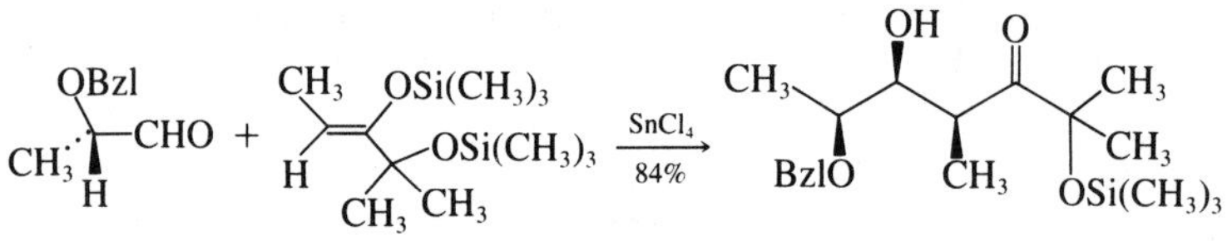

β-Chelation can also result in significant diastereoselectivity. Thus the reaction of an achiral β-hydroxy aldehyde catalyzed by TiCl₄ results in *syn*- and *anti*-adducts in the ratio 94:6 (equation III). Finally, a chiral β-alkoxyaldehyde when complexed with TiCl₄ can

(III) BzlO(CH₂)₂CHO +

94:6

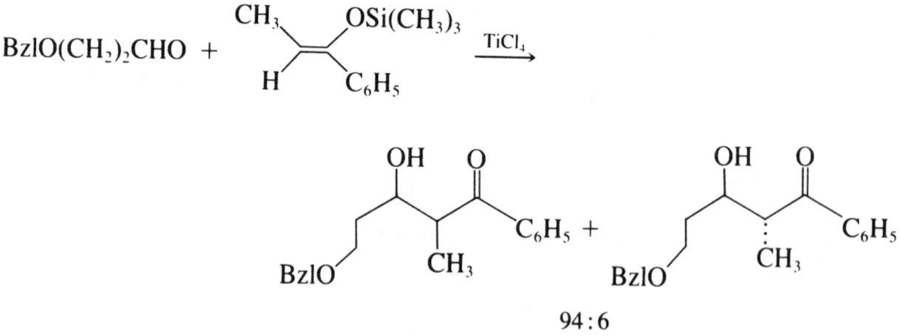

(IV)

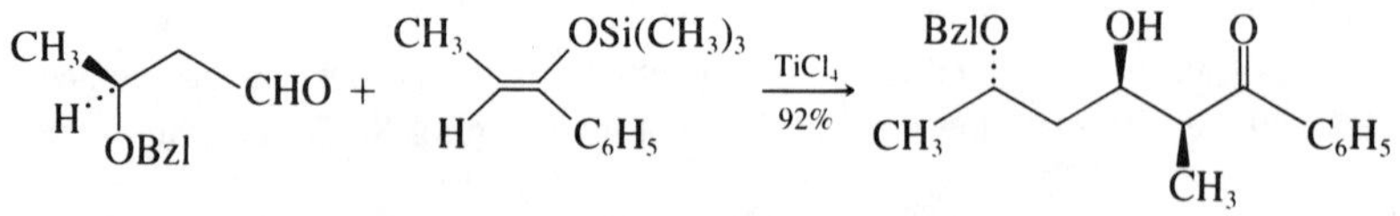

undergo aldol additions with high 1,3-asymmetric induction and high *syn*-chelation control (equation IV).

Intramolecular ene reaction.[3] A key step in a new route to anthracyclinones such as γ-citromycinone (**4**) is a regioselective intramolecular ene reaction of the unsaturated aldehyde **1** to give **2** in 93% yield. Sharpless oxidation [$(CH_3)_3COOH$, $VO(acac)_2$] of **2** is regioselective, giving the desired epoxide **3** as the only isomer. Synthesis of **4** is

1 **2**

3

4

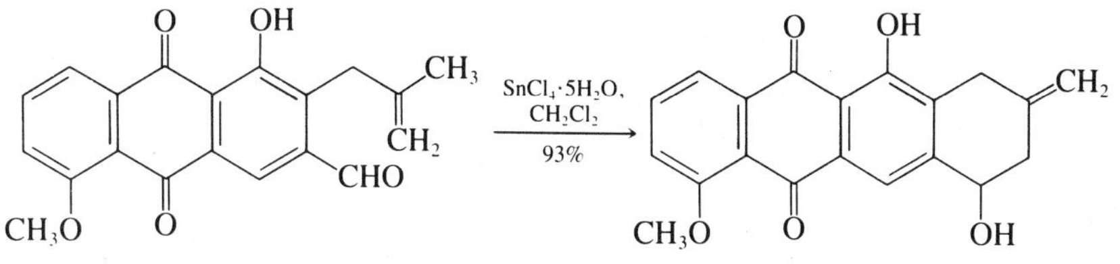

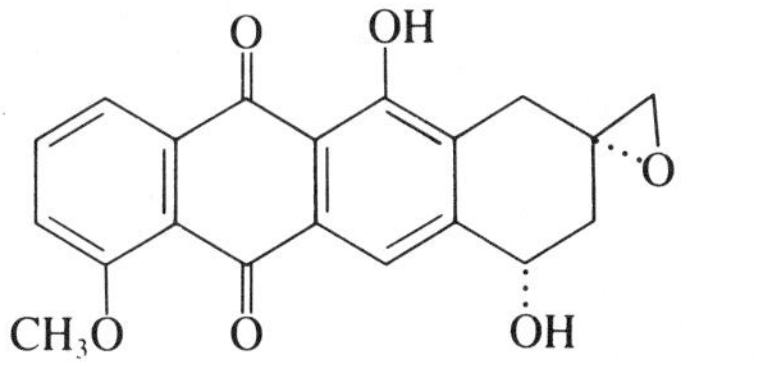

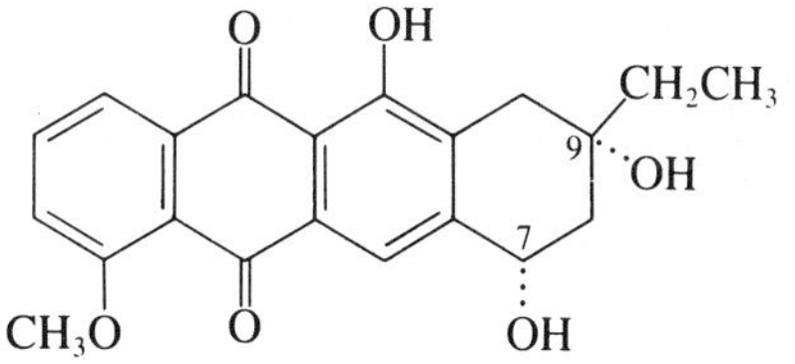

completed by conversion of the epoxide function to an ethylcarbinol group by reaction with methylcopper (prepared from CH_3Li and CuCN, 84% yield).

β-Chloro-α,β-enones.[4] These useful intermediates can be prepared stereoselectively by hydrochlorination of allenic ketones with $CH_3HN—NHCH_3·2HCl$ (**1**) in DMF or with tin(IV) chloride in benzene. In both cases, the (E)-isomer of the product predominates.

Examples:

1, DMF 50–65% 95:5

$SnCl_4$, C_6H_6 40–65% 100:1

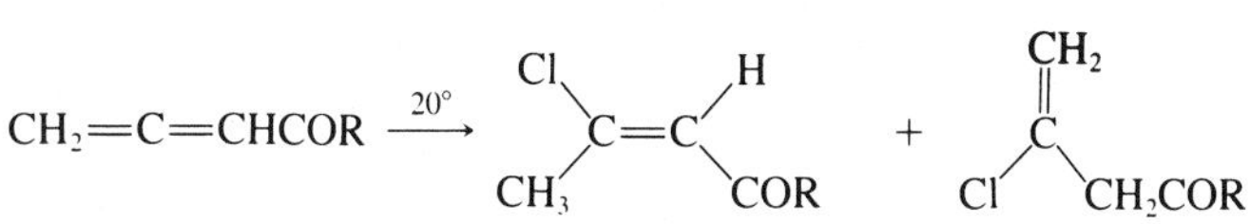

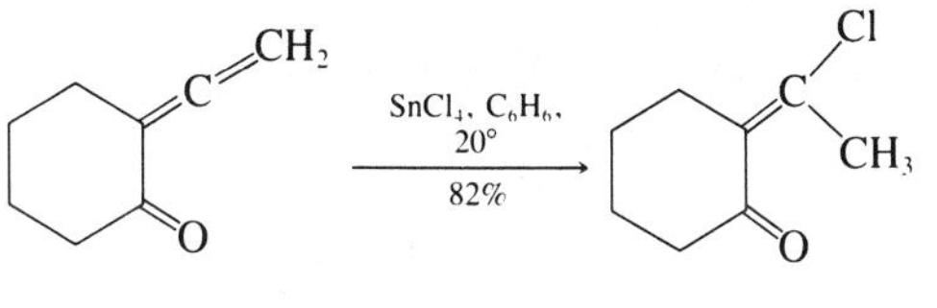

[1] E. Nakamura, K. Fukuzaki, and I. Kuwajima, *J.C.S. Chem. Comm.*, 499 (1983).
[2] M. T. Reetz, K. Kesseler, and A. Jung, *Tetrahedron*, **40**, 4327 (1984).
[3] F. M. Hauser and D. Mal, *Am. Soc.*, **106**, 1862 (1984).
[4] J.-L. Gras and B. S. Galledou, *Bull. Soc.*, II-89 (1983).

Tin(II) trifluoromethanesulfonate, 11, 525–526.

Enantioselective cross aldol reactions.[1]   A 3-acylthiazolidine-2-thione[2] can be used as an equivalent of an aldehyde or carboxylic acid in cross aldol reactions to provide β-hydroxy aldehydes or carboxylic acids. The tin enolates formed from $Sn(OTf)_2$ react with aromatic and aliphatic aldehydes with high *syn*-selectivity (equation I).

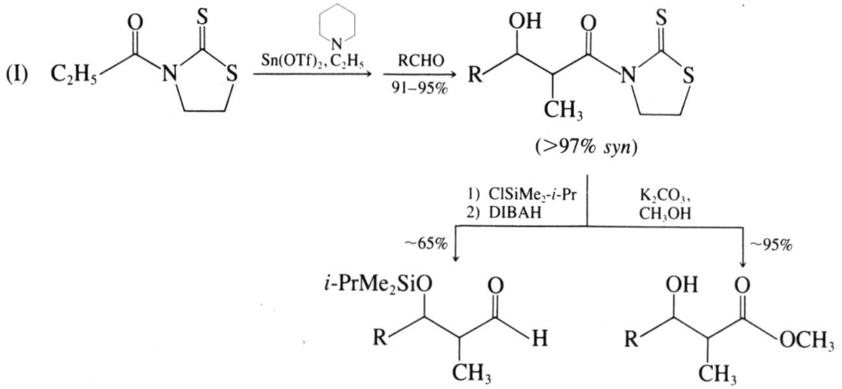

Highly enantioselective cross aldol reactions of 3-acetylthiazolidine-2-thione with aliphatic aldehydes can be effected by use of the chiral diamine **1**, derived from (S)-proline, as a ligand for the tin(II) enolate (equation II).

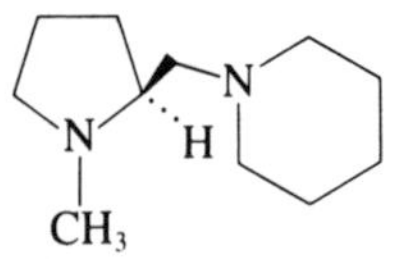

1

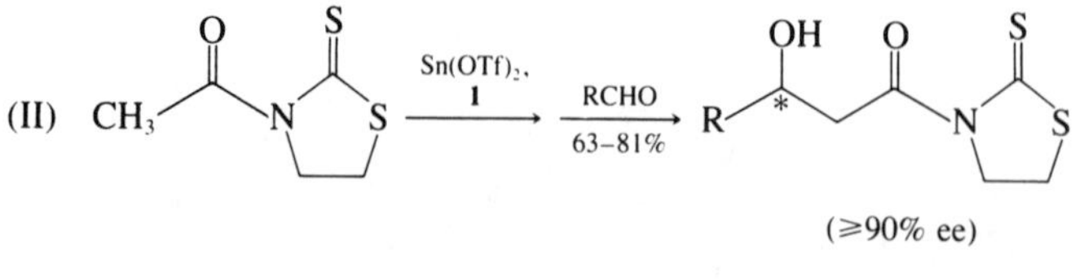

The reaction of 3-(2-benzyloxyacetyl)thiazolidine-2-thione (**2**) with an aliphatic alde-hyde mediated by tin(II) triflate affords the *syn-* and *anti-*aldols (**3**) in the ratio 3:1. This ratio is reversed in the presence of 1.2 equiv. of TMEDA. Addition of the chiral diamine

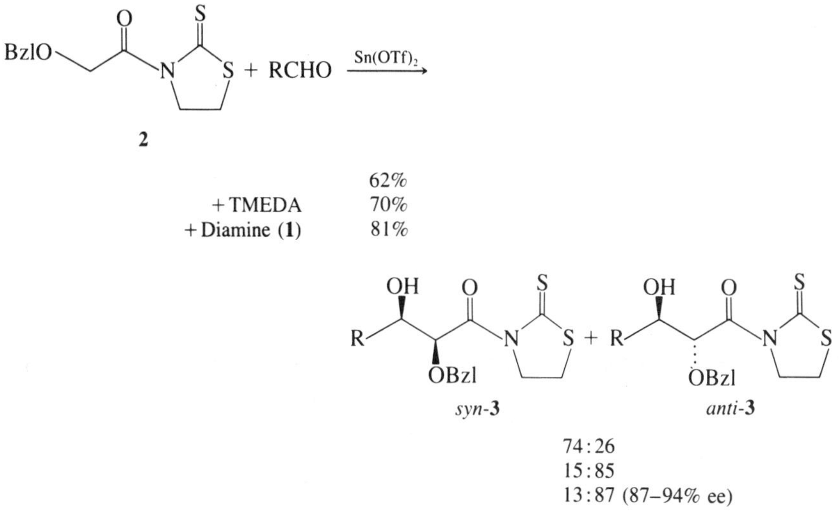

	62%
+TMEDA	70%
+Diamine (**1**)	81%

syn-**3**   anti-**3**

74:26
15:85
13:87 (87–94% ee)

1, derived from (S)-proline, has the same effect, and also results in high asymmetric induction (87–94% ee). This reversal of stereoselection is observed only with aliphatic aldehydes. Reaction of benzaldehyde with **2** catalyzed by tin(II) triflate gives mainly the *anti*-isomer in the presence or absence of a diamine (*anti/syn* ≈ 80:20).[3]

1,4-Diketones.[4]   In the presence of excess tin(II) triflate and a base (N-trimethyl-silylimidazole), β-keto aryl sulfoxides (excess) react with silyl enol ethers to provide 2-arylsulfenyl-1,4-diketones in moderate to high yield.

Example:

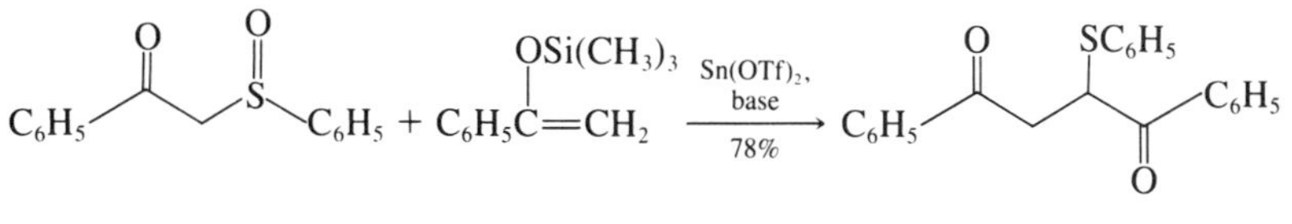

The reaction involves a Pummerer-type rearrangement to a α-thiocarbocation, which then reacts with the enol silyl ether. The products can be converted by reduction into 1,4-diketones or by oxidation to unsaturated 1,4-diketones.

[1] N. Iwasawa and T. Mukaiyama, *Chem. Letters*, 297 (1983); T. Mukaiyama, N. Iwasawa, R. W. Stevens, and T. Haga, *Tetrahedron*, **40**, 1381 (1984).
[2] E. Fujita, *Pure Appl. Chem.*, **53**, 1141 (1981).
[3] T. Mukaiyama and N. Iwasawa, *Chem. Letters*, 753 (1984).
[4] M. Shimizu, T. Akiyama, and T. Mukaiyama, *ibid.*, 1531 (1984).

Titanium(III) chloride, 2, 415; **4**, 506–508; **5**, 669–671; **6**, 587; **7**, 369; **8**, 482–483; **9**, 467; **10**, 400; **11**, 529.

Allylic pinacols.[1]   Reduction of ketones bearing an electron-withdrawing group with $TiCl_3$ in the presence of α,β-enals results in coupling to allylic pinacols in 45–85% yield. Example:

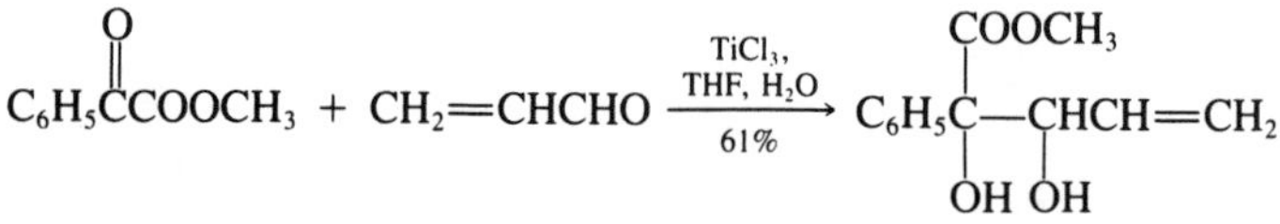

Deoxygenation of a nitrone.   Some years ago Todd and co-workers[2] noted that 1-pyrroline-1-oxides can be obtained by reductive cyclization of a γ-nitro ketone. This reaction was used recently in a synthesis of chlorins related to vitamin B_{12}.[3] Thus the nitro ketone **1** on reduction with zinc and acetic acid gives the pyrroline N-oxide **2** in high yield. The N-oxide is stable to triphenylphosphine or trimethyl phosphite, but is reduced to the imine **3** in high yield by $TiCl_3$.

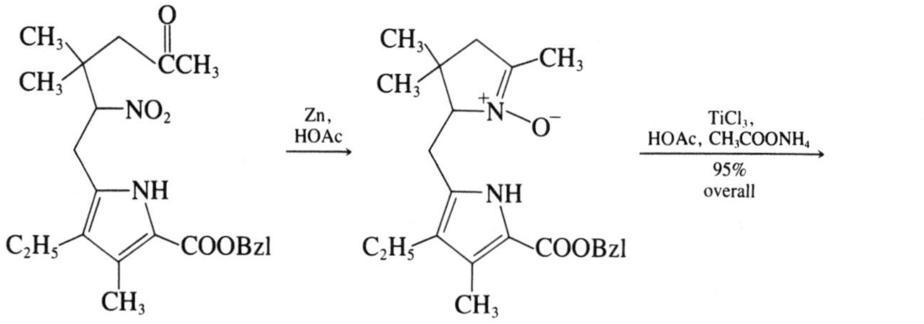

[1] A. Clerici and O. Porta, *J. Org.*, **48**, 1690 (1983).
[2] R. Bonnett, V. M. Clark, A. Giddey, and A. R. Todd, *J. Chem. Soc.*, 2087 (1959).
[3] A. R. Battersby, C. J. R. Fookes, and R. J. Snow, *J.C.S. Perkin I*, 2725 (1984).

Titanium(III) chloride–Lithium aluminum hydride, **6**, 588–589; **7**, 369–370; **9**, 468; **10**, 401; **11**, 529.

Cycloalkanones.   The synthesis of cycloalkenes by intramolecular coupling of a diketone with a Ti(0) reagent (**8**, 483) has been extended to a cycloalkanone synthesis. Thus reaction of a keto ester with $TiCl_3/LiAlH_4$ in the presence of triethylamine results in cycloalkanones in moderate to high yield.[1]

Examples:

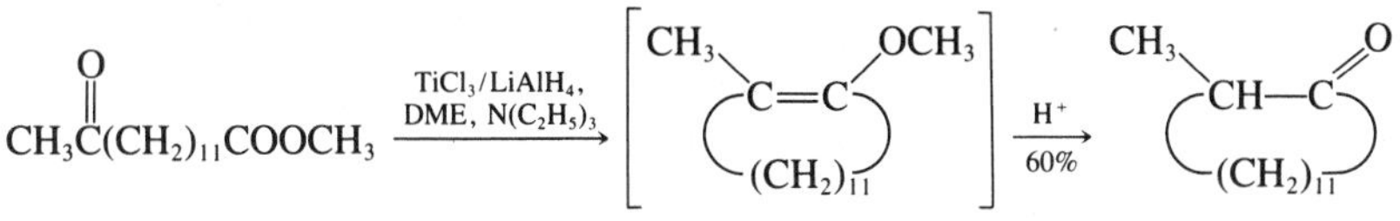

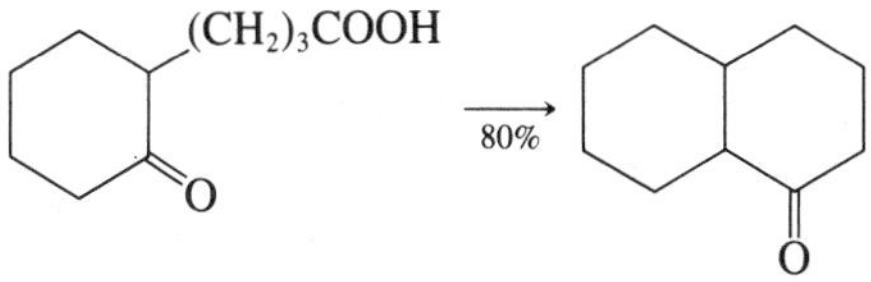

This cycloalkanone synthesis furnishes a short synthesis of isocaryophyllene (**1**), the (Z)-isomer of natural caryophyllene. In this example, the double bond isomerizes during the keto cyclization.[2]

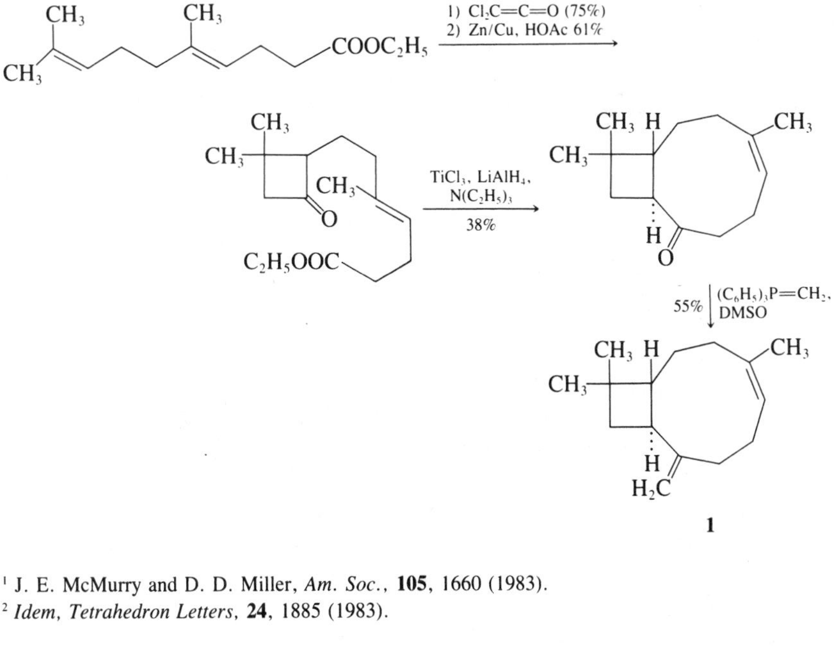

[1] J. E. McMurry and D. D. Miller, *Am. Soc.*, **105**, 1660 (1983).
[2] *Idem, Tetrahedron Letters*, **24**, 1885 (1983).

Titanium(IV) chloride, 1, 1169–1171; **2**, 414–415; **3**, 291; **4**, 507–508; **5**, 671–672; **6**, 590–596; **7**, 370–372; **8**, 483–486; **9**, 468–470; **10**, 401–403; **11**, 529–533.

Stereoselective additions to chiral α- and β-alkoxy aldehydes. Lewis-acid-catalyzed additions of enol silyl ethers to chiral α-alkoxy or β-alkoxy aldehydes can proceed with high 1,2- and 1,3-asymmetric induction. Moreover, the sense of induction can be controlled by the Lewis acid.[1] Thus BF_3, which is nonchelating, can induce diastereo-

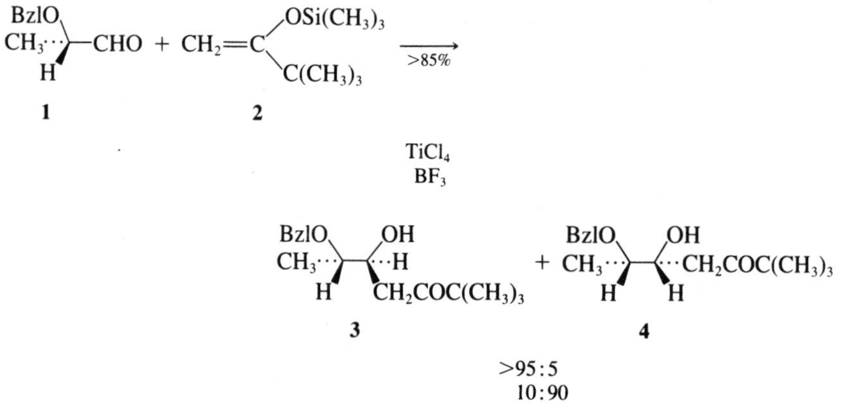

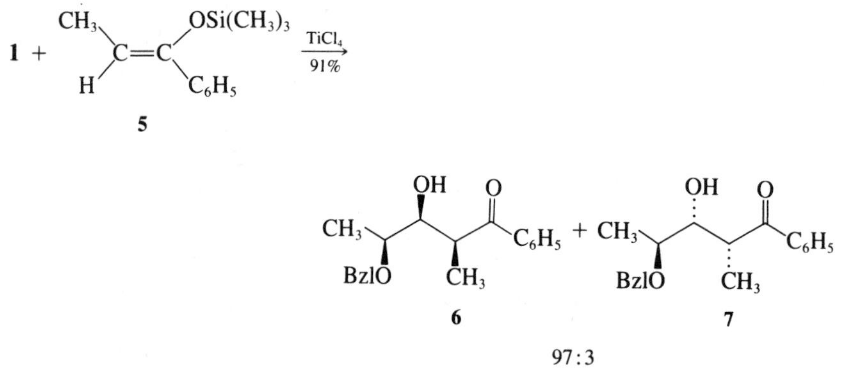

selectivity opposite to that obtained with $TiCl_4$ or $SnCl_4$.[2] An example is the reaction of α-benzyloxypropanal (**1**) with the enol silyl ether **2** to form mainly one (**3** or **4**) of the two possible diastereomers. The aldol addition of **1** with the (Z)-enol silyl ether **5** catalyzed by $TiCl_4$ or fluoride ion results in essentially only one of the two possible *syn*-diastereomers.

The stereoselectivity of additions to chiral β-benzyloxy-α-methylpropanal can also be controlled by the Lewis acid catalyst (equation I).

(I)

$$\underset{\text{BzlO}}{\overset{\text{CH}_3}{\text{H}\cdots\text{C}}}\text{-CHO} + \text{CH}_2{=}\overset{\text{OSi(CH}_3)_3}{\underset{\text{C(CH}_3)_3}{C}} \longrightarrow$$

2

$$\xrightarrow[\text{BF}_3]{\text{TiCl}_4}$$

$$\underset{\text{BzlO}\qquad\text{CH}_2\text{COC(CH}_3)_3}{\overset{\text{CH}_3\qquad\text{OH}}{\text{H}\cdots\text{C}\cdots\text{H}}} + \underset{\text{BzlO}\qquad\text{H}}{\overset{\text{CH}_3\qquad\text{OH}}{\text{H}\cdots\text{C}\cdots\text{CH}_2\text{COC(CH}_3)_3}}$$

95:5
88:12

High 1,3- and 1,4-asymmetric induction is possible in addition reactions to chiral β-alkoxy aldehydes.[3] The reaction with allylsilanes catalyzed by $TiCl_4$ shows high diastereofacial selectivity (equation II). Essentially only one *syn*-adduct of two possible diastereomers is obtained in reactions with silyl enol ethers (equation III).

(II)

$$\underset{\text{BzlO}}{\overset{\text{CH}_3}{\diagup}}\diagdown\diagup\text{CHO} + \text{CH}_2{=}\text{CHCH}_2\text{Si(CH}_3)_3 \xrightarrow[90\%]{\text{TiCl}_4}$$

8

$$\underset{\text{CH}_3}{\overset{\text{BzlO}\quad\text{OH}}{\diagup}}\diagdown\diagup\diagdown\text{CH}_2\text{CH}{=}\text{CH}_2 + \underset{\text{CH}_3}{\overset{\text{BzlO}\quad\text{OH}}{\diagup}}\diagdown\diagup\diagdown\text{CH}_2\text{CH}{=}\text{CH}_2$$

95:5

(III) **8** +

$$\underset{\text{H}}{\overset{\text{CH}_3}{C}}{=}\underset{\text{C}_6\text{H}_5}{\overset{\text{OSi(CH}_3)_3}{C}} \xrightarrow[90\%]{\text{TiCl}_4}$$

5

$$\underset{\text{CH}_3\qquad\qquad\text{CH}_3}{\overset{\text{BzlO}\quad\text{OH}\quad\text{O}}{\diagup\diagdown\diagup\diagdown}\text{C}_6\text{H}_5} + \underset{\text{CH}_3\qquad\qquad\text{CH}_3}{\overset{\text{BzlO}\quad\text{OH}\quad\text{O}}{\diagup\diagdown\diagup\diagdown}\text{C}_6\text{H}_5}$$

92:8

Sakurai reaction (**7**, 371).[4]   Blumenkopf and Heathcock[5] have examined the stereochemistry of the Sakurai reaction and of the di-*n*-alkyl cuprate addition as applied to 4- and 5-methylcyclohexenones. The stereoselectivity of the former reaction can be higher than that of the latter reaction, and can even be in a different sense.

Examples:

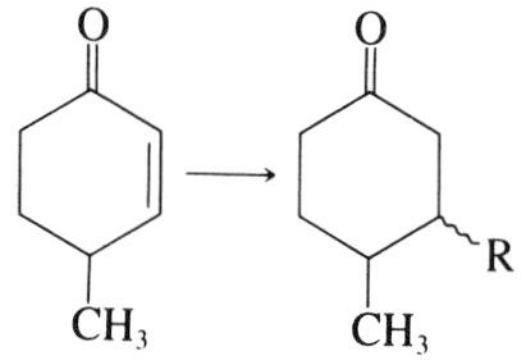

$(CH_3)_3SiCH_2CH=CH_2$, $TiCl_4$ 76% (*trans*/*cis* = 32:68)
n-C_3H_7MgBr, CuI   78% (*trans*/*cis* = 80:20)

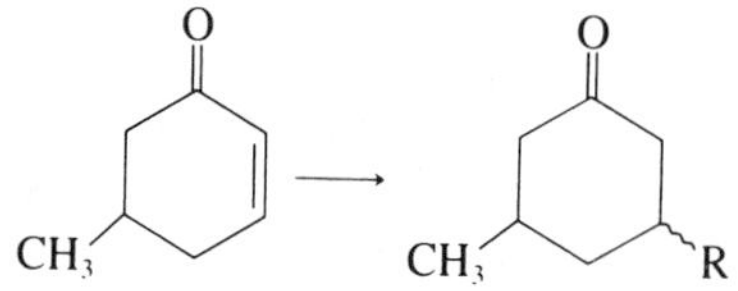

$(CH_3)_3SiCH_2CH=CH_2$, $TiCl_4$ 83% (*trans*/*cis* = >98:2)
n-C_3H_7MgBr, CuI   81% (*trans*/*cis* = 93:7)

These results can be explained in terms of an interplay of stereoelectronic and steric factors. Steric factors are evidently more important in the reaction of bulky cuprate clusters than in the Sakurai reaction. Thus stereoelectronic factors predominate in the allylsilane reaction. Similar effects are observed in conjugate additions to substituted cycloheptenones.

Intramolecular Sakurai reaction (**7**, 371).[6]   The intramolecular conjugate addition of an allylsilane group to an α,β-enone can be effected with high stereoselectivity. Thus

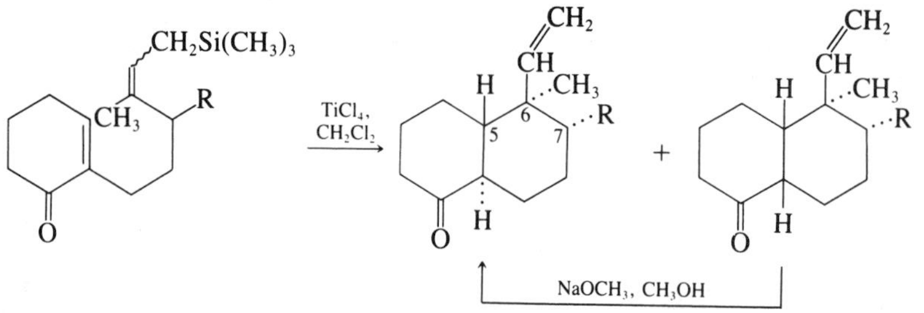

1, R = H (E/Z = 3:2)	90%	**2**	3:2	**3**
4, R = CH_3 (E/Z = 5:4)	80%	**5**	3–4:1	**6**

in the presence of TiCl$_4$, **1** cyclizes to a 3:2 mixture of **2** and **3** with the same configuration at two of the three new asymmetric centers. The stereoselectivity is even greater in cyclization of **4** to give **5** as the major product. Since **6** is readily isomerized to **5**, this reaction can result in high diastereoselection at four contiguous carbon atoms.

Aldol condensation of **1**-*naphthol with pyruvates.*[7] 1-Naphthol undergoes an aldol-type condensation with α-keto esters in the presence of TiCl$_4$. Efficient asymmetric in-

duction (de as high as 92%) can be effected in the condensation with menthyl pyruvate. The major product undergoes *in situ* epimerization at 25°, particularly in the presence of TiCl$_4$, to a 30:70 mixture. Lower asymmetric induction is observed with SnCl$_4$, ZrCl$_3$, and BCl$_3$.

syn-Homoallylic alcohols.[8]   (E)-Crotyl- and (E)-cinnamyltrimethylsilane add to aldehydes in the presence of TiCl$_4$ to form *syn*-homoallylic alcohols with >93% selectivity (equation I). The corresponding (Z)-allylsilanes show only moderate *syn*-selectivity.

(R^1 = CH$_3$, C$_6$H$_5$)

Aldol coupling of enol silyl ethers and ketals (**6**, 599).[9]   The reaction of the silyl ether **1** with dimethyl or dibenzyl ketals of 2-, 3-, or 4-methylcyclohexanone results in at least 92% selectivity for equatorial attack (equation I). The degree of equatorial attack

R = CH$_3$
R^1 = CH$_3$, CH$_2$C$_6$H$_5$

(e, 92–95%)

is considerably higher than that observed in the same reaction with the free ketones or in the reaction of Grignard reagents with the ketones (~70–80% equatorial attack).

Trichlorotitanium enolates.   These Ti enolates are formed generally in high yield by reaction of enol silyl ethers with TiCl$_4$ at 20–35° in CH$_2$Cl$_2$.[10] The corresponding reaction with SnCl$_4$ results in α-trichlorostannyl ketones.[11] The stability of these Ti enolates varies from a few seconds to a few days, depending on the structure. As in the case of other titanium enolates, these trichlorotitanium enolates are *syn*-selective in aldol reactions.[12]

Diastereoselective Mannich reaction.[13]   Mannich bases can be prepared by addition of a lithium dialkylamide to a nonenolizable aldehyde to form a lithium alkoxide. Trans-metallation provides a trichlorotitanium alkoxide, which reacts with a lithium enolate to form a Mannich base.

Example:

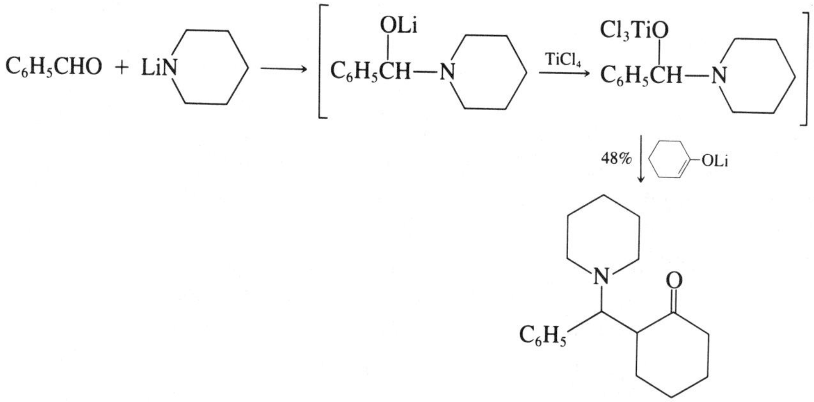

(78% ds)

The Mannich bases prepared in this way from the enolate of cyclohexanone are formed with a diastereoselectivity of 66–84%, but the configuration is not known.

α-Hydroxy amides.[14]   In the presence of 1 equiv. of TiCl$_4$, isocyanides react with acetals to form α-alkoxy amides (equation I).

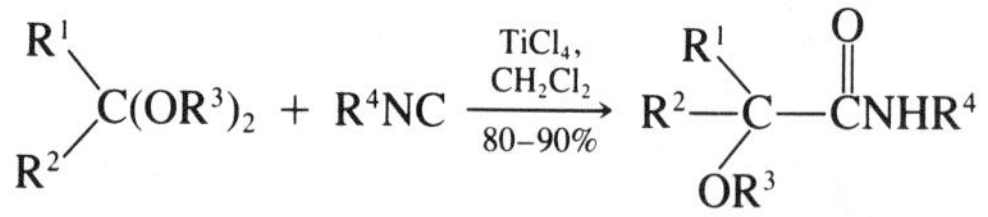

The actual reagent may be an N-alkyl(trichlorotitanio)formidoyl chloride (**1**), known to be formed readily by insertion of an isocyanide into a Ti—Cl bond of TiCl$_4$.[15] Indeed, the preformed reagent (**1**) from methyl isocyanide reacts directly with aldehydes or ketones to form α-hydroxy amides (equation II).[16]

(II) $CH_3NC + TiCl_4 \xrightarrow{CH_2Cl_2,}$ $CH_3N{=}C{\overset{\displaystyle Cl}{\underset{\displaystyle TiCl_3}{\big\langle}}}$ $\xrightarrow{R^1COR^2}$

$$\left[\; R^1\;{\overset{\displaystyle OTiCl_3}{\underset{\displaystyle R^2}{\big\rangle}}}C{-}\;\overset{\displaystyle N}{\underset{\displaystyle \underset{Cl}{|}}{C}}{=}\overset{\displaystyle}{\underset{\displaystyle CH_3}{\big\rangle}}\; \right]$$

1

$$65\text{–}95\% \downarrow H_3O^+$$

$$R^1\;{\overset{\displaystyle OH}{\underset{\displaystyle R^2}{\big\rangle}}}C{\overset{}{-}}\;\underset{\displaystyle \underset{O}{\|}}{C}NHCH_3$$

α,β-*Enals* and -*enones*.[17] The TiCl$_4$-catalyzed reaction of lithium trialkyl(1-alkynyl)borates with methyl vinyl ketone (**10**, 402) has been extended to a similar reaction with orthoesters, which results in allylic boranes that are oxidized to α,β-enones and -enals (equation I).

(I) $[R_3BC{\equiv}CR^1]Li^+ + R^2C(OR^3)_3 \xrightarrow[\substack{50\text{–}100\%}]{\substack{1)\ TiCl_4 \\ 2)\ H_2O_2,\ OH^-}} R_2C{=}C{\overset{\displaystyle R^1}{\underset{\displaystyle COR^2}{\big\langle}}}$

Titanium homoenolates of alkyl propionates.[18] Treatment of 1-alkoxy-1-trimethylsilyloxycyclopropanes (**1**)[19] with TiCl$_4$ results in moderately air-sensitive, thermally stable, purple solids formulated as **2**, but probably dimeric. They are oxidized by air to alkyl β-hydroxypropionates in good yield.

$$\triangleright\!\!<{\overset{\displaystyle OSi(CH_3)_3}{\underset{\displaystyle OR}{\big\langle}}} \xrightarrow[\substack{70\text{–}89\%}]{\substack{TiCl_4,\ CH_2Cl_2, \\ 25°}} Cl_3TiCH_2CH_2COOR \xrightarrow{O_2} HOCH_2CH_2COOR$$

1, R = CH$_3$, C$_2$H$_5$,   **2**
     CH(CH$_3$)$_2$

Ketones do not react with **2**. Surprisingly, **2** is inert to benzoyl chloride and to the more reactive complex with AlCl$_3$. Aliphatic aldehydes react with **2** to form γ-hydroxy esters in high yield.

Example:

$$C_6H_5CH_2CH_2CHO \xrightarrow[\substack{70\text{–}80\%}]{2} C_6H_5CH_2CH_2\overset{\displaystyle OH}{\overset{|}{C}}HCH_2CH_2COOCH(CH_3)_2$$

1,4-Diketones; γ-keto esters.[20]   TiCl$_4$ or SnCl$_4$ activates nitroolefins for Michael addition with silyl enol ethers to form intermediate silyl nitronates, which are hydrolyzed to 1,4-diketones.

Example:

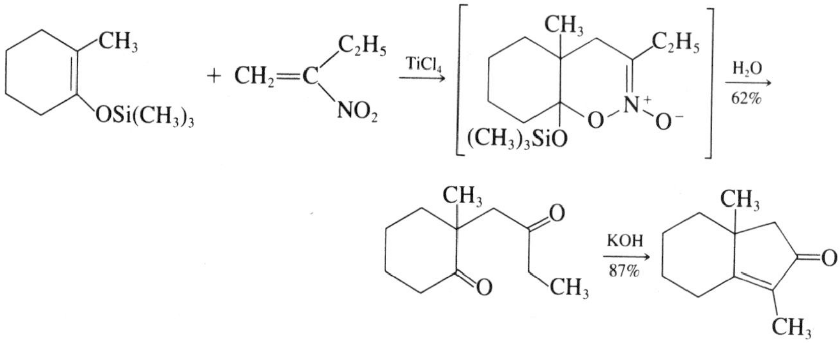

A similar reaction of nitroolefins with ketene silyl acetals provides γ-keto esters; but in this case only TiCl$_4$ is an effective catalyst. Yields are low unless titanium(IV) iso-propoxide is added to suppress hydrolysis of the ketene silyl acetals.

Example:

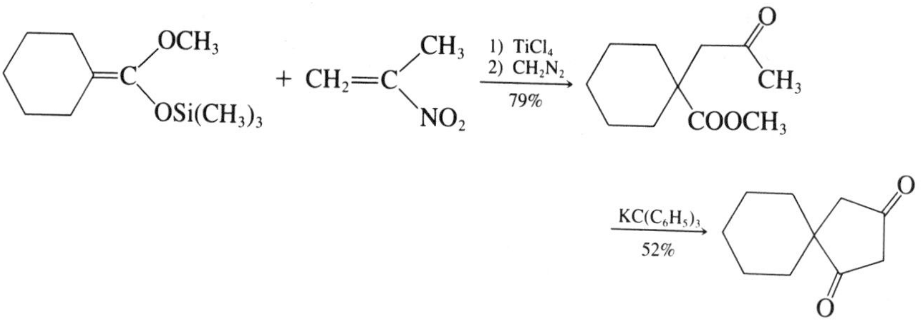

Enamines.   A few years ago White and Weingarten[21] reported that TiCl$_4$ can serve as the catalyst and water scavenger in the synthesis of enamines. The method is particularly useful in the case of enamines of acyclic ketones.[22] Yields can be improved by addition of the ketone to a preformed complex between the amine and TiCl$_4$; the ratio of amine to TiCl$_4$ should be at least 5:1.

Thioacetalization.[23]   Aldehydes and ketones react with alkylthiols and alkanedithiols in the presence of catalytic amounts of TiCl$_4$ in CHCl$_3$ at 28° to form thioacetals and -ketals in >90% yield. The reaction is satisfactory even with readily enolizable carbonyl compounds.

Cleavage of γ-lactols.[24]   γ-Lactols are generally more stable than the acyclic hydroxy aldehyde tautomers. However, they can be converted into the corresponding ring-opened hydroxy thioacetals by reaction with 1,2-ethanedithiol (or 1,3-propanedithiol) catalyzed by TiCl$_4$. Use of a variety of other Lewis acid catalysts (BF$_3$ etherate, AlCl$_3$) results in oxathioacetals (a), which are converted into the hydroxy thioacetals in the presence of TiCl$_4$.

Examples:

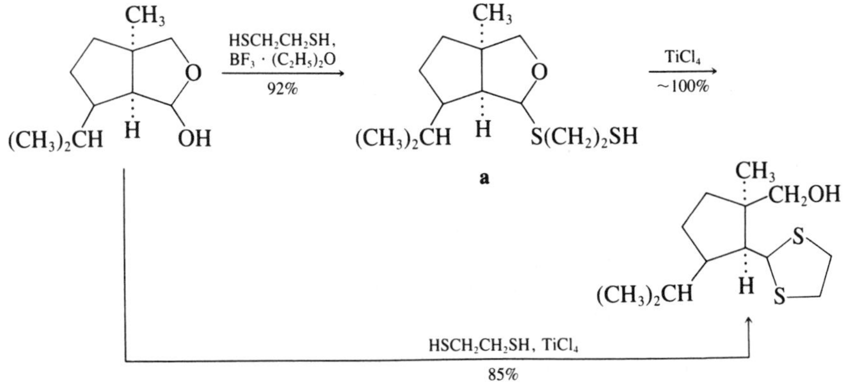

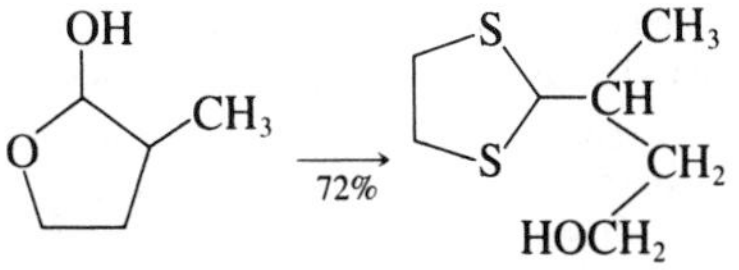

Deprotection of acetals.[25]   Acetals, ketals, and dioxolanes are converted to the carbonyl compounds by TiCl$_4$ in ether, usually in yields of 75–90%.

Intramolecular Mukaiyama aldol condensation.[26]   This reaction can be used to obtain six-, seven-, and eight-membered rings.[27] Thus the reaction of the *cis*-dioxolane **1a** with TiCl$_4$ (1–2 equiv.) gives **2a** as the exclusive product. The isomeric *trans*-dioxolane **1b** under similar conditions gives a 1:1 mixture of **2a** and **2b** (72% yield). No cyclization products are obtained with SnCl$_4$ or ZnCl$_2$.

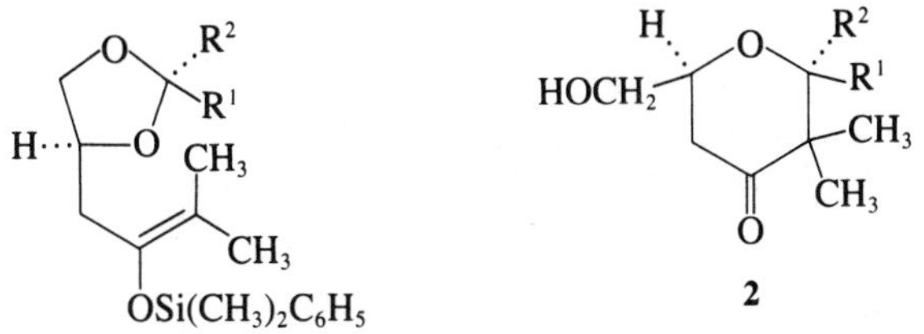

1a, R^1 = CH$_2$CH$_2$CH$_3$, R^2 = H
1b, R^1 = H, R^2 = CH$_2$CH$_2$CH$_3$

A further example is the cyclization of **3** with TiCl$_4$ to give **4, 5a,** and **5b** in the ratio 2:3:3, respectively (88% combined yield). Only **5a** and **5b** are formed with SnCl$_4$ (23% yield). The formation of an eight-membered ring is striking, particularly in the absence of high dilution, and is attributed to coordination of the silyl enol ether and acetal oxygens to the Ti catalyst.

3

4

5a, R^1 = H, R^2 = CH$_3$
5b, R^1 = CH$_3$, R^2 = H

[1] M. T. Reetz, *Angew. Chem., Int. Ed.,* **23**, 556 (1984).
[2] M. T. Reetz and K. Kesseler, *J.C.S. Chem. Comm.,* 1079 (1984).
[3] M. T. Reetz and A. Jung, *Am. Soc.,* **105**, 4833 (1983).
[4] H. Sakurai, *Pure Appl. Chem.,* **54**, 1 (1982).
[5] T. A. Blumenkopf and C. H. Heathcock, *Am. Soc.,* **105**, 2354 (1983).
[6] T. Tokoroyama, M. Tsukamoto, and H. Iio, *Tetrahedron Letters,* **25**, 5067 (1984).
[7] O. Piccolo, L. Filippini, L. Tinucci, E. Valoti, and A. Citterio, *Helv.,* **67**, 739 (1984).
[8] T. Hayashi, K. Kabeta, I. Hamachi, and M. Kumada, *Tetrahedron Letters,* **24**, 2865 (1983).
[9] E. Nakamura, Y. Horiguchi, J. Shimada, and I. Kuwajima, *J.C.S. Chem. Comm.,* 796 (1983).
[10] E. Nakamura, J. Shimada, Y. Horiguchi, and I. Kuwajima, *Tetrahedron Letters,* **24**, 3341 (1983).
[11] E. Nakamura and I. Kuwajima, *Chem. Letters,* 59 (1983).
[12] *Idem, Tetrahedron Letters,* **24**, 3343 (1983).
[13] D. Seebach, C. Betschart, and M. Schiess, *Helv.,* **67**, 1593 (1984).
[14] T. Mukaiyama, K. Watanabe, and M. Shiono, *Chem. Letters,* 1457 (1974).
[15] B. Crociani, M. Nicolini, and R. L. Richards, *J. Organometal. Chem.* **101**, C1 (1975).
[16] M. Schiess and D. Seebach, *Helv.,* **66**, 1618 (1983).
[17] S. Hara, H. Dojo, and A. Suzuki, *Chem. Letters,* 285 (1983).
[18] E. Nakamura and I. Kuwajima, *Am. Soc.,* **105**, 651 (1983).
[19] Preparation: K. Rühlmann, *Synthesis,* 236 (1971).
[20] M. Miyashita, T. Yanami, T. Kumazawa, and A. Yoshikoshi, *Am. Soc.,* **106**, 2149 (1984).
[21] W. White and H. Weingarten, *J. Org.,* **32**, 213 (1967).
[22] R. Carlson, A. Nilsson, and M. Stromqvist, *Acta Chem. Scand.,* **37B**, 7 (1983).
[23] V. Kumar and S. Dev, *Tetrahedron Letters,* **24**, 1289 (1983).
[24] P. G. Bulman-Page, R. A. Roberts, and L. A. Paquette, *ibid.,* **24**, 3555 (1983).
[25] G. Balme and J. Goré, *J. Org.,* **48**, 3336 (1983).
[26] Review: T. Mukaiyama, *Org. React.,* **28**, 203 (1982).
[27] G. S. Cockerill and P. Kocienski, *J.C.S. Chem. Comm.,* 705 (1983).

Titanium(IV) chloride–Diethylaluminum chloride.

[6 + 2] *Cycloadditions.*[1] This system, (C$_2$H$_5$)$_2$AlCl/TiCl$_4$ (20:1), induces [6 + 2] cycloaddition of cycloheptatriene with 1,3-butadiene, norbornadienes, and alkynes.

Examples:

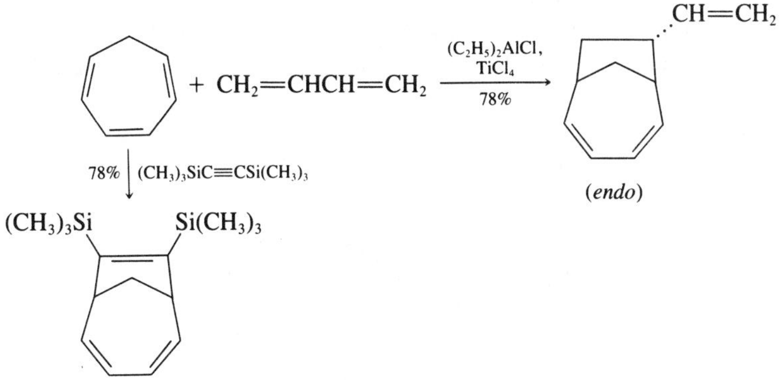

[1] K. Mach, H. Antropiusová, P. Sedmera, V. Hanuš, and F. Tureček, *J.C.S. Chem. Comm.*, 805 (1983).

Titanium(IV) chloride–Magnesium amalgam, 7, 373–374; **11**, 534.

Reduction of NO₂.[1]   This Ti(II) reagent reduces aromatic and aliphatic nitro compounds to amines in THF/*t*-butyl alcohol at 0° in yields of 85–95%. Halo, cyano, and ester groups are not reduced.

[1] J. George and S. Chandrasekaran, *Syn. Comm.*, **13**, 495 (1983).

Titanium(IV) chloride–Titanium(IV) isopropoxide.

2-Methyl-2-alkenenitriles **(3)**.[1]   The aldol-type reaction of aldehydes with the ketenimine **1** catalyzed by MgBr₂ or BF₃ etherate is nonstereoselective. However, use of a catalyst composed of TiCl₄ and Ti[OCH(CH₃)₂]₄ in the ratio 1:3 results into a stereocontrolled reaction to give **2**, which is converted into **3** by treatment with BF₃ etherate.

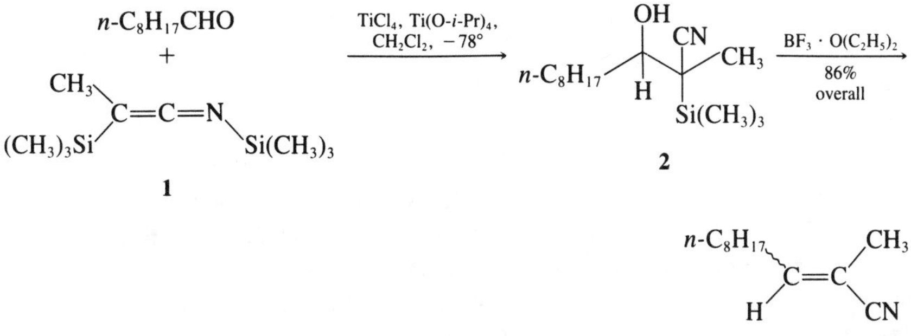

[1] H. Okada, I. Matusda, and Y. Izumi, *Chem. Letters*, 97 (1983).

Titanium(IV) isopropoxide, 10, 404–405; **11,** 528–529.

Reaction with α-hydroxy epoxides. This regioselective reaction provides a key step in a synthesis of a unique furyl sesquiterpene, pleraplysillin-1 (**5**), isolated from a marine sponge. Thus treatment of **1** with Ti(O-*i*-Pr)$_4$ results in the enediol **2**, which is converted

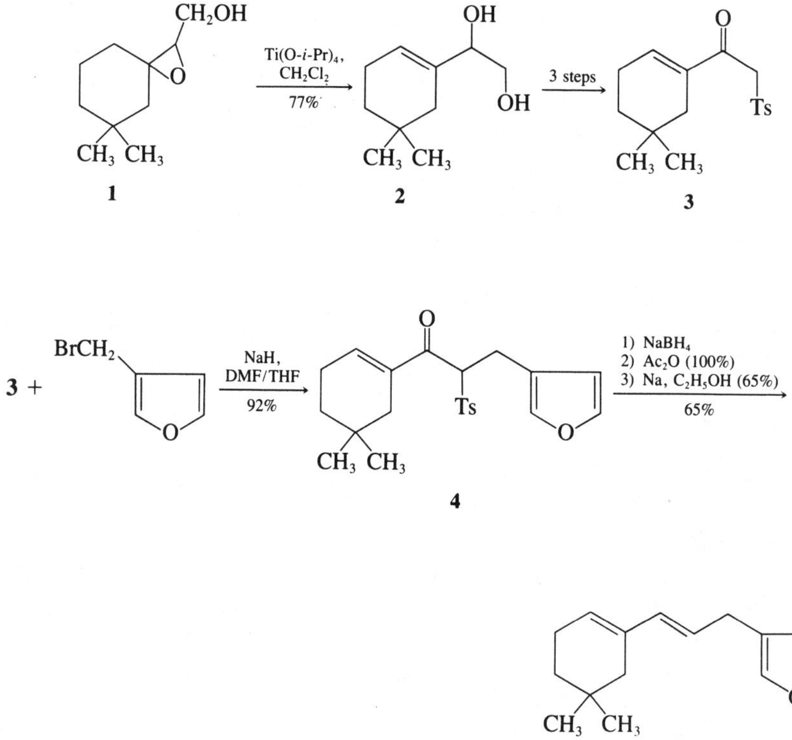

to **3**. Coupling of **3** with 3-furylmethyl bromide results in **4**. Conversion of **4** to the diene **5** gives mainly the desired (E)-isomer. Chromatography provides pure (E)-**5**.[1]

Regioselective cleavage of α,β-epoxy alcohols.[2] In the presence of Ti(O-*i*-Pr)$_4$ (1 equiv.), a variety of nucleophiles react with **1** regioselectively by cleavage at C$_3$. No reaction occurs in the absence of Ti(O-*i*-Pr)$_4$ except in the case of C$_6$H$_5$SNa.

Regioselective cleavage of α,β-epoxy acids and amides.[3] In the presence of 1

$$CH_3(CH_2)_2 \overset{O}{\diagup\!\!\!\diagdown} \diagdown_{OH} \quad \xrightarrow[\text{Nu}]{\text{Ti(O-}i\text{-Pr)}_4,}$$

1

$$\underset{\underset{OH}{\vdots}}{CH_3(CH_2)_2} \overset{Nu}{\diagup} \diagdown_{OH} \quad + \quad CH_3(CH_2)_2 \overset{OH}{\diagup}\underset{Nu}{\diagdown}_{OH}$$

2   **3**

Nu = (CH$_3$)$_3$SiN$_3$	74%	14:1
= (CH$_3$)$_2$CHOH	88%	100:1
= (CH$_3$)$_3$CCOOH	59%	100:1
= C$_6$H$_5$SNa	68%	9:1

equiv. of Ti(O-i-Pr)$_4$, amines and thiophenol react with α,β-epoxy acids and amides regioselectively at C$_3$.

Examples:

$$n\text{-C}_7H_{15} \overset{O}{\diagup\!\!\!\diagdown} \overset{O}{\overset{\|}{C}}\!\!\diagdown_{OH} \xrightarrow[\text{HNR}_2]{\text{Ti(O-}i\text{-Pr)}_4} n\text{-C}_7H_{15}\overset{NR_2}{\underset{OH}{\diagup}}\overset{O}{\overset{\|}{C}}\!\!\diagdown_{OH} + n\text{-C}_7H_{15}\overset{OH}{\underset{NR_2}{\diagup}}\overset{O}{\overset{\|}{C}}\!\!\diagdown_{OH} :$$

R = CH$_2$CH=CH$_2$ 83%   20:1

$$n\text{-C}_7H_{15} \overset{O}{\diagup\!\!\!\diagdown} \overset{O}{\overset{\|}{C}}\!\!\diagdown_{NHCH_2C_6H_5} \xrightarrow[95\%]{\substack{\text{C}_6\text{H}_5\text{SH,}\\ \text{Ti(O-}i\text{-Pr)}_4}}$$

$$n\text{-C}_7H_{15}\overset{SC_6H_5}{\underset{OH}{\diagup}}\overset{O}{\overset{\|}{C}}\!\!\diagdown_{NHCH_2C_6H_5} + n\text{-C}_7H_{15}\overset{OH}{\underset{SC_6H_5}{\diagup}}\overset{O}{\overset{\|}{C}}\!\!\diagdown_{NHCH_2C_6H_5}$$

20:1

Regioselective Peterson reaction. Aldehydes react with the anion of an (α-alkoxy)allyltrimethylsilane (**1**) at both α- and γ-positions. Addition of HMPT favors reaction at the γ-position, whereas addition of Ti(O-i-Pr)$_4$ (1 equiv.) results in exclusive reaction at the α-position to give an (E)-1,3-dienol ether (**2**). These products are readily hydrolyzed to vinyl ketones (**3**).[4]

Example:

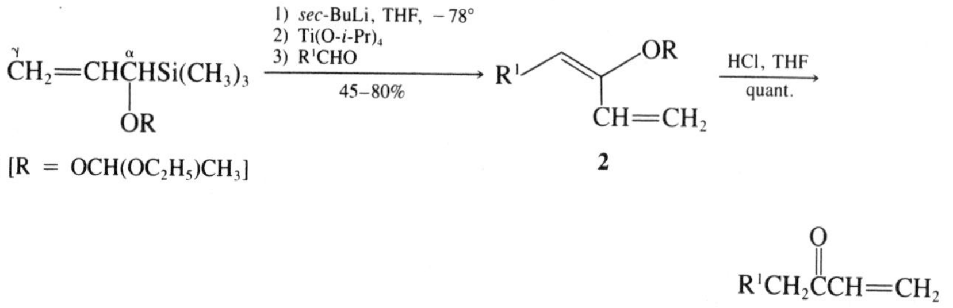

1) *sec*-BuLi, THF, $-78°$
2) Ti(O-*i*-Pr)$_4$
3) R'CHO
45–80%

HCl, THF
quant.

$$CH_2=CHCHSi(CH_3)_3$$

1 [R = OCH(OC$_2$H$_5$)CH$_3$]

2

$$R^1CH_2CCH=CH_2$$

3

This three-carbon homologation was used in a synthesis of the diterpene ($-$)-aplysin-20 (**4**) from nerolidol (equation I).[5]

(I)

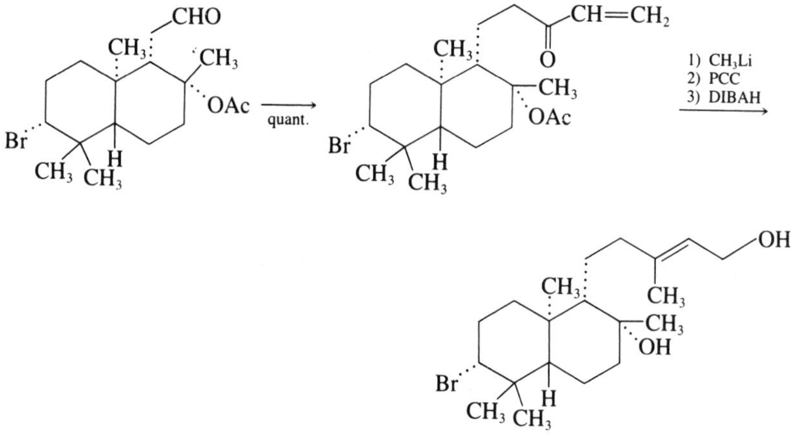

1) CH$_3$Li
2) PCC
3) DIBAH

4

[1] Y. Masaki, K. Hashimoto, Y. Serizawa, and K. Kaji, *Bull. Chem. Soc. Japan*, **57**, 3476 (1984).
[2] M. Caron and K. B. Sharpless, *J. Org.*, **50**, 1557 (1985).
[3] J. M. Chong and K. B. Sharpless, *ibid.*, **50**, 1560 (1985).
[4] A. Murai, A. Abiko, N. Shimada, and T. Masamune, *Tetrahedron Letters*, **25**, 4951 (1984).
[5] A. Murai, A. Abiko, and T. Masamune, *ibid.*, **25**, 4955 (1984).

Titanocene methylene–Zinc iodide complex, Cp$_2$TiCH$_2$·ZnI$_2$ (**1**). The reagent is prepared by reaction of Cp$_2$TiCl$_2$ with CH$_2$(ZnI)$_2$ in THF. It is related to the Tebbe reagent (**8**, 83–84; **10**, 87–88), but is easier to prepare.

Methylenation.[1]   The reagent reacts with ketones at $-10°$ to give essentially quantitative yields of the corresponding methylene derivatives. The reagent effects coupling of benzyl halides to bibenzyl (85% yield).

[1] J. J. Eisch and A. Piotrowski, *Tetrahedron Letters*, **24**, 2043 (1983).

p-**Toluenesulfinyl chloride,** p-$CH_3C_6H_4S(O)Cl$. Mol. wt. 175.05, b.p. 113–115°/35 mm. Preparation.[1]

α,β-*Unsaturated thiolactams.*[2]   Thiolactams can be dehydrogenated by reaction with this sulfinyl chloride followed by mild acid treatment (equation I). The reaction may involve attack on sulfur, rather than carbon, followed by a 1,4-elimination. The generality of this dehydrogenation has not been established. It is useful for indirect dehydrogenation

of lactams in cases where phenylselenenylation of the enolate fails, as has been observed with some indole alkaloids. The lactam is converted into the thiolactam with Lawesson's reagent (**8**, 327; **9**, 50); after dehydrogenation, reaction with triethyloxonium tetrafluoroborate followed by alkaline hydrolysis provides the α,β-unsaturated lactam.

[1] F. Kurzer, *Org. Syn. Coll. Vol.*, **4**, 937 (1955).
[2] P. Magnus and P. Pappalardo, *Am. Soc.*, **105**, 6525 (1983).

p-**Toluenesulfonic acid, 1**, 1172–1178; **4**, 508–510; **5**, 673–675; **7**, 374–375; **8**, 488–489; **9**, 471–472; **11**, 535.

Nitrile oxides.[1]   This acid is an effective catalyst for dehydration of primary nitro compounds (RCH_2NO_2) to nitrile oxides, which can be trapped with dipolarophiles to form five-membered heterocycles. The reaction is carried out in refluxing mesitylene.

Examples:

$$CH_3OOCCH_2NO_2 \quad + \quad \underset{O}{\overset{O}{\|}} NC_6H_5 \quad \xrightarrow[70\%]{TsOH, \Delta} \quad$$ (structure)

$$C_6H_5COCH_2NO_2 \quad + \quad C_6H_5C\equiv CH \quad \xrightarrow{46\%} \quad$$ (structure)

Protection of amines.[2]   The 4-methoxybenzyloxycarbonyl group of amino acids can be removed with CF_3COOH or, preferably, with TsOH in acetonitrile at 20–25°. Other solvents are not useful. This method is also useful for removal of a Boc group.

[1] T. Shimizu, Y. Hayashi, and K. Teramura, *Bull. Chem. Soc. Japan*, **57**, 2531 (1984).
[2] H. Yamada, H. Tobiki, N. Tanno, H. Suzuki, K. Jimpso, S. Ueda, and T. Nakagome, *ibid.*, 3333 (1984).

***p*-Toluenesulfonylhydrazine, 1**, 1185–1187; **2**, 419–423; **3**, 293; **4**, 571–572; **5**, 678–681; **6**, 598–560; **7**, 375–376; **8**, 489–493; **9**, 472–473; **11**, 537.

Tosylhydrazones of keto esters.   Acid-sensitive α- and β-keto esters can be converted into the tosylhydrazones in 30–80% yield by use of neutral alumina as catalyst.[1] Yields are 5–10% when BF_3 etherate is used as catalyst.[2]

[1] P. Vinczer, L. Novak, and C. Szantay, *Syn. Comm.*, **14**, 281 (1984).
[2] P. A. Grieco and M. Nishizawa, *J. Org.*, **42**, 1717 (1977).

***p*-Tolylsulfinylacetic acid, 10**, 405–407; **11**, 538.

3-p-*Tolylsulfinyl*-4-*aryl*-β-*lactams*.[1]   *p*-Tolylsulfinylacetic acid (**1**), activated with

$$p\text{-}CH_3C_6H_4\underset{*}{\overset{O}{\overset{\|}{S}}}CH_2COOH \; + \; \underset{R^3}{\overset{R^1}{\underset{\|}{\overset{|}{CH}}}} \xrightarrow[30-45\%]{DMF} \; C_7H_7$$ (structure **2**)

N,N'-carbonyldiimidazole, condenses with aryl aldimines to form the β-lactams **2** in moderate yield. The reaction affords only the two possible *trans*-diastereomers.

[1] G. Guanti, L. Banfi, E. Narisano, and S. Thea, *J.C.S. Chem. Comm.*, 861 (1984).

(R)-(+)-(α-*p*-Tolylsulfinyl)-N,N-dimethylacetamide (1).

p-CH$_3$C$_6$H$_4$—S—...—N(CH$_3$)$_2$

(R)-**1** (α_D + 194.7°, m.p. 63–64°)

The reagent is obtained by reaction of (S)-(−)-menthyl *p*-toluenesulfinate (**10**, 405) with α-lithio-N,N-dimethylacetamide (83% yield).

Chiral β-hydroxy amides.[1]   A metal enolate of (R)-**1** reacts with aldehydes to form adducts (**2**) that are desulfurized by Na/Hg to optically active hydroxy amides (**3**). The extent and the sense of chiral induction depends on the metal enolate. Use of *n*-butyllithium

C$_7$H$_7$—S—...—N(CH$_3$)$_2$

R   OH

2

OH   O

R—*—N(CH$_3$)$_2$

3

results in low to moderate enantioselectivity to give (R)-(+)-**3**. Use of *t*-butylmagnesium bromide results in much higher and opposite enantioselectivity to give (S)-(−)-**3** in 85–99% ee. In both cases, the enantioselectivity decreases with an increase in the size of R. The favorable effect of magnesium is ascribed to chelation, which favors formation of the (Z)-enolate.

[1] R. Annuziata, M. Cinquini, F. Cozzi, F. Montanari, and A. Restelli, *J.C.S. Chem. Comm.*, 1138 (1983).

(R)-(+)-3-(*p*-Tolylsulfinyl)propionic acid,

C$_7$H$_7$—S—...—OH
O

(1), α_D + 180°, m.p.

133°. The reagent is prepared by reaction of the anion of (R)-(+)-menthyl *p*-tolyl sulfoxide with lithium bromoacetate.

Optically pure butenolides.[1]   The dianion of **1** reacts with aldehydes to give mainly the two diastereomeric lactones **2** and **3** in the approximate ratio 60:40. These are separated by flash chromatography and converted into the optically pure butenolides **4** and **5** by pyrolysis.

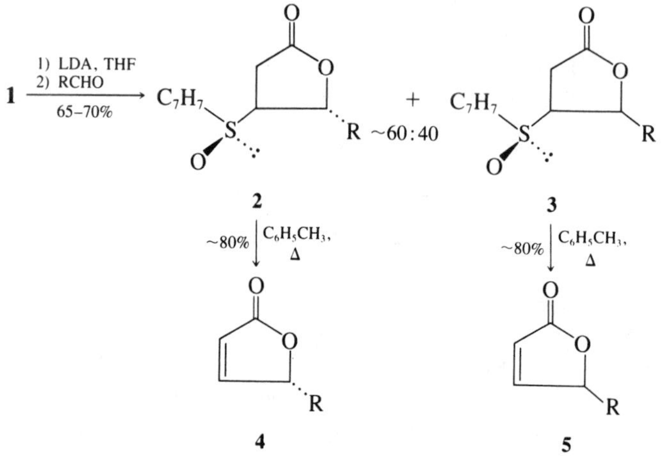

[1] P. Bravo, P. Carrera, G. Resnati, and C. Ticozzi, *J.C.S. Chem. Comm.*, 19 (1984).

(S)-(+)-*p*-Tolyl *p*-tolylthiomethyl sulfoxide (1), 9, 474; 10, 408–409.[1]

Chiral α-methoxy aldehydes.[2]   The anion of (S)-**1** reacts with benzoyl chloride to give two products (**2**) in the ratio 99:1. LiAlH₄ reduction of **2a, 2b** gives only two α-tolylthio-β-hydroxy sulfoxides (**3a,b**) rather than the expected four diastereomers. Evidently the LiAlH₄ reduction is highly regiospecific. Either **2** or **3** can be separated by crystallization or chromatography and converted into optically pure α-methoxy aldehydes (**4**, equation I).

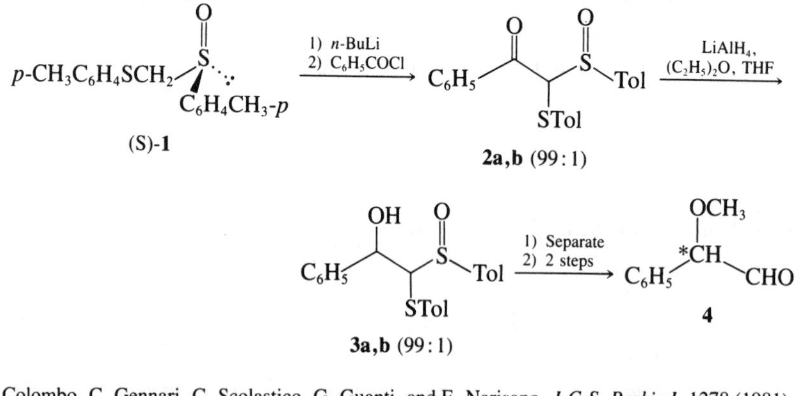

[1] L. Colombo, C. Gennari, C. Scolastico, G. Guanti, and E. Narisano, *J.C.S. Perkin I*, 1278 (1981).
[2] G. Guanti, E. Narisano, L. Banfi, and C. Scolastico. *Tetrahedron Letters*, **24**, 817 (1983).

Tosylmethyl isocyanide, 4, 514–516; **5**, 684–685; **6**, 600; **7**, 377; **8**, 493–494; **10**, 409; **11**, 539.

21-*Hydroxy-20-ketosteroids*.[1] C_{17}-Ketosteroids (**1**) undergo a Knoevenagel-type condensation with TosMIC to provide 17-(isocyanotosylmethylene)steroids (**2**). Phase-transfer catalyzed alkylation of **2** with formaldehyde gives an oxazoline, which is hydro-

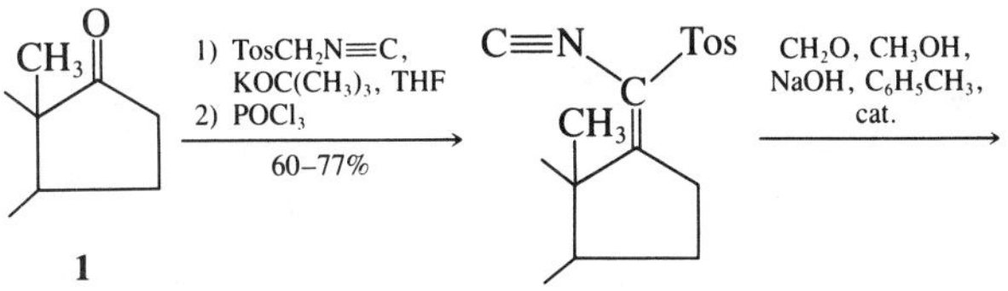

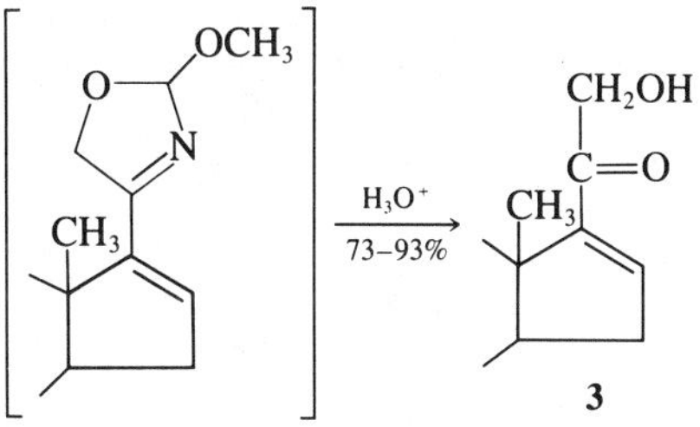

lyzed to a Δ^{16}-21-hydroxy-20-ketopregnane (**3**). The transformation of **2** to **3** is a modification of a general method for homologation of a ketone to an α,β-enone (equation I).[2] Alkylation of the anion of **5** is highly regiospecific, and proceeds in high yield with the more reactive halides, but only in moderate yield with secondary halides.

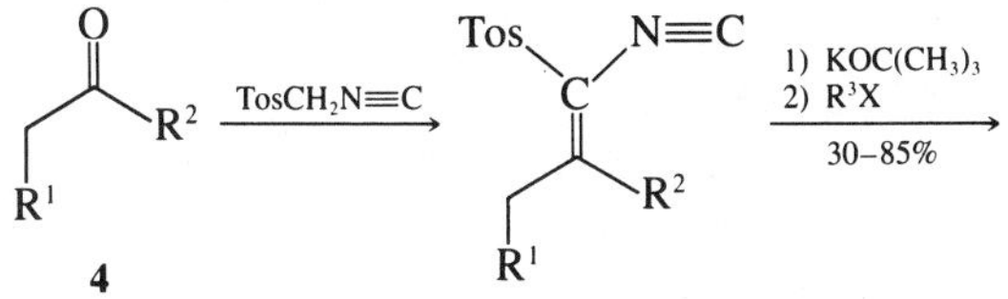

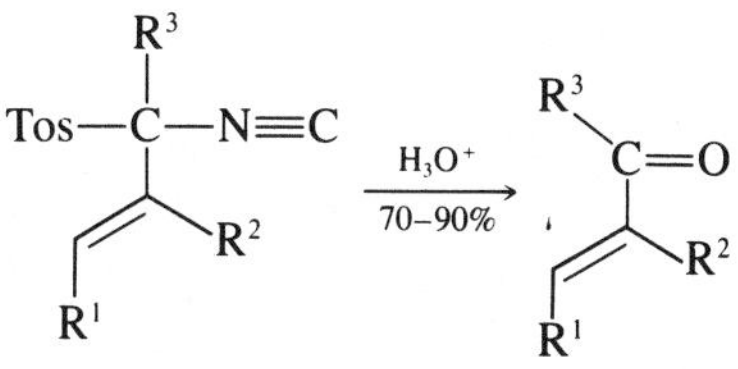

[1] D. van Leusen and A. M. van Leusen, *Tetrahedron Letters*, **25**, 2581 (1984).
[2] J. Moskal and A. M. van Leusen, *ibid.*, 2585 (1984).

Trialkylaluminums.

Asymmetric pinacol-type rearrangement.   Chiral β-mesyloxy tertiary alcohols undergo a stereospecific pinacol-type rearrangement in the presence of triethylaluminum (excess) to furnish optically pure α-aryl or α-vinyl ketones (equation I).[1] The reaction is particularly

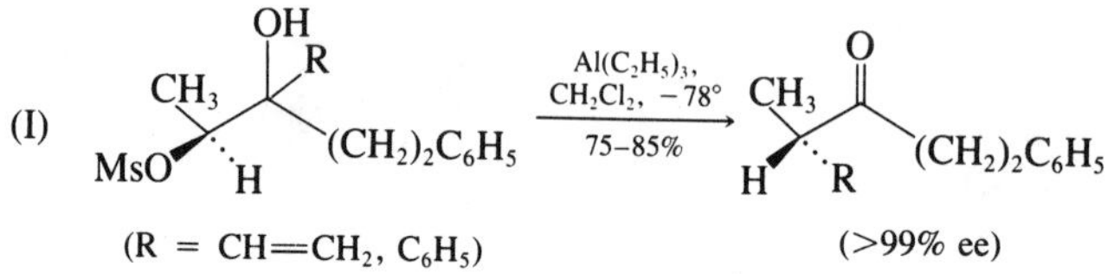

$$(R = CH=CH_2, C_6H_5)$$

$$(>99\% \; ee)$$

useful for preparation of optically pure α-methyl-β,γ-unsaturated ketones by migration of an alkenyl group, which occurs with retention of the alkene geometry (equation II).[2] These ketones are reduced by lithium tri-*sec*-butylborohydride with high *anti*-selectivity.[3]

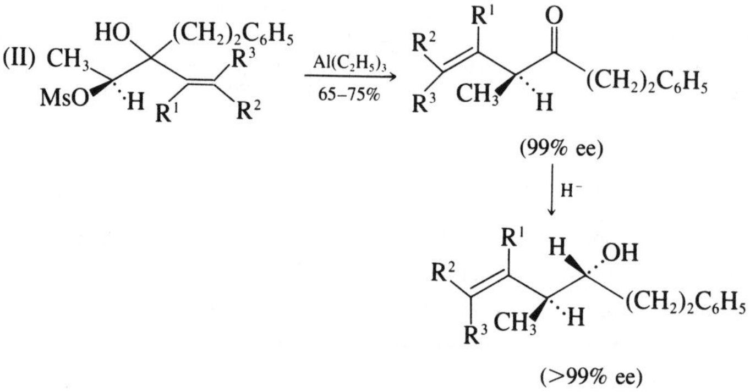

This rearrangement can be applied to chiral α-mesyloxy ketones by *in situ* reduction with DIBAH followed by addition of triethylaluminum or diethylaluminum chloride. The resulting aldehyde is reduced as formed to an optically active 2-aryl- or 2-alkenylpropanol.[4] Example:

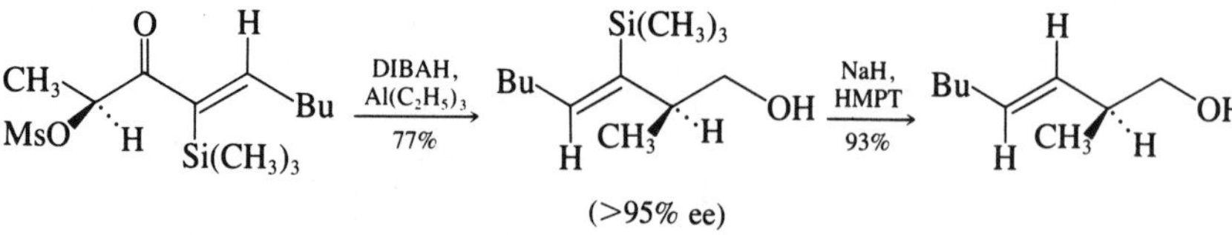

$$(>95\% \; ee)$$

The trimethylsilyl group promotes the rearrangement, and can be removed when desired by reaction with a catalytic amount of NaH in HMPT.[5]

Chiral α-methyl aldehydes.[6]   Reaction of optically active 2,3-epoxy alcohols with $Al(CH_3)_3$ results in a mixture of two diols that are not separable by conventional chromatography. However, the 1,2-diol is oxidized by $NaIO_4$ to a chiral α-methyl aldehyde, which is easily separated from the 1,3-diol.

Examples:

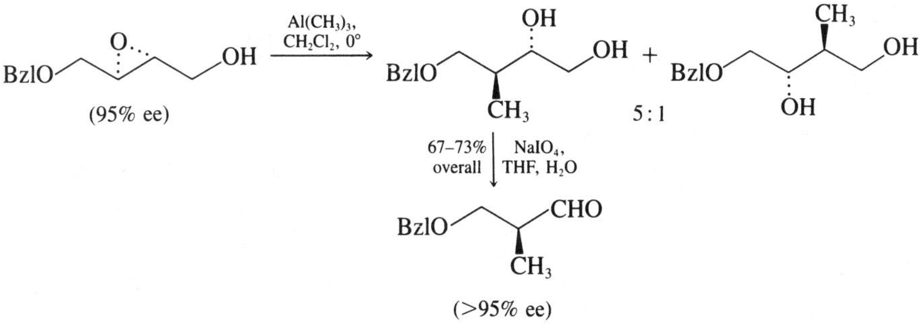

[1] K. Suzuki, E. Katayama, and G. Tsuchihashi, *Tetrahedron Letters*, **24**, 4997 (1983).
[2] *Idem, ibid.*, **25**, 1817 (1984).
[3] *Idem, ibid.*, **25**, 2479 (1984).
[4] K. Suzuki, E. Katayama, T. Matsumoto, and G. Tsuchihashi, *ibid.*, **25**, 3715 (1984).
[5] F. Sato, Y. Tanaka, and M. Sato, *J.C.S. Chem. Comm.*, 165 (1983).
[6] W. R. Roush, M. A. Adam, and S. M. Peseckis, *Tetrahedron Letters*, **24**, 1377 (1983).

Tri-*n*-butylcrotyltin, CH_3CH=$CHCH_2SnBu_3$ (**1**), **10**, 411; **11**, 542–543.

Addition to aldehydes.[1]   The allylic tin reagent reacts with aldehydes in the presence of BF_3 etherate to give preferentially the *syn*-homoallylic alcohol regardless of the geometry of the double bond (**10**, 411). When carried out in the absence of a catalyst but under high pressure, the reaction results mainly in the *anti*-isomer.

The BF_3-etherate-catalyzed reaction of **1** with the aldehyde **2** provides a short stereoselective synthesis of the Prelog-Djerassi lactonic acid **3**.

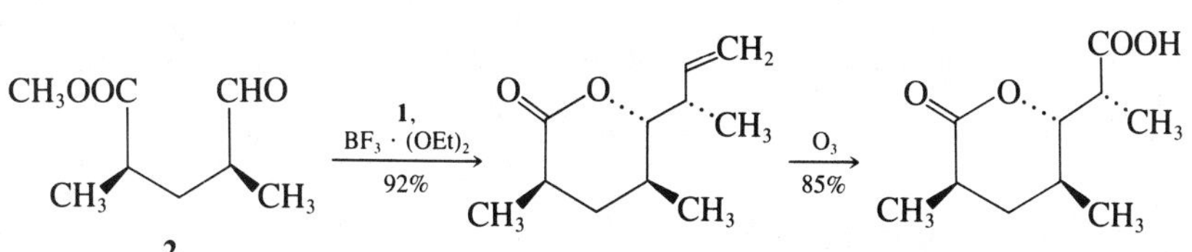

Use of $AlCl_3$/isopropanol as the Lewis acid catalyst can change the regioselectivity in the reaction with benzaldehyde and linear aldehydes to furnish the linear adducts as the major products (equation I). Presumably the catalyst is $AlCl_2[OCH(CH_3)_2]$.

(I) **1** + RCH_2CHO $\xrightarrow[\sim 70\%]{\substack{AlCl_3, \\ (CH_3)_2CHOH}}$ RCH_2 ... CH_3 + RCH_2 ... CH_3

$$(E/Z = 9:1) \qquad 90\text{–}95:10\text{–}5$$

Reaction with glyoxylates.[2]   The reaction of **1** with glyoxalates **2** is *syn*-selective and the selectivity increases with increasing steric bulk of the ester group (equation I).

(I) **1** + $\overset{O}{\overset{\|}{HCCOOR}}$ $\xrightarrow{CH_2Cl_2,}$ H_2C ... $\overset{CH_3}{COOR}$ + H_2C ... $\overset{CH_3}{COOR}$

$$\mathbf{2}$$

$$R = CH_3 \qquad\qquad 25:75$$
$$R = CH(CH_3)_2 \qquad 10:90$$

This reaction of **1** with the optically active glyoxylate ester of 8-phenylmenthol provides an enantioselective synthesis of verrucarinolactone (**3**, equation II).

(II) R^*O ... $\overset{O\quad CH_3}{\underset{OH}{}}$ CH_2 $\xrightarrow[70\%]{BH_3 \cdot S(CH_3)_2,}$ R^*O ... $\overset{O\quad CH_3}{\underset{OH}{}}$ OH

$$(R^* = \text{8-phenylmenthyl})$$

$\xrightarrow{TsOH, CH_2Cl_2}$ + *cis*-epimer

$$9:1$$

$$\mathbf{3}\ (\alpha_D - 8.8°, 91\%\ ee)$$

[1] Y. Yamamoto, H. Yatagai, Y. Ishihara, N. Maeda, and K. Maruyama, *Tetrahedron*, **40**, 2239 (1984).
[2] Y. Yamamato, N. Maeda, and K. Maruyama, *J.C.S. Chem. Comm.*, 774 (1983).

Tri-*n*-butylphosphine–Diphenyl disulfide, 6, 602.

Reduction of oximes.[1]   The reagent reduces ketoximes under anhydrous conditions to imines, which can undergo several useful transformations as well as the usual hydrolysis

to ketones. The imine can be reduced to the amine with NaBH₃CN or converted into an
amino nitrile by reaction with HCN.

Example:

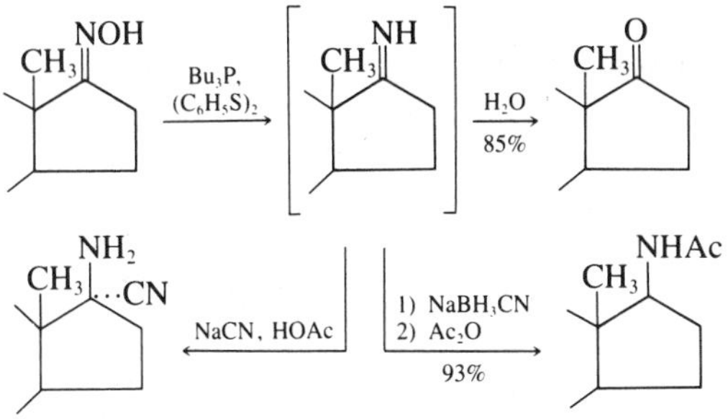

Reduction of sec-nitro compounds.[2]   The reagent also reduces secondary nitroalkanes to imines, which can be trapped with hydrogen cyanide as before to furnish α-
amino nitriles (70% yield).

Similar reduction of a 1,⁴-nitro ketone results in an intramolecular cyclization to a
pyrrole (equation I).

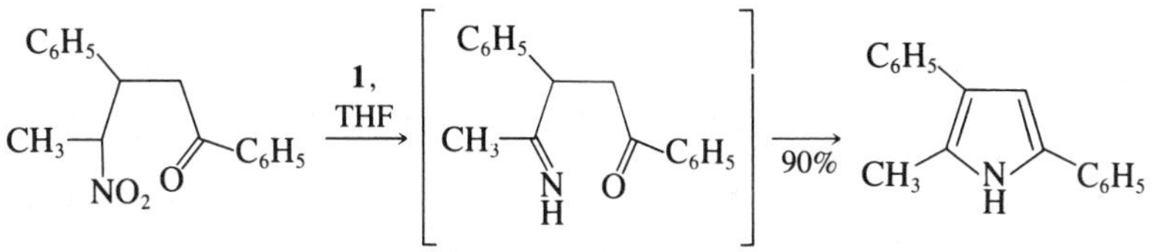

[1] D. H. R. Barton, W. B. Motherwell, E. S. Simon, and S. Z. Zard, *J.C.S. Chem. Comm.*, 337
(1984).
[2] D. H. R. Barton, W. B. Motherwell, and S. Z. Zard, *Tetrahedron Letters*, **25**, 3707 (1984).

Tri-*n*-butyltin fluoride, 11, 544. Supplier: Alfa.

Selective silyl/stannyl exchange.[1]   Regioselective aldol condensation of methyl ketones can be effected by generation of the α-stannyl derivative via silyl–stannyl exchange

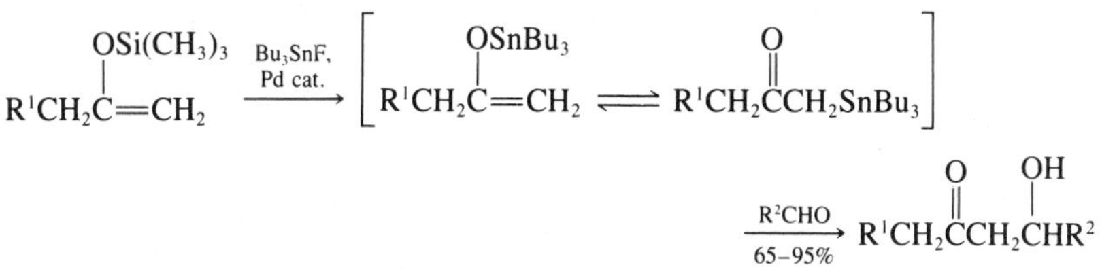

(equation I). Selective aldol reactions of a bis-silyl enol ether at the site of the methyl ketone are also possible by this exchange reaction.

Selective desilylation of silyl enol ethers.[2]   The silyl enol ethers of methyl ketones undergo desilylation on treatment with Bu_3SnF (1 equiv.). The reaction is markedly accelerated by a palladium catalyst, particularly $PdCl_2[P(o\text{-}CH_3C_6H_4)_3]_2$. The rate of desilylation is markedly decreased by steric congestion around the double bond. Thus highly selective desilylation is possible. The relative rate decreases in the following order:

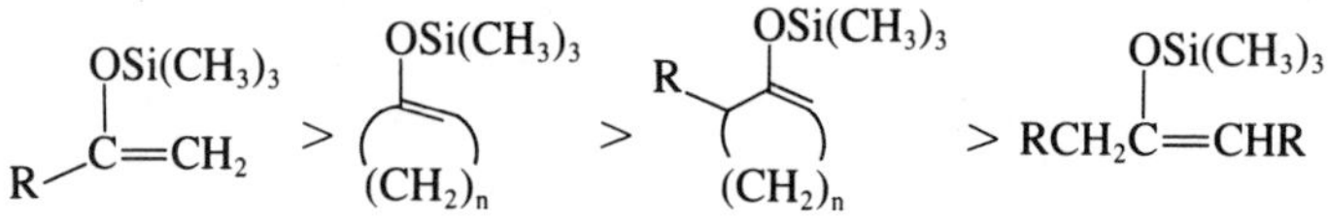

Example:

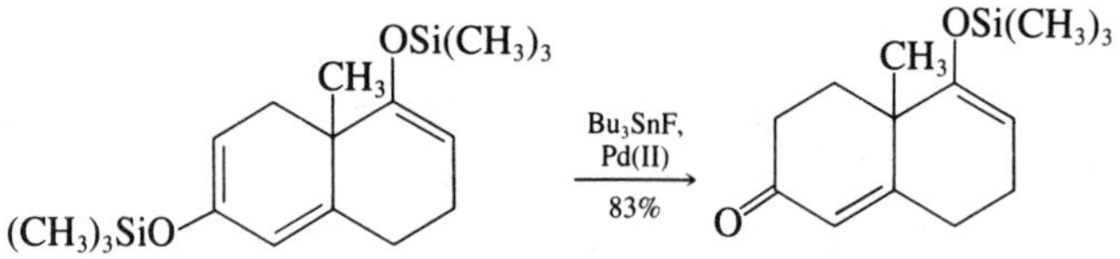

[1] H. Urabe and I. Kuwajima, *Tetrahedron Letters,* **24,** 5001 (1983).
[2] H. Urabe, Y. Takano, and I. Kuwajima, *Am. Soc.,* **105,** 5703 (1983).

Tri-*n*-butyltin hydride, 1, 1192–1193; **2,** 424; **3,** 294; **4,** 518–520; **5,** 518–520; **6,** 604; **7,** 379–380; **8,** 497–498; **9,** 379–380; **10,** 411–413; **11,** 545–551.

>*CHOH* → >*CH₂*.   A variety of thiocarbonyl derivatives of secondary alcohols are reductively cleaved by Bu_3SnH.[1] The S-methyl xanthate is a convenient derivative for use in this reaction because of convenience and low cost.[2]

Example:

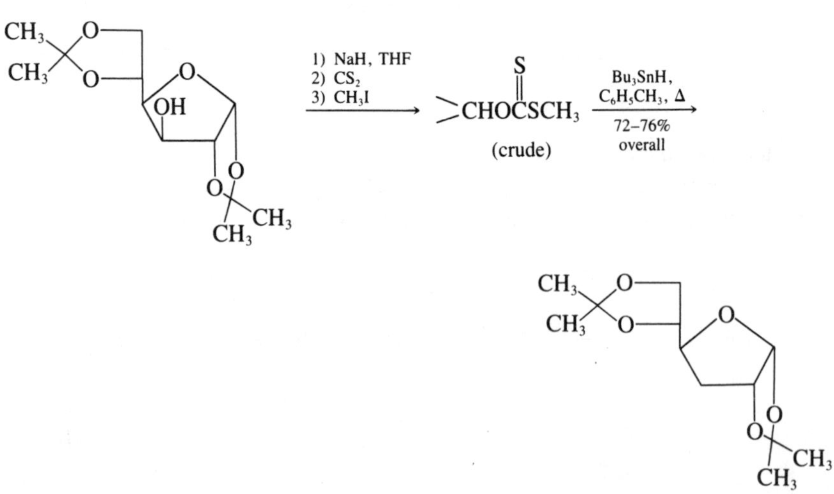

Reduction of nitro groups (**11**, 547).³ Although secondary nitro groups do not generally undergo hydrodenitration on treatment with Bu₃SnH and AIBN, benzylic and allylic nitro compounds and α-nitro ketones and esters are reduced satisfactorily. Even unactivated secondary nitro compounds can be reduced in moderate yield under forcing conditions: a large excess of reductant and elevated temperatures. This reduction is useful for reduction of a medium-sized nitro lactone (**2**) obtained by ring expansion of a nitro alcohol (**1**).

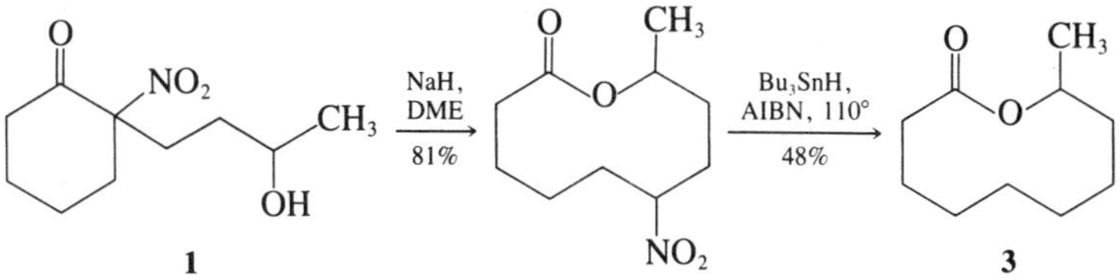

*Tropones.*⁴ 2,4,6-Trimethylphenol (**1**) reacts with dichloro- or dibromocarbene (halo-genoform and phase-transfer catalyst) to provide the dienones **2** and **3**, which on reaction with Bu₃SnH are both converted into 2,4,7-trimethyltropone (**4**).

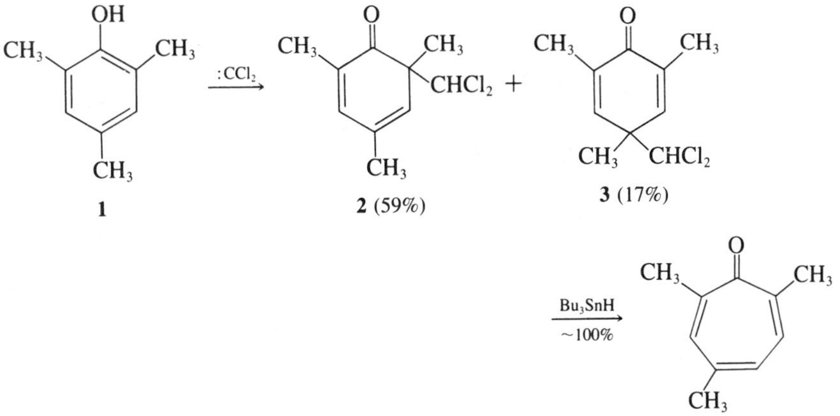

Even phenols substituted by *t*-butyl groups can undergo this transformation, but in lower yield.

*Coupling of alkyl halides with alkenes.*⁵ The radical generated from an alkyl halide by tri-*n*-butyltin hydride (AIBN, *hν*) undergoes coupling with electron-deficient alkenes. In the case of alkyl iodides, only catalytic amounts of the tin reagent are necessary when NaBH₄ is present for regeneration from the tin halide. Xanthates also undergo this coupling.

Examples:

$$(CH_3)_3CI + H_2C=CHCN \xrightarrow[87\%]{\substack{n\text{-}Bu_3SnH, \\ AIBN}} (CH_3)_3CCH_2CH_2CN$$

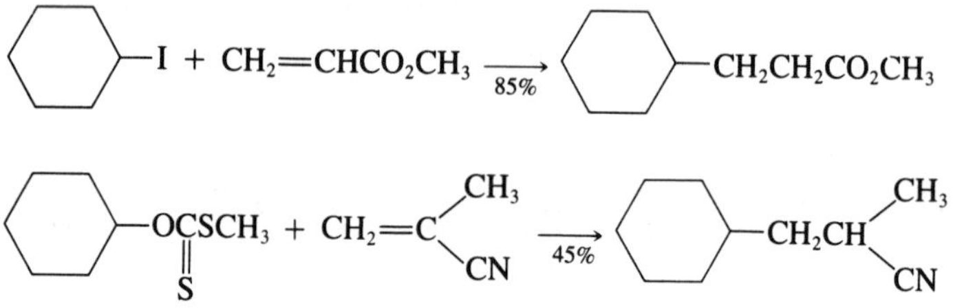

***C-Glycosides.*[6]**   Axial C-glycopyranosides are obtained exclusively by photolysis of a mixture of an α-D-glycopyranosyl bromide, acrylonitrile, and tributyltin hydride. The diastereoselectivity may involve an anomeric effect or rapid trapping of the initial radical before inversion can occur.

Example:

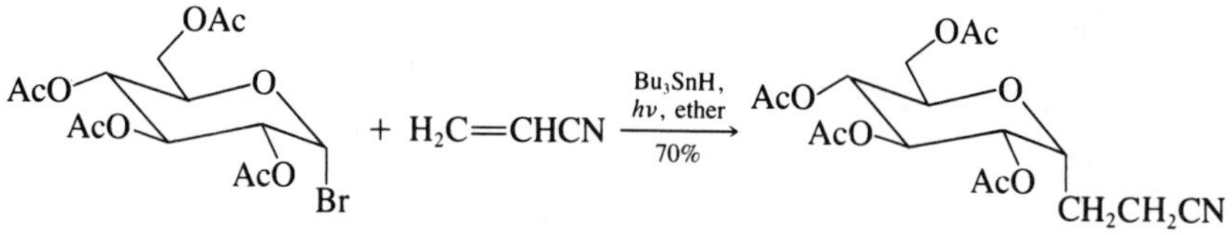

***Cyclization of bromo acetals.*[7]**   Reduction of the unsaturated bromo acetal **1** with Bu₃SnH (AIBN) gives the cyclic acetal **2**, which is converted on Jones oxidation to the *trans*-lactone **3**.

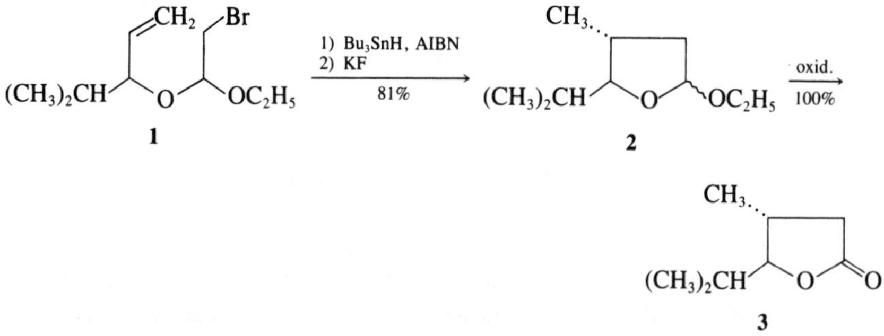

This free-radical cyclization is particularly useful for formation of *cis*-bicyclic systems.
Examples:

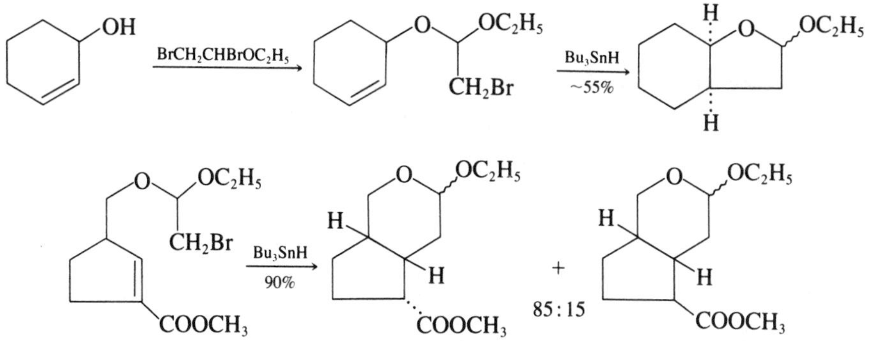

α- and β-Methylene-γ-butyrolactones.[8]   The radical cyclization of bromoacetals to
γ-butyrolactones (**11**, 545–546) has been extended to a related synthesis of methylene-
γ-butyrolactones.
Examples:

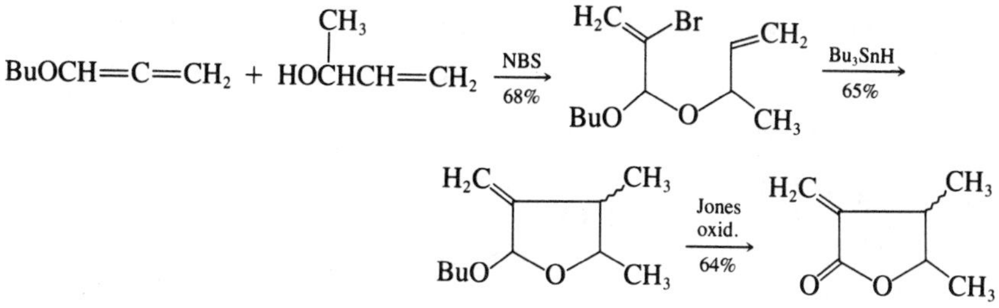

(*trans*/*cis* = 93:7)

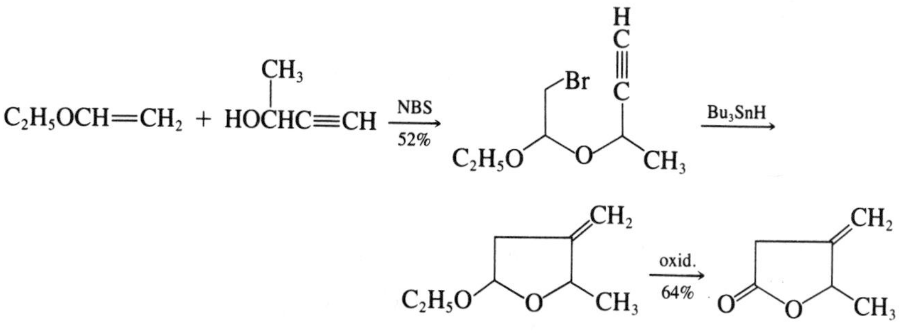

β-Hydroxy-γ-butyrolactones.[9]   The vinyl ether bromoacetal **1** undergoes cyclization
to **2** on reduction with tri-*n*-butyltin hydride in the presence of AIBN to give the cyclic

acetal ether **2**, which is oxidized to the lactone **3**. In contrast, radical cyclization with cobaloxime(I) (**11**, 135–136) gives **4** with only traces of **2**.

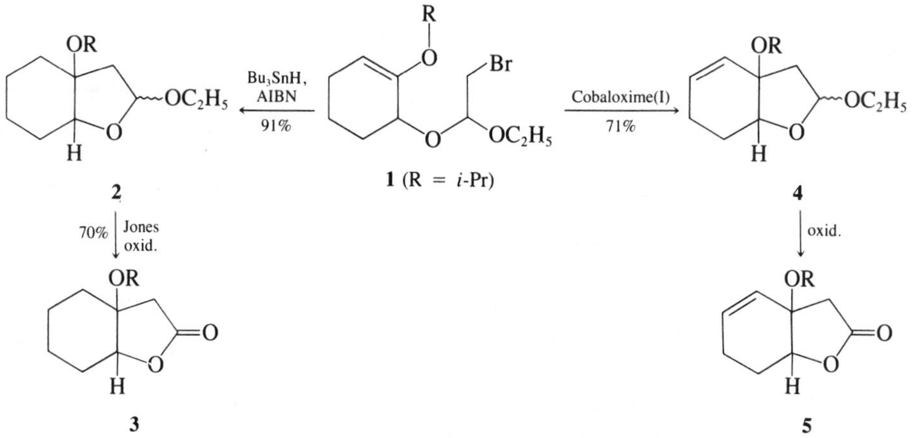

Double cyclizations to butenolides and furanes.[10]   Radicals can undergo intramolecular addition to triple bonds when separated by three carbons. This strategy can be used for synthesis of butenolides (equation I) and β-substituted furanes (equation II).

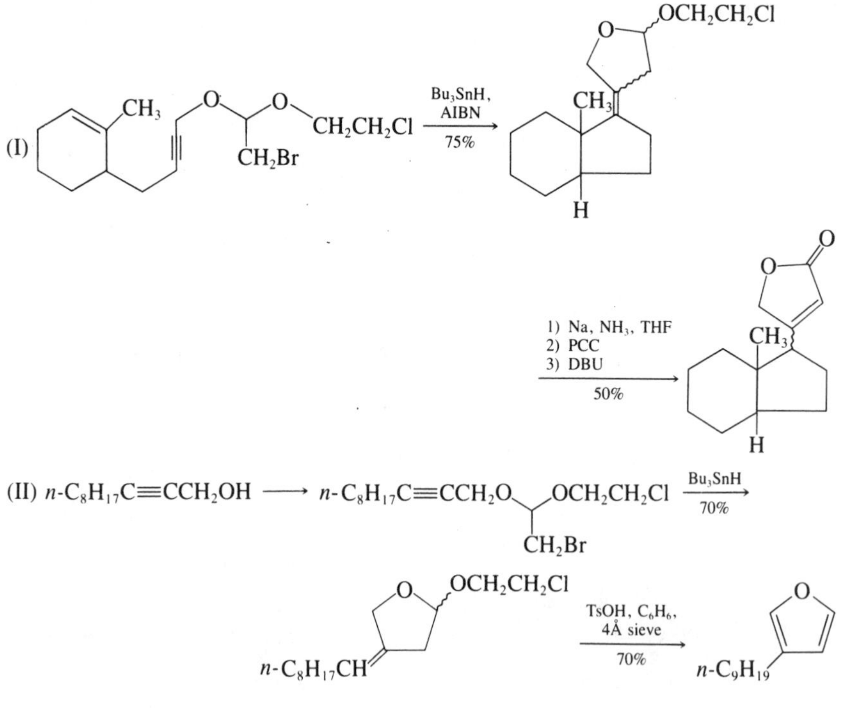

Cyclization of vinyl bromides.[11]   Fused and bridged ring systems can be prepared by intramolecular conjugate addition of vinyl radicals to an α,β-enone group. The vinyl radical is generated preferably by reaction of vinyl bromides with tri-*n*-butyltin hydride (AIBN initiation), and cyclization proceeds most readily when a five-membered ring is formed. Yields are improved by dilution. Benzene is the optimal solvent.

Examples:

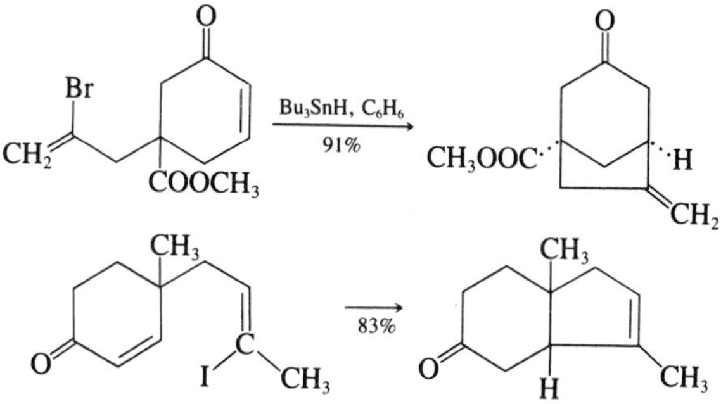

1,3-Diols.[12]   Intramolecular radical cyclization of the allylic silylmethyl ether **1** followed by oxidation of the Si—C bond with H_2O_2 (this volume) results in a 1,3-diol (**2**).

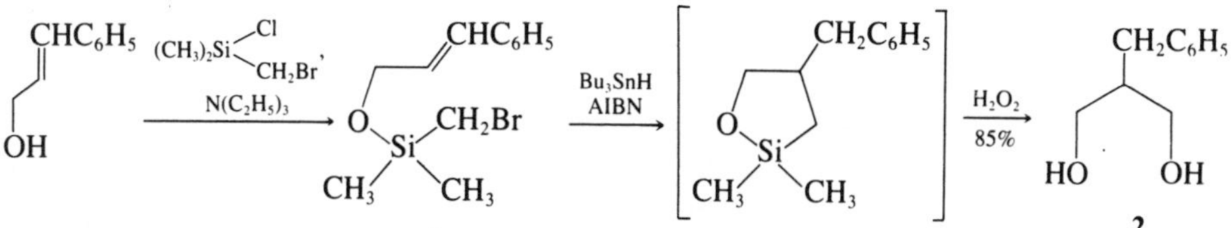

The process can provide a useful stereoselective route to 1,3-diols.

Examples:

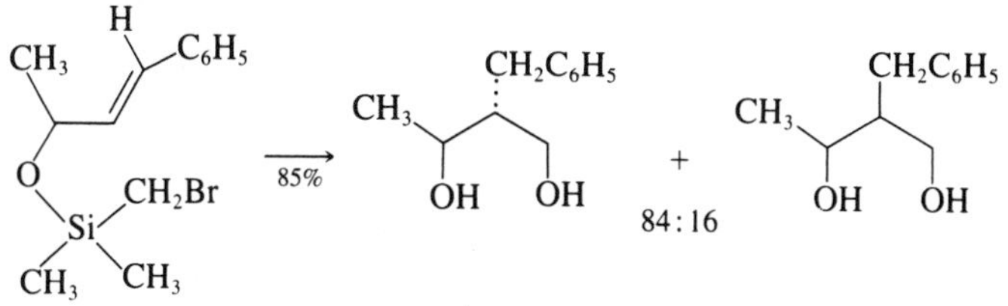

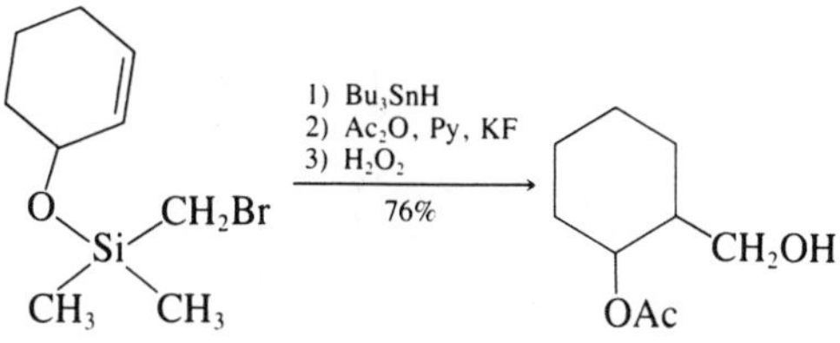

Radical cyclization to perhydroindanes.[13]   A new route to this ring system from benzoic acids involves reductive alkylation to provide an unsaturated acid such as **1** followed by iodolactonization to **2**. The radical obtained on reduction of **2** with Bu₃SnH cyclizes mainly to a *cis*-fused perhydroindane (**3**) and a perhydronaphthalene (**4**). In this

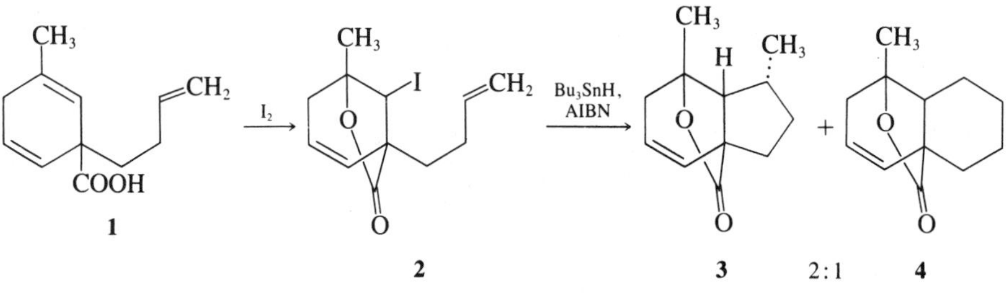

and related examples cyclization to a five-membered ring is favored over cyclization to a six-membered ring.

Pyrrolizidines (11, 546).[14]   Intramolecular cyclization of an α-acylamino radical to an allene provides a route to the pyrrolizidine alkaloid supinidine (**4**), outlined in Scheme (I).

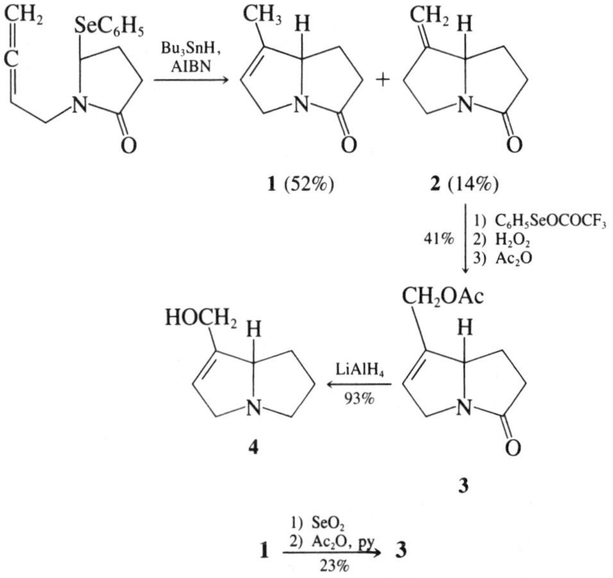

Scheme (I)

[1] D. H. R. Barton and W. B. Motherwell, *Pure Appl. Chem.*, **51**, 15 (1981).
[2] S. Iacono and J. Rasmussen, *Org. Syn.*, submitted (1983).
[3] N. Ono, H. Miyake, and A. Kaji, *J. Org.*, **49**, 4997 (1984).
[4] M. Barbier, D. H. R. Barton, M. Devys, and R. S. Topgi, *J.C.S. Chem. Comm.*, 843 (1984).
[5] B. Giese, J. A. González-Goméz, and T. Witzel, *Angew. Chem., Int. Ed.*, **23**, 69 (1984).
[6] B. Giese and J. Dupuis, *Angew. Chem., Int. Ed.*, **22**, 622, 753 (1983); *idem, Org. Syn.*, submitted (1984).
[7] G. Stork, R. Mook, Jr., S. A. Biller, and S. D. Rychnovsky, *Am. Soc.*, **105**, 3741 (1983).
[8] O. Moriya, M. Okawara, and Y. Ueno, *Chem. Letters*, 1437 (1984).
[9] M. Ladlow and G. Pattenden, *Tetrahedron Letters*, **25**, 4317 (1984).
[10] G. Stork and R. Mook, Jr., *Am. Soc.*, **105**, 3720 (1983).
[11] M. V. Marinovic and H. Ramanathan, *Tetrahedron Letters*, **24**, 1871 (1983).
[12] H. Nishiyama, T. Kitajima, M. Matsumoto, and K. Itoh, *J. Org.*, **49**, 2298 (1984).
[13] C.-P. Chuang and D. J. Hart, *ibid.*, **48**, 1782 (1983).
[14] D. A. Burnett, J.-K. Choi, D. J. Hart, and Y.-M. Tsai, *Am. Soc.*, **106**, 8201 (1984).

Tri-*n*-butyltinlithium, $(n\text{-}C_4H_9)_3SnLi$ (1), **8**, 495–497; **10**, 413–414; **11**, 551–552. The reagent can be prepared by reaction of $(n\text{-}C_4H_9)_3SnCl$ with excess granular lithium in THF.

Desulfonylation.[1]   The reagent undergoes conjugate addition to α,β-unsaturated sulfones. The resulting β-tri-*n*-butyltin sulfones undergo β-elimination to alkenes in the presence of silica gel or when heated in xylene.

Examples:

$$C_6H_5SO_2CH{=}CH(CH_2)_2C_6H_5 \xrightarrow{1,\ THF} C_6H_5SO_2CH_2\underset{\underset{Sn(C_4H_9)_3}{|}}{C}H(CH_2)_2C_6H_5$$

$$\xrightarrow[\substack{85\% \\ overall}]{SiO_2} CH_2{=}CH(CH_2)_2C_6H_5$$

$$C_6H_5SO_2CH{=}CHC_{10}H_{21}\text{-}n \xrightarrow[67\%]{\substack{1)\ \mathbf{1} \\ 2)\ CH_3I}} C_6H_5SO_2\underset{\underset{CH_3}{|}}{C}H{-}\overset{\overset{Sn(C_4H_9)_3}{|}}{C}H(CH_2)_9CH_3$$

$$(E/Z = 1.6:1)$$

$$\xrightarrow{SiO_2} CH_3CH{=}CH(CH_2)_9CH_3$$

$$(E/Z = 1:1.1)$$

[1] M. Ochiai, T. Ukita, and E. Fujita, *J.C.S. Chem. Comm.*, 619 (1983).

Tri-*n*-butyltinlithium–Diethylaluminum chloride.

Reformatsky-type reaction.[1]   Reaction of an α-bromo aldehyde, ketone, or ester with the reagent (**1**) formed from Bu_3SnLi and $(C_2H_5)_2AlCl$ provides an aluminum enolate that reacts with a ketone or aldehyde to afford a β-hydroxy aldehyde or ketone. The yields are generally higher in a reaction catalyzed by $Pd[P(C_6H_5)_3]_4$.

Examples:

$$C_2H_5OCCH_2Br + C_6H_5CH{=}CHCHO \xrightarrow[75\%]{} C_2H_5OCCH_2CHCH{=}CHC_6H_5$$

[1] S. Matsubara, N. Tsuboniwa, Y. Morizawa, K. Oshima, and H. Nozaki, *Bull. Chem. Soc. Japan*, **57**, 3242 (1984).

Tri-*n*-butyltin trifluoromethanesulfonate, $Bu_3SnOSO_2CF_3$ (**1**). The triflate, m.p. 41–43°, is prepared by reaction of tri-*n*-butyltin oxide with triflic anhydride (93% yield) or by reaction of Bu_3SnH with CF_3SO_3H.

(Z)-Disubstituted alkenes.[1] The (Z)-vinylstannane **2** is readily prepared by hydroalumination of propargyl alcohol followed by stannylation with tri-*n*-butyltin triflate. Another route to (Z)-vinylstannanes is shown in equation (II). The bifunctional **2** is useful for iterative coupling to (Z)-polyolefins. Thus **3** is converted into the vinyl Grignard

$$\text{(I)} \quad HC{\equiv}CCH_2OH \xrightarrow[80\%]{\substack{1)\ \text{LiAlH}_4 \\ 2)\ Bu_3SnOTf\ (1)}} Bu_3Sn\diagup{=}\diagdown CH_2OH$$

2

$$\text{(II)} \quad CH{\equiv}CH + (n\text{-}C_5H_{11})_2CuLi \longrightarrow \left(C_5H_{11}\diagup{=}\diagdown\right)_2CuLi \xrightarrow[\substack{90\% \\ \text{overall}}]{1} C_5H_{11}\diagup{=}\diagdown SnBu_3$$

3

reagent **4** by successive reaction with *n*-BuLi and $MgBr_2$. This product in the presence of Li_2CuCl_4 (**4**, 163–164; **5**, 226) couples with the acetate (**5**) of **2** to provide the (Z,Z)-1,4-diene **6**. This diene is then coupled with **5** by an identical procedure to give the (Z,Z,Z)-triene **7** in 89% yield. The triene was used for a synthesis of 5,6-dehydroarachidonic acid.

$$C_5H_{11}\diagdown MgBr + AcO\diagdown\diagup SnBu_3 \xrightarrow[89\%]{\substack{Li_2CuCl_4, \\ THF}} C_5H_{11}\diagdown\diagup\diagdown SnBu_3$$

4   **5**   **6**

$$\xrightarrow[89\%]{5}$$

7

(Z)-Disubstituted cyclopropanes.[2]   The (Z)-vinylstannane **2** (above) reacts with the Simmons-Smith reagent to give the *cis*-1,2-disubstituted cyclopropane **3** in 70% yield. The product is useful for preparation of optically active cyclopropanes.

3

Hydrostannation.[3]   This triflate is a very effective catalyst for the reduction of aldehydes and ketones with Bu_3SnH at room temperature in benzene or dichloroethane. Yields (85%–98%) are generally superior to those obtained with Bu_3SnH and AIBN or Bu_3SnH and CH_3OH. The stereochemistry is similar to that observed with Bu_3SnH–AIBN. Aldehydes can be reduced with high selectivity in the presence of a methyl ketone. Esters and ketals are not reduced.

In the presence of $Pd[P(C_6H_5)_3]_4$ or $Cl_2Pd[P(C_6H_5)_3]_2$ this system can reduce α,β-enones to saturated ketones, but generally $ZnCl_2$ is a more effective catalyst than **1**.

[1] E. J. Corey and T. M. Eckrich, *Tetrahedron Letters*, **25**, 2419 (1984).
[2] *Idem, ibid.*, **25**, 2415 (1984).
[3] Y. T. Xian, P. Four, F. Guibé, and G. Balavoine, *Nouv. J. Chim.*, **8**, 611 (1984).

Tri-μ-carbonylhexacarbonyldiiron, 1, 259–260; **2**, 139–140; **3**, 101; **4**, 157–158; **5**, 221–224; **6**, 195–198; **7**, 110–111; **8**, 498–499; **9**, 477–478.

1,2,4-Triazines.[1]   Fused 1,2,4-triazines are formed regioselectively by cocyclization of adiponitrile and a nitrile catalyzed by $Fe_2(CO)_9$ (equation I).

$$(I)$$

Deoximiation.[2] Oximes, oximic ethers, or oxime O-acetates can be converted into the corresponding ketones by reaction with $Fe_2(CO)_9$ in CH_3OH at 60°. This reaction proceeds particularly readily and in good yield with oxime acetates.

[1] E. R. F. Gesing, U. Groth, and K. P. C. Vollhardt, *Synthesis,* 351 (1984).
[2] M. Nitta, I. Sasaki, H. Miyano, and T. Kobayashi, *Bull. Chem. Soc. Japan,* **57**, 3357 (1984).

Trichloroacetonitrile, 1, 1194–1195; **5**, 686; **6**, 604–605; **7**, 381–382.

cis-Oxyamination.[1] A trichloromethyl imidate such as **1**, prepared by reaction of the corresponding allylic alcohol with CCl_3CN (NaH), on treatment with iodonium di-*sym*-collidine perchlorate (**10**, 212–213; **11**, 269) is converted into an oxazoline (**2**). Reduction of **2** with tri-*n*-butyltin hydride provides methyl N-acetylristosaminide (**3**). The overall sequence provides an attractive route to *cis*-hydroxyamino sugars.

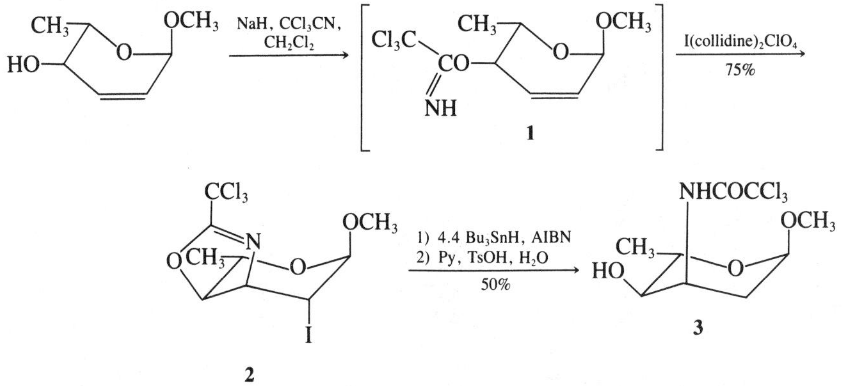

α-Amino acids.[2] α-Amino acids can be prepared from a primary allylic alcohol by rearrangement of the trichloroacetimidate (**6**, 604–605) to a protected primary allylic amine, which is oxidized by ruthenium tetroxide (**11**, 462–463, Sharpless conditions) to a protected α-amino acid.

Example:

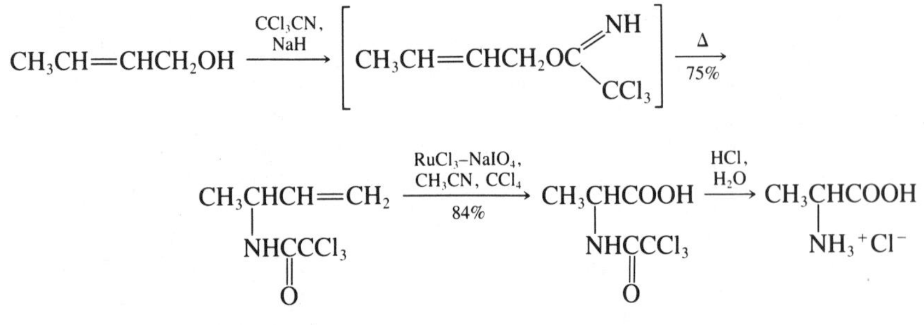

[1] H. W. Pauls and B. Fraser-Reid, *J. Org.*, **48**, 1392 (1983).
[2] S. Takano, M. Akiyama, and K. Ogasawara, *J.C.S. Chem. Comm.*, 770 (1984).

Trichloro(methyl)silane–Sodium iodide, 11, 553–554. This *in situ* equivalent of iodotrimethylsilane is also effective for cleavage of esters and lactones, selective conversion of tertiary and benzylic alcohols into iodides, dehalogenation of α-halo ketones, deoxygenation of sulfoxides, and conversion of dimethyl acetals to carbonyl compounds.[1]

[1] G. A. Olah, A. Husain, B. P. Singh, and A. K. Mehrotra, *J. Org.*, **48**, 3667 (1983).

Triethyl phosphonoacetate, 1, 1216–1217; **6**, 612.

Wittig-Horner reaction; α,β-unsaturated esters.[1] The reaction of triethyl phosphonoacetate (**1**) with aldehydes can be carried out in water using potassium carbonate at 20° or potassium hydrogen carbonate at 100°.

Examples:

$$(C_2H_5O)_2\overset{\overset{O}{\|}}{P}CH_2COOC_2H_5 + HCHO \xrightarrow[77\%]{K_2CO_3, H_2O, 20°} H_2C=C\overset{CH_2OH}{\underset{COOC_2H_5}{\diagdown}}$$

1

$$\textbf{1} + CH_3CH=CHCHO \xrightarrow[82\%]{} CH_3CH=CHCH=CHCOOC_2H_5$$

(E)

Similar conditions can be used in the reaction of diethyl 2-ketoalkanephosphonates (**2**) with aldehydes for a synthesis of α,β-unsaturated ketones.

Example:

$$(C_2H_5O)_2\overset{\overset{O}{\|}}{P}CH_2COCH_3 + n\text{-}C_7H_{15}CHO \xrightarrow[87\%]{K_2CO_3, H_2O, 20°} n\text{-}C_7H_{15}CH=CH\overset{\overset{O}{\|}}{C}CH_3$$

2   (E)

[1] J. Villieras and M. Rambaud, *Synthesis,* 300 (1983).

Triethylsilyl perchlorate.

Silylketene ketals.[1] Reaction of esters with triethylsilyl perchlorate and diisopropylethylamine or 2,2,6,6-tetramethylpiperidine provides (Z)-silylketene ketals selectively (equation I). The selectivity is higher than that obtained by deprotonation with LDA/HMPT followed by silylation.

$$(I) \quad R^1CH_2COR^2 + (C_2H_5)_3SiOClO_3 \longrightarrow \underset{H}{\overset{R^1}{\diagdown}}C{=}C\underset{OR^2}{\overset{OSi(C_2H_5)_3}{\diagup}}$$

$$(Z/E = 92{-}99 : 8{-}1)$$

[1] C. S. Wilcox and R. E. Babston, *Tetrahedron Letters*, **25**, 699 (1984).

Triethyl 1,2,4-triazine-3,5,6-tricarboxylate (1).
  Preparation:[1]

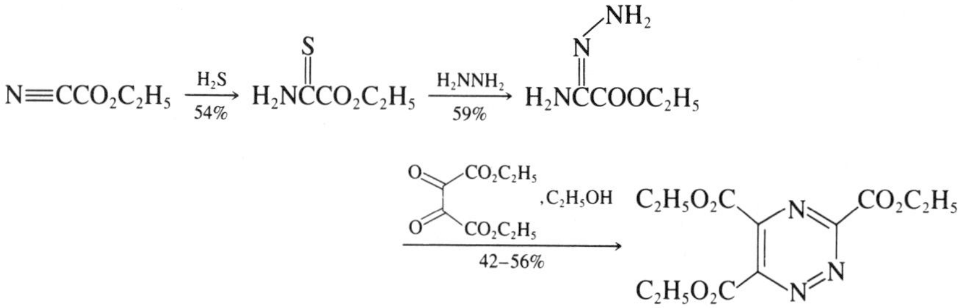

$$N{\equiv}CCO_2C_2H_5 \xrightarrow[54\%]{H_2S} H_2N\overset{S}{\overset{\|}{C}}CO_2C_2H_5 \xrightarrow[59\%]{H_2NNH_2} H_2N\overset{\overset{\displaystyle N{\diagup}NH_2}{\|}}{C}COOC_2H_5$$

Pyridine synthesis (**10**, 49).   This triazine undergoes ready cycloaddition with pyrrolidine enamines or electron-rich alkenes with loss of N_2 and aromatization to give pyridine derivatives (equation I).[2]

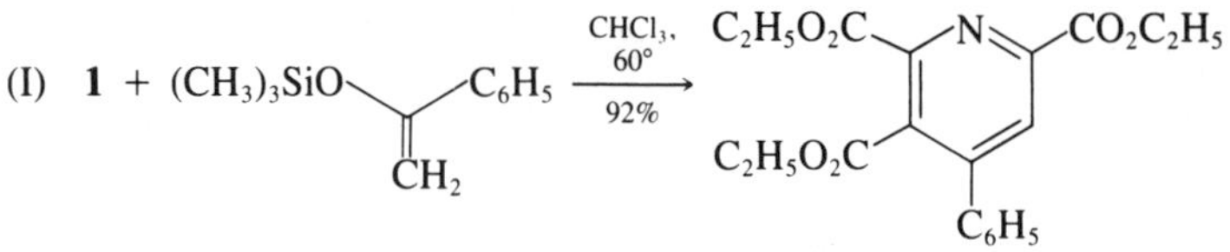

This reaction was investigated particularly as a regioselective route to the pyridyl biaryl CD ring system of the antitumor antibiotic streptonigrin (**5**). Thus the reaction of the 1,2,4-triazine **2** with the morpholino enamine **3** gives the desired tetracyclic adduct **4** together with some of the adduct formed with reverse regioselectivity. This product was converted in several steps into a known precursor of the natural product.

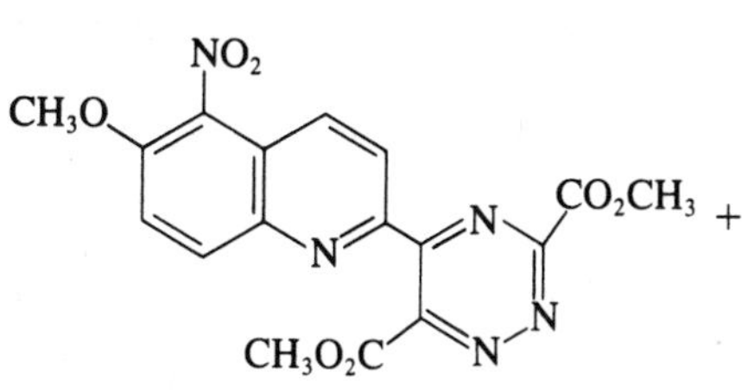

2

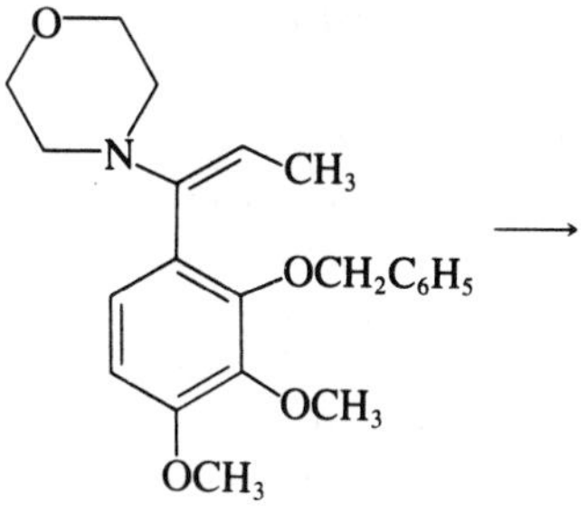

3

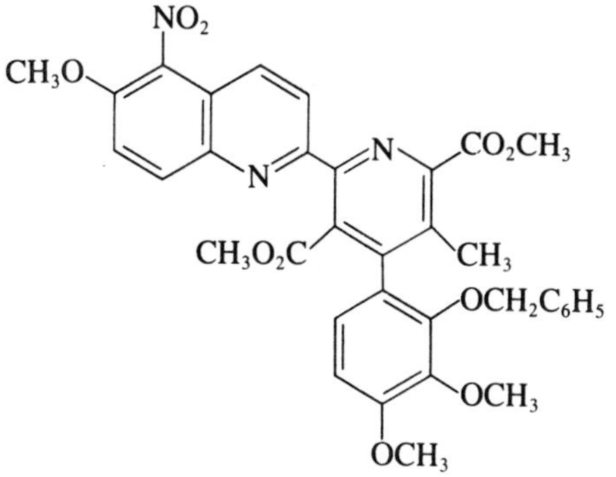

4

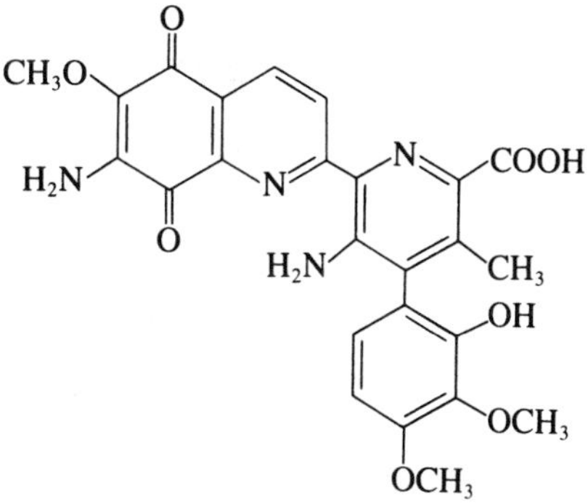

5

[1] D. L. Boger, J. S. Panek, and M. Yosuda, *Org. Syn.*, submitted (1984).
[2] D. L. Boger and J. S. Panek, *J. Org.*, **48**, 621 (1983); *idem, Am. Soc.*, **106**, 5745 (1985).

Trifluoroacetic acid, 1, 1219–1221; **2**, 433–434; **3**, 305–308; **4**, 530–532; **5**, 695–700; **6**, 613–615; **7**, 388–389; **8**, 503; **9**, 483; **10**, 418; **11**, 557–559.

Modified Curtius rearrangement. Acyl azides, preferably prepared *in situ* under phase-transfer conditions by reaction of acid chlorides and NaN_3, are converted to trifluoroacetamides by reaction with CF_3COOH in refluxing CH_2Cl_2 solution. These products are cleaved to primary amines under mild conditions (equation I).[1]

$$(I) \quad RCOCl + NaN_3 \xrightarrow[\text{H}_2\text{O, CH}_2\text{Cl}_2]{n\text{-Bu}_4\text{NBr,}} \left[RCON_3 \right] \xrightarrow[\text{80–97\%}]{\substack{\text{CF}_3\text{COOH,} \\ \text{CH}_2\text{Cl}_2}} RNHCOCF_3 \xrightarrow[\text{H}_2\text{O, CH}_3\text{OH}]{\text{K}_2\text{CO}_3,} RNH_2$$

[1] J. R. Pfister and W. E. Wymann, *Synthesis*, 38 (1983).

Trifluoroacetic anhydride, 1, 1221–1226; **3**, 308; **5**, 701; **6**, 616–617; **7**, 389–390; **9**, 484.

Pummerer rearrangement.[1]   Thioanisoles are readily prepared, but the drastic conditions required for deprotection limit their use as a protecting group for thiophenols. The conversion can be carried out readily by a three-step procedure without purification in high yield. Oxidation with 1 equiv. of *m*-chloroperbenzoic acid affords the essentially pure sulfoxide, which undergoes a regiospecific Pummerer rearrangement in neat trifluoroacetic anhydride at 40° to afford a hemithioacetal trifluoroacetate. This product is readily hydrolyzed by methanol and triethylamine to the arylthiol (equation I).

(I)

[1] R. N. Young, J. Y. Gauthier, and W. Coombs, *Tetrahedron Letters*, **25**, 1753 (1984).

Trifluoroacetyl nitrate, CF_3COONO_2 (**1**). The reagent is prepared *in situ* from ammonium nitrate (NH_4NO_3) and trifluoroacetic anhydride.

Nitration and oxidation.   This nitrate (**1**) is a more efficient and versatile nitrating agent than the well-known acetyl nitrate, generated *in situ* from acetic anhydride and nitric acid. Oxidation, rather than nitration, is observed with phenols, which are converted mainly into diphenoquinones or *o*- or *p*- benzoquinones, depending on the substitution.[1]

The reagent has recently been used to effect hydroxylation of an intermediate (**2**) in a synthesis of anthracylines. The reaction of **1** with **2** results in hydroxylation and nitration to give **3** and **4** in the ratio 3:2. The nitro compound (**4**) is converted into **3** in 75% yield by hydrogenation and diazotization.[2]

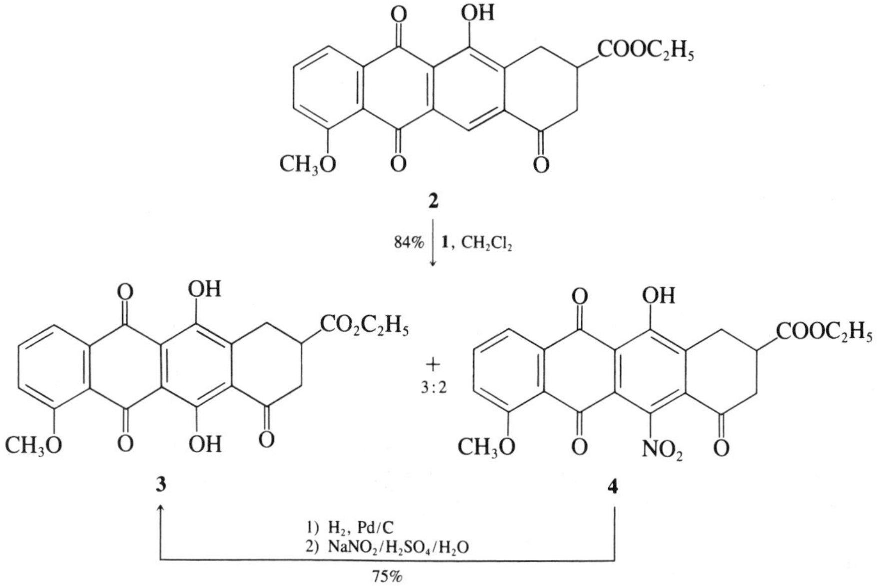

2

84% $\Big|$ **1**, CH$_2$Cl$_2$

+

3:2

3   **4**

1) H$_2$, Pd/C
2) NaNO$_2$/H$_2$SO$_4$/H$_2$O

75%

Nitration of enol acetates; α-nitro ketones.[3]   This reagent is more useful than acetyl nitrate (Ac$_2$O/HNO$_3$) for nitration of enol acetates of cyclohexanones, particularly those of 2-alkylcyclohexanones, which can undergo ring cleavage. However, yields are low unless the 2-substituent is a primary alkyl group.

Example:

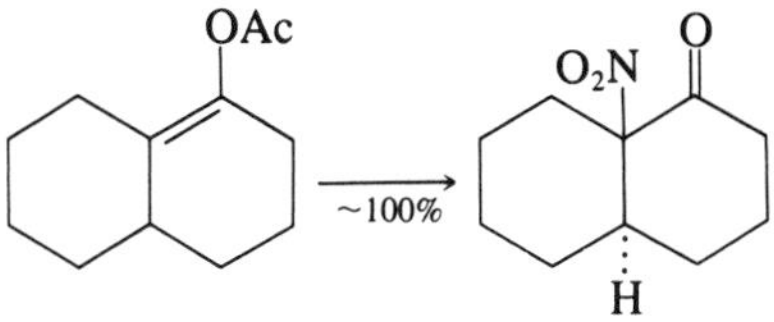

~100%

[1] J. V. Crivello, *J. Org.*, **46**, 3056 (1981).
[2] R. K. Boeckman, Jr., and S. H. Cheon, *Am. Soc.*, **105**, 4112 (1983).
[3] P. Dampawan and W. W. Zajac, Jr., *Synthesis*, 545 (1983).

Trifluoromethanesulfonic acid, 4, 533; **5**, 701–702; **6**, 617–618; **8**, 504; **9**, 485.

Intramolecular cyclization of tetraenes.[1]   The tetraene **1** does not cyclize when heated at 140°, but cyclizes rapidly to **3** in high yield in the presence of this sulfonic acid at $-23°$. The tetraene **2** does undergo thermal cyclization to a $2:1$ mixture of **4** and **5**, but acid-catalyzed cyclization is more efficient. Although the radical cation tris(p-bromo-

1, R = CH$_3$	88%	**3**
2, R = C(CH$_3$)$_3$	88%	**4** 65:23 **5**

phenyl)aminium hexachlorostibnate (**10**, 452) catalyzes the cyclization of **1** and **2**, the evidence suggests that it is functioning merely as a proton source. This tetraene cyclization is considered to involve protonation to produce an allylic cation, which undergoes intramolecular addition to the diene to form a new allylic cation, which then undergoes a second cyclization and loss of a proton.

Cyclodehydration of arylpropionic and -butanoic acids.[2]   This cyclization to 1-indanones and 1-tetralones can be effected in a one-pot reaction via the mixed anhydride formed with triflic acid (**4**, 533–534).

Example:

Review.[3]   This review covers not only trifluoromethanesulfonic acid (triflic acid), but also the anhydride, trimethylsilyl triflate, triflate salts, and alkyl and vinyl triflates (232 references).

[1] P. G. Gassman and D. A. Singleton, *Am. Soc.*, **106**, 6085 (1984).
[2] B. Hulin and M. Koreeda, *J. Org.*, **49**, 207 (1984).
[3] P. J. Stang and M. R. White, *Aldrich. Acta*, **16**, 15 (1983).

Trifluoromethanesulfonic anhydride, 4, 533–534; **5**, 702–705; **6**, 618–620; **7**, 390; **10**, 919–920; **11**, 560–562.

Aminodeoxy sugars.[1]   These sugars are prepared readily by reaction of primary or secondary sugar triflates with dry gaseous ammonia in absolute chloroform at 50° or in 1,2-dichloroethane at 70°. This method is more convenient than the usual method using azidodeoxy sugars.

[1] A. Malik, N. Afza, M. Roosz, and W. Voelter, *J.C.S. Chem. Comm.*, 1530 (1984).

2,4,6-Triisopropylbenzenesulfonyl hydrazine, 4, 535; **5**, 706; **7**, 392; **9**, 486–488; **10**, 422–423; **11**, 563–566.

Methylenecycloalkanes.[1]   The vinyl anion (**a**) generated from the trisyl hydrazone (**1**) of an ω-chloroalkyl methyl ketone with *sec*-butyllithium (2 equiv.) and TMEDA

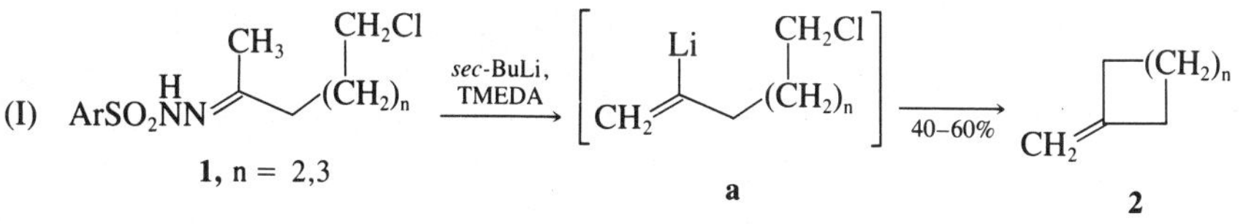

cyclizes to a methylene cycloalkane (**2**, equation I). Cyclization to methylenecycloheptane (n = 4) is effected most efficiently (70% yield) with *t*-butyllithium as base and Li_2CuCl_4 as catalyst. The method also provides a route to bicyclic systems.

Example:

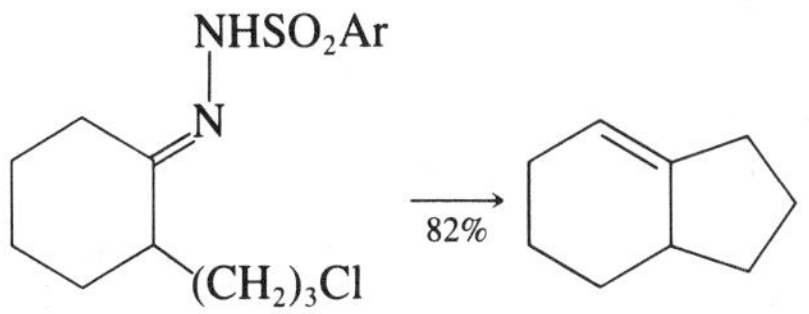

[1] A. R. Chamberlin and S. H. Bloom, *Tetrahedron Letters,* **25**, 4901 (1984).

Trimethylamine N-oxide, 1, 1230–1231; **2**, 434; **3**, 309–310; **6**, 624–625; **7**, 392; **8**, 507; **9**, 488–489; **10**, 423.

Pyrrolidines.[1]   LDA (THF, −78°) converts trimethylamine N-oxide into an ylide, $CH_2=N^+(CH_3)CH_2Li^-$, which undergoes [3 + 2] cycloaddition with simple alkenes to form pyrrolidines.

Examples:

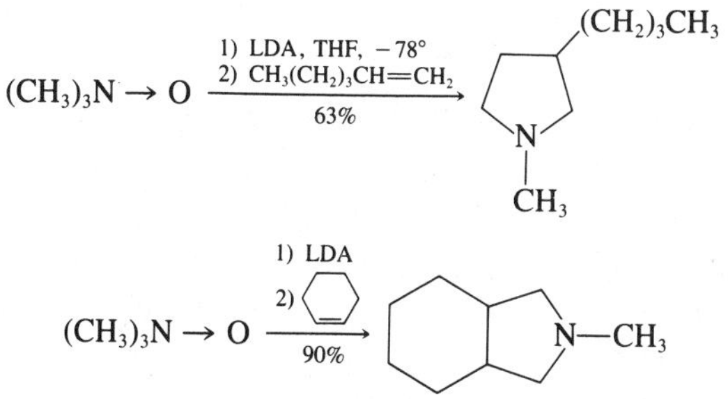

[1] R. Beugelmans, G. Negron, and G. Roussi, *J.C.S. Chem. Comm.,* 31 (1983).

4,4,6-Trimethyl-1,3-oxathiane, (structure with CH$_3$ groups) **(1).**

Stereoselective synthesis of oxathiane tertiary carbinols (**8**, 508–509).[1] This reagent is now preferred over the isomeric 4,6,6-trimethyl-1,3-oxathiane (**8**, 508–509) for a two-step, highly stereoselective synthesis of oxathiane tertiary carbinols (**3**, equation I), which can be cleaved to enantiomerically pure tertiary α-hydroxy aldehydes. The first step, acylation with retention of configuration, gives the more stable equatorial adduct **2**

(I) **1** $\xrightarrow[\text{3) oxid.}]{\substack{\text{1) }n\text{-BuLi}\\ \text{2) RCHO}}}$ (structure **2**) $\xrightarrow{\text{R'M}}$ (structure **3**)

with high stereoselectivity. The diastereoselectivity in the addition of Grignard reagents is generally high (80–100%), and is higher for the 4,4,6-system of **1** than for the 4,6,6-isomer. The stereoselectivity accords with Cram's rule in a chelated system. It depends to some extent on the organometallic reagent, the solvent, and the temperature. By proper choice of conditions, one diastereomer of **3** can usually be obtained in >95% yield.

[1] E. L. Eliel and S. Morris-Natschke, *Am. Soc.,* **106**, 2937 (1984).

N,N,P-Trimethyl-P-phenylphosphinothioic amide, C$_6$H$_5$PCH$_3$ (with S double bond and N(CH$_3$)$_2$) **(1), 11,** 569–570.

Methylenation.[1] In a synthesis of the two guaianolide sesquiterpenes dehydrocostus lactone (**4**) and estafiatin (**6**), usual methods for methylenation of the ketone group of **2** resulted only in β-elimination. However, use of the anion of **1** effected the desired reaction, albeit in modest (15%) yield.

[1] J. H. Rigby and J. Z. Wilson, *Am. Soc.,* **106**, 8217 (1984).

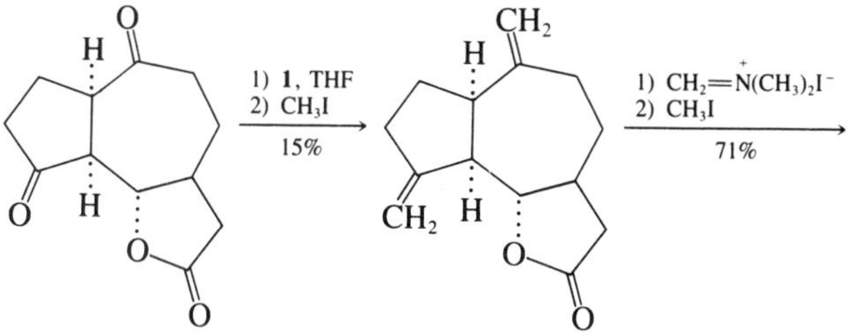

Trimethyl(phenylthio)silane, 10, 426–427.

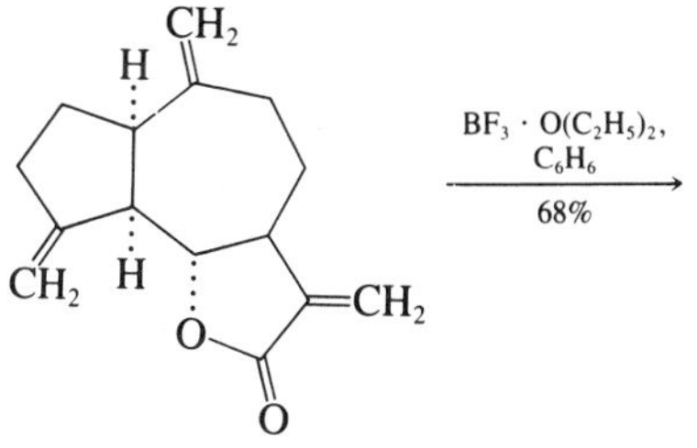

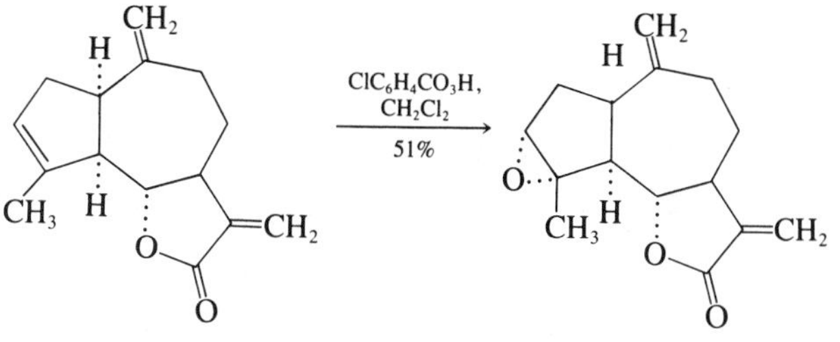

O-Glycosides.   Thioglycosides, prepared by Hanessian's method or with $C_6H_5SSi(CH_3)_3$ and trimethylsilyl triflate, can be converted into O-glycosides by reaction with NBS in the presence of the hydroxy component. The stereochemical outcome (α/β ratio) is independent of the stereochemistry of the thioglycoside, but is controlled by the solvent used. Reactions in CH_2Cl_2 usually result in α/β ratios of 1:1 to 2:1; the same reactions in CH_3CN result in α/β ratios of 3:1 to 9:1. This methodology is equally applicable to intramolecular glycosidation.[1]

Examples:

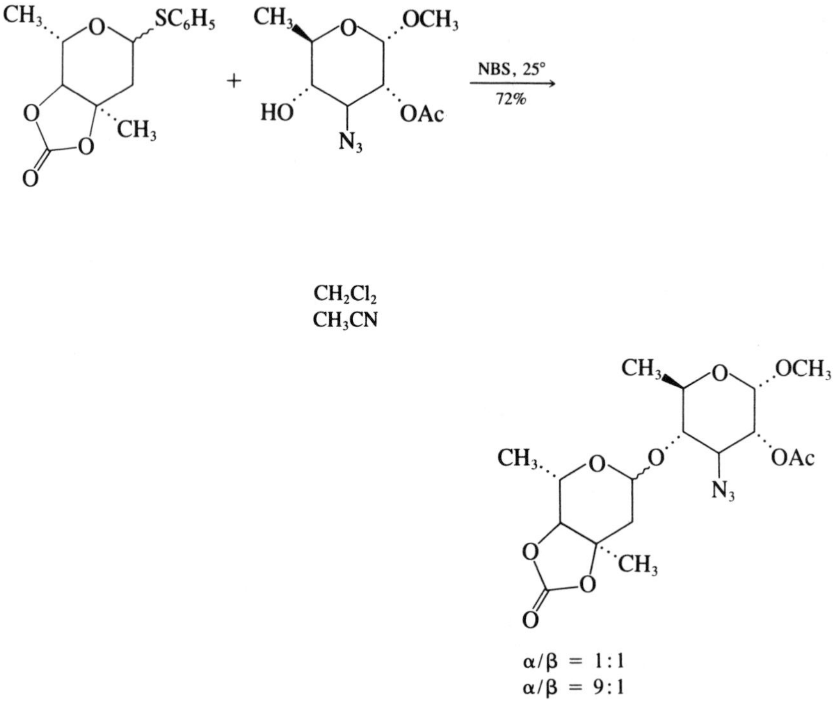

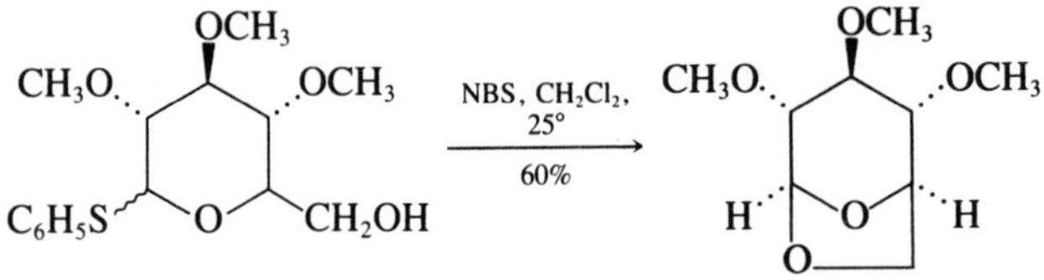

$$CH_2Cl_2$$
$$CH_3CN$$

$$\alpha/\beta = 1:1$$
$$\alpha/\beta = 9:1$$

[1] K. C. Nicolaou, S. P. Seitz, and D. P. Papahatjis, *Am. Soc.*, **105**, 2430 (1983).

Trimethyl phosphite, 1, 1233–1235; **2**, 439–441; **3**, 515–516; **4**, 591–592; **5**, 717; **7**, 393–399; **9**, 490–491; **11**, 570.

Reductive phosphorylation of **p-**benzoquinones (**1**, 1233).[1]   This reaction is generally improved by use of $P(OCH_3)_3/ClSi(CH_3)_3$, which provides *p*-hydroxyphenyl phosphates after methanolysis. Moreover, C-phosphorylation occurs to a minor extent with this modification. This reductive phosphorylation coupled with selective cleavage of the phosphoryl group with $BrSi(CH_3)_3$ (**9**, 74) provides a route to monoprotected hydroquinones from monosubstituted *p*-benzoquinones.

Example:

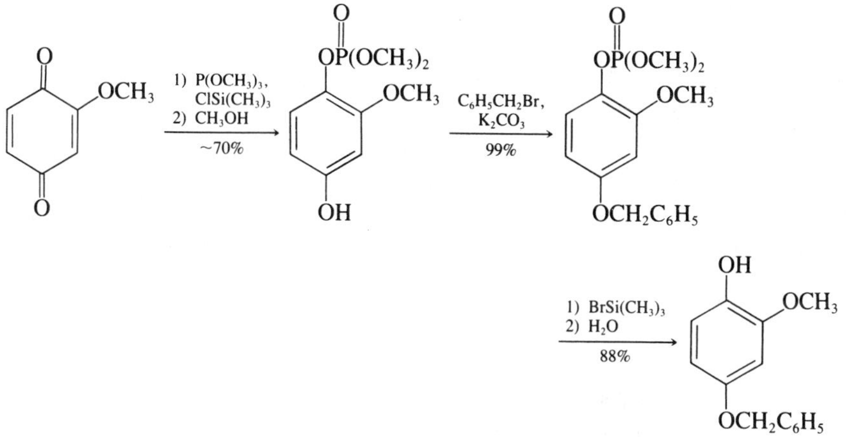

[1] R. O. Duthaler, P. A. Lyle, and C. Heuberger, *Helv.*, **67**, 1406 (1984).

Trimethylphosphonoglycolate, $(CH_3O)_2\overset{\overset{O}{\|}}{P}CH(OH)COOCH_3$ **(1).** The phosphonate is prepared by reaction of dimethyl phosphite with methyl glyoxalate (82% yield).[1,2]

α-Keto esters.[2] Protected derivatives of **1** are useful for conversion of aldehydes to the two-carbon homologated α-keto esters. The most useful derivative is the trichloro-*t*-butyloxy carbonate (**8**, 499) because of its ready reductive elimination with zinc (equation I). This process is useful in the case of base-sensitive and, in particular, chiral aldehydes.

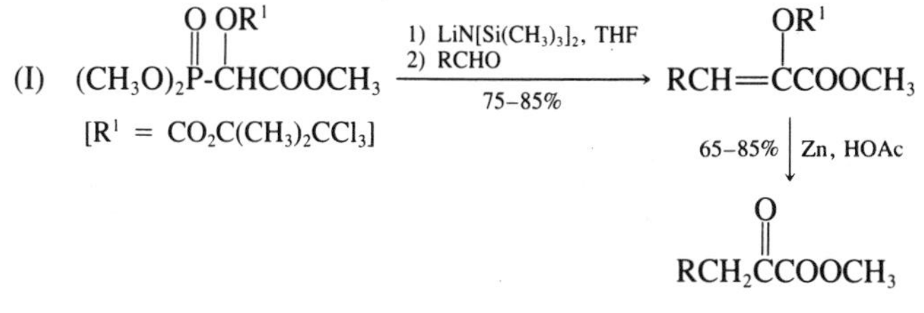

[1] E. Nakamura, *Tetrahedron Letters*, **22**, 663 (1981).
[2] D. Horne, J. Gaudino, and W. J. Thompson, *ibid.*, **25**, 3529 (1984).

4-Trimethylsilyl-2-butene-1-ol, $HOCH_2CH=CHCH_2Si(CH_3)_3$ (**1**).
Preparation:

$$HC\equiv CCH_2Si(CH_3)_3 \xrightarrow[76\%]{\substack{1)\ BuLi \\ 2)\ (CH_2O)_n}} HOCH_2C\equiv CCH_2Si(CH_3)_3 \xrightarrow[68\%]{LiAlH_4} \mathbf{1}$$

Protection of carboxylic acids.[1]   Esters of this alcohol are converted in the presence of catalytic amounts of $Pd[P(C_6H_5)_3]_4$ into butadiene and trimethylsilyl esters, which are readily hydrolyzed by water or an alcohol. This protecting group is thus useful for protection of highly functionalized and sensitive acids. The same procedure can be used for deprotection of carbonates or carbamates containing this unit.

[1] H. Mastalerz, *J. Org.*, **49**, 4092 (1984).

Trimethylsilyldiazomethane, 10, 431; **11**, 573–574, 580. A convenient synthesis involves reaction of trimethylsilyl trifluoromethanesulfonate with an ethereal solution of diazomethane and diisopropylethylamine (74% yield).[1]

O-Methylation.[2]   Phenols are converted into methyl ethers by reaction with this reagent and ethyldiisopropylamine in CH_3OH/CH_3CN. Under the same conditions, readily enolizable ketones are converted into enol methyl ethers.

[1] M. Martin, *Syn. Comm.*, **13**, 809 (1983).
[2] T. Aoyama, S. Terasawa, K. Sudo, and T. Shioiri, *Chem. Pharm. Bull. Japan*, **32**, 3759 (1984).

Trimethylsilyllithium, 7, 400; **9**, 493; **11**, 575.

Allyltrimethylsilanes.   Trimethylsilyllithium reacts with primary (E)-allylic halides to form the corresponding allyltrimethylsilanes. Reaction of the reagent prepared from $(CH_3)_3SiLi$ and CuI in HMPT with these halides results in the isomeric 3-trimethylsilyl-1-alkene.[1]

Example:

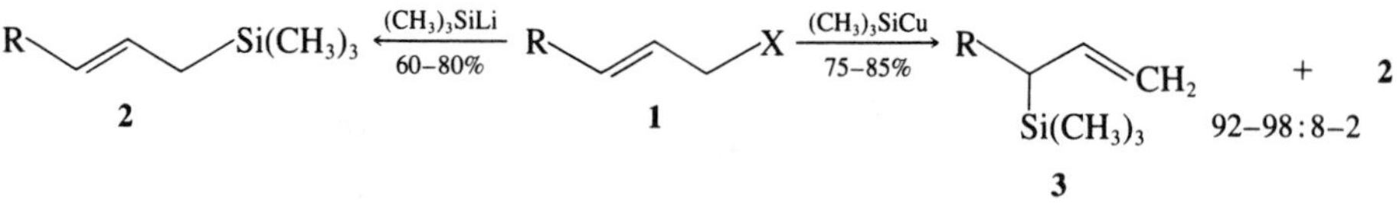

[1] J. G. Smith, S. E. Drozda, S. P. Petraglia, N. R. Quinn, E. M. Rice, B. S. Taylor, and M. Viswanathan, *J. Org.*, **49**, 4112 (1984).

Trimethylsilylmethyl azide, $(CH_3)_3SiCH_2N_3$ (**1**). B.p. 58–61°/80 mm. The azide is obtained in high yield by reaction of NaN_3 with $(CH_3)_3SiCH_2Cl$ in HMPT or DMF. It is a useful, safe substitute for methyl azide, b.p. 20°, which is explosive.

[3 + 2] Cycloaddition.[1]   The azide reacts with acetylenes to form 1(H)-trimethyl-silylmethyltriazoles (equation I).

(I) 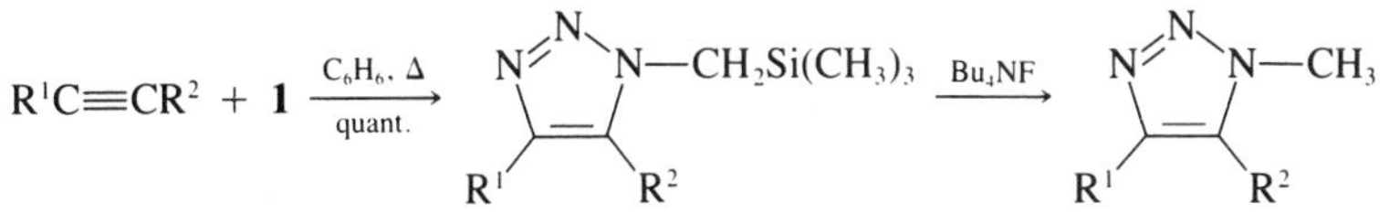 $R^1C{\equiv}CR^2$ + 1 $\xrightarrow[\text{quant.}]{C_6H_6,\ \Delta}$

Amination.[2]   The azide reacts with aryl Grignard reagents at room temperature to form an intermediate, probably a triazene, that is hydrolyzed to an aniline in 70–90% yield. Aryllithium compounds can also be used, but yields are generally lower.

[1] O. Tsuge, S. Kanemasa, and K. Matsuda, *Chem. Letters*, 1131 (1983).
[2] K. Nishiyama and N. Tanaka, *J.C.S. Chem. Comm.*, 1322 (1983).

2-Trimethylsilylmethyl-1,3-butadiene (Isoprenyltrimethylsilane) (1), 9, 493–494; **10,** 432–433.
   Preparation:

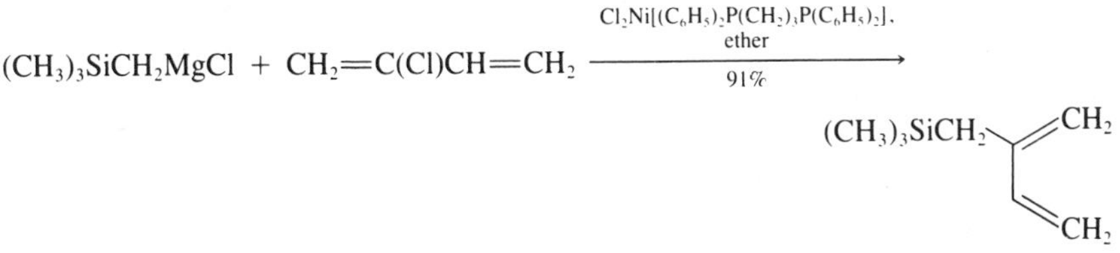

$(CH_3)_3SiCH_2MgCl$ + $CH_2{=}C(Cl)CH{=}CH_2$ $\xrightarrow[91\%]{\substack{Cl_2Ni[(C_6H_5)_2P(CH_2)_3P(C_6H_5)_2].\\ \text{ether}}}$

1, b.p. 69–70°/80 mm

Isoprenylation of carbonyl compounds.[1]   In the presence of $TiCl_4$, **1** can react as an allylsilane with acetals and acid chlorides (**7,** 370–371) to give isoprenylated compounds in generally satisfactory yields. Yields with aldehydes tend to be low. Isoprenylation of carbonyl groups is best effected with catalysis with tetrabutylammonium fluoride or iodotrimethylsilane (**10,** 216).
   Examples:

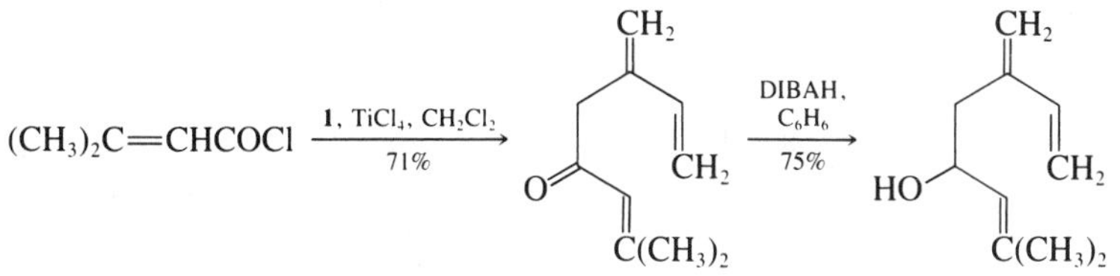

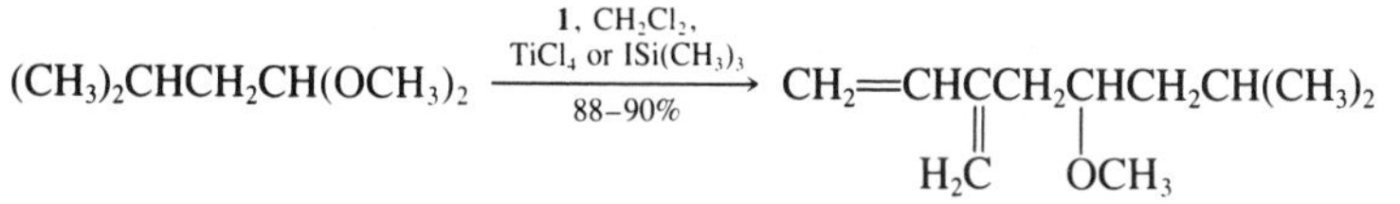

$(CH_3)_2C{=}CHCOCl$ $\xrightarrow[71\%]{1,\ TiCl_4,\ CH_2Cl_2}$

$(CH_3)_2CHCH_2CH(OCH_3)_2$ $\xrightarrow[88-90\%]{\substack{1,\ CH_2Cl_2,\\ TiCl_4\ or\ ISi(CH_3)_3}}$ $CH_2{=}CHCCH_2CHCH_2CH(CH_3)_2$

Diels-Alder reactions (**10**, 432–433).[1]   The "*para*" regioselectivity of Diels-Alder reactions of **1** is improved dramatically by catalytic amounts of aluminum chloride. Examples:

$$CH_2=CH-COOCH_3 \quad \xrightarrow[C_6H_6]{1} \quad (CH_3)_3SiCH_2 \cdots COOCH_3 \quad + \quad (CH_3)_3SiCH_2 \cdots COOCH_3$$

	80°	58%	84:16
	AlCl₃, 50–60°	72%	99:1

$$ \text{(cyclohexenone)} \longrightarrow (CH_3)_3SiCH_2\cdots \quad + \quad (CH_3)_3SiCH_2\cdots $$

	135°	17%	84:16
	AlCl₃, 80°	70%	99:1

The products can be hydrodesilylated by reaction with HCl in CH_3OH or with CsF or KF in aqueous DMSO or DMF. The former reaction occurs with almost complete shift of the bond to give *exo*-methylene products. The latter results in *endo*- and *exo*-methylene compounds in the ratio ~85:15. These reactions provide attractive routes to various terpenes.

[1] H. Sakurai, A. Hosomi, M. Saito, K. Sasaki, H. Iguchi, J. Sasaki, and Y. Araki, *Tetrahedron*, **39**, 883 (1983).

Trimethylsilylmethylmagnesium chloride, 5, 724–725; **6**, 636–637; **10**, 433–434.

α-*Trimethylsilyl ketones.* α-Silyl ketones can be prepared by reaction of α,β-enals with this Grignard reagent (**1**) followed by a rhodium-catalyzed isomerization of the resulting allylic alcohol (equation I).[1]

$$(I) \quad R^1 \underset{OH}{\overset{R^2}{\diagdown}} Si(CH_3)_3 \quad \xrightarrow[65-80\%]{\substack{HRh[P(C_6H_5)_3]_3, \\ \text{dioxane}, 105°}} \quad R^1 \underset{O}{\overset{R^2}{\diagdown}} Si(CH_3)_3$$

These ketones are deprotonated at both α-positions by various lithium amide bases, but are selectively deprotonated α to the silyl group by 1-trimethylsilylhexyllithium (**2**),

$$CH_2=CHSi(CH_3)_3 + n\text{-BuLi} \quad \xrightarrow[-78°]{THF,} \quad BuCH_2\overset{Li}{\underset{}{C}}HSi(CH_3)_3$$

2

prepared from vinyltrimethylsilane and n-BuLi. Enolates obtained with this base undergo a crossed aldol reaction with aldehydes followed by a Peterson-type *syn*-elimination to provide (E)-α,β-unsaturated ketones selectively (equation II).

$$\text{(II)} \quad R^1CH_2\overset{\displaystyle \|}{\underset{\displaystyle O}{C}}CH_2Si(CH_3)_3 \xrightarrow[-78°]{2,\ THF,} R^1CH_2C{=}CHSi(CH_3)_3 \xrightarrow{R^2CHO}$$
$$\underset{OLi}{|}$$

$$\left[R^1CH_2\overset{\displaystyle \|}{\underset{\displaystyle O}{C}}\overset{\displaystyle Si(CH_3)_3}{\overset{\displaystyle |}{C}}\underset{\displaystyle \underset{OLi}{|}}{CH}R^2 \right] \xrightarrow[75-90\%]{} R^1CH_2\overset{\displaystyle \|}{\underset{\displaystyle O}{C}} \diagup\!\!\!\diagdown R^2$$

The synthesis of (E)-7-methyl-4-octene-3-one (**3**) is a typical example of this carbonyl olefination reaction.[2]

$$CH_2{=}CHCHO \xrightarrow[56\%]{\begin{array}{l}\text{1) } \mathbf{1}\\ \text{2) } H_2O\\ \text{3) } HRh[P(C_6H_5)_3]_4\end{array}} CH_3CH_2\overset{\displaystyle \|}{\underset{\displaystyle O}{C}}CH_2Si(CH_3)_3 \xrightarrow[75\%]{\begin{array}{l}\text{1) } \mathbf{2}\\ \text{2) } (CH_3)_2CHCH_2CHO\end{array}}$$

$$(CH_3)_2CHCH_2 \diagup\!\!\!\diagdown \overset{\displaystyle }{\underset{\displaystyle O}{\diagdown}} CH_2CH_3$$

3

[1] S. Sato, H. Okada, I. Matsuda, and Y. Izumi, *Tetrahedron Letters*, **25**, 769 (1984).
[2] I. Matsuda, H. Okada, S. Sato, and Y. Izumi, *ibid.*, **25**, 3879 (1984).

Trimethylsilylmethylpotassium, $KCH_2Si(CH_3)_3$. This base is prepared by reduction of bis(trimethylsilylmethyl)mercury with potassium sand.[1] It is comparable in reactivity to n-BuLi/$KOC(CH_3)_3$ for allylic and benzylic metallation.[2]

Metallation of 7-dehydrocholesterol.[3]   Although 7-dehydrocholesterol has three different allylic positions, deprotonation of the potassium alcoholate **1** with this base followed by carboxylation and subsequent esterification results in the diastereomeric esters **2** as the major products. On equilibration of the mixture, the major isomer (probably the β-isomer) isomerizes to the minor isomer and a new ester believed to be **3**. The selective deprotonation at C_{14} probably results mainly from steric factors.

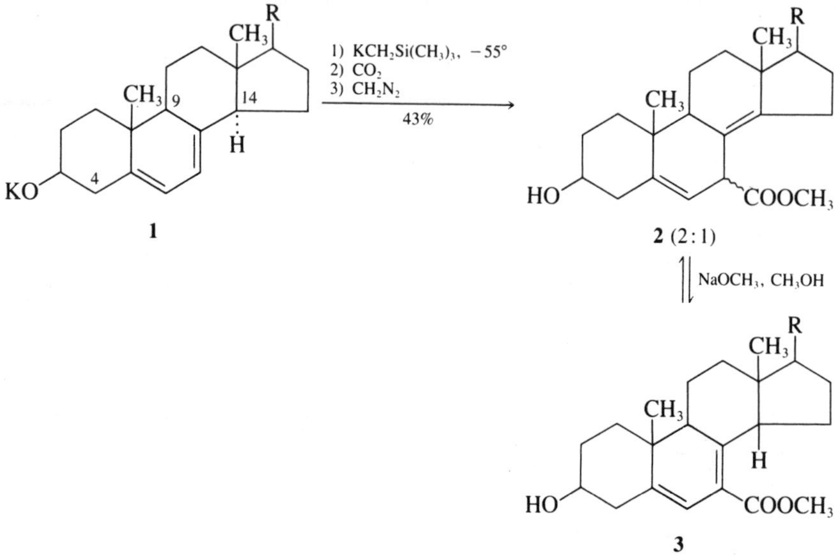

1

2 (2:1)

3

NaOCH₃, CH₃OH

1) KCH₂Si(CH₃)₃, −55°
2) CO₂
3) CH₂N₂

43%

[1] A. J. Hart, D. H. O'Brien, and C. R. Russell, *J. Organometal. Chem.*, **72**, C19 (1974).
[2] J. Hartmann and M. Schlosser, *Helv.*, **59**, 453 (1976).
[3] E. Moret and M. Schlosser, *Tetrahedron Letters*, **25**, 1449 (1984).

Trimethylsilylmethyl trifluoromethanesulfonate, 10, 434–436; **11**, 582.

Pyridinium methylides.[1] Trimethylsilylmethyl triflate converts pyridines into N-(trimethylsilylmethyl)pyridinium triflates in high yield. On treatment with cesium fluoride these products are converted into nonstabilized pyridinium methylides, which can be trapped by cycloaddition to dimethyl acetylenedicarboxylate to form indolizines.

Example:

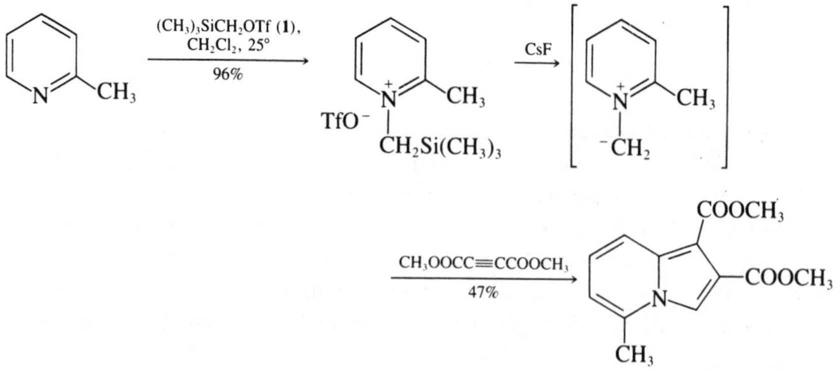

A pyridinium methylide can undergo a novel reaction with an electron-deficient alkene to provide the next higher homolog in which the original double bond is saturated and a methylene group is added. This reaction fails with hindered alkenes, and apparently requires DME as solvent. A [3 + 2] cycloaddition may be the initial step.

Examples:

(50%)   (12%)

[1] O. Tsuge, S. Kanemasa, S. Kuraoka, and S. Takenaka, *Chem. Letters,* 279, 281 (1984).

Trimethylsilyl polyphosphate (PPSE), **10**, 437; **11**, 427.

Alkyl iodides.[1]   Many primary, secondary, and benzylic alcohols are converted into alkyl iodides by reaction with sodium iodide in the presence of PPSE. The reaction proceeds with inversion in the case of secondary alcohols. Yields of 80–98% are often obtainable, but isomerization or elimination can be important side reactions, particularly in the cases of allylic alcohols, 1,2-diols, and tertiary alcohols.

[1] T. Imamoto, T. Matsumoto, T. Kusumoto, and M. Yokoyama, *Synthesis,* 460 (1983).

Trimethylsilyl trifluoromethanesulfonate, 8, 514–515; **9**, 497–498; **10**, 438–441; **11**, 584–587. The triflate can be generated *in situ* by reaction of CF_3SO_2OH with a slight excess of $Si(CH_3)_4$. Methane is formed as the by-product.[1] Another method[2] is formulated in equation (I).

(I) CF_3SO_2OH + ... $\xrightarrow{40°}$ $CF_3SO_2OSi(CH_3)_3$ + ...

(98%)

α-Glycosidation of anthracyclinones.[3]   Anthracyclinones occur in nature as anthracyclins, glycosides of amino sugars. Glycosidation of the aglycones with the 1-halo derivatives of amino sugars by the Koenigs-Knorr reaction or with silver triflate proceeds in low yield and with low stereoselectivity. Optically pure (+)-4-demethoxydaunorubicin (**2**) has been obtained in high yield by reaction of N-trifluoroacetyl-1,4-di-O-*p*-nitroben-

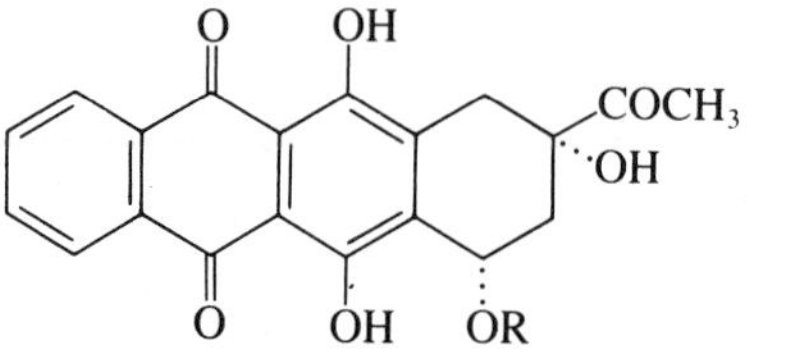

1, R = H
2, R = L-daunosamine (3)

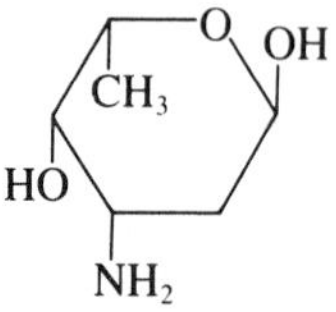

3

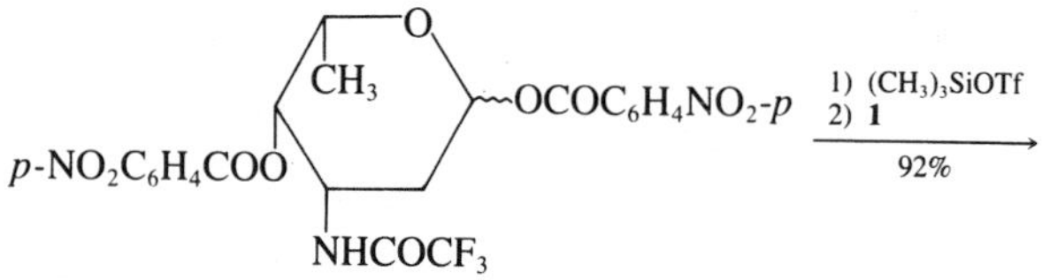

4

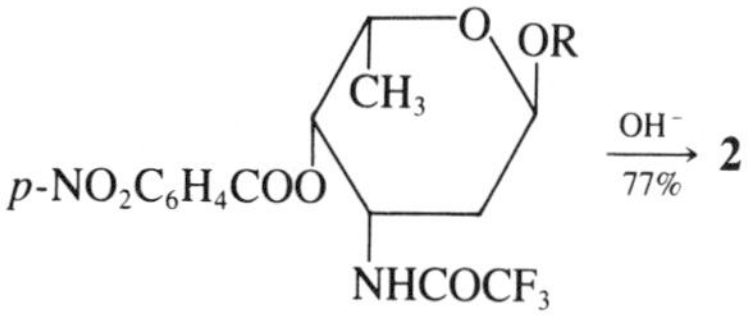

5

zoyl-L-daunosamine (**4**) with trimethylsilyl triflate and **1** to provide the α-glycoside **5**, followed by alkaline hydrolysis of the protecting groups.

C-Allylated glycopyranosides.[4] α-Glycopyranosides or α-glycopyranosyl chlorides are converted stereoselectively into α-C-allylated glycopyranosides on reaction with allylsilanes catalyzed by trimethylsilyl triflate. Iodotrimethylsilane is less effective as the catalyst.

Example:

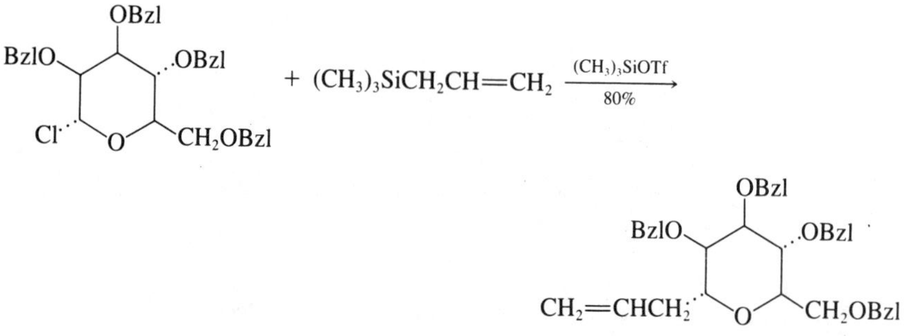

(α, 100%)

Glucuronidation.[5]   Glucuronides can be prepared in moderate yield by reaction of an aliphatic alcohol with methyl 2,3,4-tri-O-acetylglucopyranuronate (**1**) in the presence of $(CH_3)_3SiOTf$ at 0°. β-Glucuronides (**2**) are usually formed exclusively or preferentially with nucleophilic aglycones. Weakly nucleophilic aglycones (*e.g.*, phenols) do not react under these conditions.

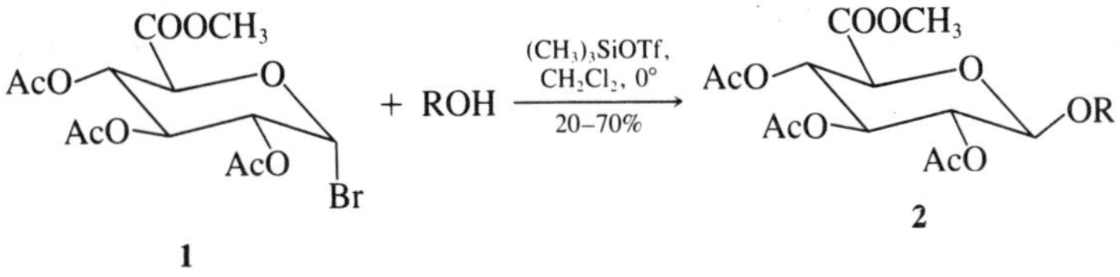

Δ³-Piperidines.[6]   The reaction of various carbon nucleophiles with the 4-hydroxy-1,2,3,4-tetrahydropyridine **1** catalyzed by Lewis acids results in an S_N2'-like reaction to give a 2-substituted-Δ³-piperidine. Trimethylsilyl triflate is a generally effective catalyst, but $SnCl_4$ and $TiCl_4$ are also useful. The reaction with ethanol or thiophenol, however, proceeds without rearrangement.

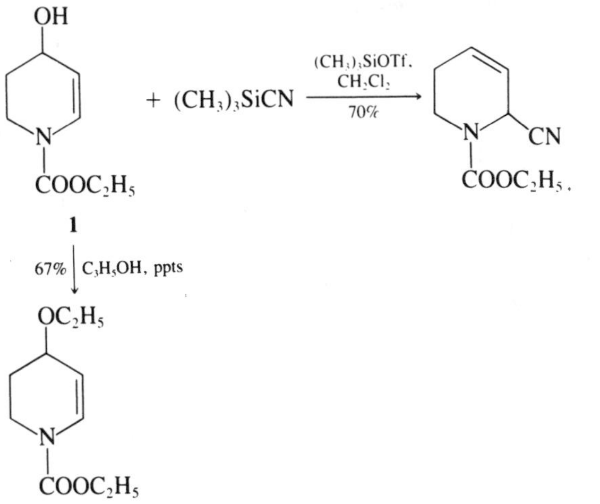

The reaction is reminiscent of the Lewis acid-catalyzed reaction of glycals with nucleophiles to give 2,3-unsaturated glycopyranosides (equation I).[7]

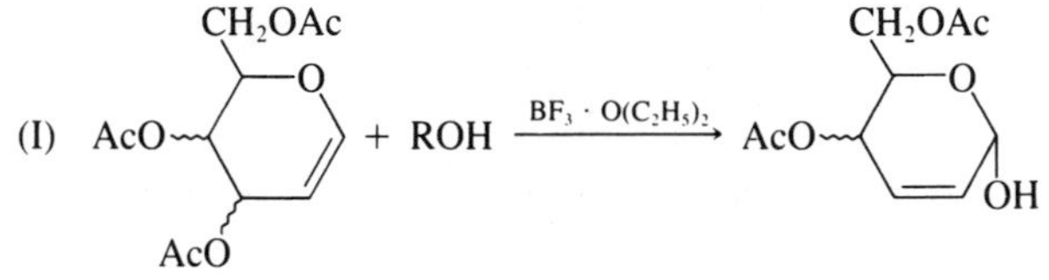

Trimethylsilyl enol ethers.[8]   Kinetic enol ethers can be obtained by trimethylsilyl triflate-catalyzed rearrangement of β-keto silanes, which can be prepared in high yield by reaction of acid chlorides with trimethylsilylmethylcopper.

Example:

$$RCOCl + ClMgCH_2Si(CH_3)_3 \xrightarrow[\text{ether}]{\text{CuI,}} RCCH_2Si(CH_3)_3 \xrightarrow[70-95\%]{\substack{1)\ (CH_3)_3SiOTf,\ pentane \\ 2)\ N(C_2H_5)_3}} RC=CH_2$$

α-Halovinyl sulfides.[9]   α-Halo sulfoxides are converted into α-halovinyl sulfides by reaction with trimethylsilyl triflate (2.2 equiv.) and triethylamine at 0°.

Example:

$$C_6H_5SCHCH(CH_3)_2 \xrightarrow[92\%]{\substack{(CH_3)_3SiOTf, \\ N(C_2H_5)_3}} C_6H_5S\diagdown C=C(CH_3)_2$$

Silyl-directed Beckmann fragmentation.[10]   A trimethylsilyl group in the β- or β'-position of cyclic ketoxime acetates controls the regio- and stereochemistry of the double bond formed on Beckmann fragmentation catalyzed by trimethylsilyl triflate.

Examples:

[1] M. Demuth and G. Mikhail, *Synthesis*, 827 (1982); *idem, Tetrahedron*, **39**, 991 (1983).

[2] M. Ballester and A. L. Palomo, *Synthesis*, 571 (1983).

[3] Y. Kimura, M. Suzuki, T. Matsumoto, R. Abe, and S. Terashima, *Chem. Letters*, 501 (1984).

[4] A. Hosomi, Y. Sakata, and H. Sakurai, *Tetrahedron Letters*, **25**, 2383 (1984).

[5] B. Fischer, A. Nudelman, M. Ruse, J. Herzig, H. E. Gottlieb, and E. Keinan, *J. Org.*, **49**, 4988 (1984).

[6] A. P. Kozikowski and P. Park, *ibid.*, 1674 (1984).

[7] R. J. Ferrier, *J. Chem. Soc.*, 5443 (1964).

[8] Y. Yamamoto, K. Ohdoi, M. Nakatani, and K. Akiba, *Chem. Letters*, 1967 (1984).

[9] R. D. Miller and R. Hässig, *Syn. Comm.*, **14**, 1285 (1984).
[10] H. Nishiyama, K. Sakuta, N. Osaka, and K. Itoh, *Tetrahedron Letters*, **24**, 4021 (1983).

Trimethylstannyllithium, 9, 449–500. Preparation.[1,2]

Disubstituted butadienes.[2] 2,3-Bis(trimethylstannyl)-1,3-butadiene (**1**) can be prepared by reaction of 2,3-dichlorobutadiene or 1,4-dichloro-2-butyne with 2 equiv. of trimethylstannyllithium (equation I). 2,3-Disubstituted butadienes can be prepared from **1** in one step by reaction of the dianion formed with methyllithium at $-78°$ with 2 equiv.

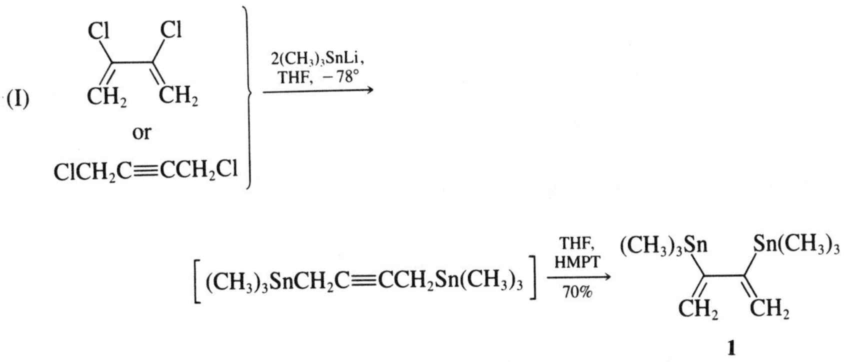

(I)

of an electrophile. Introduction of two different electrophiles is possible by a two-step process.

Examples:

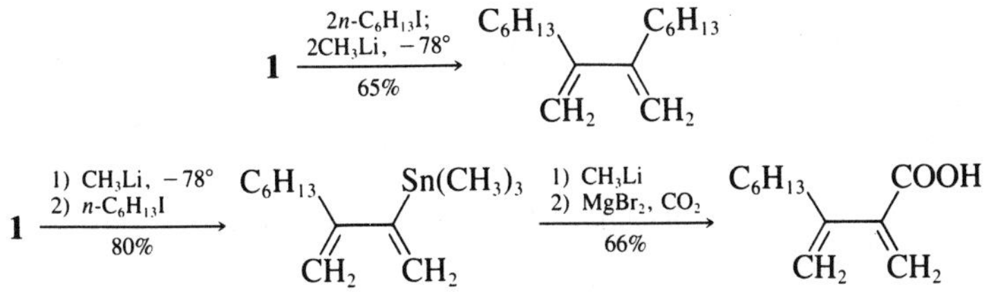

Reaction of **1** with excess methyllithium at $-50°$ results in a dianion of 2-butyne that reacts with electrophiles to give 1,4-disubstituted 2-butynes.

[1] C. Tamborski, F. E. Ford, E. J. Soloski, *J. Org.*, **28**, 237 (1963).
[2] H. J. Reich, K. E. Yelm, and I. L. Reich, *ibid.*, **49**, 3438 (1984).

Trimethylstannylmethyllithium, $(CH_3)_3SnCH_2Li$ (**1**). See also triphenylstannylmethyllithium (**9**, 509).

Preparation:

$$CH_2I_2 + SnBr_2 + CH_3MgI \xrightarrow[48\%]{} (CH_3)_3SnCH_2Sn(CH_3)_3 \xrightarrow{CH_3Li} 1$$

Methylenation.[1] The reagent reacts with carbonyl compounds to form adducts that are converted into 1-alkenes in 65–95% yield on treatment with silica gel (equation I). Methylenation of *cis*-1-decalone can be effected in 91% yield without any epimerization

in the presence of HMPT. Methylenation with the Wittig reagent proceeds in 32% yield to give a 60:40 mixture of *cis*- and *trans*-isomers.

The reagent reacts with α-chloro ketones to give allyl alcohols (equation II).

The reagent reacts with oxiranes (**2**) in the presence of HMPT to form γ-hydroxy-alkylstannanes (**3**), which cyclize to cyclopropanes (**4**) when treated with BF_3 etherate or triphenylphosphine (equation III).

[1] E. Murayama, T. Kikuchi, K. Sasaki, N. Sootome, and T. Sato, *Chem. Letters*, 1897 (1984).

Triphenylbismuth diacetate, $(C_6H_5)_3Bi(OCOCH_3)_2$ (**1**). Mol. wt. 558.37, m.p. 162°. The reagent is prepared by reaction of triphenylbismuth carbonate with acetic acid.

Monophenylation of diols.[1] Isolated hydroxyl groups are inert to **1**, but diols are converted into monophenyl ethers on reaction with **1** in refluxing CH_2Cl_2. Yields are high in reactions with *vic*-diols, but fall to 40–50% when the hydroxyl groups are separated by three or four methylene groups. In the reactions of *t*-butylcyclohexanediols with **1**, the axial epimer reacts much more rapidly than the equatorial one. Preferential phenylation of axial hydroxy groups of pyranosides is also observed.

[1] S. David and A. Thieffry, *J. Org.*, **48**, 441 (1983).

Triphenylmethyl perchlorate (Trityl perchlorate), 1, 1256–1258.

α-Glucosides.[1]   1-O-Acyl-β-D-glucoses react with alcohols in the presence of trityl perchlorate to give α-glucosides preferentially. The highest α-selectivity is obtained with 1-O-bromoacetyl-β-D-glucose derivatives, obtained by acylation of the anomeric hydroxyl group with bromoacetyl chloride catalyzed by potassium fluoride.[2] The complete sequence is shown in formula (I).

(I)

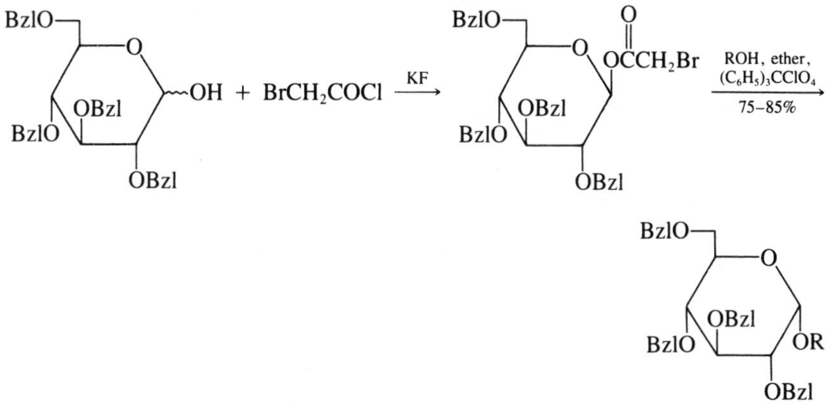

$(\alpha/\beta = 92{-}96{:}8{-}4)$

A similar reaction with 1-O-acetyl-2,3,5-tribenzyl-β-D-ribofuranose affords β-ribosides exclusively (equation II). However, α-ribosides are formed preferentially if lithium perchlorate and 4 Å molecular sieves are present.

(II)

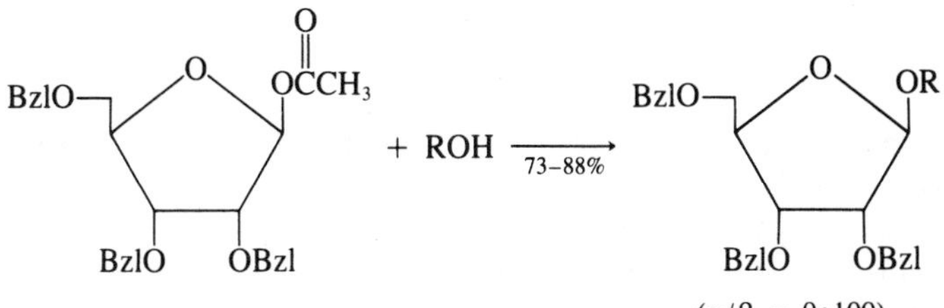

$(\alpha/\beta = 0{:}100)$

Aldol-type reactions.[3]   Trityl perchlorate catalyzes an aldol-type reaction between silyl enol ethers and acetals or ketals to give β-alkoxy ketones. The yields are comparable to those obtained with $TiCl_4$ (**6,** 594). The *syn*-aldol is formed predominantly ($\sim$4:1).

---

[1] T. Mukaiyama, S. Kobayashi, and S. Shoda, *Chem. Letters,* 907 (1984).
[2] S. Shoda and T. Mukaiyama, *ibid.,* 861 (1982).
[3] T. Mukaiyama, S. Kobayashi, and M. Murakami, *Chem. Letters,* 1759 (1984).

Triphenylmethylpotassium, 8, 524; **9**, 502. This base can be prepared by reaction of $(C_6H_5)_3CH$ with KH in diglyme containing some DMSO.[1]

Dehydrohalogenation.[2] This base is superior to DBU for dehydrohalogenation of secondary alkyl bromides or iodides. The reaction can be conducted at 0° in DME, ether, benzene, or diglyme. Secondary alkyl chlorides undergo slow dehydrohalogenation under the same conditions, and primary alkyl halides are completely stable.

[1] J. W. Huffman and P. G. Harris, *Syn. Comm.*, **7**, 137 (1977).
[2] D. R. Anton and R. H. Crabtree, *Tetrahedron Letters*, **24**, 2449 (1983).

Triphenylmethyl tetrafluoroborate, 1, 1256–1258; **2**, 459; **4**, 565–567; **6**, 657; **7**, 414–415; **8**, 524–525; **9**, 512; **10**, 455.

Cyclodehydrogenation; chromenes.[1] Reaction of *o*-(3,3-dimethylallyl)phenols with trityl tetrafluoroborate results in dehydrogenation and cyclization to chromenes. DDQ has usually been used for this purpose.

Example:

[1] P. Barua, N. C. Barua, and R. P. Sharma, *Tetrahedron Letters*, **24**, 5801 (1983).

Triphenylphosphine, 1, 1238–1247; **2**, 443–445; **3**, 317–320; **4**, 548–550; **5**, 725; **6**, 643–644; **7**, 403–404; **11**, 588.

α-*Alkoxyphosphonium bromides; enol ethers.*[1] These salts (2) are prepared by reaction of an enol ether (1) with triphenylphosphine hydrobromide at $0 \rightarrow 20°$ in CH_2Cl_2 in 85–90% yield. The ylides formed on deprotonation undergo Wittig reactions to give enol ethers (3), which are suitable for further transformations. An example is the synthesis of a C-glycoside (4).

—$N_3 \rightarrow$ —NH_2. Some time ago Staudinger[2] reported this transformation by reaction of an azide with $P(C_6H_5)_3$ to form an iminophosphorane, $RN{=}P(C_6H_5)_3$, which is hydrolyzed to RNH_2 and $(C_6H_5)_3P{=}O$. This reaction has received little attention, but recent work indicates that it is generally useful, and is compatible with epoxy, acetal, nitro, and ester groups. Yields are 80–95% (seven examples).[3]

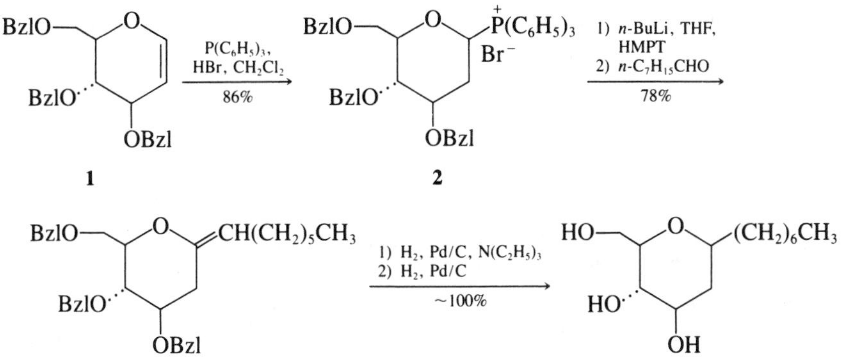

1,5-Dicarbonyl compounds.[4] Michael addition of α-nitro ketones to methyl vinyl ketone (**1a**) or acrylaldehyde (**1b**) catalyzed by $P(C_6H_5)_3$ provides the adducts **2**, which are reduced by Bu_3SnH to 1,5-dicarbonyl compounds (**3**). Additions to the enal proceed more rapidly than ones to the enone.

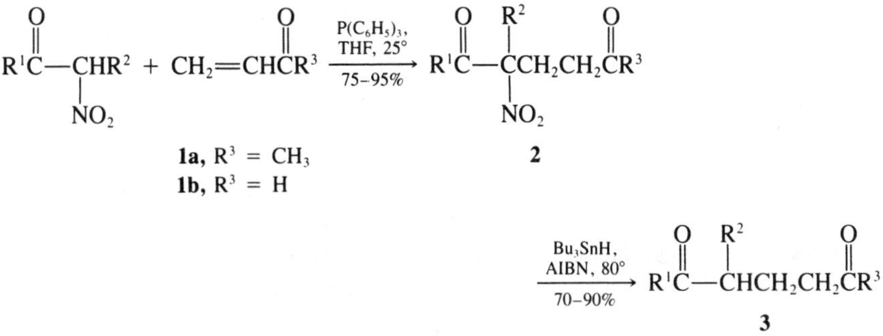

[1] J. B. Ousset, C. Mioskowski, Y.-L. Yang, and J. R. Falck, *Tetrahedron Letters*, **25**, 5903 (1984).
[2] H. Staudinger and J. Meyer, *Helv.*, **2**, 635 (1919).
[3] M. Vaultier, N. Knouzi, and R. Carrié, *Tetrahedron Letters*, **24**, 763 (1983).
[4] N. Ono, H. Miyake, and A. Kaji, *J.C.S. Chem. Comm.*, 875 (1983).

Triphenylphosphine–Carbon tetrachloride, 1, 1247; **2**, 445; **3**, 320; **4**, 551–552; **5**, 727; **6**, 644–645; **7**, 404; **8**, 516; **9**, 503; **10**, 447–448; **11**, 588.

Cyclodehydration. The reaction of this reagent with 1,4-diols results mainly in cyclodehydration to tetrahydrofuranes. Extent of formation of cyclic ethers decreases in order of the ring sizes: 5 > 6 > 4 ≈ 7. The reaction with 1,2-diols or 1,3-diols results in chlorohydrins as the major products.[1] If the reaction is conducted in the presence of solid K_2CO_3 in CCl_4 (reflux), epoxides become the major products formed from 1,2-diols.

Alkyl substitution favors epoxide formation, as does an increase in the concentration of triphenylphosphine above 1 molar equiv.[2]

[1] C. N. Barry and S. A. Evans, Jr., *J. Org.*, **46**, 3361 (1981).
[2] *Idem, Tetrahedron Letters*, **24**, 661 (1983).

Triphenylphosphine–Diethyl azodicarboxylate (DEAD), **1**, 245–247; **4**, 553–555; **5**, 727–728; **6**, 645; **7**, 404–406; **8**, 517; **9**, 504–506; **10**, 448–449; **11**, 589–590.

Macrolactonization with inversion.[1]   Lactonization of the optically active seco-acid **1** with $P(C_6H_5)_3$ and DEAD followed by hydrolysis of the acetonide group gives the cyclic dilactone colletodiol (**2**), formed with inversion of configuration at the hydroxyl-bearing carbon. Lactonization of **1** with 2,4,6-trichlorobenzoyl chloride and triethylamine (**9**, 478–479) furnishes 6-epicolletodiol after deprotection.

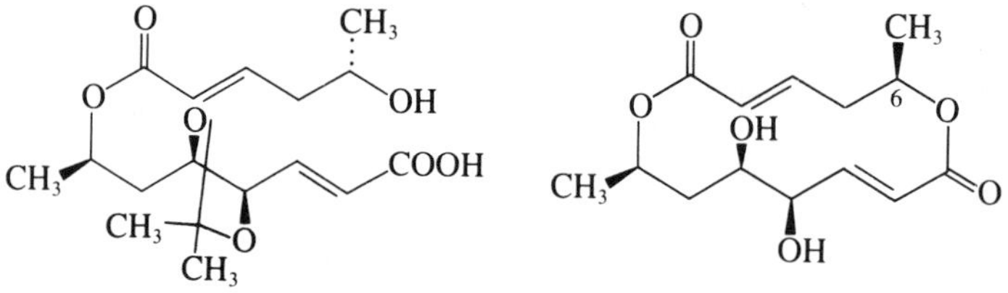

Alkyl halides.[2]   Primary, secondary, and allylic alcohols are converted into halides by a Mitsunobu reaction (**5**, 728) with an added zinc halide. Yields are generally >70%, and are highest with zinc chloride. No other metal salts were found to be useful.

Azetidines.[3]   Reaction of the salt (**1**) of a β-amino alcohol with the Mitsunobu reagent produces the azetidine **2** (50% yield).

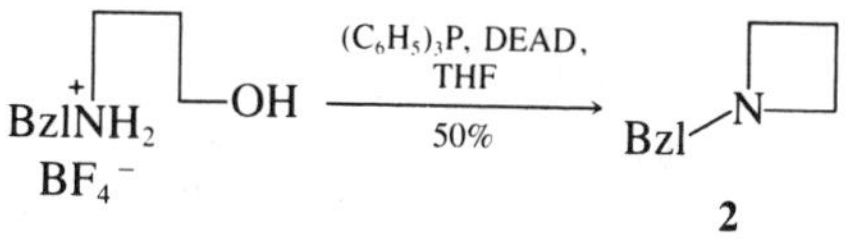

The $(C_6H_5)_3P$–CCl_4 reagent is also effective for this cyclization. Example:

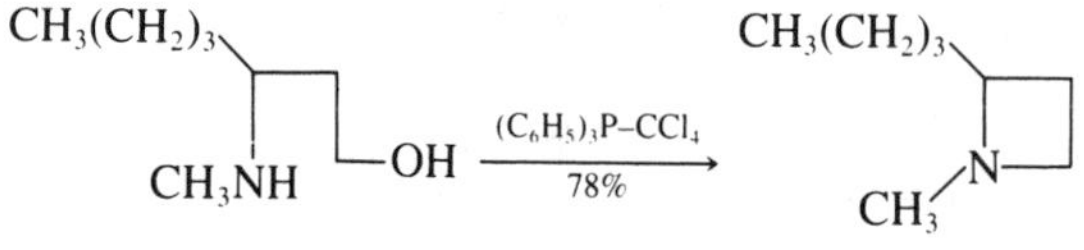

Decarboxylative dehydration of β-hydroxy carboxylic acids (**9**, 504).[4]   Study of the stereochemistry of this reaction indicates that *syn*-β-hydroxy carboxylic acids are usually converted into (E)-alkenes (equation I). In contrast, the *anti*-isomers are converted

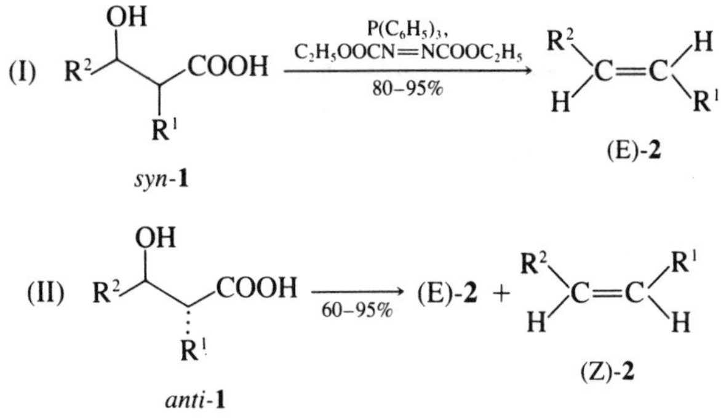

into (E)- and/or (Z)-**2**, depending on the nature of R^1 and R^2. Formation of (Z)-alkenes is favored when R^1 is a phenyl group and R^2 is an alkyl group; formation of (E)-alkenes is favored when R^1 is an alkyl group and R^2 is a phenyl group, particularly one substituted by electron-releasing groups.

This reaction can be useful for synthesis of an alkene that is difficult to obtain in pure form by a Wittig reaction, such as (Z)-asarone (**3**).

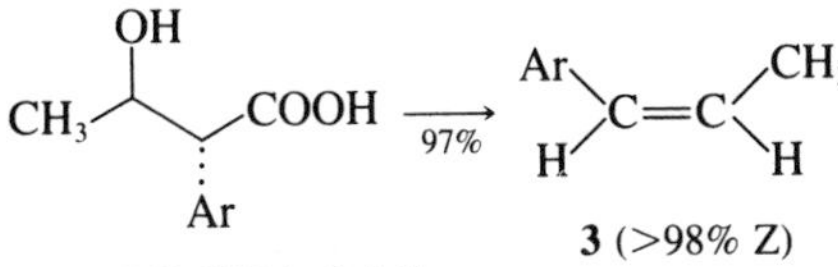

[Ar = $C_6H_2(CH_3)_3$-2,4,5]

β-Lactams (**10**, 447–448; **11**, 589).[5]   Cyclization of β-hydroxy aryl amides to β-lactams by a Mitsunobu reaction is stereospecific. Thus the protected derivative **1** of phenylserine cyclizes to **2** as the only β-lactam and an aziridine. The configuration at C_2 of **1** is retained, whereas the configuration at C_3 is inverted.

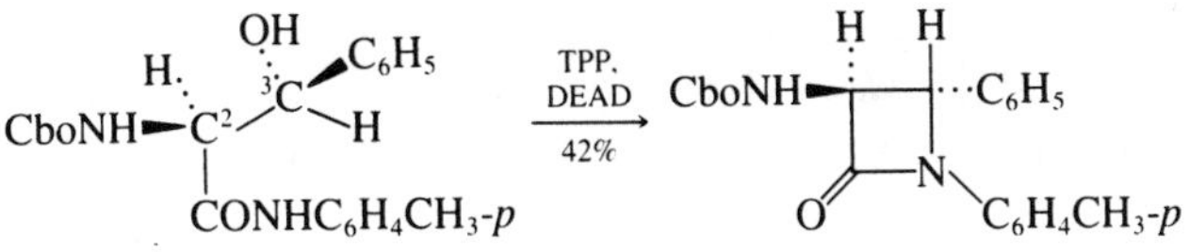

[1] H. Tsutsui and O. Mitsunobu, *Tetrahedron Letters*, **25**, 2159, 2163 (1984).
[2] P.-T. Ho and N. Davies, *J. Org.*, **49**, 3027 (1984).

[3] P. G. Sammes and S. Smith, *J.C.S. Chem. Comm.*, 682 (1983).
[4] J. Mulzer and O. Lammer, *Angew. Chem., Int. Ed.*, **22**, 628 (1983).
[5] A. K. Bose, M. S. Manhas, D. P. Sahu, and V. R. Hegde, *Can. J. Chem.*, **62**, 2498 (1984).

Triphenylphosphine dihalides, 1, 1247–1249; **6**, 646–648; **7**, 407.

ROH → RX.   β-Naphthol (**1**) is converted into 2-bromonaphthalene (**2**) in 92–94% yield by reaction with triphenylphosphine and bromine in acetonitrile at 70°.[1]

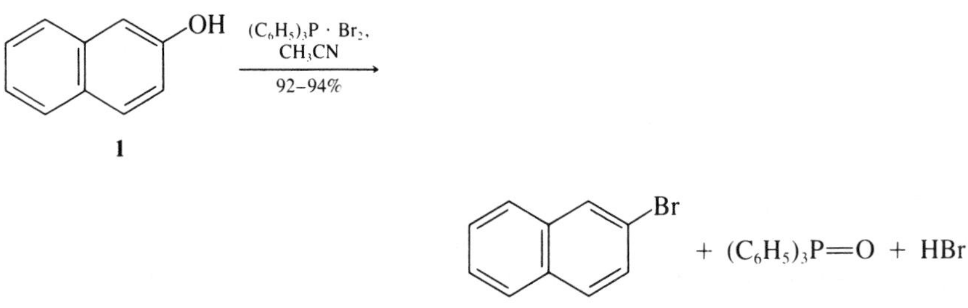

This method is recommended for conversion of cyclopropylcarbinyl alcohols into the corresponding halides without formation of homoallylic halides via ring cleavage.[2]

Example:

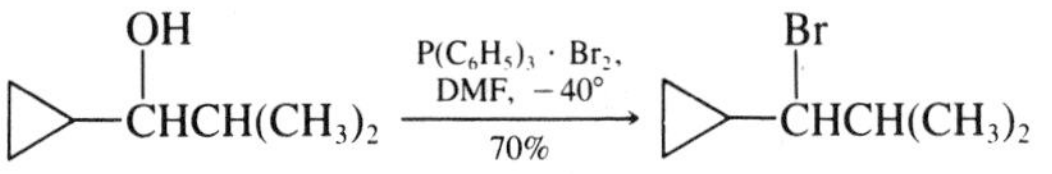

The conversion of primary hydroxyl groups of carbohydrates into chloro groups proceeds in higher yield with a reagent obtained from $(C_6H_5)_3PCl_2$ and imidazole (1:2). Imidazole may function in part as a scavenger for chlorine. Addition of imidazole also is useful for conversion of hydroxyl groups into chloro groups with triphenylphosphine–carbon tetrachloride.[3]

Deoxygenation of epoxides.[4]   Trisubstituted steroidal epoxides undergo deoxygenation to the unsaturated steroids by reaction with $P(C_6H_5)_3$ and I_2 in CH_2Cl_2 (70–100% yield).

Halohydrins.[5]   The reaction of a slight excess of I_2, Br_2 or Cl_2 and of $P(C_6H_5)_3$ in CH_2Cl_2 with an epoxide results in a halohydrin in generally high yield. The orientation depends in part on the bulkiness of the halide ion, but the halogen ion generally attacks the less substituted carbon atom.

Examples:

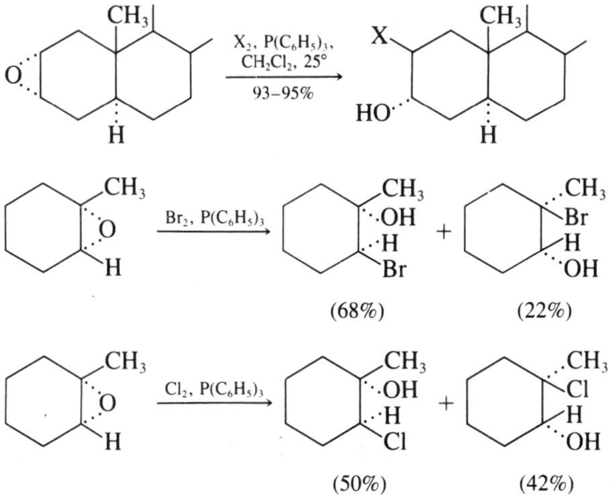

Ketone synthesis.[6]   Lithium carboxylates react with $(C_6H_5)_3PCl_2$ to form an acyloxy-phosphonium salt (**a**), which reacts with Grignard reagents to provide ketones in generally high yield. If a free carboxylic acid is used, a tertiary amine is required to neutralize the

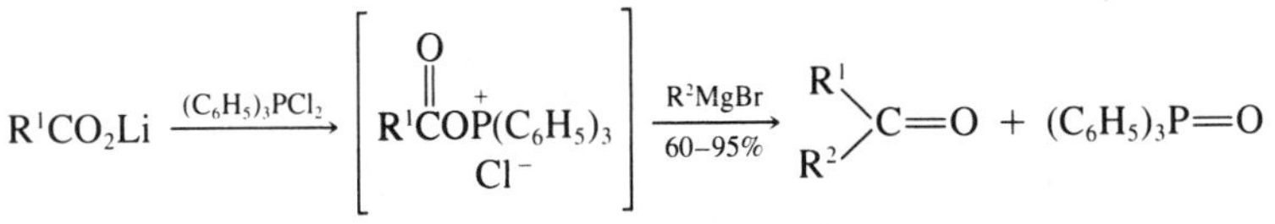

a

hydrogen chloride liberated. Formation of an acyl chloride is not involved, since tertiary alcohols are not obtained as by-products.

[1] N. Miyata, H. Yagi, and D. M. Jerina, *Org., Syn.*, submitted (1984).

[2] R. T. Hrubiec and M. B. Smith, *J. Org.*, **49**, 431 (1984); M. B. Smith and R. T. Hrubiec, *Org. Syn.*, submitted (1981).

[3] P. J. Garegg, R. Johansson, and B. Samuelsson, *Synthesis*, 168 (1984).

[4] Z. Paryzek and R. Wydra, *Tetrahedron Letters*, **25**, 2601 (1984).

[5] G. Palumbo, C. Ferreri, and R. Caputo, *ibid.*, **24**, 1307 (1983).

[6] T. Fujisawa, S. Iida, H. Uehara, and T. Sato, *Chem. Letters*, 1267 (1983).

Triphenyltin hydride, 1, 1250–1251; **2**, 448; **3**, 324; **4**, 559; **5**, 734; **6**, 649; **8**, 521–522; **10**, 451–452.

 Cyclization of cyano and acetylenic radicals.[1]   The radical generated by reduction $[(C_6H_5)_3SnH, AIBN]$ of the imidazolylthiocarbonyl derivative of an alcohol (**11**, 550) can undergo intramolecular cyclization with a suitably located nitrile group to provide cyclic ketones. The required δ-hydroxy nitriles are conveniently obtained by condensation of the pyrrolidine enamine of a cycloalkanone with acrylonitrile followed by reduction and acylation with thiocarbonyldiimidazole.

Example:

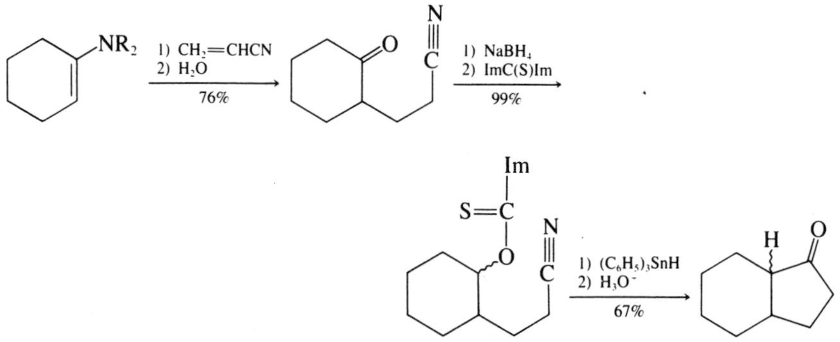

A similar cyclization can be effected starting with δ- or ε-hydroxy acetylenes, but the yield is low with terminal acetylenes.

Example:

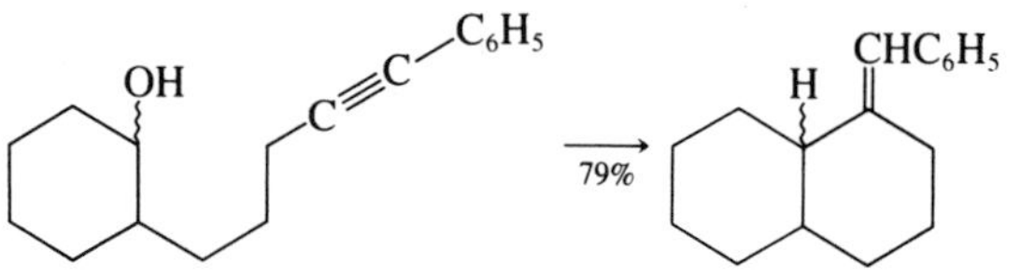

[1] D. L. J. Clive, P. L. Beaulieu, and L. Set, *J. Org.*, **49**, 1313 (1984).

Tris(acetonitrile)tricarbonyltungsten, $(CH_3CN)_3W(CO)_3$ (**1**). Preparation.[1]

Allylic alkylations.[2] This complex in combination with 2,2'-bipyridyl (bpy) catalyzes nucleophilic alkylation of allylic acetates and carbonates, but is less active than molybdenum or palladium catalysts. The displacement occurs with retention of configuration, as with Mo and Pd catalysts. However, alkylation occurs almost entirely at the more substituted end of the allylic group, regardless of the nucleophile.

Examples:

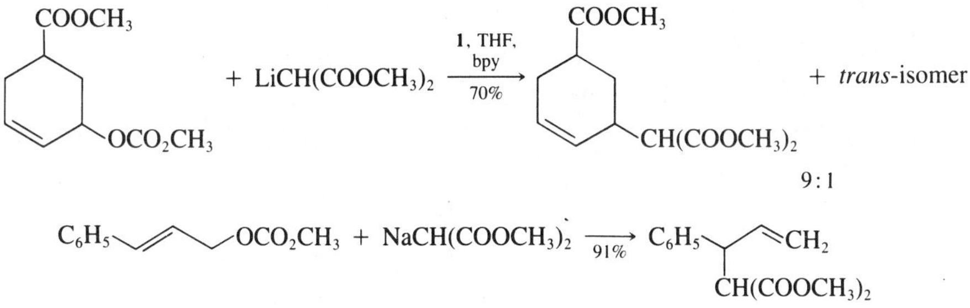

+ *trans*-isomer

9:1

[1] J. W. Faller, D. A. Haitko, R. D. Adams, and D. F. Chodosh, *Am. Soc.*, **101**, 865 (1979).
[2] B. M. Trost and M.-H. Hung, *ibid.*, **105**, 7757 (1983).

Tris(acetylacetonate)iron(III), $Fe(acac)_3$. Mol. wt. 353.18, m.p. 179°, air stable. Suppliers: Alfa, Strem.

Ketone synthesis.[1]   This iron complex is an efficient catalyst for the reaction of a wide variety of Grignard reagents with aromatic or aliphatic acyl chlorides to form ketones in 70–90% yield.

[1] V. Fiandanese, G. Marchese, V. Martina, and L. Ronzini, *Tetrahedron Letters*, **25**, 4805 (1984).

4,4′,4″-Tris(4,5-dichlorophthalamido)trityl bromide (CPTrBr),

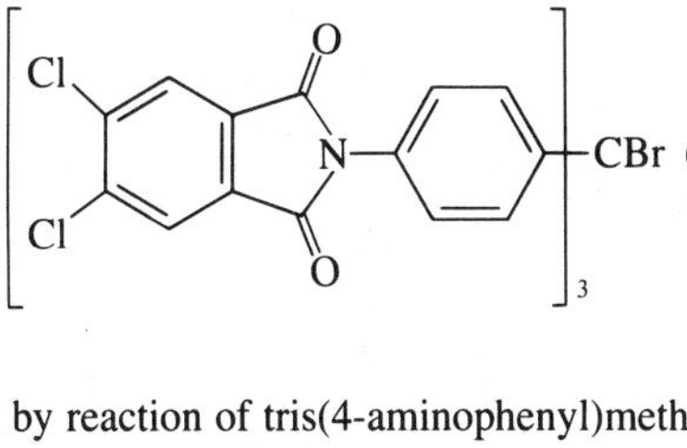

The reagent is prepared by reaction of tris(4-aminophenyl)methanol (pararosaniline) with 4,5-dichlorophthalic anhydride followed by bromination with acetyl bromide.

Protection of primary alcohols.[1]   This trityl bromide in the presence of $AgNO_3$ in DMF reacts selectively with the primary hydroxyl group of carbohydrates and nucleosides. The CPTr ether group is stable to aqueous pyridine and 80% acetic acid, but is selectively cleaved by hydrazine in pyridine–acetic acid. This protecting group is useful in the synthesis of oligoribonucleotides.

[1] M. Sekine and T. Hata, *Am. Soc.*, **106**, 5763 (1984).

Tris(dimethylamino)sulfonium difluorotrimethylsilicate (TASF), **10**, 452–453; **11**, 590.

Michael addition of ketene silyl acetals.[1]   TASF catalyzes the conjugate addition of ketene silyl acetals to enones in THF at room temperature. A similar addition can be effected without a catalyst in a polar solvent, acetonitrile at 55° (ref. 2) or CH_3NO_2 at 25°, in the case of some less hindered ketene silyl acetals. The addition shows no diastereoselection. The adducts can be alkylated to provide 2,3-disubstituted cycloalkanones as a 1:1 mixture of two diastereomers (both probably *trans*).

Examples:

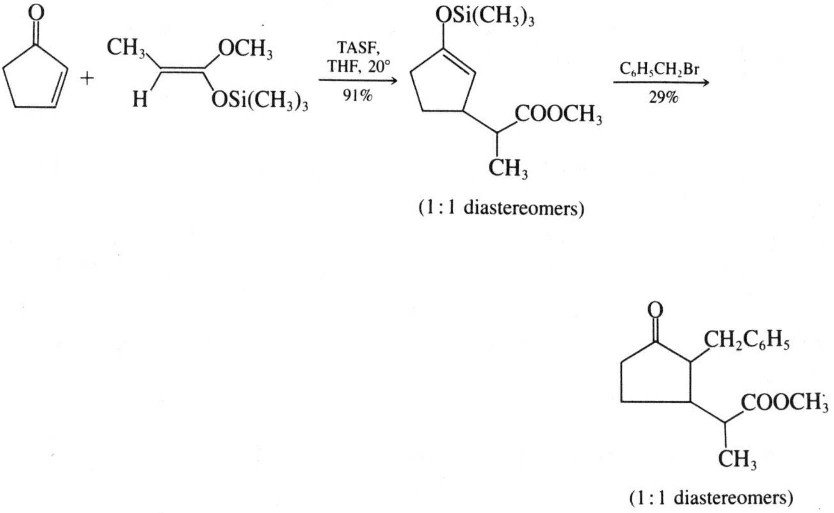

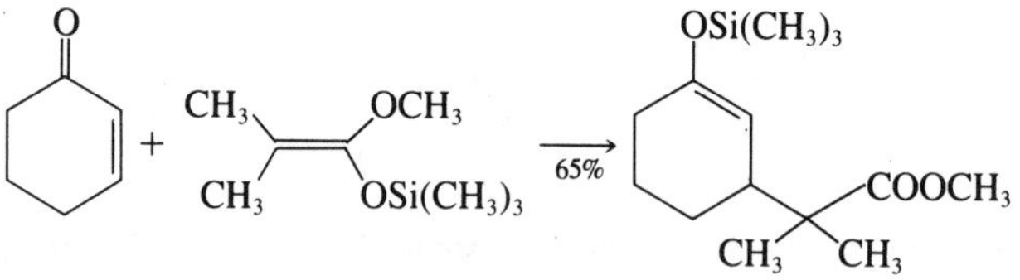

(1:1 diastereomers)

(1:1 diastereomers)

α-Nitroaryl carbonyl compounds.[3]   In the presence of 1 equiv. of TASF, nitrobenzene reacts with the silyl ketene acetal **1** to give the Michael-type adduct **a**, which is oxidized by Br_2 or DDQ to the α-nitrophenyl ester **2**. Similar reactions with less hindered ketene acetals and silyl enol ethers give mixtures of *ortho-* and *para*-adducts, but exclusive *ortho*-addition is possible with *para*-substituted nitrobenzenes.

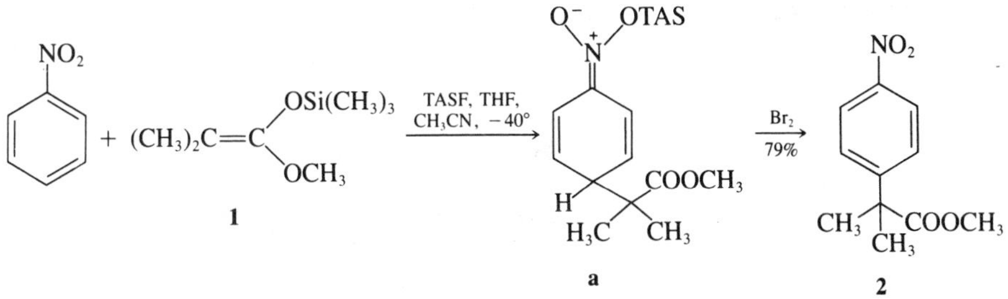

Example:

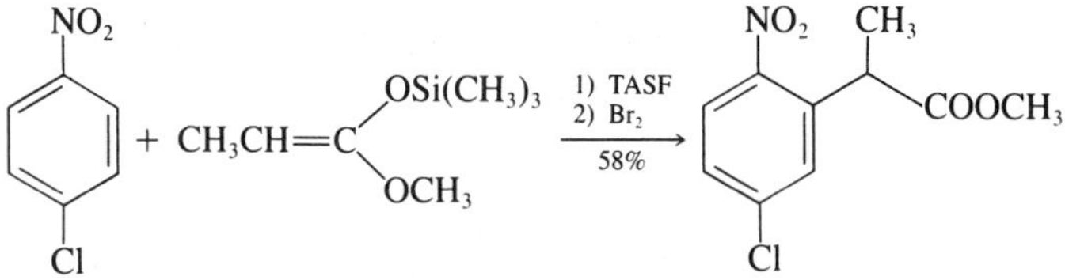

[1] T. V. RajanBabu, *J. Org.*, **49**, 2083 (1984).
[2] Y. Kita, J. Segawa, J. Haruta, H. Yasuda, and Y. Tamura, *J.C.S. Perkin I*, 1099 (1982).
[3] T. V. RajanBabu and T. Fukunaga, *J. Org.*, **49**, 4571 (1984).

Tris(6,6,7,7,8,8,8-heptafluoro-2,2-dimethyl-3,5-octanedionato)europium, Eu(fod)₃ (**1**). Mol. wt. 1037.50, m.p. 203–207°. Supplier: Aldrich.

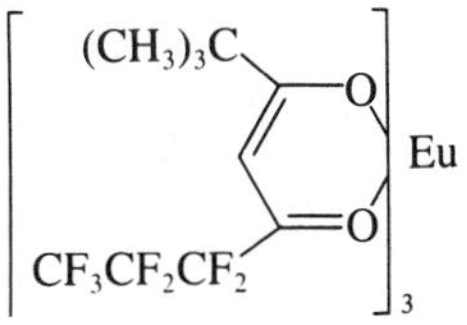

1

Diels-Alder reaction of aldehydes with activated dienes.   This lanthanide shift reagent can function as a Lewis acid catalyst in the cyclocondensation of 1-methoxy-3-trimethylsilyloxy-1,3-diene (2) with aromatic aldehydes, and permits isolation of the initial

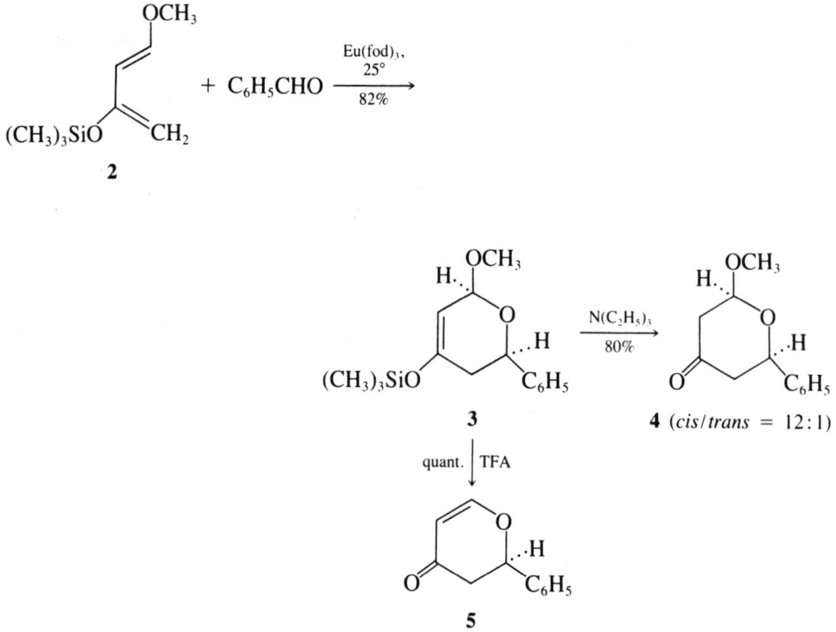

labile *endo*-adduct (3), which is readily convertible into either 4 or 5. A similar cyclocondensation of the diene 6 with either aromatic or aliphatic aldehydes results in the enol ether (7) with three chiral centers.[1]

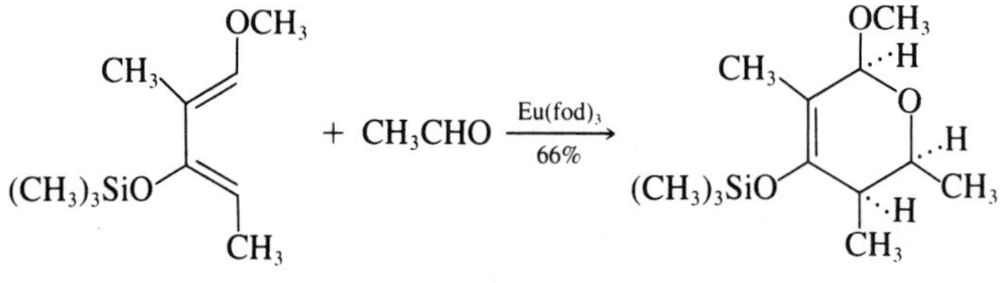

The cycloaddition reaction is applicable to various derivatives of 3-trimethylsilyloxy-1,3-diene, including 1-alkyl derivatives, thus providing stereoselective routes to various derivatives of 4-pyranone.[2]

Use of the chiral europium(III) derivative, tris[3-(heptafluoropropylhydroxymethylene)-*d*-camphorato]europium(III) [Eu(hfc)$_3$], as the catalyst results in moderate asymmetric induction (18–36% ee).[1,3]

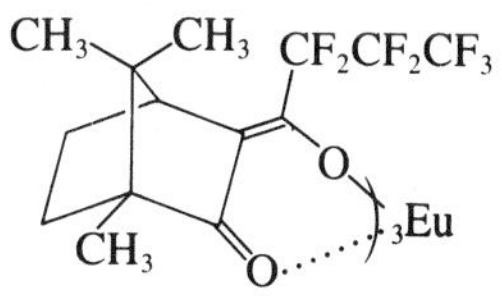

Eu(hfc)$_3$

Cycloadditions with enol ethers.[4]   The Diels-Alder reaction of vinyl ethers with α,β-unsaturated aldehydes proceeds at room temperature when catalyzed by the related lanthanide Yb(fod)$_3$. An example is the reaction of crotonaldehyde with ethyl vinyl ether (equation I).

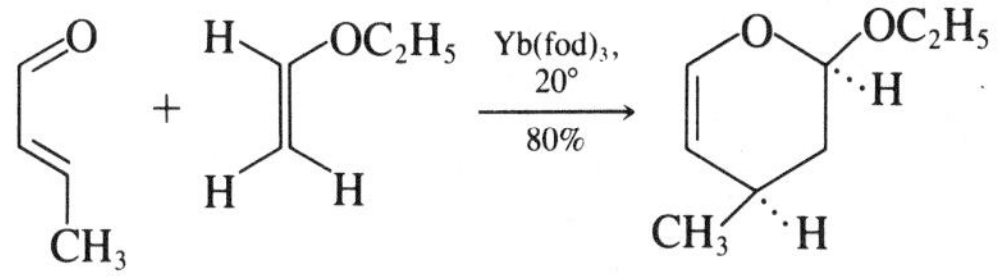

[1] M. Bednarski and S. Danishefsky, *Am. Soc.*, **105**, 3716 (1983).
[2] S. Danishefsky, D. F. Harvey, G. Quallich, and B. J. Uang, *J. Org.*, **49**, 392 (1984).
[3] M. Bednarski, C. Maring, and S. Danishefsky, *Tetrahedron Letters*, **24**, 3451 (1983).
[4] S. Danishefsky and M. Bednarksi, *ibid.*, **25**, 721 (1984).

Tris(tribenzylideneacetylacetone)tripalladium(chloroform), Pd$_3$(TBAA)$_3$·CHCl$_3$ (**1**). Preparation.[1]

Cyclization of 1,3-diene monoepoxides to δ-lactones.[2]   The (Z)-ester **2** is cyclized in the presence of **1** and a phosphite ligand to the lactones **3** and **4** in the ratio 92:8, whereas the isomeric (E)-ester **2** is cyclized mainly to the lactone **4**. The two lactones are epimeric at C_{15}. The lactone **3** was converted into 11-deoxy PGE_1.

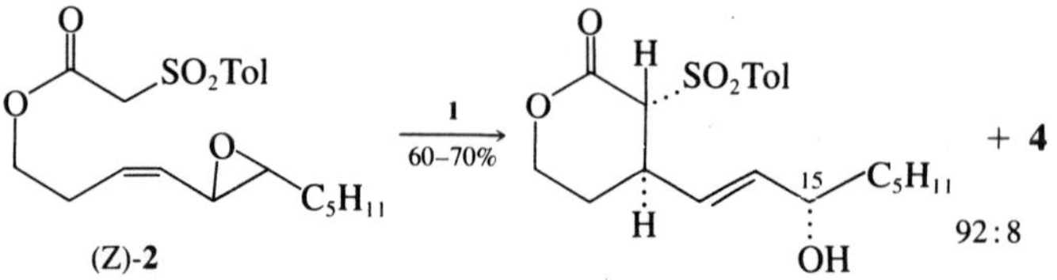

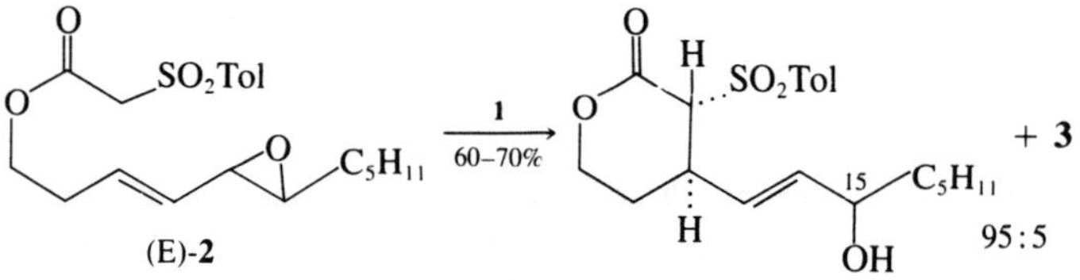

[1] Y. Ishii, S. Hasegawa, S. Kimura, and K. Itoh, *J. Organometal. Chem.*, **73**, 411 (1974).
[2] T. Takahashi, H. Kataoka, and J. Tsuji, *Am. Soc.*, **105**, 147 (1983).

V

(S)-Valine *t*-butyl ester.

Asymmetric alkylation of α-alkyl-β-keto esters.[1] The sense of asymmetric alkylation of the chiral enamines prepared from (S)-valine *t*-butyl ester with α-alkyl-β-keto esters is markedly controlled by the solvent. Thus alkylation of the anion **1**, prepared with LDA in toluene, with CH₃I in the presence of HMPT (1 equiv.) results in (R)-**2** in 99% ee, whereas alkylation in THF (2 equiv.) results in (S)-**2** in 92% ee. The effect of HMPT is general for a variety of electrophiles; depending on the electrophile, dioxolane

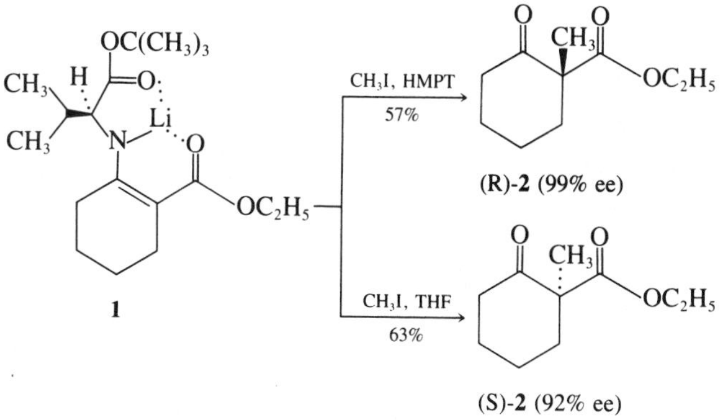

or trimethylamine can be as effective as THF. The same solvent effect is observed with acyclic β-keto esters.

[1] K. Tomioka, K. Ando, Y. Takemasa, and K. Koga, *Am. Soc.*, **106**, 2718 (1984); *idem, Tetrahedron Letters*, **25**, 5677 (1984).

L-Valinol, (CH₃)₂CHCHCH₂OH (**1**). Mol. wt. 103.17, α_D + 14.6°. Supplier: Aldrich.
                    |
                   NH₂

Chiral quaternary carbon centers.   Meyers *et al.*[1] have reported an enantioselective synthesis of chiral α,α-disubstituted-γ-keto acids (**6**) via the lactam **3** prepared from L-valinol (**1**) and the γ-keto acid **2**. Alkylation of **3** with primary alkyl halides gives mainly the *endo*-isomer (**4**). Alkylation of **4** also proceeds with *endo*-selectivity to give **5** with a quaternary center. This product is cleaved to give the γ-keto ester **6** in >95% enantiomeric purity. The order of introduction of the alkyl groups determines the configuration. The stereoselectivity is lower, and can even be reversed, when a secondary alkyl halide is used.

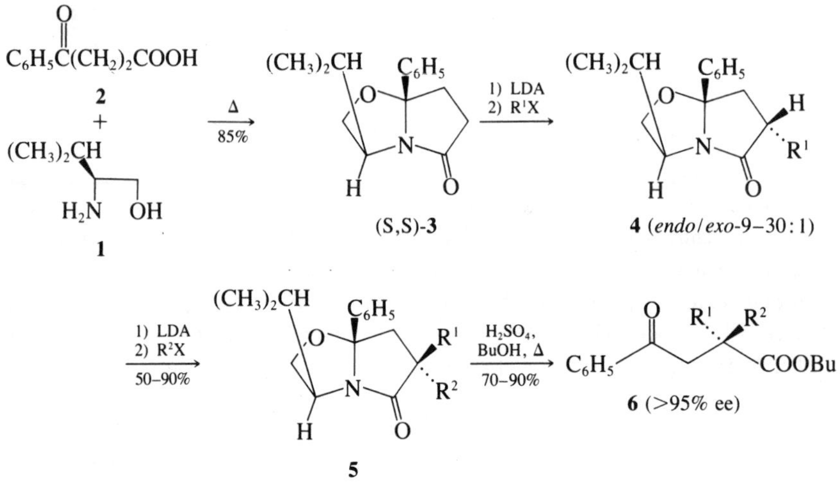

[1] A. I. Meyers, M. Harre, and R. Garland, *Am. Soc.*, **106**, 1146 (1984).

Vilsmeier reagent, 1, 284–298; **2,** 154; **3,** 116; **4,** 186; **5,** 251; **6,** 220; **7,** 422–424; **8,** 529–530; **9,** 514–515; **10,** 457.

Ketoximes from $R^1CH_2NO_2$ and R^2MgBr.[1]   Reaction of the Vilsmeier reagent with the lithium nitronate of a nitroalkane activates the α-position to attack by Grignard reagents to form ketoximes (equation I).

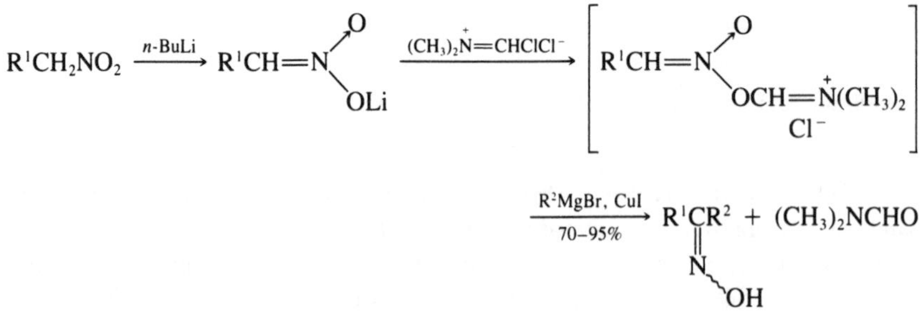

Selective reduction of —COOH to —CH₂OH.[2]   Chemoselective reduction of carboxylic acids is possible by *in situ* conversion to the carboxymethyleneiminium salt by reaction with the Vilsmeier reagent (DMF and oxalyl chloride). This salt is then reduced with NaBH₄ (2 equiv.) to the alcohol (equation I). Various functional groups are tolerated: bromo, cyano, ester, and C≡C (even when conjugated to COOH).

$$\text{(I)} \quad \text{RCOOH} \xrightarrow[\text{THF, CH}_3\text{CN}]{\overset{+}{\text{ClCH}}=\overset{+}{\text{N}}(\text{CH}_3)_2\text{Cl}^-,} \left[\underset{\displaystyle \text{RCO}-\text{CH}=\overset{+}{\text{N}}(\text{CH}_3)_2}{\overset{\displaystyle \text{O}}{\|}} \quad \text{Cl}^- \right] \xrightarrow[\substack{\text{DMF} \\ 88-97\%}]{\text{NaBH}_4,} \text{RCH}_2\text{OH}$$

Alkyl chlorides (**7**, 422).   The iminium salt obtained from N,N-diphenylbenzamide and oxalyl chloride is somewhat more effective than the salt obtained from DMF for conversion of primary or secondary alcohols into chlorides. The reaction occurs at room temperature and with complete inversion in the case of secondary alcohols. Yields are generally >85%.[3]

Quinoline synthesis.   Ethyl anilinobutenoates (**1**), prepared as shown, are converted into 3-carboethoxyquinolines (**2**) on reaction with the Vilsmeier reagent.[4]

[1] T. Fujisawa, Y. Kurita, and T. Sato, *Chem. Letters,* 1537 (1983).
[2] T. Fujisawa, T. Mori, and T. Sato, *ibid.,* 835 (1983).
[3] T. Fujisawa, S. Iida, and T. Sato, *ibid.,* 1173 (1984).
[4] D. R. Adams, J. N. Domínguez, and J. A. Pérez, *Tetrahedron Letters,* **24**, 517 (1983).

Vinyl acetate, 1, 1271.

N-Acylindoles.[1]   Reaction of N-phenylhydroxamic acids (**1**) with vinyl acetate catalyzed by Li₂PdCl₄ results in N-acylindoles (**2**) via a hetero-Cope rearrangement of an intermediate N-phenyl-O-vinylhydroxylamine (**a**). Acylation of indoles results in reaction at both nitrogen and C_3.

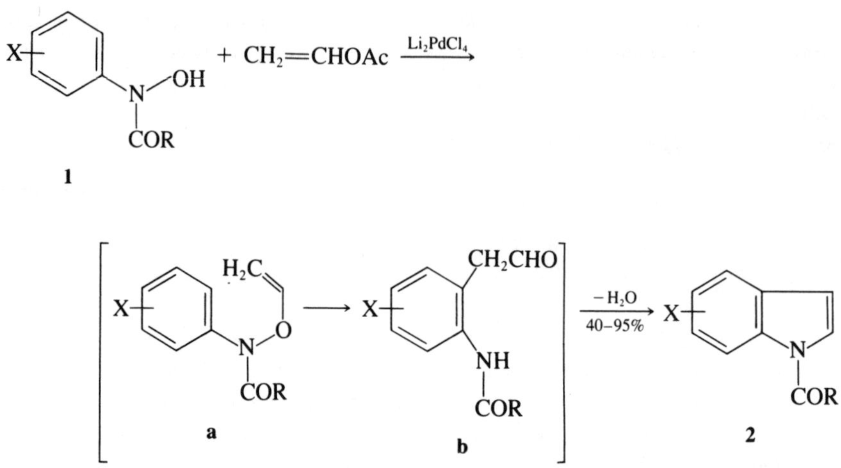

[1] P. Martin, *Helv.*, **67**, 1647 (1984).

Vinyltrimethylsilane, **9**, 498–499; **10**, 444. Suppliers: Aldrich, Fluka.

  α,β-*Unsaturated aldehydes*.   Homologation of an aldehyde to an α,β-unsaturated aldehyde can be effected by reaction of the derived nitrone with vinyltrimethylsilane to form an isoxazolidine, which undergoes acid-catalyzed fragmentation to an (E)-α,β-unsaturated aldehyde.[1]

  Example:

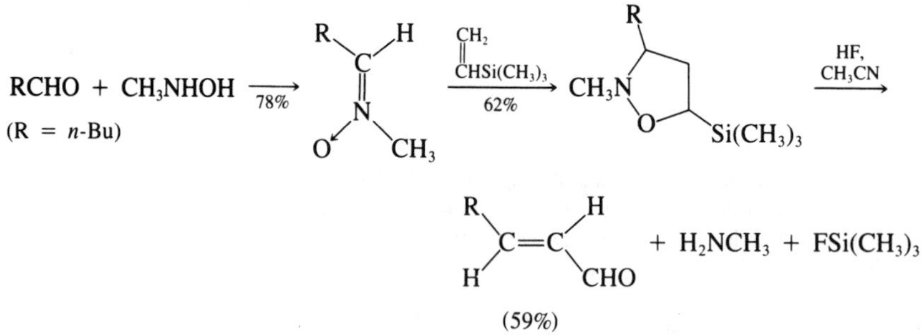

[1] P. DeShong and J. M. Leginus, *J. Org.*, **49**, 3421 (1984).

Z

Zinc, 1, 1276–1284; **2**, 459–462; **3**, 334–337; **4**, 574–577; **5**, 753–756; **6**, 672–675; **7**, 426–428; **8**, 532; **10**, 459; **11**, 598–599.

β-Keto esters.[1] β-Keto esters are obtained in good yield by slow addition of α-bromo esters to nitriles in the presence of activated zinc dust in refluxing THF (equation I). An excess of the ester is necessary, because some undergoes self-condensation rather

$$(I) \quad R^1OOCCHR^2 + N\equiv CR^3 \xrightarrow[70-95\%]{Zn, \, THF} R^1OOC \overset{R^2}{\diagup}\diagdown R^3 \xrightarrow[80-85\%]{H_3O^+} R^1OOCCHCR^3$$

(with Br substituent on the left carbon; the central product has NH₂ group; the right product has R^2 substituent and $=O$)

than addition. This synthesis was originally reported by Blaise,[2] but has seen little use because the yields are generally low, particularly in the reaction of α-bromo acetates.

Reformatsky reaction of 4-bromocrotonate.[3] The regioselectivity of the Reformatsky reaction of ethyl 4-bromocrotonate with cyclopentanone (equation I) is markedly dependent on the solvent and catalyst. α-Addition is favored in ether, whereas γ-addition

(I) (cyclopentanone) + $BrCH_2CH=CHCO_2C_2H_5 \xrightarrow{90-95\%}$

(α) product with $CH_2=CH$ and $CHCO_2C_2H_5$ and HO

(γ) product with $CH_2CH=CHCO_2C_2H_5$ and HO

Zn/Cu(HOAc), ether	100:0
Zn, benzene	25:75

is favored in benzene, cyclohexane, or THF. The Zn/Cu couple is somewhat more α-selective than Zn, particularly when activated with HCl or HOAc.

β-Lactams. Some time ago, Gilman and Specter[4] reported that α-bromo esters and Schiff bases undergo a Reformatsky-type reaction in the presence of zinc activated by a trace of iodine to form β-lactams in moderate yield. This reaction is markedly improved by ultrasonic irradiation (**11**, 599) and by use of dioxane as solvent (equation I).[5] Surprisingly, β-lactams are not obtained from α-bromopropionic esters under these conditions.

$$
(I)\quad
\begin{array}{c}
CH_2Br \\
| \\
COOCH_3
\end{array}
+
\begin{array}{c}
\diagup Ar^1 \\
\| \\
N \\
\diagdown Ar^2
\end{array}
\xrightarrow[\substack{70-95\%}]{\substack{Zn,\ I_2, \\ dioxane}}
\begin{array}{c}
Ar^1 \\
\square \\
O \quad N \diagdown Ar^2
\end{array}
$$

[1] S. M. Hannick and Y. Kishi, *J. Org.*, **48**, 3833 (1983).

[2] E. E. Blaise, *Compt. Rend.*, **132**, 478 (1901).

[3] L. E. Rice, M. C. Boston, H. O. Finklea, B. J. Suder, J. O. Frazier, and T. Hudlicky, *J. Org.*, **49**, 1845 (1984).

[4] H. Gilman and M. Speeter, *Am. Soc.*, **65**, 2255 (1943).

[5] A. K. Bose, K. Gupta, and M. S. Manhas, *J.C.S. Chem. Comm.*, 86 (1984).

Zinc–Acetic acid, 5, 757–758.

α-Amino ketones.[1]   Acyl nitriles, available by reaction of acid chlorides with CuCN or CNSi(CH$_3$)$_3$, are reduced by zinc–acetic acid and excess acetic anhydride to N-acetyl derivatives of α-amino ketones.

Example:

$$
CH_3OOC(CH_2)_2COCl \xrightarrow[70-80\%]{\substack{CuCN, \\ CH_3CN}} CH_3OOC(CH_2)_2\overset{\overset{\displaystyle O}{\|}}{C}CN \xrightarrow[83\%]{\substack{Zn/HOAc, \\ Ac_2O}}
$$

$$
CH_3OOC(CH_2)_2\overset{\overset{\displaystyle O}{\|}}{C}CH_2NHAc
$$

[1] A. Pfaltz and S. Anwar, *Tetrahedron Letters*, **25**, 2977 (1984).

Zinc–Chlorotrimethylsilane, 5, 714; **6**, 628; **7**, 429–430; **8**, 532–533; **10**, 98.

Cyclopentanols.[1]   Cyclic ketones bearing a δ,ε-unsaturated side chain at the α-position undergo five-membered ring annelation to cyclic pentanols on reaction with zinc and chlorotrimethylsilane in the presence of 2,6-lutidine. The reaction probably involves reduction of the keto group to an α-trimethylsilyloxy radical, which adds to the unsaturated center. The bicyclic products have a *cis*-ring fusion.

Examples:

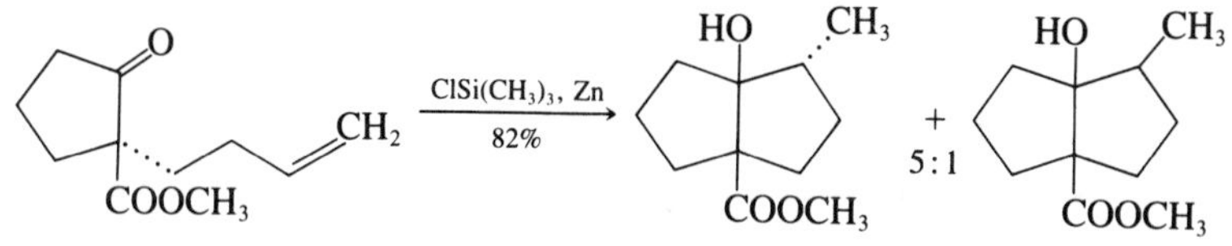

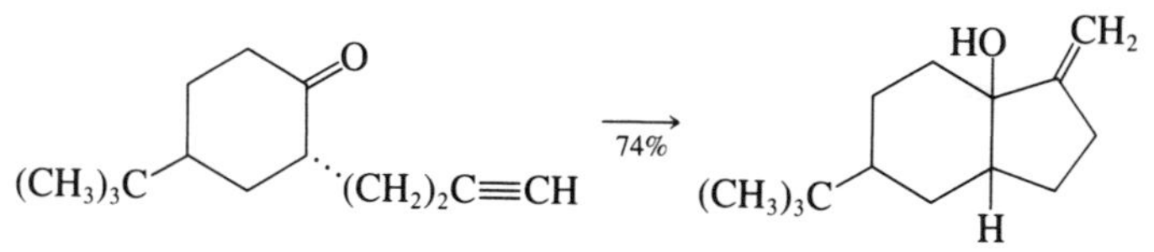

Reduction of epoxides.[2] Epoxides are reduced in CH_2Cl_2 by these two reagents. Unsymmetrical epoxides are reduced with some regioselectivity favoring the less substituted alcohol.

Example:

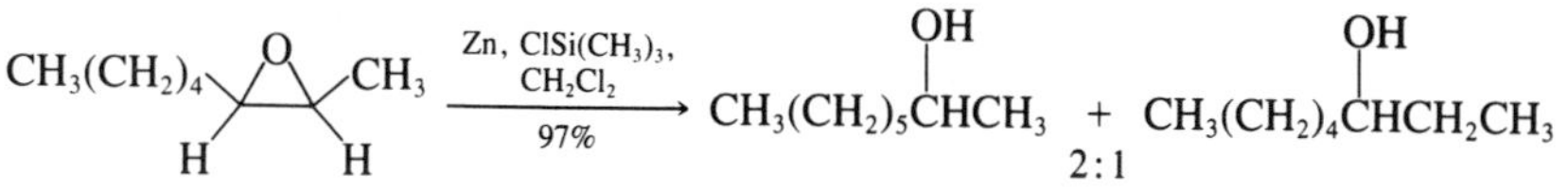

[1] E. J. Corey and S. G. Pyne, *Tetrahedron Letters*, **24**, 2821 (1983).
[2] Y. D. Vankar, P. S. Arya, and C. T. Rao, *Syn. Comm.*, **13**, 869 (1983).

Zinc–Copper couple, 1, 1292–1293; 5, 758–760; 7, 428–429; 8, 533–534; 10, 459; 11, 599. A highly active couple can be prepared by addition of DMF to $CuSO_4 \cdot 5H_2O$ and Zn dust (1:10). The mixture blackens and is ready for use after cooling and dilution with ether.

Deiodination.[1] This couple is recommended for deiodination of the 1,4-diiodide **1** to the 1,3-diene **2**.

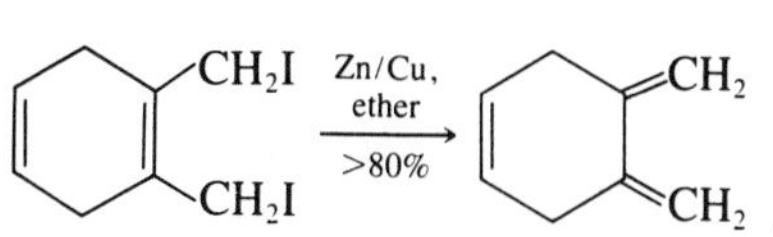

Reduction of alkenes.[2] Alkenes substituted by one or two electronegative substituents can be reduced by the Zn/Cu couple of Smith and Simmons (**1**, 1292) in refluxing methanol in yields of >85%. Some typical alkenes and the yield of the corresponding alkane are formulated.

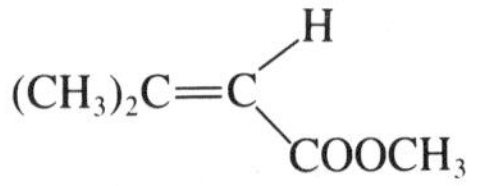

(98%)

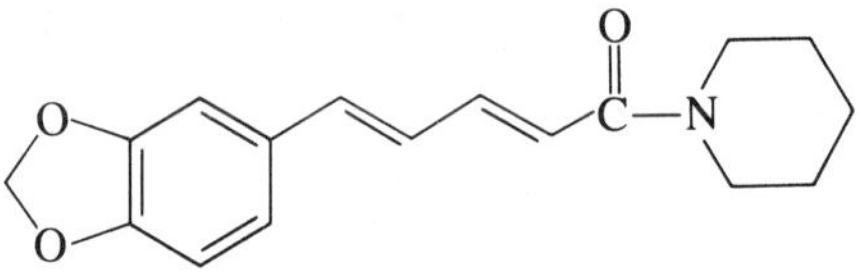

(tetrahydro derivative, 98%)

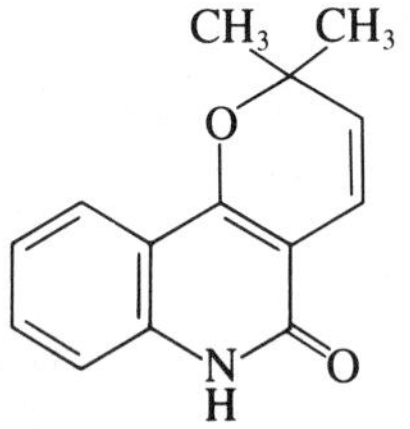

(dihydro derivative, 85%)

[1] R. O. Angus, Jr., and R. P. Johnson, *J. Org.*, **48**, 273 (1983).
[2] B. L. Sondengam, Z. T. Fomum, G. Charles, and T. MacAkam, *J.C.S. Perkin I*, 1219 (1983).

Zinc–1,2-Dibromoethane.

Reduction of alkynes.[1]   Zinc powder activated by 1,2-dibromoethane reduces conjugated diynes to (Z)-enynes in ≥70% yield. Triple bonds conjugated with an aryl group are also reduced (80–90% yield). A more active, but less selective, reagent is obtained by activation with dibromoethane and then with CuBr and LiBr. This activated zinc reduces monoacetylenic compounds substituted by OH, NR_2, and OR groups to the (Z)-alkenes.

[1] M. H. P. J. Aerssens and L. Brandsma, *J.C.S. Chem. Comm.*, 735 (1984).

Zinc–Graphite, Zn–Gr (1).

This metal is obtained by reaction of anhydrous $ZnCl_2$ with potassium–graphite suspended in THF (exothermic reaction).

This reagent is useful for preparation of Reformatsky reagents and for zincation of allylic bromides.[1]

Examples:

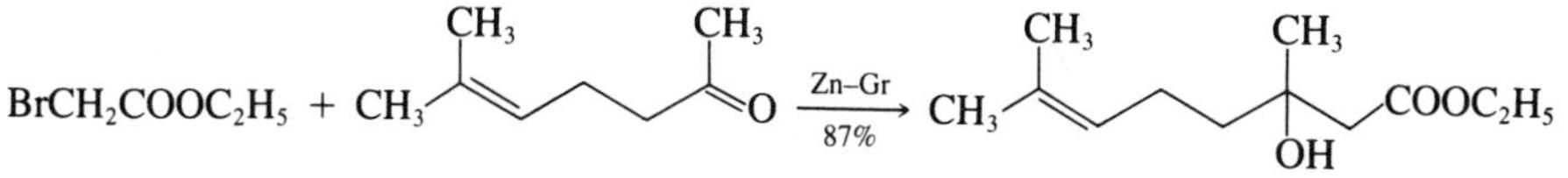

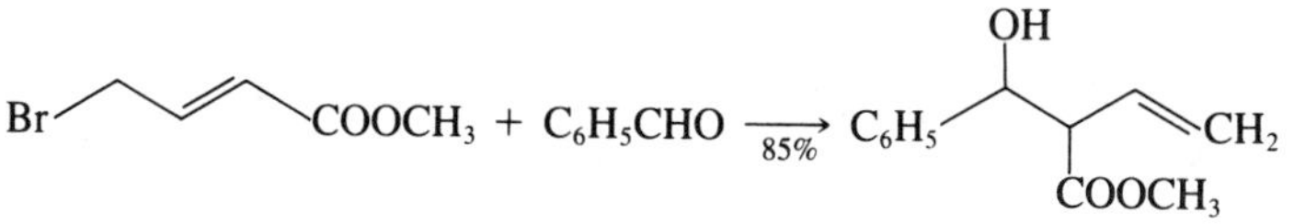

$$BrCH_2COOC_2H_5 + CH_3\text{[structure]}O \xrightarrow[87\%]{Zn-Gr} CH_3\text{[structure]}COOC_2H_5$$

$$Br\text{[structure]}COOCH_3 + C_6H_5CHO \xrightarrow{85\%} C_6H_5\text{[structure]}CH_2 \\ COOCH_3$$

(syn/anti = 45:55)

$$CH_2{=}\overset{COOSi(CH_3)_3}{\underset{}{C}}{-}CH_2Br \quad + \quad \text{[cyclohexanone]} \xrightarrow{90\%} \text{[spiro lactone]} {=}CH_2$$

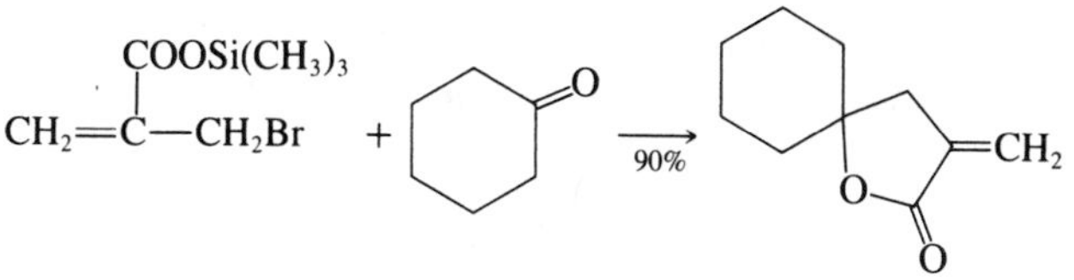

[1] G. P. Boldrini, D. Savoia, E. Tagliavini, C. Trombini, and A. Umani-Ronchi, *J. Org.*, **48**, 4108 (1983).

Zinc–Silver couple, 5, 760–761; 9, 519; 10, 460.

o-Quinodimethanes.[1] Zinc–Ag couple is superior to activated zinc for generation of the *o*-quinodimethane **2** from the α,α'-dibromo-*o*-xylene **1**. Cycloaddition of **2** with alkenes provides an attractive route to occidol sesquiterpenes.

Example:

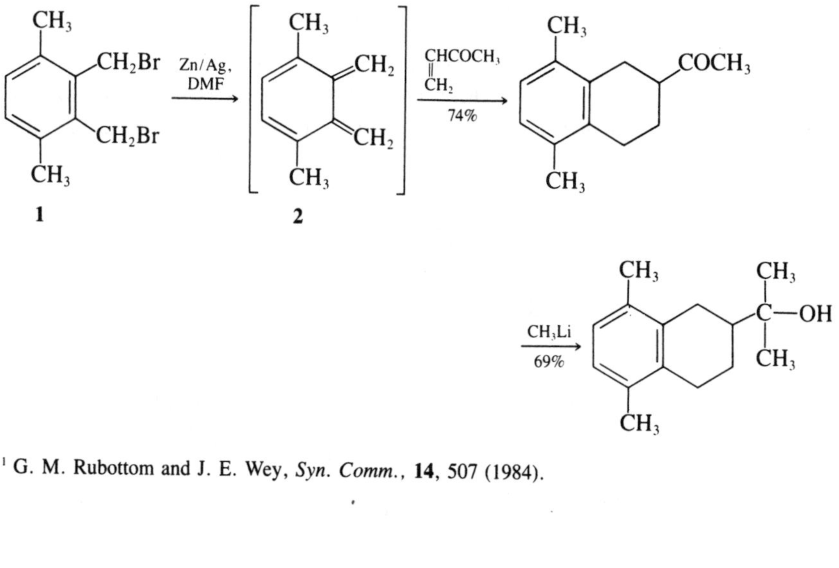

[1] G. M. Rubottom and J. E. Wey, *Syn. Comm.*, **14**, 507 (1984).

Zinc borohydride, 3, 337–338; **5**, 761–762; **10**, 460–461; **11**, 599–600.

Stereoselective reduction of β-keto amides.[1]   α-Alkyl-β-keto amides (**1**) are reduced with high *syn*-selectivity by Zn(BH$_4$)$_2$ or by the combination of NaBH$_4$ and Zn(ClO$_4$)$_2$ (equation I).

(I)

R^1 ⟶ Zn(BH$_4$)$_2$, ether, −78° / 85–97% ⟶ R^1 (OH, CH$_3$) NHR^2 + R^1 (OH, CH$_3$) NHR^2

~98 : 2

1

Stereoselective reduction of acyclic ketones.   α-Hydroxy ketones (**1**) are reduced by Zn(BH$_4$)$_2$ selectively to the *anti,vic*-diol **2**.[2] In contrast, the *t*-butyldiphenylsilyloxy derivative of **1** is reduced selectively to the *syn,vic*-diol **3** by sodium bis(2-methoxy-ethoxy)aluminum hydride.[3]

R^1 (OH) R^2, O   ⟶   R^1 (OH) R^2, OH   +   R^1 (OH) R^2, OH

1   **2**   **3**

Zn(BH$_4$)$_2$ ~85 : 15
LiAlH$_4$ ~80 : 20

R^1 (OSi(C$_6$H$_5$)$_2$C(CH$_3$)$_3$) R^2, O   —1) SMEAH / 2) F$^−$→ **2** + **3**

~15 : 85

These two stereoselective reductions also apply to reduction of an α-hydroxy-β-methyl ketone (**4**) to give *anti*- and *syn*-**5**. These diols were used to prepare the four possible diastereomers of a 1,3-dimethyl-2-hydroxy unit frequently found in polyether antibiotics.

CH$_3$ (O, CH$_3$, OH) OC(CH$_3$)$_3$ —Zn(BH$_4$)$_2$→ CH$_3$ (OH, CH$_3$, OH) OC(CH$_3$)$_3$

4   *anti*-**5** (25 : 1)

1) *t*-Bu(CH$_3$)$_2$SiCl
2) SMEAH
⟶
CH$_3$ (OH, CH$_3$, OH) OC(CH$_3$)$_3$

syn-**5** (30 : 1)

Each diol was converted into the corresponding α- and β-epoxides, which were opened regioselectively by 2-lithio-1,3-dithiane. The complete sequence from *anti*-5 is formulated in equation (I).[4]

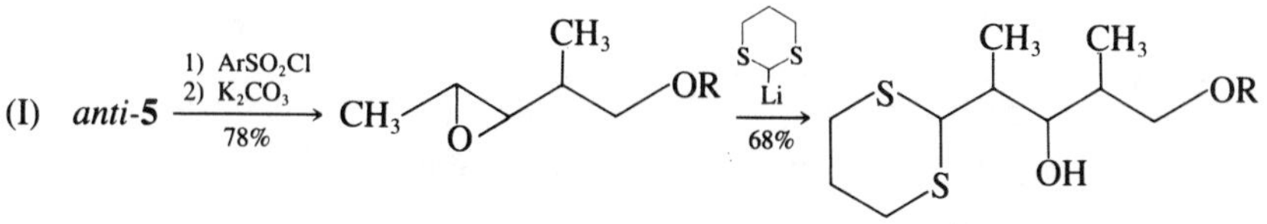

Stereoselective reduction of chiral 2-alkyl-3-keto amides.[5]   The chiral propionamide (**1**) derived from *trans*-2,5-bis(methoxymethoxymethyl)pyrrolidine[6] undergoes stereoselective acylation of the enolate in the presence of $ZnCl_2$ to give 2-alkyl-3-oxo amides (**2**). These products undergo reduction with zinc borohydride to give *syn*-2-alkyl-3-hydroxy amides (**3**).

Example:

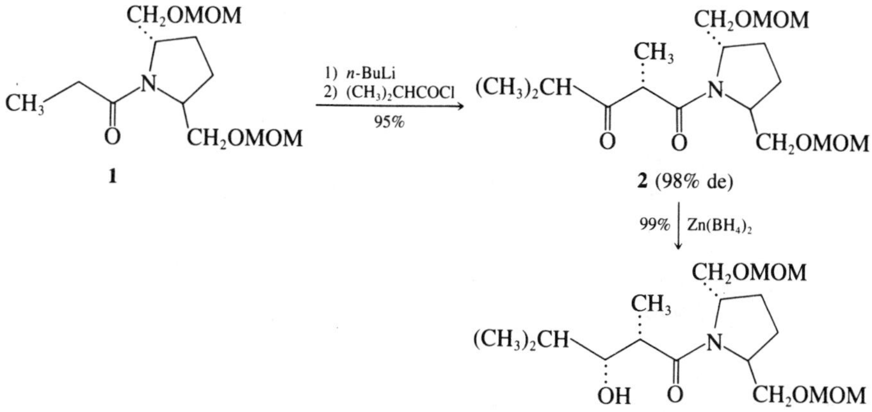

Dehalogenation.[7]   Tertiary and benzylic chlorides and bromides are reduced to the corresponding hydrocarbons by $Zn(BH_4)_2$ in ether. Yields are generally >90%. Vinylic and allylic halides are not reduced.

[1] Y. Ito and M. Yamaguchi, *Tetrahedron Letters,* **24**, 5385 (1983).
[2] T. Nakata, M. Fukui, and T. Oishi, *ibid.,* 2657 (1983).
[3] T. Nakata, T. Tanaka, and T. Oishi, *ibid.,* 2653 (1983).
[4] T. Nakata, M. Fukui, H. Ohtsuka, and T. Oishi, *Tetrahedron,* **40**, 2225 (1984).
[5] Y. Ito, T. Katsuki, and M. Yamaguchi, *Tetrahedron Letters,* **25**, 6015 (1984).
[6] Y. Kawanami, Y. Ito, T. Kitagawa, Y. Taniguchi, T. Katsuki, and M. Yamaguchi, *ibid.,* 857 (1984).
[7] S. Kim, C. Y. Hong, and S. Yang, *Angew. Chem., Int. Ed.,* **22**, 562 (1983).

Zinc borohydride–Dimethylformamide, $Zn(BH_4)_2 \cdot 1.5$ DMF. This complex is obtained as a stable solid by addition of DMF to an ethereal solution of $Zn(BH_4)_2$.

Reduction of ketones.[1] Reduction of one carbonyl group requires 2 moles of this complex because B_2H_6 is also released. Aliphatic ketones are reduced readily, but aromatic ketones are reduced much more slowly, and α,β-enones even more slowly. Consequently some selective reductions are possible.

A similar $Cd(BH_4)_2 \cdot 1.5$DMF can be prepared and appears to have similar properties. This reagent may be generally useful for reduction of aroyl chlorides to aryl aldehydes.

[1] B. J. Hussey, R. A. W. Johnstone, P. Boehm, and I. D. Entwistle, *Tetrahedron,* **38**, 3769 (1982).

Zinc chloride, 1, 1289–1292; **2**, 464; **3**, 338; **5**, 763–764; **6**, 676; **7**, 430; **8**, 536–537; **9**, 522–523; **10**, 461–462; **11**, 602–604.

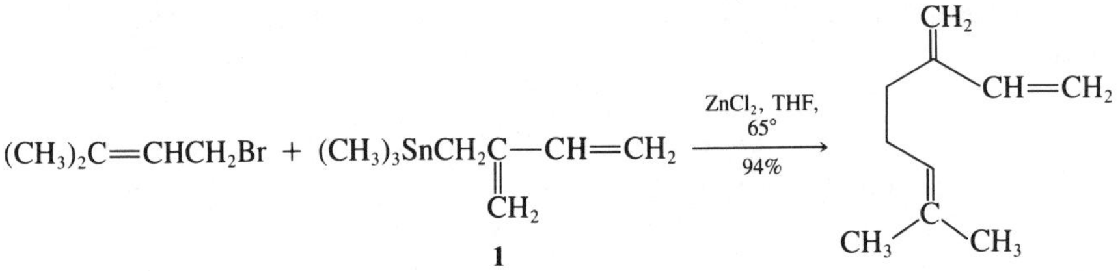

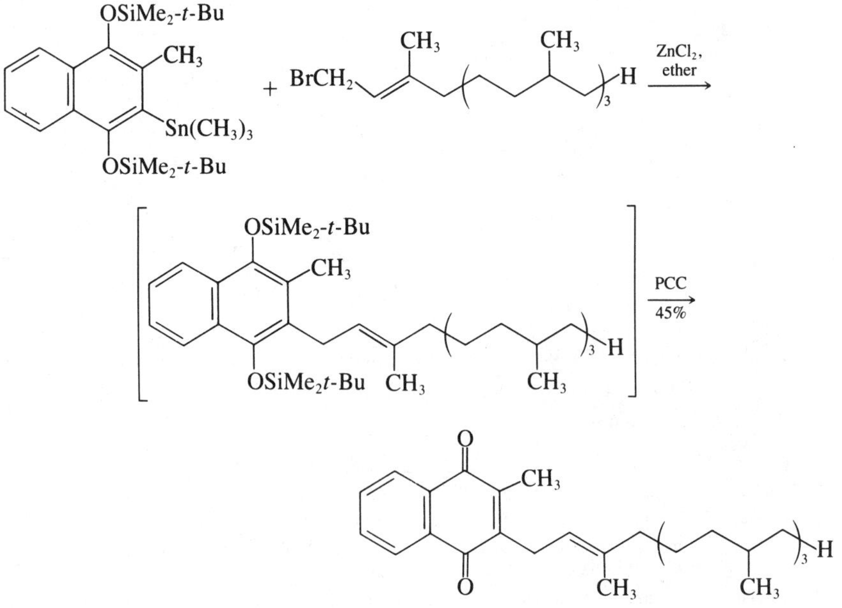

Cross-coupling of allyl bromides with allyltins (*cf.* **10**, 26).   The coupling of prenyl bromide with the tin reagent **1** in the presence of 10% $ZnCl_2$ gives myrcene (**2**) in high yield. This coupling also provides a synthesis of vitamin K_1 (**3**).[1]

Acylation of aryl and alkenyl halides.[2]   The zinc salts of enol ethers and allenic ethers couple with aryl and alkenyl halides in the presence of several Pd(0) catalysts: $Pd[P(C_6H_5)_3]_4$ or bis(dibenzylideneacetone)palladium with added phosphine ligands.

Examples:

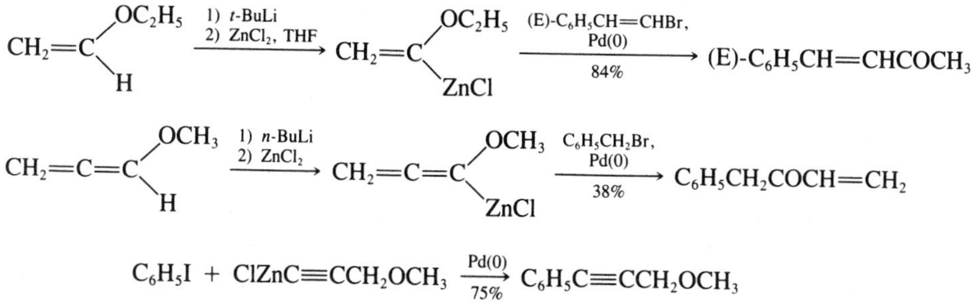

$$C_6H_5I \ + \ ClZnC\equiv CCH_2OCH_3 \ \xrightarrow[75\%]{Pd(0)} \ C_6H_5C\equiv CCH_2OCH_3$$

Stereoselective α-alkylation of ketones.[3]   This reaction can be effected by reaction of silyl enol ethers with benzyl acetates complexed with $Cr(CO)_3$ in the presence of $ZnCl_2$ (1 equiv.). This methodology is particularly useful because only the adduct *anti* to the metal is obtained. Use of an optically active chromium complex such as **1** results in 100% stereoselective alkylation.

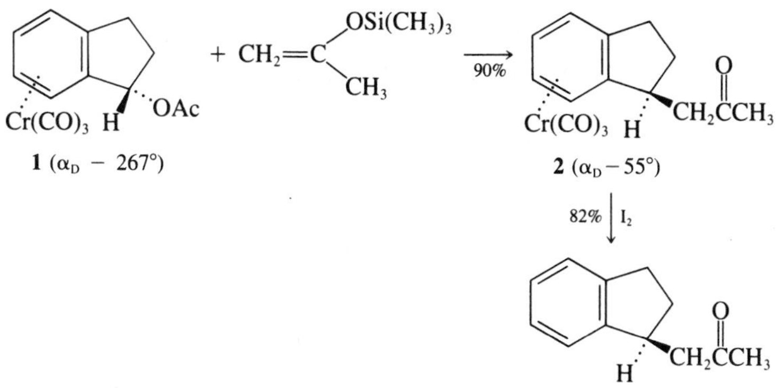

Allyl cations (**10**, 461–462).   Allyl cations undergo [2 + 2] cycloaddition with alkenes to afford cyclobutanes.[4]

Example:

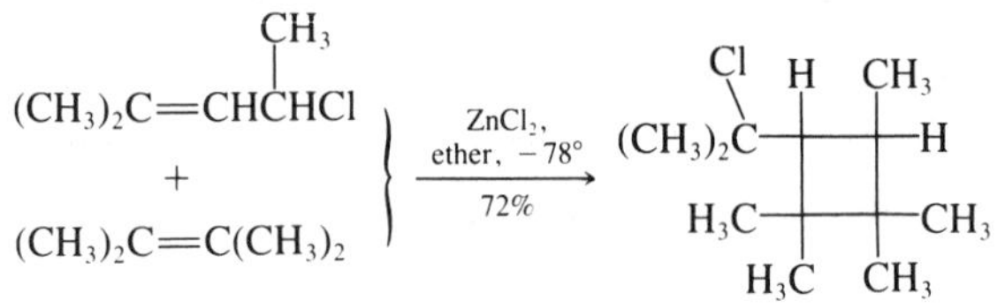

[1] J. P. Godschalx and J. K. Stille, *Tetrahedron Letters*, **24**, 1905 (1983).
[2] C. E. Russell and L. S. Hegedus, *Am. Soc.*, **105**, 943 (1983).
[3] M. T. Reetz and M. Sauerwald, *Tetrahedron Letters*, **24**, 2837 (1983).
[4] H. Klein, G. Freyberger, and H. Mayr, *Angew. Chem., Int. Ed.*, **22**, 49 (1983).

Zinc iodide, 1, 1299; **5**, 765; **10**, 462–463; **11**, 604–605.

Sulfides.[1]   Benzylic or allylic alcohols react with thiols in the presence of zinc iodide (0.5 equiv.) in 1,2-dichloroethane at 20° to form sulfides in yields generally of 80–90%.

[1] Y. Guindon, R. Frenette, R. Fortin, and J. Rokach, *J. Org.*, **48**, 1357 (1983).

Zinc permanganate, $Zn(MnO_4)_2$.
Preparation:

$$3\ BaMnO_4 + ZnO + 3H_2SO_4 \rightarrow Zn(MnO_4)_2 + MnO_2 + 3BaSO_4 + 3H_2O$$

Oxidations.[1]   Zinc permanganate supported on silica gel is useful for oxidations in organic solvents, usually $CHCl_3$ or CH_2Cl_2. Workup involves merely filtration and removal of the solvent. Under comparable conditions, $Zn(MnO_4)_2$ is more reactive than $KMnO_4$; $Mg(MnO_4)_2$ exhibits intermediate reactivity.

$Zn(MnO_4)_2$ effects several known oxidations such as oxidation of alkynes to α-diketones, cyclic ethers to lactones, and cycloalkanones to dicarboxylic acids. The paper reports one novel transformation: Oxidation of a cyclohexene to a ketol in moderate yield (36%), equation (I).

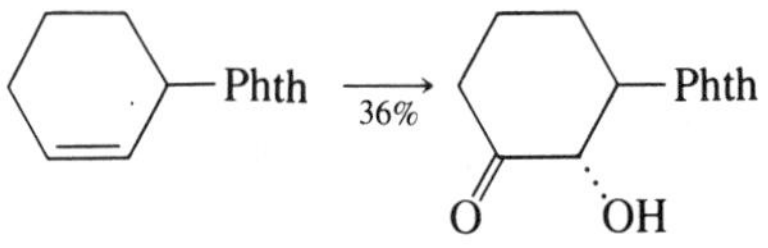

[1] S. Wolfe and C. F. Ingold, *Am. Soc.*, **105**, 7755 (1983).

Zinc trifluoromethanesulfonate, $Zn(OTf)_2$. The salt is prepared by reaction of zinc carbonate in CH_3OH with triflic acid at 23° → reflux (98% yield). Supplier: Fluka.

Thioketalization.[1] Zinc (or magnesium) triflate is an efficient catalyst for thioketalization with 1,2-ethanedithiol, particularly in the cases of acid-sensitive ketones and highly hindered ketones (*e.g.*, camphor ethane dithioketal, 99% yield). Δ^4-Cholestenone is converted into the corresponding dithioketal cleanly with zinc triflate, but into a 4:1 mixture of the Δ^4- and Δ^5-cholestenone dithioketals with magnesium triflate. In general, zinc triflate is more effective than magnesium triflate.

The ethylene thioketal of the unstable, acid- and base-sensitive ketone **1** can be obtained in 85% yield with zinc triflate as catalyst in CH_2Cl_2 at 23°.[2]

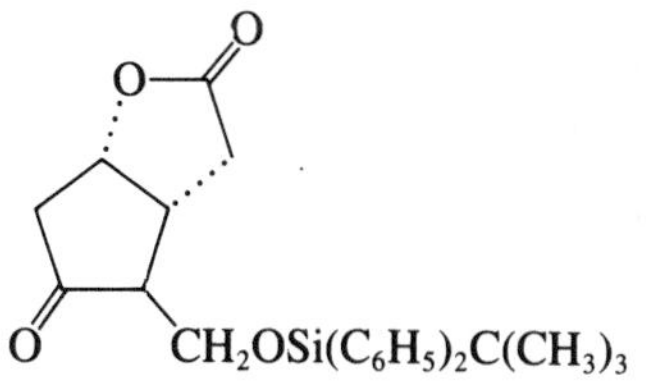

1

[1] E. J. Corey and K. Shimoji, *Tetrahedron Letters*, **24**, 169 (1983).
[2] *Idem, Am. Soc.*, **105**, 1662 (1983).

Zirconium carbene complexes, $Cp_2Zr\begin{smallmatrix}CHCH_2R\\L\end{smallmatrix}$ (**1**).

Preparation:[1]

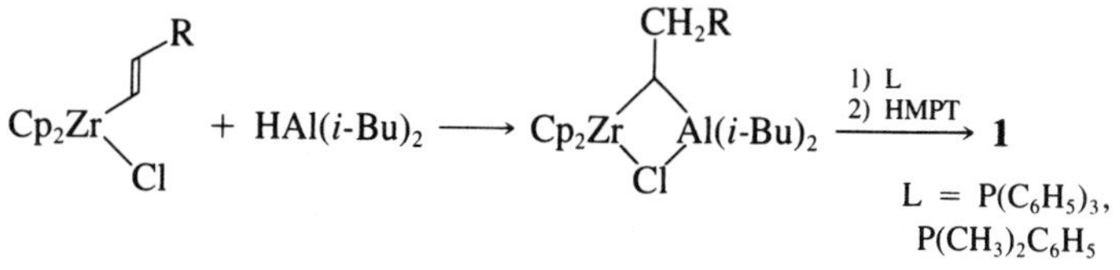

Wittig-type olefination. These complexes convert esters to vinyl ethers and ketones to olefins in 78–100% yield. They are somewhat less reactive than the corresponding Ti complexes.[1] In the olefination of aldehydes, ketones, and thioketones, the (Z)-isomer is formed with moderate selectivity. However, the (E/Z)-selectivity can be reversed by use of the imine or imidate.[2]

Example:

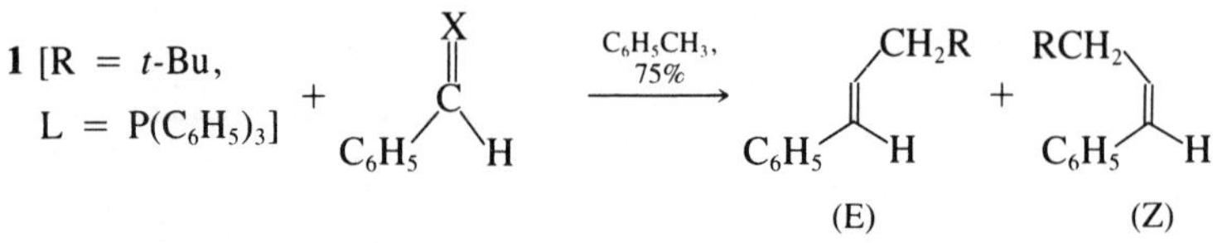

X		
X = O	76%	(E/Z = 0.50)
X = NCH$_3$	88%	(E/Z = 2.06)
X = NC(CH$_3$)$_3$	16%	(E/Z = 4.32)

[1] F. W. Hartner, Jr., and J. Schwartz, *Am. Soc.*, **103**, 4979 (1981); F. W. Hartner, Jr., J. Schwartz, and S. M. Clift, *ibid.*, **105**, 640 (1983).
[2] S. M. Clift and J. Schwartz, *ibid.*, **106**, 8300 (1984).

INDEX OF REAGENTS ACCORDING TO TYPES

ACETONYLATION: Acetoacetic acid. 2-Methoxypropene.

ALDOL CONDENSATION:
Aceto(carbonyl)cyclopentadienyl(triphenylphosphine)iron. 3-Acylthiazolidine-2-thiones. Bis(cyclopentadienyl)titanacyclobutanes. Bromomagnesium diisopropylamide. Cerium(III) chloride. Dichlorophenylborane. Dimethylphenylsilyllithium. Ethylene chloroboronate. Ketene bis(trimethylsilyl)ketals. Mandelic acid. Norephedrine. Potassium fluoride–Alumina. (S)-(—)-Proline. Tetra-n-butylammonium fluoride. Tin(IV) chloride. Tin(II) trifluoromethanesulfonate. Titanium(IV) chloride. Tri-n-butyltin fluoride. Trityl perchlorate.

ALLYLIC ACETOXYLATION: Bromine–Silver acetate. Palladium(II) trifluoroacetate.

ALLYLIC AMINATION: Bis(methoxycarbonyl)sulfur diimide.

ALLYLIC OXIDATION: t-Butyl hydroperoxide–Chromium carbonyl. 2,3-Dichloro-5,6-dicyanobenzoquinone.

ALLYLIC REARRANGEMENT: Bis(trimethylsilyl)peroxide-Vanadyl acetylacetonate.

AMINATION: Azidomethyl phenyl sulfide. Trimethylsilylmethyl azide.

ASYMMETRIC REDUCTION OF KETONES:
(S)-(—)-2-Amino-3-methyl-1,1-diphenylbutane-1-ol. (S)-4-Anilino-3-methylamino-1-butanol. 2,2-Dihydroxy-1,1-binaphthyl. (—)-N-Methylephedrine. B-3-Pinanyl-9-borabicyclo[3.3.1]nonane.

BAEYER-VILLIGER REACTION: Permaleic acid.

BAKER-VENKATORAMAN REARRANGEMENT: Potassium t-butoxide-t-Butyl alcohol.

BARBIER CYCLIZATION: Samarium(II) iodide.

BECKMANN FRAGMENTATION: Trimethylsilyl trifluoromethanesulfonate.

BECKMANN REARRANGEMENT: Alkylaluminum halides.

BENZOYLATION: Benzoyl trifluoromethanesulfonate.

BIRCH REDUCTION: Calcium–Amines.

BROMOCYCLIZATION: Bromonium dicollidine perchlorate.

BROMINATION: 1,3-Dibromo-5,5-dimethylhydantoin.

BROMOLACTONIZATION: Bromine.

BUCHNER REACTION: Rhodium(II) acetate.

CHLORINATION: Chlorine oxide. N-Chlorotriethylammonium chloride.

CLAISEN ESTER CONDENSATION: Dichlorobis(trifluoromethanesulfonate)-titanium(IV).

CLAISEN REARRANGEMENT: Lithium hexamethyldisilylazide. Lithium methylsulfinylmethylide. Organoaluminum compounds. Sodium methylsulfinylmethylide.

CLEAVAGE OF:
Acetals: N-Bromosuccinimide. t-Butyl hydroperoxide. Dimethylboron bromide. Nafion-H. Organocopper reagents. Organotitanium reagents. Titanium(IV) chloride.

Allylic alcohols: t-Butyl alcohol–Bisoxobis(2,4-pentanedionate)molybdenum.

N,N-Dimethylhydrazones: m-Chloroperbenzoic acid. Ferric nitrate/K10 Bentonite.

Disilanes: Tetra-n-butylammonium fluoride.

Enamines: Oxygen, singlet.

Epoxides: Cyanotrimethylsilane. Dilithium tetrabromonicklelate. Lithium borohydride. Organoaluminum reagents. Sodium hydrogen telluride.

Ethers: Aluminum chloride–Sodium iodide. Aluminum iodide. Bis(isopropylthio)boron bromide. Boron trichloride. Boron trifluoride etherate. Bromotrimethylsilane. 2-Chloro-1,3,2-dithioborolane. Chlorotrimethylsilane–Acetic anhydride. Chlorotrimethylsilane–Sodium iodide. Dimethylboron bromide. Pyridinium p-toluenesulfonate. Sodium methaneselenolate.

579

CLEAVAGE OF (*Continued*)
Glycosyl esters: Lithium iodide.
Lactols: Titanium(IV) chloride.
Oxetanes: Boron trifluoride. Cobalt *meso*-Tetraphenylporphine.
Oximes: Tri-μ-carbonylhexacarbonyldiiron.
Phthalimides: Sodium borohydride.
Tetrahydropyranyl ethers: Alkylaluminum halides.
Thioacetals: Ferric nitrate/K10 Bentonite.
COPE REARRANGEMENT:
Bis(acetonitrite)dichloropalladium.
CRIEGEE REARRANGEMENT: Ozone.
CURTUIS REARRANGEMENT: Trifluoroacetic acid.
[1 + 2] CYCLOADDITION: Cyclopropenone 1,3-propanediyl ketal.
[2 + 2] CYCLOADDITION: Chlorocyanoketene. N,N-Diethylaminopropyne. Diethyl (diaomethyl)phosphonate.
[3 + 2] CYCLOADDITION: N-Alkylhydroxylamines. *t*-Butyldimethylsilyl ethylnitronate. Cyclopropenone 1,3-propanediyl ketal. Hydrogen peroxide–Sodium tungstate. N-Methylhydroxylamine. Phenyl isocyanate. Trimethylamine N-oxide. Trimethylsilylmethyl azide.
[6 + 2] CYCLOADDITION: Titanium(IV) chloride–Diethylaluminum chloride.
[2 + 2 + 2] CYCLOADDITIONS:
Dicarbonylcyclopentadienylcobalt. Dicyclopentadienylcobalt.
CYCLODEHYDRATION: Trifluoromethanesulfonic acid. Triphenylphosphine–Carbon tetrachloride.
CYCLODEHYDROGENATION: Trityl tetrafluoroborate.
CYCLOPROPENATION: Bromoform–Diethylzinc. Menthol.

DARZEM REACTION: Sodium hydride.
DEALKOXYCARBONYLATION: Propionic acid.
DEALKYLATION OF AMINES: *t*-Butyldimethylsilyl triflouromethanesulfonate. 2-Chloroethyl chloroformate.
N-DEBENZYLATION: Oxygen.
DECARBONYLATION: Chlorobis[1,3-bis(diphenylphosphine)propane]rhodium.
DECARBOXYLATION: Chlorobis[1,3-bis(diphenylphosphine)propane]rhodium.

Dimethylformamide. Organocopper reagents. 2-Pyridinethiol-1-oxide.
DECARBOXYLATIVE DEHYDRATION:
N,N-Dimethylformamide dineopentyl acetal. Triphenylphosphine–Diethyl azodicarboxylate.
DEHALOGENATION: Aluminum chloride–Ethanethiol. Sodium sulfide. Zine borohydride.
DEHYDRATION: Dimethylformamide–Thionyl chloride. Ethoxy(trimethylsilyl)acetylene. Potassium hydride.
DEHYDROBROMINATION: Lithium chloride–Hexamethylphosphoric triamide. Potassium fluoride–Alumina.
DEHYDROGENATION: Dimethyl bromomalonate. Dimethylsulfoxide–Iodine. Palladium(II) acetate. Palladium black. *p*-Toluenesulfinyl chloride.
DEHYDROHALOGENATION: Alkylaluminum halides. Triphenylmethylpotassium.
DEHALOGENATION: Diethyl phosphite.
DEHYDROGENATION: Palladium black.
DEHYDROSULFONYLATION: Potassium *t*-butoxide.
DENITRATION: 1-Benzyl-1,4-dihydronicotinamide. Lithium aluminum hydride. Tri-*n*-butyltin hydride.
DEOXIMATION: Bispyridinesilver permanganate. Raney nickel. Tri-μ-carbonylhexacarbonyldiiron.
DEOXYGENATION OF:
Alcohols: Iodotrimethylsilane. Sodium borohydride. Tri-*n*-butyltin hydride.
Epoxides: Dimethyl diazomalonate. Triphenylphosphine–Dihalides.
Nitrones: Titanium(III) chloride.
Phenols: Nonafluorobutansulfonyl fluoride.
Sulfoxides: Dimethylboron bromide.
DESELENENIATION: Nickel boride.
DESILYLATION: Boron trifluoride. Cesium fluoride. Tri-*n*-butyltin fluoride.
DESULFONYLATION: Tri-*n*-butyltinlithium.
DESULFURATION: Diphosphorus tetraiodide.
DIAZO TRANSFER: *p*-Dodecylbenzenesulfonyl azide.
DIECKMANN CONDENSATION: Dichlorobis(trifluoromethanesulfonate)-titanium(IV). Potassium.
DIELS-ALDER CATALYSTS: Alkylaluminum halides. Boron trifluoride etherate. Diaza-

dieneiron(O). Tris(6,6,7,7,8,8,8-hepta-
fluoro-2,2-dimethyl-3,5-octanedion-
ato)europium.
DIELS-ALDER DIENOPHILES: Bis(2,2,2-
trichloroethyl) azodicarboxylate. 1-Bromo-
2-chlorocyclopropene. Chromium carbene
complexes. (R)-Ethyl p-tolysulfinyl-
methylenepropionate. (S)-α-Hydroxy-β,β-
dimethylpropyl vinyl ketone. (1R)-cis-3-
Hydroxy isobornyl neopentyl ether. Oxazolidones, chiral. Phenylselenenyl benzenesulfonate.
DIELS-ALDER REACTIONS: Allylidenecy-
clopropane. 1-Bromo-2-chlorocyclopro-
pene. 2-Bromomethyl-3-(trimethylsilylmethyl)butadiene. Camphor-10-sulfonic acid.
1,3-Dimethoxy-1-trimethylsilyloxy-1,3-bu-
tadiene. 1,2-Dimethoxy-1-trimethylsily-
loxy-1,3-pentadiene. Dimethyl 1,2,4,5-
tetrazine-3,6-dicarboxylate. 1-Methoxy-
2,4-dimethyl-3-trimethylsilyloxy-1,3-buta-
diene. 6-Methoxy-3,5-hexadienoic acid. 1-
Methoxy-3-trimethylsilyloxy-1,3-buta-
diene. 4-Phenyloxazole. 2-Trimethylsilyl-
methyl-1,3-butadiene.

ENE REACTIONS: Alkylaluminum halides.
Bis(methoxycarbonyl)sulfur diimide.
Chlorine oxide. 8-Phenylmenthol. Tin(IV)
chloride.
ENOL LACTONIZATION: Silver nitrate.
EPOXIDATION: Bromomethyllithium. Chloromethyllithium. Hydrogen peroxide–Sodium tungstate. Peroxytrichloroacetimidic
acid. Potassium o-nitrobenzeneperoxysul-
fonate. Potassium peroxomonosulfate. Potassium superoxide.
ESTERIFICATION: Alkyl chloroformates. 4-
(N,N-Diemethylamino)pyridinium chlorosulfite chloride. 1,1'-Di(methylcyclopentadienyl)tin. Ethyl acetoacetate.
N,N,N',N'-Tetramethylchloroformami-
dinium chloride.

FAVORSKII RING CONTRACTION: Barium
hydroxide.
FLUORINATION: Acetyl hypofluorite. Carbonyl difluoride. N-Fluoro-N-alkyl-p-tolu-
enesulfonamides.
FORMYLATION: Dimethyl(methylthio)-
sulfonium tetrafluoroborate.

GIESE REACTION: Mercury(II) acetate. Tri-n-
butyltin hydride.
GLYCOSYLATION: Tetrafluorosilane. Tri-
methylsilyl trifluoromethanesulfonate.
GOMBERG-BACHMANN BIPHENYL SYN-
THESIS: Phase-transfer catalysts.

HALOBORATION: B-Halo-9-borabicy-
clo[3.3.1]nonene. Monoisopinocamheyl-
borane. Thexylborane.
HELL-VOLHARD-ZELINSKY HALOGENA-
TION: 7,7,8,8-Tetracyanoquinodimethane.
HOMOALDOL REACTION: Allyl-9-borabicy-
clo[3.3.1]nonane. B-Allyldiisopinocam-
pheyl-borane. 2-Butenylcarbamates. Tri-n-
butylcrotyltin.
HOSOMI-SAKURAI REACTION: Tita-
nium(IV) chloride.
HYDRATION OF ALKYNES: Mercury(II)
acetate–Nafion-H.
HYDROBORATION: Diisopinocampheylborane. Di-2-mesitylborane. Monoisopino-
campheyl-borane.
HYDROFORMYLATION: Rhodium(II) carboxylates.
HYDROGENATION CATALYSTS: Arene-
chromium tricarbonyls. (1,5-Cycloocta-
diene)(pyridine)(tricyclohexylphosphine)iridium(I) hexafluorophosphate.
Di-μ-chlorobis(1,5-hexadiene)dirhodium.
Lindlar catalyst. Palladuim(II) acetate–Sodium hydride–t-Amyl alkoxide. Rhodium
catalysts.
HYDROSILYLATION: Diethoxymethylsilane.
Dimethylphenylsilane.
HYDROSTANNATION: Tri-n-butyltin trifluoromethanesulfonate.
α-HYDROXYLATION:
Ketones: Dibenzoyl peroxide. o-Iodosylbenzoic acid. 2-(Phenylsulfonyl)-3-phenyloxa-
ziridine.
HYDROZIRCONATION: Dichlorobis(cyclopentadienyl)zirconium–t-Butylmagnesium
bromide.

INVERSION, OF ALCOHOLS: Cesium propionate. Triphenylphosphine–Diethyl azodicarboxylate.
IODOCARBAMATION: Iodine.
IODOLACTONIZATION: Iodine.

ISOMERIZATION OF:
Alkenes: Potassium 3-aminopropylamide.
ISOMERIZATION:
α,β-Unsaturated carboxylic acids: Tetrakis(triphenylphospine)palladium.

MANNICH REACTION: Titanium(IV) chloride.
MARSHALK REACTION: Benzyltrimethylammonium hydroxide.
METHYLENATION: Alkyldimesitylboranes.
μ-Chlorobis(cyclopentadienyl)(dimethylaluminum)-μ-methylene-titanium. Methylene bromide–Zinc–Titanium(IV) chloride. Organomolybdenum reagents. Titanocene methylene–Zn halide complex. N,N,P-Trimethyl-P-phenylphosphinothioic amide. Trimethylstannylmethyllithium.
METHYTHIOMETHYLENATION: N-Methylthiomethylenepiperidine.
MITSUNOBU REACTION: Triphenylphosphine–Diethyl azodicarboxylate.
MUKAIYAMA ALDOL CONDENSATION: Titanium(IV) chloride.

NAZAROV CYCLIZATION: Aluminum chloride. Benzylchlorobis(triphenylphosphine)palladium. Bis(acetonitrile)-dichloropalladium(II). Boron trifluoride etherate. Tetrakis(triphenylphosphine)palladium.
NEF REACTION: N″-(t-Butyl)-N,N,N′,N′-tetramethylguanidinium m-iodylbenzoic acid.
NITRATION: Trifluoroacetyl nitrate.
NITRO-ALKOL REACTION: Tetra-n-butylammonium fluoride.

OXIDATION, REAGENTS: Barium manganate. Benzyl(triethyl)ammonium permanganate. Bispyridinesilver permanganate. Bis(trimethylsilyl)peroxide. t-Butyl hydroperoxide. t-Butyl hydroperoxide–Benzyltrimethylammonium tetrabromooxomolybdate. t-Butyl hydroperoxide–Bisoxobis-(2,4-pentadionato)molybdenum. t-Butyl hydroperoxide–Chromium carbonyl. t-Butyl hydroperoxide–Dialkyl tartrate–Titanium(IV) isopropoxide. t-Butyl hydroper-

oxide–Vanadyl acetylacetonate. N″-(t-Butyl)-N,N,N′,N′-tetramethylguanidinium m-iodylbenzoate. m-Chloroperbenzoic acid. Collins reagent. Chromic acid. 2,3-Dichloro-5,6-dicyanobenzoquinone. Diisopropyl sulfide–N-Chlorosuccinimide. Ferric nitrate/K10 Bentonite. Hydrogen peroxide. Hydrogen peroxide-Ammonium heptamolybdate. Hydrogen peroxide–Sodium tungstate. Iodosylbenzene. o-Iodosylbenzoic acid. Manganese(III) acetate. Osmium tetroxide. Periodinane. Phenyliodine(III) diacetate. Potassium permanganate. Potassium superoxide. Pyridinium chlorochromate. Pyridinium dichromate–Acetic anhydride. Pyridinium fluorochromate. Ruthenium(IV) oxide. Silver carbonate–Celite. Sodium perborate. Tetrabutylammonium chlorochromate. 2,2,6,6-Tetramethylpiperidine nitroxyl. 2,2,6,6-Tetramethylpiperidinyl-1-oxyl. Thallium(III) nitrate. Zinc permanganate.
OXIDATIVE DECARBOXYLATION: N-Iodosuccinimide.
OXIDATIVE DESULFONYLATION: Bis(trimethylsilyl)peroxide.
OXY-COPE REARRANGEMENT: Mercury(II) trifluoroacetate. Potassium hydride. Silver(I) trifluoromethanesulfonate.

PETERSON REACTION: Alkyldimesitylboranes. Chloromethyldiphenylsilane. Ethyl trimethylsilylacetate. Lithium 1-(dimethylamino)naphthalenide. Methyldiphenylchlorosilane. Phenylsulfonyl(trimethylsilyl)-methane. Titanium(IV) isopropoxide. Trimethylsilylmethylmagnesium chloride.
PINACOL REARRANGEMENT: Trialkylaluminum reagents.
PRÉVOST REACTION: Iodine–Silver acetate.
PRINS REACTION: Alkylaluminum halides.
PROTECTION OF:
Alcohols: t-Butyldiphenylchlorosilane. 2-Benzyloxy-1-propene. Bis(isopropylthio)boron bromide. t-Butylmethoxyphenylsilyl bromide. 2,3-Dichloro-5,6-dicyano-1,4-benzoquinone. Mercury(II) chloride.4,4′,4″-Tris(4,5-dichlorophthalamido)trityl bromide. Tetramethylguanidine.
Aldehydes: Lithium morpolide. Organotitanium compounds.

Amines: *t*-Butyl benzotriazol-1-yl carbonate. *t*-Butyl 2-pyridyl carbonate. N-(Ethoxycarbonyl)phthalimide. 2-(2-Pyridyl)ethyl-*p*-nitrophenyl carbonate. Sodium borohydride. 1,1,4,4-Tetramethyl-1,4-bis(N,N-dimethylamino)disilethylene. *p*-Toluenesulfonic acid.

Carboxylic acids: Phenacyl bromide. 2-Pyridimethanol. 4-Trimethylsilyl-2-butene-1-ol.

1,3-Dienes: N-Phenyl-1,2,4-triazoline-3,5-dione.

Diols: Bis(1-methoxy-2-methyl-1-propenyloxy)dimethylsilane.

Dicarboxylic acids: Bis(1-methoxy-2-methyl-1-propenyloxy)dimethylsilane.

Thiophenols: Trifluoroacetic anhydride.

Tryptophane: Di-*t*-butyl dicarbonate.

PSCHORR CYCLIZATION: Phase-transfer catalysts.

PSCHORR PHENANTHRENE SYNTHESIS: Sodium iodide.

PUMMERER REARRANGEMENT: Trifluoroacetic anhydride.

RAMBERG-BACKLUND REACTION: Bromomethanesulfonyl bromide.

REDUCTION, REAGENTS: Bis(N-methylpiperazinyl)aluminum hydride. Borane–Dimethyl sulfide. Borane–Tetrahydrofurane. Borane-Pyridine. *n*-Butyllithium–Diisobutylaluminum hydride. Calcium–Amines. Diisobutylaluminum hydride. 8-Hydroxyquinolinedihydroboronite. Lithium aluminum hydride. Lithium 9-boratabicyclo[3.3.1]nonane. Lithium *n*-butyldiisopropylaluminum hydride. Lithium tri-*sec*-butylborohydride. Lithium triethylborohydride. Monochloroalane. Nickel boride. 2-Phenylbenzothiazoline. Potassium 9-(2,3-dimethyl-2-butoxy)-9-boratabicyclo[3.3.1]nonane. Raney nickel. Sodium bis(2-methoxyethoxy)aluminum hydride. Sodium borohydride. Sodium borohydride–Nickel chloride. Sodium borohydride–Praseodymium chloride. Sodium(dimethylamino)borohydride. Sodium hydrogen telluride. Thexyl chloroborane–Dimethyl sulfide. Tri-*n*-butylphosphine–Diphenyl disulfide. Tri-*n*-butyltin hydride. Zinc–1,2-Dibromoethane. Zinc borohydride.

REFORMATSKY REACTION: Chlorotrimethylsilane. Tri-*n*-butyltinlithium–Diethylaluminum chloride-Zinc. Zinc–Graphite.

RESOLUTION: (S)-1-Amino-2-(silyloxymethyl)pyrrolidines. 1,1′-Binaphthyl-2,2′-diyl hydrogen phosphate. *t*-Butyl hydroperoxide–Diisopropyl tartrate–Titanium(IV) isopropoxide. Di-μ-carbonylhexacarbonyldicobalt. (2S)-(2α,3aα,4α,7α,7aα)-2,3,3a,4,5,6,7,7a-Octahydro-7,8,8-trimethyl-4,7-methanobenzofurane-2-ol.

RING EXPANSION: Hydrogen peroxide.

SAKURAI REACTION: Alkylaluminum halides. Titanium(IV) chloride.

SPIROKETALIZATION: Mercury(II) chloride.

STRECKER REACTION: Cyanotrimethylsilane.

SULFENOCYCLOAMINATION: Benzenesulenyl chloride.

SYNTHESIS OF:

ACETALS: Iodotrimethylsilane.

α-ACETOXY ESTERS: Lead tetraacetate.

α-ACETOXY LACTONES: Lead tetraacetate.

ACETYLENIC KETONES: Boron trifluoride etherate. Dichloro[1,1′-bis(diphenylphosphino)ferrocene]palladium(II). Silver tetrafluoroborate.

ACYLOINS: Acyllithium reagents. Grignard reagents.

α-ACYLOXY ALDEHYDES: N-*t*-Butylhydroxylamine.

ALDEHYDES: Alkyldimesitylborane. Chloromethyltrimethylsilane. Diethyl[(2-tetrahydropyranyloxy)methylphosphonate. N,N-Dimethylchloromethyleniminum chloride. Dimethyl(methylthio)sulfonium tetrafluoroborate. Methoxy(phenylthio)methyllithium. Methylthiomethyl *p*-tolyl sulfone. 1-Phenylthio-1-trimethylsilylethylene. Tetrakis(triphenylphosphine)palladium. Thexylchloroborane-Dimethyl sulfide.

1-ALKENES: Methyldiphenyl chlorosilane. Palladium(II) acetate.

β-ALKOXY KETONES: Iodotrimethylsilane.

ALKYL HALIDES: N,N′-Carbonyldiimidazole. Dimethyl formamides. Trimethylsilylsilyl polyphosphate. Triphenylphosphine dihalides. Triphenylphosphine–Diethylazodicarboxylate–Zinc halide. Vilsmeier reagent.

ALKYNES: Sodium amalgam.

SYNTHESIS OF (*Continued*)

α,β-ALKYNYL KETONES: Boron trifluoride etherate.

ALLENES:
Bis(cyclopentadienyl)titanacyclobutanes.
2-Propynyltriphenyltin.

ALLENYLSILANES: Organocopper reagents.

ALLENYNES: Tetrakis(triphenylphosphine)palladium.

ALLYLIC ALCOHOLS: Allyltri-*n*-butyltin. Aluminum chloride. Chromium(II) chloride.

ALLYLIC AMINES: N-Alkyl-β-aminothylphosphonium bromides.
Bis(methoxycarbonyl)sulfur diimide.
Chloramine-T. N-Chlorosuccinimide.

ALLYLIC CHLORIDES: 1-Chloro-N,N,2-trimethylpropenylamine.

ALLYLSILANES: Organoaluminum reagents.

AMIDES: Borane–Trimethylamine. Methoxy(phenylthio)trimethylsilylmethyllithium.

AMINES: (S)-1-Amino-2-methoxymethyl-1-pyrrolidine. Azidomethyl phenyl sulfide. Bis(methoxycarbonyl)sulfur diimide. N,N-Bis(trimethylsilyl)methoxymethylamine. O,N-Dimethylhydroxylamine. Diphenyl phosphoroazidate. Iodosylbenzene. Sodium bis(trimethylsilyl)amide. Sodium hydrogen telluride. Triphenylphosphine.

α-AMINO ACIDS: L-(—)-Serine. Trichloroacetonitrile.

β-AMINO ACIDS: N,N-Bis(trimethylsilyl)-methoxymethylamine.

1,3-AMINO ALCOHOLS: Alkylaluminum halides. Lithium aluminum hydride.

α-AMINO KETONES: Zinc–Acetic acid.

α-AMINO NITRILES: Cyanotrimethylsilane. Methylbenzylamine.

ANHYDRIDES: Ethoxy(trimethylsilyl)acetylene. Sulfene.

ANILINES: Hydroxylamine. Silver carbonate.

ANTHRACYCLINES: Benzyltrimethylammonium hydroxide. Chromium carbene complexes. 1-Methoxy-3-trimethylsilyloxy-1,3-dienes.

ANTHRAQUINONES: 1,2-Dimethoxy-1-trimethylsilyloxy-1,3-pentadiene.

ARENE OXIDES: Potassium peroxomonosulfate.

ARYLACETIC ACIDS: Iodine-Silver nitrate.

ARYL METHYL ETHERS: Methyl trichloroacetate.

ARYL NITRILES: Thallium(III) trifluoroacetate.

2-ARYLPROPANOIC ACIDS: Sodium selenophenolate.

ARYL THIOLS: Tetraisopropylthiuram disulfide.

AZETIDINES: Triphenylphosphine–Diethyl azodicarboxylate.

α-AZIDO ETHERS: Hydrazoic acid.

AZLACTONES: Ethyl orthoformate.

BENZOFURANES; BENZOPYRANES: *sec*-Butyllithium.

BENZYL KETONES: Nickel(0).

BIARYLS: Nickel(0). Thallium(III) trifluoroacetate-Palladium(II) acetate.

BINAPHTHYLS: (1S,2S)-(+)-1-Methoxy-2-amino-3-phenyl-3-hydroxypropane.

BROMOHYDRINS: Dilithium tetrabromonickelate.

BUTENOLIDES: 1-Chloro-N,N-trimethylpropenylamine. Cyclopropenone 1,3-propanediyl ketal. Peracetic acid. Silver carbonate–Celite. (R)-(+)-3-(*p*-Tolylsulfinyl)-propionic acid.

γ-BUTYROLACTONES: Bromine–Nickel(II) alkanoates. Dichloroketene.

CARBACEPHAMS: Oxygen, singlet.

CARBAMATES: Rhodium catalysts.

β-CARBOLINES: Tetrakis(triphenylphosphine)palladium.

CARBOXYLIC ANHYDRIDES:
Ethoxy(trimethylsilyl)acetylene. Sulfene.

β-CHLORO-α,β-ENONES: Tin(IV) chloride.

α-CHLORONITRILES: Cyanotrimethylsilane.

CHROMANES: Menthyl *p*-toluenesulfinate.

CHROMENES: Trityl tetrafluoroborate.

CHROMONES: Mercury(II) acetate.

CYANOHYDRIN ETHERS: *t*-Butyl isocyanide.

CYANOMETHYL ETHERS: Diethylaluminum cyanide.

CYCLITOLS: 9-(Benzyloxy)methoxyanthracene.

CYCLOALKANONES: Titanium(III) chloride–Lithium aluminum hydride.

CYCLOALKENONES: N-Phenylketeniminyl(triphenyl)phosphorane.

CYCLOBUTANES: Zinc chloride.

CYCLOBUTANONES: 1-Bromo-1-ethoxycyclopropane. 1-Hydroxycyclopropanecarboxaldehyde.

CYCLOBUTENEDIONES: Dichloroketene.

CYCLOHEXA-2,4-DIENONES: Chromium carbene complexes.

CYCLOOCTENONES: Potassium hydride.

CYCLOPENTANONES: Aluminum chloride. Chlorotris(triphenylphosphine)rhodium. Dichloroketane. N,N-Diethylaminopropyne. 1-Hydroxycyclopropanecarboxaldehyde. Rhodium(II) carboxylates.
CYCLOPENTENES: Benzylchlorobis(triphenylphosphine)palladium(II). Cyclopropenone 1,3-propanediyl ketal. Di-μ-carbonylhexacarbonyldicobalt.
CYCLOPENTENONES: Bis(acetonitrile)dichloropalladium(II). Bis(cyclopentadienyl)diiodozirconium. Di-μ-carbonylhexacarbonyldicobalt. Organocopper reagents. Tin(IV) chloride.
CYCLOPROPANECARBOXYLATES: Bromoform–Diethylzinc. Nickel carbonyl.
CYCLOPROPANES: Cyclopropene 1,3-propanediyl ketal. Trimethylstannylmethyllithium.
CYCLOPROPYL KETONES: Bis(N-t-butylsalicyldimato)copper(II). Chloroethyl dimethylsulfonium iodide.
DEHYDROAMINO ACIDS: (Diethylamino)sulfur trifluoride–Pyridine. Oxygen, singlet.
gem-DIACETATES: Acetic anhydride–Ferric chloride.
vic-DIACETATES: Ammonium persulfate.
vic-DIAMINES: Cyanamide.
DIARYL ETHERS: Copper(I) phenylacetylide.
DIARYL SULFIDES: Dimethyl sulfoxide.
DIARYL SULFONES: Phosphorus pentoxide–Methanesulfonic acid.
1,4-DIENE-3-ONES: Aluminum chloride.
2,4-DIENE-1-ONES: B-1-Alkenyl-9-borabicyclo[3.3.1]nonanes.
cis,vic-DICHLOROALKANES: Benzeneselenenyl halides.
1,3-DIENES: Allyldiphenylphosphine. Bromomethanesulfonyl bromide. 1,3,3a,4,7,7a-Hexahydro-4,7-methanobenzo[c]thiophene 2,2-dioxide. Molybdenum carbonyl. Nickel(II) chloride–Zinc. 3-Sulfolene. Tetrakis(triphenylphosphine)palladium.
1,4-DIENES: Benzylchlorobis(triphenylphosphine)palladium(II). Bis(dibenzylideneacetone)palladium. Tetrakis(triphenylphosphine)palladium.
1,5-DIENES: Palladium(II) chloride.
DIENONES: B-1-Alkenyl-9-borabicyclo[3.3.1]nonanes.
DIENYNES: (E)-1,2-Dichloroethylene.
DIHYDROFURANONES: Mercury(II) acetate–Nafion-H.

2,3-DIHYDRO-4-PYRONES: 1-Methoxy-2,4-dimethyl-3-trimethylsilyloxy-1,3-butadiene.
1,2-DIKETONES: Acyllithium reagents.
1,4-DIKETONES: Tetrakis(triphenylphosphine)palladium. Tin(II) trifluoromethanesulfonate. Titanium(IV) chloride.
1,5-DIKETONES: Tetrakis(triphenylphosphine)palladium. Triphenylphosphine.
1,6-DIKETONES: Copper(II) tetrafluoroborate.
1,3-DIOLS: Di-μ-carbonylhexacarbonyl dicobalt. Organotitanium reagents. (2R,4R)-Pentanediol. Tri-n-butyltin hydride.
DISULFIDES: Iron(III) nitrate.
DITHIOKETALS: Zinc trifluoromethanesulfonate.
DIVINYL KETONES: Benzylchlorobis(triphenylphosphine)palladium.
1,3-DIYNES: Tetrakis(triphenylphosphine)palladium.
1,4-DIYNES: Di-μ-carbonylhexacarbonyldicobalt.
ENAMINES: N-Morpholinodiphenylphosphine oxide. Titanium(IV) chloride.
ENAMINONES: Alkylaluminum halides. Bis(acetonitrile)dichloropalladium(II).
ENEDIONES: Aluminum chloride. Oxygen, singlet.
ENOL SILYL ETHERS: Bromomagnesium diisopropylamide. t-Butyldimethylchlorosilane. t-Butyldimethylsilyl trifluoromethanesulfonate. Chlorotrimethylsilyl–Sodium iodide. Iron. Lithium t-octyl-t-butylamide.
ENOL TRIFLATES: N-Phenyltrifluoromethanesulfonimide.
1,3-ENYMES: Dimethyl(methylthio)sulfonium tetrafluoroborate. Potassium t-butoxide. Tetrakis(triphenylphosphine)palladium–Copper(I)iodide.
EPOXIDES: t-Butyldimethylsilyl hydroperoxide. Hydrogen bromide–Acetic acid. Lithium tri-sec-butylborohydride. Tellurium(IV) chloride. Triphenylphosphine–Carbon tetrachloride.
α, β-EPOXY ALCOHOLS: t-Butyldimethylsilyl hydroperoxide–Mercury(II) trifluoroacetate.
ETHYLENE KETALS: 2,4,6-Collidinium p-toluenesulfonate.
FLAVONES: Dimethyl sulfoxide–Iodine.
FURANES: Cyclopropenone 1,3-propanediyl ketal. Dichlorobis(cyclopentadienyl)-titanium. (α-Formylethylidene)triphenyl-

SYNTHESIS OF, FURANES *(Continued)*
phosphorane. Palladium(II)chloride.
4-Phenyloxazole.

GLUCURONIDES: Trimethylsilyl trifluoromethanesulfonate.

GLYCOSIDES: Tetrafluorosilane. Trimethyl-(phenylthio)silane. Triphenylmethyl perchlorate.

C-GLYCOSIDES: Allyltri-*n*-butyltin. Benzeneselenenyl halides. (Diethylamino)sulfur trifluoride. Lithium naphthalenide. Tri-*n*-butyltin fluoride. Triphenylphosphine.

GLYCOSYL FLUORIDES: (Diethylamino)sulfur trifluoride. Pyridinium poly(hydrogenfluoride).

α-HALO CARBOXYLIC ACIDS: Pyridinium poly(hydrogen fluoride). 7,7,8,8-Tetracyanoquinodimethane.

HALOMETHYL KETONES: Bromomethyllithium, Chloromethyllithium.

HOMOALLYLIC ALCOHOLS: B-Allyldiisopinocampheylborane. Allyl phenyl selenide. Allyltri-*n*-butyltin. Chromium(II) chloride. Crotyltrimethylsilane. Diethyl(tributylstannyl)aluminum. (2R,4R)-Pentanediol. Pinacol chloromethaneboronate. Tin–Aluminum. Titanium(IV) chloride.

HOMOALLYLIC AMINES: N-Sulfinyltoluenesulfonamide.

HOMOALLYLIC ETHERS: Iodotrimethylsilane.

HOMOPROPARGYLIC ALCOHOLS: Organozinc reagents.

trans-Hydrindenones: Alkylaluminum halides.

α-HYDROXY ALDEHYDES: Hexahydro-4,4,7-trimethyl-4H-1,3-benzothiin. Potassium hydroxide. 4,4,6-Trimethyl-1,3-oxathiane.

α-HYDROXY AMIDES: Organotitanium reagents. Titanium(IV) chloride.

β-HYDROXY AMIDES: (S)-(—)-Menthyl *p*-toluenesulfinate. (R)-(α-*p*-Tolylsulfinyl)-N,N-dimethylamide.

α-HYDROXY AMINES: Trichloroacetonitrite.

β-HYDROXY CARBOXYLIC ACIDS: (R)-2-Acetoxy-1,1,2-triphenylethanol. Ketene bis(trimethylsilyl) ketals. Titanium(IV) chloride.

γ-HYDROXY CARBOXYLIC ACIDS: Hexahydro-4,4,7-trimethyl-4H-1,3-benzothiin.

α-HYDROXY-β-DIKETONES: Tetra-*n*-butylammonium fluoride.

α-HYDROXY ESTERS: Borane–Dimethyl sulfide.

β-HYDROXY ESTERS: Aceto(carbonyl)cyclopentadienyl(triphenylphosphine)iron. (S)-(—)-Menthyl *p*-toluenesulfinate.

γ-HYDROXY ESTERS: Titanium(IV) chloride.

α-HYDROXY- β-KETO ESTERS: Lithium diisopropylamide.

α-HYDROXY KETONES: *t*-Butylhydrazine. *o*-Iodosylbenzoic acid. Samarium(II) iodide.

β-HYDROXY KETONES: Sodium hydrogen telluride.

HYDROXYLAMINES: Dibenzoylperoxide.

21-HYDROXY-20-KETOSTEROIDS: Tosylmethyl isocyanide.

5-HYDROXYPYRONES: Bromine–Methanol.

γ-HYDROXY-α,β-UNSATURATED NITRILES: α-(Phenylsulfinyl)acetonitrile.

IMIDES: 1,1′-Di(methylcyclopentadienyl)tin.

1-INDANONES: Trifluoromethanesulfonic acid.

INDOLES: Organoaluminum reagents.

INDOLIZINES: Trimethylsilylmethyl trifluoromethanesulfonate.

α-IODO ALDEHYDES: Iodine–Pyridinium chloro chromate.

α-IODOCARBOXYLIC ACIDS: Iodine–Copper(II) acetate.

α-IODO KETONES: Iodine–Pyridinium chlorochromate.

β-IODO KETONES: Tetraethylammonium iodide.

ISOALLOXAZINES: Oxygen.

ISOCOUMARINS: Dimethylsulfoxonium methylide. Thallium(III) trifluoroacetate.

KETENE-O,S-ACETALS: Methoxy(phenylthio)trimethylsilylmethyllithium.

KETENE DIETHYL KETALS: Diethyoxymethyldiphenylphosphine oxide.

KETENE THIOACETALS: Lithium iodide.

α-KETO ALDEHYDES: Nafion-H.

α-KETO ESTERS: *t*-Butyl hydroperoxide. Cyanotrimethylsilane. 1,2-Diethoxy-1,2-disilyloxyethane. Trimethylphosphonoglycolate.

β-KETO ESTERS: *t*-Butylacetothioacetate. Methyl cyanoformate. Zinc.

γ-KETO ESTERS: Cyclopropenone 1,3-propane diyl ketal. 1-Ethoxy-1-trimethylsilyloxycyclopropane. Methyl α-chloro-α-phenylthioacetate. Titanium(IV) chloride.

δ-KETO ESTERS: (S)-1-Amino-2-methoxyme-
thylpyrrolidine.

KETONES: Benzylchlorobis(triphenylphosphine)palladuim. 1-Chloro-N,N,2-tri-
methylpropenylamine. Dimethylaluminum
methyl selenoate. Organocopper
reagents.Sodium benzeneselenoate.Tetrakis-
(triphenylphosphine)palladuim. Triphenylphosphine dichloride. Tris(acetylacetonate)-
iron(III).

α-KETO NITRILES: *t*-Butyl hydroperoxide.

KETOXIMES: Vilsmeier reagent.

LACTAMS: Benzeneselenyl halides. Di-*n*-bu-
tyltin oxide.

β-LACTAMS: Aceto(carbonyl)cyclopenta-
dienyl-(triphenylphosphine)iron. Benzenesulfenyl chloride. Chlorodicarbonylrho-
dium(I) dimer. 2-Chloro-1-methylpyridin-
ium iodide. Dimethylformamide–Thionyl
chloride. Ketene bis(trimethylsilyl)-
ketals. *p*-Tolylsulfinylacetic acid. Triphenylphosphine–Diethyl azodicarboxylate. Zinc.

LACTONES: Mercury(II) acetate. Sodium
bromite.

γ-LACTONES: Benzeneselenenyl halides. Bromine–Nickel(II) alkanoaotes. 2-Chloro-1-
methylpyridinium iodide. Dichloroketone.
Iodine. Iodoacetoxy(tri-*n*-butyl)tin. Man-
ganese(II) acetate. Menthyl *p*-toluenesul-
finite. Silica.

δ-LACTONES: Iodotrimethylsilane. Oxygen,
singlet. Rhodium(II) carboxylates.
Tris(tribenzylideneacetylacetone)-
tripalladium.

MACROLIDES: Chloromethyl methyl sulfide.
Dialkylboryl trifluoromethanesulfonate.
Di-*n*-butyltin oxide. Dichloroketene. 4-Di-
methylamino-3-butyne-2-one. Ketenyli-
dene triphenylphosphorane. Tetra-*n*-butyl-
ammonium fluoride. 1,1,6,6-Tetra-*n*-butyl-
1,6-distanna-2,5,7,10-tetraoxycyclodecane.
Tetrakis(triphenylphosphine)-
palladium. N,N,N,N-Tetramethylchloro-
formamidinium chloride.

α-METHOXY ALDEHYDES: (S)-*p*-Tolyl *p*-tolyl-
thiomethyl sulfoxide.

α-METHYL ALDEHYDES: Trimethylaluminum.

N-METHYLAMINO ACIDS: Methyl trifluoromethanesulfonate.

α-METHYLENE-γ-BUTYROLACTONES: Iodotrimethylsilane. Tri-*n*-butyltin hydride.

METHYLENECYCLOALKANES: 2,4,6-Triisopro-
pylbenzenesulfonyl hydrazine.

α-METHYLENECYCLOBUTANONES: Chloro-
[(trimethylsilyl)methyl]ketene.

METHYLENECYCLOPENTANES: 4-Chloro-2-
lithio-1-butene.

α-METHYLENECYCLOPENTANONES: α-Methoxyallene.

5-METHYLENE-2(5H)-FURANONES: Pyridinium
dichromate.

α-METHYLENE KETONES: Bromomethanesulfonyl bromide.
Chloro[(trimethylsilyl)methyl]ketene. Formaldehyde. Titanocene methylene–Zinc
halide.

α-METHYLENELACTONES: Ethyl formate.

δ-METHYLENE-δ-LACTONES: Mercury(II) oxide.

3-METHYLENETETRAHYDROFURANES: 4-Chloro-
2-lithio-1-butene.

METHYL KETONES: Acetoacetic acid. Bis(cyclopentadienyl)titanacyclobutanes. μ-Chlo-
robis(cyclopentadienyl)(dimethylaluminum)-μ-methylenetitanium. Chlorotrimethylsilane. (—)Norephedrine.

NAPHTHOQUINONES: Chlorotris(triphenyl phosphine)cobalt.

NITRILE OXIDES: Phenyl isocyanate. *p*-Tolu-
enesulfonic acid.

NITRILES: Cyanotrimethylsilane.

NITROALKANES: Lithium tri-*sec*-butylborohy-
dride. Ozone.

NITROALKENES: Nitromethane.

α-NITRO KETONES: Pyridinium chlorochromate. Trifluoroacetyl nitrate.

NITRONES: Hydrogen peroxide–Sodium tungstate. N-Methylhydroxylamine.

NITROSOALKENES: O-*t*-Butyldimethylsilyl-
hydroxylamine.

C-NUCLEOSIDES: Benzeneselenenyl halides.

1,5-OCTADIENES: Bis(1,5-cyclooctadiene)-
nickel(0).

ORTHO ESTERS: 3-Methyl-3-hydroxymethyl-
oxetane.

OXETANES: Dimethyloxosulfonium methylide.
Dimethyl N(*p*-tosyl)sulfoximine.

PENTAFULVALENES: Copper(II) chloride.

PEPTIDES: O-Benzotriazolyl-N,N,N′,N′-tetra-
methyluronium hexafluorophosphate. 1-(3-
Dimethylaminopropyl)-3-ethylcarbodi-
imide. Di(N-succinimidyl)oxalate. Tetra-

SYNTHESIS OF, PEPTIDES (*Continued*)
methylguanidine. 1,3-Thiazolidine-2-thione.

PHENANTHRENEQUINONES: Potassium–Graphite.

PHENOLS: 2-*t*-Butyl-1,3,2-dioxaborolane. Chromium carbene complexes.

PHLOROGLUCINOLS: Malonyl dichloride.

PHTHALIDES: Thallium(III) trifluoroacetate.

PINACOLS: Samarium(II) iodide. Tetra-*n*-butylammonium fluoride.

Δ¹-PIPERIDEINES: Diisobutylaluminum hydride.

Δ³-PIPERIDINES: Trimethylsilyl trifluoromethanesulfonate.

1,3-POLYOLS: Organocopper compounds.

PYRANES: Palladium(II) chloride.

PYRANONAPHTHOQUINONES: Chlorotris(triphenylphosphine)cobalt.

PYRIDINES: Triethyl 1,2,4-triazine-3,5,6-tricarboxylate.

PYRIDINIUM METHYLIDES: Trimethylsilylmethyl trifluoromethanesulfonate.

PYRIDONES: Dicarbonylcyclopentadienylcobalt.

α-PYRONES: Iron-carbene complexes.

PYRROLIDINES: Copper(I) trifluoromethanesulfonate. 1,4-Dichloro-1,1,4,4-tetramethyldisilylethylene. Trimethylamine oxide.

Δ¹-PYRROLINES: Diisobutylaluminum hydride.

PYRROLIZIDINES: Tri-*n*-butyltin hydride.

o-QUINODIMETHANES: Cesium fluoride. Chromium(II) chloride. Zinc–Silver couple.

QUINOLINES: Diisopropyl peroxydicarbonate. Vilsmeier reagent.

QUINONES: Chromium carbene complexes.

1-SILYL-1-ALKENES: Organomagnesium reagents.

SILYLKETENE KETALS: Triethylsilyl perchlorate.

α-SILYL KETONES: *n*-Butyllithium–Potassium *t*-butoxide. Trimethylsilylmagnesium chloride.

SPIRO[2,3]HEXANES: Cyclobutylidene.

SPIROLACTONES: *t*-Butyl isocyanide.

SPIROKETALS: N,N-Dimethylhydrazines. Potassium hydride.

SUCCINIMIDES: Molybdenum carbonyl.

TETRAHYDROFURANES: Iodine. Lead tetraacetate. Silver(I) nitrate. 2,4,4,6-Tetrabromo-2,5-cyclohexadienone.

TETRAHYDROPYRIDINES: Mercury(II) acetate.

1-TETRALONES: Trifluoromethanesulfonic acid.

THIONO ESTERS: Hydrogen sulfide.

THIOPEPTIDES: 2,4-Bis(4-methoxyphenyl)-1,3,2,4-dithiaphosphetane-2,4-disulfide.

1,2,4-TRIAZINES: Tri-μ-carbonylhexacarbonyldiiron.

TROPONES: Tri-*n*-butyltin hydride.

β,γ-UNSATURATED ACETALS: Ethyl diazoacetate.

α,β-UNSATURATED ALDEHYDES: Iodosylbenzene–Boron trifluoride etherate. 3-Methylthio-2-propenyl *p*-tolyl sulfane. Palladium(II) acetate. Thiophenol. Titanium(IV) chloride. Vinyltrimethylsilyl.

β,γ-UNSATURATED ALDEHYDES: Diethyl[(2-tetrahydropyranyloxy)methylphosphonate.

γ,δ-UNSATURATED ALDEHYDES: (E)-(Carboxyvinyl)trimethylammonium betaine.

β,γ-UNSATURATED AMINES: Cyanotrimethylsilane.

α,β-UNSATURATED ESTERS: Di-*n*-butyltelluronium carboethoxymethylide. Methyl bis(trifluoroethyl)phosphonoacetate. Methyldiphenylchlorosilane. Triethyl phosphonoacetate.

β,γ-UNSATURATED ESTERS: Organocopper reagents. Potassium hexamethyldisilazide.

γ,δ-UNSATURATED ESTERS: Reformatsky reagent.

α,β-UNSATURATED KETONES: Allyl chloroformates. *t*-Butyl hydroperoxide–Chromium carbonyl. Organolithium compounds. Oxygen, singlet. Palladium(II) acetate. N-Phenyl keteniminyl(triphenyl)phosphorane. Proline. Tetrakis(triphenylphosphine)palladium. Titanium(IV) chloride. Tosylmethyl isocyanide.

β,γ-UNSATURATED KETONES: Dichlorobis(tri-*o*-tolylphosphine)palladium(II). Pyridine.

δ,ε-UNSATURATED KETONES: Bis(benzonitrile)-dichloropalladium.

α,β-UNSATURATED δ-LACTONES: Pyridinium chlorochromate.

α,β-UNSATURATED NITRILES: Diethyl phosphorocyanidate.

α,β-UNSATURATED THIOLACTAMS: *p*-Toluenesulfinyl chloride.

VINYL ETHERS: Methoxymethyl phenyl thioether.

VINYL HALIDES: B-Halo-9-borabicyclo-[3.3.1]nonene. Hydrazine. 1,1,3,3-Tetramethylguanidine.

VINYLSILANES: Chlorotrimethylsilane–Sodium. Dichlorobis(cyclopentadienyl)titanium. Diethyl[dimethyl(phenyl)silyl]aluminum.

VINYL SULFIDES: Iodotrimethylsilane.

VINYL SULFONES: Copper(II) acetate. Phenylsulfonyl(trimethylsilyl)methane. Se-Phenyl p-tolueneselenosulfonate.

TEBBE REAGENT: μ-Chlorobis(cyclopentadienyl)(dimethylaluminum) μ-methylenetitanium.

THIONATION: 2,4-Bis(4-methoxyphenyl)-1,3,2,4-dithiaphosphetane-2,4-disulfide.

TRANSESTERIFICATION: Pyridinium m-nitrobenzenesulfonate. Titanium(IV) alkoxides.

ULLMANN DIARYL ETHER SYNTHESIS: Copper(I) phenylacetylide.

WACKER OXIDATION: Phase-transfer catalysts.

WITTIG-HORNER REAGENTS: Carbomethoxyethyl(diphenyl)phosphine oxide. Diethoxymethyldiphenyl phosphine oxide. Diethyl [(2-tetrahydropyranyloxy)-methyl]-phosphonate. Methyl bis(trifluoroethyl)-phosphonoacetate. N-Morpholinodiphenyl-phosphine oxide. Triethylphosphonoacetate.

WITTIG-HORNER REACTION: Lithium chloride–Diisopropylethylamine. Potassium carbonate–18-crown-6. Tetramethylethylenediamine.

WITTIG REAGENTS: N-Alkyl-β-aminoethyl-phosphonium bromide. α-Formylethylidenetriphenylphosphorane. N-Phenyl-keteniminyl(triphenyl)phosphorane.

WITTIG REARRANGEMENT: n-Butyllithium.

WITTIG-TYPE REACTIONS: Chromium carbene complexes. Di-n-butyltelluronium carboethoxymethylide. Zirconium carbene complexes.

WURTZ-FITTIG REACTION: Chlorotrimethylsilane–Sodium.

AUTHOR INDEX

Cortes, D. A., 480
Cossy, J., 61
Cottrell, D. A., 121
Couturier, D., 36
Cox, D. P., 462
Cozzi, F., 298, 509
Crabtree, R. H., 152, 550
Cram, D. J., 312, 383
Crass, G., 193, 480
Craven, B. M., 11
Crawford, R. J., 465
Crich, D., 417
Criegee, R., 245, 366
Crimmins, M. T., 350
Crisp, G. T., 474
Crivello, J. V., 531
Crociani, B., 502
Croudace, M. C., 167
Crouse, G. D., 390
Crumbie, R. L., 233
Cruz de Maldonado, V., 322
Cuiffreda, P., 479
Cummins, C. H., 95
Curran, D. P., 422

Dabescat, F., 86
Daggett, J. U., 14
Dale, J. A., 282
Dampawan, P., 531
Danheiser, R. L., 350
Danhmen, W., 205
Daniels, K., 344
Daniewski, W. M., 61
Danishefsky, S., 303, 313, 391, 561
Danishefsky, S. J., 313
Dao, G. M., 251
Dappen, M. S., 86
Daub, J. P., 282
Daugherty, B. W., 329
D'Auria, M., 256
Dave, L. D., 202
Dave, V., 224
David, S., 548
Davidson, S. K., 186
Davies, N., 553
Davies, S. G., 36
Davis, F. A., 393
Davis, J. T., 70, 249, 277
Davis, P., 381
Davis, P. D., 46
Dawe, R. D., 195

Dawson, J. W., 179
Dax, S. L., 108
De, B., 324
De Bernardis, F., 179
de Groot, A., 317
de Kaifer, C. F., 322
De Loach, J. A., 350
De Lombaert, S., 183
Debost, J. L., 318
Dedericks, E., 205
Dehasse-Delombaert, C. G., 125
Deiko, S. A., 484
Dekoker, A., 125
deLasalle, P., 278
Della Pergola, R., 449
Delorme, D., 79, 396
Demuth, M., 546
Denis, J. N., 219
Denis, P., 53
Denmark, S. E., 86, 283, 441, 451
Depaye, N., 231
Després, J.-P., 178
Desai, R. C., 109, 158
Deshmukh, M. N., 94
DeShong, P., 566
Desmaële, D., 183
Desmond, R., 25
Dess, D. B., 379
Detar, D. F., 450
Dev, S., 502
Devant, R., 3
DeVos, M. J., 295
Devys, M., 523
Deyo, D., 389
Di Fabio, R., 418
Di Furia, F., 94
DiBattista, P., 449
Diercks, R., 157
Dieter, J. W., 465
Dieter, R. K., 350, 465
Dietsche, W., 182
Differding, E., 125
Ditrich, K., 101
Doboszewski, B., 420
Docken, A. M., 200
Dötz, K. H., 136
Doi, E., 485
Dojo, H., 236, 502
Dolence, E. K., 412
Dolle, R. E., 185, 476
Dolling, U. H., 381

SUBJECT INDEX